T5-CUR-667

TEXTBOOK OF ANATOMY AND PHYSIOLOGY

THE MACMILLAN COMPANY
NEW YORK • CHICAGO
DALLAS • ATLANTA • SAN FRANCISCO
LONDON • MANILA
IN CANADA
BRETT-MACMILLAN LTD.
GALT, ONTARIO

TEXTBOOK *of* ANATOMY *and* PHYSIOLOGY

DIANA CLIFFORD KIMBER
CAROLYN E. GRAY, A.M., R.N.

REVISED BY

CAROLINE E. STACKPOLE, M.A.
LATE ASSOCIATE IN BIOLOGY, TEACHERS COLLEGE, COLUMBIA UNIVERSITY

LUTIE C. LEAVELL, R.N., M.A., M.S.
PROFESSOR OF NURSING EDUCATION
TEACHERS COLLEGE, COLUMBIA UNIVERSITY

THIRTEENTH EDITION

NEW YORK · THE MACMILLAN COMPANY · 1959

PRINTED IN THE UNITED STATES OF AMERICA

Fifth Printing 1959

Earlier editions: Text-book of Anatomy and Physiology for Nurses *by Kimber, copyright, 1893, by Macmillan and Co., and copyright, 1902, by The Macmillan Company;* Text-book of Anatomy and Physiology for Nurses *by Kimber and Gray, copyright, 1909, 1914, 1918, by The Macmillan Company;* Text-book of Anatomy and Physiology *by Kimber and Gray, copyright, 1923, 1926, by The Macmillan Company;* Textbook of Anatomy and Physiology *by Kimber and Gray, copyright, 1931, by The Macmillan Company;* Textbook of Anatomy and Physiology *by Kimber, Gray, and Stackpole, copyright, 1934, 1938, 1942, by The Macmillan Company;* Textbook of Anatomy and Physiology *by Kimber, Gray, Stackpole, and Leavell, copyright, 1948, by The Macmillan Company.*

PREFACE TO THE THIRTEENTH EDITION

In an era of extensive research and rapid change in many physiological concepts, textbook revision is not an easy task. This is especially true when materials must be highly selected to meet the needs of beginning students. For example, in the area of hormones "knowledge is just at the beginning" in relation to the adrenal cortex, especially the adrenal-pituitary relationships in response to stress. The interrelationships of other hormonal activities to body function are still not completely understood. However, in the light of present-day accepted hormonal functions, the chapter on hormones has been revised; but highly controversial material has not been included.

In response to many requests, the discussion of epithelial tissues has been placed with that of the other tissues in Chapter 4, and an effort has been made to clarify functions. The chapters on cells have been enlarged and emphasis has been placed on the cell, its function, and its liquid environment. The chapter on muscles has been completely revised to emphasize functional relationships. A new chapter has been added on water and electrolyte balance and buffer action.

Each chapter has been scrutinized for clarity. Outdated ideas have been replaced by new concepts, and physiological relationships have been brought into better focus. The ever-present question of how best to organize learning experiences and whether anatomy and histology should precede physiology have been again considered. We still believe that sometimes there are advantages in studying anatomy and physiology together, each giving immediate meaning to the other. Sometimes the study of anatomy can well precede the study of physiology, a clear understanding of function requiring detailed knowledge of structure as a whole. Again, in some instances there are advantages in studying function first, the physiology illuminating anatomy at every point, the anatomical knowledge thus gained making the physiology in its wider aspects more clear. Actually what is needed for a real understanding of either anatomy or physiology is a study of each in terms of the other, or an appreciation of the interlocking relationships of structure and function. Physiology has been elaborated and reinforced, but in many instances it is integrated with anatomical structure to clarify points of issue.

Many of the illustrations have been replaced by new ones, and the number has been increased from 360 to 388. As in previous editions, cross reference to diagrams and their captions has been retained.

The general organization of the chapters is in five units, as indicated in the Table of Contents.

This text can be supplemented by the *Laboratory Manual in Anatomy and Physiology* by the authors. There is also a set of Kodachrome (Medichrome) slides, Series MH2, selected or made by the authors for the Clay-Adams Company of New York City.

Grateful acknowledgment is made to teachers and students who have made suggestions for arrangement of material and also for content. We are indebted to Miss Mary Lorenc and Miss Frances Russell, who made many of the new illustrations. Other acknowledgments are made in the text.

We wish to express our indebtedness to Mr. W. Holt Seale and his staff of the Medical Department of The Macmillan Company, and especially to Miss Barbara Russell, who has been most helpful in handling the editorial details of this thirteenth edition.

Teachers College
Columbia University
New York City
June, 1955

Caroline E. Stackpole
Lutie C. Leavell

PREFACE TO THE EIGHTH EDITION

In a previous revision of this textbook, the aim was stated as follows: "It represents an attempt to describe in as simple a manner as possible the phenomena of life and the principal conclusions which have been reached as to their interdependence and cause." This aim has dominated the present revision.

A further aim has been to include the material outlined in the course in anatomy and physiology in the curriculum published by the National League of Nursing Education. It would be flattering to think these aims have been achieved. It is highly probable that teachers and students will discover mistakes as well as many places where improvement is needed.

It is encouraging to be asked to incorporate new material, and an attempt has been made to grant the requests that have come to the publisher and author. Obviously it is difficult to do this and keep the size of the book within reasonable limits. Two hundred fifty-five illustrations account for many pages.

A large proportion of the illustrations have been redrawn, while 30 new ones have been added.

In a text where much space is given to explanations of scientific terms and a very complete index is provided, it is a question whether a complete glossary is necessary. Students should be helped to form the habit of using the index. It is highly probable that the index will be more helpful than any glossary, because the limitations of space make a glossary as complete as the index an impracticable ideal.

The advisability of placing a list of reference readings at the end of each chapter has been considered. The principal sources of reference material are the scientific journals and textbooks for advanced students. It is doubtful whether the students for whom this book is intended are sufficiently prepared to profit by the study of scientific journals, and to list chapter or page references in advanced textbooks does not commend itself. New editions of such advanced texts are constantly coming out, and chapter or page references that were correct at the time of publication may easily be incorrect a few months later. Hence the bibliography on p. 605, which represents a minimum of necessary reference books.

Nothing is more encouraging than the growth of reference libraries in some schools of nursing. Many of them have outgrown the bibliography. Others are struggling to get along with so few reference books that the marvel is not that some students fail but that the majority do so well.

It is desirable to stimulate an open-minded questioning attitude and to help students to realize that physiology is a growing subject constantly adding to its knowledge, and discarding old theories or reinterpreting them in the light of experimental results. An appreciation of, and respect for, knowledge gained by careful detailed experiments can best be taught in connection with laboratory work. To teach anatomy and physiology by any but the laboratory method seems inexcusable. The fact that in many schools the laboratory method is used makes the question why all schools do not use it all the more pertinent.

Having been a student and teacher of anatomy and physiology for many years, the number of students and coworkers to whom I am indebted is very large. It is a pleasure to acknowledge this indebtedness and express my appreciation of it. In connection with this revision I am especially indebted to Caroline E. Stackpole, A.M., Associate in Biology, Teachers College, New York, who has read and criticized my manuscript; Dr. George W. Corner of Rochester University, who helped with the preparation of Chaps. 3 and 24; and Dr. R. J. E. Scott, who has again made the index. The representatives of The Macmillan Company, particularly Mr. J. Norris Myers and his staff, and Mr. J. P. Smith of the Manufacturing Department, have been most co-operative and helpful. The authors whom I have consulted and the various publishers who have granted me permission to use illustrations from their books have been most generous.

City Hospital
New York City
September, 1931

Carolyn E. Gray

SUGGESTIONS FOR ILLUSTRATIVE MATERIALS

There is much truth in the statement that "one picture is often worth a thousand words." Increasing emphasis is placed on visual aids in teaching, and we believe that the learner can grasp and understand the problem at hand if she can see, handle, and read about it at the same time.

Equipment for laboratory work need not be elaborate; in fact, the simpler it is the better it is for beginning students.[1]

The black-and-white lantern slides, Kodachrome slides, and microscope slides are of great value in clarifying structures and in many instances physiological relationships.

Films and film strips have their rightful place in all learning situations, but need to be selected with care.

Charts and models are of value and can be used in many instances.

Whenever practical, the students themselves should be used as subjects.

Sources of Materials. Available from the National League for Nursing, 2 Park Avenue, New York 16, N. Y., are bibliographies that include books, pamphlets, and audiovisual aids of all kinds. Volume I includes materials for the sciences. In the section on anatomy and physiology are included books of all kinds for the student and teacher, as well as films, slides (of all types), charts and models, descriptions, and addresses where they can be purchased. Suggestions for how to use the students as subjects and for using fresh materials of all kinds are included. Sources of all kinds of materials are carefully listed. These bibliographies are kept up to date by supplements.

The Clay-Adams Company, 141 East 25 Street, New York, N. Y., has several series of Kodachrome (Medichrome) slides. The Stackpole-Leavell selection, Series MH2, consisting of about 140 Kodachrome (Medichrome) slides (2x2) on anatomy, histology, and physiology, was especially prepared for use with this textbook. The syllabus SL/K explains each slide in detail.

[1] C. E. Stackpole and L. C. Leavell, *Laboratory Manual in Physiology,* 2nd ed. The Macmillan Company, New York, 1948. See Appendix for equipment needed.

TABLE OF CONTENTS

UNIT I

The Body as a Whole—Structural and Functional Relationships and Organization

CHAPTER 1

ANATOMY, PHYSIOLOGY, HYGIENE
THE ANATOMICAL POSITION, BODY REGIONS

GENERAL STRUCTURE OF THE HUMAN BODY

A knowledge of anatomy and physiology is essential as a basis for healthful living or *hygiene.* It is equally essential in attempting to understand the pathological states that engage the attention of students of the science of medicine in their efforts to forestall departure from normal, through *preventive hygiene* or *prophylaxis* and *preventive medicine,* and to bring sick individuals back to normal anatomical and physiological conditions through professional diagnosis and skillful medical and nursing care, which may be called *remedial medicine.*

Both anatomy and physiology, along with other classes of organized knowledge concerning living matter, are divisions of the great science of *biology.*

Anatomy is concerned with structure. Physiology is concerned with function. Anatomy describes the structure, but usually the function is referred to. Physiology discusses the work done by the structure, but usually in terms of the anatomy; to put it another way, an understanding of anatomy is incomplete without physiology, and an understanding of physiology is incomplete without anatomy.

Anatomy belongs to that group of biological sciences known as *morphology*—the group which deals with structure and spatial relationships, the way bodies are built, the kinds of material used, and the architecture of the entire organism in action.

Gross anatomy is the science of macroscopic structure, that which can be seen with the unaided eye. Anatomy may be considered from many points of view. As in all the biological sciences, the focus of thought may be on plant anatomy useful to agriculturists, etc., animal anatomy useful to veterinarians, etc., or on the human being alone, in which case the study is *human anatomy,* or it may be on comparing the structures of animals with each other and with the human structure, which is *comparative anatomy.* There is also *systemic anatomy* with a point of view slightly different from that of *regional anatomy,* the former giving attention to the systems of the body, the latter stressing regions, such as the abdomen, the thorax, the arm, or the head. Great advances of skill in physical diagnosis, brought about by the use of such methods as measurement, palpation, radiography, transillumination, radioactive isotope studies, fluoroscopy, etc., the use of kymograph records (basal metabolism, etc.), the use of electrograms such as electroencephalograms, electrocardiograms, and sphygmograms made evident the need of regional anatomy and the conceptions it gives of the effects that changing conditions in one organ exert upon other organs, as, for instance, when it becomes distended through altered blood and lymph supply and presses on nearby organs. If the focus of study is related to the embryonic and later development of body structures, the science is called *developmental anatomy.*

Again, anatomy may be considered from the viewpoint of activity. *Functional anatomy,* the study of the interaction of organ upon organ as each continually changes in shape, size, temperature, pressure, and in other respects, is a dynamic rather than a static subject. Today, more than hitherto, anatomy considers the structure of the living, moving parts of the living organism.

Pathological anatomy has to do with diseased structures, their location, and their regional effects. The term *pathology* may be used in connection with anatomy, histology, physiology, etc., and is concerned with deviation from normal structures and functions.

Histology stresses minute or microscopic structure, structure that can be seen only with the aid of lenses, and often, therefore, is defined as microscopic anatomy of microanatomy. It makes clear the structure and activities of cells, the arrangement of cells in tissues, and the manner in which tissues are built into organs.

This may be studied as *normal histology,* dealing with healthy cells, tissues, and organs, and as *pathological histology,* dealing with the same parts when they are diseased. Early studies in histology were chiefly limited to the shapes and appearances of the cells and tissues in the manner of *anatomical histology;* but modern histology, in making use of microscopic study of living cells, tissues, organs and organisms, tissue cultures, micromanipulation of living tissues, etc., is largely *physiological histology* and gives much attention to cell and tissue development and activities. *Comparative histology* stresses a comparison of cells and tissues of animals and plants with each other as well as with human cells and tissues.

Physiology is the science of function and activity. It shows what cells, tissues, and organs do individually, in relation to each other, and in the integrated behavior of the organism as a unit. The investigation of these organic modes of behavior splits into many branches. There are *plant physiology, animal physiology, human physiology,* and *comparative physiology,* the latter using information concerning the bodily working of plants and animals as compared with each other and with the human organs and systems. Another aspect is *general*[1] or *cellular physiology,* a term applied to the individual cells themselves, as they live out on a small scale all the activities that characterize the larger systemic units of the organism—such activities as respiration, excretion, absorption of food, and movement. An understanding of cells is important in interpreting the vital processes that take place in plant or animal organisms. Two other aspects are routine physiology and emergency physiology. *Routine physiology* is concerned with minute and always changing adjustments that have to be undertaken in and outside the body while the factors of the environment are within adjustable range. *Emergency physiology* is concerned with the adjustment of the activities of the tissue cells when conditions surrounding them get beyond their adaptive span and become dangerously harmful, as they become, for instance, when suddenly climate is extremely dry or wet or hot or cold, when water to drink is lacking, or when such notable events as general or local disturbances in circulation, fever, bruises, injuries, or infections happen within the body.

Embryology in its widest sense means the science of growth from the one-cell stage to the adult but frequently is restricted to mean the period of growth and development before birth. This period is followed by the postnatal development of infancy, childhood, adoles-

[1] Many physiologists consider cellular physiology as only a part of general physiology and would define general physiology as the study of physiological principles which are general in the sense of being common to all living things.

cence, and early, middle, and late maturity. Development can be studied as an anatomical subject but recently is being studied much more vigorously from the standpoint of physiology.

Hygiene is defined as organized knowledge concerning the application of the principles of physiology to living in health and is divided into many kinds of hygiene. Examples of two important divisions are *external,* or *exterofective, hygiene,* by which is meant the preparation of the environment for the individual, and *internal,* or *interofective, hygiene,* by which is meant the healthful reaction of the individual to environment.

External hygiene covers such matters as pure foods and the scientific selection and preparation of the diet, ventilation, cleanliness in regard to all supplies and wastes, the conditions of workrooms and living rooms in regard to light, heat, humidity, body posture, and opportunities afforded to change body position at will, noise, color schemes, quality of surfaces, etc. It considers a plan of daily living for individual, family, or group that gives interesting work, change of environment and varied activities, leisure, and recreation, and that does away with fatigue, worry, emotional friction, and other harmful conditions such as lack of opportunity for growth and development in an occupation chosen on the basis of the individual's fitness for it. It assumes self-induced, dynamic, constructive participation in social problems connected with individual and group living. A requisite of a balanced life is the clear consciousness of individual responsibility and its challenging activation in definite doing.

The purpose of external hygiene, in other words, is to create an environment in which an individual can "live most and serve best."

Internal hygiene embodies knowledge of the adequate handling of supplies and wastes inside the body. This includes the transformation of the periodic intake of supplies and outgo of wastes into the state of constancy or *homeostasis* found in the body fluids and brought about through the body's ability to distribute, store, and redistribute according to the varying general and local needs of the body cells in their steady state of rapid flux.

These sciences and others, including psychology, sociology, etc., can be grouped as the *biological sciences,* having to do with living things, as contrasted with the *physical sciences,* represented by physics, mathematics, and the like. The branches of science are all closely related and overlap.

Science is one thing, but for purposes of study, it is broken into separate sciences for convenience and to emphasize different or more items than would be possible if considered as a whole.

GENERAL STRUCTURE

An anatomical characteristic of all *vertebrate animals* is their structure as a tube known as the *body wall,* enclosing a tube known as the *viscera,* the cavity[2] between the two tubes being the *body cavity,* or *celom* (Figs. 378 and 379).

Cavities. The *celom,* or *body cavity,* is a *ventral cavity,* and it is enclosed by the body wall. This wall is composed of skin, connective tissues, bone, muscles, and serous membrane. In mammals, during embryonic life this cavity becomes subdivided by a dome-shaped, musculomembranous partition, the *diaphragm,* into the thoracic and abdominal cavities. The pericardial cavity is also developed embryologically from the celom.

1. *The thoracic cavity,* or *chest,* contains the trachea, the bronchi, the lungs, the esophagus, nerves, the heart, and the great blood and lymph vessels connected with the heart. It also contains lymph nodes and the thymus gland.

The thoracic cavity is lined with pleura which divides it into right and left *pleural cavities,* each containing a lung. The other thoracic organs lie in the *mediastinum* between these pleural cavities (Figs. 1 and 5).

2. *The abdominal cavity* contains the stomach, liver, gallbladder, pancreas, spleen, kidneys, and small and large intestines.

The abdominal cavity is lined with the peritoneum enclosing the *peritoneal cavity.* The kidneys are described as being retroperitoneal, lying, as they do, behind the peritoneum.

The *pelvic cavity* is that portion of the abdominal cavity lying below an imaginary line drawn across the prominent crests of the hip-bones. It is more completely bounded by bony walls than the rest of the abdominal cavity. It is divided by a narrow bony ring, the pelvic inlet, into the greater, or false, pelvis above and the lesser, or true, pelvis below. The greater, or false, pelvic cavity is the lower

[2] The term *cavity* is used here in different senses. In one sense it refers to the body cavity and its parts, thoracic and abdominal; in another sense it refers to dorsal and ventral cavities; and in a still more general sense it includes the more open cavities, such as the orbital, nasal, and buccal cavities.

part of the peritoneal cavity and contains parts of the organs listed for the abdominal cavity. The lesser, or true, pelvis contains the bladder, rectum, and some of the reproductive organs. The true pelvis is, in general, below the peritoneum. Study Figs. 1, 2, and 3.

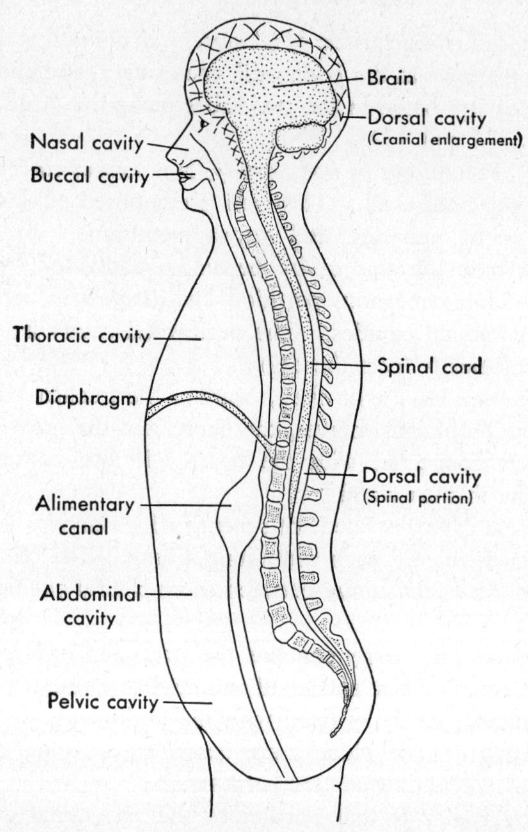

Fig. 1. Diagram of a longitudinal section to show dorsal and ventral body cavities.

The *dorsal cavity* is within the *dorsal body wall*. It contains the brain and spinal cord. The dorsal cavity is a continuous bony cavity formed by the cranial bones and the vertebrae, and it is lined by the meninges of the brain and spinal cord.

A survey of the skeleton shows small cavities in the skull, in addi-

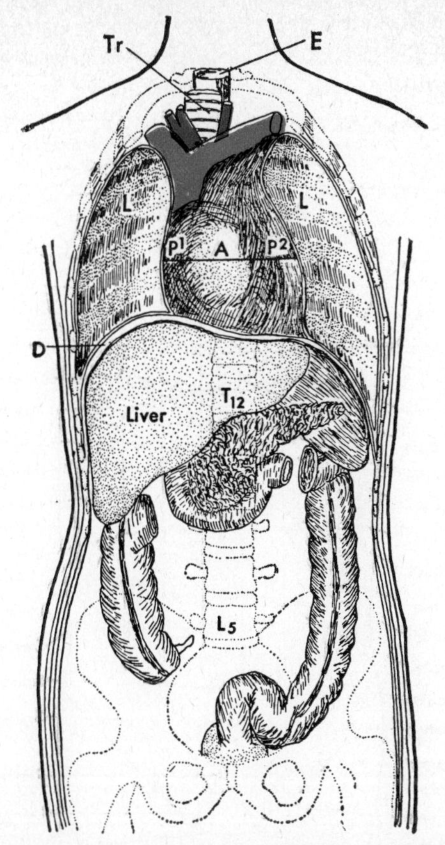

Fig. 2. Diagram to show thoracic and abdominal cavities, seen from the front. Line *A* indicates the mediastinum extending from P^1 to P^2, the mediastinal pleurae; *D*, diaphragm; *E*, esophagus; *L*, lung; L_5, fifth lumbar vertebra; T_{12}, twelfth thoracic vertebra; *Tr*, trachea; blue, superior vena cava; red, innominate and left common carotid arteries.

tion to the cranial cavity, which, for the sake of simplicity in study, are included here.

The *orbital cavities* contain the eyes, the optic nerves, the muscles of the eyeballs, and the lacrimal apparatus.

The *nasal cavity* is filled in with the structures forming the nose (pp. 482–83).

The *buccal cavity*, or *mouth cavity*, contains the tongue and teeth.

(Text continued on p. 13.)

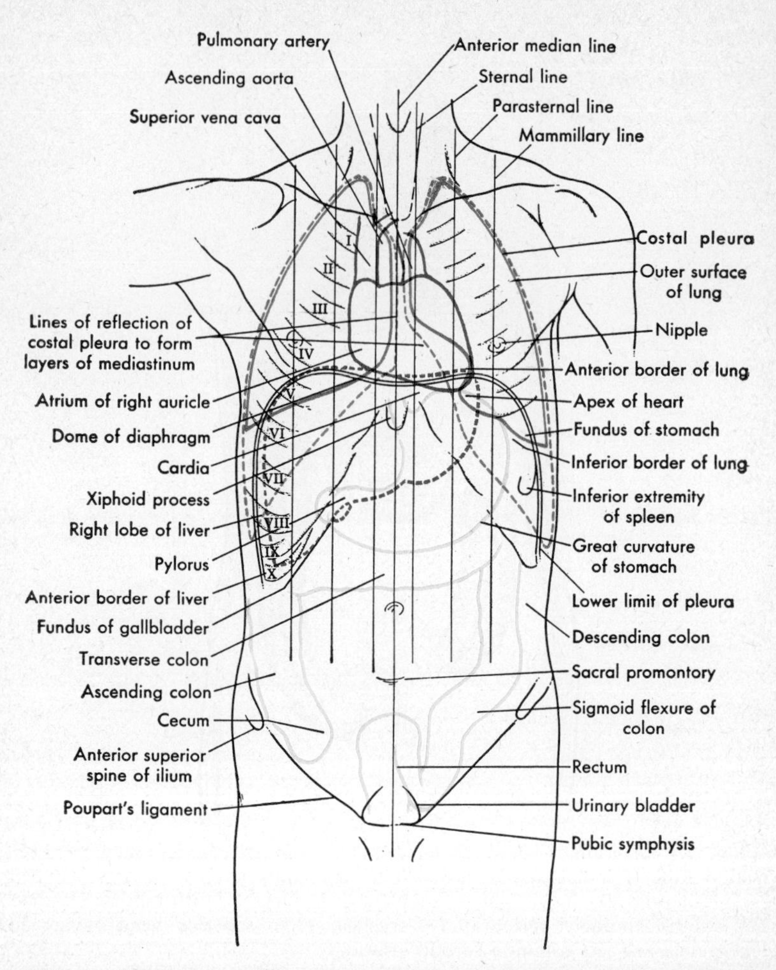

Fig. 3. Projection outlines of the thoracic and abdominal organs on the anterior surface of the trunk. Continuous red, outline of heart, superior vena cava, ascending aorta, pulmonary artery; continuous blue, two lungs; dotted blue, boundaries of pleural cavities; dotted red, liver and fundus of gallbladder; yellow, stomach and parts of large intestine; black, dome of diaphragm and lower edge of spleen. (Toldt.)

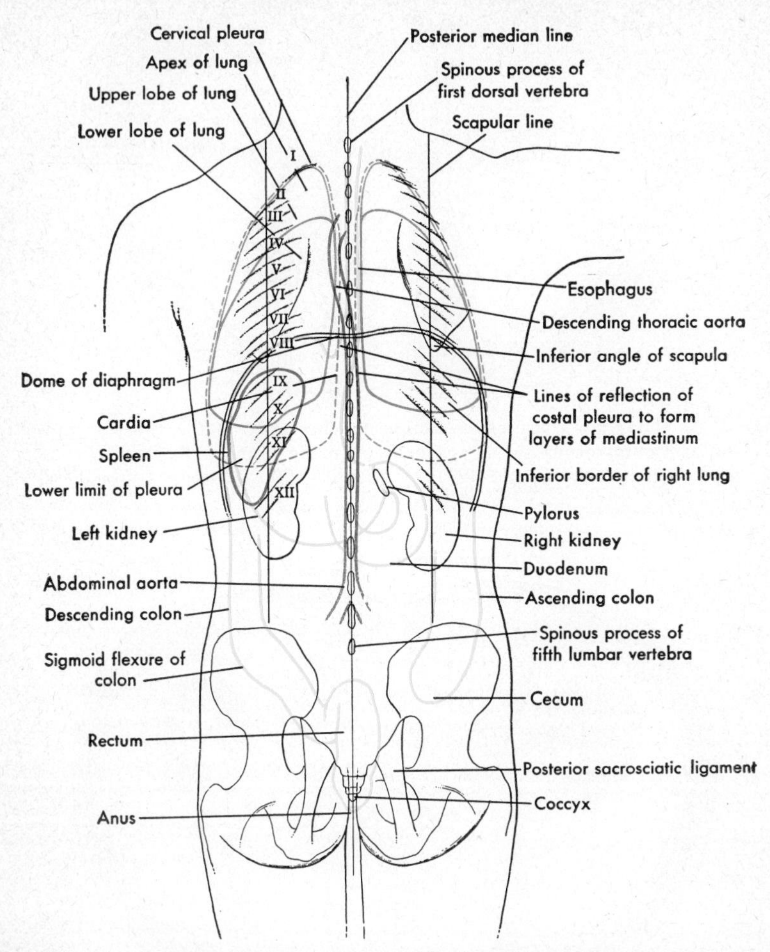

Fig. 4. Projection outlines of the thoracic and abdominal organs on the posterior surface of the trunk. Red, outline of descending thoracic aorta, abdominal aorta, and spleen; continuous blue, two lungs; dotted blue, boundaries of pleural cavities; yellow, stomach, duodenum, parts of large intestine; black, dome of diaphragm and two kidneys. (Toldt.)

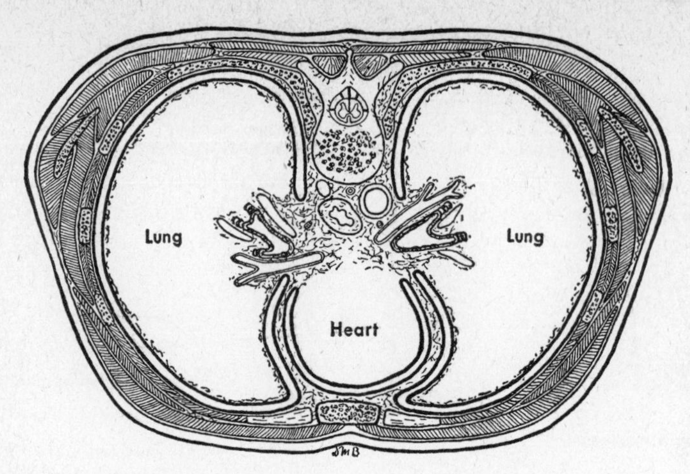

Fig. 5. Diagram of a cross section of the body in the thoracic region. The mediastinum occupies the space between the lungs and extends from the sternum to the vertebrae (p. 497). (Adapted from Toldt.)

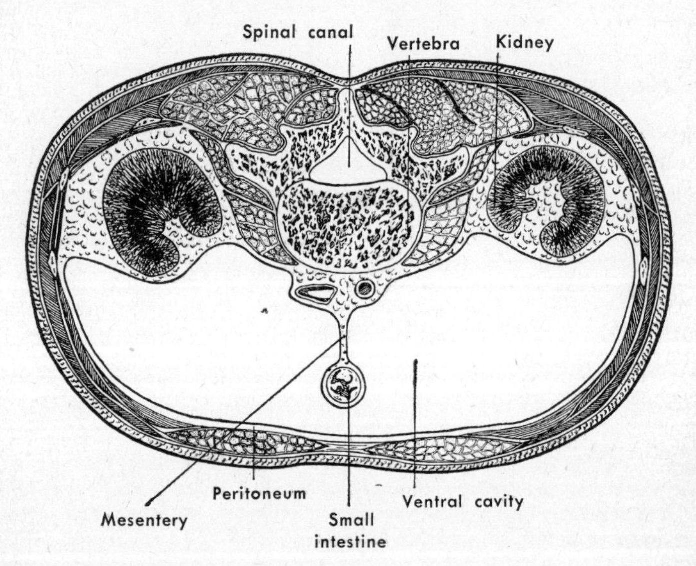

Fig. 6. Diagrammatic transverse section of the body to show abdominal part of ventral cavity and spinal part of dorsal cavity.

BODY REGIONS

Owing to the fact that man walks erect and the majority of other mammals go on all fours, confusion sometimes arises in the use of terms which describe corresponding parts of man and other animals. To avoid this confusion, anatomists have given these terms arbitrary significance.

Fig. 7. In the anatomical position man's vertebral column extends up at practically a right angle to the floor.

The anatomical position. In describing the body, anatomists always consider it as being in the erect position with the face toward the observer, the arms hanging at the sides, and the palms of the hands turned forward. All references to location of parts assume the body to be in this position.

Textbooks of human anatomy use both *dorsal* and *posterior* for the side containing the backbone, and *ventral* or *anterior* for the opposite side. Comparative anatomists call the head end of an animal or man *anterior,* the opposite end *posterior,* the side containing the backbone *dorsal* and the opposite side *ventral.* The head end is spoken of as *cranial* or *superior,* and the opposite end as *caudal* or *inferior.*[3] A part above another part is described as superior to it. A part below another is said to be inferior.

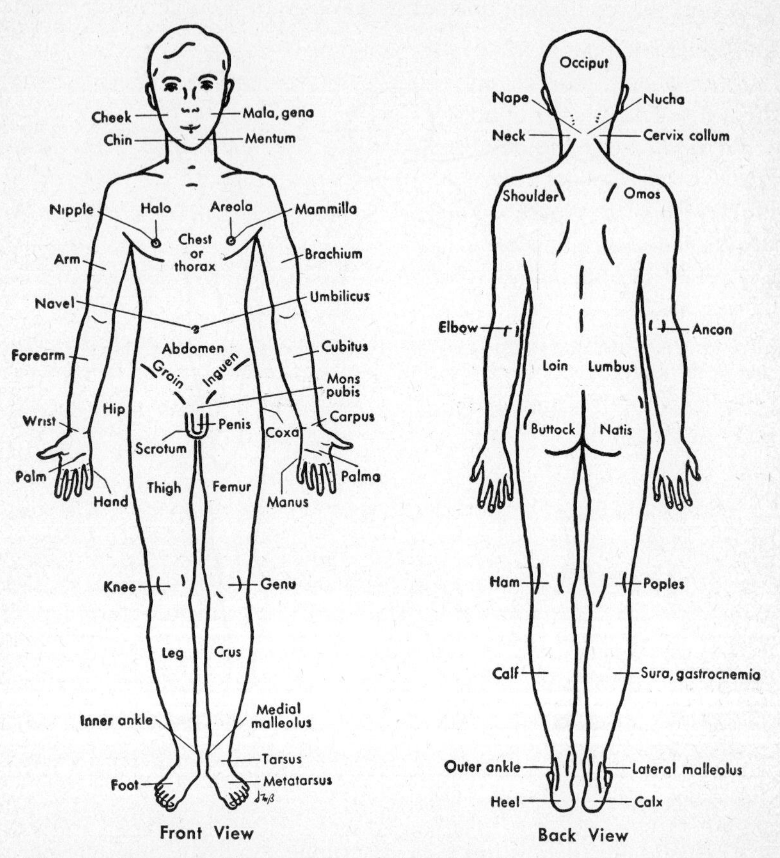

Fig. 8. The anatomical position and regional names. Latin on one side and English on the other side.

General anatomical terms. *Sagittal plane* is the dorsoventral plane dividing the body into right and left sides. It is usually used in the sense of a *midsagittal plane* dividing the body into right and left halves. A *coronal* or *frontal plane* divides the body into ventral

[3] See also *Developmental Anatomy,* L. B. Arey, 1954, p. 9.

and dorsal parts. A *transverse plane* divides the body into cranial and caudal parts. The parts nearest the midsagittal plane are *medial* (mesial); those farthest from this plane are *lateral.* A *horizontal plane* is parallel to the horizon, and the term should be used only in relating the individual to his environment. If the term is used solely in relation to the individual, it indicates a transverse plane, since the individual is supposed to be in the anatomical position.

Internal and *external* are reserved almost entirely for describing the walls of cavities or of hollow viscera.

Proximal is used to describe a position near the origin of any part. *Distal* is used to describe a position distant, or farthest away, from the source of any part. Thus we speak of the proximal end or of the distal end of a finger.

Parietal (Latin, *paries,* a wall) is used to describe the walls enclosing the body cavity or surrounding the organs.

Visceral (Latin, *viscus,* an organ) is applied to the organs within the body cavities.

In anatomy the term *periphery* means the outside or surface of a body or an organ.

SUMMARY

Biological Sciences Deal with Living Things (Science is organized or classified knowledge)

- Morphology, or structure
  - Anatomy or macroscopic structure
  - Histology or microscopic structure
- Physiology, or function
- Development, or growth
  - Morphological
  - Physiological
  - Embryological or early growth and development

Anatomy

- Plant
- Animal
- Human
- Comparative
- Systemic
- Regional
- Functional, or physiological
- Developmental
- Pathological

Histology

- Anatomical
- Functional, or physiological
- Pathological
- Comparative

- **Physiology**
  - Plant
  - Animal
  - Human
  - Comparative
  - General, or cellular
  - Routine
  - Emergency
- **Embryology**
  - Morphological
  - Physiological
- **Hygiene**
  - Definition
  - External, *exterofective*
  - Internal, *enterofective*
- **HUMAN BODY**
  - **Ventral cavity**
    - (1) **Thoracic cavity**
      - Esophagus, trachea, lungs, heart, blood and lymph vessels, thymus gland
    - Right and left **pleural cavities** lined with pleura; each contains a lung
    - The *mediastinum* between cavities contains thymus, trachea, bronchi, lymph nodes and vessels, nerves, esophagus, heart, and large blood vessels
    - The *diaphragm* separates the thoracic and abdominal cavities
    - (2) **Abdominal cavity**
      - Stomach, spleen, pancreas, liver, gallbladder, kidneys, large and small intestines
      - **Pelvic cavity** (Lower portion abdominal cavity)
        - (1) Greater, or false, pelvis
        - (2) Lesser, or true, pelvis: bladder, rectum, some of the reproductive organs
    - **Peritoneal cavity,** lined with peritoneum, contains organs as above, except kidneys, which lie between peritoneum and spinal column—retroperitoneal in position—and organs of true pelvis, which lie, in general, below the peritoneum
  - **Dorsal cavity**
    - (1) **Cranial cavity**—brain
    - (2) **Spinal canal**—spinal cord
  - **Facial aspect of skull**
    - (1) **Orbital cavities**
      - Eyes, optic nerves, muscles of the eyeballs, lacrimal apparatus
    - (2) **Nasal cavity**—structures forming the nose
    - (3) **Buccal cavity**
      - Tongue, teeth

Body Regions	**Anatomical position**	Definition	Superior
		Dorsal	Inferior
		Ventral	Cranial
		Anterior	Caudal
		Posterior	
	Anatomical terms	Sagittal and midsagittal plane	
		Coronal (frontal) plane	
		Transverse plane	
		Medial (mesial, mesal, mesad)	
		Lateral	

CHAPTER 2

THE BODY AS AN ORGANIZED WHOLE
SYSTEMS, ORGANS, TISSUES, CELLS

The cell is the structural and physiological, as well as the developmental, unit of the body. It is desirable, however, to begin the study of the body with an analysis of it into its component parts—the systems, organs, tissues, unit patterns, and cells.

A system is an arrangement of organs closely allied to one another and concerned with the same functions.

The skeletal system consists of the bones of the body and the connective tissues which bind them together. The main functions are support, protection, and motion.

The muscular system consists of the striped muscle trunks, (e.g., biceps muscle) and the unstriped muscles (e.g., the muscle coats of the stomach), and its main function is to cause movement by contracting.

The nervous system consists of the brain, spinal cord, ganglia, nerve fibers, and their sensory and motor terminals, e.g., motor end plates on striped muscle. These are grouped into two integrated systems—the *somatic* and the *visceral systems.* The main functions are to correlate the afferent nerve impulses in the sensory centers and to co-ordinate the nerve impulses in the motor centers, thus acquainting the organism with the environment and integrating the nerve impulses into appropriate or adaptive responses. The nervous system contains centers for sensation, emotion, and thinking.

The vascular, or **circulatory, system** consists of the heart, the blood vessels and blood, and the lymphatic vessels and lymph. The main function is to distribute body fluids steadily to all the cells, maintaining the tissue fluid which bathes them in a state of constancy (homeostasis, stable state) in all respects at all times.

The endocrine system, or system of ductless glands, includes the thyroid gland, parathyroids, thymus, adrenals, pituitary body, pineal body, and portions of the glands with ducts, such as the islands of Langerhans in the pancreas, portions of the ovaries and testes, and portions of the liver. There is much evidence to support the theory that every cell is a ductless gland. The members of this system contribute to the body fluids specific substances which affect the activity of cells, as do all organs, tissues, and cells.

The respiratory system consists of the nose, pharynx, larynx, trachea, bronchi, and lungs. The main functions are to provide oxygen and get rid of excess carbon dioxide.

The digestive system consists of the alimentary canal and the accessory glands, i.e., the salivary glands, the pancreas, and the liver. The main functions are to receive, digest, and absorb food and eliminate some wastes.

The excretory system consists of the urinary organs, i.e., the kidneys, ureters, bladder, urethra, and also the respiratory and digestive systems and the skin. The main function is to eliminate the waste products that result from cell activity.

The reproductive system consists of the testes, seminal ducts, seminal vesicles, penis, urethra, prostate, and bulbourethral glands in the male; the ovaries, uterine tubes, uterus, vagina, and vulva in the female.

All these systems are closely interrelated and dependent on each other. While each forms a unit especially adapted for the performance of some function, that function cannot be performed without the co-operative activity of the other systems; for instance, the skeleton does not support unless assisted by the muscular, nervous, circulatory, and other systems. It is the function of the body fluids and the nervous system to integrate the work of the systems.

An organ is a member of a system and is composed of tissues associated in performing some special function for which it is especially adapted. Systems are made up of *organs* with a corresponding division of labor and special adaptation of the organ to its particular share of the work of the system. For example, the urinary system consists of the following organs: (1) two kidneys, which secrete the urine by filtering under control waste materials from the blood; (2) two ureters, ducts which convey the urine from the kidneys to the bladder; (3) the bladder, a reservoir for the reception of urine; and

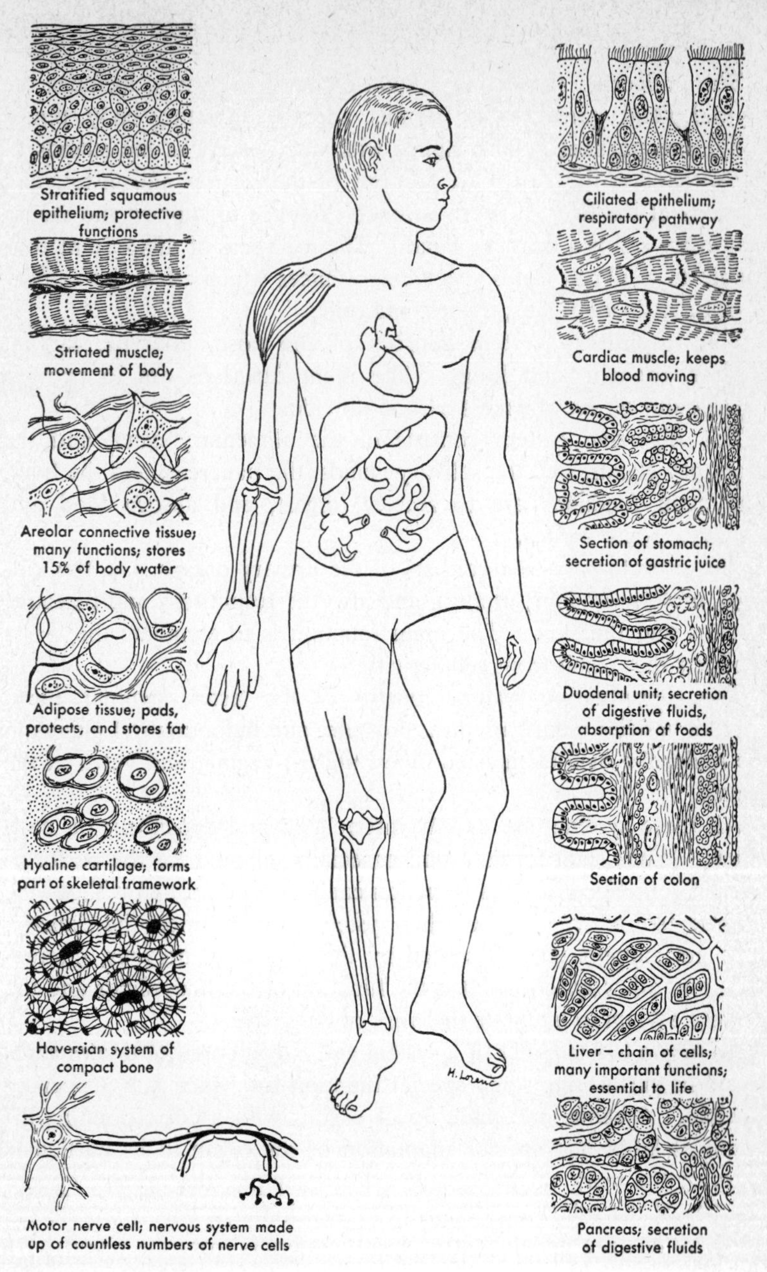

Fig. 9. Diagrams showing some unit patterns and tissue relationships of the body as a whole.

(4) the urethra, a tube through which the urine passes from the bladder and is finally voided. The interdependence of these organs is obvious, and differences in structure suggest differences in function in the work of the system.

Tissues. The organs can be analyzed into component *tissues*. For example, the stomach is composed of columnar epithelial tissue, smooth muscle tissue, connective tissue, and serous tissue.

Microscopic study of tissues reveals the fact that tissues are made up of smaller units, or cells. Each tissue is a group of cells with more or less intercellular material. The intercellular material varies in amount and in composition and in many cases determines the nature of the tissue, as, for instance, in the case of bone.

Unit pattern. Study of the tissues which compose the organs shows that the tissues are arranged in an orderly way. Some of these tissue arrangments stand out clearly on microscopic examination because the *pattern* which they make is unique; hence, they have received names such as *lung* or *pulmonary unit, nephron* of kidney (*renal tubule*), *chain of cells* of the liver, *reflex arc* of the nervous system, *Haversian system* of bone, etc. Sometimes the special arrangement of tissues is not so well known, partly because the units are not so clearly observable under the microscope or are difficult to name, e.g., a villus and a gland of the duodenum and the portion of the wall behind them. This may be called the *duodenal unit*. These units are thought to be both structural and physiological, and it is suggested that each be called a *unit pattern,* or *functional unit*.

A *unit pattern,* or *functional unit,* can be defined approximately as the smallest aggregate of cells which, when repeated many times, composes an organ. It can be said that these unit patterns are simple, minute, and repeated a vast number of times to form the organ. If the liver is studied in this way, it will be seen that the lobules (smallest macroscopic units) are composed of *chains of cells* with their definite supply paths of blood, lymph, and nerve fibers. This arrangement gives an *enormous area,* for the volume of cells and body fluid concerned, over which the cells and circulatory fluids can be brought into diffusion relations and shows an orderly arrangement of blood and lymph tubes and nerve fibers throughout the organ.

These units can be studied as to structure (the shape, size, kinds, and arrangement of cells composing them, their grouping in the organ, orderly supply of blood and lymph vessels and nerve fibers, etc.). Some have been dissected out from the organs of lower organisms and can be worked with singly in experiments. Some have been cultured on nutrient media outside the bodies of animals, and their activities have been observed and studied.

Some of these units are probably better known in terms of their physiology, as, for instance, the reflex arc. Most of these reflex arcs are too long to be traced easily with the microscope, but at the same time they are too fine to be seen with the naked eye. They were known as functional units before they were *seen* as unit patterns. Most reflex arcs are today traced by function rather than by the microscope (e.g., the exact distribution of most of the *individual fibers* of the sciatic nerve has not been worked out objectively).

Each *functional unit* can be studied in terms of its structure (anatomy or histology) and also in terms of its activities (physiology).

These unit patterns, or functional units, share in common certain structures of the organ and system as a whole; for instance, the millions of *chains of cells* of the liver share in the passageway to the outside, their secretion eventually entering the hepatic duct on its way to the duodenum.

The tissues which compose these functional units are in turn composed of cells and intercellular material which the cells have made. This leads to a primary generalization of biology, the *cell theory,* which states that the cell is the unit of structure and the unit of function as well as the unit of development of the organism.

The directional forces bringing about this high degree of structural organization or structural integration are at present the basis of much experimental work. Recent new ways of studying cells and their unit pattern aggregations and the directional forces concerned in structural and functional integration are bringing the cell theory up for keen study once more.[1]

Cells are the physiological and structural units of the body. It is therefore necessary to understand their activities and structure. Low down in the scale of life are simple animals consisting of one cell. These unicellular animals carry on the biological functions which are essential to life. These biological functions are support, moving, breathing, digesting, collecting and distributing, excreting, irritability to environment, and reproducing. The ameba, a typical one-cell animal, can be observed under the microscope carrying on these

[1] *Biological Symposia,* Vol. I, 1940.

functions. The life of each individual cell in the body is dependent on its ability to carry on these biological functions.

Higher in the scale of life are animals that consist of a greater number of cells. The human being may be described as a multicellular animal consisting of an enormous number of cells and inter-

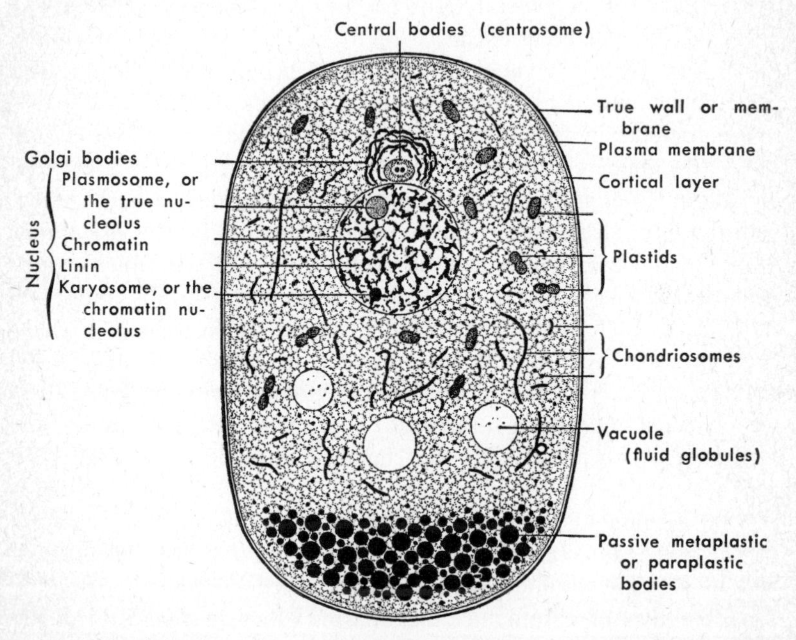

Fig. 10. Diagram of a generalized cell. The chromatin is also called *basichromatin* since it stains with basic dyes; the linin is called *acidichromatin* since it stains with acid dyes. The *Golgi bodies*[2] are usually found near the nucleus. *Plastids* are areas of special chemical activity leading to the production of special substances, e.g., starch grains in plant cells. (E. B. Wilson, *The Cell in Development and Heredity*. Courtesy of The Macmillan Company.)

cellular material which the cells have made. In multicellular animals, individual cells are often remote from air, food, and the excreting organs and must rely upon the circulating fluids to carry oxygen and food to them and waste matters from them. The systems of the body represent an adaptation and specialization of groups of cells to carry on the biological activities for the body as a whole. A comparison of the biological activities of a unicellular animal with

[2] Camillo Golgi, Italian anatomist (1844–1926).

the biological activities of the cells of multicellular animals and the functions performed by the systems of the body shows them to be essentially the same.

Structure of the cell. A cell[3] is a minute portion of living substance edged by a cell membrane, which is thin and delicate.

Protoplasm is a general term for living substance. The protoplasm of the cell body exclusive of the nucleus is called *cytoplasm* or the *cytosome.* It varies in extent and in appearance in different cells, being sometimes homogeneous, sometimes alveolar, and sometimes granular in structure. The ground substance or *hyaloplasm* (Fig. 10) appears structureless under the microscope but is seen to contain minute granules of various sizes (colloidal) when viewed with the ultramicroscope. In the hyaloplasm are granules of various shapes and sizes, which can be seen easily with the microscope. Both the cytoplasm and the nucleus of which protoplasm is composed have been much studied, chemically, by the application of various dyes and also by the techniques applied in micromanipulation[4] or micrurgy. Early studies in the field of cytology were concerned with the chemical morphology (structure) of cells. Recent studies are increasingly concerned with the chemical physiology or activity of cells.

That portion of the protoplasm composing the *nucleus* is called *karyoplasm.* It consists of a fine network of material called *linin,* and in this are granules of a material called *chromatin.* Besides these granules of chromatin other masses called *nucleoli* are found. By studying the behavior of cells or parts of cells deprived of a nucleus and comparing this behavior with that of similar cells containing a nucleus, it has been demonstrated that the nucleus is essential for producing chemical changes on which the nutrition and functioning of the cells depend, for instigating the process by which cells multiply, and for transmitting characteristics from parents to children. In a cell which is about to divide, the chromatin becomes aggregated into bodies called *chromosomes.* The number of chromosomes is constant for the members of a species but varies for different species.

[3] The word *cell* (Latin *cella,* "a cavity") was used by Robert Hooke (1665) to designate the minute cavities separated by solid walls observed in cork.

[4] R. Chambers, "The Micromanipulation of Living Cells" in *The Cell and Protoplasm.* Publication No. 14 of the American Association for the Advancement of Science.

The term *central bodies* is applied to structures, usually double, which are associated with many cell activities, particularly cell division.

In the meshwork of protoplasm various passive bodies are often suspended, such as food granules, pigment bodies, drops of oil, etc.

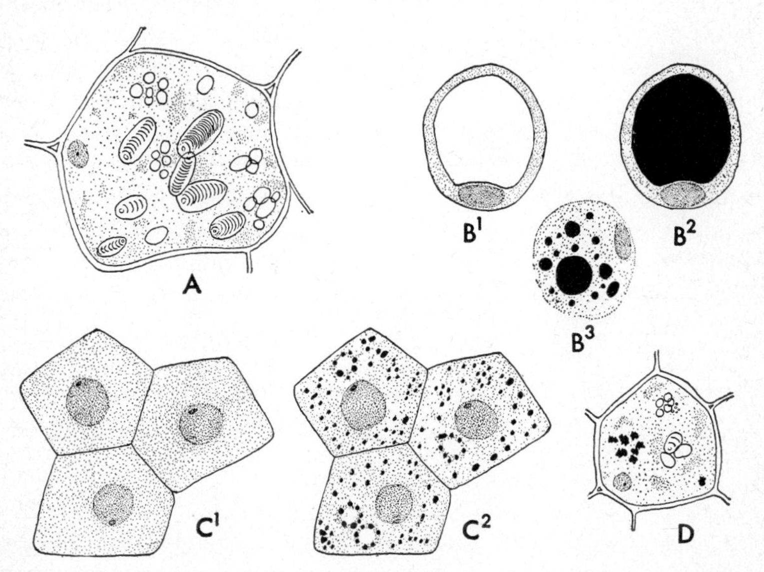

Fig. 11. Diagram of cells showing metaplasm. In *A* and *B* the metaplasm is visible under the microscope. *A* is a potato cell and contains starch grains. The large grains show concentric markings; the small starch grains show no markings. *B* shows adipose cells. In B^1 the white center indicates fat; in B^2 the fat has been made black with osmic acid; B^3 is a cell not yet filled with fat. The droplets coalesce as they increase in number. In *C* and *D* the metaplasm is practically invisible until dyes reveal it. The fine dots indicate protoplasm. C^1 shows liver cells containing glycogen (invisible); C^2 shows the same liver cells after staining with Best's carmine. *D* shows glucose made visible with Benedict's solution. In C^2 and *D* dark spots indicate glycogen.

These may represent reserve food material, or waste matters, and are nonliving substances collectively designated as *metaplastic* or *paraplastic* substances or *metaplasm.*

Physiology of cells. The general term *metabolism* summarizes the activities each living cell must carry on. The changes which involve the building up of living material within the cell have received the general name of *anabolic* changes, or *anabolism;* those which involve the breaking down of such material into other and simpler products are known as *catabolic* changes, or *catabolism;* the sum of all the anabolic and catabolic changes which are proceeding within the cell is spoken of as the *metabolism* of the cell. These chemical changes are always more marked as the activity of the cell is hastened by warmth, electrical or other stimulation, the action of certain drugs, etc.

The activities exhibited by cells are:

1. *Support.* Each cell gains support by the cohesiveness of its own structure and by its cell membrane.

2. *Respiration.* Each cell coming in contact with oxygen absorbs it. During this absorption some of the cell contents are oxidized,

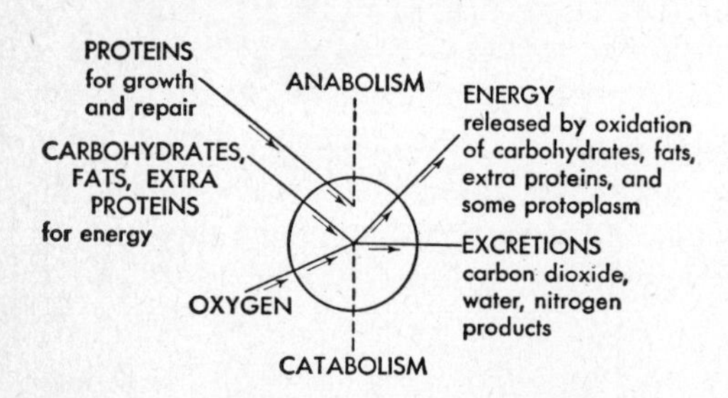

Fig. 12. Diagram showing metabolic processes. The cell is shown as a circle.

and as a result of this oxidation, energy is liberated, and carbon dioxide, a waste product, is formed and given off. This is known as cellular respiration. The purpose of respiration by the body is to convey oxygen from the environment to each individual cell and to remove from the cell the excess carbon dioxide and convey it to the environment. This is accomplished by the respiratory movements and by the circulating body fluids.

3. *Use of foods.* Each cell can absorb and convert into its own protoplasm certain materials (foods) that are nonliving; in this way the protoplasm may undergo repair, and the cell may grow. Foods may also be oxidized to yield energy and to regulate body functions.

4. *Motion.* This is exhibited in two forms and is due to the ability of protoplasm to contract.

a) AMEBOID MOVEMENT consists in the pushing outward by the cell of protoplasmic processes, called *pseudopodia.* These pseudopodia may be slowly retracted, they may be bent, or the contents of the cell may flow slowly or rapidly into them and change both the shape and position of the cell. By a repetition of this process the cell may move slowly about, so that an actual locomotion takes place, e.g., ameboid or pseudopodial movement of white blood cells. (See Fig. 14*A*, A^1, A^2.)

b) CILIARY MOVEMENT is the whipping motion exhibited by short, microscopic, filamentous processes called *cilia,* which project from the surface of some epithelial cells.

5. *Circulation.* This consists of a "streaming"[5] of the protoplasm within the cell. By this means nutritive material and oxygen may be distributed gradually to all parts of the protoplasm, and the waste substances are gradually brought to the surface of the cell for elimination.

6. *Excretion.* Each cell is able to discharge waste substances.

7. *Irritability* is that property which enables a cell to respond to stimuli (change in its environment). If the stimulus received by a cell is strong enough, it is conducted throughout the protoplasm of the cell, and the cell responds. The response may take the form of an increase of some kind of activity, such as motion, growth, etc., or it may take the form of a decrease of these activities. If the response is an increased activity, the protoplasm is said to be *excited;* if the response is a decreased activity, the protoplasm is said to be *inhibited.*

8. *Cell division.* As do all living organisms, each cell grows and produces other cells. As the cells of the body are constantly wearing out and leaving the body in the excretions, the need for constant reproduction of cells is apparent. Cells usually divide by indirect cell division, called also *karyokinesis* or *mitosis.* In mitosis the nucleus passes through a series of changes, illustrated in Fig. 13. Chromatin is that part of the nucleus which becomes colored when certain basic dyes are applied to the cell. It will be noted that the chromatin, which at first (*A*) exists as granules in the nucleus, becomes in (*C*) definite bodies, the chromosomes. These chromosomes divide lengthwise (*E*), and the half-chromosomes thus formed move to the poles of the spindles (*F*) and (*G*), so that when the cell divides each daughter cell has a longitudinal half of each chromo-

[5] In plant cells and protozoa, where this streaming is sometimes conspicuous, it is often called *cyclosis.*

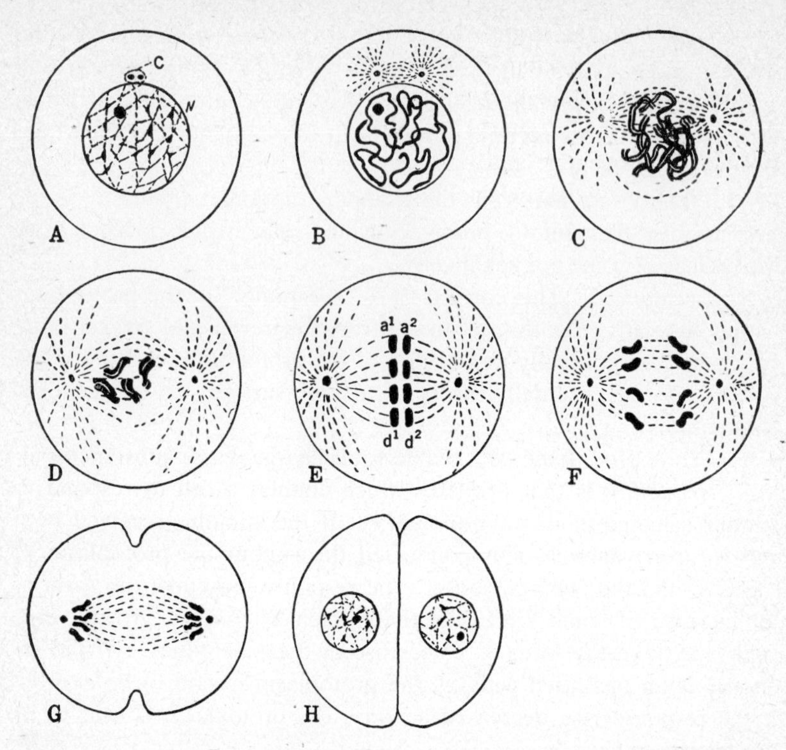

Fig. 13. Diagrams to show mitosis. The cell is presumed to have four chromosomes. *A,* resting cell with nucleus (*n*) and centrosome (*c*); a nucleolus and network of chromatin are shown in the nucleus. *B,* spindle fibers (*s*) forming; a chromatin thread, or *spireme,* is breaking into chromosomes. *C,* nuclear membrane disappearing, *chromosomes* shown dividing *lengthwise* into halves. *D,* chromosomes shorter and thicker, staining power increased. *E,* chromosomes arranged on the *spindle,* the two halves of each chromosome opposite each other as at a^1 and a^2, d^1 and d^2, etc. *F,* chromosomes moving toward the poles of the spindle. *G,* cell beginning to divide. *H,* cell division complete. *A–C* called *prophase; C–D* called *metaphase; E–G* called *anaphase; G–H* called *telophase.* (Modified from *The Cell in Development and Heredity,* by E. B. Wilson, The Macmillan Company.)

some. The chromosomes in the daughter cells (*H*) are changed into granules, so that the daughter cells are like the parent cell except in size. Chromosomes contain the *genes,* which are considered to be determiners for inherited traits. The determining effects of the genes probably depend on their chemical composition, the chemical composition of the chromosomal complex, and the relative location of genes in this complex. Thus, traits said to be inherited according to the Mendelian[6] laws are related to the chromosomes.

Some cells are thought to divide directly, or by *amitosis.* In direct division the cell elongates, the nucleus and cytoplasm become constricted in the center, and the cell divides, forming two cells which grow to the size of the original cell.

SUMMARY

System. An arrangement of organs, closely allied and concerned with the same function

Systems found in the human body:

Skeletal	Endocrine
Muscular	Respiratory
Nervous	Digestive
Vascular, or circulatory	Excretory
Reproductive	

Organ. A physiological unit composed of two or more tissues associated in performing some special function

Tissue. A group of cells with more or less intercellular material

Functional Unit, or Unit Pattern	Definition	Smallest aggregate of cells (with intercellular material) which when repeated many times composes an organ
	Names	Lung or pulmonary unit Nephron Chain or cord of cells Haversian system Reflex arc etc.
	No names for others	Example . . . villus and gland of duodenum with portion of wall behind them (duodenal unit) etc.
	Distribution of blood and lymph tubes and nerve fibers	In orderly way following unit pattern, giving great diffusion area between cells and body fluids for volumes concerned
	Anatomy and physiology of functional unit	They can be studied as to structure—anatomy They can be studied as to their activities—physiology

[6] Gregor Johann Mendel (1822–1884), an Austrian monk, was the first to demonstrate experimentally the laws of inheritance, now called Mendelian laws.

The Cell Theory
- Since intercellular material is thought to be made by the cells, a fundamental generalization in biology is that the cell is the unit of structure, of function, and of development of the body. The name *unit pattern*, or *functional unit*, is used for the smallest aggregate of cells characteristic of any organ

Cell. The structural and physiological unit of the body

Ameba. An example of a unicellular animal which lives because it is able to carry on the biological activities of support, breathing, digesting, collecting and distributing, moving, excreting, irritability, and reproducing

Comparable Activities
- Biological activities of ameba
- Functions of the systems in human body
- Life activities of cells

Cell Anatomy
- Protoplasm
  - Cytoplasm
    - Central bodies
    - Golgi bodies
    - Plastids
    - Chondriosomes
    - Vacuoles
  - Nucleus
    - Plasmosome, or true nucleolus
    - Chromatin (basichromatin)
    - Linin (acidichromatin)
    - Karyosome, or chromatin nucleolus
- Metaplasm
  - Foods
  - Wastes
  - Other storage products

Cell Physiology
- **Metabolism**—life activities of cells
  - Anabolism—building-up process
  - Catabolism—breaking-down process
- (1) Support
- (2) Respiration
  - Provides oxygen for oxidation
  - Liberates heat
  - Removes excess carbon dioxide
- (3) Use of foods
  - Materials for protoplasm manufacture
  - Foods may also be oxidized to yield energy and regulate body processes
- (4) Motion
  - *a*) Ameboid movement
  - *b*) Ciliary movement
- (5) Circulation—streaming of the protoplasm in the cell
- (6) Excretion—discharge of waste substances
- (7) Irritability—ability to respond to stimuli and to *conduct* the stimuli throughout the cell
- (8) Cell division
  - *a*) Indirect division, mitosis, or karyokinesis
  - *b*) Direct division, or amitosis

CHAPTER 3

PHYSIOLOGY OF THE CELL

The physiological activities of cells can be grouped into (1) the relationships between cells and the surrounding tissue fluid and (2) the relationships between the parts of cells.

1. *Relationships between cells and tissue fluid* include (*a*) the transfer of supplies, mainly by diffusion, from blood circulating in capillaries to tissue fluid and then into the cells in which the supplies are used and (*b*) the transfer of wastes, mainly by diffusion, from the cells to the tissue fluid and then to blood, circulating in capillaries, in which the wastes are carried away. The relationships between the cells and their environment are concerned with the selective permeability of the cell membranes, their areas in relation to the volumes of cell substance, and the relative concentrations of particles of diffusible substances within and outside the cells. Hence, the shape, size, and arrangement of the cells and tissue fluid are significant.

2. *Relationships between the parts of cells* include cell organization for the use of these supplies by the cells for needed energy, storage, repair, and growth with the formation of by-products or wastes which must be eliminated.

It is obvious that these two groups of activities are very closely related. The building of supplies into products (e.g., glucose into glycogen) keeping their concentrations (e.g., glucose) relatively low in cells thereby controls to some extent the rate of diffusion into the cell. The formation of wastes (e.g., CO_2) keeps the concentration of these wastes relatively high in the cell as compared to tissue fluid and thereby to some extent controls the rate of diffusion out of the cell. These two processes are mutually dependent. They are therefore discussed somewhat together in this chapter.

Differences in cells. Cells differ in form, size, appearance, chemical composition, and activities.

It is assumed that the marked difference in the appearance, etc., of cells is an expression of a chemical difference, which in turn is an evidence of difference in function or activity.

Each cell in a multicellular animal has a dual function: (*a*) to

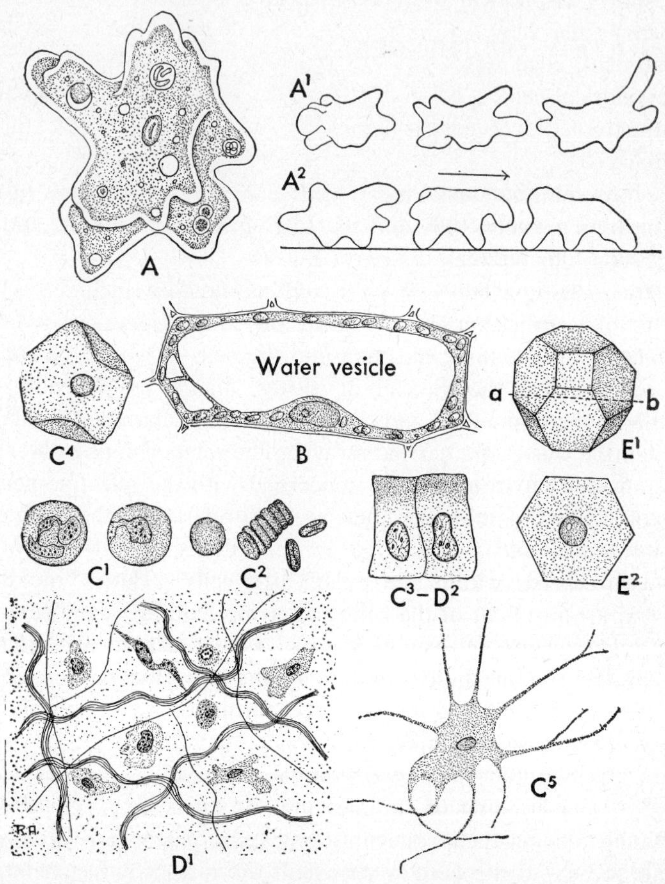

Fig. 14. Diagrams of cells of many shapes as they would appear under the microscope. Description in text. The cells are not drawn to any scale.

carry on the activities on which its own life depends, and (*b*) by specialization in some activities to assist in the special work of the tissue of which it forms a part.

Form or shape of cells. Figure 14 shows cells of various shapes. *A* is a "shapeless" cell—the ameba. At the right, outlines of an ameba in pseudopodial motion are shown, in A^1 from above and in A^2 from the side; the straight line represents the *edge* of a microscope slide. *B* shows a typical plant cell with a cell wall of cellulose giving it fixed shape, with somewhat angled corners.

C^1 shows a spherical white blood cell, at left seen in face view, at right in profile view. C^2 shows at left a face view of a red blood cell; at right, edge views of a few red cells. C^3 shows side views of two cylindrical cells. C^4 shows a surface view of a flat (squamous) cell, the folded-over edge showing its thickness. C^5 shows a very irregular cell.

D^1 shows an abundance of intercellular material between cells (as seen in areolar connective tissue); D^2 shows two cells with very little intercellular material between the cells (represented by the black line).

E^1 shows the shape which soft living cells (e.g., fat cells) assume when lightly pressed together (as in adipose connective tissue); E^2 shows how the microscope would show this cell if focused at plane *a . . . b*.

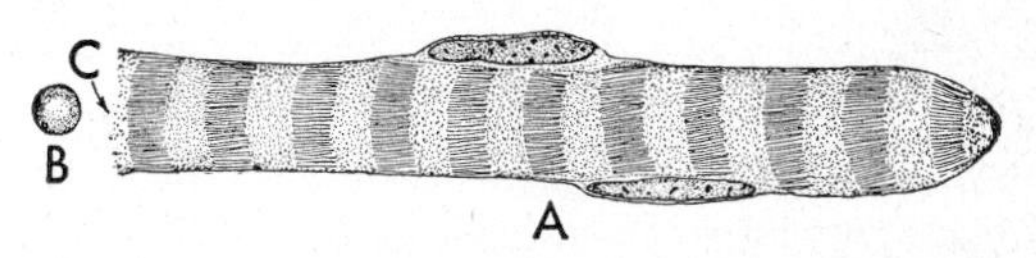

Fig. 15. Size of cells. *A*, voluntary muscle cell magnified in width 200 times and represented as cut off at *C*. At this magnification it would be about 200 in. long. *B*, red blood cell, also magnified about 200 times.

As a generalization it can be said that for a *given volume* spherical cells have least surface area, and that the more irregular the cell, the greater is its surface area. For example, irregular connective-tissue cells and nerve cells expose an enormous surface area to body fluids (for a given volume of cell contents) as compared with spherical white blood cells, cylindrical or columnar cells, or the disk-shaped red blood cells, etc.

The surface area–volume ratio in cells of various shapes is conspicuous as it concerns nerve cells, which contain relatively little metaplasm (e.g., stored food) and may have very long, exceedingly fine processes (fibers); this means an enormous area in relation to the volume of protoplasm concerned. This surface area is in contact with circulating lymph in the perineural lymph spaces, and hence the protoplasm of the cell has abundant opportunity to get supplies (e.g., food) in relation to its need for supplies. Or consider a *piece* of nerve fiber 25 mm long and less than 0.01 mm in diameter in terms of surface area–volume ratio! This is one of the many items in favor of the membrane theory of the nerve impulse (p. 236).

Size of cells. The factors which determine the sizes of cells are unknown, though much experimental work has been done in this field, and several theories have been proposed. However, cells, with few exceptions, are minute. The average diameter of a red blood cell is about 0.0075 mm, and it is about one-fourth as thick. A striped muscle cell may be an inch or more long, but the diameter is seldom over 0.05 mm. The total length of the peripheral and central processes of a spinal sensory nerve fiber may reach from the toe to the medulla. In a tall person, then, it may be more than 4 or 5 ft long, but the diameter of the nerve fiber would probably be less than 0.01 mm. If all the processes of this nerve cell, divested of their sheaths, were wrapped closely around the cell body, the total volume of the cell would still be minute. Figure 16 shows that for a given volume of protoplasm, the greater the number of cells composing it, the greater will be the surface area.[1] As a generalization it can be said that the protoplasm of the body is so partitioned in minute cells as to expose to the tissue fluid around the cells a truly enormous surface area for the volume of protoplasm itself.

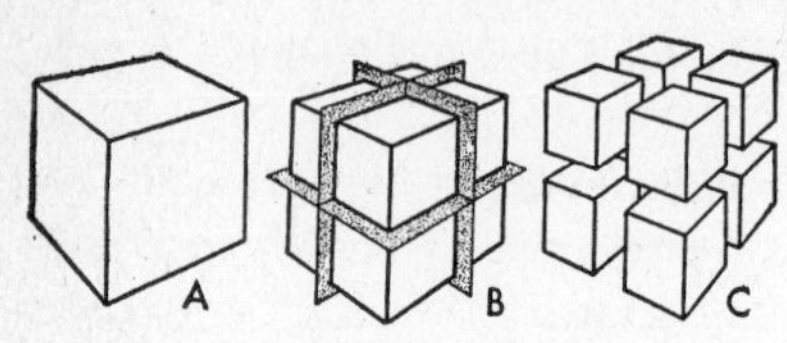

Fig. 16. Diagrams to illustrate increased proportion of total surface area to total volume on fragmenting an object. *A* is a cube 2 in. on a side. *B* shows planes in which it may be cut to produce the eight cubes shown in *C*. Each of these eight cubes is 1 in. on a side. The total area of *A* is 24 sq in. The total area of the eight cubes in *C* is 48 sq in. The volume of material in *A* and *C* is the same, 8 cu in.

The ratio of nuclear volume to cytoplasmic volume is in many cells somewhat the same, the volume of the nucleus being from about $\frac{1}{25}$ to $\frac{1}{50}$ the volume of the cytoplasm.[2] In young cells with a high metabolic rate, the nucleus is relatively large. In position the nucleus is usually near the most active (chemically) part of the cell. These are indications of the fact that "the activities of the cell can only be carried on within certain small limits of size." Since the

[1] *Specific surface* (ratio of surface area to volume) is high.

[2] C. A. Kofoid, "Cell and Organism" in *The Cell and Protoplasm*. Publication No. 14 of the American Association for the Advancement of Science.

surface area of the nucleus "increases only in the ratio of the square while the cell volume of cytoplasm and the nucleus increase in the ratio of the cube," the area of contact between nucleus and cytoplasm does not keep pace with their growth in volume, and hence the "efficiency of the nucleus" in relation to cell activities decreases

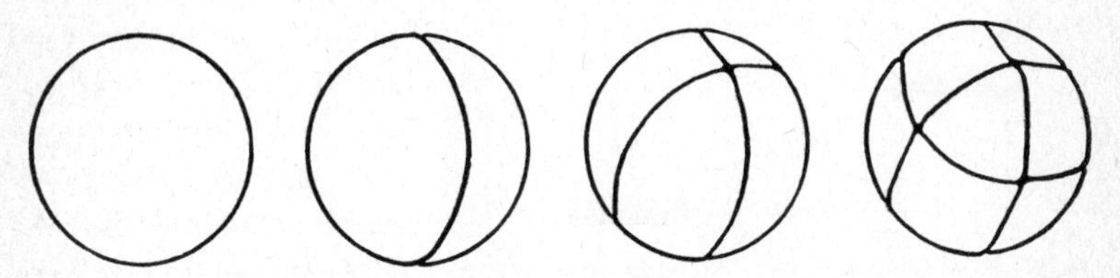

Fig. 17. Animal cells* are more nearly spheres than cubes. A group of eight small cells having the same volume as one large cell will expose to body fluids very much more diffusion area than the large cell.

*$4\pi R^2$ = area of a sphere $\frac{4}{3}\pi R^3$ = volume of a sphere

markedly. This may be one of the important factors in limiting the sizes of cells.

The constituents of protoplasm. Chemical analysis of the human body has shown that it contains the following *chemical elements:*

Element	Symbol		Percent
Oxygen	(O)	Form 96% of total weight of body	65.0%
Carbon	(C)		18.0
Hydrogen	(H)		10.0
Nitrogen	(N)		3.0
Sulfur	(S)		0.25
Calcium	(Ca)		2.2
Phosphorus	(P)		0.8–1.2
Potassium	(K)		0.35
Chlorine	(Cl)		0.15
Sodium	(Na)		0.15
Magnesium	(Mg)		0.05
Iron	(Fe)		0.004
Iodine	(I)		0.00004
Silicon	(Si)	Very minute amounts	
Fluorine	(F)		

Perhaps *traces* of Cu, Mn, Zn, Co, Ni, Ba, Li, etc., spoken of as trace elements.

In the human body free oxygen, hydrogen, and nitrogen have been found in the blood and intestines, but the bulk of these elements, as well as all of the others, exists in the form of complex compounds which are constituents of the cells and body fluids. The compounds are divided into two classes, organic and inorganic.

The *organic compounds* found in protoplasm are proteins, carbohydrates, and fats. Proteins, wherever found in living organisms, possess the same chemical elements and are therefore fundamentally similar. This similarity is also found in carbohydrates and fats. In addition to this universal similarity, the proteins, carbohydrates, and fats of each species of plant or animal possess distinctive characteristics. The essential chemical elements are present in varying absolute quantities, in varying relative quantities, and in varying combinations. The essential elements in proteins are carbon, hydrogen, oxygen, and nitrogen, and in some also sulfur and phosphorus; in carbohydrates and fats the essential elements are carbon, hydrogen, and oxygen. The body ingests these proteins, carbohydrates, and fats. They are split up into simpler compounds (i.e., carbohydrates into simple sugars, fats into glycerin and fatty acids, proteins into amino acids), which are carried by the blood and lymph to the cells, where they are converted into carbohydrates, fats, and proteins having the distinctive characteristics of those found in the human body. These may then be stored in the cells, or utilized for making protoplasm or for liberating energy, as indicated in Fig. 12. A more complete discussion will be found in Chaps. 18 and 19.

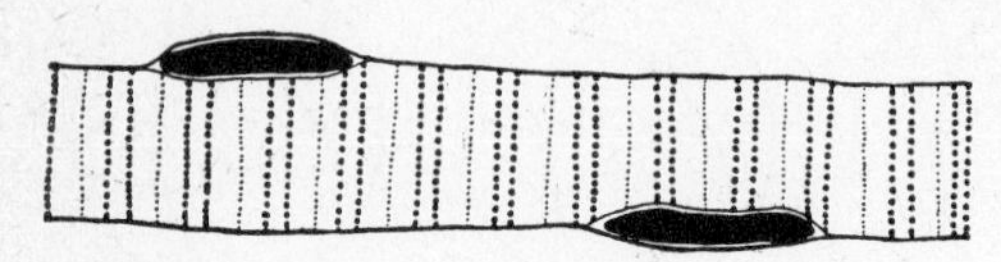

Fig. 18. Diagram of a striated muscle cell which has been incinerated, as seen under the microscope, highly magnified. The black dotted lines indicate salt deposits. The water and organic material have been burned off.

The *inorganic compounds* exist in cells partly as dissolved salts and partly in combination with the organic compounds. In chemical analyses of cells, the *mineral elements* remain either wholly or largely in the ash when the cells are incinerated; hence, they are grouped as the *ash constituents.* Only small amounts of these elements are needed, but they are essential parts of cells.

Water is the most abundant constituent of cells and constitutes about two-thirds of body weight and more than 80 per cent of nonbony body weight.

It is difficult to obtain the normal water content of an isolated living cell. One estimate gives it at from 85 to 92 per cent of the weight of the cell. It is evident that water is by far the predominant constituent of protoplasm. As will be seen later, this is true also of the body fluids—blood, lymph, and tissue fluid—which differ from protoplasm in many respects to only a slight degree.

Following are some of the characteristics of water that make it significant as the most abundant constituent of all body cells:

1. The solvent power of water is great. Nothing can compare with water in relation to the kinds of substances which can be dissolved in it and the great and varied concentrations that can be obtained.

2. The ionizing power of water is high; hence, the great number and many kinds of solute molecules yield large numbers of varied ions.

3. Water has a high heat capacity (high specific heat). This means that it can hold more heat with less change of temperature than most substances; hence, the heat produced by cell metabolism makes comparatively little change in the temperature of the cell.

4. The heat-conducting power of water is high (for liquids). This means that the heat produced in the cells can pass to body fluids even if the temperature of the cell is barely above that of the "lymph around the cell" (p. 44), which likewise can hold this heat with comparatively little rise in temperature and pass it on with little change in temperature to the blood and finally to the skin.

5. The latent heat of evaporation of water is high. Owing to the high latent heat of evaporation, practically a maximum of heat is taken from the skin for the evaporation of perspiration—about 0.5 large calorie per gram of water evaporated.

6. Another significant characteristic of water as a constituent of the cell is its high surface tension. Because of this any immiscible liquid with which it comes in contact must expose to it the minimum of surface; but since substances which dissolve in water lower its surface tension, since dilute concentrations especially have a tendency to lower surface tensions, and because of the great solvent capacity of water, the surface tension of water can be lowered greatly and by a great variety of substances. This permits the area of contact of the immiscible liquids to be enormously increased. Also it is known that any dissolved substances which do lower surface tension will accumulate or be *adsorbed* at the surface of contact of the immiscible

liquids. Protoplasm is often described as an emulsion[3] of colloidal particles of proteins, etc. (e.g., lipins) in water or the reverse. These colloidal particles are the smallest particles that can take other substances into solution; hence, as a generalization it may be said that for the volume of material concerned there is a maximum interfacial area of material that can take other substances in solution, between the phases of the emulsion composing protoplasm, and therefore a maximum area over which solutes can be concentrated.

Sometimes in experimental work cells behave as *sols*, that is, as emulsions in which the continuous phase is water, the colloidal

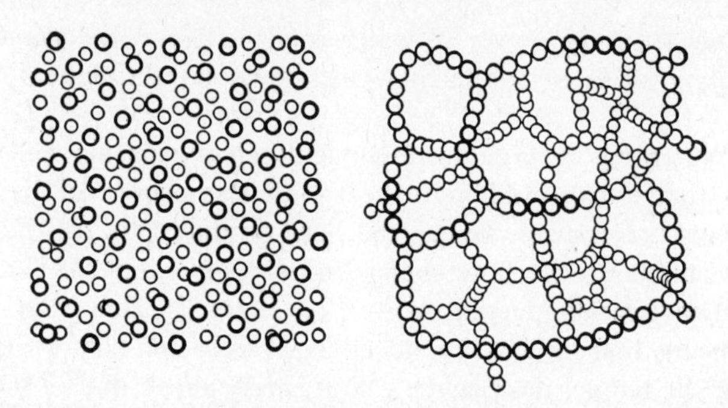

Fig. 19. Colloidal particles of the sol state (*left*) are separated from one another like islands in a lake. Such a sol is therefore a continuous liquid and flows easily. When the particles stick together in interlacing strands in the gel state (*right*), the mass is more comparable to lakes separated by strips of land. Such a gel, then, is a spongy solid and, though soft, holds its shape. (Courtesy of Dr. Ralph W. Gerard and Harper & Brothers.)

particles of proteins and lipoids being dispersed in it. Sometimes they behave as *gels*, in which the protein molecules form networks enclosing areas of water between them. It has been said of protoplasm that its characteristics are like those of a reversible sol-gel colloidal system. "Hence to speak of protoplasm as a liquid or as a solid has little meaning."[4] (Fig. 19.)

It can be seen that these characteristics of water lend great stability

[3] An emulsion is a mixture of immiscible liquids. A colloidal particle is a particle whose diameter is between 0.1 μ and 1 mu. An emulsion whose dispersed particles are 0.1 μ to 1 mu in diameter would be a colloidal emulsion.

[4] R. Chambers, "The Micromanipulation of Living Cells" in *The Cell and Protoplasm*. Publication No. 14 of the American Association for the Advancement of Science.

to the cell but at the same time allow many diverse chemical changes to take place, varied products being made and unmade with minute changes in kinds and concentration of ions, temperature, etc. These characteristics of water form the basis of a chemical equilibrium always tending to completeness but never quite reaching it, because from the "lymph bathing the cell" (tissue fluid) small quantities of these varied substances in dilute solution are always entering the cell, and small quantities of varied cellular products in dilute solution are always leaving the cell, the whole bringing about the smooth, slowly acting dynamic equilibrium characteristic of the physiology of cells. The cell is to be regarded as a highly organized or integrated unit engaged in ceaseless chemical activities. These activities are dependent on the continuous reception of substances from the so-called *internal environment* (tissue fluid) and the continuous elimination of substances to this tissue fluid. The circulatory liquids continually bring substances from the supply organs which obtain them from the *external environment* and continuously take eliminated substances to the eliminating organs for final elimination to the external environment.

These activities of the cell eventually release energy for the use of the cell in its constructive activities such as food storage, repair, growth, etc., this energy coming from the substances diffusing from the tissue fluid. It is conspicuous that destruction of an organized cell (by mechanical crushing, etc.) interferes with its ability to carry on these constructive chemical activities. Hopkins has said this organized and integrated protoplasmic cell "interposes a barrier and dams up a reservoir" of energy "which provides potential for its own remarkable activities."[5]

The substances entering and leaving the cell do so by the process spoken of as *diffusion*. This process may be described as follows.

Diffusion. This term is applied to the spreading or scattering of molecules of gases or liquids.[6] When two gases are brought into contact, the continual movement of the molecules of gas will soon produce a uniform mixture. If a solution of salt is placed in a receptacle and a layer of water poured over it, there will be a mingling of salt molecules and water molecules, producing a solution of

[5] Sir Frederick Gowland Hopkins, "Some Chemical Aspects of Life," *Science*, 1933.

[6] Diffusion takes place slowly also in solids.

uniform composition. Such a mingling would also occur if two solutions of different salts were brought into similar relations with each other.

Osmosis. The usual definition of osmosis is the diffusion of solvent particles, such as water molecules, through a membrane. If a saline solution and water are separated by a membrane permeable to water, the water molecules will pass through the membrane to the salt solution, thereby raising the level of the latter. Theoretically, molecules of liquid are constantly in motion, a permeable membrane offering no resistance to their passage, and therefore the movement

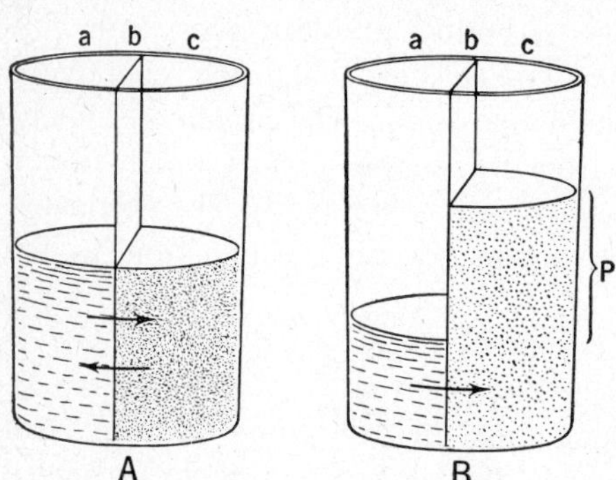

Fig. 20. Diagram to illustrate osmosis. *a*, chamber containing water; *c*, chamber containing a solution of salt in water; *b*, membranous partition between the two chambers. In *A*, *b* is permeable to all substances; in *B*, the membrane *b* is semipermeable—impermeable to salt particles but permeable to water particles. *P*, difference of level of the liquids when the chambers *a* and *c* are separated by a semipermeable membrane, measures the osmotic pressure, the pressure that prevents the rest of the water in *a* from entering *c*.

of water particles is in both directions; but the water molecules will travel in greater numbers per unit of time from the place where their concentration is highest to where it is lowest. If the concentration of the water particles on both sides of the membrane is the same, in a given time equal numbers of particles will travel in each direction, and equilibrium will be reached. If the concentrations of water molecules are kept constant by the addition of water molecules on one side and the removal of water molecules on the other, equilibrium will be prevented, and the movement of the water molecules from a constant source to a constant place of disappearance will be continuous.

In physiology the osmotic characteristics of different solutions are often determined by the way in which they affect the red cells of the blood. In other words, their effect is compared with that of the blood serum. If red cells are subjected to contact with any fluid other than normal serum, they may remain unchanged or they may shrink or swell. If they remain unchanged, the solution is said to have the same osmotic characteristics as the blood serum and is called *isotonic* or *isosmotic*. If they shrink, the solution has higher osmotic characteristics than that of the blood serum and is called *hypertonic* or *hyperosmotic*. If they swell, the solution has lower osmotic characteristics than that of the blood serum and is called *hypotonic* or *hyposmotic*.

Sometimes the word *dialysis* is used for the diffusion of molecules of the soluble constituents (solutes) through a permeable membrane. If two saline solutions of unequal concentration are separated by a membrane permeable to salt, a greater number of salt molecules will pass from the more concentrated solution over to the less concentrated per unit of time. Not only salt molecules but particles (ions, atoms, atom groups, molecules, small molecular aggregates) of many substances in solution may pass to and fro through membranes, so that two liquids separated by a permeable membrane and originally unlike in composition may, by the action of diffusion, come to have the same composition. In the body, diffusion of soluble constituents from the points of greater to those of less concentration and from constant source to constant place of disappearance is continuous. A cell membrane may be called a *diffusion membrane*. The cell or plasma membrane is delicate in structure and is thought to be composed of lipoprotein molecules with tiny channels of water molecules between them. These channels form ultramicroscopic pores through which the smallest molecules, such as water and other molecules, can pass readily.

Figure 21 gives a good idea of a cell membrane as concerns its relation to diffusion. The diffusing particles move through the membrane in solution in the water between the molecules of protein or lipoid material; or they may possibly move through the membrane in solution in the protein or in the lipoid material, depending upon their ability to go into solution in these substances.

Particles of substances which have a higher concentration outside the membrane than inside the membrane will diffuse in greater numbers per unit of time *from outside in* (if they can move through the membrane). In general these will be supplies for the cell.

Particles of substances which have a higher concentration inside the cell than outside will diffuse in greater numbers per unit of time *from inside out* (if they can move through the membrane). In general these will be cell wastes.

It is possible that there are at least two items in the selectivity of cells for the substances that enter them from the body fluids. First, the permeability of the membrane itself: in general only small particles can diffuse through the membrane.[7] Many of the cases where the reverse seems to be true show that there are other factors, such as chemical interaction with the membrane itself, etc., to be considered. Second, as brought out by Hopkins, selection of materials

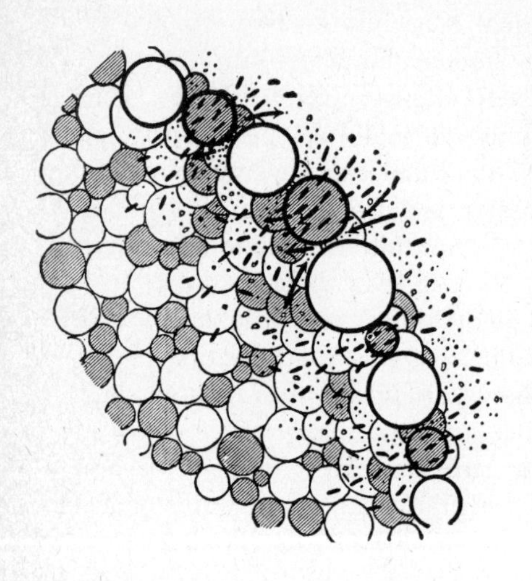

Fig. 21. Diagram of a cell membrane, cut in the plane of the paper. Lipoid particles are shaded, protein ones not. Continuous water channels, through the "pores" between particles, permit small molecules to diffuse between the outside and inside of the cell. Larger molecules which dissolve only in water are blocked by the pores, but those large ones which are soluble in fats can dissolve into the lipoid particles of the membrane, mix with the molecules which compose the particle, and finally emerge again on the inside as well as on the outside of the membrane. (Courtesy of Dr. Ralph W. Gerard and Harper & Brothers.)

in tissue fluid which enter the cells is primarily related to the permeability of the cell membrane but of *"deeper significance in that selection"* is the system of specific enzymes (catalysts) in the cell.

Substances which diffuse into the cell and rapidly undergo chemical changes, thereby *disappearing as such* in the cell, will permit a great number of particles to diffuse in before equilibrium with the tissue fluid outside is attained. This cell membrane has apparently selected certain substances; but it is the *catalysts* (enzymes) and organization within the cell which have prevented a rise in concentration of the diffusible substance and therefore caused the apparent selectivity.

[7] Reports of experiments by J. V. Osterhout, *Science,* Dec. 2, 1938, May 12, 1939.

It is the cell and not the membrane alone which has *selected* certain substances.

The study of the physical and chemical activities going on in cells is difficult. Although certain constituents of cells (e.g., enzymes) have been isolated and studied, protoplasm in all its physical and chemical characters has never been duplicated in the laboratory, and crushed cells do not behave physically or chemically as do living cells. This makes it seem as if *cell organization* itself is one of the chief items to be considered in cell physiology. There is continuously going on a complicated series of successive and interlocking changes. Attention is called to some of these in subsequent chapters, in relation to metabolism, respiration, etc. A series of probable chemical events concerned with contraction of a muscle shows how, in parallel series of chemical changes, energy is obtained by some of these changes to be used for others. (See p. 142.)

Cell studies today are devoted to the chemistry that takes place within the cell rather than to its structure or anatomy. The rapid development of chemistry as a science, the number of research workers concerned, the array of ultramicroscopic methods of measuring with an amazingly high degree of accuracy the very minute components of cells, indicate that knowledge of cell activities will develop rapidly.

Studies of enzyme activity, vitamin activity, hormone activity, gene activity, and their relations to each other, the relations of carbohydrate metabolism to fat and protein metabolism, the use of oxygen by cells, the sol-gel state, and surface tension phenomena, etc., are gathering results that are dependable in interpreting the physiology of cells.[8]

THE CELLS AND TISSUE FLUID

All cells lie in a liquid environment called tissue fluid. This fluid serves as the only medium of exchange between blood plasma and the cells. Substances needed by cells for maintenance, growth, and repair diffuse from the plasma to the tissue fluid and on into the

[8] David E. Green, *Currents in Biological Research.*
"Reconstruction of Chemical Events in Living Cells" by David E. Green in *Perspectives in Biochemistry,* edited by Joseph Needham and David E. Green, Cambridge University Press.

cell. The products of cell metabolism or other cell activity diffuse into the tissue fluid and enter either the blood or lymph capillaries.

Sources of tissue fluid. The walls of the capillaries are thin, and some of the fluid passes out into the spaces between the tissue cells. *Tissue fluid* is derived from the plasma of the blood mainly by diffusion and filtration. There is sometimes assumed an active secretory process on the part of the endothelial cells of the capillaries.

Points of view differ in regard to classifying the liquids concerned in the exchange of material between the blood and the tissue cells. Anatomists in general claim that the lymph vessels form a closed system and suggest that the name *lymph* be applied to the fluid within the vessels only and that the fluid outside of the vessels, in the tissue spaces, be called *tissue fluid.* Usually the term *lymph* is applied to the tissue fluid in general, i.e., to the fluids in the tissue spaces and to the excess drained off in the lymph tubes. In this discussion the name *lymph* is restricted to include only the fluids found in the lymph vessels, in the different serous spaces of the body (e.g., the pericardial, pleural, and peritoneal cavities), and in the spaces of the cerebrum, spinal cord, eyes, ears, and joints. Lacteals are lymph tubes in the small intestine. During digestion, they are filled with *chyle,* a milk-white fluid composed mainly of emulsified fat.

The name *tissue fluid* covers all fluids *not* in the blood vascular system, the lymph vascular system, the great spaces of the body, or the cells themselves. The body fluids may be grouped as follows:

Body Fluids	(1) Cellular fluid	Protoplasm is sometimes thought of as a *sol* (liquid), sometimes as a *gel* (semisolid)
	(2) Tissue fluid, also called It occupies the area around all cells	"Lymph around cells" Tissue lymph Matrix lymph Interstitial lymph Extravascular lymph Extracellular lymph Intercellular lymph etc.
	(3) Blood plasma	Blood circulatory system
	(4) Lymph	Lymph lacteals Lymph capillaries, nodes, ducts, sinuses
	(5) Filling serous spaces	Pericardial fluid Pleural fluid Peritoneal fluid Cerebrospinal fluid

Body Fluids	(6) Fluids in spaces	Endolymph and perilymph of internal ear Fluid of eyes Synovial fluids and fluids of bursae, sheaths

Formation of tissue fluid. Tissue fluid is formed from plasma by the process of diffusion and filtration. There is difference in pressure within the blood capillary and in the tissue spaces surrounding the capillary. For instance, at the arterial end of the capillary the hydrostatic pressure is about 30 mm Hg and in the tissue spaces surrounding the capillary the pressure is much lower. Since the pressure is highest within the capillary, fluid and other substances are driven from the capillary into the tissue spaces. Another force that must be considered is the protein osmotic pressure formed by the plasma proteins, which act as a "pulling" force to hold fluids within the vessels as well as to "attract fluids in." This opposing force prevents undue loss of fluid from the capillaries.

At the venous end of the capillary, hydrostatic pressure is about 15 mm Hg. This means that the differences in pressure within the capillary and in the tissue spaces is not as great as at the arterial end. It is also conceivable that as blood moves through the capillary network and fluid is lost to the tissue spaces, plasma protein concentration is slightly raised, and hence the "pulling force" is increased so that water and crystalloids re-enter the capillaries readily, but colloids, along with water and crystalloids, enter the lymph capillaries. Increases in hydrostatic pressure within the capillaries from any cause will interfere with returns of substances to the lymphatics or capillaries and will result in excess accumulation of tissue fluid or edema (p. 448).

The composition of tissue fluid is similar to that of blood plasma. It is a colorless or yellowish fluid possessing an alkaline reaction, a salty taste, and a faint odor. When examined under the microscope, it is seen to consist of cells floating in a clear liquid. Its resemblance to the plasma is indicated in the table below. In consequence of the varying needs and wastes of different tissues at different times, both the tissue fluid and blood must vary in composition in different parts of the body. But the loss and gain are so fairly balanced that the average composition is pretty constantly maintained. The composition of the fluids in the serous, cerebrospinal, and synovial cavities, and that of the fluids of the eye and ear, vary.

Comparison of Blood and Tissue Fluid

Blood	*Tissue Fluid*
Specific gravity about 1.055	Specific gravity varies between 1.015 and 1.023
Contains erythrocytes	May contain a few erythrocytes
Contains white cells	Granulocyte count lower, lymphocyte count higher
Contains blood platelets	Does not contain blood platelets
A high content of blood proteins	A lower content of blood proteins
A low content of waste products	A higher content of waste products
A high content of nutrients	A lower content of nutrients
Normally—clots quickly and firmly	Clots slowly, and clot is not firm
Relatively high in colloidal protein	Relatively low in colloidal protein; globulin practically absent

Water, sugar, salts, same concentration in both

Function of the tissue fluid. The tissue fluid bathes all cells of the body. It delivers to the cells the material they need to maintain functional activity and picks up and returns to the blood the products of this activity. These products may be simple waste or materials capable of being made use of by some other tissue. There is thus a continual interchange going on between the blood and the tissue fluid. This interchange is effected by means of *diffusion.*

The tissue fluid becomes altered by the metabolic changes of the tissues which it bathes. There are three different fluids separated by the moist membranes which form the walls of the blood vessels and lymphatics—the blood inside the capillary walls, the tissue fluid in the tissues outside the walls of the blood vessels, and the lymph in the lymph channels. Some of the constituents of the blood pass into the tissue fluid; some of the constituents of the tissue fluid pass into the blood directly, and some into the lymphatics. The capillary walls and walls of the lymph vessels are permeable and allow water molecules (solvent) and molecules of substances in solution (solute) to pass through. Diffusion of this kind is dependent on differences in concentration of diffusible particles of any substance on the two surfaces of the diffusion membranes.

Tissue fluid is also closely related to physiological integration. As cells go into activity, varying needs must be met in relation to supplies and the products of metabolism.

The table on p. 47 gives some results of experiments related to exchange of substances through capillary walls in voluntary muscle. Lines *A, B, C, D* give data concerning muscle in various states of muscular effort from (*A*) muscle at rest, requiring less exchange of materials between blood and tissue fluid and between tissue fluid and cells, to (*D*) muscle in which a maximum circulation has been established.

ILLUSTRATION OF PHYSIOLOGICAL INTEGRATION

		a	b	c	d	e	f	g
		O_2 consumption assumed per minute, Vols. % of the tissue	*Number of capillaries per mm²* *cross section*	*R*	*2r*	*T_O-T_R in % of atmosphere*	*Total surface of capillaries in 1 cu cm muscles, cm²*	*Total capacity of capillaries, Vols. % of the tissue*
A	Rest	0.5	31*	100	3.0	6.3	3	0.02
		1.0	85	61	3.0	3.4	8	0.06
		3	270	34	3.8	2.5	32	0.3
B	Massage	5	1,400	15	4.6	0.6	200	2.8
C	Work	10	2,500	11	5.0	0.4	390	5.5
D	Maximum circulation	15	3,000	10	8	0.25	750	15

A, radius of cylinder of cells and tissue fluid physiologically related to a capillary; 2*r*, diameter of average capillary—*R* and 2*r* measured in μ; cm², sq cm; mm², sq mm; T_O-T_R, difference in oxygen pressure in a capillary and in the surrounding tissue fluid; *, estimated. (A Krogh, *The Anatomy and Physiology of Capillaries,* 1929, p. 271, 269 especially. Courtesy of Yale University Press.)

It will be noted in this table that as use of oxygen increases, the number of open capillaries in the muscle increases, the capillary diameter increases, the total area of capillary wall increases, the volume of blood in the muscle increases, the distance of the farthest cell from a capillary decreases, and the difference in oxygen pressure inside and outside the capillary decreases. There are also other changes, as increased temperature, velocity of blood flow, etc.

If, as indicated in the table, there is increased cellular metabolism (oxidation, etc.) when muscular effort is increased, there is produced in the cells an increased quantity of metabolites (CO_2, etc.) to be eliminated. Thus the chemical activity of the muscle cells may be a factor, and probably the chief factor, controlling the amount and distribution of blood through the muscle. This is an automatic control—the graded need to get and to give off brings about the graded means (variable blood flow) to do so accurately. If maximum blood supply does not bring sufficient oxygen, etc., and remove metabolites fast enough, oxygen hunger, etc., followed by fatigue, results.

This automatic control for optimum distribution of blood in relation to muscular effort involves adjustment of pulse rate, pulse volume, general and local peripheral resistance, respiratory rate, respiratory volume, and, in fact, an adjustment of all body functions. This control is brought about by the effects of variations in chemical equilibrium on the tissues themselves, or by these same effects and the nervous system on distant organs concerned.

This indicates the relative functions of blood plasma and tissue fluid. The blood brings (and takes away) substances. The volume of blood per minute in the muscle, the capillary area for diffusion, etc., are constantly varied. To the muscle cells the blood is the source kept relatively high in concentration because continually in motion, etc. Or in case of wastes it is the place of disappearance kept relatively low.

In the cells the supplies are used and wastes produced.

The tissue fluid stands between the two. It is a fluid that moves very slowly and is separated both from the blood and from the cell contents, which it closely resembles chemically and physically, by diffusion membranes. In the cell the rate of change of chemical equilibria, controlled by catalysts, constantly uses supplies and produces wastes; the blood constantly brings supplies and carries off wastes. The tissue fluid regulates this transfer and makes it possible for large amounts of substances to be transported and used with relatively small differences in concentration of soluble constituents in any of the body fluids.

This, together with the fact that blood returning from all tissues is *mixed in the heart* and hence all tissues receive the same blood, is probably the basis of all *physiological integration.*

For cell arrangements in gland formation and secretion, see Chap. 15.

SUMMARY

- **Physiology of Cells** (Two groups of activities)
  - (1) Relationships between cells and tissue fluid
  - (2) Relationships between the parts of cells
- **Difference in Cells**
  - Size
    - In general, microscopic in size
  - Arrangement
    - Very little intercellular material between cells, e.g., epithelium
    - A great deal of intercellular material between cells, e.g., areolar connective tissue
- **Constituents of Protoplasm**
  - Bulk of chemical elements found in body exist in form of compounds
    - Organic compounds contain carbon
      - Carbohydrates
      - Fats
      - Proteins
    - Inorganic compounds—contain no carbon
  - Proteins contain
    - Carbon
    - Hydrogen
    - Oxygen
    - Nitrogen
    - Sulphur
    - Phosphorus
  - Carbohydrates contain
    - Carbon
    - Hydrogen
    - Oxygen
  - Fats contain
    - Carbon
    - Hydrogen
    - Oxygen
  - Ash constituents
    - Sulphur
    - Phosphorus
    - Chlorine
    - Sodium
    - Potassium
    - Calcium
    - Magnesium
    - Iodine
    - Iron
    - Silicon
    - Traces of others
  - Water
    - Most abundant constituent of
      - Cells
      - Intercellular material
    - More than two-thirds of weight of body
    - More than 80% of nonbony body weight
    - Estimated 85–92% of weight of cell
    - Some physical characteristics giving its physiological significance
      - Solvent power is high
        - As to kinds of solutes
        - As to variable and high concentration of solutes
      - Ionizing power is high
      - Surface tension is high
      - Specific heat is high
      - Thermal conductivity (for liquids) is high
      - Latent heat of evaporation is high
      - etc.

Constituents of Protoplasm
- Cell organization
  - Cells may act like liquids (sols)
  - Cells may act like semisolids (gels)
  - Body cells organized close to line of demarcation between sol and gel

Diffusion
- Definition
  - Diffusion
  - Osmosis
  - Dialysis
- Relationships (If experiments are standardized and conditions kept the same throughout)
  - Relationship between area of membrane and amount of a substance diffused
  - Relationship between time duration and amount of a substance diffused
  - Relationship between difference in concentration of a diffusible substance on the two sides of the membrane and amount of this substance diffused
  - Relationship between temperature change and amount of a substance diffused
  - etc.

Cell Metabolism (Chemical activities in cells)
- The place of disappearance of supplies diffusing into it; the source of wastes, diffusing out
- In relation to slow continuous diffusion
- Relations of cell, water, colloids, adsorption, enzymes to cell activities
- Protoplasm has never been isolated or synthesized with characteristics of living cells
- Organization of protoplasm exceedingly complex
- Activities of cells are studied by isolation or synthesis of known cellular constituents and their physical and chemical characteristics, e.g., vitamins, hormones, etc.
- Cell organization as an emulsion of colloidal particles thought to be a chief factor in orderly progression of chemistry in the cell
- An illustration of orderly progression of chemical changes shown on p. 150 in relation to contraction of a muscle
- Many other illustrations of orderly progression of chemical changes will be found throughout the text

The Cells and Body Fluids
- (1) Cellular fluid
  - Protoplasm is sometimes thought of as a sol (liquid), sometimes as a gel (semisolid)
- (2) Tissue fluid, also called
  It occupies the area around all cells
  - "Lymph around cells"
  - Tissue lymph
  - Matrix lymph
  - Interstitial lymph
  - Extravascular lymph
  - Extracellular lymph
  - Intercellular lymph
  - etc.

The Cells and Body Fluids	(3) Blood plasma	Blood circulatory system
	(4) Lymph	Lymph lacteals Lymph capillaries, nodes, ducts, sinuses
	(5) Filling serous	Pericardial fluid Pleural fluid
The Cells and Tissue Fluid	**Location**	Surrounding all cells Occupies the spaces of areolar connective tissue and all other tissues
	Source	Diffused blood plasma Diffused cellular materials Cell membrane is a two-way filtering and diffusion membrane between all cells and tissue fluid, including capillary cells
	Composition	Chemically and physically similar to blood plasma and protoplasm
	Function	Go-between for blood and lymph and cells

CHAPTER 4

THE TISSUES OF THE BODY

Microscopic anatomy refers to the study of any structure under the microscope. Histology limits microscopic anatomy to the study of tissues. It is concerned with structural characteristics of cells and groups of cells as arranged to form tissues. The structure and function of the tissues are closely related. It is structure that accomplishes function, or function that determines the tissue. An understanding of structure will clarify function and help to build the foundation for physiology.

The kinds of tissues of which the body is formed are few in number, and some of these, although apparently distinct, have so much in common in their structure and origin that only four distinct tissues are usually recognized:

1. The epithelial tissues
2. The connective tissues
3. The muscular tissues
4. The nerve tissues

For convenience and to avoid repetition, muscular tissues are placed with the muscles and the nervous tissues are placed with the nervous system.

Epithelial tissue is composed of cells held together by cell cement or by intercellular fluid with a surface boundary of cell cement. The cells are generally arranged so as to form a membrane covering the external surfaces and lining the internal parts of the body; hence it is called a boundary tissue. It is devoid of blood vessels and is nourished by the fluid which passes to the cells by way of the intercellular substance. The general functions of epithelial tissues are protection, excretion, secretion, absorption, and the reception of stimuli.

Epithelial tissues may be classified in various ways. A simple classification is as follows:

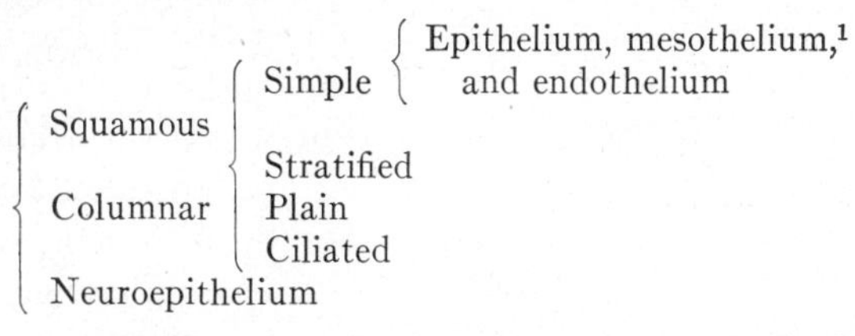

Squamous epithelium is so called because the cells on the free surface are flattened, scalelike, and fitted together to form a mosaic.

Simple squamous epithelium consists of one layer of flat cells. It forms smooth surfaces and secretes serum to lubricate them. It is found lining the alveoli of the lungs, in the crystalline lens of the eye, and in the membranous labyrinth of the inner ear. These tissues are derived from the ectoderm and entoderm[2] and are called *epithelium.*

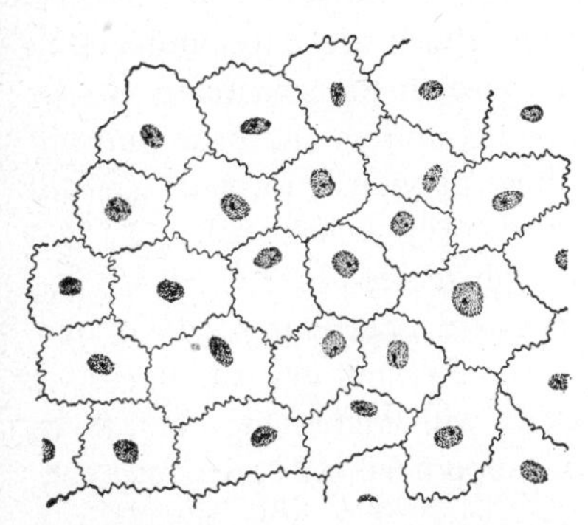

Fig. 22. Simple squamous epithelium, surface view.

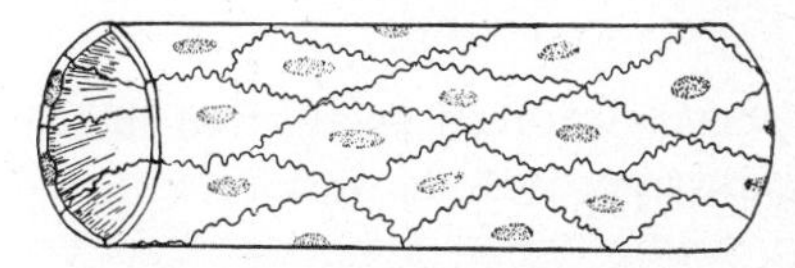

Fig. 23. Diagram of endothelium, surface and sectional views. Seen in the wall of a capillary.

A tissue similar in structure is found lining the heart and the blood and lymph vessels, and forming the capillary networks. This tissue is derived from the mesoderm and is called *endothelium.* Mesothelium is the name given to this tissue where it lines the serous cavities.

Another tissue somewhat similar in structure is called *mesenchyme* and is found lining the perilymph chambers of the ear, the cavities of the eyeball, and the spaces between the dura and the pia of the brain and cord.

[1] Mesothelial tissue lining the blood and lymph circulatory system is frequently called endothelium. Endothelium is, therefore, a geographical area of mesothelium. P. Smith, *Bailey's Textbook of Histology,* 1953.

[2] Often spelled *endoderm.*

Stratified squamous epithelium consists of several layers of cells which differ in shape. As a rule the cells of the deepest layer are cylindrical, the next manysided, and those nearest the surface are flattened and scalelike. The deeper cells of a stratified epithelium are separated from one another by a system of channels, which are bridged across by numerous protoplasmic threads. They are continually multiplying by cell division, and as the new cells which are thus produced in the deeper parts increase in size, they compress and push outward those previously formed. In this way cells which were at first deeply seated are gradually shifted outward, becoming dehydrated, shrinking, and growing harder as they are forced away from their contact with underlying body fluids and approach the surface. The older superficial cells are being continually rubbed off, and new ones continually rise up to replace them. Hence, when injured, this tissue has great capacity for repair.

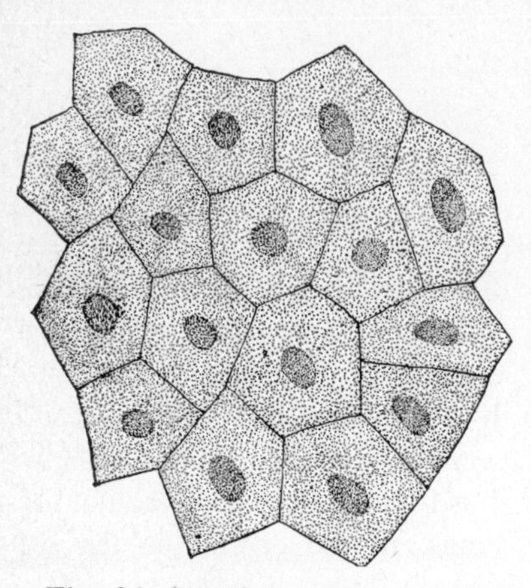

Fig. 24. Stratified squamous epithelium, surface view.

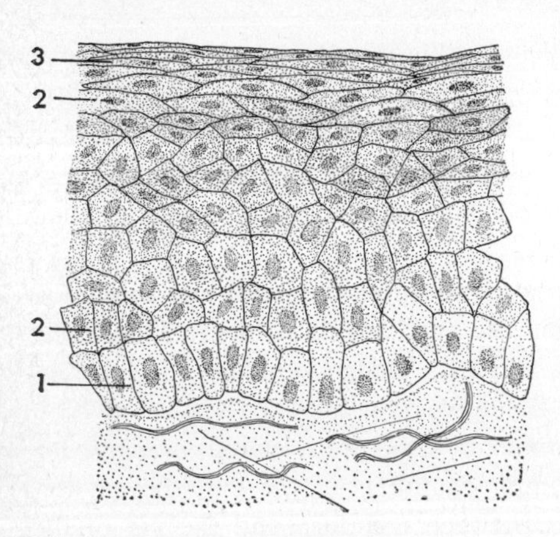

Fig. 25. Stratified squamous epithelium, sectional view. Seen in the skin. Note areolar connective tissue underneath it. *1*, cylindrical layer; *2*, transitional layers; *3*, flat, or squamous, layers of cells.

Function. Stratified squamous epithelium is a protective tissue. It protects the body in many ways. It covers the body, forming the epidermis, and is found wherever the ectoderm folds in from the outside, e.g., mouth, nose, and anus. It prevents loss of body fluids and contains structures for the reception of stimuli.

Transitional epithelium. This tissue is somewhat like stratified

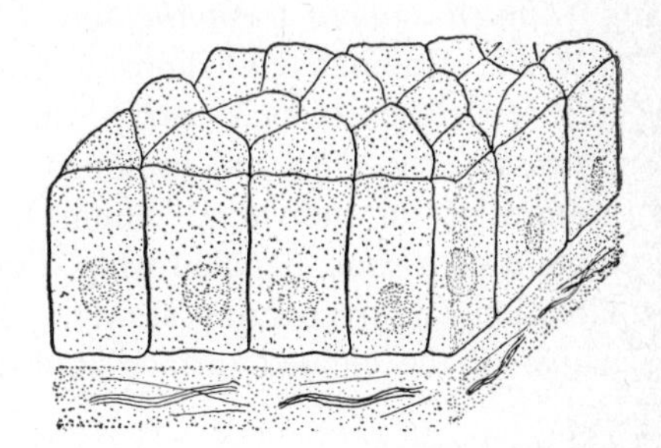

Fig. 26. Simple columnar epithelium, sectional and surface views. The columnar cells are relatively short and wide.

squamous epithelium, as it is composed of several layers of cells; however, there are usually only five or six layers. The deepest cells are polyhedral in shape, while the superficial cells are more flattened. It permits marked distension and relaxation, thereby adjusting the membrane to the content of the organ. It is found in the pelvis of the kidney, ureters, bladder, and part of the urethra. When the bladder is empty the five or six layers of cells are evident, but when it is distended there are only two or three layers of cells.

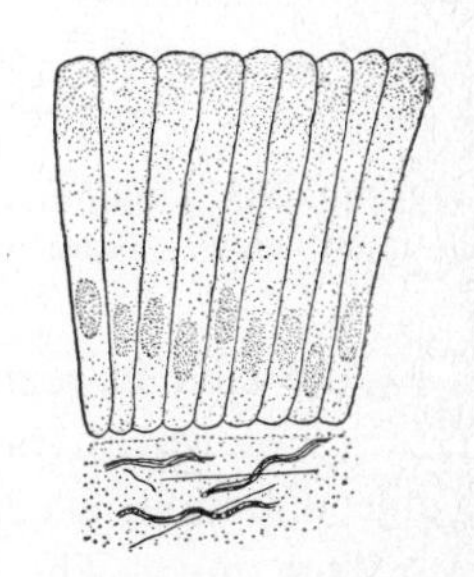

Fig. 27. Columnar epithelium, sectional view, showing tall, slender cells.

Columnar epithelium. In plain columnar epithelium the cells have a cylindrical shape and are set upright on the surface which they cover. Epithelium consisting of only one layer of prismatic cells constitutes the simple columnar variety. Columnar epithelium is found in its most characteristic form lining the stomach, small and large intestines, and digestive glands.

The chief functions of columnar epithelium are the secretion of digestive fluids and absorption of digested food and fluids.

Modifications of columnar epithelium result in three types of cells —goblet cells, ciliated columnar epithelial cells, and neuroepithelial cells.

When these cells assume a peculiar chalice form, resulting from an accumulation of mucoid secretion, they are called *goblet* cells and may be regarded as unicellular glands. Goblet cells are numerous in the mucosa of the small intestine and are especially abundant in the large intestine. The mucus which they secrete protects the lining from abrasion and in many other ways.

Ciliated columnar epithelium. In some localities the free surface of columnar epithelium is provided with microscopic, threadlike processes which carry on ciliary motion. These minute processes, the *cilia,* are prolongations of the cell protoplasm. The motion of an individual cilium may be compared to the lashlike motion of a short-handled whip, the cilium being rapidly bent in one direction and recovering slowly. The motion does not involve the whole of the ciliated surface at the same moment but is performed

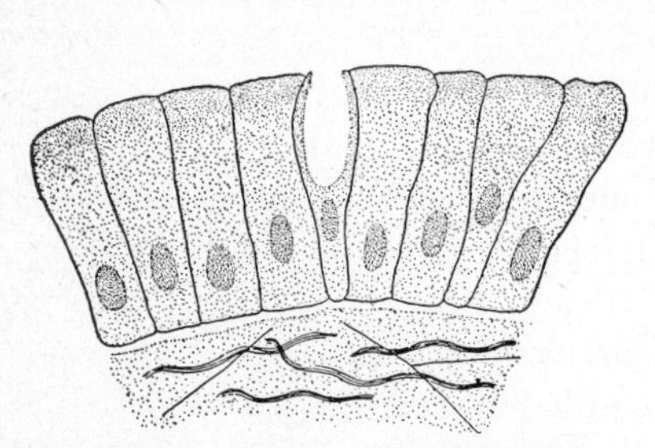

Fig. 28. Columnar epithelium, sectional view. Note one goblet cell.

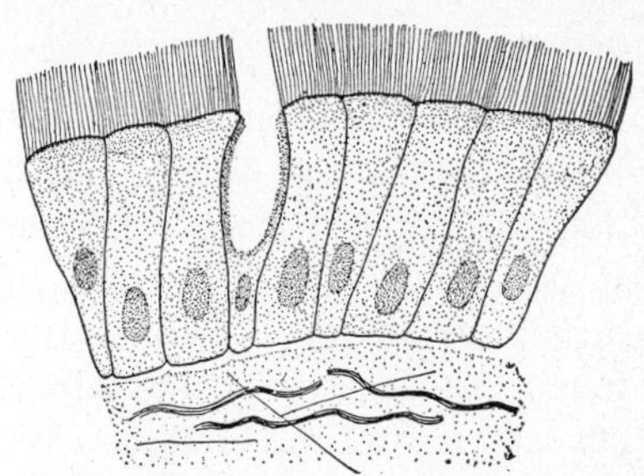

Fig. 29. Ciliated columnar epithelium, sectional view. Note one goblet cell.

by the cilia in regular succession, giving rise to the appearance of a series of waves traveling along the surface. Since they all move in one direction, a current of much power is produced.

Function. The function of cilia is motion, to impel secreted fluids and other materials along the surfaces from which they extend and to prevent the entrance of foreign matter into cavities.

Ciliated epithelium is found in the respiratory tract from the nose to the end of the bronchial tubes (with the exception of the pharynx and vocal cords), in the uterine tubes and the upper part of the uterus in the female, and in the efferent ducts of the testes in the male. In the trachea and larger bronchi, the internal auditory canal, and the efferent ducts of the testes in the male, the shape and position of the ciliated cells are changed by pressures. Most of the cells are narrow and tall. Between these are many cells that do not reach the surface. These cells function in the repair of epithelium in these areas. Numerous mucus-secreting goblet cells are found in ciliated epithelium.

Neuroepithelium (sensory) contains the end organs of nerve fibers. For example, in the epithelium lining the nose two kinds of cells develop—olfactory and supporting cells. The olfactory cells send their axons into the brain, where they come into relation with other neurons in the olfactory tract. The olfactory epithelium is thus a neuroepithelium; its sensory cells are nerve cells. They are derived from the ectoderm. The cells of the organ of Corti, the taste buds, and the retina are other examples of neuroepithelium.

Functions. The function of the sensory cells is sensation, and the term *supporting* describes the function of the others.

MEMBRANES

The word *membrane* in its widest sense is used to designate any thin expansion of tissues. In the commonest sense, the word *membrane* is used to denote an enveloping or a lining layer made up of tissues. Epithelial tissues form membranes.

Classification of membranes. The chief membranes of the body are classified as serous, synovial, mucous, and cutaneous.

Serous membranes are thin, transparent, strong, and elastic. The surfaces are moistened by a self-secreted fluid called *serum.* They consist of simple squamous epithelium and a layer of areolar connective tissue which serves as a base. Since the epithelium is derived from mesoderm, it is called *mesothelium.*

Serous membranes are found (1) lining the body cavities and covering the organs which lie in them, (2) lining the blood and lymph vascular system, and (3) forming the fascia bulbi and part of the membranous labyrinth of the ear.

1. *Lining the body cavities and covering the organs which lie in them.* With one exception, these membranes form closed sacs, one part of which is attached to the walls of the cavity which it lines—the *parietal* portion—while the other is reflected over the surface of the organ or organs contained in the cavity and is named the *visceral* portions of the membrane. In this way the viscera are not contained within the sac but are really placed outside of it; and some of the organs (e.g., lung) may receive a complete, while others (e.g., kidney) receive only a partial, investment.

The free surface of a serous membrane is smooth and lubricated; in this group the free surface of one part is applied to the corresponding free surface of some other part, only a small quantity of fluid being interposed between the surfaces. The organs situated in a cavity lined by a serous membrane, being themselves also covered by it, can thus glide easily against the walls of the cavity or upon each other, their motions being rendered smoother by the lubricating fluid.

This class of serous membranes includes (*a*) the two *pleurae,* which cover the lungs and line the chest (Figs. 3, 5), (*b*) the *pericardium,* which covers the heart and lines the outer fibrous pericardium (Fig. 5), (*c*) the *peritoneum,*[3] which lines the abdominal cavity and covers its contained viscera and the upper surface of some of the pelvic viscera (Fig. 6).

2. *Lining the vascular system.* This is the internal coat of the heart, blood vessels, and lymph vessels.

3. *Forming the fascia bulbi and part of the membranous labyrinth of the ear.*

a) Between the pad of fat in the back of the orbit and the eyeball is a serous sac—the fascia bulbi—which envelops the eyeball from the optic nerve to the ciliary region and separates the eyeball from the bed of fat on which it rests.

b) The membranous labyrinth of the ear has somewhat the same general form as the bony cavities in which it is contained.

Function. The function of the membranes is mainly protective; and this protection is accomplished in many ways, as, for example, by secreting serum which covers its surface, supplying the lubrication for organs as they move over each other.

Synovial membranes are membranes associated with the bones and muscles. They consist of an outer layer of fibrous tissue and an inner layer of areolar connective tissue with loosely arranged collagenous and elastic fibers, connective-tissue cells, and fat cells. There is no definite cellular layer at the inner surface. Synovial membranes secrete *synovia,* a viscid, glairy fluid that resembles the white of egg.

They are divided into three classes: (1) articular, (2) mucous sheaths, and (3) bursae mucosae.

[3] The peritoneal cavity in the female is an exception to the rule that serous membranes form perfectly closed sacs, since it communicates with the uterine (*Fallopian*) tubes at their ovarian ends.

1. *Articular synovial membranes* line the articular capsules of the freely movable joints.

2. *Synovial sheaths* (mucous sheaths) are elongated closed sacs which form sheaths for the tendons of some of the muscles, particularly the flexor and the extensor muscles of the fingers and toes. They facilitate the gliding of the tendons in the fibro-osseous canals.

3. *Synovial bursae* (bursae mucosae) are simple sacs interposed to prevent friction between two surfaces which move upon each other. They may be subcutaneous, submuscular, subfacial, or subtendinous. The large bursa situated over the patella is an example of a subcutaneous bursa. Similar, though smaller, bursae are found over the olecranon, the malleoli, the knuckles, and other prominent parts.

The function of synovial membranes is the same as that of serous membranes. Both the serous and the synovial membranes are derived from the mesoderm.

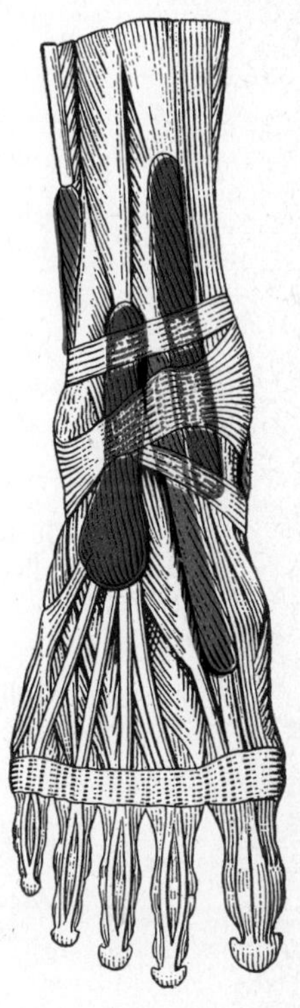

Fig. 30. The anterior annular ligament of the ankle and the synovial membranes of the tendons beneath it. Artificially distended. (Gerish.)

The mucous membranes may be grouped in two great divisions: (1) gastropulmonary and (2) genitourinary. Mucous membranes secrete *mucus,* a watery fluid containing *mucin* (a glycoprotein), salts, etc.

1. The *gastropulmonary* mucous membrane lines the alimentary canal, the air passages, and the cavities communicating with them. It commences at the edges of the lips and nostrils, extends through mouth and nose to the throat, and is continued throughout the length of the alimentary canal to the anus. At its origin and termination it is continuous with the external skin. It also extends throughout the trachea, bronchial tubes, and air sacs. From the interior of the nose the membrane extends into the frontal, ethmoid, sphenoid, and maxillary sinuses, also into the lacrimal passages, becoming the conjunctival membrane over the fore part of the eyeball and inside of the eyelids, on the edges of which it meets the

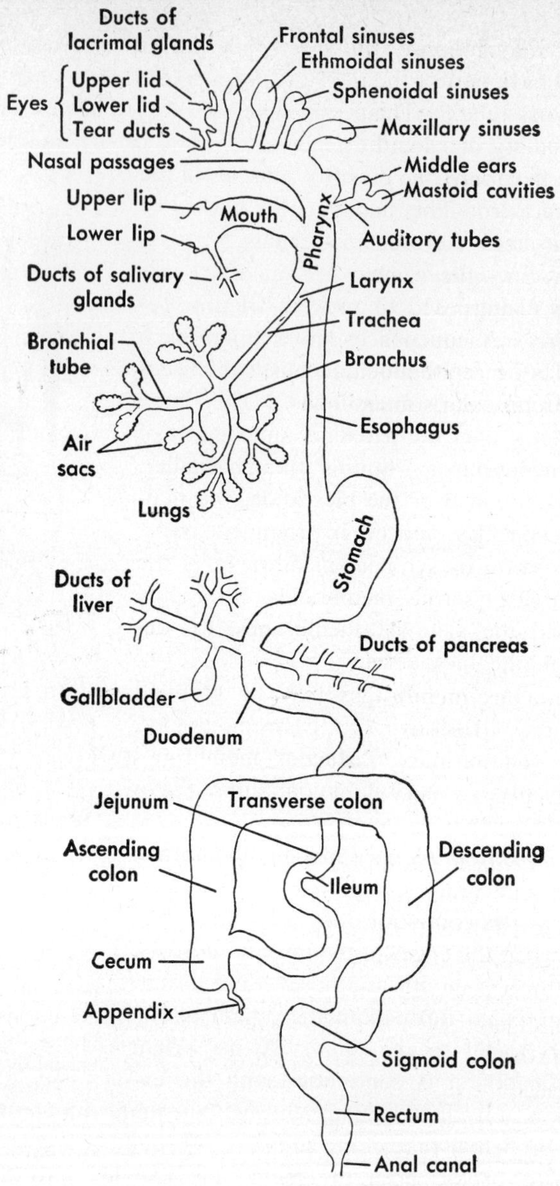

Fig. 31. Diagram showing continuity of the gastropulmonary mucous membrane.

skin. A prolongation of this membrane extends on each side of the upper and back part of the pharynx, forming the lining of the auditory (Eustachian) tube.[4] This membrane also lines the salivary, pancreatic, and biliary ducts and the gallbladder.

2. The *genitourinary* mucous membrane lines the bladder and the urinary tract from the interior of the kidneys to the orifice of the urethra; it lines the ducts of the testes, epididymis, and seminal vesicles; it lines the vagina, uterus, and uterine (Fallopian)[5] tubes. A study of Figs. 31, 32 and 33 will make this plain.

Structure. A mucous membrane is usually composed of four layers of tissue: epithelium, basement membrane, areolar connective tissue, and muscularis mucosae.

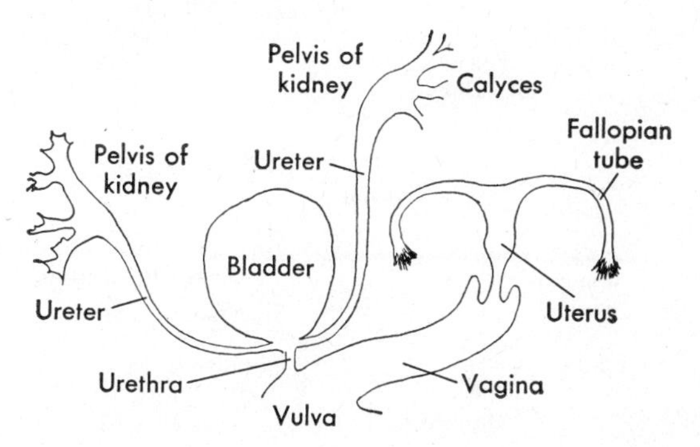

Fig. 32. Diagram showing continuity of mucous membrane in the female genitourinary pathway.

1. *Epithelium* is the surface layer. It may be stratified squamous, as in the throat; columnar, as in the stomach and intestine; or ciliated, as in the respiratory tract.

2. The *basement membrane*, if present, consists of a layer of flattened cells, the differentiated edge of the underlying connective tissue, or a secretion of the epithelial cells.

3. The *stroma* (*corium*) is composed of either areolar or lymphoid connective tissue and contains blood vessels.

4. The *muscularis mucosae* consists of a thin layer of muscular tissue. This layer is not always present, e.g., in the trachea.

The mucous membranes are attached to the parts beneath them by areolar connective tissue, in this case called *submucous* connective

[4] Bartolomeo Eustachio, Italian anatomist (1500–1574).
[5] Gabriel Falloppius, Italian anatomist (1523–1562).

tissue. This differs greatly in quantity as well as in consistency in different parts. The connection is in some cases close and firm, as in the cavity of the nose. In other instances, especially in cavities subject to frequent variations in capacity, like the esophagus and the stomach, it is lax. When such a cavity is narrowed by contraction of its outer coats, the mucous membrane is thrown into folds, or *rugae*, which disappear again when the cavity is distended. In certain parts the mucous membrane forms permanent folds that cannot be effaced, and these project conspicuously into the cavity which it lines. The best example of these folds is seen in the small intestine,

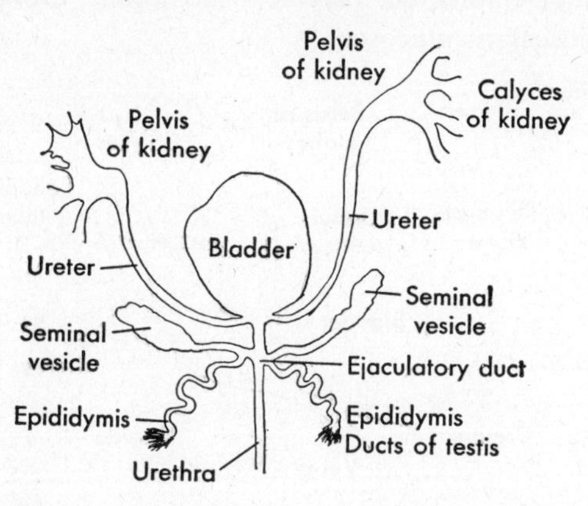

Fig. 33. Diagram showing continuity of mucous membrane in the male genitourinary pathway.

where they are called *circular folds* (valvulae conniventes), which increase the area of absorbing surface for the products of digestion. In some locations the free surface of mucous membrane contains minute glands or is covered with papillae or villi.

Functions of mucous membranes. The functions of mucous membranes are protection, support of blood vessels and lymphatics, and provision of a large amount of surface for secretion and absorption.

A mucous membrane protects by forming a lining for all the passages that communicate with the exterior, i.e., those passages subject to contact with foreign substances which are introduced into the body and with waste materials which are expelled from the body. The mucus secreted is a thicker and more viscid liquid than either serum or synovia and, by coating the surface, lessens the irritation from food, waste materials, or secreted substances. The cilia of the respiratory tract also assist in protection. They keep up an incessant motion and thus carry mucus toward the outlet of these passages. Dust and foreign materials usually become entangled in the mucus and are forced out with it.

The redness of mucous membranes is due to their abundant supply of blood. The small blood vessels which convey blood to the mucous membranes divide in the submucous tissues and send smaller branches into the corium, where they form a network of capillaries just under the basement membrane. The lymphatics also form networks in the corium and communicate with larger vessels in the submucous tissue below.

The projections of mucous membrane, such as the circular folds, covered with glands and villi as they are, increase enormously the surface area of the membrane for secretion and absorption and enable the membrane to carry more blood vessels and lymphatics (Figs. 298 and 299).

Cutaneous membrane refers to membrane which covers the body and is commonly spoken of as skin. It is a complex structure and has several functions in addition to serving as a protective covering for the deeper tissues lying beneath it. It will be more fully considered in Chap. 22.

THE CONNECTIVE TISSUES

The connective tissues differ in appearance but are alike in that the cellular elements are relatively few and the intercellular material is relatively abundant. They serve to connect and support the other tissues of the body, and with few exceptions they are highly vascular. The intercellular substance determines the physical characteristics of the tissue. The connective tissues may be classified as follows:

A. Embryonal
B. Connective tissue proper
   1. Areolar
   2. Adipose
   3. Liquid
   4. Reticular
   5. Elastic
   6. Fibrous
C. Cartilage
   1. Hyaline
   2. Fibrous
   3. Elastic
D. Bone

A. Embryonal tissue. In the development of connective tissue from the mesoderm, the cells unite to form a network. The cytoplasm increases rapidly, with a resulting differentiation into cells and a semifluid intercellular substance. This embryonal tissue is abundant in the embryo. It represents a stage in the development of connective tissue. Normally it does not occur in the adult but is found in connective-tissue repair after injuries and in certain pathological growths. Embryonal tissue, in which the ground substance is rich in mucin, is called *mucous connective tissue.* In the mucin, or ground substance, are found irregular branching cells and small bundles of white fibrils. Wharton's jelly,[6] in the umbilical cord, and the vitreous body of the eye are mucous connective tissues.

B. Connective tissue proper.

1. Areolar[7] connective tissue is composed of cells separated from one another by a semifluid ground substance, or *matrix.* Lying in this ground substance is an irregular network of silvery-white collagenous fibers and yellow elastic fibers. These fibers are the predominant characteristic of the tissue. The microscopic appearance of this tissue is shown in Fig. 34. Note the semiliquid ground substance, the cells, the fibrous fibers, and the elastic fibers lying in it.

A white collagenous fiber is composed of an albuminoid protein called collagen. Collagen, when boiled in water, yields gelatin. A collagenous fiber is composed of many minute wavy fibrils lying parallel and held together by cement. The fibrils frequently separate into groups extending in different directions. These fibrils are flexible but possess great tensile strength. Elastic fibers are homogeneous, branch freely, and are composed of a protein called elastin.

This tissue is soft and pliable, and when fresh it is transparent.

The cells of connective tissue. The cells found in connective tissue include:

1. THE FIBROBLASTS. These are most numerous and are large, flat branching cells with many processes. They are thought to play a part in the final formation of the collagenous fibers.

2. THE HISTIOCYTES OR MACROPHAGES. These are irregularly shaped cells with processes. They are also found in the sinusoids

[6] Thomas Wharton, English anatomist (1614–1673).

[7] Areolar means containing minute spaces. At one time these were thought to be empty spaces. Modern techniques have shown the spaces to be filled with fluid, called *tissue fluid.* (See p. 44 and Fig. 34.)

of the liver, lymph organs, and bone marrow. The histiocytes are phagocytic and function in normal physiological processes. They also have great phagocytic capacity under such conditions as inflammatory processes.

3. THE PLASMA CELLS. These are small round or irregular-shaped cells and are found in greatest numbers in the connective tissues of the membranes of the alimentary canal and great omentum.

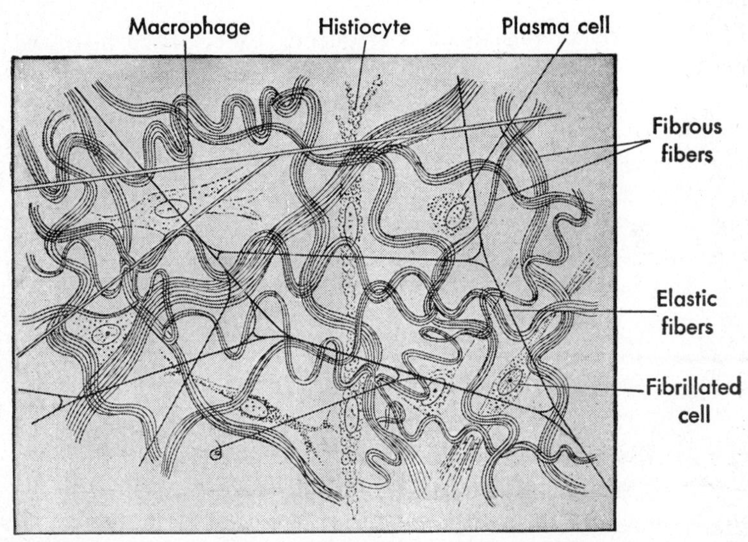

Fig. 34. Areolar connective tissue from a young rabbit. (Highly magnified.) The fibrous fibers are in wavy bundles; the elastic fibers form an open network. The ground substance, or matrix, in which the cells and fibers lie in tissue fluid—here shaded. Fifteen per cent of the body fluid lies in this tissue. (Schäfer.)

They are probably derived from the lymphocytes, but their function is not understood.

4. THE MAST CELLS in areolar connective tissue are most numerous along blood-vessel beds. Recent evidence strengthens the idea that an anticoagulant is formed by these cells which resembles heparin.

5. BLOOD CELLS. Lymphocytes, neutrophils, and eosinophils from the blood and lymph wander in and out of this tissue.

Function. Areolar connective tissue serves to connect the various tissues of an organ. Nerves, blood vessels, lymph vessels, and cells lie in it. It is found under the skin (subcutaneous), under the mucous membranes (submucous), and filling in the spaces around the blood vessels and nerves. Moreover, it is continuous throughout the body, and from one region it may be traced into any other, however distant—an interesting fact in medicine, as in this way water, other fluids, and pus, effused into the areolar connective tissue, may spread far from the spot where they were first introduced or deposited.

The matrix of this tissue is frequently called tissue fluid and in most areas of the body is spoken of as the *internal environment.*

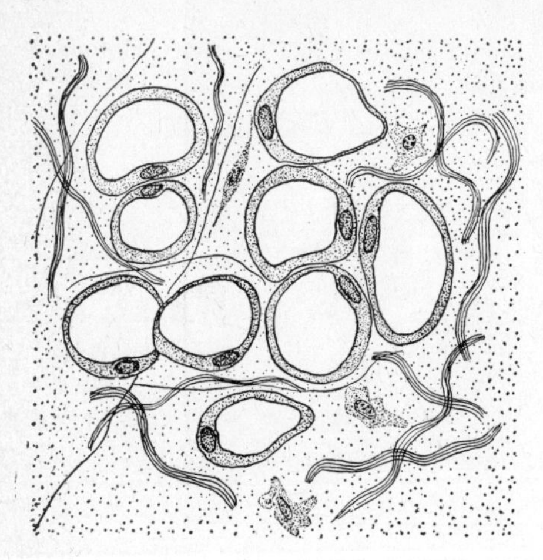

Fig. 35. Diagram of a thin section of adipose connective tissue. (Highly magnified.) Note adipose cells, matrix, fibrous fibers, elastic fibers, connective tissue. The white centers of the adipose cells represent fat. Excess fat adds to the work of the heart.

Speaking generally, it is this tissue that delivers supplies from the blood to the cells and delivers wastes from the cells to the blood and lymph. It stores water, salts, and glucose temporarily.

2. Adipose connective tissue is areolar connective tissue in which many of the cells are filled with fat. Adipose tissue exists very generally throughout the body, accompanying the still more widely distributed areolar connective tissue in most parts in which the latter is found. It is found chiefly:

Underneath the skin, in the subcutaneous layer.

Beneath the serous membranes or in their folds, e.g., omentum.

Collected in large quantities around certain internal organs, especially the kidneys, helping to hold them in place.

Covering the base and filling up furrows on the surface of the heart.

As padding around the joints.

In the marrow of the long bones.

Function. Adipose tissue has many functions, among which are: (1) to constitute an important reserve food, which when needed can be returned to the cells by the blood and oxidized, thus producing energy—adipose tissue stores more calories of energy-producing substance than any other tissue in the body for volume concerned; (2) to serve as a jacket or covering under the skin and, being a poor conductor of heat, to reduce the loss of heat through the skin; (3) to support and protect various organs, e.g., the kidneys; and (4) to serve to fill up spaces in the tissues, thus supporting delicate structures such as blood vessels and nerves.

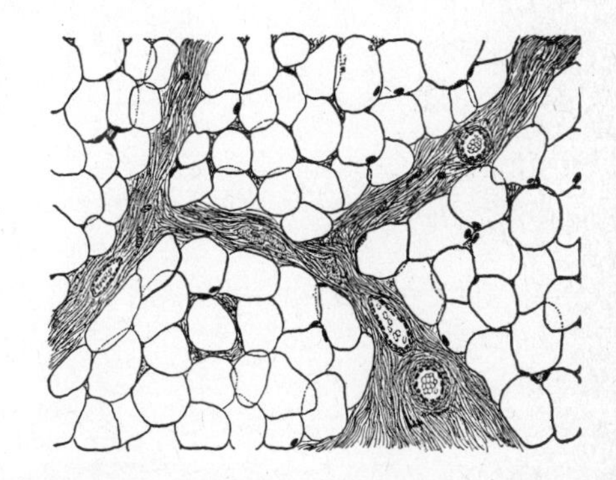

Fig. 36. Diagram of a thin section of adipose connective tissue. (Low magnification.) Note fibers close together in strands containing blood tubes and a few cells. The adipose cells are crowded together.

3. Liquid tissues. Blood and lymph may be classified as liquid tissue. They consist of cells and an intercellular substance, which is liquid. (See Chap. 10.)

4. Reticular[8] tissue. Reticular tissue is a variety of areolar tissue with a network, or reticulum, of fibrous fibers. The cells are thin and flat and are wrapped around the fibers. It can also be described as a meshwork of stellate cells forming minute cavities in which lymph cells are found.

Lymphoid or adenoid[9] tissue is reticular tissue in which the meshes of the network are occupied by lymph cells. This is the most common variety of reticular tissue.

Function. Reticular tissue forms a supporting framework in the lymph nodes, in bone marrow, and in muscular tissue. It is also

[8] Reticulum (Latin *reticulum,* "a small net").

[9] Adenoid (Greek *aden,* "a gland," and *eidos,* "resemblance").

present in the spleen, in the mucous membrane of the gastrointestinal tract, and in the lungs, liver, and kidneys.

5. Elastic connective tissue is areolar connective tissue in which the elastic fibers predominate. It consists of a ground substance containing cells with a few fibrous fibers and a predominance of elastic fibers which branch freely. They give it a yellowish color.

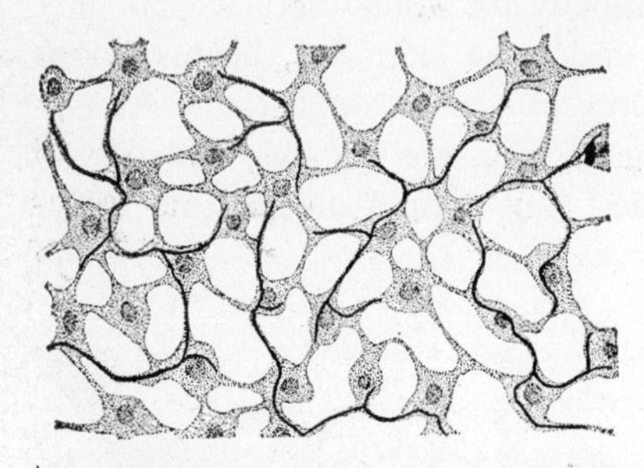

Fig. 37. Diagram of reticular tissue as seen in a thin section of a bit of lymph node. (Highly magnified.) This tissue forms the framework of lymph tissue.

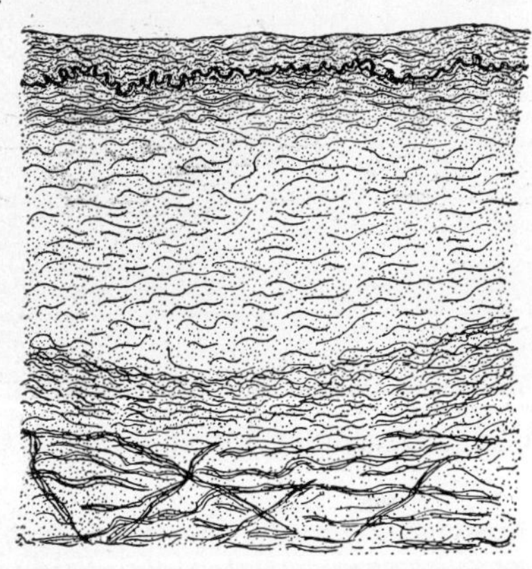

Fig. 38. Elastic connective tissue as seen in a thin section of the wall of a large artery. (Highly magnified.) The elastic fibers appear as short, black lines.

Function. Elastic tissue is extensile and elastic. It is found:

Entering into the formation of the lungs and uniting the cartilages of the larynx.

In the walls of the arteries, the trachea, bronchial tubes, and vocal folds.

In a few elastic ligaments and between the laminae of adjacent vertebrae (ligamenta flava, ligamentum nuchae).

6. Fibrous connective tissue is areolar connective tissue in which the fibrous fibers predominate. It consists of a ground substance in which there are cells and wavy, collagenous fibers, which cohere very closely and are arranged side by side in bundles which have an undulating outline. The matrix between the bundles contains cells arranged in rows, but the cells are not a prominent feature of this tissue.

Fibrous tissue is silvery-white, strong, and tough, yet perfectly pliant; it is almost devoid of extensibility and is very sparingly supplied with nerves and blood vessels.

Function. Fibrous connective tissue is part of the supporting framework of the body. It forms:

1. *Ligaments,* strong flexible bands, or capsules, of fibrous tissue that help to hold the bones together at the joints.

2. *Tendons* or *sinews,* white glistening cords or bands which serve to attach the muscles to the bones.

3. *Aponeuroses,* flat, wide bands of fibrous tissue which connect one muscle with another or with the periosteum of bone.

4. *Membranes* containing fibrous connective tissue found investing and protecting different organs of the body, e.g., the heart and the kidneys.

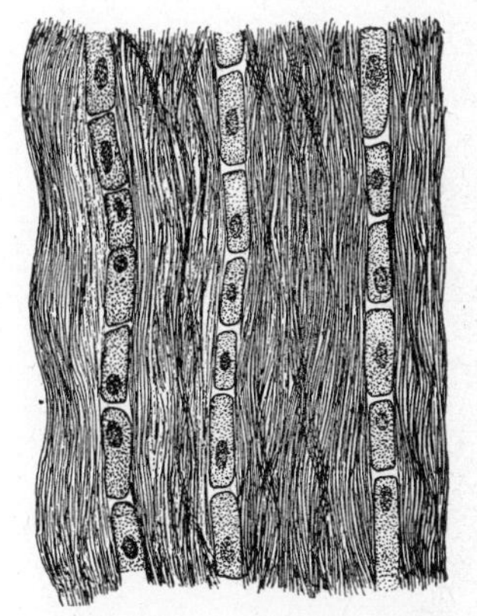

Fig. 39. Fibrous connective tissue as seen in a thin, longitudinal section of a tendon. (Highly magnified.) Fibrous fibers and cells lying in a matrix (white).

5. *Fasciae.* The word *fascia* (Latin) means a band or bandage. It is most frequently applied to sheets of fibrous connective tissue which are wrapped around muscles and serve to hold them in place. Fasciae are divided into two groups, (*a*) superficial fascia and (*b*) deep fasciae.

(*a*) SUPERFICIAL FASCIA, composed of subcutaneous areolar connective tissue, forms a nearly continuous covering beneath the skin. It varies in thickness and usually permits free movement of the skin on the subjacent parts.

Infection of the superficial fascia is called *cellulitis.* In this loose, areolar connective tissue infection spreads readily, and it is difficult to keep it from extending to surrounding areas.

(*b*) DEEP FASCIAE are sheets of white, fibrous tissue, enveloping and binding down the muscles, also separating them into groups.

The term *fasciae* usually designates the *deep fasciae*. Subcutaneous areolar tissue is rarely called by the name *fascia,* though it is correctly classed as such.

C. Cartilage, the well-known substance called *gristle,* is firm, tough, and flexible. When a very thin section is examined with a microscope, it is seen to consist of groups of cells in a mass of intercellular substance called the matrix. According to the texture of the intercellular substance, three principal varieties can be distinguished: (1) hyaline, or true, cartilage; (2) fibrocartilage; (3) elastic cartilage.

1. Hyaline cartilage. Comparatively few cells lying in fluid spaces or lacunae are embedded in an abundant quantity of intercellular substance. This substance appears glassy or homogeneous but is made up of collagenous fibers forming a feltlike mass. These fibers are similar to those found in fibrous connective tissue.

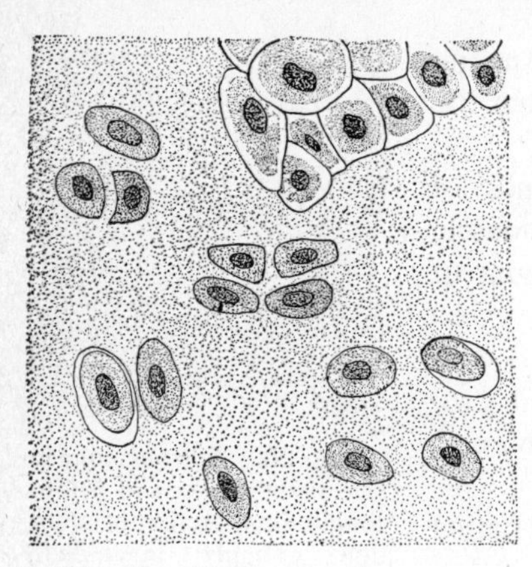

Fig. 40. Hyaline cartilage as seen in a thin section. (Highly magnified.) Some cells are shrunken to show that they lie in fluid spaces. Matrix is stippled.

Hyaline cartilage covers the ends of the bones in the joints, forming articular cartilage; it forms the ventral ends of the ribs as the costal cartilages.

In these situations the cartilages are in immediate connection with bone and may be said to form part of the skeleton; hence they are frequently described as skeletal cartilages. It also enters into the formation of the ears, nose, larynx, trachea, bronchi, and bronchial tubes.

Function. In covering the ends of the bones in the joints, the articular cartilages provide the joints with a thick, springy coating which gives ease to motion. In forming part of the bony framework of the thorax, the costal cartilages impart flexibility to its walls.

In the embryo a type of hyaline cartilage, known as embryonal cartilage, forms the matrix in which most of the bones are developed.

2. Fibrocartilage. The intercellular substance is pervaded with bundles of white fibers, between which are scattered cartilage cells. It closely resembles fibrous tissue.

Fibrocartilage is found joining bones together, the most familiar instance being the flat round disks or symphyses of fibrocartilage connecting the bodies of the vertebrae and the symphysis pubis between the pubic bones. In these cases the part in contact with the bone is always hyaline cartilage, which passes gradually into fibrocartilage. It forms the interarticular cartilage of other joints. In the center of the intervertebral disks there is a softened mass called

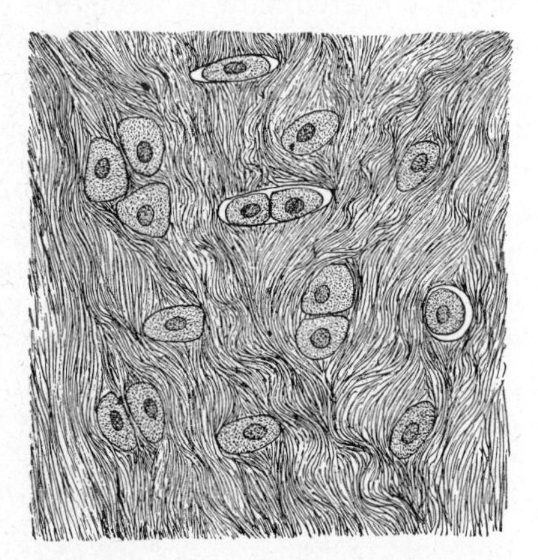

Fig. 41. Fibrous cartilage as seen in a thin section. (Highly magnified.)

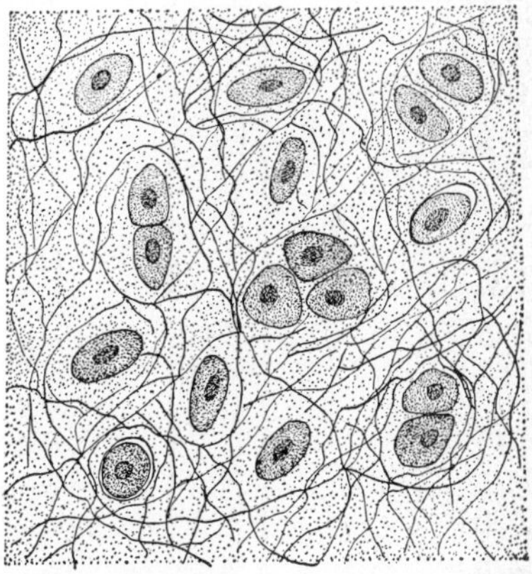

Fig. 42. Elastic cartilage as seen in a thin section. (Highly magnified.)

the nucleus pulposus. Herniation of this mass may occur into the spinal canal.

Function. Fibrocartilage serves as a strong, flexible connecting material between bones and is found wherever great strength combined with a certain amount of rigidity is required.

3. Elastic cartilage. The intercellular substance is pervaded with elastic fibers which form a network. In the meshes of the network the cartilage cells are found. This form of cartilage is found in the epiglottis, cartilages of the larynx, auditory tube, and external ear.

Function. It strengthens and maintains the shape of these organs and yet allows a certain amount of change in shape.

Cartilage is not supplied with nerves and very rarely with blood vessels. *Perichondrium,* a moderately vascular fibrous membrane, covers and nourishes cartilage except where it forms articular surfaces. Perichondrium also functions in the repair process of injured cartilage. When injured, the area is invaded by the perichondrial tissue, which is gradually changed into cartilage. Regeneration of cartilage is slow and may not take place in some instances.

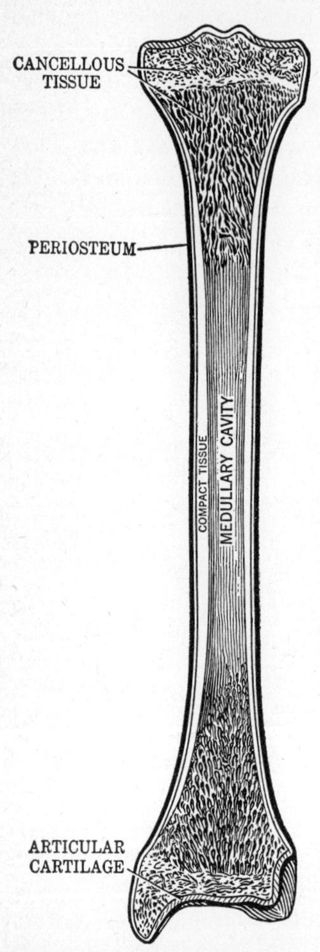

Fig. 43. Longitudinal section of a long bone. Note the cancellous structure at the ends of the bone. (Gerrish.)

D. Bone, or osseous tissue, is connective tissue in which the intercellular substance is rendered hard by being impregnated with mineral salts, chiefly calcium phosphate and calcium carbonate. The mineral salts, or inorganic matter, constitute about two-thirds of the weight of bone. The organic matter, consisting of cells, blood vessels, and cartilaginous substance, constitutes about one-third. The inorganic matter can be dissolved out by soaking a bone in dilute acid, or the organic matter may be driven off by heat. In both cases the shape of the bone will be preserved. Bone freed from inorganic matter is called *decalcified*. It is a tough, flexible, elastic substance, so free from stiffness that it can be tied in a knot. Bone free from organic matter is white and so brittle that it can be crushed in the fingers.

Structure of bone. On sectioning a bone, it will be seen that in some parts it consists of slender fibers and lamellae which form a structure resembling latticework, whereas in others it is dense and close in texture, appearing like ivory. There are two forms of bony tissue: (1) the *cancellous,* or *spongy;* and (2) the *dense,* or *compact.*

All bone is porous, the difference between the two forms being a matter of degree. The *compact* tissue has fewer spaces and is always found on the exterior of a bone, whereas the *cancellous* has larger cavities and more slender intervening bony partitions and is found in the interior of a bone. The relative quantity of these two kinds of tissue varies in different bones and in different parts of the same bone, depending on the need for strength or lightness. The shafts of the long bones are made up almost entirely of compact tissue, except that they are hollowed out to form a central canal, the medullary canal, which is lined by a vascular tissue called the medullary membrane.

Marrow is of two distinct kinds, red and yellow.

RED MARROW consists of a small amount of connective tissue that acts as a support for a large number of blood vessels; a large number of marrow cells, or *myelocytes,* which resemble the white blood cells; a small number of fat cells; a number of cells called *erythroblasts,* from which the red blood cells are derived; and *giant cells* (*osteoclasts*) found in both kinds of marrow but more abundant in the red marrow. Red marrow is found in the articular ends of the long bones, mainly femur and humerus, and in the cancellous tissue.

YELLOW MARROW consists of connective tissue containing numerous blood vessels and cells. Most of the cells are fat cells; only a few are myelocytes. It is found in the medullary canals of the long bone and extends into the spaces of the cancellous tissue and the Haversian[10] canals. It is thought that in the adult the white cells of the blood are formed in the marrow tissue from its myelocytes.

Periosteum. All bones are covered, except at their cartilaginous extremities, by a membrane called periosteum. It consists of an outer layer of connective tissue and an inner layer of fine fibers which form dense networks. In young bones the periosteum is thick, vascular, and closely connected with the epiphyseal cartilages. Later in life the periosteum is thinner and less vascular.

Blood vessels and nerves. Unlike cartilage, the bones are plentifully supplied with blood. If the periosteum is stripped from a fresh bone, many bleeding points representing the canals (Volkmann's[11]) through which the blood vessels enter and leave the bone are seen. These blood vessels proceed from the periosteum to join the system of Haversian canals. Around the Haversian canals the lamellae are disposed, and lying between them, arranged in circles, are found the

[10] Clopton Havers, English anatomist (1650–1702).
[11] Alfred Wilhelm Volkmann, German physiologist (1800–1877).

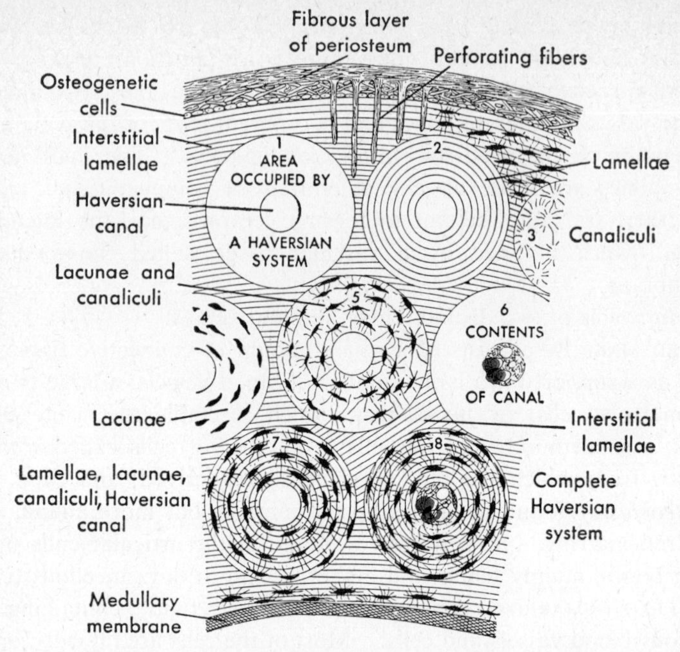

Fig. 44. Diagram of a cross section of osseus tissue. Details are drawn to a very much larger scale than the complete drawing. A small part of a *transverse section* of a long bone is shown. At the uppermost part is the periosteum, covering the outside of the bone; at the lowermost part is the medullary membrane, lining the marrow cavity. Between these is compact tissue, consisting largely of a series of Haversian systems, each being circular in outline and perforated by a central canal, left blank in the canals of five of the systems, in this illustration, and represented in color in two. The *first* circle shows the area occupied by a system. The *second* shows the layers of bony tissue, or lamellae, arranged around the central canal.

In the *third,* fine dark radiating lines represent canaliculi, or lymph channels. In the *fourth,* dark spots arranged in circles between the lamellae represent lymph spaces, or lacunae, which contain the bone cells. In the *fifth,* the central canal, lacunae, and canaliculi, which connect the lacunae with each other and with the central canal, are shown.

The *sixth* shows the contents of the canal: artery, veins, lymphatics, and areolar tissue. The *seventh* shows the lamellae, lacunae, canaliculi, and Haversian canal. The *eighth* shows a complete Haversian system.

Between the systems are interstitial lamellae, only a few of which show lacunae. The periosteum is made up of an outer fibrous layer and an inner osteogenetic layer, so called because it contains bone-forming cells, or osteoblasts. (Gerrish.)

lacunae which contain the bone cells. Radiating from one lacuna to another and toward the center are the canaliculi. Following this scheme, it will be seen that the innermost canaliculi run into the Haversian canals; thus, direct communication is established between the lymph in these canals and the cells in the lacunae surrounding the Haversian canals. The marrow in the body of a long bone is supplied by one large artery (sometimes more) called the *medullary* or *nutrient* artery, which enters the bone at the nutrient foramen, situated in most cases near the center of the body, and perforates the compact tissue obliquely. It sends branches upward and downward, which ramify in the marrow and enter the adjoining bony tissue. The twigs of these vessels anastomose with the arteries of the compact and cancellous tissue. In this way the whole substance of the bone is penetrated by intercommunicating blood tubes. In most of the flat and in many of the short spongy bones, larger apertures give entrance to vessels which pass to central parts of the bone and correspond to nutrient arteries of long bones. Veins emerge with or apart from arteries.

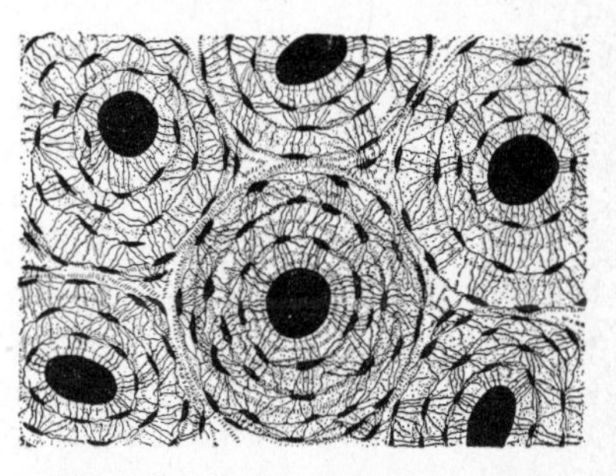

Fig. 45. Bony tissue as seen in a thin cross section of a bone. (Highly magnified.)

Lymphatic vessels have been traced into the substance of bone and accompanying the blood vessels in the Haversian canals. The periosteum is well supplied with nerves, which accompany the arteries into the bone. These nerve fibers form a plexus around the blood vessels. The nerve fibers include afferent myelinated and autonomic unmyelinated fibers.

Development of bones. In the early embryo some bones, such as those forming the roof and sides of the skull, are preformed in membrane; others, such as those of the limbs, are preformed in cartilage. Hence two kinds of ossification occur, the intramembranous and the intracartilaginous.

INTRAMEMBRANOUS OSSIFICATION. Before the cranial bones are formed, the brain is covered by inner meningeal membranes, a middle fibrous membrane, and an outer layer of skin. The fibrous membrane occupies the place of the future bone. From it periosteum and bone are formed. It is composed of fibers and cells called *osteoblasts*[12] in a matrix, or ground substance. When bone begins to form, a network of spicules radiates from a point or center of ossification. These spicules develop into fibers. Lime salts are deposited in the fibers and matrix, enclosing some of the osteoblasts in minute spaces, called lacunae. As the fibers grow out, they continue to calcify and give rise to fresh bone spicules. Thus, a network of bone is formed, the meshes of which contain the blood vessels and a delicate connective tissue crowded with osteoblasts. The bony network thickens by the addition of fresh layers of bone formed by the osteoblasts, and the meshes are correspondingly encroached upon. Successive layers of bony tissue are deposited under the periosteum, so that the bone increases much in thickness and presents the structure of compact bone on the outer and inner surfaces with a layer of soft, spongy, cancellous tissue between. The cancellous tissue between the layers, or tables, of the skull is called the *diploë*. Ossification of the bones of the skull is not complete at birth. At the site of the future union of two or more bones, membranous areas persist and are called fontanels.

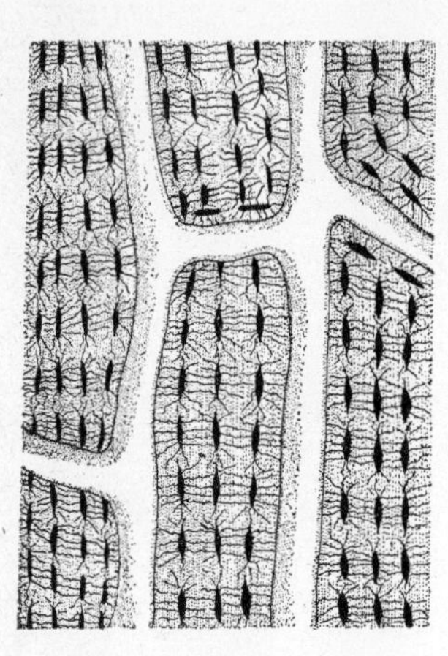

Fig. 46. Bony tissue as seen in a thin longitudinal section of a bone. (Highly magnified.) Haversian canals light in color, lacunae and canaliculi black.

INTRACARTILAGINOUS, OR ENDOCHONDRAL, OSSIFICATION. By the end of the second month the skeleton of the embryo is preformed in cartilage. Soon after this, ossification begins. The first step in intracartilaginous ossification is that the cartilage cells at the center of ossification enlarge and arrange themselves in rows. Following enlargement, lime is deposited in the matrix between the cells, first separating them and later surrounding them so that all nutriment is cut off, resulting in their atrophy and disappearance. The membrane called *perichondrium,* which covers the cartilage, assumes the character of periosteum. From this membrane grow cells which are deposited in the spaces left by the atrophy of the cartilage cells.

[12] Osteoblasts are bone-forming cells.

These two processes, destruction of cartilage cells and the formation of bone cells to replace them, continue until ossification is complete. The number of centers of ossification varies in differently shaped bones. In most of the short bones ossification begins at a point near the center and proceeds toward the surface. In the long bones, there is first a center of ossification for the body called the *diaphysis* and one or more centers for each extremity called *epiphyses*. Ossification proceeds from the diaphysis toward the epiphyses and from the epiphyses toward the diaphysis. As each new portion is ossified, thin layers of cartilage continue to develop between the diaphysis

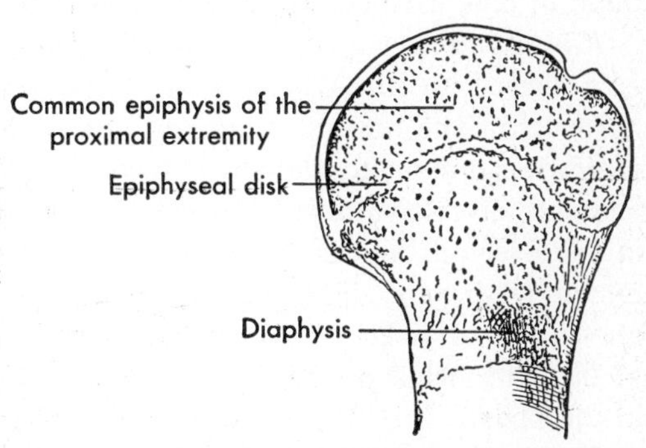

Fig. 47. The common epiphysis of the proximal extremity of the humerus from a boy of 13 years. Bone growth takes place at the epiphysis.

and epiphyses; and during the period of growth these outstrip ossification. When this ceases, the growth of a bone stops.

Ossification begins soon after the second month of intrauterine life and continues well into adult life. Such bones as the sternum, the sacrum, and the hipbones do not unite to form single bones until the individual is well beyond 21 years of age.

From this brief discussion two points of practical interest stand out: (1) the bones of the newborn infant are soft and largely composed of cartilage; (2) since the process of ossification is going on continually, the proper shape of the cartilage should be preserved in order that the shape of the future bone may be normal. Therefore it is obvious that a young baby's back should be supported, and a child should always rest in a horizontal position. The facility with which bones may be molded and become misshapen is seen in the flattening of the head of some Indians who bind a board across the top of the infant's skull, and in the bowlegs of children. However, the softness of the skeleton of a child accounts for the fact that the many jars and tumbles experienced in early life are not as injurious to the cartilaginous frame as they would be to a harder structure.

Numerous experiments have demonstrated that the proper ossification and growth of bone depend upon (1) adequate amounts of calcium and phosphorus in the food and (2) chemical substances which enable the bone cells to utilize calcium and phosphorus. The chemical substances may be vitamins derived from food or a hormone, such as thyroxin, derived from the internal secretion of the thyroid gland. Low calcium content of the blood may be caused by inadequate calcium in the diet, by poor absorption of calcium, or by too rapid excretion of it in the feces. Milk contains well-balanced proportions of calcium and phosphorus; hence the dietary rule of a quart of milk each day for every child.

Rickets. Rickets is a condition in which the mineral metabolism is disturbed so that calcification of the bones does not take place normally. The bones remain soft and become misshapen, resulting in bowlegs and malformations of the head, chest, and pelvis. Liberal amounts of calcium and phosphorus in food, and vitamin D, found in fish-liver oils and in significant amounts in egg yolk, whole milk, butter, fresh vegetables, etc., are important factors in the cure and prevention of rickets. Exposure to sunshine, especially irradiation of the skin, helps also in the optimal use of calcium and phosphorus in the body.

Three types of rickets are recognized by students[13] as due to chemical deficiency of the blood. The first, or so-called low-phosphorus rickets, is caused by a subnormal content of phosphate ions in the blood. This is the commonest type, sometimes called true rickets, and is characterized by histological changes resulting in large joints, deformed bones of the cranium, chest, and spine, and a condition in which beadlike deposits occur at the ends of the ribs. The second type is characterized by deformed bones of the head, trunk, and limbs and is frequently accompanied by tonic muscular spasms (tetany) lasting for considerable periods of time. It is sometimes called low-calcium rickets or a "rickets-like condition." The third type is one in which both calcium and phosphorus are below normal and is characterized by progressive porosity of the bones; it is sometimes spoken of as osteoporosis.

Fracture is a term applied to the breaking of a bone. It may be either partial or complete. As a result of the greater amount of organic matter in the bones of children, they are flexible, bend easily, and do not

[13] H. C. Sherman, *Chemistry of Food and Nutrition,* 1946.

break readily. In some cases the bone bends like a bough of green wood. Some of the fibers may break, but not the whole bone, hence the name *green-stick fracture.* The greater amount of inorganic matter in the bones of the aged renders the bones more brittle, so that they break easily and heal with difficulty.

Regeneration of bone. A fracture is usually accompanied by injury to the periosteum and tissues. This results in inflammation, which means that an increased amount of blood is sent to the part. The plasma and white cells from the blood exude into the tissues and form a viscid substance, which sticks the ends of the bone together and is called *callus.* Usually bone cells from the periosteum and lime salts are gradually deposited in the callus, which eventually becomes hardened and forms new bone. Occasionally the callus does not ossify, and a condition known as *fibrous union* results. The periosteum is largely concerned in the process of repair. If a portion of the periosteum is stripped off, the subjacent bone may die, whereas if a large part or the whole of a bone is removed and the periosteum at the same time is left intact, the bone will wholly or in great measure be regenerated.

SUMMARY

Classification tissues
- The epithelial tissues
- The connective tissues
- The muscular tissues
- The nerve tissues

Epithelial Tissues. Boundary tissues composed of cells and a minimum of intercellular substance

Classification of Epithelial Tissues
- Squamous
  - Simple—one layer of flat cells
    - Mesothelium and endothelium—derived from mesoderm
    - Epithelium—derived from ectoderm and entoderm
  - Stratified—several layers of cells
- Columnar
  - Plain—cylindrical cells, upright on surface
  - Ciliated—microscopic, threadlike processes at free end
- Neuroepithelium

Functions
- (1) *Protection.* Some varieties are specially modified so as to form protective membranes. Example—skin
- (2) *Motion.* This is seen in the cilia
- (3) *Absorption.* This is particularly well seen in the digestive tube
- (4) *Secretion.* Every secreting organ contains epithelial cells. Mucous and serous membranes are examples of secreting membranes
- (5) *Special sensation.* The organs of the special senses contain epithelial cells. Examples—eye, ear, nose, etc.

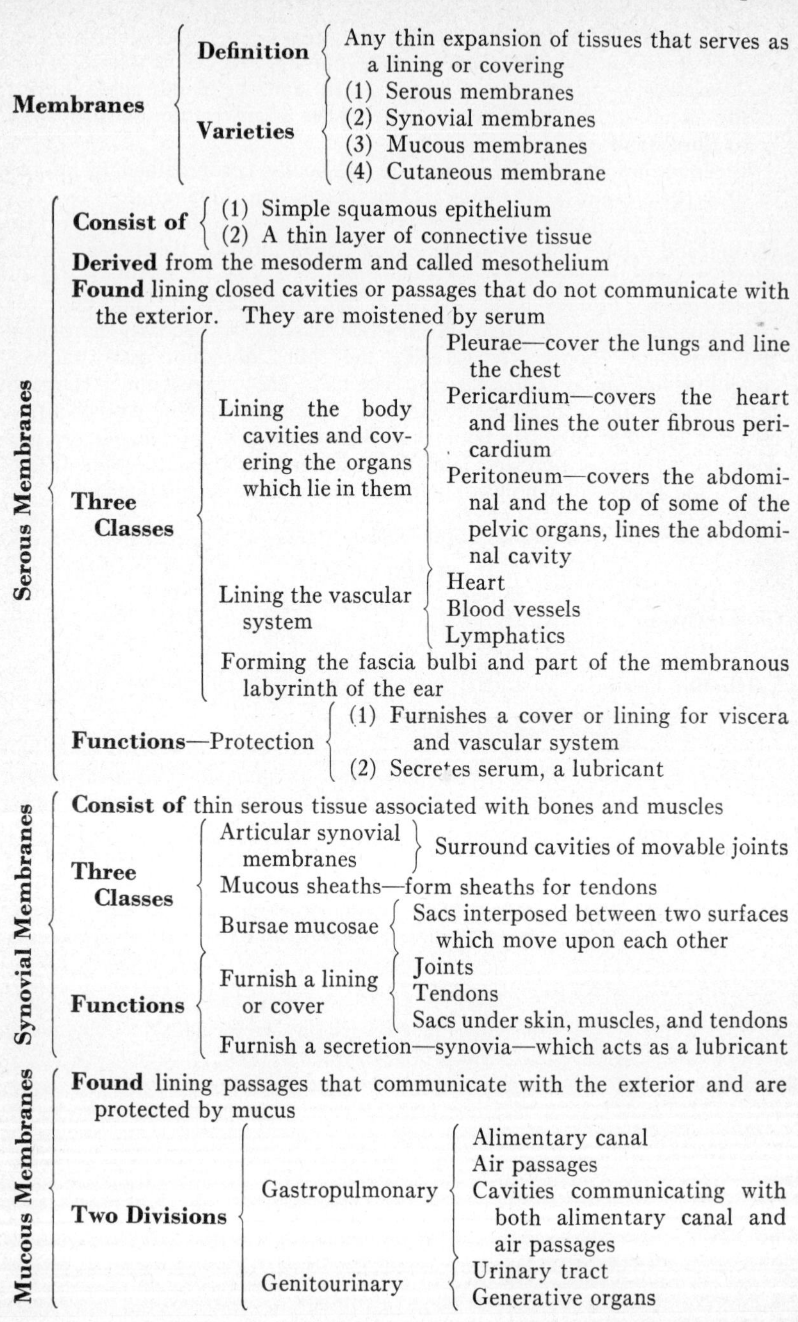

Membranes
- **Definition** — Any thin expansion of tissues that serves as a lining or covering
- **Varieties**
  - (1) Serous membranes
  - (2) Synovial membranes
  - (3) Mucous membranes
  - (4) Cutaneous membrane

Serous Membranes
- **Consist of**
  - (1) Simple squamous epithelium
  - (2) A thin layer of connective tissue
- **Derived** from the mesoderm and called mesothelium
- **Found** lining closed cavities or passages that do not communicate with the exterior. They are moistened by serum
- **Three Classes**
  - Lining the body cavities and covering the organs which lie in them
    - Pleurae—cover the lungs and line the chest
    - Pericardium—covers the heart and lines the outer fibrous pericardium
    - Peritoneum—covers the abdominal and the top of some of the pelvic organs, lines the abdominal cavity
  - Lining the vascular system
    - Heart
    - Blood vessels
    - Lymphatics
  - Forming the fascia bulbi and part of the membranous labyrinth of the ear
- **Functions**—Protection
  - (1) Furnishes a cover or lining for viscera and vascular system
  - (2) Secretes serum, a lubricant

Synovial Membranes
- **Consist of** thin serous tissue associated with bones and muscles
- **Three Classes**
  - Articular synovial membranes — Surround cavities of movable joints
  - Mucous sheaths—form sheaths for tendons
  - Bursae mucosae — Sacs interposed between two surfaces which move upon each other
- **Functions**
  - Furnish a lining or cover
    - Joints
    - Tendons
    - Sacs under skin, muscles, and tendons
  - Furnish a secretion—synovia—which acts as a lubricant

Mucous Membranes
- **Found** lining passages that communicate with the exterior and are protected by mucus
- **Two Divisions**
  - Gastropulmonary
    - Alimentary canal
    - Air passages
    - Cavities communicating with both alimentary canal and air passages
  - Genitourinary
    - Urinary tract
    - Generative organs

Mucous Membranes

- **Consist of**
  - (1) Epithelium
    - Stratified
    - Columnar
    - Ciliated
  - (2) Basement membrane, a layer of flat cells, etc.
  - (3) Stroma
    - Areolar or
    - Lymphoid tissue, which contains blood vessels
  - (4) Muscularis mucosae—thin layer of muscular tissue which is not always present
- **Projections**
  - Rugae—temporary folds
    - Esophagus
    - Stomach
  - Circular folds—permanent folds of mucous membrane found in small intestine
  - Papillae—conical processes of mucous membrane best seen on tongue. Contain blood vessels and nerves
  - Villi—tiny threadlike projections of the mucous membrane of small intestine
- **Functions**
  - Protection
    - Inside skin
    - Secretion of mucus
    - Action of cilia
  - Support for network of blood vessels
  - Absorption and Secretion — Various modifications increase the surface

Cutaneous Membrane

- Forms the skin
- Covers the body
- Serves to protect underlying tissues
- Prevents loss of body fluids
- Contains structures for the reception of stimuli
- See Chap. 22 for details of structure and function

Connective tissues. Tissues composed of cells with much intercellular substance, which is derived from the cells

Characteristics

- Cellular element at a minimum, intercellular element abundant
- Intercellular material determines characteristic of the tissue
- Serves to connect and support other tissues
- With the exception of cartilage, they are highly vascular

Varieties

- A. Embryonal
- B. Connective tissue proper
  - 1. Areolar
  - 2. Adipose
  - 3. Liquid
  - 4. Reticular
  - 5. Elastic
  - 6. Fibrous
- C. Cartilage
- D. Bone

A. Embryonal tissue. Represents a stage in development of connective tissue. Consists of cells and a primitive intercellular ground substance. When ground substance is rich in mucin, it is called mucous connective tissue

B. Connective tissue proper

Areolar tissue. Formed by interlacing of wavy bundles of fibrous fibers and some straight elastic fibers with cells lying in the ground substance.

Function. Connects, insulates, forms protecting sheaths, and is continuous throughout the whole body

Fluid matrix is called tissue fluid

Fluid matrix, often called internal environment, serves as a medium for transfer of supplies from blood and lymph vessels to cells, and wastes from cells to blood and lymph. Stores water, salts, glucose, etc.

Adipose tissue. Modification of areolar tissue, with cells filled with fat. Distribution quite general but not uniform

Function
1. Forms a reserve food to be drawn upon in time of need
2. Prevents the too rapid loss of heat
3. Serves to protect and support delicate organs

Liquid tissues. Cells in a liquid intercellular substance, e.g., blood and lymph

Reticular tissue. Areolar tissue with a network of fibrous fibers. Cells wrapped around fibers

Lymphoid tissue. Reticular tissue with meshes of network occupied by lymph cells

Function. Reticular tissue forms a supporting framework in many organs, e.g., lymph nodes, bone marrow, and muscular tissue. Reticular tissue is present in the spleen, mucous membrane of the gastrointestinal tract, lungs, liver, and kidneys

Elastic tissue. Consists of cells with few fibrous fibers and a predominance of elastic fibers

Function. It is extensile and elastic. Found in blood vessels, air tubes, larynx, vocal folds, and lungs, ligamenta flava, ligamentum nuchae

Fibrous tissue. Formed of wavy bundles of fibrous fibers only, with cells in rows between bundles; very strong and tough but pliant

Function. Is found in form of ligaments, tendons, aponeuroses, protecting sheaths, and fasciae

C. Cartilage. Cartilage consists of a group of cells in a matrix. It is firm, tough, and elastic, covered and nourished by perichondrium

Varieties
1. **Hyaline cartilage** — Articular, Costal } Skeletal
2. **Fibrocartilage**
3. **Elastic cartilage**

D. Bone, or osseous tissue. Bone is connective tissue in which the intercellular substance derived from the cells is rendered hard by being impregnated with mineral salts

BONE, OR OSSEOUS TISSUE

- **Composition**
  - *Inorganic matter about 67%*
    - Calcium phosphate
    - Calcium carbonate
    - Calcium fluoride
    - Magnesium phosphate
    - Sodium chloride
  - *Organic matter about 33%*
    - Cells
    - Blood vessels
    - Gelatinous substance
- **Varieties**
  - Cancellous, or spongy
  - Dense, or compact, like ivory
- **Canals**
  - Medullary—red and yellow marrow
  - Haversian
    - Blood vessels
    - Lymphatics
- **Haversian System**
  - Haversian canals are surrounded by lamellae, lacunae, and canaliculi
  - Lamellae—bony fibers arranged in rings around Haversian canals
  - Lacunae—hollow spaces between lamellae occupied by bone cells
  - Canaliculi—canals which radiate from one lacuna to another and toward the Haversian canals
- **Medullary Membrane.** A vascular tissue that lines the medullary canal
- **Marrow**
  - **Red** — Consists of connective tissue supporting blood vessels, myelocytes, fat cells, erythroblasts from which red blood cells are derived, and giant cells. Found in the marrow cavity at the ends of long bones and in cancellous tissue
  - **Yellow** — Contains more connective tissue and fat cells than red marrow, fewer myelocytes, few if any red cells, and fewer giant cells. White cells of blood and lymph are derived from its myelocytes. Found in the medullary canals of the long bones

Periosteum. A vascular fibrous membrane that covers the bones except at their cartilaginous extremities and serves to nourish them. Important in the reunion of broken bone and growth of new bone

Blood vessels. Twigs of nutrient artery in medullary canal anastomose with twigs from Haversian canals, and these in turn anastomose with others which enter from periosteum. Nerves accompany arteries into bone

- **Development of Bone**
  - In the embryo bones are preformed in membrane and in cartilage
  - **Ossification**
    - Intramembranous
    - Intracartilaginous, or endochondral
  - **Dependent upon**
    - Adequate amounts of calcium and phosphorus in food
    - Vitamins and hormones
- **Rickets**
  - A **disturbance** of mineral metabolism
  - **Prophylaxis or Preventive hygiene**
    - Adequate amounts of calcium and phosphorus in food
    - Vitamin D supplied by fish oils, egg yolk, milk, butter, fresh vegetables, direct sunlight
  - **Types**
    - True rickets
    - "Rickets-like condition"
    - Sometimes called osteoporosis

UNIT II

The Structural and Functional Relationships for Correlation and Coordination of External Activities

CHAPTER 5

SKELETON, BONES AND SINUSES OF HEAD, TRUNK, EXTREMITIES

The bones are the principal organs of support and the passive instruments of locomotion. They form a framework of hard material to which the skeletal muscles are attached. This framework affords attachment for the soft parts, maintains them in position, shelters them, helps to control and direct varying internal pressures, gives stability to the whole fabric, and preserves its shape. They form joints which may be movable. Here the bones act as levers for movement. Blood cells are formed in bone marrow.

The adult skeleton consists of 206 named bones.

Bone		Number	Total
Cranium		8	206
Face		14	
Ear	Malleus 2, Incus 2, Stapes 2	6	
Hyoid		1	
The spine, or vertebral column (sacrum and coccyx included)		26	
Sternum and ribs		25	
Upper extremities		64	
Lower extremities		62	

This list does not include the sesamoid[1] and Wormian[2] bones. Sesamoid bones are found embedded in the tendons covering the bones of the knee, hand, and foot. Wormian bones are small

[1] Ses′amoid (Greek *sesamon,* a "seed of the sesamum," and *eidos,* "form," "resemblance").

[2] Olaus Worm, Danish physician (1588–1654).

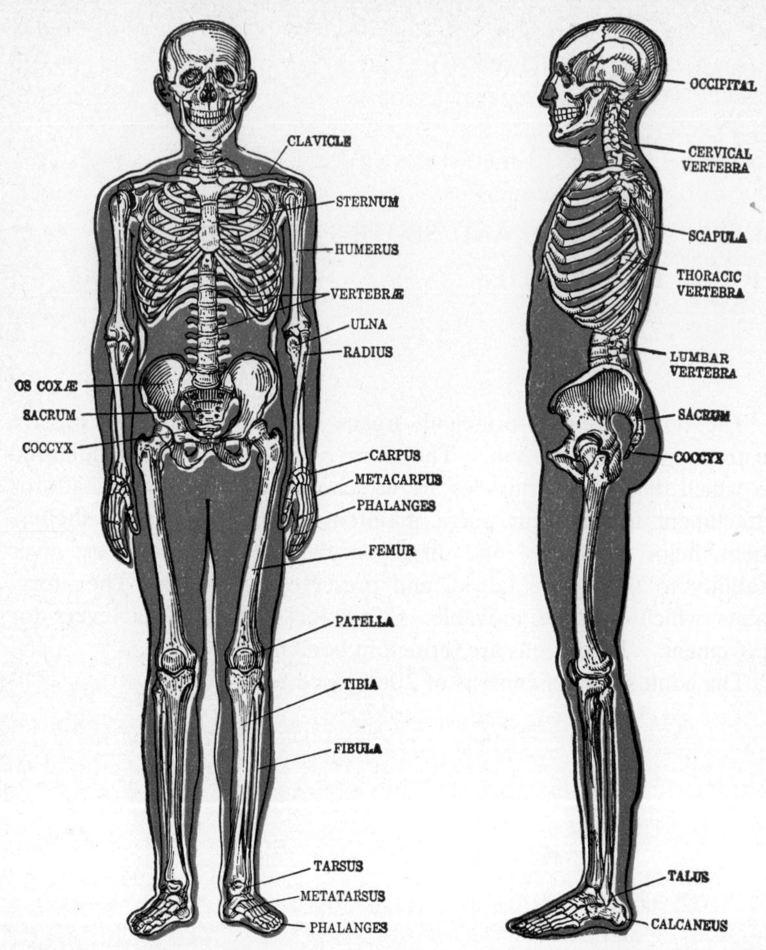

Fig. 48. The human skeleton, front view.

Fig. 49. The human skeleton, side view.

isolated bones which occur in the course of the sutures, most frequently the lambdoid suture.

Classification. The bones may be divided according to their shape into four classes: (1) *long,* (2) *short,* (3) *flat,* and (4) *irregular*.

A *long bone* consists of a shaft and two extremities. The shaft is formed mainly of compact tissue, this compact tissue being thickest in the middle, where the bone is most slender and the strain greatest, and it is hollowed out in the interior to form the *medullary canal.* The extremities are made of cancellous tissue, with only a thin coating of compact tissue, and are more or less expanded for greater convenience of mutual connection and to afford a broad surface for muscular attachment. All long bones are more or less curved, which gives them greater strength. They are found in the arms and legs, e.g., humerus.

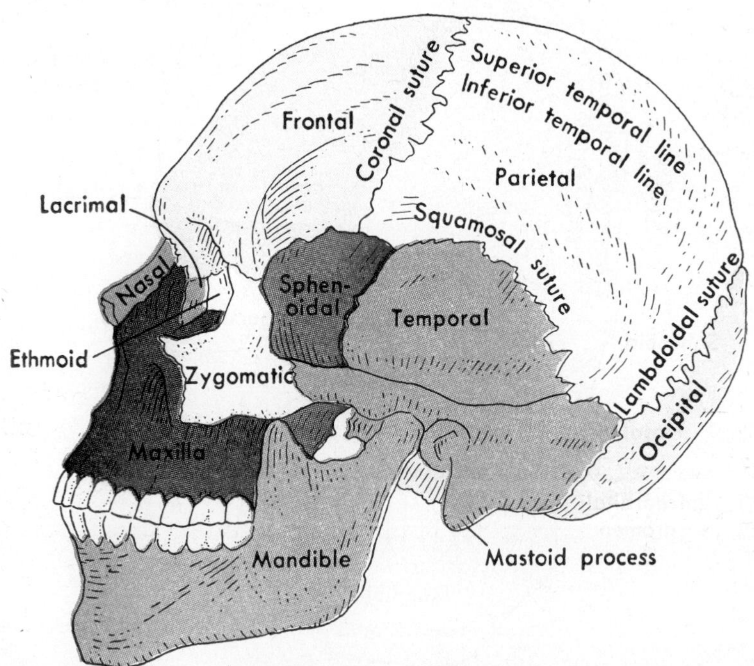

Fig. 50. Side view of skull.

The *short bones* are irregularly shaped. Their texture is spongy throughout, except at their surface, where there is a thin layer of compact tissue. The short bones are the 16 bones of the carpus, the 14 bones of the tarsus, and the 2 patellae.

Where the principal requirement is extensive protection or the provision of broad surfaces for muscular attachment, *flat bones* are found. The bony tissue expands into broad or elongated flat plates which are composed of two thin layers of compact tissue, enclosing between them a variable quantity of cancellous tissue, e.g., occipital bone.

The *irregular bones*, because of their peculiar shape, cannot be grouped under any of the preceding heads. A vertebra is a good example. The bones of the ear are so small that they are described as *ossicles*.

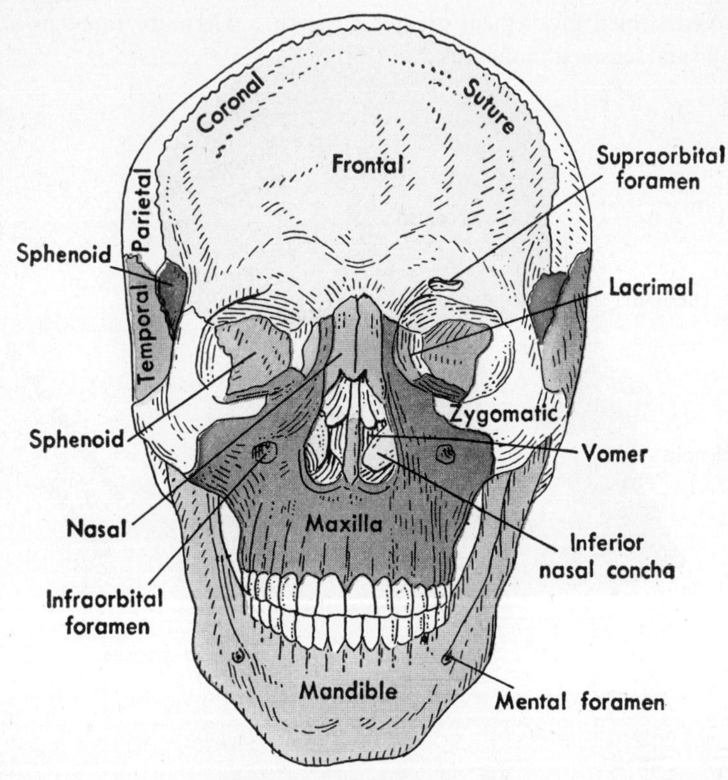

Fig. 51. Front view of skull.

Processes and depressions. The surface of bones shows projections, or *processes*, and depressions, called *fossae* or *cavities*. Qualifying adjectives or special names may be used to describe them. Both processes and depressions are classified as (1) *articular*—those serving for connection of bones to form joints, and (2) *non-articular*—those serving for the attachment of ligaments and muscles. The following terms are used:

Cavities

Fissure. A narrow slit

Foramen. A hole or orifice through which blood vessels, nerves, and ligaments pass

Meatus, or canal. A long, tubelike passageway

Sinus[3] and *antrum.* Applied to cavities within certain bones

Groove, or *sulcus.* A furrow

Fossa. A depression in or upon a bone

Processes

Process. Any marked bony prominence

Condyle. A rounded or knucklelike process

Tubercle. A small rounded process

Tuberosity. A large rounded process

Trochanter. A very large process

Crest. A narrow ridge of bone

Spine, or *spinous process.* A sharp, slender process

Head. A portion supported on a constricted part, or *neck*

DIVISIONS OF THE SKELETON

In taking up the various divisions of the skeleton, it will be considered as consisting of:

Axial Skeleton	(1) Head, or skull	Cranium; Face
	(2) Hyoid	
	(3) Trunk	Vertebrae; Sternum; Ribs
Appendicular Skeleton	(4) Upper extremities	
	(5) Lower extremities	

The head, or skull, rests upon the spinal column and is composed of the cranial and facial bones. It is divisible into *cranium,* or *brain case,* and *anterior region,* or *face.*

Bones of the Cranium

Occipital, base of skull	1	8
Parietal, crown	2	
Frontal, forehead	1	
Temporal, ear region	2	
Ethmoid, between cranial and nasal cavities	1	
Sphenoid, base of brain and back of orbit	1	

[3] The term *sinus* is also used in surgery to denote a narrow tract leading from the surface down to a cavity.

The occipital bone is situated at the back and base of the skull. The internal surface is deeply concave and presents many eminences and depressions for parts of the brain. The *foramen magnum* is a large opening in the inferior portion of the bone for the transmission of the medulla oblongata where it narrows to join the spinal cord. At the sides of the foramen magnum it has two processes called condyles, which articulate with the atlas.

The external surface is convex and presents midway between the summit of the bone and the foramen magnum a projection—the ex-

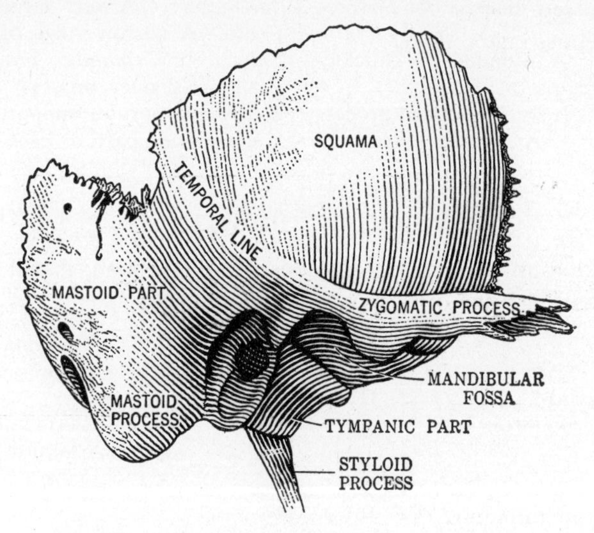

Fig. 52. The right temporal bone, outer surface.

ternal occipital protuberance—which can be felt through the scalp. From this a median ridge—the external occipital crest—leads to the foramen magnum. The protuberance and crest give attachment to the ligamentum nuchae (Fig. 74). Two curved lines extend laterally from the protuberance. To these lines and the expanded plate behind the foramen magnum—the squama—several muscles are attached.

The parietal bones, right and left, form by their union the greater part of the sides and roof of the skull. The external surface is convex and smooth; the internal surface is concave and presents eminences and depressions for lodging the convolutions of the brain, and numerous furrows for the ramifications of arteries supplying blood to the dura mater, which covers the brain.

The frontal bone forms the forehead and part of the roof of the orbits and of the nasal cavity. The arch, formed by part of the frontal bone over the eye, is sharp and prominent and is known as the supraorbital margin. Just above the supraorbital margins are hollow spaces, the *frontal sinuses*, that are filled with air and open into the nose. In the upper and outer angle of each orbit are two depressions called lacrimal fossae, in which lie the lacrimal glands, which secrete tears. At birth the bone consists of two pieces, which later become united along the middle line by a suture that runs from

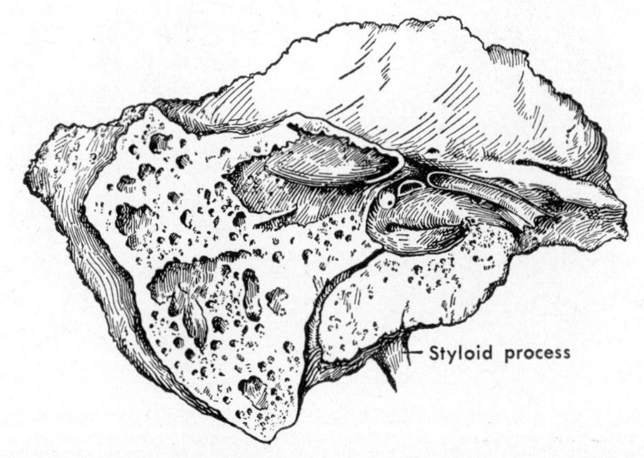

Fig. 53. Right temporal bone, sectioned to show mastoid cells.

the vertex of the bone to the root of the nose. This suture usually becomes obliterated within a few years after birth.

The temporal bones,[4] right and left, are situated at the sides and base of the skull. They are divided into five parts, the squama, the petrous, mastoid, and tympanic parts, and the styloid process.

The *squama*, a thin, expanded portion, forms the anterior and upper part of the bone. A curved line, the temporal line or supramastoid crest, runs backward and upward across its posterior part. Projecting from the lower part of the squama is the long, arched, zygomatic process, which articulates with the temporal process of the zygomatic bone.

[4] Named temporal from the Latin *tempus,* "time," as it is on the temples that the hair first becomes gray and thin.

The *petrous* portion is shaped like a pyramid and is wedged in at the base of the skull between the sphenoid and occipital bones. The internal ear, the essential part of the organ of hearing, is contained in a series of cavities in the petrous portion. Between the squamous and the petrous portions is a socket, called the mandibular fossa, for the reception of the condyle of the mandible.

The *mastoid* portion projects downward behind the opening of the meatus. It is filled with a number of connected spaces, called

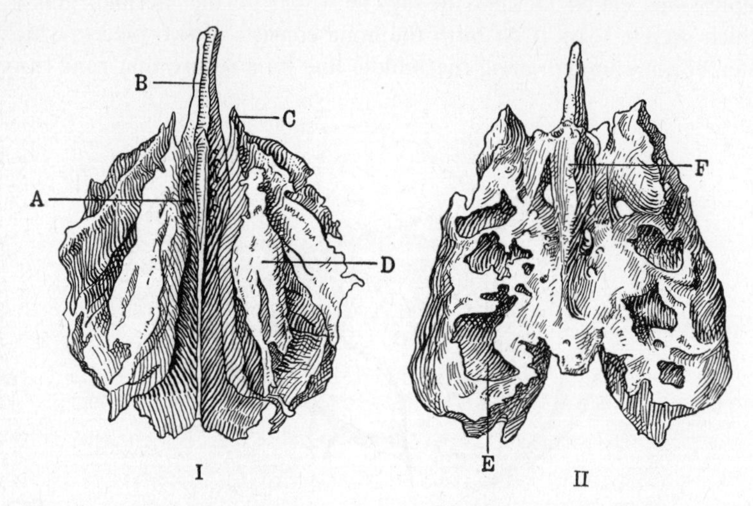

Fig. 54. The ethmoid bone. *I,* under surface; *II,* upper surface showing ethmoid cells. *A,* horizontal plate showing foramina; *B,* perpendicular plate; *C,* ala; *D,* labyrinth or lateral mass; *E,* ethmoid cell; *F,* crista galli.

mastoid cells or sinuses, which contain air and communicate with the cavity of the middle ear. Inflammation of the lining of these sinuses is known as mastoiditis. The bony partition between the mastoid cells[5] and the brain is thin. A danger in mastoiditis is that the infection may penetrate the bone, reach the meninges or coverings of the brain, and cause meningitis. The purpose in operating is to bring about external drainage to prevent the infection reaching the meninges.

The *tympanic* portion is a curved plate of bone below the squama

[5] *Cells.* Histologically, the word *cell* refers to one of the component units of the body, such as an epithelial cell. Occasionally it refers to such minute chambers as mastoid cells.

and in front of the mastoid process. It forms a part of the acoustic meatus leading to the internal ear.

The *styloid* is a slender, pointed process that projects downward from the under surface of the temporal bone. To its distal part are attached ligaments and some of the muscles of the tongue.

The ethmoid bone is a light, cancellous bone consisting of a horizontal or cribriform plate, a perpendicular plate, and two lateral masses, or labyrinths. The horizontal plate forms the roof of the nasal cavity and closes the anterior part of the base of the cranium. It is pierced by numerous foramina, through which the olfactory nerve fibers pass from the mucous membrane of the nose to the

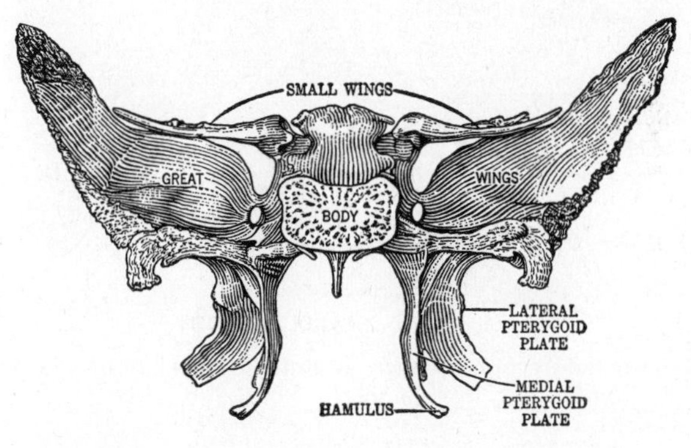

Fig. 55. The sphenoid bone, seen from behind.

olfactory bulb. Projecting upward from the horizontal plate is a smooth, triangular process called the *crista galli* (cock's comb), which serves for the attachment of the *falx cerebri* (Fig. 182). Descending from the horizontal plate is the perpendicular plate, which helps to form the upper part of the nasal septum. On either side, the lateral masses form part of the orbit and part of the corresponding nasal cavity. The lateral masses contain a number of thin-walled cavities, the ethmoidal cells or sinuses, which communicate with the nasal cavity. Descending from the horizontal plate on either side of the septum are two processes of thin, cancellous, bony tissue, the superior and middle conchae.

The sphenoid bone is situated at the anterior part of the base of the skull and binds the other cranial bones together. In form it resembles a bat with extended wings and consists of a body, two great and two small wings extending outward from the sides of the body, and two pterygoid processes which project downward. The body is joined to the ethmoid in front and the occipital behind. It contains cavities called sphenoidal sinuses, which communicate with the nasopharynx. The upper portion of the body presents a fossa

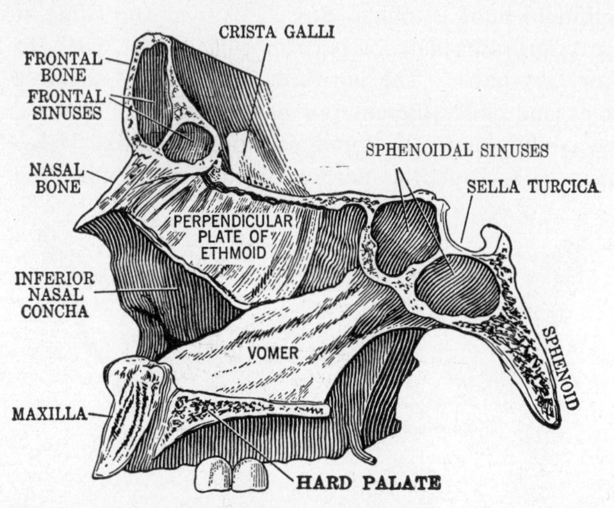

Fig. 56. Sagittal section of face, a little to the left of the middle line, showing the vomer and its relations. Note also frontal and sphenoidal sinuses. (Gerrish.)

with anterior and posterior eminences. This is called the *sella turcica*, from its resemblance to a Turk's saddle. The *hypophysis cerebri* is lodged in the sella turcica.

Bones of the Face[6]

Bone	Number	Total
Nasal	2	14
Vomer	1	
Inferior nasal concha (inferior turbinated)	2	
Lacrimal	2	
Zygomatic (malar)	2	
Palatine (palate)	2	
Maxilla (upper jaw)	2	
Mandible (lower jaw)	1	

The nasal bones are two small oblong bones placed side by side at the middle and upper part of the face, forming by their junction

[6] The BNA classifies the inferior nasal conchae, the lacrimals, the nasals, and the vomer as cranial bones.

the upper part of the bridge of the nose, the lower part being formed by the nasal cartilages.

The vomer is a single bone placed at the lower and back part of the nasal cavity, forming part of the central septum of the nasal cavity. It is thin and varies in different individuals, being frequently bent to one or the other side, thus making the nasal chambers of unequal size.

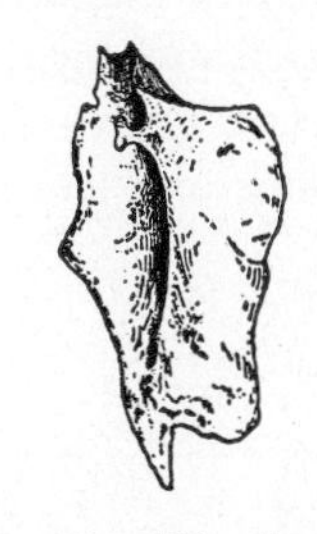

Fig. 57. Lacrimal bone.

The inferior nasal conchae (turbinated bones) are situated in the nostril on the outer wall of each side. Each consists of a layer of thin, cancellous bone curled upon itself like a scroll. They are below the superior and middle conchae of the ethmoid bone. Structural deviations and abnormal conditions of these bones and the membranes covering them are involved in some of the more common nasal abnormalities.

The lacrimal bones are situated at the front part of the inner wall of the orbit and somewhat resemble a fingernail in form, thinness, and size. They are named lacrimal because they contain part of the canal through which the tear duct runs.

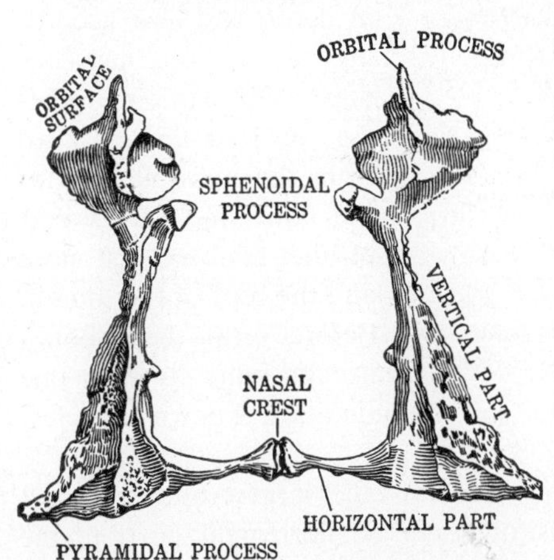

Fig. 58. The two palatine bones in there natural position, viewed from behind. (Gerrish.)

The zygomatic, or **malar, bones** form the prominence of the cheeks and part of the outer wall and floor of the orbits. A long, narrow, and serrated process of each malar bone, called the temporal process, projects backward and articulates with the zygomatic process of the temporal bone, thus forming the zygomatic arch of each side.

Palatine bones. Each one is shaped somewhat like an L and consists of a horizontal part, a vertical part, and three processes, the pyramidal, orbital, and sphenoidal processes. They are situated at the back part of the nasal cavity between the maxillae and the pterygoid processes of the sphenoid and help to form (1) the back part of the roof of the mouth, (2) part of the floor and outer wall of the nasal cavities, and (3) a very small portion of the floor of the orbit.

The maxillae, or **upper jawbones,** are two in number, right and left, and form by their union the whole of the upper jaw. Each bone assists in forming (1) part of the floor of the orbit, (2) the floor and lateral wall of the nasal cavities, and (3) the greater part of the roof of the mouth.

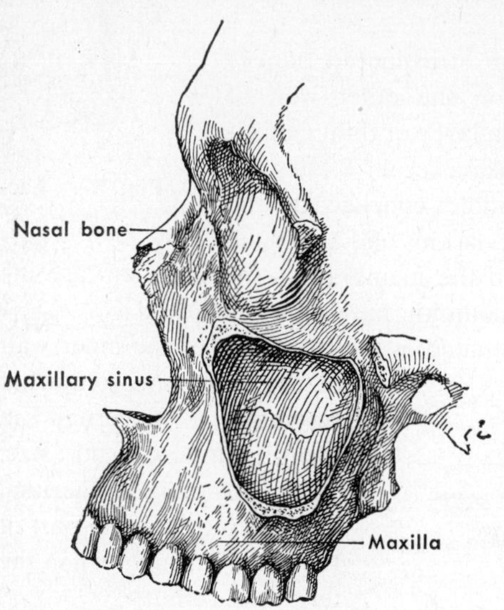

Fig. 59. The left maxilla. Outer surface cut away in part to show maxillary sinus, or antrum. (Gerrish.)

Each consists of a body and four processes. The body of the bone contains a large cavity known as the *antrum of Highmore,* or *maxillary sinus,* which opens into the nose. The alveolar process is excavated into cavities varying in depth and size according to the teeth they contain. The palatine process projects medialward from the nasal surface of the bone and forms part of the floor of the nose and the roof of the mouth. Before birth these bones usually unite to form one bone. When they fail to do so, the condition known as cleft palate results.[7]

The mandible, or **lower jawbone,** is the largest and strongest bone of the face and consists of a curved horizontal portion, the body, and two perpendicular portions, the rami. The superior or alveolar border of the body is hollowed out into cavities for the reception of the teeth. Each ramus has a condyle which articulates with the mandibular fossa of the temporal bone and a coronoid

[7] In its simplest form cleft palate is a divided uvula. In a more severe form the cleft extends through the soft palate, the posterior part of the hard palate may be involved, or the cleft may extend through the maxilla between the teeth. It may affect one or both sides and may be complicated with a cleft in the lip (harelip or cleft lip).

process which gives attachment to the temporal muscle and some of the fibers of the buccinator. The deep depression between these two processes is called the mandibular notch. The mental foramen, which is just below the first molar tooth, serves as a passageway for the inferior dental nerve, which is a terminal branch of the mandibular nerve, which in turn is a branch of the fifth, or trigeminal, nerve. Branches of the inferior dental nerve supply the molar and premolar teeth of the lower jaw.

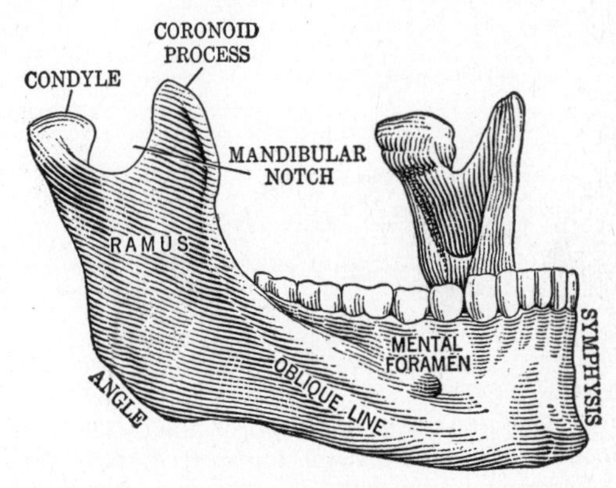

Fig. 60. The mandible, viewed from the right and a little in front. (Gerrish.

At birth the mandible consists of two parts, which join at the symphysis in front and form one bone, usually during the first year. It undergoes several changes in shape during life, due mainly to the first and the second dentition and to the loss of teeth in the aged with the subsequent absorption of that part of the bone which contained them.

The skull as a whole. The cranium is a firm case, or covering, for the brain. Four[8] of the eight bones which form this bony covering are flat bones and consist of two layers of compact tissue, the outer one thick and tough, the inner one thinner and more brittle. The base of the skull is much thicker and stronger than the walls and roof; it presents a number of openings, or foramina, for the passage of the cranial nerves, blood vessels, and other structures.

The bones of the cranium, most of which are thin and flat, begin to develop in early fetal life. Ossification of these bones is gradual and takes place from ossification centers, generally near the center of the future completed bones. Ossification is not complete at birth; hence, membrane-filled spaces are found between the bones. These spaces are called fontanels.

The fontanels. At birth there may be many of these fontanels. The shape and location of six of them are quite constant.

[8] The occipital, two parietals, and the frontal.

The *anterior*, or *bregmatic*, *fontanel* is the largest and is a lozenge-shaped space between the angles of the two parietal bones and the two segments of the frontal bone. Normally this fontanel closes at about 18 months of age.

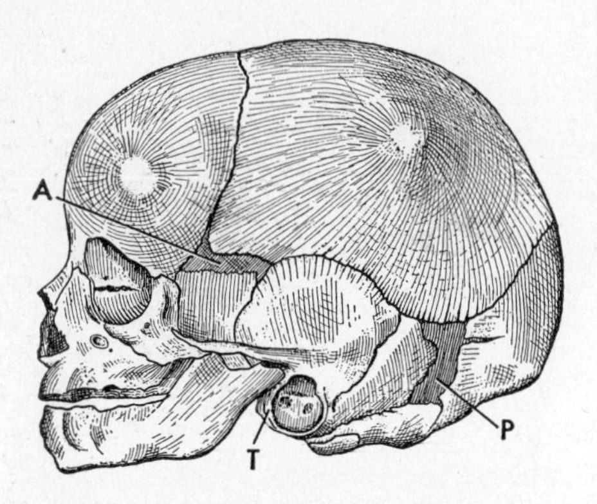

Fig. 61. Skull of newborn infant, side view. *A,* anterolateral fontanel; *P,* posterolateral fontanel; *T,* tympanic ring. (Toldt.)

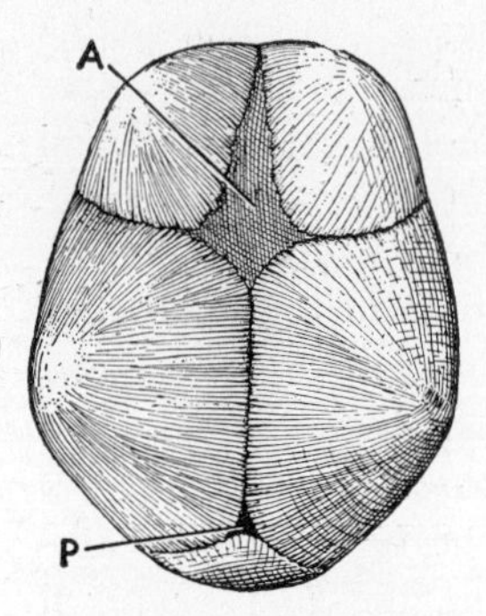

Fig. 62. Same, seen from above. *A,* anterior fontanel; *P,* posterior fontanel. (Toldt.)

In abnormal conditions the fontanel may close much earlier or much later. In cases of retarded brain growth, called microcephalus, it closes early. In hydrocephalus the increased pressure may cause it to remain open. In rickets and cretinism which are not yielding to treatment it may not close until much later.

The *posterior,* or *occipital, fontanel* is much smaller in size and is a triangular space between the occipital and two parietal bones. Usually this closes by an extension of the ossifying process a few months after birth.

There are two *anterolateral*, or *sphenoidal*, *fontanels* at the junction of the frontal, parietal, temporal, and sphenoid bones. They are quite small and usually close by the third month after birth.

There are two *posterolateral*, or *mastoid*, *fontanels* at the junction of the parietal, occipital, and temporal bones. They decrease in size but usually do not close entirely until the second year.

The membranous tissue between the cranial bones at the sutures and fontanels allows more or less overlapping during labor, thus reducing the diameters of the skull. This is called *molding* and accounts for the elongated shape of the head of a newborn infant, particularly if the labor has been long.

The sinuses of the head. Four air sinuses communicate with each nasal cavity: the frontal and the ethmoidal open into the nasal cavity, the sphenoidal opens into the nasopharynx, and the maxillary, or antrum of Highmore, opens on the lateral wall of each nasal passage. The mucous membrane which lines the nose also lines

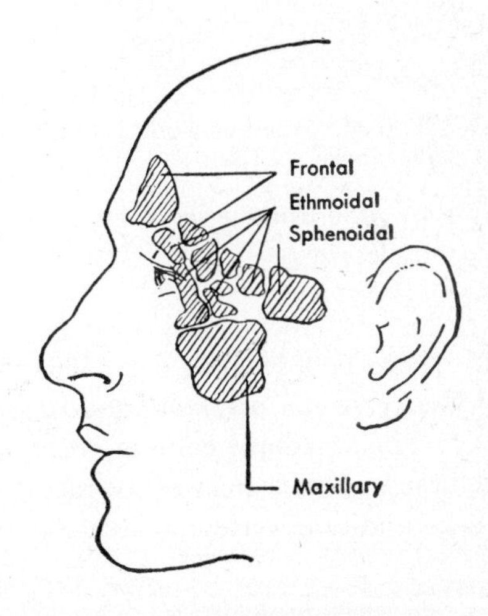

Fig. 63. Sinuses projected to the surface of the face. All except the maxillary sinus are near the central line of the skull when viewed from the front.

See also Fig. 53 for mastoid cells; Fig. 54 for ethmoid cells; Fig. 56 for sphenoidal and frontal sinuses; Fig. 59 for maxillary sinuses.

these sinuses, and inflammation of this membrane may extend into any of them, causing *sinusitis*. The mastoid cells are comparable to the sinuses and are lined by an extension of the same mucous membrane that lines the sinuses.

At birth the skull is proportionately larger than other parts of the skeleton, and the facial portion is small. The small size of the maxillae and mandible, the noneruption of the teeth, and the small size of the sinuses and nasal cavities account for the smallness of the face. With the eruption of the first teeth there is an enlargement of the face and jaws. This enlargement is much more pronounced after the eruption of the second set of teeth. Usually the skull becomes thinner and lighter in old age, but occasionally the inner table hypertrophies, causing an increase in weight and thickness (pachycephalia). The most noticeable feature of the old skull is the decrease in the size of the maxillae and mandible, resulting from the loss of the teeth and the absorption of the alveolar processes.

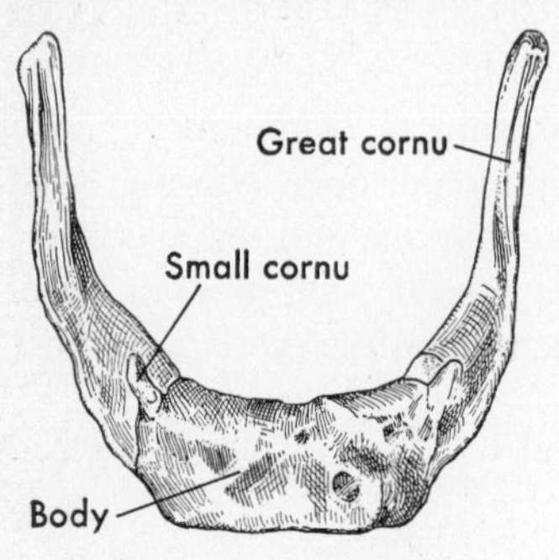

Fig. 64. The hyoid bone, seen from above. (Toldt.)

The hyoid bone is shaped like a horseshoe and consists of a central part called the body and two projections on each side called the greater and lesser cornua. It is suspended from the styloid processes of the temporal bones and may be felt in the neck just above the laryngeal prominence (Adam's apple). It supports the tongue and gives attachment to some of its numerous muscles.

THE TRUNK

The bones which enter into the formation of the *trunk* consist of the *vertebrae, sternum,* and *ribs.*

The vertebral column is formed of a series of bones called vertebrae and in a man of average height is about 71 cm long (28 in.). In youth the vertebrae are 33 in number:

Cervical, in the neck	7	Movable, or true, vertebrae
Thoracic, or dorsal, in the thorax	12	
Lumbar, in the loins	5	
Sacral, in the pelvis	5	Fixed, or false, vertebrae
Coccygeal, in the pelvis	4	

In the three upper portions of the spine the vertebrae are separate and movable throughout life. Those found in the sacral and coccygeal regions are firmly united in the adult, so that they form two bones, five entering into the upper bone, or sacrum, and four into the terminal bone, or coccyx; because of their union the number of vertebrae in the adult is 26.

The vertebrae differ in size and shape, but in general their structure is similar. Seen from above, as in Fig. 69, they consist of a body from which two short, thick processes, called the pedicles, project backward, one on each side. to join with the laminae which unite posteriorly and form the vertebral, or neural, arch. This arch encloses the spinal foramen. Each vertebra has seven processes: four articular, two to connect with bone above, two to connect with bone below; two transverse, one at each side where the pedicle and

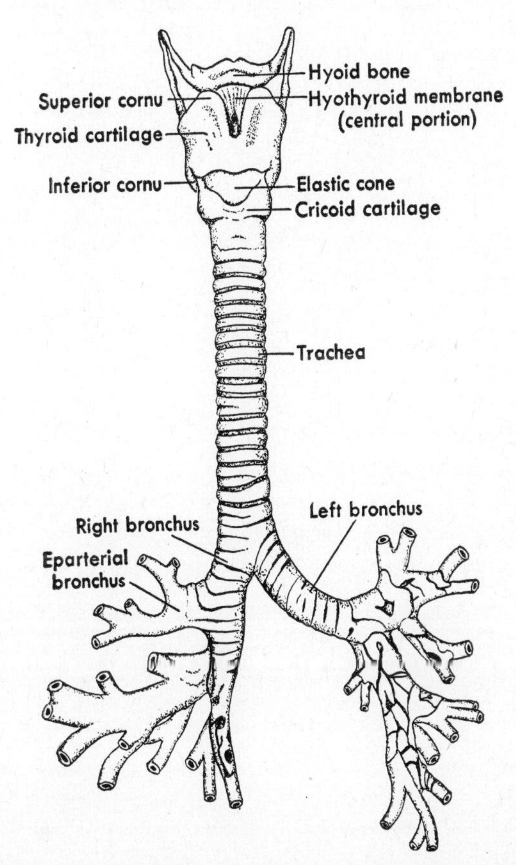

Fig. 65. Larynx and trachea, front view. Shows position of hyoid bone in relation to larynx. (Courtesy of Dr. Carl C. Francis and C. V. Mosby Company.)

lamina join; and one spinous process, projecting backward from the junction of the laminae.

Cervical vertebrae. The bodies of the cervical vertebrae are smaller than the thoracic, but the arches are larger. The spinous

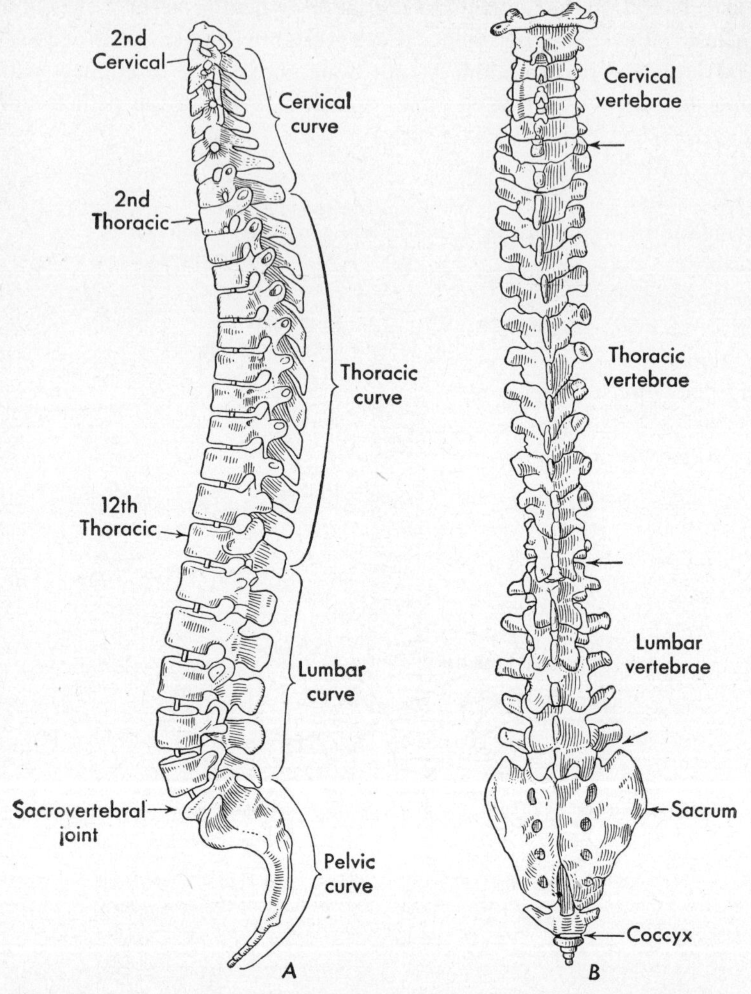

Fig. 66. The vertebral column. *A,* left lateral view showing curves; *B,* dorsal view.

processes are short and are often cleft in two, or bifid. Each transverse process is pierced by a foramen (foramen transversarium) through which nerves, a vertebral artery, and a vein pass.

The first and second cervical vertebrae differ considerably from the rest. The first, or *atlas*, so named from supporting the head, is a bony ring consisting of an anterior and posterior arch and two bulky lateral masses. Each has a superior and inferior articular surface. Each superior surface forms a cup for the corresponding condyle of the occipital bone and thus makes possible the backward and forward movements of the head. The bony ring is divided into an anterior and posterior section by a transverse ligament. The posterior section of this ring contains the spinal cord, and the anterior, or front, section contains the bony projection which arises from the upper surface of the body of the second cervical vertebra, the *epistropheus,* or *axis.* This bony projection, the *odontoid* process, forms a pivot; and around this pivot the atlas rotates when the head is turned from side to side, carrying the skull, to which it is firmly articulated.

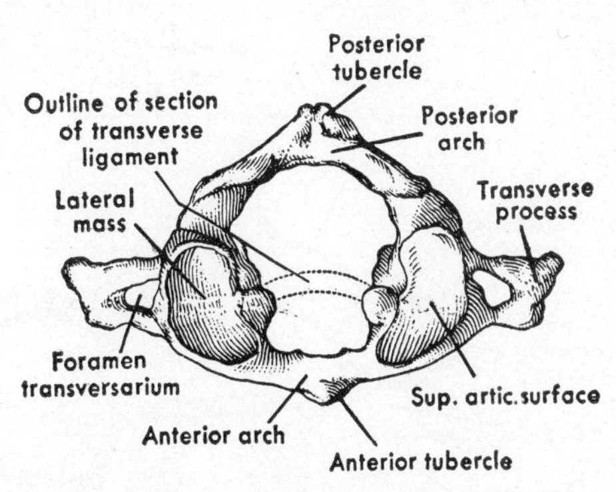

Fig. 67. The atlas, or first cervical vertebra. (After Gray's *Anatomy.*)

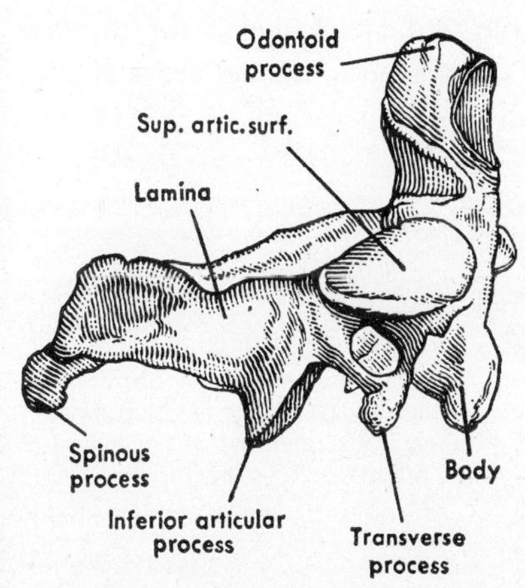

Fig. 68. The epistropheus, or axis, seen from the right side. (After Gray's *Anatomy.*)

Thoracic, or dorsal, vertebrae. The bodies of the thoracic vertebrae are larger and stronger than those of the cervical and have a facet or demifacet for articulation with the heads of the ribs. The transverse processes are longer and heavier than those of the cervical, and all except those of the eleventh and twelfth vertebrae have facets for articulation with the tubercles of the ribs. The spinous processes are long and are directed downward.

Lumbar vertebrae. The bodies of the lumbar vertebrae are the largest and heaviest in the whole spine. The processes are short, heavy, and thick.

Structure of vertebral column. The bodies of the vertebrae, which are piled one upon another, form a strong, flexible column for the support of the cranium and trunk and provide articular surfaces for the attachment of the ribs. The arches form a hollow cylinder for the protection of the spinal cord. Viewed from the side, the vertebral column presents four curves, which are alternately convex and concave. The two concave ones, named thoracic and pelvic, are called primary curves because they exist in fetal life and are designed for the accommodation of viscera. The two convex ones, named cervical and lumbar, are called secondary, or compensatory, curves because they are developed after birth. The cervical curve begins its development

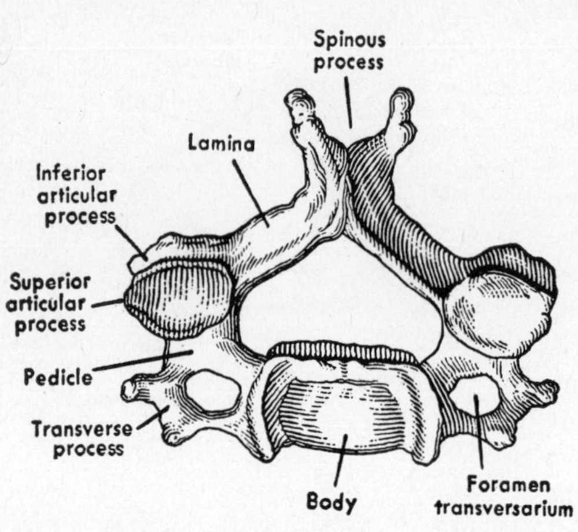

Fig. 69. A cervical vertebra, viewed from above. (After Gray's *Anatomy.*)

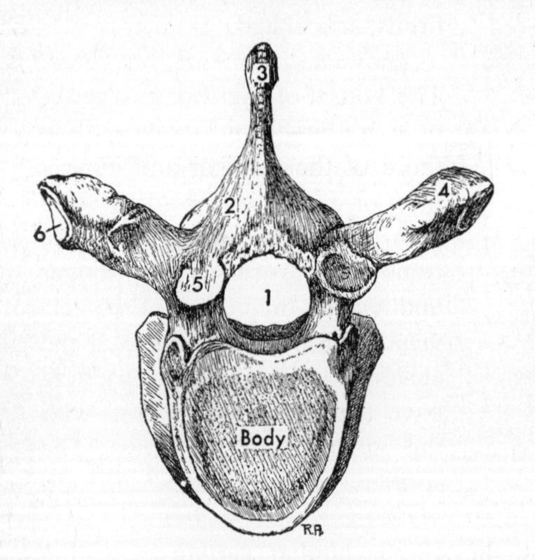

Fig. 70. Sixth thoracic vertebra, seen from above. *1,* spinal foramen; *2,* lamina; *3,* spinous process; *4,* transverse process; *5,* superior articular process; *6,* facet for tubercle of rib. (Toldt.)

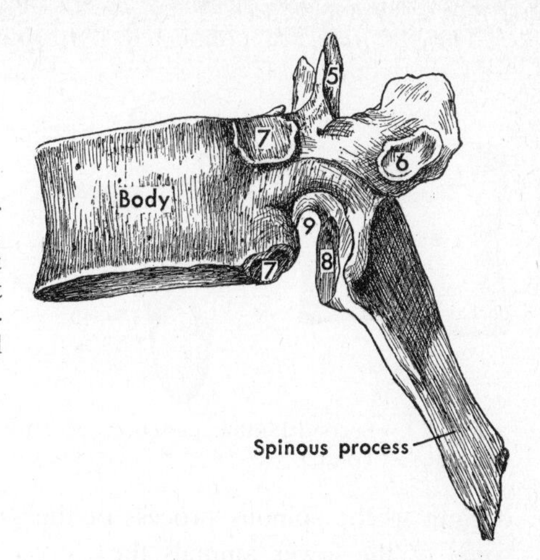

Fig. 71. Sixth thoracic vertebra, seen from the left side. *5,* superior articular process; *6,* facet for tubercle of rib; *7,* demifacet for head of rib; *8,* inferior articular process; *9,* inferior vertebral notch. (Toldt.)

when the child is able to hold up his head (at about 3 or 4 months) and is well formed when he sits upright (at about 19 months). The lumbar develops when the child begins to walk (from 12 to 18 months).

The joints between the bodies of the vertebrae are slightly movable, and those between the arches are freely movable. The *bodies* are connected (1) by disks of fibrocartilage placed between the vertebrae; (2) by the *anterior longitudinal ligament,* which extends along the anterior surfaces of the bodies of the vertebrae from the axis to the sacrum; and (3) by the *posterior longitudinal ligament,* which is inside the vertebral canal and extends along the posterior surfaces of the bodies from the axis to the sacrum.

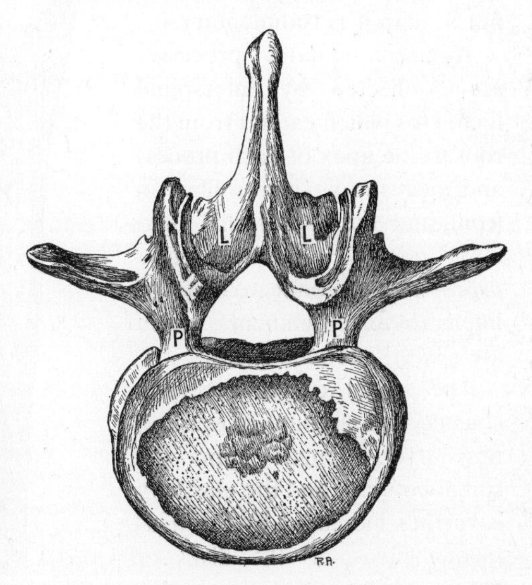

Fig. 72. Second lumbar vertebra, seen from above. *P,* pedicle; *L,* lamina. (Toldt.)

The *laminae* are connected by broad, thin ligaments called the *ligamenta flava* (*ligamenta subflava*).

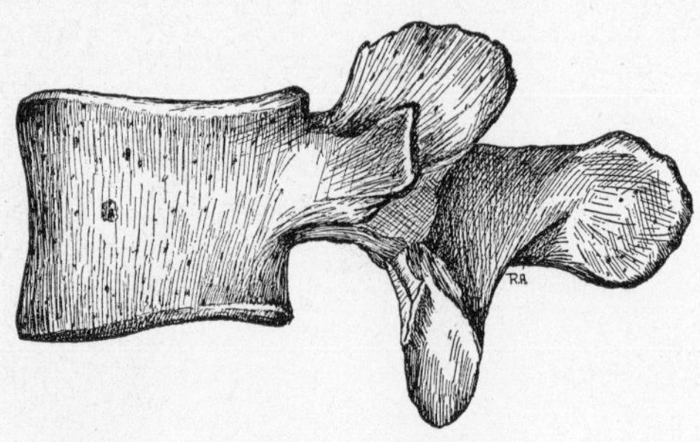

Fig. 73. Second lumbar vertebra, seen from left side. (Toldt.)

The *spinous processes* are connected at the apices by the supraspinal ligament, which extends from the seventh cervical to the sacrum. It is continued upward as the *ligamentum nuchae,* which extends from the protuberance of the occiput to the spinous process of the seventh cervical vertebra. In some of the lower animals the ligamentum nuchae serves to sustain the weight of the head, but in man it is rudimentary.

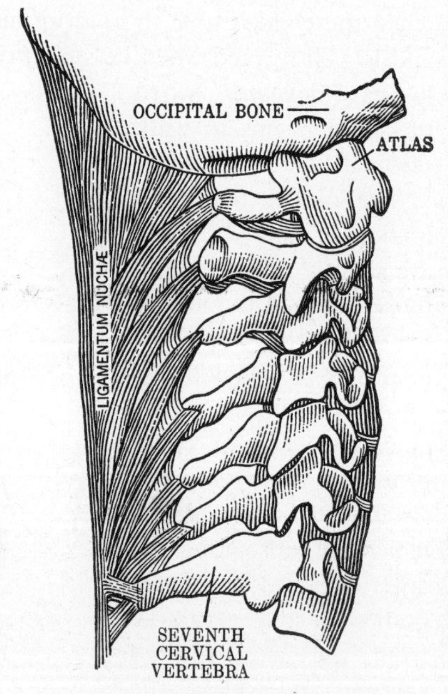

Fig. 74. The ligamentum nuchae, seen from the right side. (Henle.)

Adjacent spinous processes are connected by interspinal ligaments which extend from the root to the apex of each process and meet the ligamenta flava in front and the supraspinal ligament behind. The *transverse processes* are connected by the intertransverse ligaments, which are placed between them.

The spinal curves confer a considerable amount of springiness and strength upon the spinal column, and the elasticity is further increased by the ligamenta flava and the disks of fibrocartilage. These pads also mitigate the effects of concussion arising from falls or blows. The vertebral column is freely movable, being capable of bending forward freely, backward and from side to side less freely.[9] In the

[9] Certain exercises increase the flexibility of the spine to a marked degree.

cervical and thoracic regions a limited amount of rotation is possible.

Posture. The weight of the body should rest evenly on the two hip joints. A perpendicular dropped from the ear should fall through shoulder, hip, and ankle (Fig. 75). In this position the chest is up, the head erect, the lower abdominal muscles are retracted, and the body is well balanced and functioning efficiently.

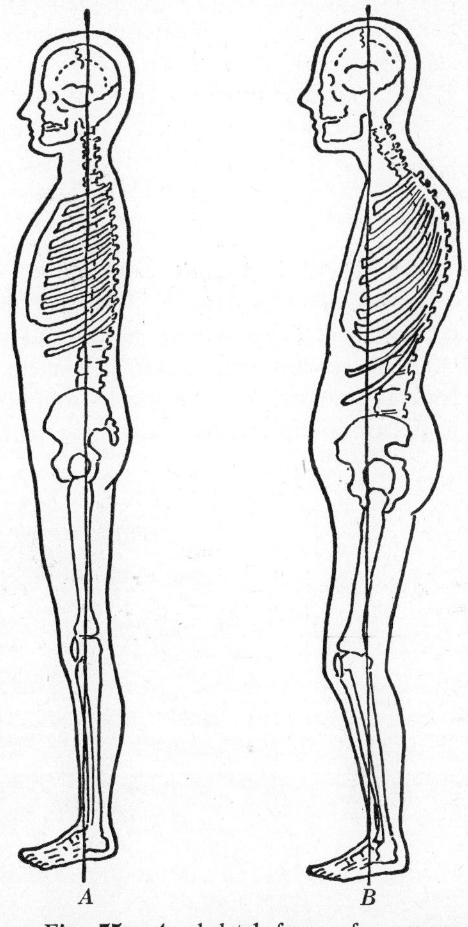

Fig. 75. *A*, skeletal form of a person with good body mechanics; *B*, skeletal form of a person with poor body mechanics. (Courtesy of the Children's Bureau, U.S. Department of Health, Education, and Welfare.)

As a result of postural habits,[10] injury, or disease, the normal curves may become exaggerated and are then spoken of as *curvatures.* If the thoracic curve is exaggerated, it is called *kyphosis* or humpback; if the exaggeration is in the lumbar region, it is called *lordosis* or hollow back. If the curvature is lateral, i.e., toward one side, it is called *scoliosis.* Lateral curvature is usually to the right side, because even in normal people there is often a slight curve toward the right in the thoracic region.

It occasionally happens that the laminae of a vertebra do not unite and a cleft is left in the arch (*spina bifida*). As a result the membranes and the spinal cord itself may protrude, forming a tumor on the child's back. This most often occurs in the lumbosacral region, though it may occur in the thoracic or cervical region.

[10] See publications Nos. 164 and 165 of the Children's Bureau of the U.S. Department of Health, Education, and Welfare.

The sacrum is formed by the union of the five sacral vertebrae. It is a large triangular bone situated like a wedge between the coxal bones and is curved upon itself in such a way as to give increased capacity to the pelvic cavity.

The coccyx is usually formed of four small segments of bone and is the most rudimentary part of the vertebral column.

THORAX

The thorax is a bony cage formed by the sternum and costal cartilages, the ribs, and the bodies of the thoracic vertebrae. It is cone-shaped, being narrow above and broad below, flattened from before backward and shorter in front than in back. In infancy the chest is rounded and the width from shoulder to shoulder and the depth from the sternum to the vertebrae are about equal. With

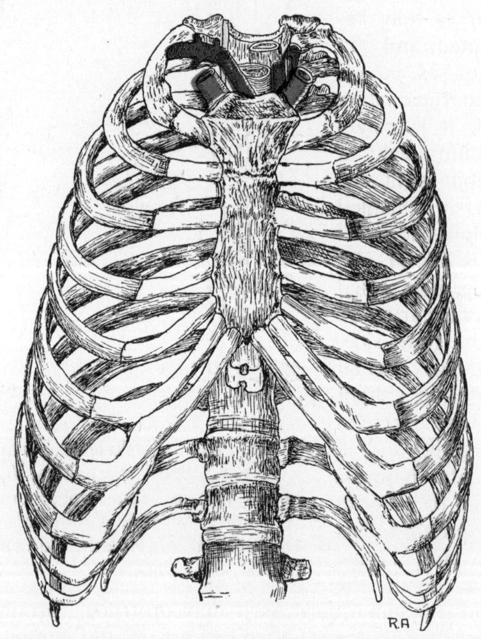

Fig. 76. Bones of thorax, seen from the front. Lying in the superior aperture of the "thoracic basket," note: esophagus (close to vetebral column), trachea, vagus nerves (yellow), arteries (red), veins (blue). (Toldt.)

growth the width increases more than the depth. The thorax supports the bones of the shoulder girdle and upper extremities and contains the principal organs of respiration and circulation.

The sternum, or **breastbone,** is a flat, narrow bone about 6 in. long, situated in the median line in the front of the chest. It develops as three separate parts. The upper part is named the *manubrium;* the middle and largest part, the *body,* or *gladiolus;* the lowest portion, the *ensiform,* or *xiphoid, process.* On both sides of the manubrium and body are notches for the reception of the sternal ends of the upper seven costal cartilages. The xiphoid process has no ribs attached to it but affords attachment for some of the abdominal muscles.

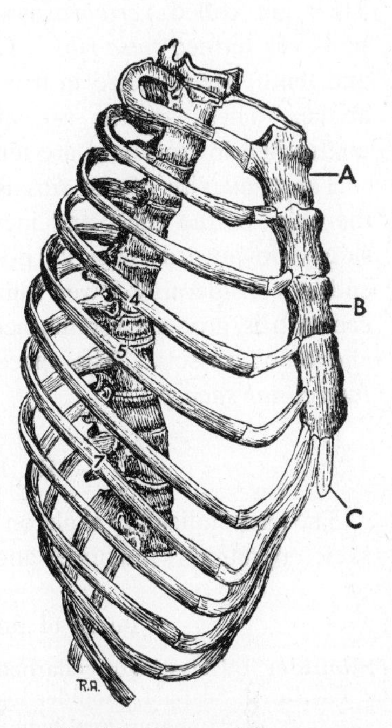

Fig. 77. Bones of thorax, seen from right side. Between the fourth and fifth ribs, note articulation of the head and tubercle of the seventh rib. *A,* manubrium; *B,* body showing articular notches of ribs and lines of union of parts of body; *C,* xiphoid process.

At birth the sternum consists of several unossified portions, the body alone developing from four centers. Union of the centers in the body begins at about puberty and proceeds from below upward until at about 25 years of age they are all united. Sometimes by 30 years of age, more often after 40, the xiphoid process becomes joined to the body. In advanced life the manubrium may become joined to the body by bony tissue. Posture, activity (play, work), and dietary hygiene have much to do with shaping the sternum and the thoracic cavity.

The ribs (costae), 24 in number, are situated 12 on each side of the thoracic cavity. They are elastic arches of bone consisting of a body, or shaft, and two extremities, the posterior, or vertebral, and the anterior, or sternal. Each rib is connected with the thoracic vertebra by the head and tubercle of the posterior extremity. The head fits into a facet formed on the body of one vertebrae or formed by the adjacent bodies of two vertebrae;[11] the tubercle articulates with the transverse processes. Strong ligaments surround and bind these articulations but permit slight gliding movements.[12]

The anterior extremities of each of the first seven pairs are connected with the sternum in front by means of the costal cartilages. They are called *vertebrosternal* or *true ribs*. The remaining five pairs are termed *false ribs*. Of these, the upper three, eighth, ninth, and tenth, are attached in front to the costal cartilages of the next rib above. These are the *vertebrochondral ribs*. The two lowest are unattached in front and are termed *floating* or *vertebral ribs*.

The convexity of the ribs is turned outward, giving roundness to the sides of the chest and increasing the size of its cavity; each rib slopes downward from its posterior attachment, so that its sternal end is considerably lower than its vertebral. The lower border of each rib is grooved for the accommodation of the intercostal nerves and blood vessels. The spaces left between the ribs are called the *intercostal spaces*.

THE EXTREMITIES

The appendicular skeleton consists of the appendages of the skeleton, namely, the upper and lower extremities.

Bones of the Upper Extremities

	Bone	Number	Total
Shoulder Girdle	Clavicle, or collarbone	2	64
	Scapula, shoulder blade	2	
Upper Limb	Humerus, arm bone	2	
	Ulna, elbow bone	2	
	Radius, small bone of forearm	2	
	Carpus, wrist (ossa carpi)	16	
	Metacarpus, body of hand	10	
	Phalanges, 2 in thumb, 3 in each finger	28	

The two clavicles and the two scapulae form the *shoulder girdle*, which is incomplete in front and behind. The clavicles articulate with the sternum in front but behind the scapulae are connected to

[11] The heads of the first, tenth, eleventh, and twelfth ribs each articulate with a single vertebra. The heads of the remaining ribs articulate with facets formed by the bodies of two adjacent vertebrae.

[12] In the eleventh and twelfth ribs the articulation between the tubercle and the adjacent transverse process is missing.

the trunk by muscles only. The shoulder girdle serves to attach the bones of the upper extremities to the axial skeleton.

The clavicle, or collarbone, is a long bone with a double curvature, which is placed horizontally at the upper and anterior part of the thorax, just above the first rib. It articulates with the sternum by its inner end, which is called the sternal extremity. Its outer, or acromial, end articulates with the scapula. In the female, the clavicle is generally less curved, smoother, shorter, and more slender than in the male. In those persons who perform considerable manual labor, which brings the muscles connected with this bone into constant action, it acquires considerable bulk.

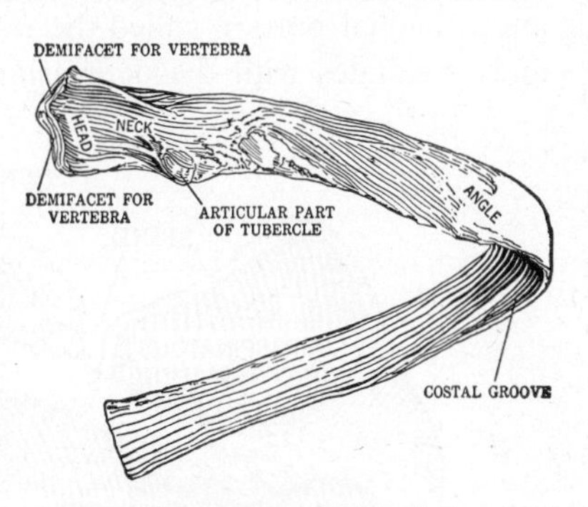

Fig. 78. A central rib of the right side, viewed from behind. (After Gray's *Anatomy*.)

The scapula, or shoulder blade, is a large, flat, triangular bone between the second and seventh ribs at the back of the thorax. It is unevenly divided on its dorsal surface by a prominent ridge, the spine of the scapula, which terminates in a large triangular projection called the *acromion process*, which articulates with the clavicle. Below the acromion process, at the head of the shoulder blade, is a shallow socket, the *glenoid cavity,* which receives the head of the humerus.

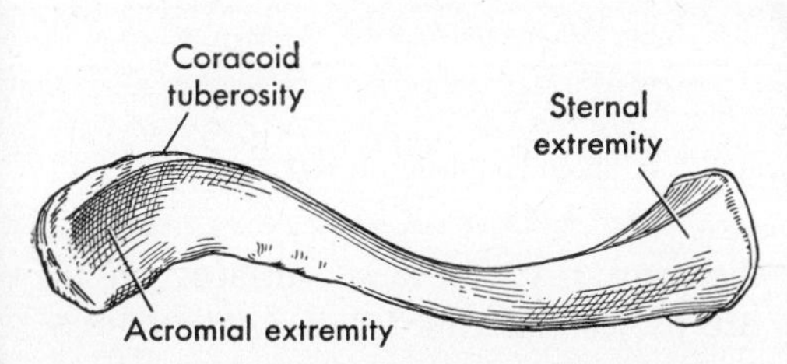

Fig. 79. The right clavicle, seen from above.

The humerus, or arm[13] **bone,** is the longest and largest bone of the upper extremity. Its upper end consists of a rounded head joined to the shaft by a constricted neck and of two eminences, called the *greater* and *lesser tubercles,* between which is the intertubercular (bicipital) groove. The constricted neck above the tubercles is

[13] Anatomically, the word *arm* is reserved for that part of the upper limb which is above the elbow; between the elbow and wrist is the forearm; below the wrist are the hand and fingers.

called the *anatomical neck*, and that below the tubercles the *surgical neck*, because it is so often fractured. The head articulates with the glenoid cavity of the scapula. The lower end of the bone is flattened from before backward and ends below in an articular surface which is divided by a ridge into a lateral eminence called the *capitulum* and a medial portion called the *trochlea*. The capitulum is rounded and articulates with the depression on the head of the radius. The

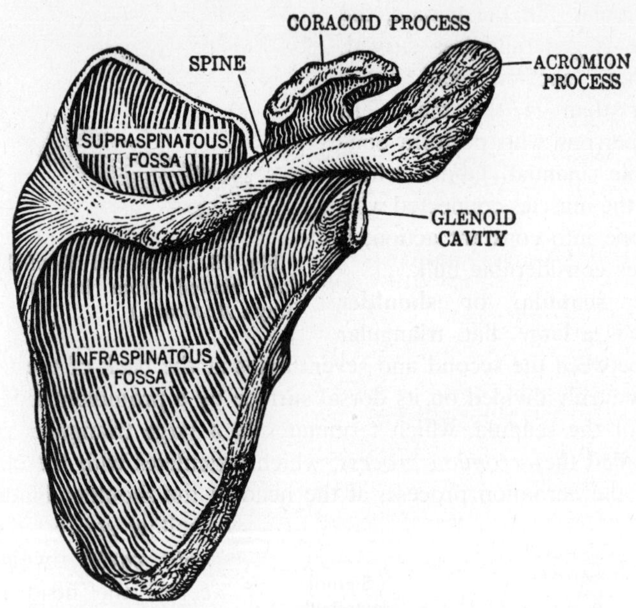

Fig. 80. The right scapula, or shoulder blade; dorsal surface.

trochlea articulates with the ulna. Above these surfaces on the lateral and medial aspects are projections called *epicondyles*.

The ulna, or elbow bone, is the large bone of the forearm and is placed at the medial side of the radius. Its upper extremity shows two processes and two concavities; the larger process forms the prominence of the elbow, called the *olecranon process*. The smaller process is called the *coronoid process*. The trochlea of the humerus fits into the semilunar notch (greater sigmoid cavity) between these two processes. The radial notch (lesser sigmoid cavity) is on the lateral side of the coronoid and articulates with the radius. The lower end of the ulna is small and ends in two eminences; the larger head articulates with the fibrocartilage disk which separates it from the wrist; the smaller is the styloid process, to which a ligament from the wrist joint is attached.

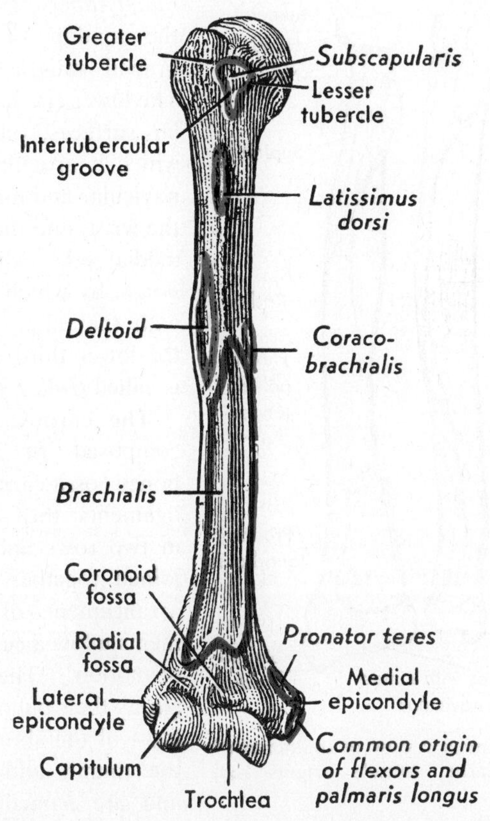

Fig. 81. The right humerus, or arm bone, ventral view. Muscle origins red, insertions blue.

The radius is placed on the lateral side of the ulna and is shorter and smaller than the ulna. The upper extremity presents a head, a neck, and a tuberosity. The head is small and rounded and has a shallow, cuplike depression on its upper surface for articulation with the capitulum of the humerus. A prominent ridge surrounds the head, and by means of this it rotates within the radial notch of the ulna. The head is supported on a constricted neck. Beneath the neck on the medial side is an eminence called the *radial tuberosity,* into which the tendon of the biceps brachii muscle is inserted. The lower end has two articular surfaces, one below, by which it articulates with the navicular and lunate bones of the wrist, and the other at the medial side, called the *ulnar notch,* by which it articulates with the ulna. Fracture of the lower third of the radius is called *Colles' fracture.*[14]

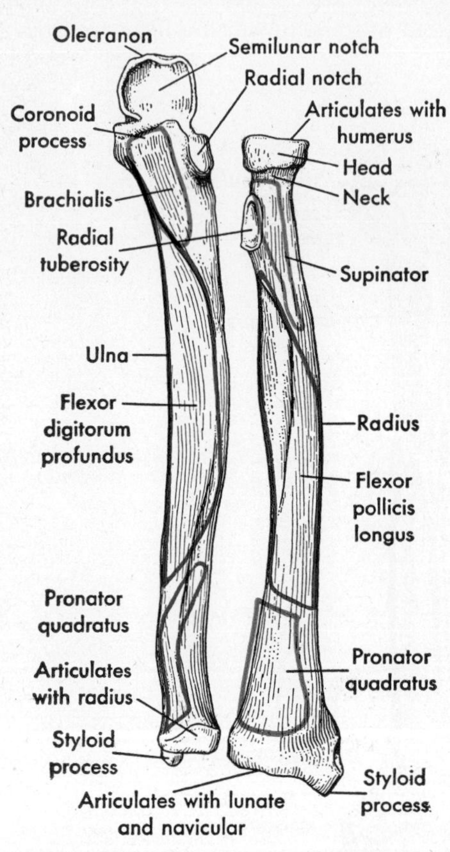

Fig. 82. Anterior view of the bones of the left forearm. Muscle origins red, insertions blue.

The carpus, or wrist, is composed of eight small bones (ossa carpi), united by ligaments; they are arranged in two rows and are closely joined together, yet by the arrangement of their ligaments allow a certain amount of motion. They afford origin by their palmar surface to most of the short muscles of the thumb and little finger and are named as follows:

Proximate, or Upper Row		**Distal, or Lower Row**		
(1) Navicular (scaphoid)	1	(5) Greater multangular (trapezium)	1	8
(2) Lunate (semilunar)	1	(6) Lesser multangular (trapezoid)	1	
(3) Triangular (cuneiform)	1	(7) Capitate (os magnum)	1	
(4) Pisiform	1	(8) Hamate (unciform)	1	

Metacarpus, or body of hand. Each metacarpus is formed by five bones (ossa metacarpalia), numbered from the lateral side.

[14] Abraham Colles, Irish surgeon (1773–1843).

The bones are curved longitudinally, convex behind, concave in front. They articulate at their bases with the second row of carpal bones and with each other. The heads of the bones articulate with the bases of the first row of the phalanges.

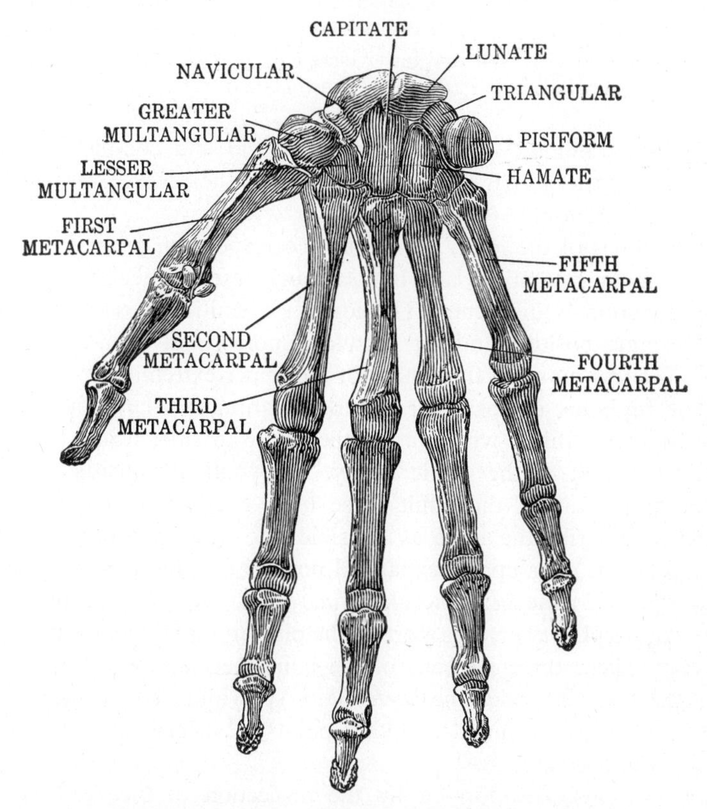

Fig. 83. The bones of the right hand, volar surface.

The phalanges are the bones of the fingers, 14 in number in each hand, three for each finger and two for the thumb. The first row articulates with the metacarpal bones and the second row of phalanges; the second row articulates with the first and third; the third articulates with the second row.

The two hipbones, which articulate with each other in front, form an arch called the *pelvic girdle*. This arch is completed behind by

Bones of the Lower Extremities

Hipbones, ossa coxae or ossa innominata		2	62
Femur, thighbone		2	
Patella, kneecap		2	
Tibia, shinbone, 2	leg	4	
Fibula, small bone of calf, 2			
Tarsus, ossa tarsi		14	
Metatarsus, sole and lower instep		10	
Phalanges, 2 in great toe, 3 in others		28	

the sacrum and the coccyx, forming a rigid and complete ring of bone called the *pelvis*. The pelvis attaches the lower extremities to the axial skeleton.

The bones of the lower extremities correspond in general to those of the upper extremities and bear a rough resemblance to them, but their function is different. The lower extremities support the body in the erect position and are therefore more solidly built, and their parts are less movable than those of the upper extremities.

The hipbone, os coxae, or os innominatum, is a large, irregularly shaped bone which, with that of the opposite side, forms the sides and front wall of the pelvic cavity. In youth it consists of three separate parts. In the adult these have become united, but it is usual to describe the bone as divisible into three portions: (1) the *ilium* (pl. *ilia*), or upper, expanded portion forming the prominence of the hip; (2) the *ischium* (pl. *ischia*), or lower strong portion; (3) the *pubis* (pl. *pubes*), or portion helping to form the front of the pelvis. These three portions of the bone meet and finally ankylose in a deep socket, called the *acetabulum* (cotyloid cavity), into which the head of the femur fits. Other points of special interest to note in the hipbones are:

1. The processes formed by the projection of the crest of the ilium in front, called the *anterior superior iliac spine* and the *anterior inferior iliac spine.* The former is a convenient landmark in making anatomical and surgical measurements.

2. The largest foramen in the skeleton, the *obturator foramen,* situated between the ischium and pubis.

3. The articulation formed by the two pubic bones in front, called the *symphysis pubis,* serving as a convenient landmark in making measurements.

The pelvis, so called from its resemblance to a basin, is strong and massively constructed. It is composed of four bones, the two hipbones forming the sides and front, the sacrum and coccyx completing it behind, and is divided by a narrowed bony ring into the greater, or false, and the lesser, or true, pelvis. The narrowed

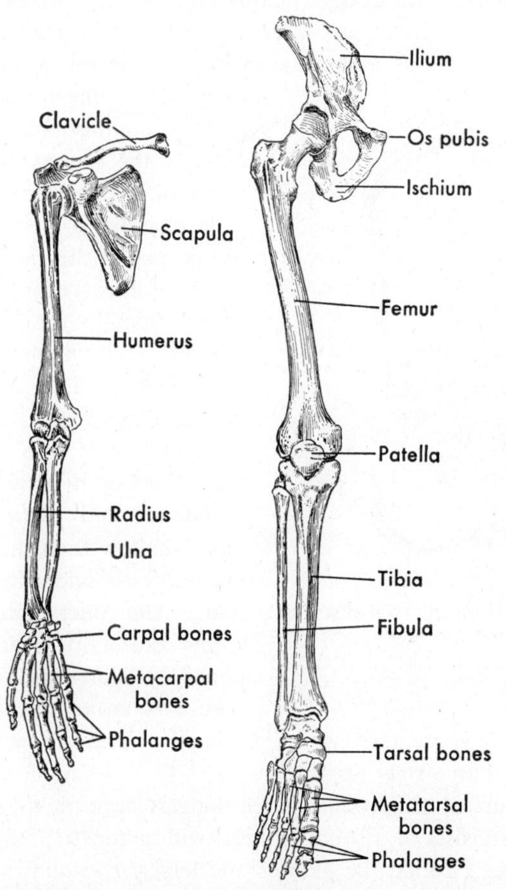

Fig. 84. Comparison of bones of the upper and lower extremities. (Toldt.)

bony ring which is the dividing line is spoken of as the *brim of the pelvis.*

The *greater pelvis* is the expanded portion situated above the brim, bounded on either side by the ilium; the front is filled in by the walls of the abdomen. The lesser pelvis is below and behind the pelvic brim, bounded on the front and sides by the pubes and ischia and behind by the sacrum and coccyx. It consists of an inlet, an outlet, and a cavity. The space included within the brim of the pelvis is called the superior aperture, or inlet; and the space below, between the tip of the coccyx behind and the tuberosities of the ischia on either side, is called the inferior aperture, or outlet. The cavity of the lesser pelvis is a short, curved canal, deeper on the posterior than on its anterior wall. In the adult it contains part of the sigmoid colon,[15] the rectum, bladder, and some of the reproductive organs. The bladder is behind the symphysis pubis; the rectum is in the curve of the sacrum and coccyx. In the female the uterus, tubes, ovaries, and vagina are between the bladder and the rectum.

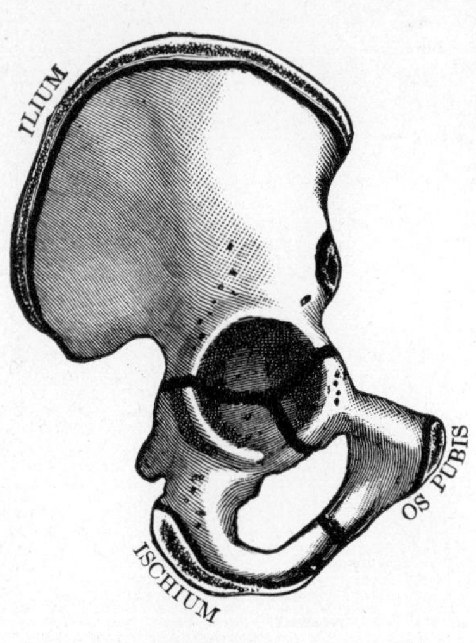

Fig. 85. Hipbone, showing the union of ilium, ischium, and os pubis in the acetabulum (cotyloid cavity). (Gerrish.)

The female pelvis differs from that of the male in those particulars which render it better adapted to pregnancy and parturition. It is more shallow than the male pelvis but relatively wider in every direction. The inlet and outlet are larger and more nearly oval, the bones are lighter and smoother, the coccyx is more movable, and the subpubic arch is greater than a right angle. The subpubic angle in a male is less than a right angle.

The femur, or thighbone, is the longest bone in the body. The upper end consists of a rounded head with a constricted neck, and of two eminences, called the greater and lesser *trochanters*. The head articulates with the cavity in the hipbone, called the acetabulum. The lower extremity of the femur is larger than the upper, is flattened from before backward, and is divided into two large eminences, or *condyles*, by an intervening notch. The condyles are called lateral and medial, and the intervening notch is the *intercondyloid fossa.*

[15] The sigmoid colon is freely movable and may be displaced into the abdominal cavity.

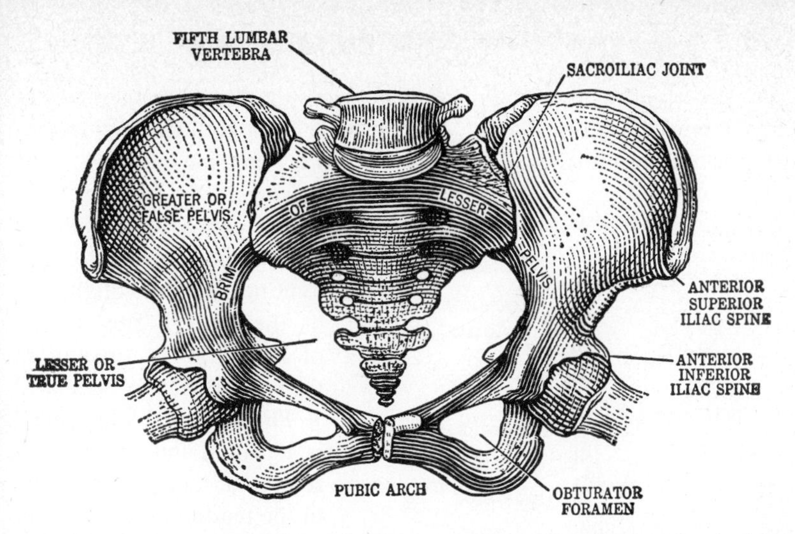

Fig. 86. The female pelvis, ventral view. The disk of fibrocartilage which unites the pubic bones is not shown in order that the end of the coccyx may be seen.

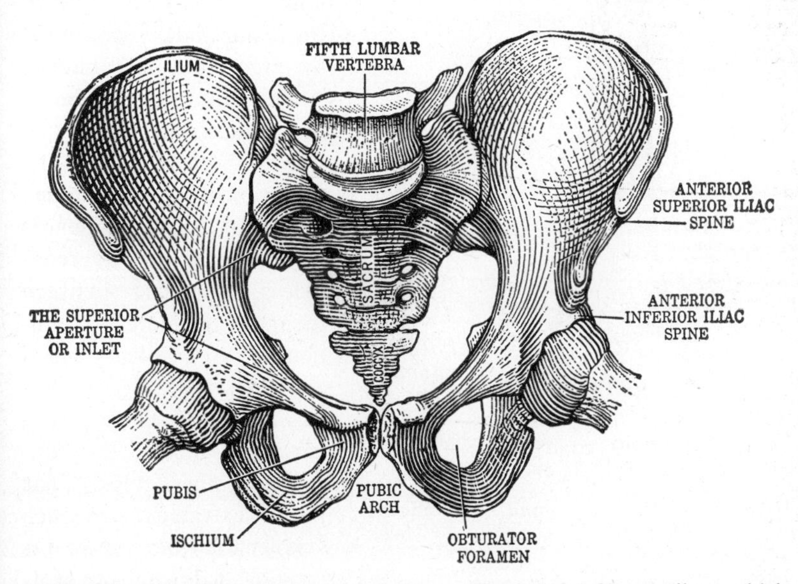

Fig. 87. The male pelvis, ventral view. The disk of fibrocartilage which unites the pubic bones is not shown in order that the end of the coccyx may be seen.

The lower end of the femur articulates with the tibia and the patella. In the erect position it is not vertical, being separated from its fellow by the entire breadth of the pelvis. The bone inclines gradually downward and inward, so as to approach its fellow below to bring the knee joint near the line of gravity of the body. The degree of inclination is greater in the female than in the male.

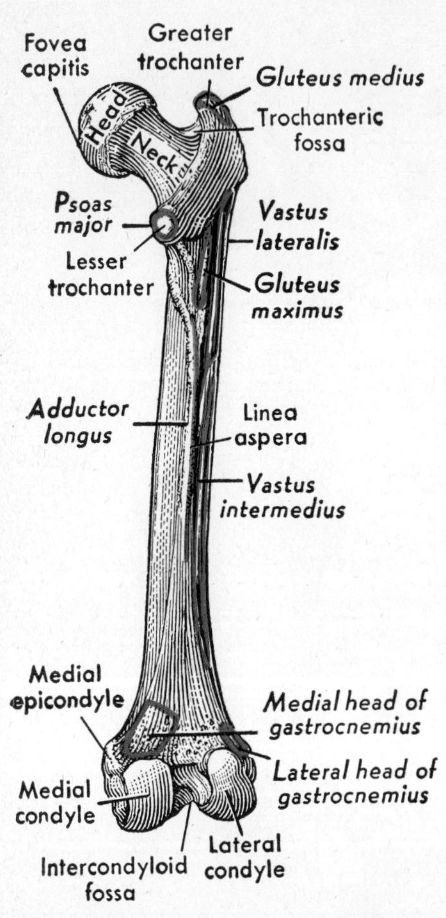

Fig. 88. The right femur, or thighbone, dorsal aspect. Muscle origins red, insertions blue.

The patella, or **kneecap,** is a small, flat, triangular, sesamoid bone developed in the tendon of the quadriceps femoris muscle and placed in front of the knee joint. It articulates with the femur and is surrounded by large, fluid-filled bursae.

The tibia, or **shinbone,** lies at the front and medial side of the leg.[16] The upper extremity is expanded into two lateral eminences with the sharp intercondyloid eminence between them. The lateral eminences are called medial and lateral condyles. The superior surfaces are concave and receive the condyles of the femur. The lower extremity is much smaller than the upper; it is prolonged downward on its medial side into a strong process, the *medial malleolus,* which

[16] Generally speaking, the lower extremity is called the leg. Anatomically, the word *leg* is reserved for that part of the lower extremity between the knee and the ankle. Above the knee is the thigh.

forms the inner prominence of the ankle. At the same lower extremity is the surface for articulation with the talus, which forms the ankle joint. The tibia also articulates with the lower end of the fibula. In the male the tibia is vertical and parallel with the bone of the opposite side, but in the female it has a slightly oblique direction lateralward, to compensate for the oblique direction of the femur medialward.

The fibula, or **calf bone,** is situated on the lateral side of the tibia, parallel with it. It is smaller than the tibia and, in proportion to its length, is the most slender of all the long bones. Its upper extremity consists of an irregular quadrant head by means of which it articulates with the tibia, but it does not reach the knee joint. The lower extremity is prolonged

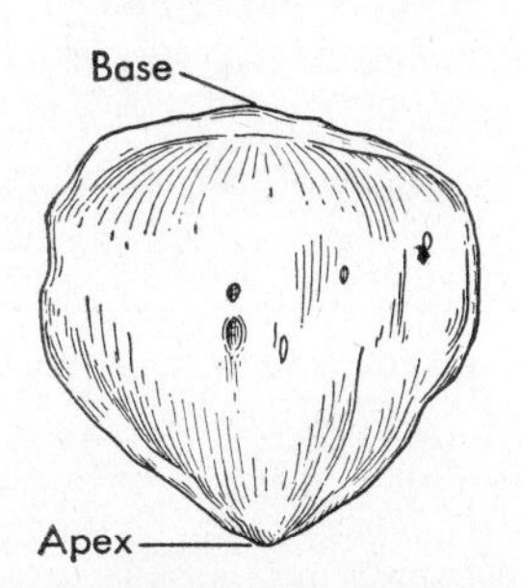

Fig. 89. The right patella, anterior surface.

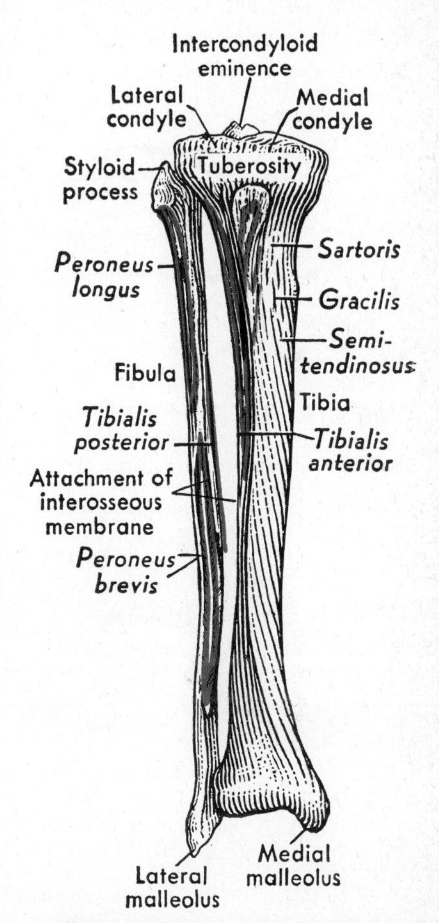

Fig. 90. The bones of the right leg, ventral surface. Muscle origins red, insertions blue.

downward into a pointed process, the *lateral malleolus,* which lies just beneath the skin and forms the outer anklebone. The lower extremity articulates with the tibia and the talus. The talus is held between the lateral malleolus of the fibula and the medial malleolus of the tibia. A fracture of the lower end of the fibula with injury of the lower tibial articulation is called a *Pott's*[17] *fracture*.

Tarsus. The seven tarsal bones, namely, the calcaneus, talus, cuboid, navicular, and the first, second, and third cuneiforms, differ from the carpal bones in being larger and more irregular. The largest and strongest of the tarsal bones is called the *calcaneus*, or

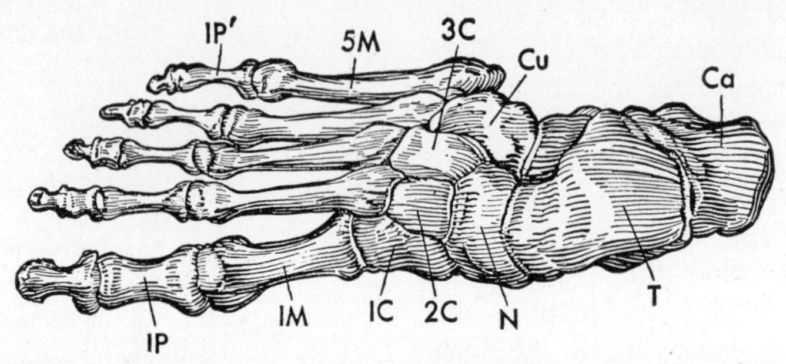

Fig. 91A. The bones of the right foot, viewed from above. *Ca,* calcaneus; *Cu,* cuboid; *1C,* first cuneiform; *2C,* second cuneiform; *3C,* third cuneiform; *1M,* first metatarsal; *5M,* fifth metatarsal; *N,* navicular; *1P,* first phalanx of hallux; *1P',* first phalanx of fifth toe; *T,* talus.

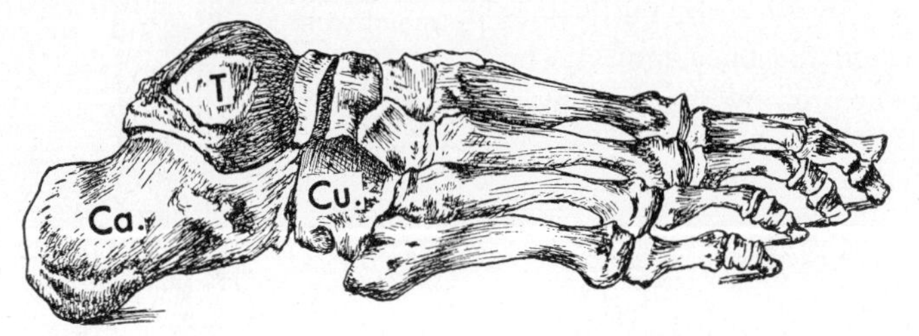

Fig. 91B. Outer side of right foot. *T,* talus or astragalus; *Ca,* calcaneus; *Cu,* cuboid. (Toldt.)

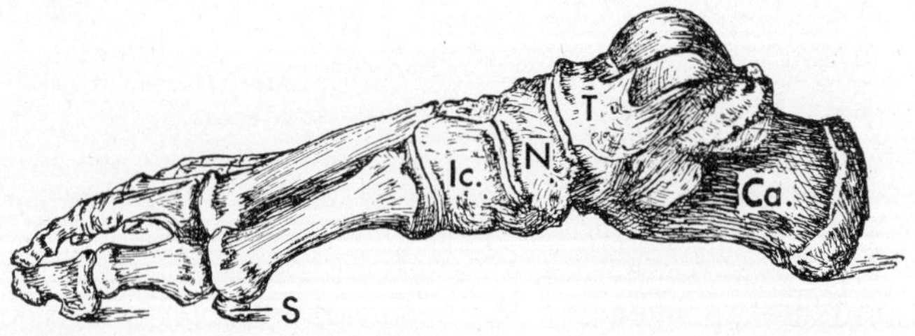

Fig. 91C. Inner side of right foot. *Ca,* calcaneus; *T,* talus or astragalus; *N,* navicular; *1C,* first cuneiform; *S,* sesamoid. (Toldt.)

[17] Percival Pott, English surgeon (1714–1788).

heel bone; it serves to transmit the weight of the body to the ground and forms a strong lever for the muscles of the calf of the leg.

The metatarsus, or **sole and instep of the foot,** is formed by five bones which resemble the metacarpal bones of the hand. Each bone articulates with the tarsal bones by one extremity and by the other with the first row of phalanges. The tarsal and metatarsal bones are so arranged that they form two distinct arches; the one running from the heel to the toes on the inner (medial) side of the foot is called the *longitudinal arch*, and the other across the foot in the metatarsal region is called the *transverse arch.*

These arches may become weakened and progressively broken down, a condition known as flatfoot. This condition is thought to be due to prenatal conditions, dietary or hormone disturbances, improper posture, weight or fatigue conditions, or the wearing of shoes ill-fitting in last or size.

Phalanges. Both in number and general arrangement they resemble those in the hand, there being two in the great toe and three in each of the other toes.

SUMMARY

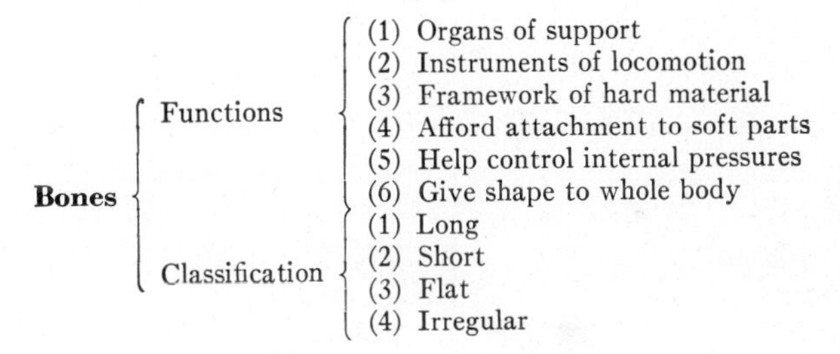

Bones
- Functions
  - (1) Organs of support
  - (2) Instruments of locomotion
  - (3) Framework of hard material
  - (4) Afford attachment to soft parts
  - (5) Help control internal pressures
  - (6) Give shape to whole body
- Classification
  - (1) Long
  - (2) Short
  - (3) Flat
  - (4) Irregular

TABLE OF THE BONES

BONES OF THE HEAD

Cranium		*Face*	
Occipital	1	Nasal	2
Parietal	2	Vomer	1
Frontal	1	Inferior nasal concha (Inf. turb.)	2
Temporal	2	Lacrimal	2
Sphenoid	1	Zygomatic (Malar)	2
Ethmoid	1	Palatine (Palate)	2
	–	Maxilla	2
	8	Mandible	1
			14

Fontanels	Anterior	1
	Posterior	1
	Anterolateral	2
	Posterolateral	2
Sinuses Opening into Nasal Cavity	Frontal	
	Ethmoidal	
	Sphenoidal	
	Maxillary	
Ear	Malleus	2
	Incus	2
	Stapes	2
		—
	Described with ear (Chap. 23)	6
	Hyoid bone in the neck	1

BONES OF THE TRUNK

		Child	Adult
Vertebrae	Cervical	7	7
	Thoracic	12	12
	Lumbar	5	5
	Sacral	5	1
	Coccygeal	4 = 33	1 = 26
Ribs			24
Sternum			1
			—
			51

BONES OF THE UPPER EXTREMITY

Clavicle	1
Scapula	1
Humerus	1
Ulna	1
Radius	1
Carpus	
Navicular (Scaphoid)	1
Lunate (Semilunar)	1
Triangular (Cuneiform)	1
Pisiform	1
Greater multangular (Trapezium)	1
Lesser multangular (Trapezoid)	1
Capitate (Os magnum)	1
Hamate (Unciform)	1
Metacarpus	5
Phalanges	14
	—
	32
	32 × 2 = 64

BONES OF THE LOWER EXTREMITY

Hipbone (Os coxae)	1
Femur	1
Patella	1
Tibia	1
Fibula	1
Tarsus	
Calcaneus (Os calcis)	1
Talus (Astragalus)	1
Cuboid	1
Navicular (Scaphoid)	1
Third cuneiform (External cuneiform)	1
Second cuneiform (Middle cuneiform)	1
First cuneiform (Internal cuneiform)	1
Metatarsus	5
Phalanges	14
	—
	31
	31 × 2 = 62

COMPARISON OF FEMALE AND MALE PELVIS

	FEMALE	MALE
Bone	Slender	Heavier and rough
Sacrum	Broad, less curved	Narrow, more curved
Symphysis	Shallow	Deeper
Major pelvis	Narrow	Wide
Minor pelvis	Shallow and wide, capacity great	Deeper and narrower, capacity less
Great sciatic notches	Wide	Narrow
Superior aperture	Oval	Heart-shaped

CHAPTER 6

ARTHROLOGY

JOINTS, OR ARTICULATIONS

The bones of the skeleton come into close contact at various points and areas of their surfaces. These contacts are called joints or articulations. The articulating surfaces of the bones are sometimes separated by a thin membrane, sometimes by strong strands of connective tissue, or fibrocartilage, and in the freely moving joints are completely separated. Strong ligaments extend over the joints or form capsules, which ensheathe them. Tendons of muscles also extend over the joints.

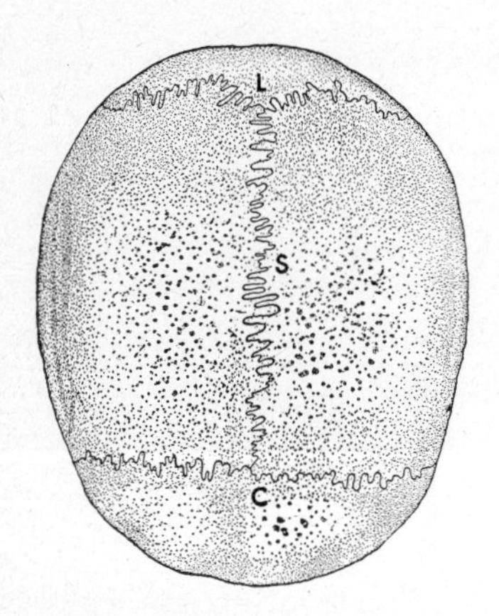

Fig. 92. A toothed, or dentated, suture, seen on the top of the skull. *L,* lambdoidal suture; *S,* sagittal suture; *C,* coronal suture.

Classification.[1] Joints are classified according to the amount of movement of which they are capable.

1. Immovable joints, or synarthroses.
2. Slightly movable joints, or amphiarthroses.
3. Freely movable joints, or diarthroses.

In all instances some softer substance, i.e., cartilage or fibrous tissue, is placed between the bones, uniting them or covering the opposed surfaces.

Immovable joints, or synarthroses. The bones are connected by fibrous tissue or cartilage. There are four varieties. These are

[1] A more complete classification of joints is given in the summary at the end of this chapter.

listed in the summary on p. 134. The bones of the cranium and the facial bones (with the exception of the lower jaw) have their adjacent surfaces applied in close contact, with only a thin layer of fibrous tissue between their margins. In most of the cranial bones union is by a series of interlocked processes and indentations which form the *sutures.* The three most important sutures are: (1) *coronal,* uniting the frontal and parietal bones; (2) *lambdoid,* uniting the parietal and occipital bones; and (3) *sagittal,* which begins at the base of the nose, extends along the middle line on the top of the crown, separates the fontal bone into two parts and the parietal bones from each other, and ends at the posterior fontanel. That portion of the sagittal suture which separates the frontal bone into two parts is often called the *frontal suture.* The junction of the sagittal and coronal sutures is called the *bregma;* the junction of the sagittal and lambdoid sutures is called the *lambda.*

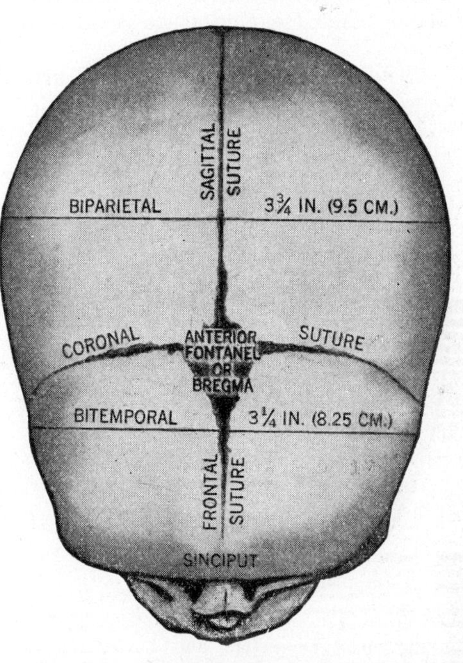

Fig. 93. Diameters and landmarks of the fetal skull, upper surface. (Edgar.)

Slightly movable joints, or amphiarthroses, include two varieties: (1) the symphysis and (2) the syndesmosis.

Symphysis. In this form of articulation the bony surfaces are joined together by broad, flattened disks of fibrocartilage, as in the articulations between the bodies of the vertebrae. These intervertebral disks being somewhat compressible and extensile, the spine can be moved to a limited extent in every direction. In the pelvis the articulations between the two pubic bones (symphysis pubis) (Fig. 75) and between the sacrum and ilia (sacroiliac articulation) are slightly movable. The pubic bones are united by a disk of fibrocartilage and by ligaments. In the sacroiliac articulation the sacrum is united more closely to the ilia, the articular surfaces being covered by cartilage and held together by ligaments.

The fibrocartilage between these joints (symphysis pubis and sacroiliac) becomes thickened and softened during pregnancy and allows a certain limited motion which is essential to a normal birth.

Syndesmosis. In this type of articulation the bony surfaces are united by an interosseous ligament, as in the lower tibiofibular articulation.

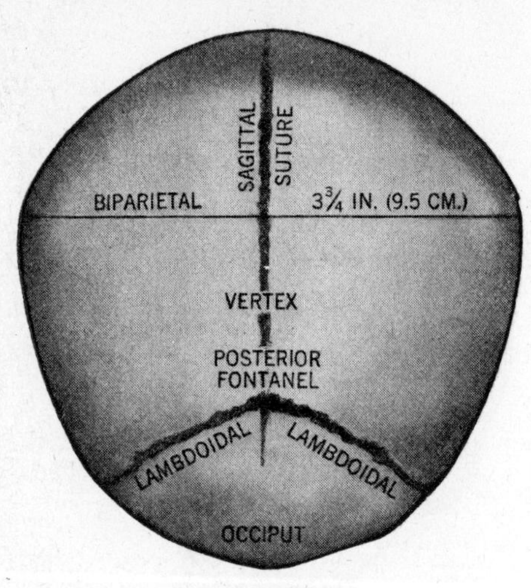

Fig. 94. Diameters and landmarks of the fetal skull, posterior surface. (Edgar.)

Freely movable joints, or **diarthroses,** include most of the joints in the body. The adjacent ends of the bones are covered with hyaline cartilage and surrounded by a more or less perfect fibrous capsule (articular capsule) which is strengthened by ligaments and lined with synovial membrane.

These joints have been classified as follows:

1. *Gliding joints* permit gliding movement only, as in the joints between the carpal bones of the wrist, between the tarsal bones of the ankle, and between the articular processes of the vertebrae. The articular surfaces are nearly flat, or one may be slightly concave, the other slightly convex. This type of movement is to some extent common to all movable joints.

2. *Hinge joints* allow angular movement in one direction, like a door on its hinges. The articular surfaces are of such shape as to permit motion in the forward and backward plane. These movements are called flexion and extension and may be seen in the joint between the humerus and ulna, in the knee and ankle joints, and in the articulations of the phalanges.

3. *Condyloid joints* admit of an angular movement in two directions. When an oval-shaped head, or condyle, of bone is received

into an elliptical cavity, it is said to form a condyloid joint, e.g., the wrist joint. Movements permitted in this form of a joint include flexion, extension, adduction, abduction, and circumduction but no axial rotation.

4. *Saddle joints* are like condyloid joints in providing for angular movement in two planes, but the structure is different. The articular surface of each of the articular bones is concave in one direction and convex in another, at right angles to the former. The metacarpal bone of the thumb is articulated with the greater multangular bone of the carpus by a saddle joint. The movements at these joints are the same as in condyloid joints.

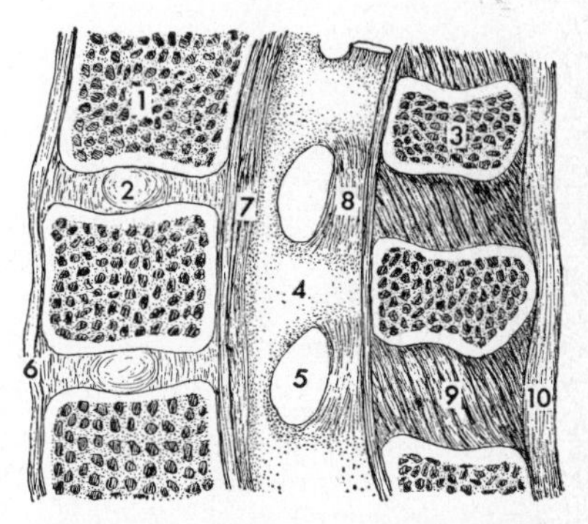

Fig. 95. Diagram of two intervertebral symphyses, seen in a longitudinal section through three segments of the vertebral column. *1,* body of vertebra; *2,* intervertebral cartilaginous disk; *3,* spinous process; *4,* spinal canal; *5,* intervertebral foramen; *6,* anterior longitudinal ligament; *7,* posterior longitudinal ligament; *8,* ligamentum flavum; *9,* interspinous ligament; *10,* supraspinous ligament.

5. *Pivot joints* are joints with a rotary movement in one axis. In this form a ring rotates around a pivot, or a pivotlike process rotates within a ring being formed of bone and cartilage. In the articulation of the axis and atlas, the front of the ring is formed by the anterior arch of the atlas and the back by the transverse ligament. The odontoid process of the axis forms a pivot, and around this pivot the ring rotates, carrying the head with it. In the proximal articulation of the radius and ulna, the head of the radius rotates within the ring formed by the radial notch of the ulna and the annular ligament. The hand is attached to the lower end of the radius, and the radius, in rotating, carries the hand with it; thus, the palm of the hand is alternately turned forward and backward. When the palm is turned forward or upward, the attitude is called *supination;* when backward or downward, *pronation.*

6. *Ball-and-socket joints* have an angular movement in all directions and a pivot movement. In this form of joint a more or less rounded head lies in a cuplike cavity, as the head of the femur in

the acetabulum and the head of the humerus in the glenoid cavity of the scapula. The shoulder joint is the most freely movable joint in the body.

Movement. Bones thus connected are capable of four different kinds of movement, which rarely occur singly but usually in combinations which produce great variety.

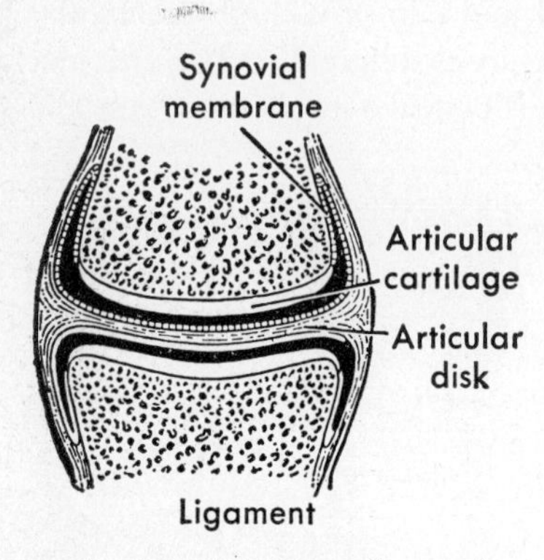

Fig. 96. Diagram of a section of a movable joint with an articular disk of cartilage. (After Gray's *Anatomy*.)

1. *Gliding movement* is the simplest kind of motion that can take place in a joint, one surface moving over another without any angular or rotatory movement. The costovertebral articulations permit a slight gliding of the heads and tubercles of the ribs on the bodies and transverse processes of the vertebrae.

2. *Angular movement* occurs only between long bones, and by it the angle between two bones is either increased or diminished. It includes flexion, extension, abduction, and adduction.

a) *Flexion.* A limb is flexed when it is bent, e.g., bending the arm.

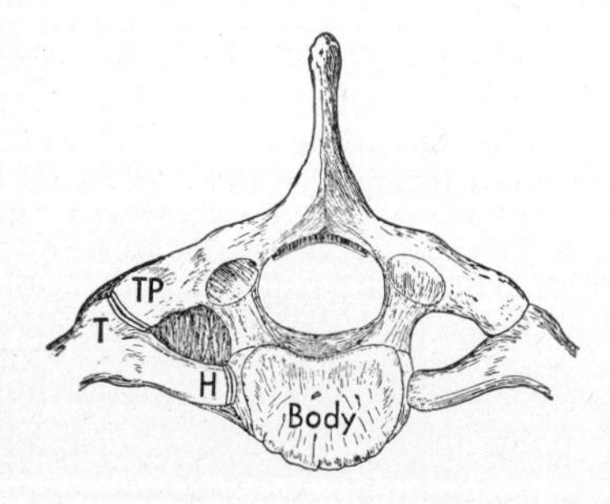

Fig. 97. Diagram of gliding joints between head of a rib and body of a vertebra, and also between the tubercle of a rib and the transverse process of a vertebra. *H,* head of rib; *T,* tubercle of rib; *TP,* transverse process.

b) *Extension.* A limb is extended when it is straightened out, e.g., straightening the arm; hence, the reverse of flexion.

c) *Abduction.* This term means drawn away from the middle line of the body, e.g., lifting the arm away from, or at right angles to the body.

d) *Adduction.* This term means brought to, or nearer, the middle line of the body, e.g., bringing the arm to the side of the body.

Both abduction and adduction have a different meaning when

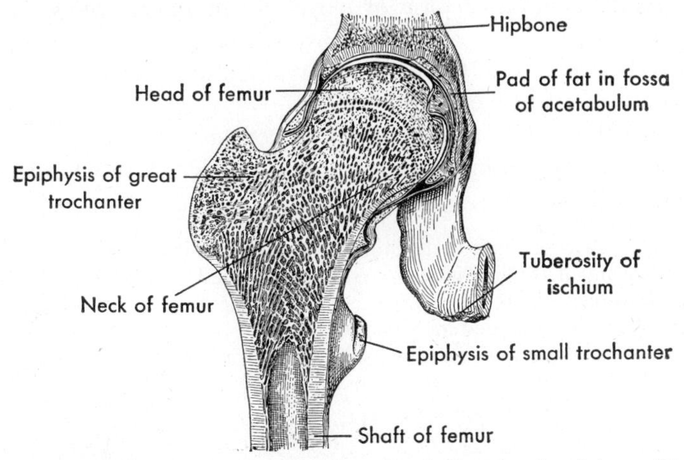

Fig. 98. Diagram of hip joint. Section of a ball-and-socket joint. (Toldt.)

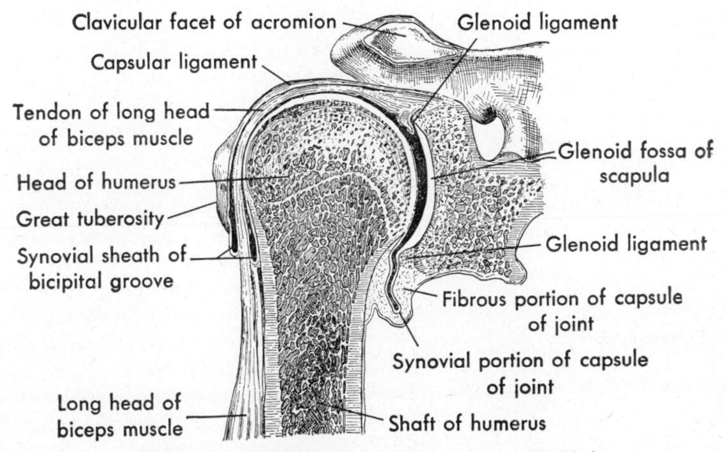

Fig. 99. Diagram of shoulder joint. (Toldt.)

used with reference to the fingers and toes. In the hand abduction and adduction are referred to an imaginary line drawn through the middle finger, and in the foot to an imaginary line drawn through the second toe.

3. *Circumduction* means that form of motion which takes place between the head of a bone and its articular cavity, when the bone is made to circumscribe a conical space by rotation around an imaginary axis, e.g., swinging the arms or legs.

4. *Rotation* means a form of movement in which a bone moves around a central axis, often an imaginary one, without undergoing any displacement from this axis, e.g., rotation of the atlas around the odontoid process of the axis.

Complete rotation, as of a wheel, is not possible in any joints of the body, as such motion would tear asunder the vessels, nerves, muscles, etc.

Sprain. A wrenching or twisting of a joint accompanied by a stretching or tearing of the ligaments or tendons is called a sprain.

Dislocation. If, in addition to a sprain, the bone of a joint is displaced, the injury is called a dislocation.

Ankylosis. Immobility and consolidation of a joint.

SUMMARY

Joints or Articulations—connections existing between bones

- **Immovable Joint, or Synarthrosis**
  - Bones are connected by fibrous tissue or cartilage
    - (1) *Sutures.* Articulations by processes and indentations interlocked together
      - *True sutures*
        - Sutura dentata—toothlike, e.g., sutures between parietal bones
        - Sutura serrata—sawlike, e.g., sutures between two portions of frontal bone
        - Sutura limbosa—in addition to interlocking, the articular surfaces are beveled and overlap, e.g., suture between parietal and frontal bones
      - *False sutures*
        - Sutura squamosa—scalelike, e.g., suture between the temporal and parietal bones
        - Sutura harmonica—simple apposition of rough surfaces, e.g., articulations between the maxillae
    - (2) *Schindylesis.* A thin plate of bone is received in a cleft or fissure of another bone, e.g., the reception of the vomer in the fissure between the maxillae and between the palatine bones

Immovable Joint, or Synarthrosis
- Bones are connected by fibrous tissue or cartilage
  - (3) *Gomphosis.* A conical process fits into a socket, e.g., roots of teeth into the alveoli of the maxillae and mandible
  - (4) *Synchondrosis.* Temporary joint. Cartilage between bones ossifies in adult life, e.g., between the occipital and sphenoid

Slightly Movable Joint, or Amphiarthrosis
- Bones are connected by disks of cartilage or interosseous ligaments
  - (1) *Symphysis.* The bones are united by a plate or disk of fibrocartilage of considerable thickness
  - (2) *Syndesmosis.* The bony surfaces are united by an interosseous ligament, as in the lower tibiofibular articulation

Movable Joint, or Diarthrosis
- (1) Hyaline cartilage covering adjacent ends of the bones
- (2) Fibrous capsule strengthened by ligaments
- (3) Synovial membrane lining fibrous capsule
  - (1) *Arthrodia.* Gliding joint; articulates by surfaces which glide upon each other
  - (2) *Ginglymus.* Hinge, or angular, joint; moves backward and forward in one plane
  - (3) *Condylarthrosis.* Condyloid joint; ovoid head received into elliptical cavity
  - (4) *Reciprocal Reception.* Saddle joint; articular surfaces are concavoconvex
  - (5) *Trochoides.* Pivot joint; articulates by a process turning within a ring or by a ring turning around a pivot
  - (6) *Enarthrosis.* Ball-and-socket joint; articulates by a globular head in a cuplike cavity

Movement
- (1) Gliding movement
- (2) Angular
  - Flexion
  - Extension
  - Adduction
  - Abduction
- (3) Circumduction
- (4) Rotation

CHAPTER 7

MUSCULAR TISSUE, PHYSIOLOGY OF CONTRACTION, LEVERS, SKELETAL MUSCLES

Motion is an important activity of the body which is made possible by special development of the function of contractility in muscle tissue. Motion in this sense includes not only movements of the entire body or parts of the body from place to place, but those of breathing, the beating of the heart, movements of the parts of the alimentary canal and its glands, and movements of the other viscera, including those of the blood and lymph tubes.

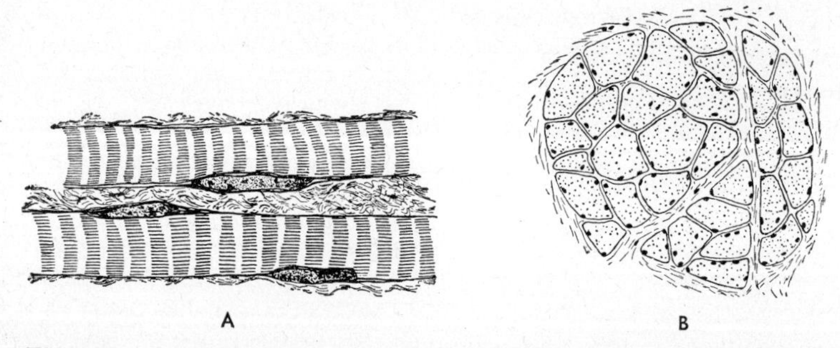

Fig. 100. Striated muscle cells. *A,* parts of two cells seen in longitudinal section showing sarcolemma, sarcoplasm, nuclei, cross striations. Note that each cell lies in connective tissue. *B,* three bundles (fasciculi) of cells seen in cross section. Note sarcoplasm, nuclei, connective tissue between cells and around fasciculi.

All physiological activities are closely related to motion brought about by contraction of muscle. Consider the effect of the contraction (change in shape) of the enormous number of muscle cells so intimately related as they are to all the other tissue cells of the body; for instance in supplying gland cells with the requisite raw materials for the manufacture of their secretion, in making it possible for them to empty this secretion into cavities or fluids of the body, in stirring up the tissue fluid and moving it about in relation to supplying all cells, say nerve cells or the muscle cells themselves, with a continuous source of supplies at low concentration, and removing their wastes constantly.

MUSCULAR TISSUE

In the human, muscular tissue constitutes 40 to 50 per cent of the body weight.

Special characteristics of muscle tissue are irritability (excitability), contractility, extensibility, and elasticity. *Irritability*, or *excitability*, is the property of receiving stimuli and responding to them. All cells possess this property. The response of any tissue to stimulation is to perform its special function, which, in the case of muscular tissue, is contraction. *Contractility* is the property which enables muscles to change their shape and become shorter and thicker. This property is characteristic of all protoplasm but is more highly developed in muscular tissue than in any other. *Extensibility* of a living muscle cell means that it can be stretched, or extended; and *elasticity* means that it readily returns to its original form when the stretching force is removed.

Types of muscle tissue. Muscle tissue is composed, as is every other tissue, of cells and intercellular substance. The cells become elongated and are termed *fibers*. The intercellular substance consists of a small amount of cement, which holds the cells to the framework of areolar connective tissue in which they are embedded.

Muscular tissue may be classified according to its structure or location:

Structure	Location
Striated, or cross-striped	Skeletal
Nonstriated, or smooth	Visceral
Indistinctly striated	Cardiac

Striated, or cross-striped, muscular tissue is called striated because of the parallel cross stripes, or *striae*, which characterize its microscopic appearance. It is called *skeletal* because it forms the muscles which are attached to the skeleton, *somatic* because it helps to form the body wall, and *voluntary* because the movements accom-

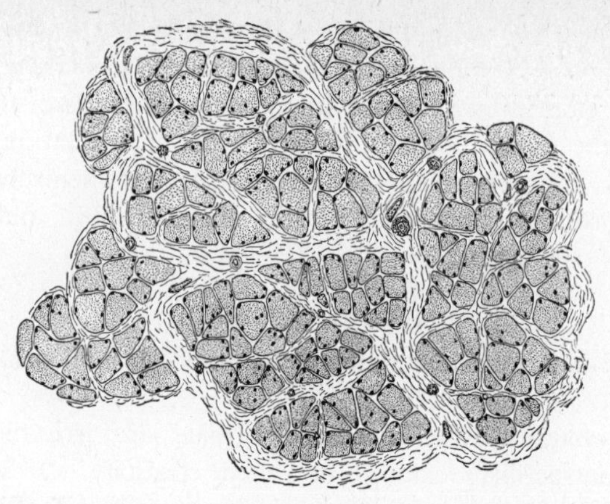

Fig. 101. Cross section of a bit of muscle trunk. The fibers are surrounded by connective tissue. Bundles of fibers are surrounded by wider areas of connective tissue. Primary bundles form secondary bundles surrounded by still wider areas of connective tissue.

plished are in most instances under conscious control. It is composed of spindle-shaped fibers, or cells, which may be 1 to 40 mm long and from 0.01 to 0.15 mm in diameter. These fibers consist of a tubular sheath, or *sarcolemma,* which encloses a soft, contractile substance. On the inner surface of the sarcolemma nuclei are seen. Microdissection shows cross-striped fibrils (myofibrils) which are closely packed together and run lengthwise through the entire fiber. Fluid material, called *sarcoplasm,* surrounds the *fibrils.* The individual fibers show variation in length, width, number of nuclei, and in the relative amount of fibrils and sarcoplasm present in them. Two types of fibers are found, *red* and *white.* Probably each skeletal muscle contains some fibers of each type; but, in general, rapid movements are carried out

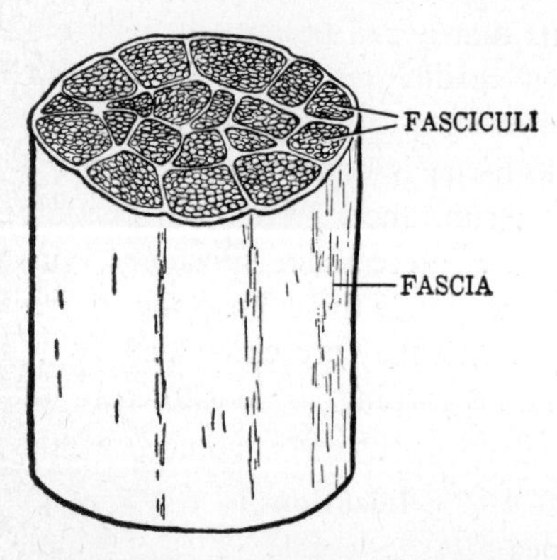

Fig. 102. A section of voluntary muscle trunk composed of fasciculi covered with fascia. Note that fasciculi are composed of cells and connective tissue and are surrounded by connective tissue.

by muscles in which white fibers predominate, while the slower movements are executed by those in which red fibers predominate.

The muscle fibers lie closely packed, forming primary bundles, or *fasciculi*. Areolar connective tissue forms a supporting framework, penetrating between the fibers and surrounding the small bundles, grouping them into larger bundles. It also surrounds the larger bundles and forms a covering for the entire muscle trunk, which is called a *fascia*. This connective tissue carries an intricate network of blood and lymph tubes and nerves, so that every muscle fiber is supplied with nerve endings and surrounded by tissue fluid (Fig. 155).

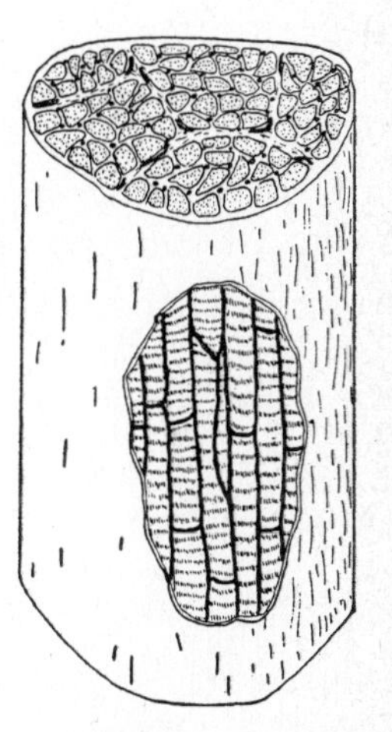

Fig. 103. A bundle of striated muscle fibers. At the top the fibers are shown cut across, each fiber being surrounded with connective tissue. The bundle of cells is wrapped with fascia. In one area the fascia is cut off to show the muscle fibers and a part of the capillary network surrounding the fibers.

Each skeletal muscle is a separate organ having its own sheath of connective tissue, called *epimysium*. The muscles vary in length from about 1 mm to nearly 60 cm (24 in.) and are diverse in form. In the trunk they are broad, flattened, and expanded, forming the walls of the cavities which they enclose. In the limbs they form more or less elongated spindles or straps. A typical muscle consists of a body and two extremities.

ORIGIN AND INSERTION. Skeletal muscles in general pass over joints, some of which are movable, some immovable. If the joint is movable, it is convenient to speak of the origin and insertion of the muscle. The *origin* is the end attached to the relatively less movable bone, and the *insertion* is the attachment to a bone moving in the ordinary activity of the body. The origin alone is fixed in a small number of muscles such as those of the face, many of which are attached by one end to a bone and by the other to skin. A muscle increases in diameter as it contracts lengthwise, pulling the attachments at each end nearer to each other.

Tendons and *aponeuroses* attach muscles to bones. As the end of a muscle is approached, the connective-tissue framework of the muscle increases in quantity and usually extends beyond the muscle fibers as a dense white cord, or *tendon,* or as a flattened tendon, or *aponeurosis.* The tendons are cablelike in that they are exceedingly strong, inextensible, and at the same time flexible.

Fasciae, as previously stated (p. 69), are sheets of connective tissue by which most muscles are closely covered. They not only envelop and bind down the muscles but also separate them into groups. Such groups are named according to the parts of the body where they are found—cervicle fascia, thoracic fascia, abdominal fascia, pelvic fascia, etc. Individual fasciae are frequently given the names of the muscles which they envelop and bind down, as pectoral fascia, etc. In the vicinity of the wrist and ankle, parts of the deep fascia become blended into tight transverse bands, or *annular ligaments,* which serve to bind down the tendons close to the bones.

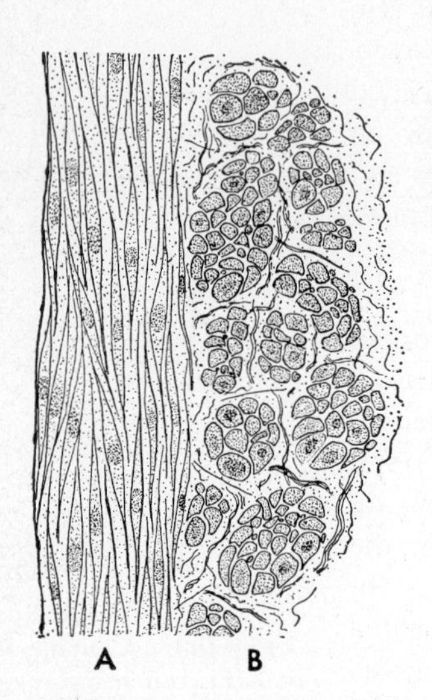

Fig. 104. Longitudinal section of a visceral organ to show smooth muscle tissue. *A,* in longitudinal section; *B,* in cross section. Note that there is a little connective tissue between the cells and that the cells are arranged in bundles.

It is important to realize the continuity of the connective tissues of the body. Tendons, ligaments, and fasciae blend with periosteum; tendons and fasciae serve as ligaments; tendons lose themselves in fasciae; and tendons of some muscles serve as fasciae for others.

The function of skeletal muscle is to operate the bones of the body, thereby producing motion.

Skeletal muscle enables one by body gesture or facial expression to show pleasure or displeasure; the ability to speak, to hear, to see are all dependent upon muscle activity.

Nonstriated, or **smooth, muscular tissue** is so called because it does not exhibit cross stripes, or striae. It is called *visceral* because it forms the muscular portion of the visceral organs. The cells are usually arranged in two main layers: the inner, thick, *circular coat* and the outer, thin, *longitudinal coat.* The motions caused by it are involuntary; usually we are not conscious of them. This tissue is composed of spindle-shaped cells each containing a single large nucleus. The cells are 0.015 to 0.5 mm long and 0.002 to 0.02 mm in diameter, and the coats are surrounded by a network of connective tissue which carries lymph, blood tubes, and nerves. In contrast with the skeletal muscle, smooth muscle contracts more slowly but possesses greater extensibility and maintains a state of contraction, or tonus (p. 142), even when all nerve connections are severed. The cells of a smooth muscle coat seem to be connected by fibrils which extend from cell to cell, and therefore the smooth muscle coat acts more like a striped muscle fiber than like a muscle (muscle trunk).[1]

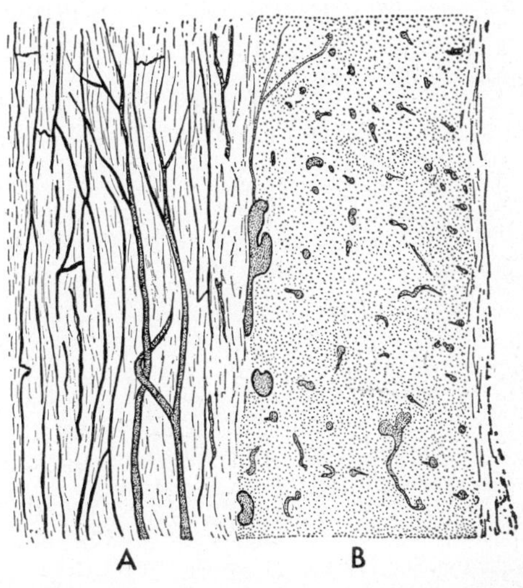

Fig. 105. An injected longitudinal section of the wall of a visceral organ to show smooth muscle tissue. *A*, in longitudinal section; *B*, in cross section. Note that the blood vessels run in the same general direction as the muscle fibers. In *B*, the blood vessels are in groups lying between the muscle bundles rather than between the muscle cells. Outside of organ is at left in Figs. 104 and 105.

Functions of smooth muscle. The walls of all visceral organs, including blood tubes, contain smooth muscle cells. It is the function of these cells to produce such changes in shape and size as occur in these organs. An example is peristalsis of the alimentary canal.

Cardiac muscle tissue forms the heart. Transverse striations are less distinct than in skeletal muscles and the cells are smaller. Each cell has a large, centrally placed nucleus. The cells are roughly quadrangular in shape, arranged end to end forming a continuous protoplasmic sheet, or *syncytium,* more or less cut up by unique dark-stained bands. These bands are not found in the hearts of children, and their function is not known. The cells are surrounded by a close network of blood and lymph capillaries. The

[1] Emil Bozler, "Physiological Evidence for the Syncytial Character of Smooth Muscle," *Science,* 1937.

rhythm of the heartbeat is presumably preserved by means of this "network" nature of cardiac muscle over which impulses may pass in any direction, the whole acting as one great muscle cell rather than as a muscle trunk.

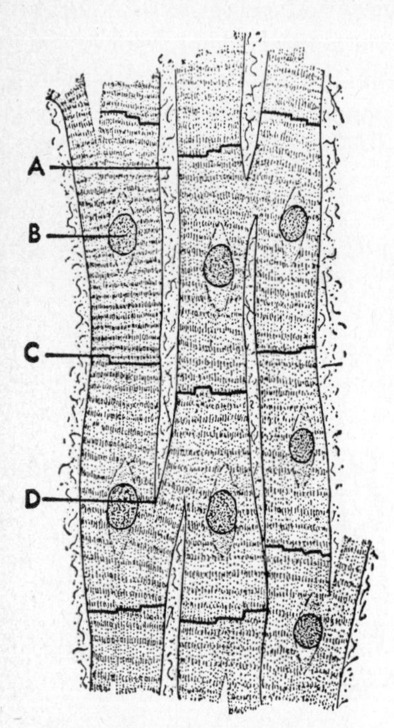

Fig. 106. Diagram of a thin section of cardiac muscle tissue. *A,* connective tissue between cells; *B,* nucleus; *C,* dark-stained bands; *D,* branching of a cell.

PHYSIOLOGY OF CONTRACTION

Tone is a property of muscle whereby a steady, partial contraction varying in degree is maintained. The fundamental mechanism whereby *tone,* or *tonus,* is produced is not fully understood, but physiologically it is known to be due to nerve impulses. By means of tonic contraction in skeletal muscles, posture is maintained for long periods with little or no evidence of fatigue. Absence of fatigue is brought about mainly by means of different groups of muscle fibers contracting in relays, giving alternating periods of rest and activity for a given muscle-fiber group. In man the antigravity skeletal muscles (retractors of neck, extensors of the back, etc.) exhibit the highest degree of tonus. In an unconscious person the body collapses if these antigravity muscles are completely relaxed. During sleep, tone is at a minimum.

The tone of skeletal muscles gives them a certain firmness and maintains a slight, steady pull upon their attachments; it also functions in the maintenance of a certain pressure upon the contents of the abdomen. Both the rapidity and the smoothness of movement are dependent upon it. In fractures the overriding of the broken ends of the bone is often due to the contractions of the muscles because of this property of tonicity.

Both visceral and cardiac muscle exhibit tonus even when isolated from the nervous system. This is probably due to the plexuses of nerve cells and fibers distributed through them. The maintenance of normal blood pressure is partly dependent upon the tone of the muscles in the walls of the small arteries. Likewise healthy digestion is dependent upon the tone of the muscles of the stomach and intestines. Although the tone of visceral and cardiac muscles is inherent in them, it is probably also under chemical control similar to that for striped muscle. The fact that tonus varies according to the general condition of health, and under certain conditions such as during abnormally high concentration of carbon dioxide disappears altogether, indicates the influence of environmental conditions upon it. It is possible that tonic changes are the result of variation in hydrogen ion concentration acting directly upon the muscle fibers.

Muscle contraction is concerned with the tone of the protoplasm of muscle cells; first the tone of individual muscle cells themselves, then the integrated tone of muscle cells in individual muscles. This tone is maintained by the chemical and physical composition of tissue fluids and by the nervous system. The cerebellum makes the final adjustments needed for *muscle groups* to act together, though muscle tone is maintained through centers in the spinal cord or brain.

Excitation (stimulation). A muscle is excitable because the muscle fibers composing it are excitable. All protoplasm possesses the property of *excitability*. Any force which affects this excitability is called a *stimulus*. Physiologically, a stimulus represents a change in the environment of the muscle cells. Protoplasm also possesses the property of *conductivity*, and when stimulated at one point the response may travel through the cell. The response is the specialized one which is characteristic for the tissue stimulated; in muscles it is contraction. Normally, the muscles (muscle trunks, as gastrocnemius) are stimulated by impulses conveyed to them by nerve fibers. These stimuli may serve to inhibit muscular activity as well as to excite it. Various other forms of stimuli, such as mechanical, thermal, chemcial, and electrical, are used in experimental work with muscles and are called *artificial stimuli*. A common source of stimulation is electricity because it is available and convenient, easily controlled as to strength and speed of application, and least destructive of the tissues stimulated. While muscles are within the body or when isolated from the body, they can be excited by artificial stimuli applied to their nerves or to the muscles directly.

Muscles are supplied with two types of nerve fibers—*sensory* fibers, conveying to the central nervous system the state of contraction of the muscle, and *motor* fibers, conveying impulses from the central nervous system to the muscles, controlling their contraction. If a motor nerve is severed or the center in the brain or cord is damaged, no stimulus is carried to the muscle, and its function is lost. When a sensory nerve is severed, no stimuli are carried from the sensory end organ to the central nervous system, and sensation is thereby lost.

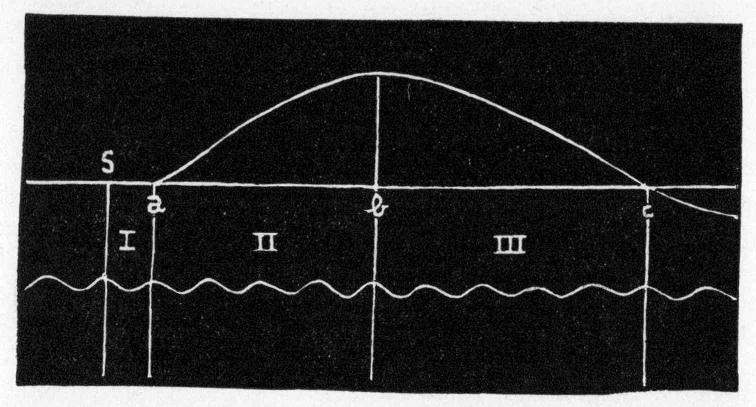

Fig. 107. Diagram curve of a simple contraction of a muscle. *S* indicates the time when the stimulus enters the muscle. Phase I (*S* to *a*), the *latent period,* in isolated frog muscle lasts about 0.01 seconds. Phase II (*a* to *b*), the *period of contraction,* lasts about 0.04 second. Phase III (*b* to *c*), the *period of relaxation,* lasts about 0.05 second. From crest to crest of the wavy line below represents an interval of 0.01 second. The apparatus for recording a muscle contraction is shown in Fig. 108.

Conditions of contraction. Skeletal muscles contract quickly and relax promptly. In sharp contrast to this, the contractions of visceral muscle coats develop slowly, are maintained for some time, and fade out slowly. The contraction of a skeletal muscle is the result of stimuli discharged by the nerve fibers innervating it. If one of these contractions is analyzed, it will be found that there is a brief period after the muscle is stimulated before it contracts. This is called the *latent period* and is followed by a *period of contraction,* which in turn is followed by a *period of relaxation* (Fig. 107). It has been demonstrated experimentally that if electrical stimuli are applied to a muscle, as for instance the gastrocnemius muscle of the frog, and the contractions are recorded on a moving drum, the contractions will vary in strength (in height on the drum record) depending on certain factors: (1) the strength of the stimulus, (2) the speed of application of the stimulus, (3) the duration of the stimulus, (4) the weight of the load, and (5) the temperature. In general, the stronger the stimulus up to a certain maximum, the stronger the contraction will be, i.e., the greater the number of single cells which

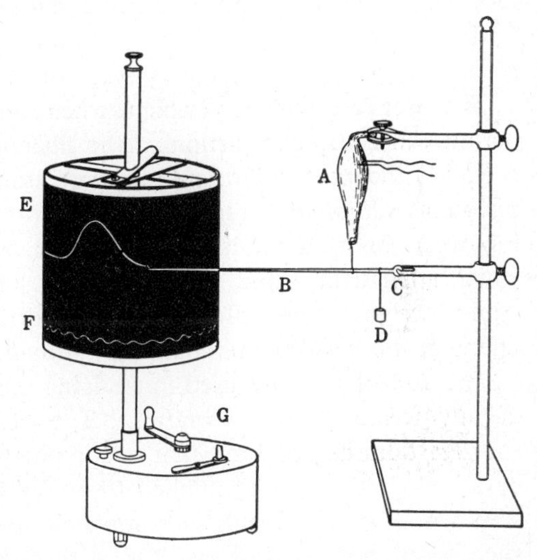

Fig. 108. Record made by a contracting muscle. *A*, muscle extending from clamp to writing lever; *B*, writing lever, hinged at *C* and counterweighted at *D*; *E*, revolving smoked drum, on which the point of the writing lever draws a line; *F*, time record—from crest to crest of a wave may be 0.01 second; *G*, fan for regulating speed. A muscle attached to the lever draws a line on the moving surface. When the muscle contracts, the lever is raised; and when the muscle relaxes, the lever falls. The drum is called a kymograph.

will contract. Strongest contractions result from stimuli of moderate duration. Some load is necessary in order to get the best response, but increase of load beyond the optimum decreases the height of contraction. Muscles do their best work at a certain optimum temperature. For man this is about 37° C (98.6° F) body temperature. If the temperature is raised much above this, the muscle loses its excitability and becomes functionally depressed, entering finally the state of heat rigor, i.e., a condition of permanent shortening.

Response to stimuli. Laboratory studies are frequently conducted to investigate the physiology of muscle cells under varying conditions such as temperature. In these studies an electric current serves to stimulate the muscle cell directly or through the nerve fiber to the cell and may therefore be said to take the place of the natural stimulus to muscle cells within the body.

It has been found in such experiments that if a stimulus applied to a single muscle fiber is strong enough to produce a response, it will give a contraction which is maximal, no matter what the strength of the stimulus. This is known as the *all-or-none law*. This does not mean that the excitability of the cell cannot be changed, but merely that it must be accomplished by other means than an increase in this stimulus. In other words, each muscle fiber gives a maximal response or none at all under the conditions of the experiment. Fatigue and varying conditions of nutrition may alter the cell's response, but increasing the strength of the stimulus will not do so.

The weakest stimulus which when applied over a reasonable period will cause contraction of the fiber under specified conditions is called the *minimal stimulus*. Any stimulus weaker than this is known as *subminimal* (*subliminal*). Two subminimal stimuli (each too weak in itself to cause contraction of the muscle fiber) may, when applied in rapid succession, by their combined forces be equivalent to a minimal stimulus and the cell will respond. This phenomenon is called *summation of stimuli*.

The unit of measure used in studying the excitability or irritability of any tissue is the *chronaxie*. The chronaxie of a cell is the shortest duration of time that a stimulus of twice minimal (rheobasic) strength must be applied to evoke a response.

When a muscle trunk is stimulated to contract many times in succession, the contractions for a time become progressively increased in extent, resulting in a record which shows a staircase effect. This effect is probably determined by an increase in irritability brought about by metabolic wastes formed during the first few contractions. As these waste products increase, irritability is decreased, and the contractions diminish progressively in extent until fatigue develops to a point where the muscle fails to respond. From this staircase phenomenon it is judged that activity of muscles is at first physiologically beneficial to the extent that irritability of muscular tissue is thereby increased.

If a second stimulus occurs during the apex of contraction from the first, the resulting height of contraction will be maximal. This is known as *summation of contractions*. If, however, the second stimulus occurs during a certain period of time after the accomplishment of the first contraction, there will be no second contraction. During this exceedingly brief lapse of time, known as the *absolute refractory period*, the muscle will not respond to any stimulus, however strong.

This is followed by the *relative refractory period,* often called the *period of depressed excitability,* during which the muscle gradually regains its irritability.

If stimuli are applied to a muscle in such rapid succession that each occurs before the fibers have relaxed from the one preceding, the cells will remain in a state in which no relaxation is apparent. Such sustained contraction is called *tetanus.* Probably all voluntary acts, even the simplest, have tetanic contractions as their basis, and they

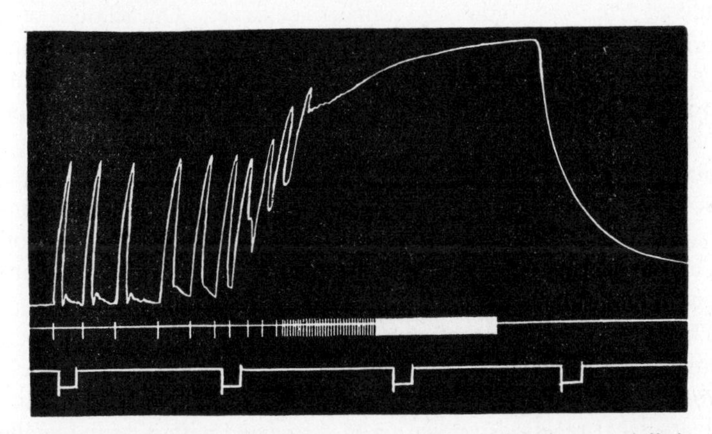

Fig. 109. Tetanus in skeletal muscle. When stimuli are applied to a muscle at a gradually increasing rate of frequency (indicated by signal), the individual muscle twitches blend together so that, when the stimuli are applied in very rapid succession, a smooth, sustained contraction results. Ordinary muscle movements are of this tetanic nature. Note that the height of contraction is greater in tetanus than in a single twitch. (A. J. Carlson and V. Johnson, *The Machinery of the Body.* Courtesy of the University of Chicago Press.)

are especially to be noted in such continued muscular work as holding the body erect or in carrying a load. Postural tonus is believed to be the result of a slight state of tetanic contraction of skeletal muscles due to fiber summation.

Types of contractions. When a muscle trunk contracts and a weight is lifted, the muscle becomes shorter and thicker, but its tone remains the same. Since the tone of the fibers is not altered, such contractions are called *isotonic.* If the muscle is compelled to contract against some weight which it cannot lift, the tension in the fibers increases, but the muscle length remains unaltered. Since the length of the fibers is unchanged, such contractions are called *isometric.*

Contraction of a skeletal muscle is usually of the isotonic type, but complexities of muscular activity involve the coordinated development of both isometric and isotonic contractions in the different fibers of a muscle trunk.

Contraction in skeletal muscles. The height of contraction of a skeletal muscle is in direct proportion to the strength of the stimulus applied. This is not a contradiction of the "all-or-none" law. It is explained by the fact that voluntary muscle cells are separate units insulated from each other by connective tissue. On account of environmental conditions the minimal stimulus of these separate fibers may vary. Thus, the minimal stimulus of a skeletal muscle trunk is one which evokes contraction from a single fiber; the maximal stimulus is one which will cause the contraction of every fiber present. A skeletal muscle trunk varies in this respect from the heart muscle, which responds as a unit to any stimulus which will contract a single cardiac cell. It will be recalled that cardiac muscle is a "network" of cells and that the rhythmic action of the heart is brought about by the contraction of these cells in unison.

The immediate cause of muscular contraction is not known, although there is evidence that *diffusion* and physical characteristics of *colloidal* conditions such as *surface tension* may help to explain it. The chemical reactions known to be associated with it seem to follow contraction rather than to produce it. There are two phases in the contraction of muscular fibers: the *anaerobic,* or *contractile,* phase, in which oxygen is not required, and an *aerobic,* or *recovery,* phase, for which oxygen is necessary and in which the muscle is restored to its previous state with regard to its glycogen supply. In the anaerobic phase, energy is made available with little or no production of lactic acid and without oxidative processes entering into the reaction. Lactic acid does not, therefore, furnish directly the energy for contraction but supplies instead the energy for the recovery process, in which one-fifth of the lactic acid furnishes energy for the building up of the remaining four-fifths into glycogen.

CHEMICAL CHANGES DURING MUSCLE CONTRACTION

The active parts of striated muscle are composed of very minute fibers of a protein known as actomyosin which is linked to a compound called adenosinetriphosphate. Szent-Györgyi[2] compares muscle fibers to "chemical engines" that derive their energy from the adenosinetriphosphate (ATP). When muscle is stimulated by a nerve impulse or by an electric shock, the chemical potential energy is converted into mechanical energy. ATP stores chemical energy and contains in the links binding phosphate groups together about 11,000 cal of free energy which forms the source of energy for muscle contraction. When muscle fibers contract, waves of excitation pass over the muscle fibers, ionic concentrations of K^+, Na^+, and Cl^- change, and ATP releases its phosphate-bond energy. This causes a rearrangement of the ATP molecules. Phosphocreatine, which is also present in the muscle fibers, functions in the restoration of the high-energy phosphate bonds of ATP. Oxygen is not utilized in these reactions.

During the recovery stage which follows, glycogen is utilized to provide energy for the resynthesis of phosphocreatine. Oxygen is used in the changes necessary to carry to completion the various chemical steps necessary to form carbon dioxide and water. Four-fifths of the lactic acid formed in the chain of chemical activity is synthesized to glycogen and the one-fifth is oxidized to carbon dioxide and water and heat is liberated. The liver aids in converting lactic acid to glycogen.

If sufficient oxygen is not available, the formed lactic acid accumulates and oxidative processes are not carried to completion. This is called oxygen debt.

Not all of the chemical changes that take place in muscle fibers during activity are completely understood. There are many enzymes, coenzymes, and enzyme systems that have specific functions in relation to chemical activity. Vitamins are needed to carry certain chemical processes to completion, as for instance the B-complex group and carbohydrate metabolism. The vitamins in some instances function as coenzymes. The reversibility of reactions is believed to be dependent upon a system of enzymes within the muscle cells.

During muscle contraction heat is formed, called (1) initial heat, which is produced during contraction; and (2) delayed heat, which is produced during the recovery stage. Delayed heat is produced

[2] A. Szent-Györgyi, *Nature of Life.* Academic Press, Inc., New York, 1945.

more slowly than initial heat, but there is very little difference in the amount of heat formed.

It is in the recovery process that glycogen, the ultimate source of energy, is used up, lactic acid and carbon dioxide are formed, and most of the heat is liberated.

Summary of Some of the Chemical Changes Taking Place in Muscle Cells in Relation to Contraction

1. Adenosinetriphosphate → adenosine diphosphate (ADP) + phosphoric acid + energy (for contraction)
2. Phosphocreatine + ADP → creatine + ATP + energy (for resynthesis of ATP)
3. Glycogen → lactic acid + energy (for resynthesis of phosphocreatine)
4. Lactic Acid
   - $\frac{4}{5}$ reformed into glycogen
   - $\frac{1}{5}$ oxidized to carbon dioxide + water + energy (for resynthesis of glycogen)

Fatigue and exercise. In muscles undergoing contraction, the first effect of the formation of carbon dioxide and lactic acid is to increase irritability; but if a muscle is continuously stimulated, the

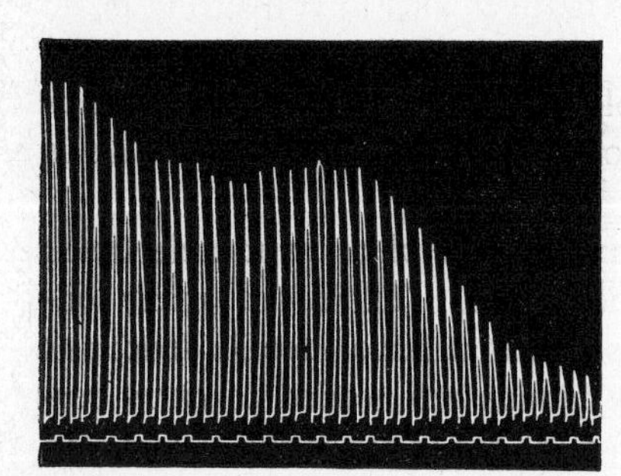

Fig. 110. Record of muscular fatigue of finger. A weight is lifted by a finger by means of a cord which runs over a pulley. The writing lever is attached to the weight and writes on a moving drum. The apparatus is called an ergograph.

strength of contraction becomes progressively less until the muscle refuses to respond. This is true fatigue and is caused, in part at least, by the toxic effects of waste products (carbon dioxide, acid phosphate, lactic acid) which accumulate during exercise. The loss of nutritive materials may also be a factor in fatigue, but recent conceptions of the phenomenon stress the accumulation of wastes. In moderate exercise the system is able to get rid of these waste products readily. After prolonged contractions a period of rest may be necessary to furnish opportunity for the blood to carry the fatigue substances to the excretory organs and nutritive material and oxygen to the muscle. Probably it is chiefly lactic acid which brings on fatigue by disturbing the hydrogen ion concentration of the cell fluids, thus inhibiting the enzyme action which is responsible for further breakdown of glycogen. It has been demonstrated that injection of the blood of a fatigued animal into a rested one will promptly bring on signs of fatigue. Another theory states that fatigue is due, at least in part, to the depressing influence of the potassium salts produced during muscular contraction.

If fatigue is carried on to the point of absolute exhaustion, the cells do not recover. The protein constituents of the fibers coagulate, exhibiting the phenomenon known as *rigor mortis.* Rigor mortis occurs in muscles from 10 minutes to 7 hours after death.

The body is susceptible to other fatigue than that of muscles. Most easily fatigued of all are nerve cells, next the junctions between nerves and muscle fibers, then the muscles themselves, and last of all the connecting nerve fibers. It must not be thought that the state which we ordinarily recognize in ourselves as fatigue is entirely physiological. The sense of fatigue is very complex and is often associated with various mental states. See p. 245 for discussion.

Exercise stimulates circulation and thereby brings about a change in conditions for cells in all locations throughout the body. This great stirring-up effect of exercise brings fresh blood to all the arterioles, and the local pressure as well as the lymph environment of all cells is changed. Exercise has been shown to increase the size, strength, and tone of the muscle fibers. Massage and passive exercise may, if necessary, be used as a partial substitute for exercise. While physical recreation is desirable for aiding metabolic processes, continued use of fatigued muscles is injurious if, during such conditions, the muscles exhaust their glycogen supply and utilize the protein of their own cells. Under normal conditions it is the sensation of fatigue which protects us from such extremes.

MUSCLES AND THE BONY LEVERS

Levers. Direct muscular contraction alone is not generally responsible for bodily motions. Intermediate action of bony levers is usually essential. In the body, cooperative functioning of bones and muscles forms levers. A knowledge of levers gives a basis for understanding the principles underlying good posture and the movements of the body.

A *simple lever* is a rigid rod which is free to move about on some fixed point or support called the *fulcrum.* It is acted upon at two different points by (1) the *resistance* (*weight*), which may be thought of as something to be overcome or balanced, and (2) the *force* (*effort*) which is exerted to overcome the resistance. In the body bones of varying shapes are levers, and the resistance may be a part of the body to be moved or some object to be lifted or both of these. The muscular effort is applied to the bone at the insertion of the muscle and brings about the motion or work.

For example, when the forearm is raised, the elbow is the fulcrum, the weight of the forearm is the resistance, and the pull due to contraction of the biceps muscle is the effort.

Levers act according to a law which may be stated thus: When the lever is in equilibrium, the effort times the effort arm[3] equals the resistance times the resistance arm ($E \times EA = R \times RA$).

For example, if the distance from the effort to the fulcrum is the same as the distance from the resistance to the fulcrum, an effort of 5 lb will balance a resistance of 5 lb.

Levers may be divided into three classes according to the relative position of the fulcrum, the effort, and the resistance. In levers of the *first class* the fulcrum lies between the effort and the resistance, as in a set of scales. In this type of lever the resistance is moved in the opposite direction to that in which the effort is applied. In the body when the head is raised, the facial portion of the skull is the resistance, moving upon the axis as a fulcrum, while the muscles of the back produce the effort.

In levers of the *second class* the resistance lies between the fulcrum and the effort and moves, therefore, in the same direction as that in which the effort is applied, as in the raising of a wheelbarrow. There are no levers of the second class in the body.[4]

In levers of the *third class* the effort is exerted between the fulcrum

[3] The "resistance arm" is the perpendicular distance from the fulcrum to the line of action of the resistance (weight). The "effort arm" is the perpendicular distance from the fulcrum to the line of action of the effort (force).

[4] J. P. Schaeffer, *Morris' Human Anatomy,* 1953, p. 406.

and the resistance. Levers in which the resistance arm is thus longer than the effort arm produce rapid delicate movements wherein the effort used must be greater than the resistance. The flexing of the forearm is a lever of this type, as are most of the levers of the body. The *law of levers* applies in the maintenance of correct posture. The head held erect in correct standing posture rests on the atlas as a fulcrum, with little or no muscular effort being exerted to maintain this

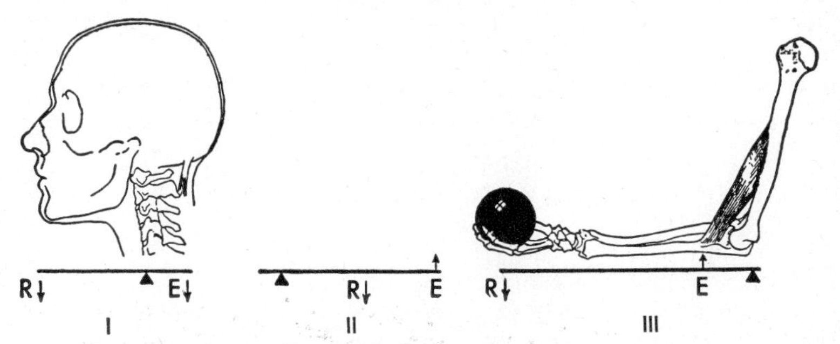

Fig. 111. Diagram of simple levers. Note insertion of muscle in relation to fulcrum and resistance. The effort is applied at the place of insertion of muscle to the bone. The ▲ represents the fulcrum; *R,* the resistance; *E,* the effort; *arrows,* the direction of motion. There are no levers of the second class in the body.

position. The head in this position is in the "line of gravity" which passes through the hip joints, knee joints, and the balls of the feet (Fig. 75). With shoulders stooped and head bent forward, constant muscular effort is exerted against the pull of gravity on the head.

SKELETAL MUSCLES

Skeletal muscles are arranged in groups with specific functions to perform: flexion and extension, external and internal rotation, abduction and adduction. For example, in flexing the elbow, several muscle groups are involved in varying degrees. The *agonists,* or prime movers, give power for flexion; the opposing group, the *antagonists,* contribute to smooth movements by their power to maintain tone yet relax and give way to movement of the flexor group. Other groups of muscles act to hold the arm and shoulder in a suitable position for action and are called *fixation* muscles. The *synergists* are muscles which assist the agonists (prime movers) and reduce undesired action

(Text continued on p. 156.)

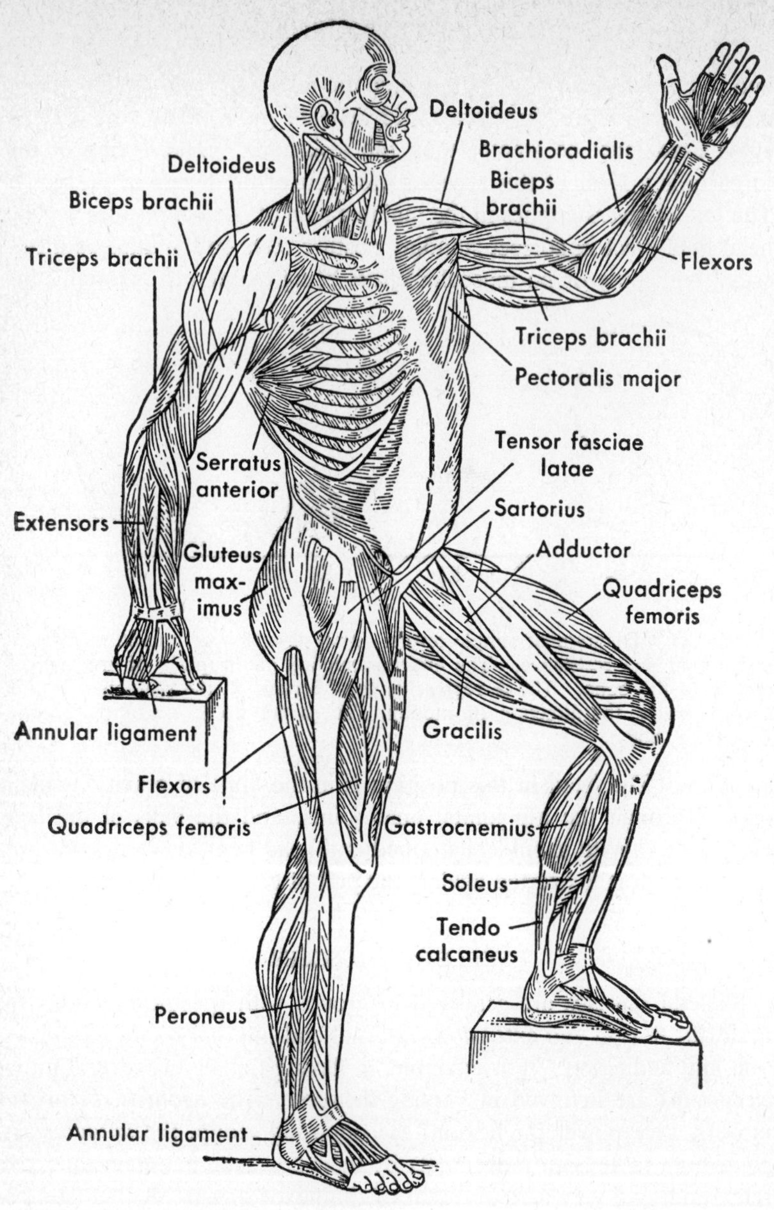

Fig. 112. Human body, showing muscles. (Right side and front.) (Courtesy of William Wood & Company.)

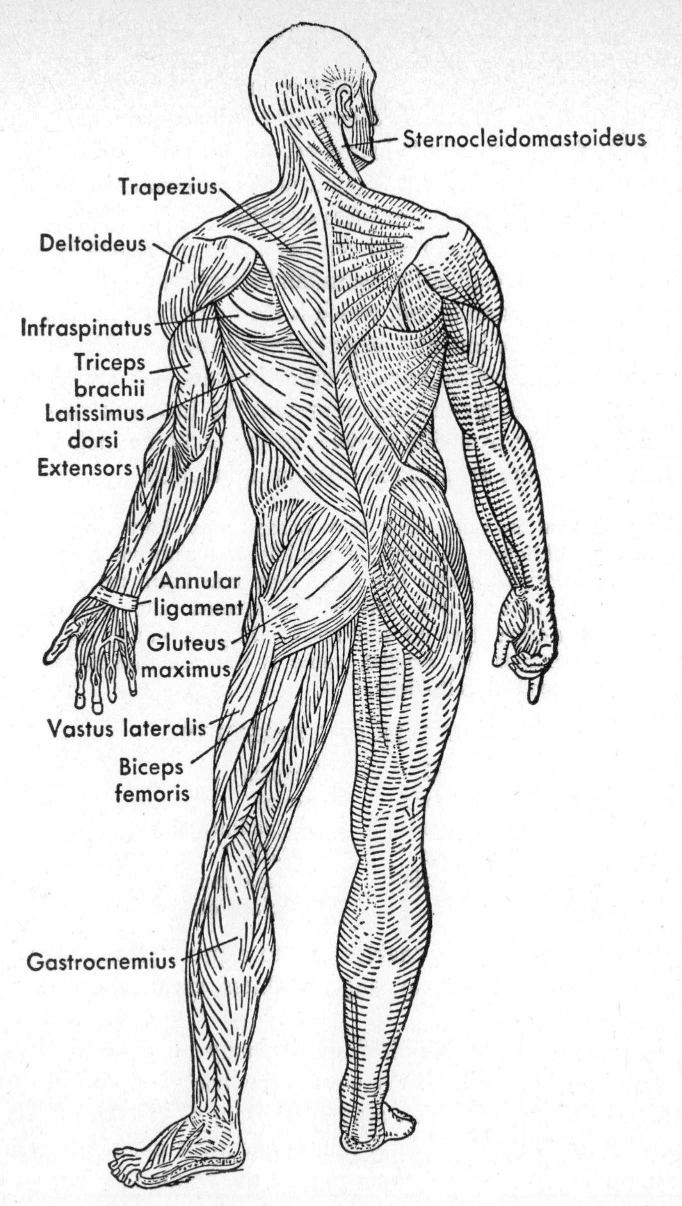

Fig. 113. Human body, showing muscles. (Back.) (Courtesy of William Wood & Company.)

or unnecessary movement. Activity of these opposing muscle groups is coordinated in relation to degree of tension exerted. When tension of the flexor muscles is increased, the tone of the extensor muscles is decreased; movement is controlled and position is maintained against varying degrees of pressures or pulls.

Each muscle has motor and sensory nerve fibers and is well supplied with arteries, capillaries, veins, and lymphatics.

Each muscle has an *origin,* which is the more fixed attachment and serves as basis of action, and an *insertion,* or the movable attachment.

Almost all the skeletal muscles occur in pairs. A few single muscles situated in the median line represent the fusion of two muscles. Only a few of the over 600 muscles of the body are included in this chapter. They are arranged in relation to function.

Many skeletal muscles bear two names, one Latin and the other English, e.g., obliquus externus abdominis and external abdominal oblique. Sometimes a muscle has more than one Latin name, e.g., psoas magnus and psoas major, vastus intermedius and vastus crureus. Frequently a muscle has no well-known English name, e.g., levatores costarum; sometimes the English name is the one that is best known, e.g., deltoid instead of deltoideus.

MUSCLES OF EXPRESSION

Epicranial	Platysma
Corrugator	Risorius
Buccinator	The quadrati labii
Zygomatic	Mentalis
Triangularis	The nasal muscles
	Orbicularis oris

This group of muscles enables one to express pleasure, pain, disgust, disdain, contempt, fear, anger, sadness, surprise, or other emotional states.

The epicranial, or **occipitofrontalis,** may be considered as two muscles—the occipital, which covers the back, or occiput, of the head, and the frontal, which covers the front of the skull. The two muscles are held together by a thin aponeurosis extending over and covering the whole of the upper part of the cranium. The occipital takes its origin from the occipital bone and the mastoid portion of the temporal bone and is inserted into the aponeurosis. The frontal takes its origin from the aponeurosis and is inserted into the tissues in the region of the eyebrows.

Action. The frontal portion of this muscle is the more powerful; by its contraction the eyebrows are elevated and the skin of the forehead thrown into transverse wrinkles and thereby expresses surprise. The occipital portion draws the scalp backward.

The corrugator muscle wrinkles the skin of the forehead vertically as in frowning. It has its origin from the medial end of the superciliary arch and is inserted into the skin of the forehead.

The buccinator (trumpeter's muscle) arises from the alveolar processes of the maxilla and mandible. The fibers converge toward

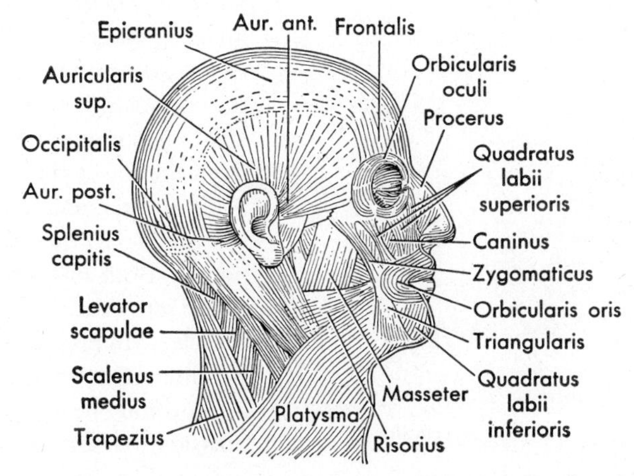

Fig. 114. Superficial muscles of head and neck. (Gerrish.)

the angle of the mouth and are inserted into the orbicularis oris. It forms the principal muscle of the cheeks and lateral wall of the oral cavity.

Action. It compresses the cheek during mastication, keeping the food under pressure of the teeth.

The zygomatic arises from the zygomatic bone and descends obliquely to its insertion into the orbicularis oris muscle. It pulls the angles of the mouth upward and backward as in smiling or laughing.

The triangularis is a broad, flat muscle which arises from the oblique line of the mandible. The fibers converge and are inserted in the orbicularis oris muscle. It depresses the angle of the mouth.

The platysma (broad sheet muscle) arises from the skin and fasciae, covering the pectoral and deltoid muscles, and is inserted in the mandible and muscles about the angle of the mouth.

Action. It depresses the mandible and draws down the lower lip and angle of the mouth and wrinkles the skin of the neck.

The risorius has its origin in the fascia over the masseter muscle and passes horizontally forward and is inserted into the skin at the angle of the mouth. When contracted strongly it gives an expression of strain and tenseness.

The quadratus labii superioris is a thin quadrangular muscle with three heads. The angular head arises in the upper part of the maxilla and passes obliquely downward, dividing into two slips. One of these is inserted into the greater alar cartilage and skin of the nose and the other into the orbicularis oris muscle. The infraorbital head arises from the lower margin of the orbit; the zygomatic head has its origin in the zygomatic bone. Both of these are inserted into the orbicularis oris muscle. It elevates the upper lip. Contraction of the infraorbital head gives expression of sadness. When the whole muscle contracts it gives an expression of disdain and contempt.

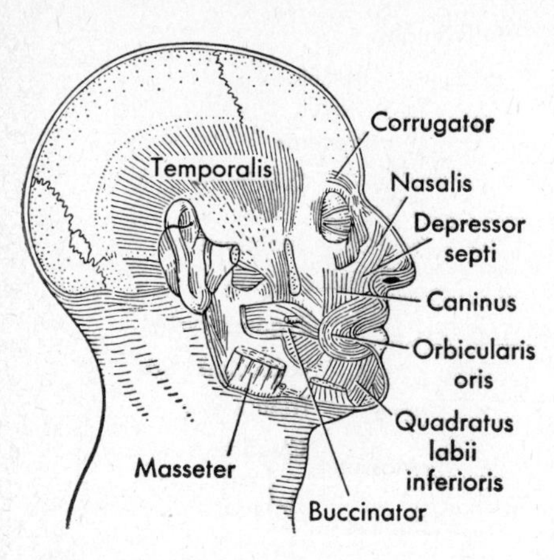

Fig. 115. Temporal and deep muscles about the mouth.

The quadratus labii inferioris is a small quadrilateral muscle which extends from the oblique line of the mandible upward and medialward and is inserted into the skin of the lower lip and the orbicularis oris muscle. It draws the lower lip downward, as in the expression of irony.

The mentalis is a short, thick muscle which has its origin in the incisive fossa of the mandible and is inserted into the skin of the chin. The mentalis protrudes the lip and wrinkles the skin of the chin, which expresses disdain or doubt.

The nasal muscles vary in size and strength in different individuals, and one or more may be absent.

The nasalis, depressor septi, and the posterior and anterior dilator naris are small muscles about the nasal openings which *constrict and enlarge the apertures of the nares.* The procerus covers the nasal bones and lateral nasal cartilage and is inserted into the skin over the lower part of the forehead between the eyebrows. It draws down the medial angle of the eyebrows and produces transverse wrinkles over the bridge of the nose.

The orbicularis oris (ring-shaped muscle of the mouth) consists of numerous layers of muscular fibers which surround the opening of the mouth and pass in different directions. Some of the fibers are derived from other facial muscles which are inserted into the lips; some fibers of the lips pass in an oblique direction from the under surface of the skin through the thickness of the lips to the mucous membrane; other fibers connect the maxillae and septum of the nose above with the mandible below.

Action. It causes compression and closure of the lips in various ways, e.g., tightening the lips over the teeth, contracting them, or causing pouting or protrusion of one or the other.

MOVEMENT OF THE EYE AND LIDS

Movement of the eyeball is controlled by six muscles. The orbit contains seven muscles; six of them are attached to the eyeball, arranged in three opposing pairs.

The four recti muscles, called respectively superior, inferior, medial, and lateral, arise at the apex of the orbital cavity. Each muscle passes forward in the position which its name indicates and is inserted into the eyeball.

The two oblique muscles are called, respectively, superior and inferior. The superior oblique arises from the apex of the orbit, courses forward to the upper and inner angle of the orbit, where it passes through a ring of cartilage; then it bends at an acute angle, passes around the upper part of the eyeball, and is inserted between the superior and lateral recti. The inferior oblique arises from the orbital plate of the maxilla and passes around the under portion of the eyeball to its attachment between the superior and lateral recti.

Action. The four recti acting singly turn the corneal surface of the eye upward, downward, inward, or outward, as their names suggest. The action of the two oblique muscles is somewhat complicated, but their general tendency is to roll the eyeball on its axis. These muscles do not act singly but rather cooperatively and with a high degree of coordination.

The levator palpebrae superioris (lifter of the upper lid) arises from the apex of the orbit, passes forward, and is inserted into the tarsal cartilage of the upper lid.

Action. It raises the upper lid and opens the eye.

The orbicularis oculi (circular muscle of the eye) arises from the nasal portion of the frontal bone, from the frontal process of the maxilla, and from a short, fibrous band, the medial palpebral (tarsal)

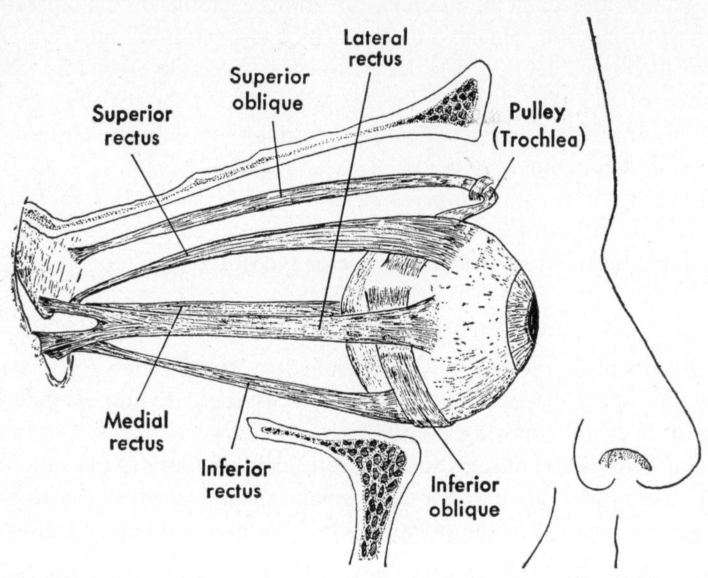

Fig. 116. Extrinsic muscles of the eyeball in the right orbit. Note tendinous insertions of superior and inferior oblique muscles between superior and lateral recti.

ligament. It spreads lateralward, forming a broad, thin layer which occupies the eyelid, and is inserted at the union of the upper and lower lids at the outer side of the eye. This is called the palpebral portion. A broader, thicker part, called the orbital portion, surrounds the circumference of the orbit and spreads over the temple and downward on the cheek; its fibers form a complete ellipse, the upper ones being inserted in the frontalis muscle.

Action. It serves as a sphincter muscle of the eyelids. The action of the palpebral portion is involuntary. It closes the lids gently as in sleep or blinking. The orbital portion is under the control of the will. The action of the entire muscle is to close the lids forcibly, drawing the parts toward the inner angle and tightening the brow. It is the antagonist of the levator palpebrae superioris.

MUSCLES CONCERNED WITH MASTICATION

The muscles of mastication are the masseter (chewing muscle), the temporal (temple muscle), the internal pterygoid, and the external pterygoid.

The masseter arises from the zygomatic process and adjacent portions of the maxilla and is inserted into the angle and lateral surface of the ramus of the mandible.

The temporal (temporalis) arises from the temporal fossa of the skull and the deep surface of the temporal fascia by which it is covered. It is inserted into the coronoid process of the mandible.

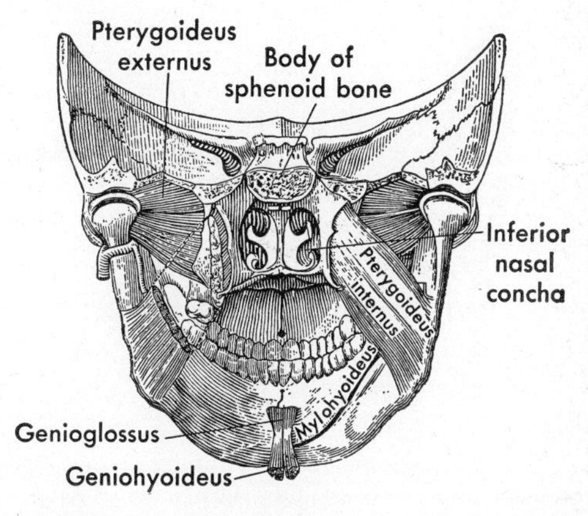

Fig. 117. Pterygoid muscles, viewed from behind, the back portion of the skull having been removed. (Gerrish.)

The internal pterygoid (pterygoideus internus) arises from the medial surface of the lateral pterygoid plate, the pyramidal process of the palatine bone, and the tuberosity of the maxilla. The fibers pass downward, lateralward, and backward, to be inserted into the ramus of the mandible.

The external pterygoid (pterygoideus externus) is a short, thick muscle which arises by two heads, an upper from the zygomatic surface of the great wing of the sphenoid and a lower from the lateral surface of the pterygoid plate. The fibers extend backward and are inserted in front of the neck of the condyle of the mandible and into the articular disk of the joint between the temporal and mandible bones.

Action. The masseter, temporal, and internal pterygoid raise the mandible against the maxillae. The posterior fibers of the temporal retract the mandible. The external pterygoid assists in opening the mouth, and the internal and external pterygoids acting together cause the lower jaw to protrude, so that the lower teeth are projected in front of the upper. The internal and external pterygoids of one side produce lateral movements of the jaw such as take place during the grinding of food.

Movement of the tongue. The muscles of the tongue are concerned with speaking, mastication, and swallowing. They are divided into right and left paired groups by a fibrous septum which is attached below to the hyoid bone. In each side there are two sets of muscles; the *extrinsic* have their origin outside the tongue, and the *intrinsic* are contained within it. Two extrinsic muscles are the genioglossus and the styloglossus.

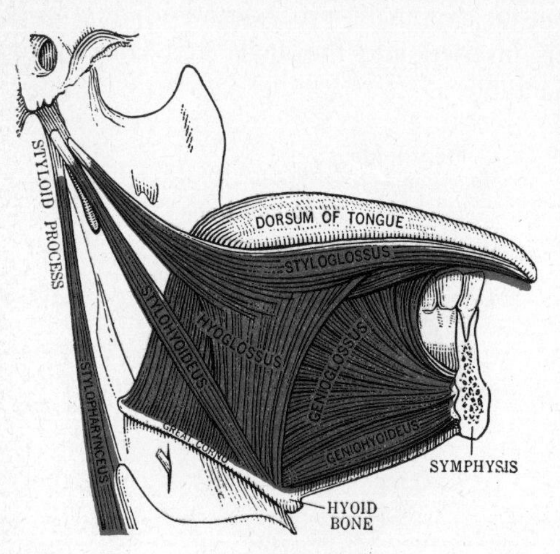

Fig. 118. Some of the muscles of the tongue, viewed from the right side. (After Gray's *Anatomy*.)

The genioglossus arises by a short tendon from the inner surface of the mandible at the symphysis and spreads out in a fanlike form. It is attached by a thin aponeurosis to the hyoid bone, and the fibers are inserted the whole length of the under surface of the tongue, in and at the side of the mid-line.

Action. It thrusts the tongue forward, retracts it, and also depresses it.

The styloglossus has its origin in the styloid process of the temporal bone and is inserted in the whole length of the side and under part of the tongue.

Action. It draws the tongue upward and backward.

During general anesthesia these, together with all the other muscles, become relaxed, and it is necessary to press the angle of the lower jaw upward and forward in order to prevent the tongue from falling backward and obstructing the larynx.

MOVEMENT OF THE HEAD

Flexion	**Extension**
Sternocleidomastoid	Splenius capitis
	Semispinalis capitis
	Longissimus capitis

The atlanto-occipital articulation permits flexion and extension of the head. Movement occurs between the condyles of the occipital bone and the superior articular surfaces of the atlas, and a backward and forward movement occurs as in nodding of the head. When movement takes place at the vertical axis the head is rotated to the right when the muscles on the right side contract, and rotated to the left when those on the opposite side contract.

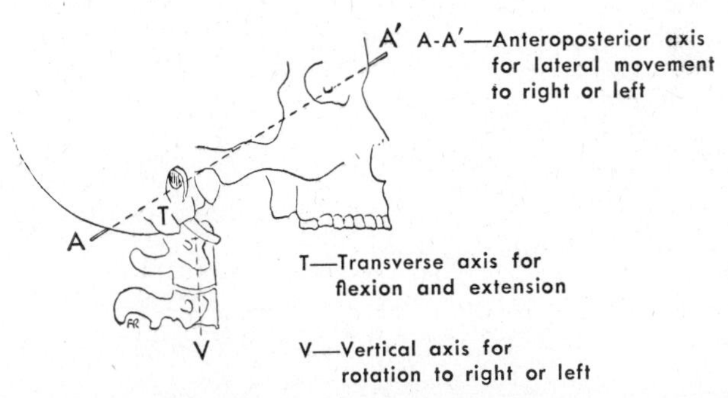

Fig. 119. Diagram showing axis of movement of skull on vertebral column.

The **sternocleidomastoid** (sternocleidomastoideus) muscle is named from its origin and insertion. It arises by two heads from the upper part of the sternum and the inner border of the clavicle and is inserted by a strong tendon into the mastoid portion of the temporal bone. This muscle is easily recognized in thin persons by its forming a cordlike prominence obliquely situated along each side of the neck.

Action. When one muscle acts alone, it draws the head toward the shoulder of the same side and rotates the head, pointing the chin upward to the opposite side. Both muscles acting together flex the head in a forward direction, and in forced inspirations they assist in elevating the thorax. If one of these muscles is either abnormally contracted or paralyzed, deformity called *torticollis,* or wry neck, results.

The splenius capitis arises from the lower half of the ligamentum nuchae and from the spinous process of the seventh cervical vertebra and the upper three or four thoracic vertebrae. The fibers extend upward and lateralward and are inserted into the outer part of the occipital bone and into the mastoid process of the temporal bone, under the sternocleidomastoid muscle.

Action. When both muscles act together, the head is pulled backward in extension. Acting alone, the head is rotated to the same side.

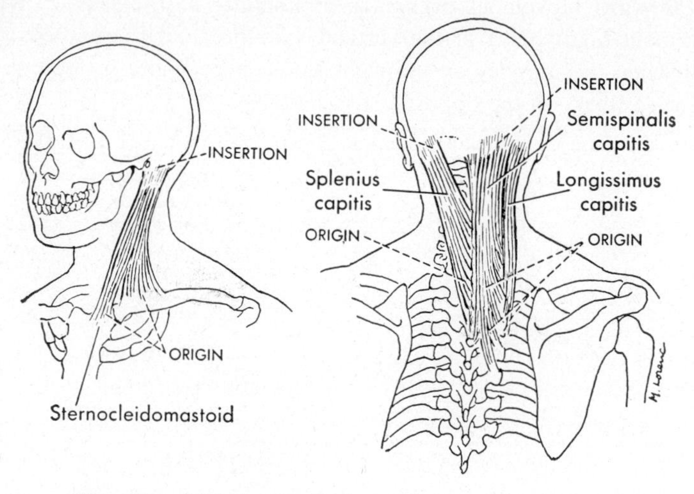

Fig. 120. Muscles for flexion and extension of the head.

The semispinalis capitis arises from the transverse processes of the upper six thoracic vertebrae and from the articular processes of the four lower cervical vertebrae and is inserted between the superior and inferior nuchal lines of the occipital bone.

Action. When both muscles contract, the head is extended; acting alone, the head is rotated toward the same side.

The longissimus capitis arises by tendons from the transverse processes of the upper four thoracic vertebrae and is inserted into the posterior margin of the mastoid process.

Action. When both muscles contract, the head is extended; acting alone, the head is bent to the same side and the face is rotated toward that side.

Also see trapezius muscle for action on the head.

MOVEMENT OF THE VERTEBRAL COLUMN

Flexion	**Extension**
Quadratus lumborum	Sacrospinalis

Forward and backward movement of the spine is limited in the thoracic region, but movement is free in the lumbar region, particularly between the fourth and fifth lumbar vertebrae.

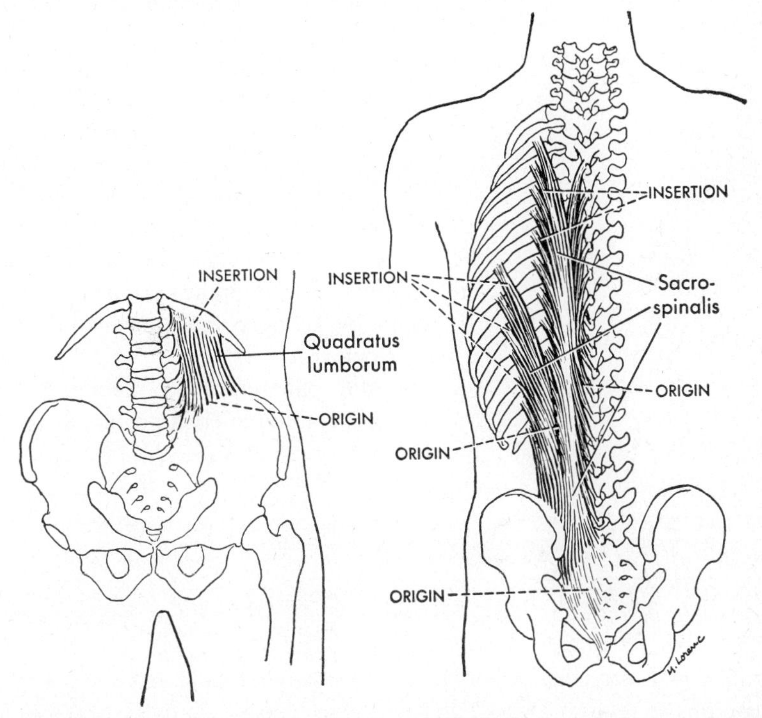

Fig. 121. Muscles producing movement of vertebral column in the lumbar region.

The quadratus lumborum is a rectangular muscle which forms part of the posterior wall of the abdominal cavity. It arises from the posterior part of the crest of the ilium and the iliolumbar ligament and is inserted into the lower border of the twelfth rib and upper four lumbar vertebrae.

Action. Acting together, the two muscles flex the spine at the lumbar vertebrae.

The sacrospinalis (erector spinae) arises from the lower and posterior part of the sacrum, from the posterior portion of the iliac crests, and from the spines of the lumbar and the lower two thoracic vertebrae. The fibers form a large mass of muscular tissue, which splits in the upper lumbar region into three columns, namely, a lateral, the *iliocostalis,* an intermediate, the *longissimus,* and a medial, the *spinalis.* Each of these consists from below upward of three parts.

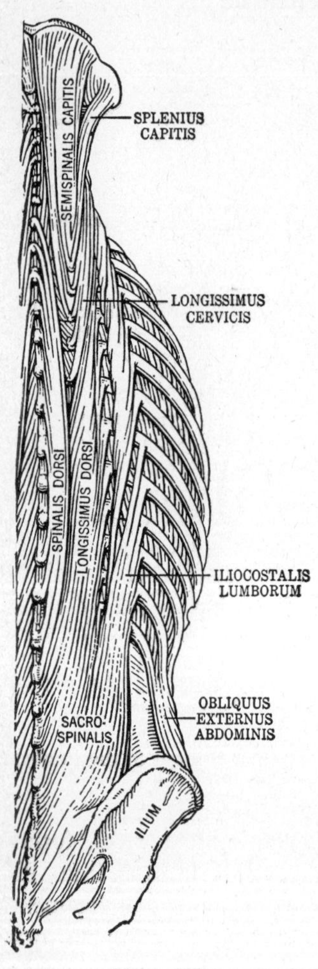

Fig. 122. Diagram to show more details of some of the muscles of the back.

These muscles are attached to the ribs and vertebrae at different levels all the way up the back to the occipital bone and the mastoid process of the temporal bone. As the muscle climbs up the back, it does not relinquish one foothold before it establishes another. The result is not merely a continuity of structure but an overlapping, as one segment begins back of the insertion of the segment below it.

Action. It serves to maintain the vertebral column in erect posture against gravity.

MOVEMENT OF THE SHOULDER GIRDLE

Elevation

Levator scapulae
Rhomboideus major
Rhomboideus minor
Trapezius (upper fibers)

Depression

Pectoralis minor
Subclavius
Trapezius (lower fibers)

The levator scapulae arises from the upper four or five cervical vertebrae. It is inserted into the vertebral border of the scapula between the medial angle and the root of the spine.

Action. As the name suggests, it lifts the angle of the scapula.

The rhomboideus major arises from the spines of the first four or five thoracic vertebrae and is inserted into the vertebral border of the scapula between the root of the spine and the inferior angle.

The rhomboideus minor arises from the lower part of the ligamentum nuchae and from the spinous processes of the last cervical and the first thoracic vertebrae. It is inserted into the vertebral border of the scapula at the root of the spine.

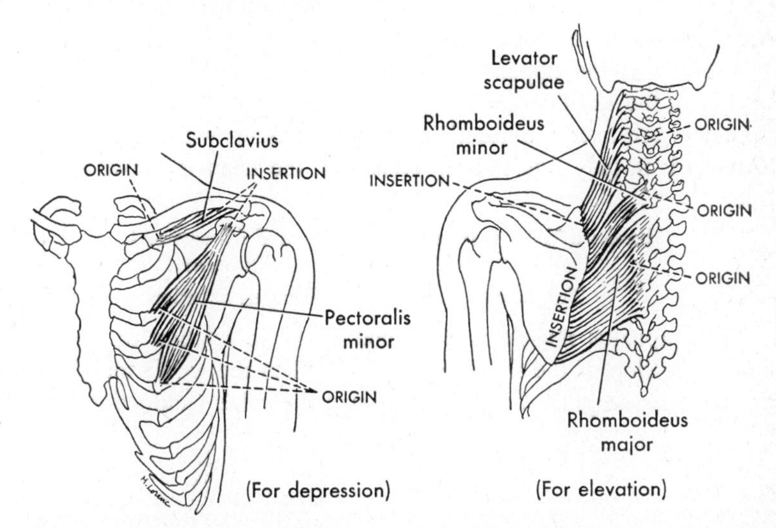

Fig. 123. Opposing muscle groups that elevate and depress the shoulder girdle.

Action. The rhomboidei carry the inferior angle of the scapula backward and upward and thus produce a slight rotation.

The pectoralis minor is underneath and entirely covered by the pectoralis major. It arises from the upper margins and outer surfaces of the third, fourth, and fifth ribs near their cartilages and is inserted into the coracoid process of the scapula.

Action. It depresses the point of the shoulder and rotates the scapula downward. In forced inspiration the pectoral muscles help in drawing the ribs upward and expanding the chest.

The subclavius arises from the junction of the first rib and its cartilage and is inserted into a groove on the under surface of the clavicle.

Action. The subclavius depresses the shoulder, i.e., carries it downward and forward.

The trapezius, so called because right and left together make a large diamond-shaped sheet, arises from the occipital bone, the ligamentum nuchae, and the spinous processes of the seventh cervical and the 12 thoracic vertebrae. From this extended line of origin the fibers converge to their insertion in the clavicle, the acromion process, and the spine of the scapula. It is a very large muscle and covers the other muscles of the upper part of the back and neck, also the upper portion of the latissimus dorsi.

Action. If the upper end is fixed, the shoulder is raised, as in shrugging the shoulder or carrying weights on the shoulder. If the shoulders are fixed, contractions of both muscles will draw the head backward; if only one muscle contracts, the head is drawn to that side.

Backward Movement (*adduction*)	**Forward Movement** (*abduction*)
Trapezius	Serratus anterior

The **trapezius** muscle is described above in relation to elevation and depression of the shoulder girdle. When the whole muscle acts together the scapulae are drawn toward the spine (adduction). At

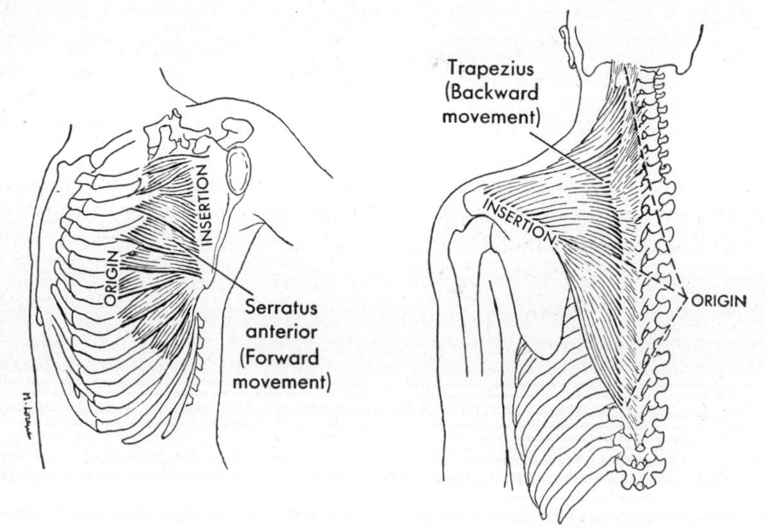

Fig. 124. Opposing muscles that abduct and adduct the shoulders.

the same time the scapulae are rotated, raising the point of the shoulder or the glenoid cavity upward.

The **serratus anterior** (serratus magnus) arises from the outer surfaces and superior borders of the upper eight or nine ribs and from the intercostals between them. The fibers pass upward and backward and are inserted in various portions of the ventral surface of the vertebral border of the scapula.

Action. Since its origin is on the chest wall, it moves the scapula forward away from the spine (abduction), as in the act of pushing. It also moves the scapula downward and inward toward the chest wall.

MUSCLES HAVING OPPOSING ACTION AT THE JOINTS

Movement of the Humerus

The shoulder joint is a ball-and-socket joint held by ligaments and tendons. The head of the humerus fits into the shallow glenoid cav-

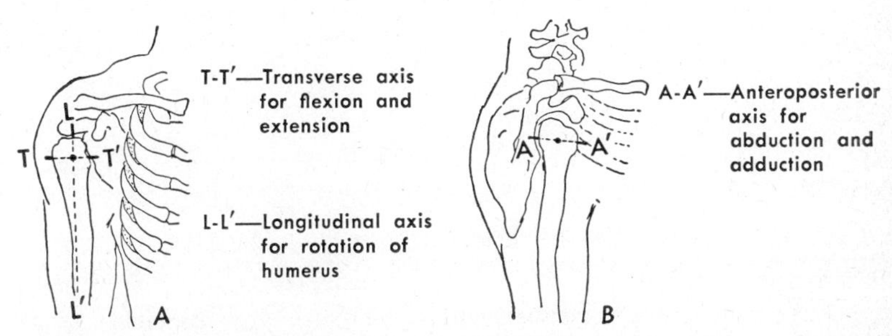

Fig. 125. Axis of movement of shoulder joint. *A*, anterior view; *B*, lateral view.

ity of the scapula. This permits free movement of flexion, extension, abduction, adduction, and rotation.

Flexion	**Extension**
Coracobrachialis	Teres major
Abduction	**Adduction**
Deltoid Supraspinatus	Pectoralis major
External rotation	**Internal rotation**
Infraspinatus Teres minor	Latissimus dorsi

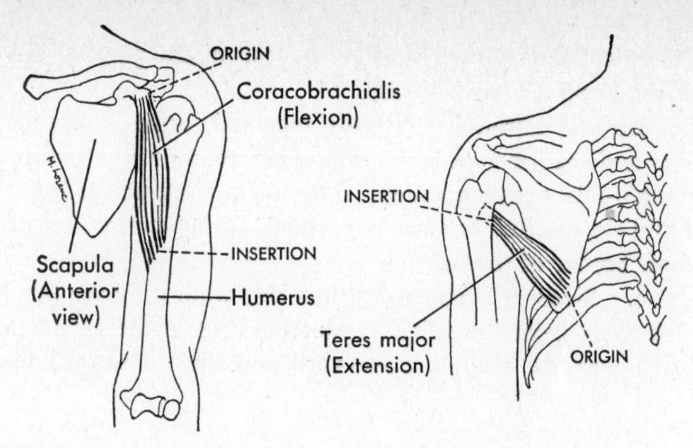

Fig. 126. Opposing muscles that flex and extend the humerus.

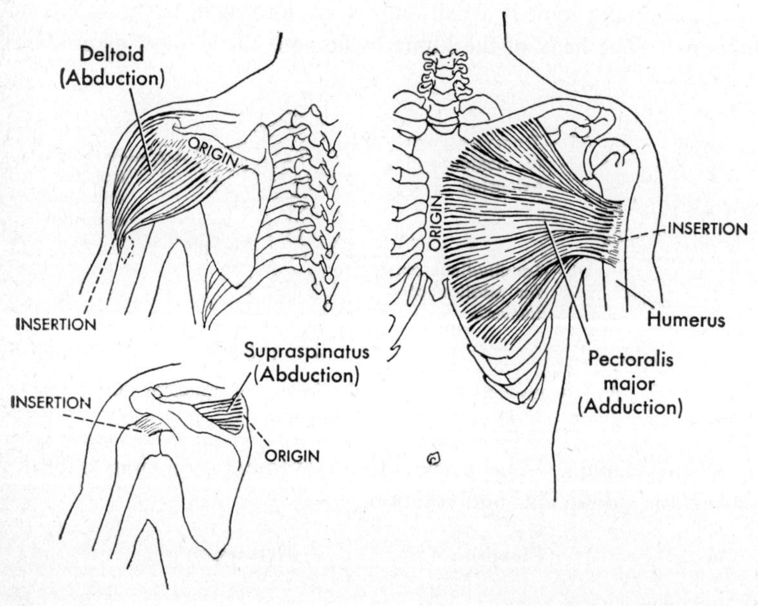

Fig. 127. Muscles having opposing action for abduction and adduction of the humerus.

The coracobrachialis is located at the upper and medial part of the arm. It has its origin on the coracoid process of the scapula and is inserted into the middle and medial surface of the humerus.

Action. It carries the arm forward, as in flexion. It also assists in adduction of the arm.

Teres major is a thick, flat muscle that has its origin on the dorsal side of the axillary border of the scapula and is inserted into the crest of the lesser tubercle of the humerus.

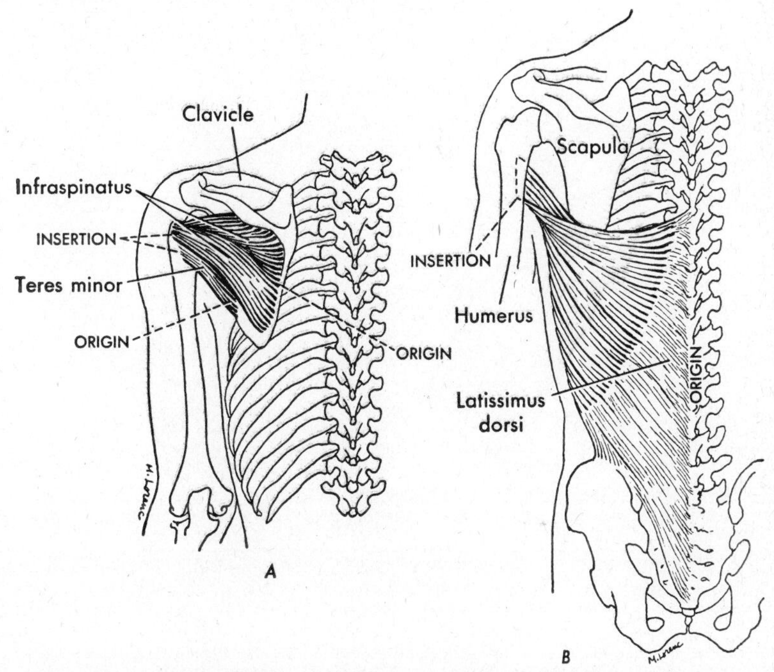

Fig. 128. Muscles having opposing action for rotation of humerus. *A,* muscles for external rotation; *B,* muscles for internal rotation.

Action. It extends the humerus, drawing it downward. It also helps to adduct and rotate the arm medially.

The deltoid. The deltoid is a thick triangular muscle which covers the shoulder joint. It arises from the clavicle, the acromion process, and the spine of the scapula and is inserted into the lateral side of the body of the humerus.

Action. Abduction—raises the arm from the side, so as to bring it at right angles to the trunk.

The supraspinatus muscle has its origin from the fossa above the spine of the scapula. The muscle passes over the shoulder joint and is inserted into the highest facet of the greater tubercle of the humerus.

Action. It assists the deltoid in abduction of the arm.

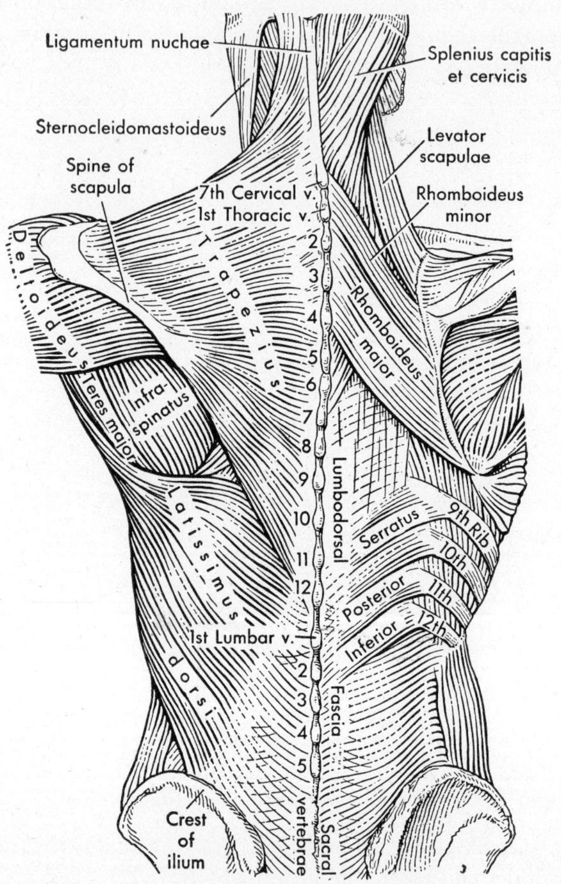

Fig. 129. Diagram showing the anatomical relationships of some of the muscles of the shoulders and back. (After Gray's *Anatomy.*)

The pectoralis major muscle. The pectoralis major is a large fan-shaped muscle that covers the upper and front part of the chest. It arises from the anterior surface of the sternal half of the clavicle, the anterior surface of the sternum, the cartilages of the true ribs, and the aponeurosis of the external oblique. The fibers converge and form a thick mass, which is inserted by a flat tendon into the crest of the greater tubercle of the humerus.

Action. If the arm has been raised by the abductors, the pectoralis major, acting with other muscles, draws the arm down to the side of the chest. Acting alone, it adducts and draws the arm across the chest, and also rotates it inward.

The infraspinatus muscle has its origin from the infraspinatus fossa on the back of the scapula and is inserted into the middle facet of the greater tubercle of the humerus.

Action. It rotates the humerus outward.

The teres minor is a long, narrow muscle that has its origin from the axillary border of the scapula and is inserted into the lowest facet of the greater tubercle of the humerus.

Action. It functions with the infraspinatus to rotate the humerus outward.

The latissimus dorsi is a flat, broad muscle that has its origin from a broad aponeurosis which is attached to the spinous processes of the lower six thoracic vertebrae, the spinous processes of the lumbar vertebrae, the spine of the sacrum, the posterior part of the crest of the ilium, and from the outer surface of the lower four ribs. Its fibers converge and form a flat tendon which is inserted into the bottom of the intertubercular groove of the humerus.

Action. It rotates the arm inward. It also extends and adducts the humerus.

Movement at the Elbow

The elbow is a compound hinge joint.

The trochlea of the humerus articulates with the semilunar notch of the ulna, permitting flexion and extension, and the articulation between the humerus and radius permits a rolling movement which supinates and pronates the hand. The strength and security of the elbow is due to arrangement of bones rather than to tendons and ligaments.

Movement between the Humerus and Ulna

Flexion	**Extension**
Brachialis	Triceps
also	
Biceps brachii	
Brachioradialis	

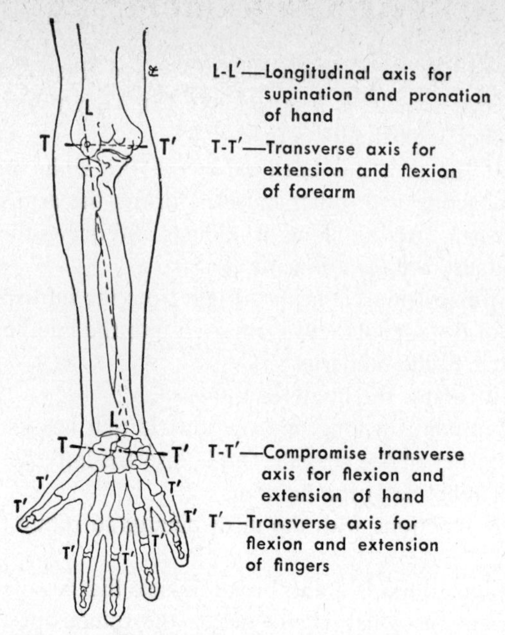

Fig. 130. Diagram showing axes of movement for elbow, hand, and fingers.

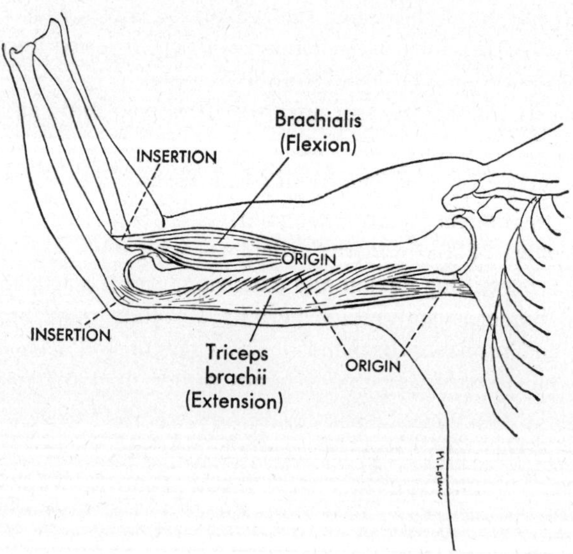

Fig. 131. Opposing muscles which flex and extend the elbow.

The brachialis muscle has its origin on the lower half of the anterior surface of the humerus and is inserted into the tuberosity of the ulna and by tendinous bands into the coronoid process.

Action. It is a strong flexor of the forearm.

The triceps (triceps brachii) arises by three heads, the long head from the infraglenoid tuberosity of the scapula and the lateral and medial heads from the posterior surface of the body of the humerus, the lateral head above the medial. The muscle fibers terminate in

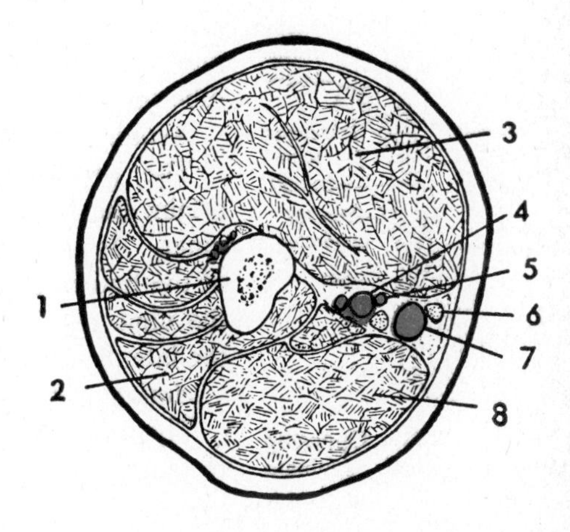

Fig. 132. Transverse section through the middle of the right upper arm. *1,* humerus; *2,* brachial muscle; *3,* triceps muscle; *4,* brachial artery; *5,* companion vein; *6,* ulnar nerve; *7,* basilic vein; *8,* biceps muscle.

two aponeurotic laminae, which unite above the elbow and are inserted into the olecranon of the ulna.

Action. The triceps is the great extensor of the forearm and is the direct antagonist of the brachialis.

Movement at the Radioulnar Joint
(Movement of the Hand)

Supination	**Pronation**
Biceps brachii	Pronator teres
Supinator (brevis)	Pronator quadratus

The biceps (biceps brachii) arises by two heads. The long head arises from a tuberosity at the upper margin of the glenoid cavity, and the short head from the coracoid process (Fig. 99). These tendons are succeeded by elongated bodies which are separate until within a short distance (7 cm) of the elbow joint, where they unite and terminate in a flat tendon, which is inserted into the tuberosity of the radius.

Action. As the biceps contracts, the radius turns outward, supinating the hand. It also flexes the forearm and its long head helps to hold the head of the humerus in the glenoid cavity, thus stabilizing the shoulder joint.

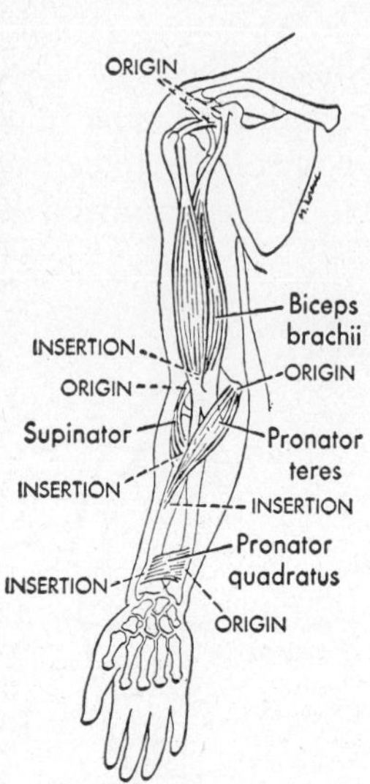

Fig. 133. Opposing muscles which supinate and pronate the hand. (Right arm.)

The supinator (brevis) muscle has its origin on the lateral epicondyle of the humerus and ridge of the ulna and is inserted into the dorsal and lateral margin of the tuberosity and oblique line of the radius.

Action. Supination of the hand.

The pronator teres has two heads of origin, the humeral or larger head and the ulnar. It arises from the medial epicondyle of the humerus and the coronoid process of the ulna. It is inserted into the middle of the lateral surface of the body of the radius.

Action. Pronation of the hand.

The pronator quadratus is a small flat, rectangular muscle extending across the lower part of the radius and ulna. It arises from the ulna and is inserted on the radius.

Action. Pronation of the hand.

Movement of the Wrist

Flexion	Extension
Flexor carpi radialis	Extensor carpi radialis longus
Flexor carpi ulnaris	Extensor carpi ulnaris

The flexor carpi radialis muscle arises from the medial epicondyle of the humerus, and its tendon is inserted into the base of the second metacarpal bone.

Action. Flexion of the hand and helps to abduct it.

The flexor carpi ulnaris muscle arises from the medial epicondyle of the humerus and the upper part of the dorsal border of the ulna. It is inserted into the pisiform bone.

Action. Flexion of the hand and helps to adduct it.

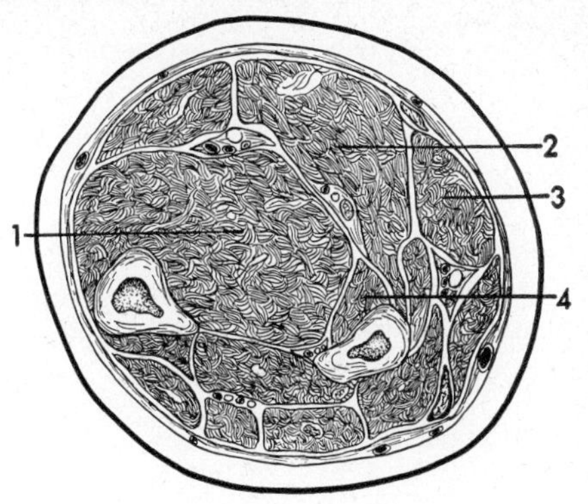

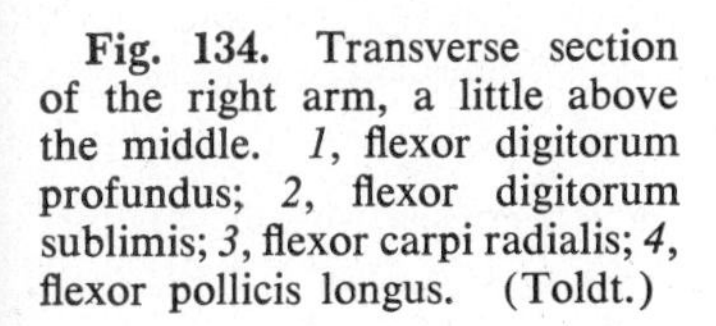
Fig. 134. Transverse section of the right arm, a little above the middle. *1*, flexor digitorum profundus; *2*, flexor digitorum sublimis; *3*, flexor carpi radialis; *4*, flexor pollicis longus. (Toldt.)

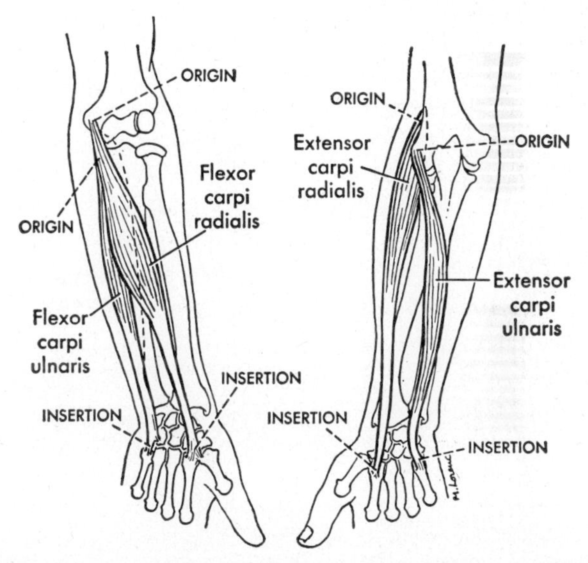

Fig. 135. Opposing groups of muscles which flex and extend the wrist.

Extensor carpi radialis longus arises from the lower third of the lateral supracondylar ridge of the humerus and is inserted on the dorsal side of the base of the radial side of the second metacarpal bone.

Action. It extends and abducts the hand.

Extensor carpi ulnaris arises from lateral epicondyle of the humerus and is inserted into the base of the fifth metcarpal bone.

Action. It extends and adducts the hand.

Movement of the Fingers
(A few of the muscles)

Flexion	**Extension**
Flexor digitorum profundus	Extensor digitorum communis
Flexor digitorum sublimis	

The flexor digitorum profundus muscle arises from the upper part of the volar and medial surfaces of the body of the ulna. Four tendons are formed which run to the tendons of the sublimis and are finally inserted into the bases of the last phalanges.

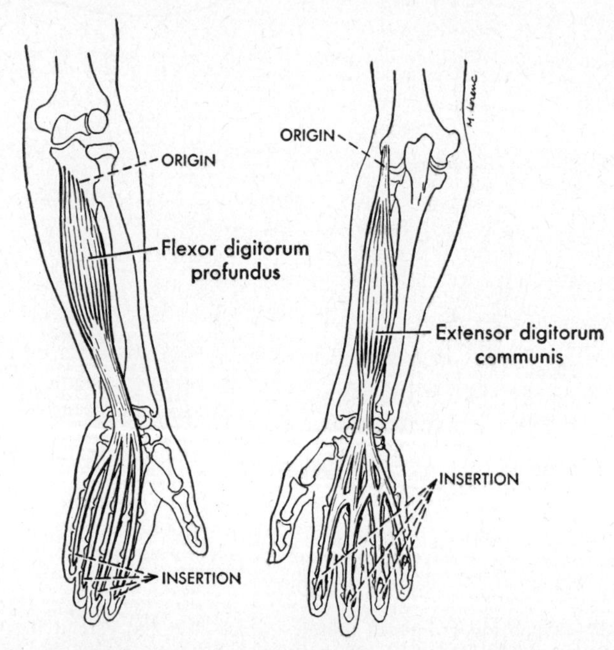

Fig. 136. Opposing muscles which flex and extend the fingers. (Left hand.)

Action. Flexes the terminal phalanx of each finger and by continued action flexes the other phalanges.

The flexor digitorum sublimis muscle arises by three heads from the epicondyle of the humerus, the coronoid process of the ulna, and the tuberosity of the radius. Four tendons are formed which are finally inserted into the sides of the second phalanx of each finger.

Action. Flexes the second phalanx of each finger, and by continued action flexes the first phalanx, and the hand. These two muscles function together in flexing the fingers.

Extensor digitorum communis muscle arises on the lateral epicondyle of the humerus and is finally inserted by four tendons into the dorsal surface of the last phalanx.

Action. Extends the phalanges and by continued action extends the wrist.

Movement of the Thumb
(A few of the muscles)

Flexion	**Extension**
Flexor pollicis longus	Extensor pollicis longus
Abduction	**Adduction**
Abductor pollicis longus	Adductor pollicis obliquus Adductor pollicis transversalis

The flexor pollicis longus muscle arises on the volar surface of the body of the radius and the interosseous membrane and is inserted into the base of the distal phalanx of the thumb.

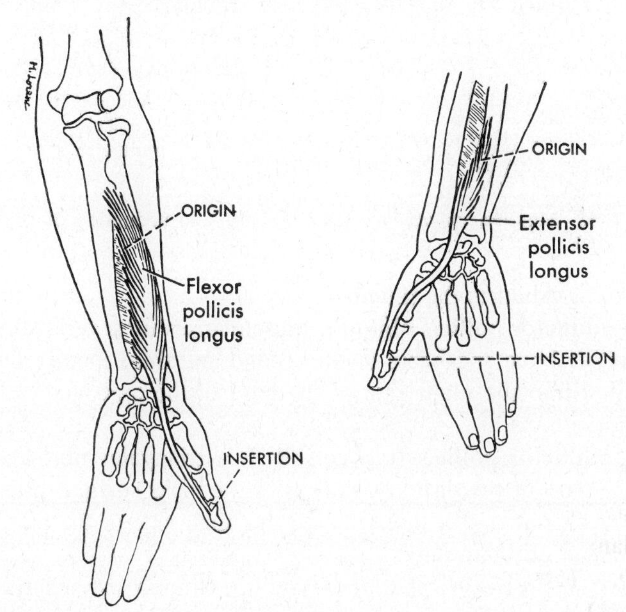

Fig. 137. Opposing muscles which flex and extend the thumbs.

Action. Flexes the thumb.

The extensor pollicis longus muscle arises on the lateral side of the middle part of the dorsal surface of the ulna and is inserted into the base of the last phalanx of the thumb.

Action. Extends the thumb.

The abductor pollicis longus muscle arises high on the lateral side of the dorsal surface of the ulna and from the middle third dorsal surface of the radius and is inserted into the radial side of the base of the first metacarpal bone.

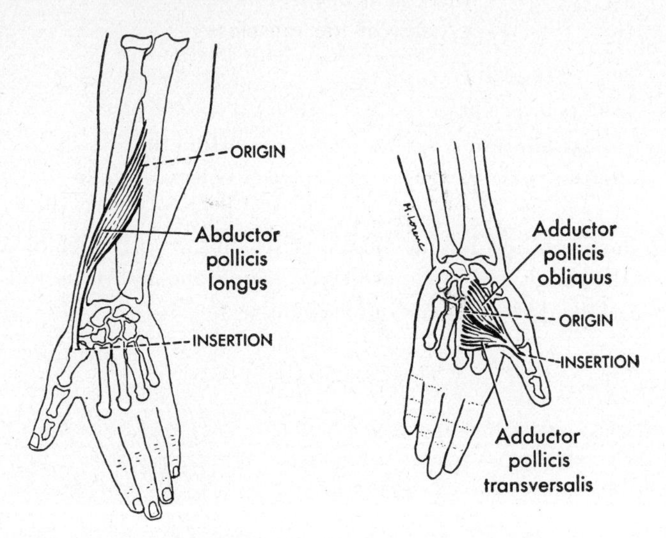

Fig. 138. Muscles concerned with adduction and abduction of the thumb. (Left hand.)

Action. Abducts the thumb.

The adductor pollicis obliquus muscle arises by several slips from the capitate bone and bases of the second and third metacarpals and is inserted into the ulnar side of the base of the first phalanx of the thumb.

The adductor pollicis transversalis is a triangular muscle arising from the base of the distal two-thirds of the volar surface of the third metacarpal bone and is inserted into the ulnar side of the base of the first phalanx of the thumb.

Action. These two muscles bring the thumb toward the palm (adduction).

MUSCLES OF RESPIRATION

Inspiration	**Expiration**
Diaphragm	Internal intercostals
External intercostals	Abdominal muscles (see p. 183)

The diaphragm is a dome-shaped, musculofibrous partition which forms the convex floor of the thoracic cavity and the concave roof of the abdominal cavity. The peripheral muscular fibers arise from the lower circumference of the thorax and are inserted into a central ten-

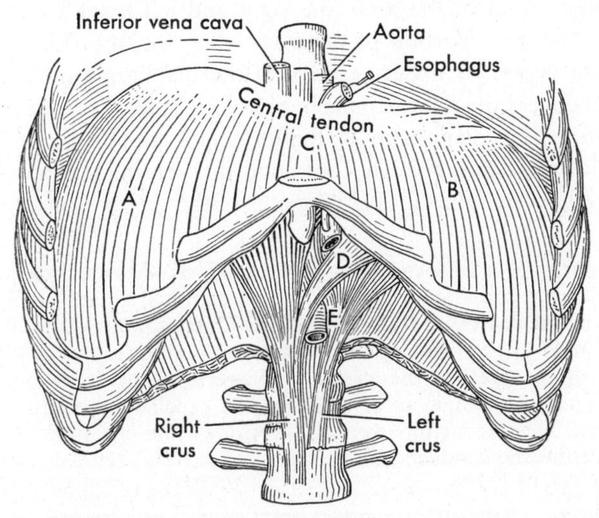

Fig. 139. Diaphragm, viewed from in front. At *A* the liver and at *B* the cardiac end of the stomach are underneath the diaphragm and push it up; at *C* the tip of the heart pushes the diaphragm down. At *D* and *E* the esophagus and aorta are shown.

don. The fibers are grouped according to their origin into three parts: the *sternal,* which arise from the back of the xiphoid process: the *costal,* which arise from the cartilages of the lower six ribs on either side; the *lumbar,* which arise from the *lumbocostal arches*[5] and the lumbar vertebrae by two pillars, or *crura.*[6] The fibers converge

[5] The *lumbocostal arches* are tendinous arches, one of which extends from the body to the transverse process of the first or second lumbar vertebra; the other extends from the transverse process of the first lumbar vertebra to the last rib.

[6] The *crura* (singular *crus*) are tendinous slips. The right crus arises from the bodies of the upper three lumbar vertebrae and the left from the bodies of the upper two.

toward the central portion, which is aponeurotic and serves for the insertion of the muscular portion.

The diaphragm has three large openings: the esophageal, for the passage of the esophagus, some esophageal arteries, and the vagus nerves; the aortic, for the passage of the aorta, the azygos vein, and the thoracic duct (strictly speaking, behind the diaphragm); the vena caval, for the passage of the inferior vena cava and some branches of the right phrenic nerves. The upper, or thoracic, surface of the diaphragm is highly arched; the heart is supported by the central tendinous portion of the arch, the right and left lungs by the lateral portions, the right portion of the arch being slightly higher than the left. The lower, or under, surface of the diaphragm is deeply concave and covers the liver, stomach, pancreas, spleen, and kidneys.

Action. The diaphragm is the principal muscle of inspiration. When the muscular portion contracts, the central tendon is pulled downward, so that the vertical diameter of the thorax is increased.

In forcible acts of expiration and in efforts of expulsion from the thoracic and abdominal cavities, the diaphragm and all the other muscles which tend to depress the ribs and those which compress the abdominal cavity concur in powerful action to empty the lungs, to fix the trunk, and to expel the contents of the abdominal viscera. Thus it follows that the action of the diaphragm is of assistance in expelling the fetus from the uterus, the feces from the rectum, the urine from the bladder, and the contents from the stomach in vomiting.

The intercostal muscles (intercostales) are found filling the spaces between the ribs. Each muscle consists of two layers, one external and one internal; and as there are 11 intercostal spaces on each side and 2 muscles in each space, it follows that there are 44 intercostal muscles. The fibers of these muscles run in opposite directions.

The external intercostals (intercostales externi) extend from the tubercles of the ribs behind to the cartilages of the ribs in front, where they end in membranes which connect with the sternum. Each arises from the lower border of a rib and is inserted into the upper border of the rib below. The direction of the fibers is obliquely downward.

The internal intercostals (intercostales interni) extend from the sternum to the angle of the ribs and are connected with the vertebral column by thin aponeuroses. Each arises from the inner surface of a rib and is inserted into the upper border of the rib below. The direction of the fibers is obliquely downward and opposite to the direction of the external intercostals.

Action. Investigators disagree as to the functions of the intercostal muscles. One authority states that the external and internal intercostals contract simultaneously and prevent the intercostal spaces from being pushed outward or drawn inward during respiration. Another classes the external ones as inspiratory and the internal ones as expiratory.

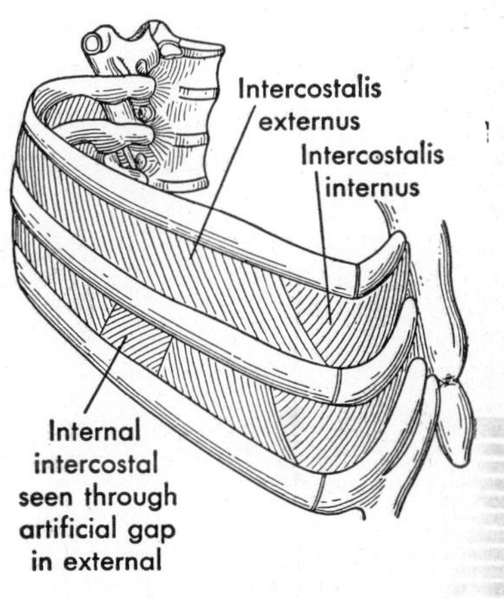

Fig. 140. Intercostal muscles in right wall of thorax.

The levatores costarum are 12 small muscles which arise from the transverse processes of the vertebrae from the seventh cervical to the eleventh thoracic. They pass obliquely downward and lateralward like the external intercostals. Each one is inserted into the outer surface of the rib, just below the vertebrae from which it takes origin.

Action. Raise the ribs, increasing the thoracic cavity (inspiratory), and rotate the vertebral column.

THE ABDOMINAL MUSCLES

The external, or descending, oblique (obliquus externus) is the strongest and most superficial of the abdominal muscles. It arises from the external surface of the lower eight ribs. The fibers from the lowest ribs pass downward and are inserted into the anterior half of the iliac crest; the middle and the upper fibers pass downward and forward and terminate in the broad aponeurosis which, meeting its fellow of the opposite side in the linea alba, covers the whole of the front of the abdomen. Between the anterior superior iliac spine and the pubic tubercle, the aponeurosis forms a thick band which is called the *inguinal ligament* (*Poupart's ligament*).

The internal, or ascending, oblique (obliquus internus) lies just beneath the external oblique. It arises from the inguinal ligament, the crest of the ilium, and the lumbodorsal fascia. The fibers are inserted into the costal cartilages of the lower six ribs, the linea alba (by means of an aponeurosis), and the crest of the pubis. At the lateral border of the rectus the aponeurosis divides into two layers, which continue forward, one in front of and the other behind the rectus muscle; they reunite at the linea alba and thus form a sheath for the rectus.

The transversus (transversalis) muscle is just beneath the internal oblique. The fibers arise from the lower six costal cartilages, the lumbodorsal fascia, the anterior three-quarters of the iliac crest, and the lateral third of the inguinal ligament. The greater part of its fibers have a horizontal direction and end in front in a broad aponeurosis, which is inserted into the linea alba and the crest of the pubic bone.

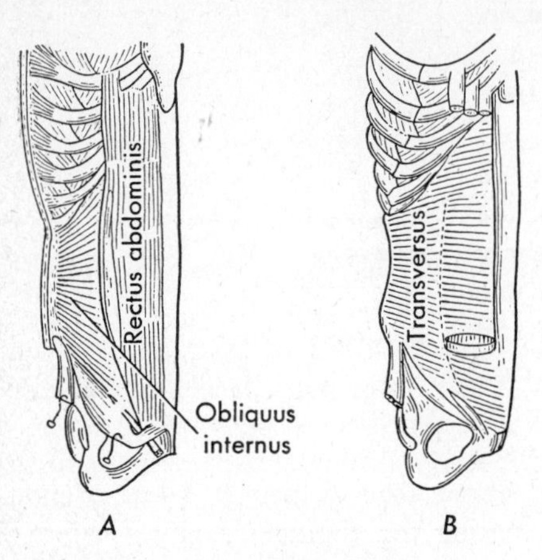

Fig. 141. Abdominal muscles. *A*, rectus abdominis and obliquus internus of right side; *B*, transversus (transversalis) of right side.

The rectus abdominis is a long, flat muscle, consisting of vertical fibers, situated at the front part of the abdomen and enclosed in the fibrous sheath formed by the aponeuroses of the internal oblique, external oblique, and transversus muscles. It arises from the pubic bone and the ligaments covering the front of the symphysis pubis and is inserted into the cartilages of the fifth, sixth, and seventh ribs. It is separated from the muscles of the other side by a narrow interval which is occupied by the linea alba.

Action of the abdominal muscles. When these muscles contract, they compress the abdominal viscera and constrict the cavity of the abdomen, in which action they are much assisted by the descent of the diaphragm.

This action assists in expiration. When the diaphragm contracts, the abdominal muscles relax; when the diaphragm relaxes, the abdominal muscles contract. When the abdominal muscles are contracted they assist in parturition, defecation, micturition, and emesis. They also bend the thorax forward. When the muscles of only one side contract, the trunk is bent toward that side.

The linea alba is a tendinous line in the middle of the abdomen formed by the blending of the aponeuroses of the two oblique and the transversus muscles of both sides. It stretches from the xiphoid process to the symphysis pubis. It is a little broader above than below; and a little below the middle it is widened into a flat, circular space, in the center of which is situated the umbilicus. The pyramidalis muscle increases the tension of the linea alba. It arises from the front of the pubis and is inserted into the linea alba halfway between the umbilicus and the pubis. It is in front of the rectus, within the same sheath.

The inguinal canal. Parallel to, and a little above, the inguinal ligament is a tiny canal about 4 cm ($1\frac{1}{2}$ in.) long, called the inguinal canal. The internal opening of the canal is called the *abdominal inguinal ring* and is situated in the fascia of the transversus muscle, halfway between the anterior superior spine of the ilium and the symphysis pubis. The canal ends in the *subcutaneous inguinal ring,* which is an opening in the tendon of the external oblique just above and lateral to the crest of the pubis. This canal transmits the spermatic cord in the male and the round ligament of the uterus in the female.

Weak places in the abdominal walls. The abdominal inguinal and the subcutaneous inguinal rings, described above, the umbilicus, and another ring situated just behind the inguinal ligament, called the *femoral ring,* are often the seat of *hernia.*

Hernia, or rupture, is a protrusion of a portion of the contents of a body cavity, and in this instance would mean a protrusion of a portion of the intestine or mesentery through one of these weak places. If it occurs in the umbilicus, it is called *umbilical hernia;* in the inguinal rings, *inguinal hernia;* and in the femoral ring, *femoral hernia.* Conditions which favor hernia are (1) lifting, coughing, etc., which greatly increase the pressure of the abdominal contents against the body wall, and (2) lack of tone of the ventral abdominal wall, which comes about in old age or as the result of illness. The inguinal canal is larger in the male than in the female; hence inguinal hernia is more common in the male than in the female. There are other abdominal hernias.

MOVEMENT OF THE FEMUR

The hip joint is a ball-and-socket joint. The head of the femur fits into the deep cup-shaped cavity of the acetabulum (Fig. 98), held by strong ligaments, yet permitting free movement.

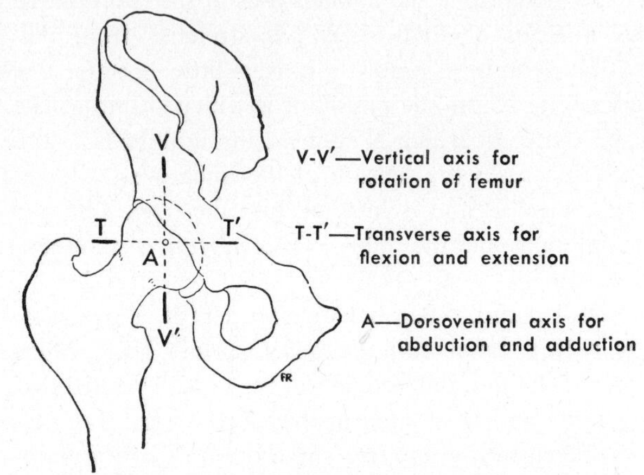

Fig. 142. Diagram showing axes of movement of the hip joint.

Flexion	**Extension**
Psoas major	Gluteus maximus
Iliacus	
Abduction	**Adduction**
Gluteus medius	Adductor magnus
Tensor fasciae latae	Adductor longus
	Adductor brevis
Outward Rotation	**Inward Rotation**
Piriformis	Gluteus minimus (anterior part)
Quadratus femoris	Gluteus medius (anterior part)
Obturators	

The psoas major (magnus) is a long, powerful muscle that arises from the bodies and transverse processes of the last thoracic and all the lumbar vertebrae with the included intervertebral cartilages. It extends downward and forward, then downward and backward, to its insertion in the small trochanter of the femur.

The iliacus arises from the iliac crest and fossa. The fibers converge and are inserted into the lateral side of the tendon of the psoas major and the body of the femur below and in front of the lesser trochanter. The relation of this muscle to the psoas major is well shown in Fig. 143.

Action. The psoas major and iliacus act as one muscle to flex the thigh on the pelvis.

The gluteus maximus is a large, powerful muscle that arises from the posterior fourth of the iliac crest, the posterior surface of the

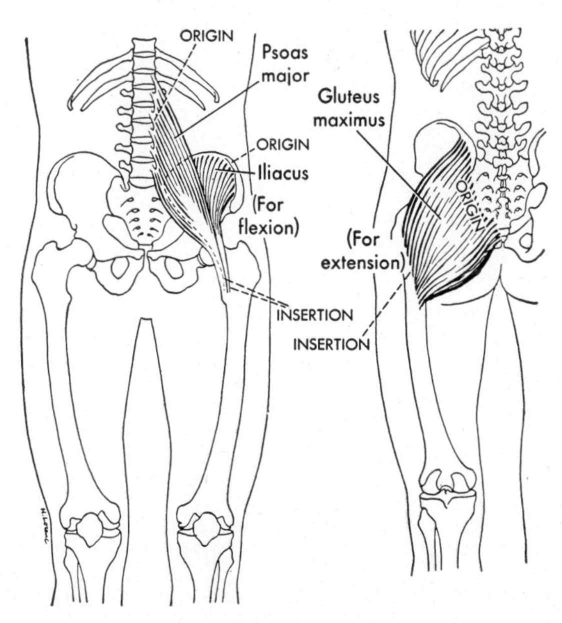

Fig. 143. Muscles having opposing action for flexion and extension of femur.

lower part of the sacrum, the side of the coccyx, and the aponeuroses of the sacrospinalis and the gluteus medius. It is inserted into the fascia lata and the gluteal ridge, a prolongation of the upper end of the linea aspera of the femur. The thigh muscles are covered by a heavy cylindrical fascia—the fascia lata—which extends from the highest margin of the thigh to the bony prominences around the knee and helps form the capsular ligament of the knee joint. It varies in thickness. In the region of the gluteus maximus and the tensor fasciae latae, it splits into two layers.

Action. It opposes the action of the iliopsoas muscle. It extends the femur and rotates it outward.

Gluteus medius and minimus lie under the maximus on the lateral side of the hip joint.

The gluteus medius arises from the outer surface of the ilium between the crest of the ilium and the posterior gluteal line and is in-

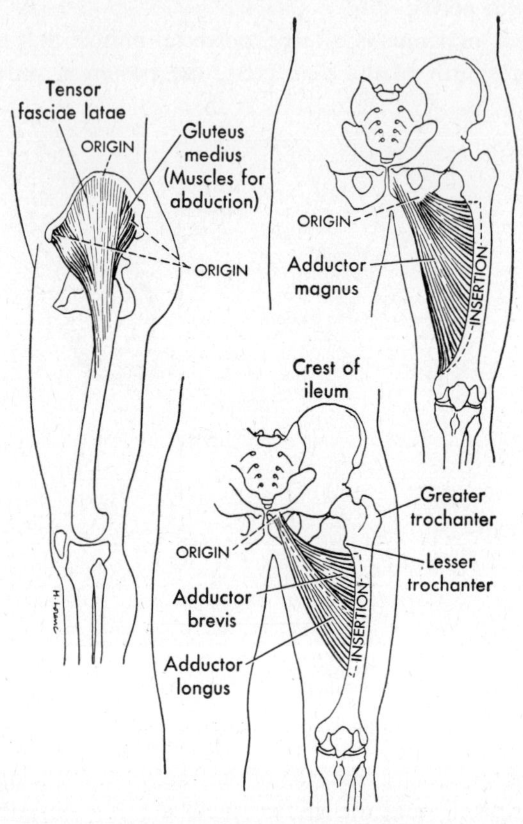

Fig. 144. Muscles having opposing action for adduction and abduction of the thigh.

serted by a strong flat tendon into the lateral surface of the great trochanter.

Action. It abducts the thigh and rotates it inward.

The gluteus minimus arises from the outer surface of the ilium and is inserted into the anterior surface of the great trochanter.

Action. It rotates the femur inward, opposing the muscles of outward rotation.

The tensor fasciae latae has its origin from the anterior outer part of the crest of the ilium and iliac spine and is inserted into the iliotibial band of the fascia lata.

Action. When the foot is lifted off the ground, it abducts, flexes, and medially rotates the thigh. When the foot is on the ground it flexes and abducts the pelvis and laterally rotates it.

Adduction. The three adductor muscles, magnus, longus and brevis, have their origin on the pubic bone and are inserted into the

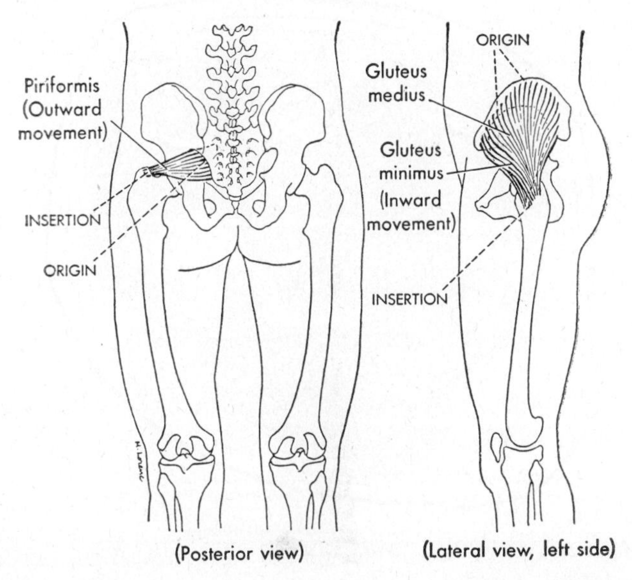

Fig. 145. Opposing muscles for rotation of the thigh. Other outward rotators are not shown.

linea aspera of the femur. The *magnus* is large and triangular in shape, has its origin on the inferior ramus of the pubis and ischium, and is inserted the full length of the linea aspera. The *longus* is triangular in shape, has its origin on the front of the pubis, and is inserted into the middle third of the linea aspera. The *brevis* arises from the outer surface of the inferior ramus of the pubis and is inserted into the upper part of the linea aspera.

Action. These muscles are powerful adductors of the femur.

The piriformis is a flat pyramid-shaped muscle that arises from the front of the sacrum and passes out of the pelvis through the great sciatic notch to be inserted into the upper border of the great trochanter.

Action. It abducts the femur and rotates it outward.

The quadratus femoris is a flat quadrilateral muscle that arises from the upper part of the tuberosity of the ischium and is inserted into the upper part of the linea quadrata of the femur.

Action. Outward rotation.

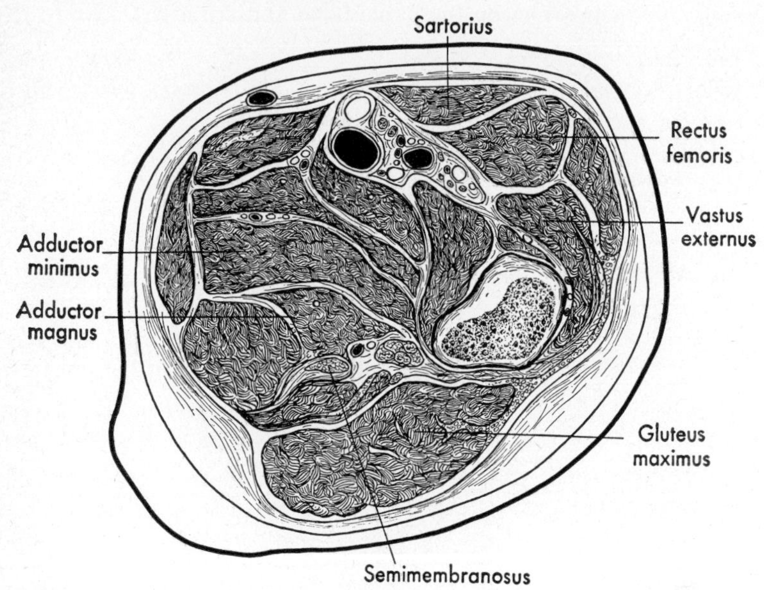

Fig. 146. Transverse section through the right thigh at the level of the small trochanter. (Toldt.)

The external obturator muscle is a flat triangular muscle that arises from the margin of bone around the obturator foramen and from the outer surface of the obturator membrane. The fibers pass backward, upward, and laterally and end in a tendon that is inserted into the trochanteric fossa of the femur.

Action. Outward rotation of the femur and supports the floor of the pelvis.

The following muscles also externally rotate the thigh:

The obturator internus arises from the inner surface of the anterolateral wall of the pelvis, the greater part of the obturator foramen, and the pelvic surface of the obturator membrane. It is inserted in the fore part of the medial surface of the greater trochanter.

Action. It brings about external rotation of the thigh.

The gemellus superior arises from the ischial spine and edge of the lesser sciatic notch. It is inserted in the fore part of the medial surface of the great trochanter.

The gemellus inferior arises from the upper and inner border of the tuberosity of the ischium. It is inserted with the gemellus superior.

Action. They act with the obturator internus.

MOVEMENT AT THE KNEE JOINT

The knee is a modified hinge joint. The condyles of the femur articulate with the condyles of the tibia. The two semilunar rings of fibrocartilage, called the lateral and medial menisci, deepen the articulation. There is also the articulation between the femur and patella. Many ligaments surround and securely hold the joint.

Flexion

Biceps femoris
Semitendinosus
Semimembranosus
Popliteus
Gracilis
Sartorius

Extension

Quadriceps femoris
 Rectus femoris
 Vastus lateralis
 Vastus medialis
 Vastus intermedius

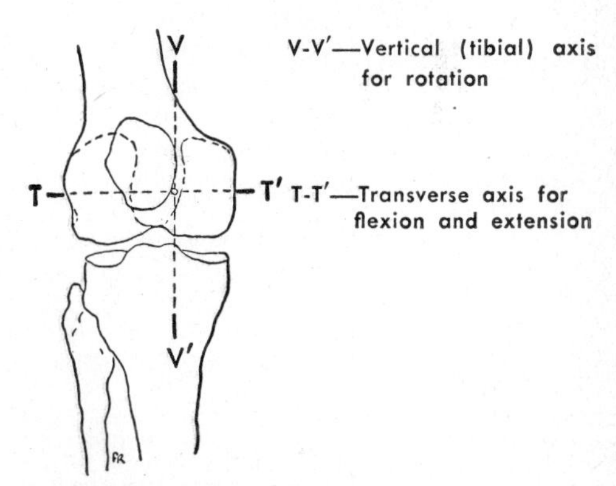

Fig. 147. Diagram showing axes of movement at the knee joint.

The biceps femoris arises by two heads, the long head from the tuberosity of the ischium and the short head from the linea aspera of the femur. It is inserted into the lateral side of the head of the fibula and the lateral condyle of the tibia.

The semitendinosus arises from the tuberosity of the ischium and is inserted into the upper part of the medial surface of the body of the tibia.

The semimembranosus arises from the tuberosity of the ischium and is inserted on the medial condyle of the tibia.

The tendons of insertion of these muscles are called the hamstrings; hence the muscles are often called the hamstring muscles.

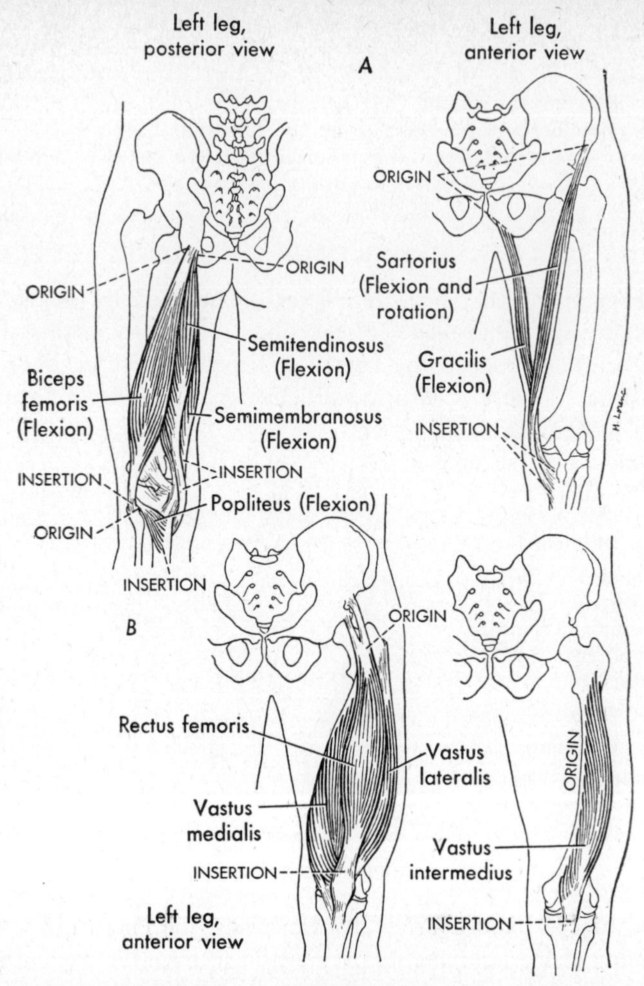

Fig. 148. *A,* muscles concerned with flexion of the knee. The gracilis also adducts the thigh and the sartorius assists in flexion and lateral rotation of the thigh. *B,* quadriceps femoris muscle for extension of the knee.

Action. They flex the leg upon the thigh and extend the thigh. When the knee is flexed, the semitendinosus and semimembranosus rotate the leg inward.

The popliteus is a thin, flat, short muscle that is located behind the knee joint.

It arises from the lateral condyle of the femur and is inserted into the posterior surface of the body of the tibia.

Action. The popliteus assists in flexing the leg upon the thigh and rotates the tibia inward.

The gracilis is a long, slender muscle on the inner side of the thigh. It arises from the symphysis pubis and the pubic arch and is inserted into the medial surface of the tibia below the condyle.

Action. It assists in flexion of the leg and in adduction of the thigh.

The sartorius is a long muscle that extends obliquely across the front of the thigh. It arises from the anterior superior spine of the ilium and is inserted into the upper part of the medial surface of the body of the tibia.

Action. It flexes the leg upon the thigh and the thigh upon the pelvis; it also rotates the thigh outward.

The quadriceps femoris (quadriceps extensor) is a four-headed muscle covering the front and sides of the thigh. Each head is described as a separate muscle.

1. *The rectus femoris* arises by two tendons, one from the anterior inferior iliac spine, the other from a groove above the brim of the acetabulum.
2. *The vastus lateralis* (vastus externus) arises by a broad aponeurosis from the great trochanter and the linea aspera of the femur.
3. *The vastus medialis* (vastus internus) arises from the medial lip of the linea aspera.
4. *The vastus intermedius* (crureus) arises from the ventral and lateral surfaces of the body of the femur.

The fibers of these four muscles unite at the *lower* part of the thigh and form a strong tendon, which is inserted into the tuberosity of the tibia. The tendon passes in front of the knee joint, and the patella is a sesamoid bone developed in it.

Action. The quadriceps femoris extends the leg upon the thigh, and the rectus portion flexes the thigh.

MOVEMENT OF THE FOOT

The ankle is a hinge joint. It is formed by the articulation of the tibia, the malleolus of the fibula, and the convex surface of the talus. Movement here gives flexion (dorsiflexion), or foot bent toward the

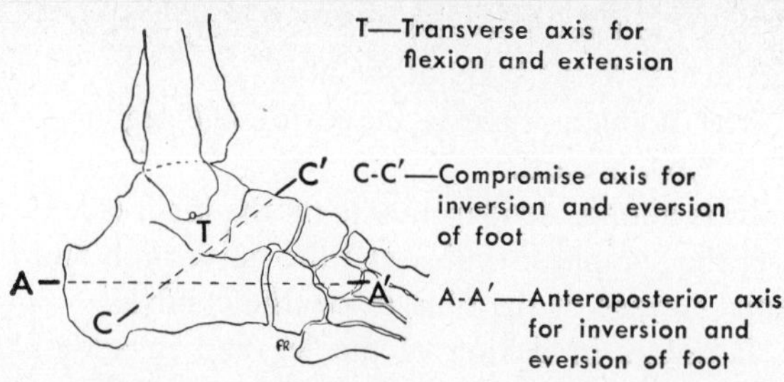

Fig. 149. Diagram showing axes of movement of the ankle joint.

anterior part of the leg, and extension (plantar flexion), or downward movement of the foot. Inversion, or turning the foot in, and eversion, or turning the foot out, also take place at the joint.

Plantar flexion	**Dorsiflexion**
Gastrocnemius	Tibialis anterior
Soleus	Peroneus tertius
Tibialis posterior	

The gastrocnemius and soleus form the calf of the leg. The gastrocnemius arises by two heads from the medial and lateral condyles of the femur. The soleus arises from the back of the head of the fibula and the medial border of the tibia. The direction of both is downward, and they are inserted into a common tendon, the tendo calcaneus (tendo Achillis), which is the thickest and strongest tendon in the body. The tendon is about 15 or 16 cm long and starts about the middle of the leg. The tendon receives muscle fibers on its anterior surface almost to the end. It is inserted into the calcaneus, or heel bone.

Action. The gastrocnemius and soleus extend or plantar flex the foot at the ankle joint, and the gastrocnemius flexes the femur upon the tibia.

The tibialis posterior (posticus) arises from the aponeurotic septum, between the tibia and the fibula, and from the adjoining parts of these two bones. It is inserted into the under surface of the navicular bone and gives off fibers which are attached to the calcaneus, the three cuneiforms, the cuboid, and the second, third, and fourth metatarsal bones.

Action. It extends the foot at the ankle joint. Acting with the tibialis anterior, it inverts the foot, i.e., turns the sole of the foot upward and medialward.

The tibialis anterior (anticus) arises from the lateral condyle and upper portion of the lateral surface of the body of the tibia and is inserted into the under surface of the first cuneiform bone and the base of the first metatarsal.

Action. It flexes the foot at the ankle joint and with the tibialis posterior raises the medial border of the foot or inverts the foot.

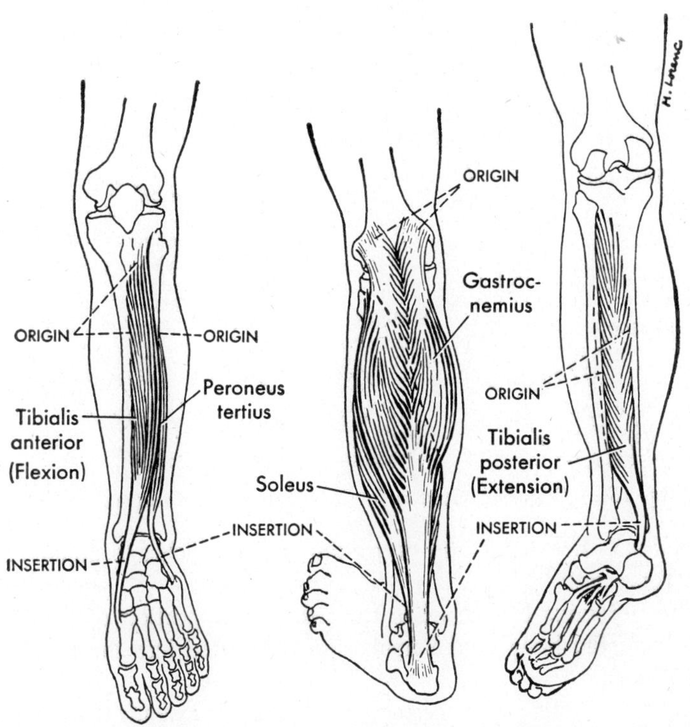

Fig. 150. Muscles having opposing action for flexion and extension of the ankle.

The peroneus tertius has its origin from the anterior part of the lower third of the fibula and is inserted on the base of the fifth metatarsal bone.

Action. It flexes and everts the foot.

The peroneus longus arises from the head and the lateral surface of the body of the fibula. The terminal tendon passes under the cuboid bone, crosses the sole of the foot obliquely, and is inserted into the lateral side of the first metatarsal and first cuneiform bones.

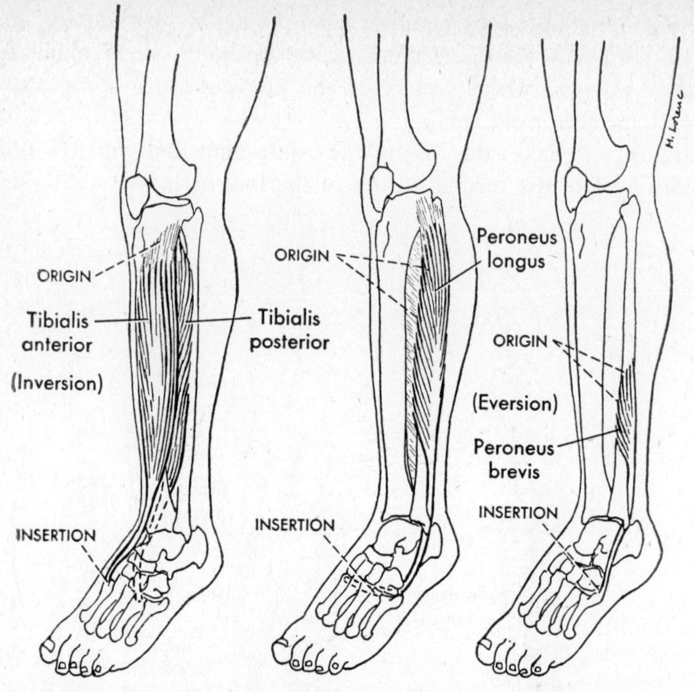

Fig. 151. Opposing muscles for inversion and eversion of the foot.

Action. It extends and everts the foot and helps to maintain the transverse arch.

The peroneus brevis arises from the lower portion of the lateral surface of the body of the fibula and is inserted into the fifth metatarsal bone.

Action. Acting with the peroneus longus, it extends the foot upon the leg and everts the foot.

The muscles of the foot. The leg muscles have important relations with the foot; hence, they are called extrinsic foot muscles. The muscles within the foot proper are called the intrinsic foot muscles. There are many small intrinsic muscles which are important in all movements of the foot and in maintaining the arches of the foot.

MOVEMENT OF THE TOES

Flexion	**Extension**
Flexor hallucis longus	Extensor hallucis longus
Flexor digitorum longus	Extensor digitorum longus

The flexor hallucis longus arises from the distal two-thirds of the posterior surface of the fibula and is inserted into the base of the last phalanx of the great toe.

The flexor digitorum longus arises from the posterior surface of the body of the tibia and is inserted by four tendons into the bases of the last phalanges of the four lesser toes.

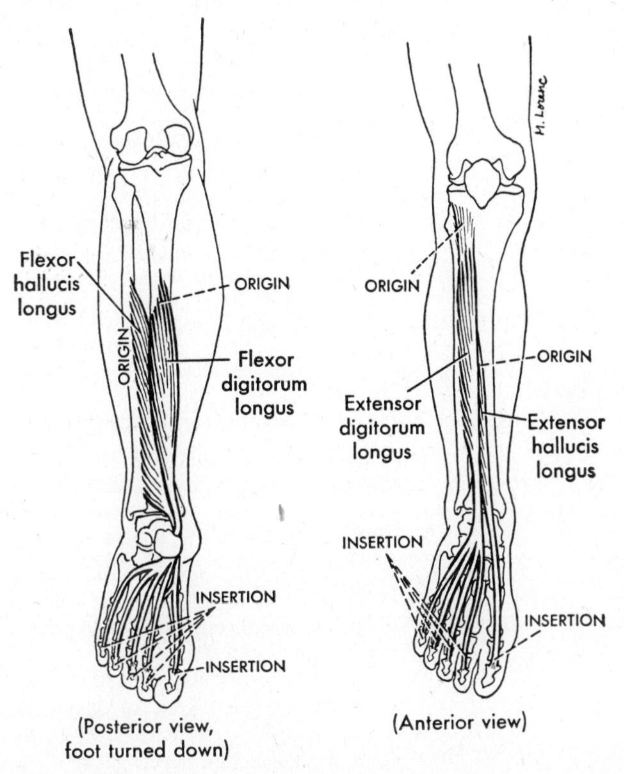

Fig. 152. Muscles having opposing action for flexion and extension of the toes.

Action. The flexor hallucis longus and the flexor digitorum longus flex the phalanges and extend the foot.

The extensor hallucis longus arises from the anterior surface of the fibula and is inserted into the distal phalanx of the great toe.

The extensor digitorum longus arises from the lateral condyle of the tibia and the upper portion of the anterior surface of the fibula.

The terminal tendon divides into four slips, which are inserted into the second and third phalanges of the four lesser toes. A fifth tendon of this muscle is sometimes described as the *peroneus tertius* muscle.

Action. The two extensor muscles extend the phalanges of the toes and flex the foot upon the leg.

SUMMARY

The movements of the body are dependent upon the contractions of muscular tissue

Characteristics of Muscular Tissue
- (1) Irritability or excitability—property of receiving stimuli and responding to them
- (2) Contractility—muscle becomes shorter, thicker, because the cells do
- (3) Extensibility—muscle can be stretched, i.e., property of individual cells
- (4) Elasticity—muscle readily returns to original shape

Classification
- (1) Striated, skeletal
- (2) Nonstriated, visceral
- (3) Indistinctly striated, cardiac

Muscular Tissue
- Cells become elongated and are called fibers
- Intercellular substance between fibers
- Connective tissue—supporting framework
- Well supplied with nerves and blood and lymph vessels

Striated { **Cross-striped** / **Skeletal** }
- (1) Marked with transverse striae
- (2) Movements accomplished by it are voluntary
- (3) Attached to skeleton
- (4) Muscle fibers are long and spindle-shaped
- (5) Connective-tissue framework carries blood vessels and nerves
- (6) Origin—more fixed attachment
- (7) Insertion—more movable attachment
- (8) Origin in periosteum of bone or intervening tendon
- (9) Insertion either by tendons or aponeuroses
- (10) Muscles closely covered by sheets of fasciae
- (11) Deep fasciae form annular ligaments in vicinity of wrist and ankle

Function. To operate the bones of the body, producing motion

Nonstriated { **Smooth**, **Visceral** }

- (1) Not marked with transverse striae
- (2) Movements accomplished by it are involuntary
- (3) Found in walls of blood vessels and viscera
- (4) Composed of spindle-shaped cells that contain one large nucleus
- (5) Connective-tissue framework carries blood vessels and nerves

Function. To cause visceral motion

Cardiac { **Striated**, **Involuntary**, **Visceral** }

- (1) Striated, but not distinctly
- (2) Not under control of will
- (3) Cells are quadrangular in shape
- (4) Cells grouped in bundles
- (5) Connective tissue forms a supporting framework

Physiology of Contraction

- **Tone**
  - Steady, partial contraction existing under normal conditions
  - Gives skeletal muscles firmness and a slight, sustained pull upon attachments
  - Both visceral and cardiac muscle exhibit tonus even when isolated from body
- **Excitation (stimulation)**
  - A property of all protoplasm
  - *Stimulus*—a change in the environment of the cell
  - *Conductivity*—a property of protoplasm by which responses are brought about
  - *Artificial stimuli*—pressure, temperature, electrical, etc.
  - Nerve fibers
    - *Sensory*, convey to central nervous system the state of contraction of a muscle
    - *Motor*, convey impulses from central nervous system to the muscles and control their contraction
- **Conditions of contraction**
  - *Latent period*—time between stimulation and contraction
  - *Period of contraction*
  - *Period of relaxation*
  - Factors influencing contraction
    - (1) Strength of stimulus—the stronger (up to a certain maximum) the stimulus, the stronger the contraction of muscle trunk
    - (2) Duration of stimulus

- **Physiology of Contraction**
  - **Conditions of contraction**
    - Factors influencing contraction
      - (3) Weight of load—increase of load decreases the height of contraction
      - (4) Temperature—optimum is 37° C (98.6° F)
  - **Response to stimuli**
    - *All-or-none law*—contraction of muscle cell is maximal or none at all for the conditions
    - *Minimal stimulus*—the weakest stimulus which will give contraction
    - *Subminimal stimulus*—any stimulus weaker than minimal
    - *Summation of stimuli*—the combined forces of subminimal stimuli which result in contraction
    - *Chronaxie*—the shortest duration of time that a stimulus of twice minimal strength must be applied to evoke a response
    - *Summation of contraction*—a maximal contraction resulting from a second stimulus occurring during apex of contraction from the first
    - *Absolute refractory period*—the time between the accomplishment of a contraction and the reception of the next stimulus
    - *Tetanus*—sustained contraction resulting from rapid succession of stimuli
  - **Types of contractions**
    - *Isotonic*—muscle shortens and thickens but its tone is not altered
    - *Isometric*—tension increases but length of muscle is unaltered
  - **Contraction in skeletal muscle**
    - Voluntary muscle cells are separate units, in contrast to heart muscle, which responds as a unit
    - Phases in contraction
      - *Anaerobic* (contractile)
        - Oxygen not required
        - Little or no lactic acid produced
      - *Aerobic* (recovery)
        - Oxygen necessary
        - Lactic acid produced and supplies energy for resynthesis of itself into glycogen. One-fifth supplies energy for resynthesis of other four-fifths
      - Cleavage of phosphocreatine thought to supply energy for contraction
      - Reversibility of these reactions thought to be due to system of enzymes within muscle cell

Physiology of Contraction

- **Fatigue and exercise**
  - Formation of carbon dioxide at first increases irritability of muscles
  - Continuous contraction brings about accumulation of waste products, causing fatigue
  - Moderate exercise aids in getting rid of these waste products
  - *Rigor mortis*—fatigue carried to point beyond possible recovery. Protein constituents of muscle fibers coagulate
  - Fatigue is complex
    - Other cells than muscles show fatigue
    - Associated with various mental states
  - Exercise brings change in conditions (new blood, etc.) for all cells of body. May increase size, strength, and tone of muscle fibers
- **Levers**
  - *Lever*—a rigid rod free to move about on some fixed point, the *fulcrum*
  - Lever acted upon by
    - *Resistance* (weight)—something to be overcome or balanced
    - *Force* (effort)—that which is exerted to overcome the resistance
  - Three types of levers
    - (1) Fulcrum lies between the effort and the resistance
      Example, raising head
    - (2) Resistance lies between fulcrum and effort
    - (3) Effort exerted between fulcrum and resistance
      Example, flexing forearm

SUMMARY: FUNCTIONALLY IMPORTANT MUSCLES

	Name of Muscle		*Origin*	*Insertion*	*Function*
Facial Expression	Epicranial, or Occipitofrontalis	Occipital	Occipital bone	Aponeurosis—top of skull	Draws the scalp backward
		Frontal	Aponeurosis—top of skull	Tissues of the eyebrows	Elevates the eyebrows, causes transverse wrinkles of forehead
	Corrugator		Medial end superciliary arch	Skin of forehead	Wrinkles the skin of the forehead vertically as in frowning
	Buccinator		Alveolar processes of maxilla and mandible	Orbicularis oris	Compresses the cheeks, brings them in contact with the teeth
	Zygomatic		Zygomatic bone	Orbicularis oris muscle	Draws angle of the mouth backward and upward as in laughing
	Triangularis		Oblique line of the mandible	Orbicularis oris muscle	Depresses angle of the mouth
	Platysma		Skin and fascia of the pectoral and deltoid muscles	Mandible and muscles about the angle of the mouth	Depresses the mandible and draws down the lower lip
	Risorius		Fascia over the masseter muscle	The skin of the angle of the mouth	Retracts angle of the mouth, produces an unpleasant grinning expression
	Quadratus labii superioris Angular head		Upper part of frontal process of the maxilla	Alar cartilage, skin of nose and upper lip	Elevates upper lip, angular head dilates the naris
	Infraorbital head		Lower margin of the orbit	Upper lip	Gives expression of sadness
	Zygomatic head		Zygomatic bone	Orbicularis oris muscle	

SUMMARY: FUNCTIONALLY IMPORTANT MUSCLES.—(*Continued*)

	Name of Muscle	*Origin*	*Insertion*	*Function*
Facial Expression (*Continued*)	Quadratus labii inferioris	Oblique line of the mandible below canine and premolar teeth	Skin of lower lip and orbicularis oris	Draws lower lip downward and lateralward as in the expression of irony
	Caninus	Canine fossa of maxilla	Orbicularis oris muscle	Produces the nasolabial furrow
	Mentalis	Incisive fossa of mandible	Skin of chin	Raises and protrudes lower lip, expresses doubt and disdain
	Orbicularis oris	Facial muscles and partition between nostrils and maxillae	Lips and mandible	Closes the lips, sphincter of mouth
	Procerus	Tendinous fibers from fascia, lower part of nasal bones	Skin over lower part of forehead between the two eyebrows; its fibers blend with those of the frontalis	Draws angle of eyebrows down and produces transverse wrinkles over bridge of nose
	Depressor septi	Incisive fossa of maxilla	Septum and ala of nose	Draws ala of nose downward, constricting aperture of naris
	Dilator naris posterior Dilator naris anterior	Margin—nasal notch of maxilla and lesser alar cartilage	Skin—margin of nostril	Enlarges aperture of naris
Movement of Eye and Lids	Superior rectus	Apex of orbit	Upper and central portion of eyeball	Rolls the eyeball upward
	Inferior rectus	Apex of orbit	Lower and central portion of eyeball	Rolls the eyeball downward
	Medial rectus	Apex of orbit	Midway on inner side of eyeball	Rolls the eyeball, to which it is attached, inward

SUMMARY: FUNCTIONALLY IMPORTANT MUSCLES.—(*Continued*)

	Name of Muscle	*Origin*	*Insertion*	*Function*
Movement of Eye and Lids (*Continued*)	Lateral rectus	Apex of orbit	Midway on outer side of eyeball	Rolls the eyeball, to which it is attached, outward
	Superior oblique	Apex of orbit	Eyeball—between superior and lateral recti	Rotates eyeball on its axis, directs cornea downward and lateralward
	Inferior oblique	Orbital plate of the maxilla	Eyeball—between superior and lateral recti	Rotates eyeball on its axis, directs cornea upward and lateralward
	Levator palpebrae superioris	Apex of orbit	Tarsal cartilage of upper lid	Elevates upper lid and opens eye
	Orbicularis oculi	Nasal portion of frontal bone, frontal process of maxilla, and a short, fibrous band, the medial palpebral ligament	Palpebral portion is inserted into lateral palpebral raphe Orbital portion surrounds orbit—upper fibers blend with the frontalis muscle	Palpebral portion closes the lids as in blinking. This action is involuntary Entire muscle closes lids forcibly
Mastication	Masseter	Zygomatic process and adjacent portions of maxilla	Ramus of mandible	Raises the mandible and closes the mouth
	Temporal	Temporal fossa	Coronoid process of mandible	Raises the mandible and closes the mouth, draws the mandible backward
	Internal pterygoid	Pterygoid plate, palatine, and maxilla	Ramus of mandible	Raises the mandible and closes the mouth
	External pterygoid	Pterygoid plate and great wing of sphenoid	Condyle of mandible	Moves the jaw forward and sideways, helps to open the mouth

SUMMARY: FUNCTIONALLY IMPORTANT MUSCLES.—(*Continued*)

	Name of Muscle	*Origin*	*Insertion*	*Function*
Movement of Tongue	Genioglossus	Symphysis of mandible	Hyoid bone and under surface of tongue	Thrusts the tongue forward, retracts it, and also depresses it
	Styloglossus	Styloid process of temporal bone	Whole length of side and under part of tongue	Draws the tongue upward and backward
Movement of Head:				
Flexion	Sternocleidomastoid	Sternum and clavicle	Mastoid portion of temporal bone	Each muscle acting alone draws the head toward shoulder of same side; both acting together flex the head on the chest or neck
Extension	Splenius capitis	Lower half ligamentum nuchae spinous processes of the seventh and upper 4 thoracic vertebrae	Outer part of occipital bone and mastoid process	When both muscles act together, head is pulled back. Acting alone, head is rotated to same side
	Semispinalis capitis	Transverse processes of upper 6 thoracic and 4 lower cervical vertebrae	Occipital bone	Both muscles acting together, head is extended. Acting alone, head is rotated to the same side
	Longissimus capitis	Transverse processes of upper 4 thoracic vertebrae	Mastoid process	Same as above
Movement of Vertebral Column:				
Flexion	Quadratus lumborum	Iliac crest and the iliolumbar ligament	Twelfth rib and transverse processes of four upper lumbar vertebrae	Acts as a muscle of inspiration by holding outer edge of diaphragm steady, and flexes the spine

SUMMARY: FUNCTIONALLY IMPORTANT MUSCLES.—(*Continued*)

	Name of Muscle	*Origin*	*Insertion*	*Function*
Extension	Sacrospinali Lateral column Iliocostalis { lumborum, dorsi, cervicis Intermediate column Longissimus { dorsi, cervicis, capitis Medial column Spinalis { dorsi, cervicis, capitis	Lower and posterior part of sacrum, posterior portion of iliac crests, spines of the lumbar and lower two thoracic vertebrae	Series of attachments to ribs and vertebrae all the way up the back to the occipital bone and mastoid process of temporal bone	Serves to maintain the vertebral column in the erect posture
Movement of Shoulder Girdle:				
Elevation	Trapezius (upper fibers)	Occipital bone, ligamentum nuchae, spinous process of the seventh cervical and the spinous processes of 12 thoracic vertebrae	Clavicle, acromion process, and spine of scapula	If upper end is fixed, shoulder is raised If shoulders are fixed, both muscles draw head backward Contraction of whole muscle retracts the scapula and braces back the shoulder
	Rhomboideus major	Spines of first four or five thoracic vertebrae	Vertebral border of scapula	Carry the inferior angle of the scapula backward and upward and thus produce slight rotation
	Rhomboideus minor	Ligamentum nuchae and spinous processes of last cervical and first thoracic vertebrae	Vertebral border of scapula	
	Levator scapulae	First four cervical vertebrae	Vertebral border of scapula	Lifts the angle of the scapula

SUMMARY: FUNCTIONALLY IMPORTANT MUSCLES.—(*Continued*)

	Name of Muscle	*Origin*	*Insertion*	*Function*
Depression	Pectoralis minor	Upper margins and outer surfaces of third, fourth, and fifth ribs	Coracoid process of scapula	Depresses the shoulder and rotates the scapula downward
	Subclavius	Junction of first rib and its cartilage	Groove on under surface of clavicle	Carries the shoulder downward and forward
	Trapezius (lower fibers)			
Backward Movement (adduction)	Trapezius	As above		
Forward Movement (abduction)	Serratus anterior	Surfaces and superior borders of upper eight or nine ribs	Various portions of the ventral surface of scapula	Carries scapula forward and raises vertebral border as in pushing; assists deltoid in raising the arm
Movement of Humerus:				
Flexion	Coracobrachialis	Coracoid process of scapula	Mesial surface of humerus	Flexes the arm at shoulder
Extension	Teres major	Dorsal surface, lower part of scapula	Crest of lesser tubercle of humerus	Adduction and rotation of arm
Abduction	Deltoideus	Clavicle, acromion process, and spine of scapula	Lateral side of the body of the humerus	Abducts the arm
	Supraspinatus	Supraspinous fossa of scapula	Greater tubercle of humerus	Abducts the arm
Adduction	Pectoralis major	Clavicle, sternum, cartilages of true ribs, and external oblique	Crest, greater tubercle of humerus	It adducts and draws the arm across the chest, also rotates it inward

SUMMARY: FUNCTIONALLY IMPORTANT MUSCLES.—(*Continued*)

	Name of Muscle	*Origin*	*Insertion*	*Function*
External Rotation	Infraspinatus	Infraspinous fossa of scapula	Greater tubercle of humerus	Outward rotation of arm
	Teres minor	Axillary border of scapula	Greater tubercle of humerus	Outward rotation of arm
Internal Rotation	Latissimus dorsi	Lower six thoracic vertebrae, lumbar and sacral vertebrae, crest of ilium, and lower three or four ribs	Intertubercular groove of humerus	Lower fibers help to depress scapula Depresses the humerus, draws it backward, and rotates it inward
	Subscapularis	Subscapular fossa of scapula	Lesser tubercle of humerus	Inward rotation of arm
Movement of the Elbow:				
Flexion	Brachialis	Lower half of front of humerus	Tuberosity of ulna and coronoid process	Flexes the forearm
	Biceps brachii	Long head from tuberosity at upper margin of glenoid cavity Short head from coracoid process of scapula	Tuberosity of the radius	Flexes the elbow and supinates the hand
	Brachioradialis	Supracondylar ridge of humerus	Styloid process of radius	Flexes the elbow joint, assists in bringing hand into supine position
Extension	Triceps brachii	Long head from infraglenoid tuberosity of scapula, lateral and medial heads from body of humerus	Olecranon of the ulna	Great extensor muscle of forearm
	Anconeus	Lateral epicondyle of humerus	Side of olecranon and dorsal surface of ulna	Assists the triceps in extending the forearm

SUMMARY: FUNCTIONALLY IMPORTANT MUSCLES.—(*Continued*)

	Name of Muscle	*Origin*	*Insertion*	*Function*
Movement of the Hand:				
Supination	Biceps brachii	As on page 208		
	Supinator (brevis)	Lateral epicondyle of humerus and radial ligament of elbow	Dorsal and lateral surfaces of body of radius	Assists the biceps in bringing hand into supine position
Pronation	Pronator teres	Humerus and ulna	Body of the radius	Rotates radius upon ulna, renders the hand prone
	Pronator quadratus	Pronator ridge on body of ulna	Volar surface of body of radius	Rotates the radius upon the ulna
Movement of Wrist:				
Flexion	Flexor carpi radialis	Medial epicondyle of humerus	Base of second metacarpal bone	Flexes and abducts the wrist
	Palmaris longus	Medial epicondyle of humerus	Transverse carpal ligament and palmar aponeurosis	Flexes the wrist joint, assists in flexing the elbow
	Flexor carpi ulnaris	Humerus and ulna	Pisiform, hamate, and fifth metacarpal	Flexor and abductor of wrist, assists in bending elbow
Extension	Extensor carpi radialis longus	Supracondylar ridge of humerus	Dorsal surface of base of second metacarpal	Extends the wrist and abducts the hand
	Extensor carpi radialis brevis	Lateral epicondyle of humerus	Dorsal surface of base of third metacarpal bone	Extends the wrist; may abduct the hand
	Extensor carpi ulnaris	Lateral epicondyle of humerus and dorsal border of ulna	Ulnar side of fifth metacarpal bone	Extends the wrist
Movement of the Fingers:				
Flexion	Flexor digitorum sublimis	Humerus, radius, and ulna	Second phalanges of the four fingers	Flexes the middle and proximal phalanges, assists in flexing the wrist and elbow
	Flexor digitorum profundus	Volar and medial surfaces of body of ulna	Bases of the last phalanges	Flexes the phalanges

SUMMARY: FUNCTIONALLY IMPORTANT MUSCLES.—(*Continued*)

	Name of Muscle	*Origin*	*Insertion*	*Function*
Extension	Extensor digiti quinti proprius	From tendon of extensor digitorum communis	Tendon of extensor digitorum communis on dorsum of first phalanx of little finger	Extends the little finger
	Extensor digitorum communis	Lateral epicondyle of humerus	Second and third phalanges of fingers	Extends the phalanges, then the wrist, finally the elbow
	Extensor indicis proprius	Dorsal surface of body of ulna	The tendon of extensor digitorum communis	Extends the index finger
Movement of Thumb:				
Flexion	Flexor pollicis longus	Volar surface of body of radius	Distal phalanx of thumb	Flexes the phalanges of thumb
Extension	Extensor pollicis longus	Dorsal surface of body of ulna	Base of last phalanx of thumb	Extends terminal phalanx of thumb
	Extensor pollicis brevis	Dorsal surface of body of radius	Base of first phalanx of thumb	Extends proximal phalanx of thumb
Abduction	Abductor pollicis longus	Dorsal surface of body of ulna	Radial side of base of first metacarpal bone	Carries thumb laterally from the palm of the hand
Adduction	Adductor pollicis obliquus	Capitate bone bases of second and third metacarpals	Ulnar side, base of first phalanx of thumb	Adduct the thumb, bring the thumb toward the palm
	Adductor pollicis transversalis	Base, distal two thirds, volar surface of the third metacarpal bone	Ulnar side, base of first phalanx of the thumb	
Muscles of Respiration:				
Inspiration	Diaphragm	Lower circumference of the thorax	A central aponeurotic tendon	Principal muscle of inspiration, modifies size of chest and abdominal cavity; aids in expulsion of substances from body

SUMMARY: FUNCTIONALLY IMPORTANT MUSCLES.—(*Continued*)

	Name of Muscle	Origin	Insertion	Function
Inspiration (*Continued*)	External intercostals	Arise from lower border of a rib	Upper border of rib below	Elevates ribs, increases anteroposterior and transverse diameters of thorax
	Levatores costarum	Transverse processes of vertebrae from seventh cervical to the eleventh thoracic	Outer surface of the rib just below vertebrae from which it arises	Act as rotators and lateral flexors of the vertebral column; may be inspiratory muscles
Expiration	Internal intercostals	Arise from inner surface of a rib	Upper border of rib below	Decrease thoracic diameters by lowering the ribs
	External, or descending, oblique	External surface of lower eight ribs	Anterior half of iliac crest and broad aponeurosis, meeting its fellow of opposite side in linea alba	Compresses the abdominal viscera
	Internal, or ascending, oblique	Inguinal ligament, crest of the ilium, and the lumbodorsal fascia	Lower six ribs, linea alba, and crest of pubis	Compresses the abdominal viscera
	Transversus	Lower six costal cartilages, lumbodorsal fascia, iliac crest, and lateral third of the inguinal ligament	Linea alba and crest of the pubis	Compresses the abdominal viscera
	Rectus abdominis	Pubic bone and ligaments covering front of symphysis pubis	Costal cartilages of fifth, sixth, and seventh ribs	Compresses the abdominal viscera
Support Pelvic Floor	Levator ani	Posterior surface of body of pubic bone, spine of ischium, and obturator fascia	Side of the coccyx and a fibrous band which extends between the coccyx and anus	Helps to form pelvic floor, constricts the lower end of rectum and vagina

SUMMARY: FUNCTIONALLY IMPORTANT MUSCLES.—(*Continued*)

	Name of Muscle	*Origin*	*Insertion*	*Function*
Support Pelvic Floor (*Cont'd*)	Coccygeus	Spine of the ischium	Coccyx and the sides of the sacrum	Helps to form pelvic floor
	Psoas minor			
	Obturator externus			
Movement of Femur:				
Flexion	Psoas major	Bodies and transverse processes of last thoracic and all the lumbar vertebrae	Small trochanter of femur	Flexes thigh on pelvis
	Iliacus	Iliac fossa	Tendon of the psoas major and body of the femur	Acts with psoas major
Extension	Gluteus maximus	Iliac crest, sacrum, side of coccyx, and aponeurosis of sacrospinalis	Fascia lata and gluteal ridge of femur	Extends the femur and rotates it outward
Abduction	Gluteus medius	Outer surface of ilium and gluteal aponeurosis covering it	Lateral surface of greater trochanter	Abduction of thigh and inward rotation
	Tensor fasciae latae	Anterior crest and spine of ilium	About one-third of way down thigh in fascia lata	Tightening of the fascia lata, abduction and inward rotation of thigh
Adduction	Adductor longus	Front of pubis	Linea aspera of femur	Adduct, flex, and rotate thigh outward
	Adductor brevis	Outer surface of inferior ramus of pubis	Linea aspera of femur	
	Adductor magnus	Inferior ramus of pubis and tuberosity of ischium	Linea aspera of femur	
Outward Rotation	Piriformis	Anterior surface of sacrum	Upper border of great trochanter	Supports floor of pelvis, rotates thigh outward

SUMMARY: FUNCTIONALLY IMPORTANT MUSCLES.—(*Continued*)

	Name of Muscle	Origin	Insertion	Function
Outward Rotation (*Continued*)	Quadratus femoris	Tuberosity of ischium	Upper part of linea quadrata	External rotation of thigh
	Obturator internus	Inner surface of antero-lateral wall of obturator foramen and obturator membrane	Fore part of medial surface of greater trochanter	External rotation of thigh
	Obturator externus	Margin of bone around the obturator foramen and obturator membrane	Tendinous insertion into trochanteric fossa	Supports floor of pelvis, external rotation of thigh
	Gemelli	Act with obturators		
Inward Rotation	Gluteus medius (anterior part)			
	Gluteus minimus (anterior part)	Outer surface of ilium	Anterior border of greater trochanter	Abduction of thigh and inward rotation
Movement of Knee Joint:				
Flexion	Biceps femoris	Tuberosity of ischium, linea aspera of femur	Head of fibula and lateral condyle of tibia	Flex the leg upon the thigh and extend the thigh
	Semitendinosus	Tuberosity of ischium	Medial surface of body of tibia	
	Semimembranosus	Tuberosity of ischium	Medial condyle of tibia	
	Popliteus	Lateral condyle of femur	Posterior surface of body of tibia	Assists in flexing leg upon thigh, rotates tibia medially
	Gracilis	Symphysis pubis and pubic arch	Medial surface of tibia below condyle	Adducts the thigh, flexes the leg
	Sartorius	Anterior superior spine of ilium	Upper medial surface of body of tibia	Flexes the leg upon the thigh and the thigh upon the pelvis
	Plantaris	Linea aspera of femur, popliteal ligament	Calcaneus	Accessory to the gastrocnemius

SUMMARY: FUNCTIONALLY IMPORTANT MUSCLES.—(*Continued*)

	Name of Muscle	*Origin*	*Insertion*	*Function*
Extension	Quadriceps femoris arises by four heads:			
	(1) Rectus femoris	Anterior inferior iliac spine and brim of acetabulum		
	(2) Vastus lateralis	Great trochanter and linea aspera of femur	Unite and form tendon which is inserted into tuberosity of tibia	Extends the leg upon the thigh; rectus portion flexes the thigh
	(3) Vastus medialis	Medial lip of linea aspera		
	(4) Vastus intermedius	Anterior and lateral surfaces of body of femur		
Movement of the Foot:				
Plantar Flexion	Gastrocnemius	Medial and lateral condyles of femur	Calcaneus, or heel bone	The gastrocnemius flexes the femur upon the tibia, the gastrocnemius and soleus together extend the foot at the ankle joint
	Soleus	Head of fibula and medial border of tibia		
	Tibialis posterior	Shaft of tibia and fibula and interosseous membrane	Under surface of navicular bone, calcaneus, three cuneiforms, the cuboid, second, third, and fourth metatarsals	Extends the foot at the ankle joint
	Peroneus longus	Head and lateral surface of body of fibula	Lateral side of first metatarsal and first cuneiform	Extends and everts the foot, helps to maintain transverse arch
	Peroneus brevis	Lateral surface of body of fibula	Fifth metatarsal bone	Extends the foot

SUMMARY: FUNCTIONALLY IMPORTANT MUSCLES.—(*Continued*)

	Name of Muscle	*Origin*	*Insertion*	*Function*
Dorsiflexion	Tibialis anterior	Lateral condyle and upper portion of body of tibia	Under surface of first cuneiform and base of first metatarsal	Flexes the foot at the ankle joint and, with the tibialis posterior, inverts the foot
	Peroneus tertius	Lower third of fibula	Base of fifth metatarsal bone	Dorsiflexes the foot
Movement of the Toes:				
Flexion	Flexor hallucis longus	Distal two-thirds of posterior of fibula surface	Base of last phalanx of great toe	
	Flexor digitorum longus	Posterior surface of body of tibia	By four tendons into last phalanges of four outer toes	Flexes the phalanges and extends the foot
Extension	Extensor hallucis longus	Anterior surface of fibula	Distal phalanx of great toe	
	Extensor digitorum longus	Lateral condyle of tibia and anterior surface of fibula	Second and third phalanges of four lesser toes	Extend the phalanges of the toes and flex the foot upon the leg
Other Muscles Functioning in Toe Movement	Extensor digitorum brevis	Calcaneus and cruciate ligament	By four tendons—phalanges of medial four toes	Extends the phalanges of the medial four toes
	Flexor digitorum brevis	Medial process tuberosity of calcaneus	Phalanges of second, third, and fourth toes	Flexes the toes
	Abductor hallucis	Medial process tuberosity of calcaneus	Tibial side of base of first phalanx, great toe	Abducts the great toe
	Abductor digiti quinti	Lateral process tuberosity of calcaneus	Fibular side of first phalanx of fifth toe	Abducts the little toe
	Quadratus platae	Calcaneus and plantar ligament	Tendon of flexor digitorum longus	Assists flexor digitorum longus in flexing the toes
	Lumbricales	Tendons of flexor digitorum longus	Dorsal surface of phalanges of lateral four toes	Extends the last phalanges of the toes and flexes the first

SUMMARY: FUNCTIONALLY IMPORTANT MUSCLES.—(*Continued*)

	Name of Muscle	*Origin*	*Insertion*	*Function*
Other Muscles Functioning in Toe Movement (*Continued*)	Flexor hallucis brevis	Cuboid and cuneiform	First phalanx of great toe	Flexes the toe
	Adductor hallucis Oblique head	Bases of second, third, and fourth metatarsal bones and ligaments	Side of base of first phalanx of great toe	Adducts and aids flexing of the great toe
	Transverse head	Plantar metatarsophalangeal ligaments of third, fourth, and fifth toes	Side of base of first phalanx of great toe	Holds head of metatarsal bones together
	Flexor digiti quinti brevis	Base of fifth metatarsal bone	Base of first phalanx of the fifth toe	Flexes the little toe, draws its metatarsal bone downward and medialward
	Interossei dorsales	Metatarsal bones and ligaments	Phalanges of the second, third, and fourth toes	Abducts these toes
	Interossei plantares	Metatarsal bones and ligaments	First phalanges of toes	Adducts the third, fourth, and fifth toes

CHAPTER 8

THE NERVOUS SYSTEM
PARTS OF THE NERVOUS SYSTEM
NEURONS, RECEPTORS
THE REFLEX ARC AND ACT

The varied activities of the body are regulated with respect to each other by the general chemical composition of body fluids, including hormones, and by the nervous system. Through the nervous system rapid coordination of the functions of widely separated cells is brought about in cooperation with the body fluids at these distant points.

It is also through this medium that acquaintance with the environment is possible.

The parts of the nervous system may be classified in many ways. A simple classification is into brain, spinal cord, and nerves (Fig. 153). Another classification follows:

- **Nervous System**
  - **Central**
    - Brain
    - Spinal cord
  - **Peripheral** (cranial and spinal nerves and end organs)
    - (1) Connections of centers in the central nervous system with the body wall by cranial and spinal nerve fibers. *The somatic,*[1] *or* **cerebrospinal,** *system*
    - (2) Connections of centers in the central nervous system with the viscera by **visceral fibers**
      - *a*) Nerve fibers from the brain and sacral spinal cord and ganglia (autonomic). *The craniosacral, or parasympathetic, system*
      - *b*) Nerve fibers from the thoracolumbar region of the spinal cord and ganglia (autonomic). *The thoracolumbar (thoracicolumbar), or sympathetic, system*

[1] Pertaining to the body, especially the body wall.

The cerebrospinal system is also known as the somatic, craniospinal, or voluntary nervous system. It includes (1) those parts of the brain which are concerned with consciousness and mental activities; (2) the parts of the brain, spinal cord, and their nerve fibers, both sensory and motor, that control the skeletal muscles; and (3) the end organs, receptors and effectors, of the body wall.

The autonomic system. The term *autonomic* means self-acting. This system includes all parts of the nervous system which innervate the plain muscular tissue, the heart, and the glands. As plain muscular tissue is found in all the viscera, this division of the nervous system is sometimes called visceral or splanchnic. It is also called involuntary because it is not under the control of the will. However, it must be thought of not as an independent system but as a closely correlated division of the whole nervous mechanism. It is suggested that the term *visceral* be used for those parts of the nervous system innervating visceral structures and that the term *autonomic* be restricted to the efferent (motor) parts of the visceral system. The autonomic system may be divided into the craniosacral and the thoracolumbar portions. The somatic and visceral systems are closely integrated both centrally and peripherally. Throughout the body efferent visceral fibers are found in all the spinal nerves and in most of the cranial nerves along with somatic fibers. The visceral afferent fibers arise from cells in the sympathetic ganglia. The higher brain centers regulate both somatic and visceral functions. Visceral reflexes may be started from afferent somatic fibers coming from any receptor, and also visceral changes may give rise to somatic acts.

Nerve tissue enters into all parts of the nervous system. Like all other tissues it is composed of cells, but the cells are differentiated from other cells in that the protoplasm is extended, often to a distance of 2 or 4 ft, to form threadlike processes, the nerve fibers. These cells are called neurons.

Properties of nerve tissue. Nerve tissue possesses marked characteristics: (1) irritability (excitability), or the power to respond to stimulation; and (2) conductivity, or the power to transmit the stimuli or nerve impulses to other cells.

Neuroglia tissue consists of cells, called glia cells, which give off numerous processes that extend in every direction and intertwine

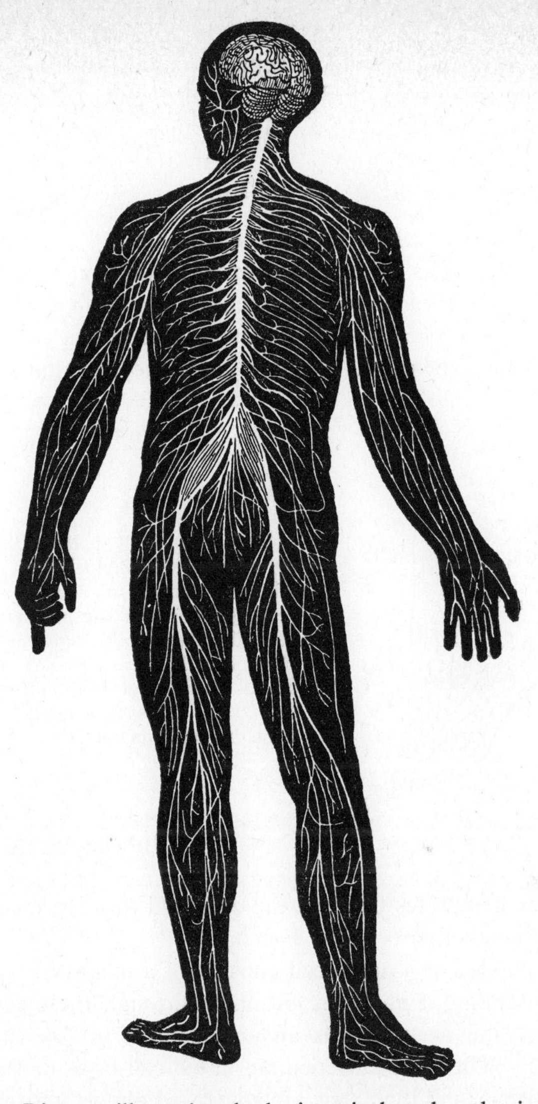

Fig. 153. Diagram illustrating the brain, spinal cord, and spinal nerves.

among the neurons, forming a supporting framework. It is found in the brain and spinal cord. Neuroglia is derived from the ectoderm.[2]

NEURONS

Nerve cells are called *neurons* (neurones).[3] Neurons develop from embryonic cells called *neuroblasts.* They vary greatly in size, shape, manner of branching, and number of processes but consist of a cell body and processes. The nervous system consists of an enormous number of neurons. Connective tissue containing blood and

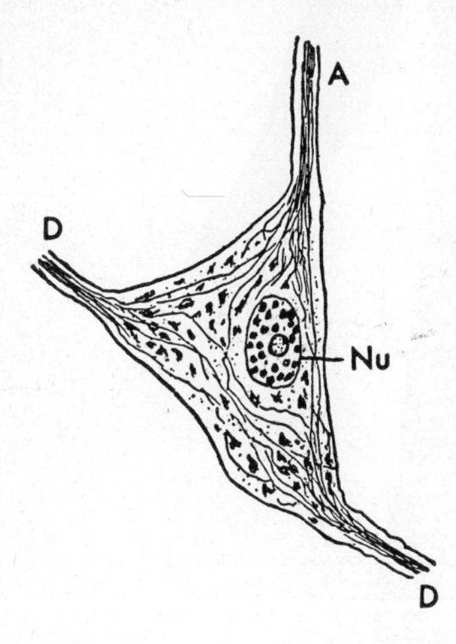

Fig. 154. Cell body of nerve cell. *A,* axon; *D, D,* dendrites. *Nu,* nucleus with contained nucleolus. The lines represent fibrils in the protoplasm of the cell; the dark spots represent Nissl, or chromophilic, granules. Note the absence of Nissl granules at the base of the axon. This area is known as the axon hillock. (Opitz.)

lymph tubes penetrates between the cells and forms protective membranes covering all parts of the system.

The cell body[4] of a nerve cell consists of a mass of granular cytoplasm surrounding a nucleus. Running through the cytoplasm and processes of the nerve cell is an arrangement of fine fibrils called *neurofibrils.* They form a reticulum in the cell body and are present in the dendrites, and the axons are composed almost entirely of them. Scattered throughout the cell body and the protoplasm of the larger dendrites is a substance which stains deeply with basic dyes, such as

[2] The microglia cells may be derived from the mesoderm.

[3] Cowdry classifies nerve cells into *primitive cells* (as in the myenteric plexus) and *neurons,* which form synapses.

[4] Also called perikaryon and cyton.

methylene blue. It is called Nissl,[5] or chromophilic, substance and is thought to represent a store of energy. The quantity is variable, depending upon the fatigue of the cell, rested cells showing a relatively greater amount. Loss of chromophilic substance (chromatolysis) occurs in fatigue, certain fevers, asphyxia, and injury to the axon. The bodies of nerve cells vary greatly in size. The granule cells of the cerebellum have a diameter of 4 or 5 μ, while the large motor cells of the ventral column of the cord may be 125 to 130 μ in diameter. The cells of the lateral columns and of the autonomic ganglia are of medium size.

The cell processes are given off from the cell body. These processes are named dendrites and axons. They differ in many ways.

From the viewpoint of structure they are called dendrites and axons.[6] From the viewpoint of function they are called afferent and efferent processes.

Dendrites (*dendrons*) are usually short and rather thick at their points of origin. They have a rough outline, diminish in caliber as they extend farther from the cell body, and branch in a treelike manner. The number of dendrites varies.

Axons in some instances attain a length equal to more than half that of the whole body. They have a smooth outline and diminish very little in caliber. They give off one or more minute branches called *collaterals*. Usually a neuron has only one axon.

The function of the neuron is to receive nerve impulses and convey them to other cells. The structure is such that normally the neuron can conduct in only one direction. Consequently each neuron possesses a distinct polarity, and the general arrangement of neurons depends in a large measure upon the connections which they establish with each other for functional purposes. The cell body is the source of energy and affords nutriment to its processes, as is evidenced by the fact that if a nerve fiber is cut, the part separated from the cell body dies. The essential function of the processes is conduction of nerve impulses either to the cell body or from it.

[5] Franz Nissl, German neurologist (1860–1919). Nissl substance is called chromophilic (color-loving) because it takes stains readily. In many nerve cells this substance appears as granules, known as the Nissl, chromophilic, or tigroid bodies.

[6] Strong and Elwyn, *Human Neuroanatomy,* 1943, p. 24. The names "axon" and "dendrite" should refer to processes having different *structural* characteristics only.

In general, processes which carry impulses to the cell body are called afferent processes, and those which carry impulses from the cell body are called efferent processes.

Classification of neurons. Neurons are classified in many ways. One classification is based on the number of processes they possess.

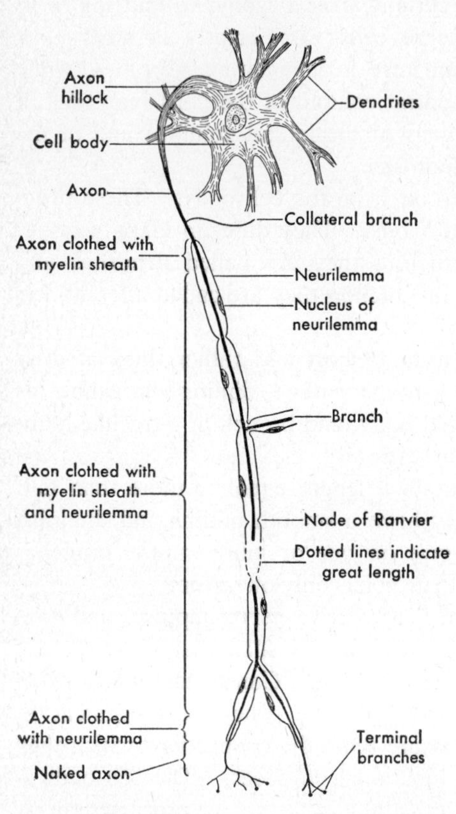

Fig. 155. Diagram of a motor neuron from ventral column of spinal cord showing cell body in detail.

1. *Bipolar* cells are somewhat oval in shape, and from the two poles nerve processes are given off. They are found in the vestibular and cochlear ganglia of the ear. In the ganglia of the dorsal roots of the spinal nerves, so-called unipolar cells give off a single T-shaped process, which rapidly divides. During the early stages of development they are bipolar cells, which gradually develop into the *unipolar* or *pseudounipolar* cells.

2. *Multipolar* cells possess numerous processes, which correspond to the number of angles or poles possessed by the cell body. It is in the large motor cells of this group that the difference between axons and dendrites is recognized. This group includes, in addition to the motor cells of the spinal cord, the pyramid-shaped cells of the cerebral cortex and the flask-shaped cells of Purkinje[7] found in the cortex of the cerebellum. The pyramidal cells give off branching dendrites from each

[7] Johannes Evangelista Purkinje, Bohemian anatomist and physiologist (1787–1869).

angle and one axon from the middle of the base. Each cell of Purkinje gives off a single axon from its base and from the apex gives off dendrites, which branch abundantly.

Another classification is based on the functions neurons perform.

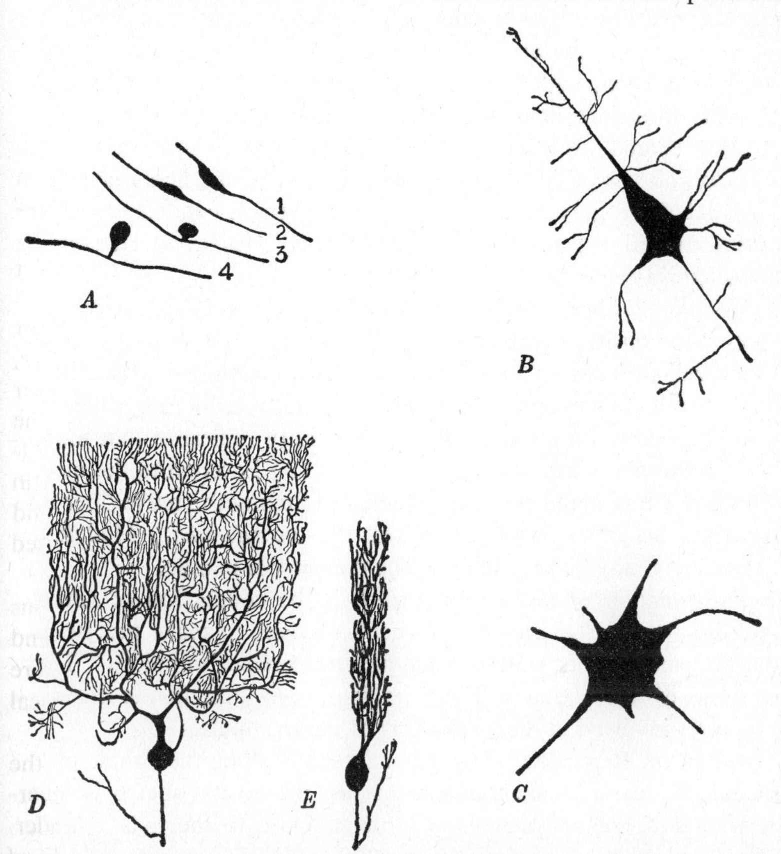

Fig. 156. Types of neurons. *A*, cell of dorsal root ganglion; *1*, *2*, *3*, *4* show how it gradually develops into a unipolar cell. *B*, pyramidal cell of cerebral cortex. *C*, motor cell of spinal cord. *D* and *E*, Purkinje cells of cerebellum; *E*, profile view.

Neurons which carry impulses from the periphery to the center are described as *afferent, receptor,* or *sensory*. The cell bodies of this type of neuron are often at some distance from their terminals. Neurons which carry impulses from the center to the periphery are described as *efferent, effector,* or *motor* if they produce motion, or *secretory* if they produce secretion. The cell bodies of this type of neuron are placed near the central end of the fiber. The effect of the impulse carried may be to excite or to lessen activity. The former are called *excitatory,* and the latter *inhibitory, neurons.* Certain neurons carry impulses from the afferent neurons to the efferent neurons and are designated as *central, connecting, internuncial, intercalated, association,* or *Golgi type II cells.*

Nerve fibers are elongated outgrowths of nerve cells. These fibers as they extend away from the cell may become surrounded with sheaths and are of two kinds: medullated, or white, fibers, and nonmedullated, or gray, fibers. Nerve fibers which conduct nerve impulses toward a center are called *afferent.* Those which conduct away from a center are called *efferent.*

Medullated fiber. Microscopic studies of a medullated nerve fiber show it to consist of three parts: (1) A central core, or *axis cylinder,* is the cell process. (2) Immediately surrounding the axis cylinder is a sheath, or covering, of a semifluid, fatty substance called the *medulla,* or *myelin sheath.* It is to this fatty substance that medullated nerve fibers owe their white color. (3) External to the myelin sheath is a thin membrane completely enveloping the nerve fiber and forming the outer covering called the *neurilemma.* Medullated nerve fibers may be very long. The diameter is minute.

The function of the myelin sheath is the conduction of nervous impulses at the rate which will enable muscles to make precise and delicate movements, rather than just to conduct impulses. There is some evidence that it plays an important part in the chemical processes involved in the production of nerve impulses.

NODES OF RANVIER. At regular intervals along the course of the medullated nerve fibers found in nerves, the myelin sheath is interrupted, and the neurilemma is brought close to the axis cylinder. These constrictions (some 80 to 200 μ apart) are the *nodes of Ranvier.*[8] It is thought that the purpose of the nodes is to allow tissue fluid to get to the fiber for nutritive purposes. Branching of the nerve fiber takes place at the nodes. In each internodal segment the neurilemma is seen to have a nucleus. These nuclei play an important part in the degeneration and regeneration of nerve fibers which are cut off from their cell bodies (Fig. 176 and p. 266).

Medullated fibers found within the brain and spinal cord differ

[8] Louis Antoine Ranvier, French histologist (1835–1922).

from those of the peripheral nerves in that the myelin sheath is not segmented and the neurilemma and nuclei are lacking.

Nonmedullated, or *amyelinated, nerve fibers* (*fibers of Remak*[9]) differ from medullated nerve fibers in the great reduction or absence of the myelin sheath, the fiber being directly invested with the neurilemma. Owing to the absence of the myelin sheath, they present a gray or yellow color. Amyelinated fibers are delicate processes of small nerve cells. Most of the cells of the sympathetic ganglia and the small cells of the cerebrospinal ganglia give rise to amyelinated fibers.

Nerve fibers may be grouped in three general groups—the fine fibers (amyelinated and slightly myelinated) having a diameter between 2 and 4 μ, the medium fibers varying in diameter from 4 to 10 μ, and the large fibers having a diameter from 10 to 20 μ.

Synapse. Each neuron is a separate and distinct unit. The fine branches of the axon of one neuron seem to interlace with the dendrites of, or lie on the surface of, another neuron, forming a synapse. At the synapse the two neurons involved come into contact. There is no protoplasmic continuity of the neurons across the synapse, but there is contact of the terminals of the axon of one neuron and the cell bodies or dendrites of other neurons. This implies something in the way of a thin "surface layer" or membrane separating the cells, which may act to raise the resistance (threshold) to nerve impulses or may set up fresh nerve impulses.

End organs are the peripheral structures related to nerve fibers. (1) They are sensory, or receptor, if associated with afferent fibers, or (2) effector, or motor, if associated with efferent fibers.

1. *Receptor end organs.* The structure of some receptor end organs is well known. The structure of others has not been adequately worked out. There are, therefore, many classifications of them. None is satisfactory, however, as some sense organs are known by name, e.g., eye, whereas others can be spoken of in terms of sensations only, e.g., "for pain."

C. J. Herrick's classification of sensory receptors follows:

Receptor End Organs	**Somatic**	**Exteroceptors**	*Distance* (teleceptors)—such as those for sight, hearing, smell, and, to some extent, temperature

[9] Robert Remak, German anatomist (1815–1865).

Receptor End Organs	**Somatic**	**Exteroceptors**	*Contact*—for taste, cutaneous sensations as touch, pressure, part of temperature, pain, chemical sensibility, etc.
		Proprioceptors	For muscular, tendon, and joint sensibility, static and equilibrium sensibility
	Visceral	**Interoceptors**	For taste, smell, hunger, thirst, respiratory, circulatory, and abdominal sensations, visceral pain, etc.

To some students, *only* the receptors of the mucosa of the alimentary canal are *interoceptors;* the receptors of the muscular and areolar parts of the alimentary wall and those of heart and blood vessels are classed as proprioceptors.

Receptor end organs may be classified as to location into those of epithelium, of connective tissue, and of muscle. The receptor end organs of connective tissue may be *free* or *encapsulated.*

Many receptor end organs are *free* or *diffuse fiber endings.* In the epidermis many of the nerve fibers lose their myelin sheaths, gradually becoming arborized into neurofibrils which ramify between the cells in varying degrees of complexity. Many of these neurofibrils show minute varicose expansions along their courses and at their ends, which are on the surfaces of cells. Free endings of this kind are found in the sclera and cornea of the eye, in the areolar connective tissue of serous and mucous membranes, on muscles and tendons, and in the periosteum of bone.

Many receptors are surrounded by capsules of connective tissue and are said to be *encapsulated.* Here the comparatively coarse nerve fiber lies in a semiliquid substance enclosed by a connective-tissue capsule. The myelin sheath is lost; the neurofibrils form a network which shows varicosities. Tactile, bulbous, articular, lamellar, and cylindrical forms are found. The corpuscles of Pacini,[10] Meissner,[11] Ruffini,[12] Krause,[13] the neuromuscular bundle (spindles), and tendon spindles (neurotendinous organs) belong to this group.

Muscle spindles are groups of a few fine, abundantly nucleated muscle fibers some 0.25 mm long and about 0.1 mm in diameter, surrounded by connective-tissue capsules chiefly found near the ends

[10] Filippo Pacini, Italian anatomist (1812–1883).
[11] Georg Meissner, German histologist (1829–1905).
[12] Angelo Ruffini, Italian anatomist (1864–1929).
[13] Wilhelm Krause, German anatomist (1833–1910).

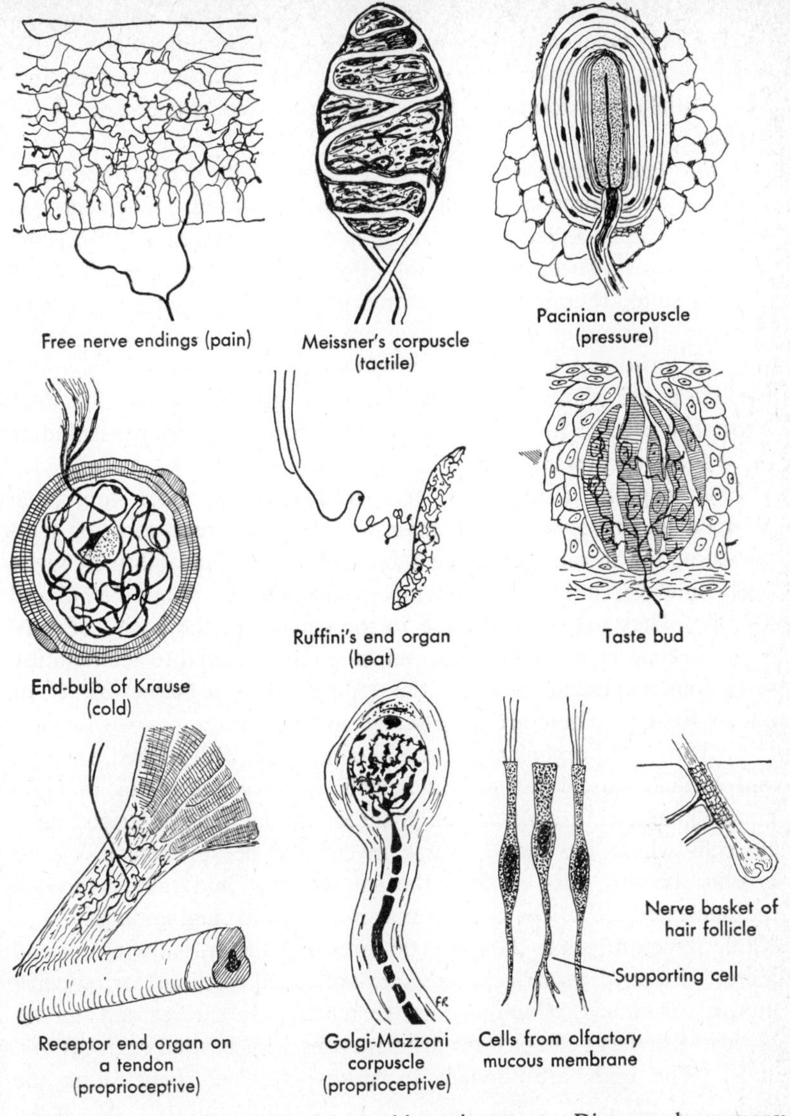

Fig. 157. Receptors acquaint us with environment. Diagram shows many kinds.

of muscle trunks. Some of the sensory nerve fibers penetrate the capsule and twine about the muscle fibers with nodular ends on the sarcolemma, the epilemmal receptors. Other sensory fiber endings (interstitial) are found in the connective-tissue framework of the muscle.

Somesthetic receptors are those concerned with general body sensibility (touch, pressure, pain, temperature, position, movement, visceral). The so-called *special sense organs* are concerned with sight, hearing, smell, taste, equilibrium. Phono, tango (touch), pressure, chemo, stato, thermo, and photoreceptors derive their names from the type of stimulus to which they are sensitive. *Kinesthetic receptors* are those concerned with muscle tone. Receptors of *superficial sensibility* are those of the exteroceptive group. Those for *deep sensibility* are the interoceptive and proprioceptive groups, including those for deep pressure.

Exteroceptors (for touch, temperature, pain, etc.) are often grouped as *protopathic* and *epicritic.* Epicritic receptors for temperature, for instance, include those related to the ability to make fairly fine discriminations in temperature between about 25° C and 40° C. They are located close to the surface of the body. *Protopathic* receptors for temperature include those related to temperature sensations experienced when the temperature is below 25° C or above 40° C. These are deep-seated in location.

2. *Motor end organs* are found in connection with striped, smooth, and cardiac muscle and glandular cells. *Motor end plates,* or myoneural junctions, are elevations of sarcoplasmic areas of striped muscle cells where the rather coarse myelinated nerve fibers lose their myelin sheaths, pass through the sarcolemma, and ramify through the nucleated elevations of sarcoplasm as hypolemmal endings. The whole structure, some 0.04 to 0.06 mm in diameter, is the motor end plate. *Visceral motor nerve endings* are called cardiomotor, visceromotor, vasomotor, pilomotor, and secretory. In cardiac and smooth muscle, the nerve fibers form arborizations with beaded surfaces and ends. The ends are thought to be hypolemmal, penetrating the surface of the cells. The motor fibers to glands ramify between the cells ending on their surfaces.[14]

[14] W. B. Cannon and A. Rosenblueth, *Autonomic Neuro-effector Systems,* 1937; p. 19 pictures a nerve ending inside a liver cell of a rabbit.

Ganglion. A collection of nerve cells outside the central nervous system is called a *ganglion.* Ganglia appear in the course of the cranial and spinal nerves and form the ganglia of the autonomic system.

Nucleus. A collection of nerve cells within the central nervous system, the fibers of which go to form one anatomical nerve or a tract within the brain or cord, is called a *nucleus,* e.g., the facial nucleus or nucleus of the facial nerve, basal nuclei, etc.

Center. A group of neurons and synapses regulating a certain function is called a *center.* It may be either a nucleus or a ganglion, as these terms refer to an anatomical entity and the term *center* refers to a functional entity. For instance, the rate of respiration is regulated by a center in the medulla oblongata, and odor is interpreted by a center in the cerebrum. The center is in communication with the organ by means of nerve fibers and end organs which are adapted to receive and transmit impulses. Usually the fibers, which connect the centers with the organs they control, do not extend all the way to the organs but terminate in masses of gray matter which serve as *relay stations,* where they form synapses with other neurons that carry their impulses onward, possibly to one or more relay stations or directly to the organs. There may be one or several relay stations. In any such series, the neuron first to be stimulated is called the *neuron of the first order,* and the succeeding neurons are called neurons of the second, third, and fourth order, etc.

Gray and white matter. The cell bodies of neurons and many of their processes and synapses are grouped together into gray matter. *Gray matter* is found in the cortex and other nuclei of the brain and in the core of the spinal cord. It composes the nuclei and ganglia and the unmyelinated nerve fibers.

The medullated processes of cell bodies are grouped together into nerves and tracts of brain and cord, forming the *white matter.* It will therefore be seen that the gray matter contains the cell bodies and the synapses where the adjusting of sensory to motor neurons takes place, forming the *collecting and distributing stations.* The white matter is made up of nerve fibers. The white matter of the brain and cord and the myelinated nerves contain many amyelinated nerve fibers. Groups of nerve fibers in the spinal cord and brain are called *tracts,* e.g., pyramidal tract, ascending tract, etc.

Nerves. A nerve fiber consists of an axis cylinder with its coverings. A bundle of these fibers enclosed in a tubular sheath is called a *funiculus.* A nerve may consist of a single funiculus or of many funiculi collected into larger bundles. Between the individual fibers is connective tissue called *endoneurium,* which serves to bind the fibers together into funiculi. Connective tissue called *perineurium* surrounds each funiculus in the form of a tubular sheath, and all the funiculi are held in a connective-tissue covering called the *epineurium.* The capillaries of the blood vessels supplying a nerve

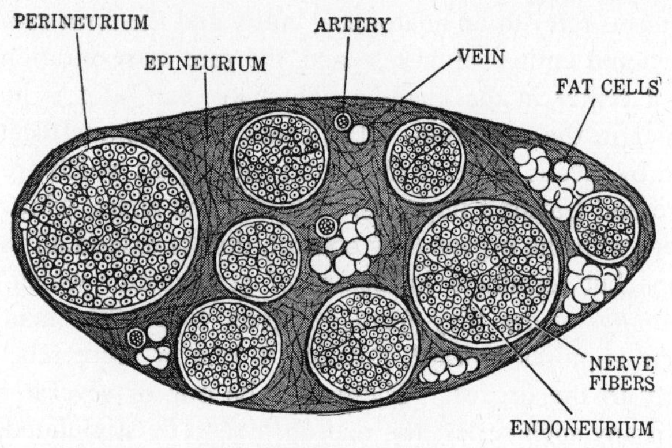

Fig. 158. Transverse section of the sciatic nerve of a cat. This nerve consists of eight bundles (funiculi) of nerve fibers. Each bundle has its own wrappings (perineurium); and all the bundles are embedded in connective tissue (epineurium) in which arteries, veins, and fat cells can be seen. See Figs. 155 and 159 for structure of a nerve fiber.

penetrate the perineurium and either run parallel with the fibers or form short transverse connecting vessels. Fine nerve fibers (vasomotor) accompany these capillaries. Considerably more than 75 per cent of a nerve is composed of nonnervous substance—more than 50 per cent of it is areolar connective tissue, blood, etc., and more than 25 per cent is myelin.

The nerves branch frequently throughout their courses, and these branches often meet and fuse with one another or with the branches of other nerves; yet each fiber always remains distinct. The nerve is thus merely an association of individual fibers which have very different activities and which function independently of one another. Most nerves are *mixed nerves* containing both sensory and motor fibers. The arrangement of nerve fibers in a nerve trunk can be seen in a cross section of a nerve (Fig. 158).

The nerve fibers in a nerve have various diameters. The sciatic nerve of the frog is said to consist of more than 8,000 fibers, more than two-thirds of which are amyelinated. The diameters of the fibers of the ventral root of the sciatic nerve range up and down from 14 μ. In general, it has been found that large fibers are heavily myelinated, originate from large cells, and extend to specialized somatic structures, e.g., motor fibers to striped muscle cells. The opposite is also true. Nerves as a rule contain all sizes of fibers.

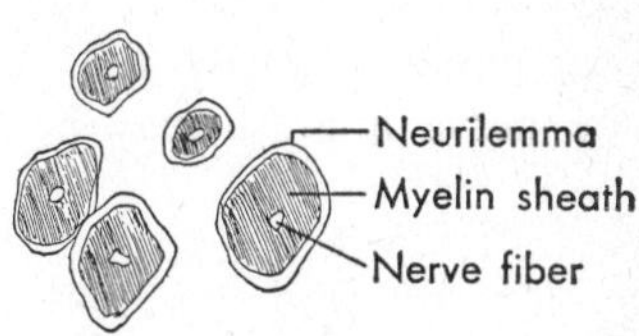

Fig. 159. Diagram showing structure of nerve fibers in cross section.

THE REFLEX ARC AND ACT

The reflex arc. The *unit pattern,* or *functional unit,* of the nervous system is the *reflex arc,* or *circuit.* The work which it does is the *reflex act.* The simple reflex arc in man consists of a sensory, a connecting, and a motor neuron with sensory and motor end organs. The number of such units in the nervous system is enormous, it being estimated that there are more than 8,000,000,000 neurons in the cerebral cortex alone. Each of these neurons is connected with many others, forming organizations of the unit patterns in bewildering numbers and complexity, linking all parts of the body together. Figure 160 shows possible types of linkages of these reflex arcs. *A* shows a unit pattern of three neurons; *B*, two reflex arcs in sequence; *E,* a reflex arc in which four connecting neurons extend between the sensory and motor neurons. *C* and *D* show how one receptor can send nerve impulses into two (or many) effectors and how two (or many) receptors can send nerve impulses into one effector. Each part of the reflex arc has its own function in carrying out the reflex act.

The reflex act. On applying an appropriate stimulus to the receptor ending of the sensory neuron, an impulse is initiated which passes along the afferent process to the cell body, thence along its efferent process, across the synapse, to the connecting neuron, along the afferent process to the cell body, thence along its efferent process and across another synapse to an afferent process of the motor neuron, then through its cell body and efferent process to the muscle or gland cell where action is produced.

The receptor receives the stimulus, transforms it to a nerve impulse, and passes it on to the nerve fiber. In the adjusting mechanism the nerve impulse may be transferred across some synapses rather

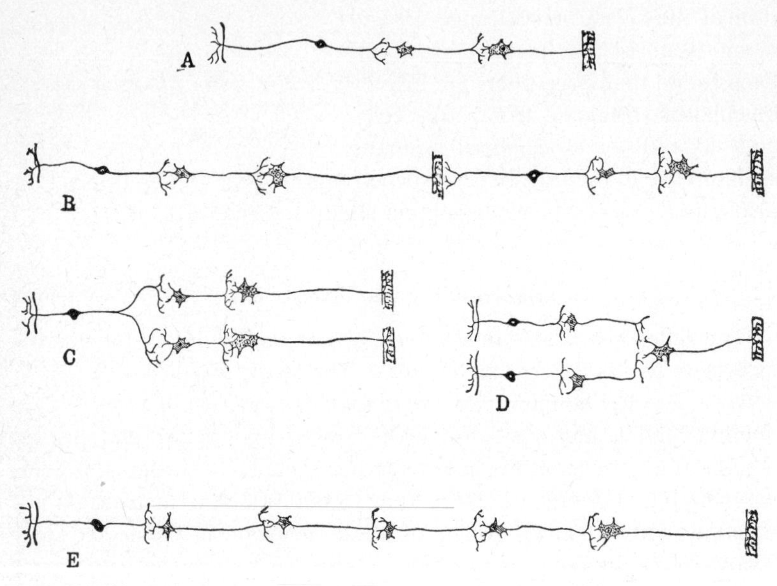

Fig. 160. Reflex arcs or circuits. *A,* reflex arc of three cells—*sensory cell body,* black; *connecting cell body,* striped; *motor cell body,* dotted. *B,* a short chain reflex of two simple reflex arcs in sequence. *C,* one sensory cell connected with two motor cells. *D,* two sensory cells connected with one motor cell. *E,* a long simple reflex arc in which there are four connecting cells between the sensory cell and the motor cell.

than others, for it must be remembered that each neuron forms synapses not simply with one neuron but often with many. When the nerve impulse reaches the effector mechanism, this mechanism transforms the impulse into a stimulus and passes it on to a muscle or gland cell. Between the simplest form of involuntary activity and the higher activities that involve consciousness, memory, or control are many types of reactions.

Reflexes may be classified in many ways:

Simple reflexes.[15] These depend upon a sensory neuron, a central neuron, and a motor neuron. Stimulation of the sensory neuron results in response by a muscle. Examples of such simple reflexes are the winking reflex caused by an object striking or appearing as if it would strike the cornea, the swallowing reaction due to food on the back of the tongue, avoiding reactions due to tickling, pricking, the application of heat or cold, etc. These simple reflexes are of value in diagnosis of diseases of the nervous system, because in general they are simple and direct, the response is rapid, they can be predicted with relative safety, persist throughout life, and can be modified or inhibited only with difficulty.

Reflexes involving the cord. Some of the reflexes that involve the spinal cord are: (1) Those concerned with withdrawal from harmful stimuli, called *flexion reflexes.* Walking involves the flexion reflexes. (2) The *extensor* or *stretch reflex,* such as the knee jerk and those concerned with posture and muscle-tone maintenance. (3) The *scratch reflex* or responses to local irritation.

Other cord reflexes may be more complex and involve more segments of the cord. In some instances cord reflexes involve the viscera, as in the reflex that empties the bladder. Afferent nerve impulses from the bladder enter the sacral and lumbar regions of the cord over internuncial neurons to the efferent neurons and the bladder is emptied. At the same time impulses reach higher brain levels. Applications of heat to an arm or leg causes dilatation of the blood vessels and may also produce sweating.

The reason a reflex may be involuntary yet conscious is that at the same time impulses continue over collaterals to internuncial neurons to the higher centers of the brain.

Reflexes involving the brain stem and cerebellum. These involve neurons extending up into the brain stem and cerebellum. These reactions are more complex, not so rigid, and more readily modified and varied, and, as a rule, involve wider regions of the body. Many

[15] Neurologists have developed special tests for such reflexes as the *wink, pupillary,* and *patellar* reflex. Patellar reflex is the name given to the jerk of the foot caused by tapping upon the patellar ligament while the leg is crossed, or suspended across the edge of a table or chair. The impulses generated are conveyed to the sciatic center and thence to the quadriceps femoris muscle, which contracts and extends the leg. The varying intensity of this reflex is indicative of the irritability of the entire nervous system. Injuries to the cord may abolish this reflex entirely. Lesions of the higher centers increase its intensity.

of the muscular coordinations concerned with walking, running, etc., are examples of reactions of this kind. The heart, blood vessels, and respiration are regulated by reflex centers in the medulla.

When the swallowing reflex is initiated, other reflexes inhibit respiration.

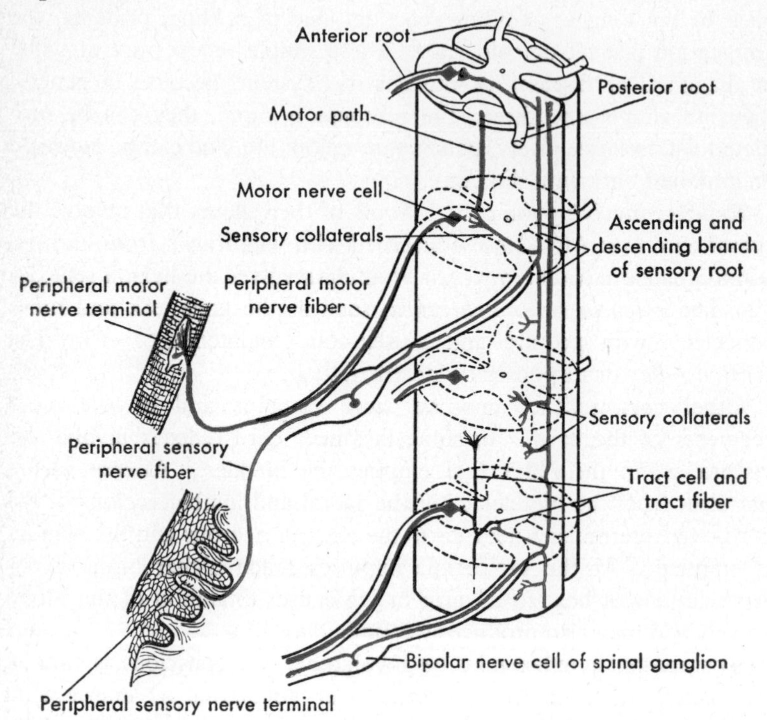

Fig. 161. Motor (red) and sensory (blue) conduction paths and reflex arcs of the spinal cord. (Toldt.)

Reflexes involving the diencephalon. In a warm environment the receptors of the skin convey impulses to the hypothalamus and the sweat glands are reflexly stimulated through efferent autonomic fibers.

Reflexes involving the cerebral cortex. In the cerebral cortex the possibilities of a wide range of connections between the receptor and effector mechanisms of the body are almost limitless. The reactions often involve many widely different regions of the body. The simple withdrawal of a finger from a prick is an example of a simple reflex.

If at the same time one cries out, there is an example of involvement of the brain stem. If in addition one deliberately dresses the wound, there is an example of a reflex involving the cerebrum.

PHYSIOLOGY OF THE REFLEX ARC

Receptors, or the ends of sensory nerve fibers. These may be simple fiber ends, or these ends may be surrounded by complex accessory structures. Receptors are sensitive to *stimuli,* which are thought of as electrochemical changes in the immediate environment of the fiber endings. These changes may be brought about by chemical or physical *factors of change.* It is the function of a receptor to convert this change in its environment into a staccatolike succession of nervous impulses in the nerve fiber. All receptors behave in the same general way as they do this.

Receptor stimuli. If nerve impulses are to be initiated at the receptors, the stimuli must be adequate and at least minimal. By an *adequate stimulus* is meant the *kind* of change to which the receptor is sensitive. For fiber terminals about the hair follicles the adequate stimulus is mechanical (bending of hairs). This is true also for receptors connected with muscle spindles, tendons, and joints. The adequate stimulus for some end organs is temperature change (for corpuscles of Krause, cold; for corpuscles of Ruffini, warmth); for some the change is brought about by chemical agents (as in taste buds, olfactory cells); for some the adequate stimulus is photochemical (as in the retina). A *minimal* (*threshold*) stimulus refers to the least change which can excite the receptor and produce a response in the effector. This varies greatly in receptor end organs. Subminimal stimuli are below threshold level, but, if repeated at the right time intervals, subminimal stimuli may undergo a summation to threshold strength and produce activity.

Intensity of stimulus is probably related to the degree of change at the receptor and is reflected in the frequency (number per second) of the impulses initiated. With increase in intensity of stimulus, the frequency of discharge from the receptor increases; also, the more rapid the increase in intensity of stimulus, the more frequent the discharges from the receptor to the nerve fiber.

Train of impulses. End organs vary greatly in relation to the speed with which they reach approximate equilibrium with their environment, or in relation to *adaptation.* The longer the period of adaptation, the longer is the train of impulses started in the nerve fibers generally by a single stimulus. Some receptors show slow adaptation with comparatively long trains (sequences) of nerve impulses issuing from them; some show rapid adaptation with short trains of impulses. There may be a correlation between the speed of adaptation and the function. The simpler the end organ, the more rapid is its adaptation and the vaguer and less discriminatory is the physiological response or conscious knowledge on the part of the individual. For instance, hairs adapt quickly, and discrimination in relation to them is slight; but muscle spindles are slow in adaptation, and hence the long train of impulses can bring about the varied tonic reflexes concerned with maintenance of posture, where discrimination is fine and actions are minutely graded. Adaptation in a receptor is slower than in a nerve fiber; hence, the nerve fiber is always ready for the nerve impulse. Adaptation of receptor refers to decreased excitability by a stimulus quite apart from previous activity.

Sensations derived from stimuli applied to receptors differ in intensity not only because of differences in the intensity of the stimulus and the rapidity with which it is applied, but also because of differences in the number of receptors stimulated. Then too, different receptors in the region will receive stimuli of different intensity. This will make a difference in the sensation; e.g., a pencil point pressed on the skin causes a conical depression with differing intensities of stimuli at base and sides of the depression. Doubtless there are other factors as well. Probably stimulation of a single receptor seldom occurs.

The nerve fiber and the nerve impulse. Nerve fibers possess the properties of *irritability* and *conductivity*. Once the nerve impulses are started from the receptor along the fiber, they are considered to be identical in character but to vary in frequency, velocity, and length of train. According to this belief the impulses carried by sensory nerve fibers, such as the optic fibers, are similar to those carried by motor nerve fibers. In one instance there is a visual sensation and in the other, contraction of a muscle. The difference in result may be due to the optic nerve fibers ending in the visual center in the cerebrum and the motor nerve fibers ending in a muscle. In this connection, physiologists use the expression *specific nerve energy* to express the fact that, when stimulated, nerve fibers give only one kind of reaction regardless of what the stimulus may be.

Normally, impulses start at the receiving end organ of a nerve fiber, but they can be induced at any part of the nerve fiber. When this happens, the brain projects the resulting sensation to the part containing the receptors of the fibers stimulated. This may explain why patients who have suffered an amputation refer the pain in the stump to the part that has been removed; the pressure of the surgical dressings or of new tissue on the nerve fibers is interpreted as a stimulus at their receptor end organs.

Of the many theories which have been advanced to explain the nerve impulse, the membrane theory is most firmly held today. Figure 162 is a well-known diagram to illustrate this theory. The method of study is by means of an applied electrical stimulus, the "current of injury" or the "action current," which is made to pass through a galvanometer, which shows the direction of the current, its strength, and its duration. When so studied, it is found that the "current of injury" and the "action current" flow from the normal or inactive portion of the nerve fiber to the injured or active portion. Similar studies of all tissues, with galvanometers sensitive to the weak bioelectric currents generated in living tissues, show weak action currents during activity.

The nerve fiber, or process of a nerve cell, is an exceedingly fine *protoplasmic thread* bounded by a semipermeable cell membrane and surrounded by tissue fluid. Or in terms of neurophysiology, "the nerve fiber is considered to be a *fluid conductor*, surrounded by a labile *surface film* maintained in a steady state through expenditure of energy derived from oxidation in the fiber."[16]

On the outside of the cell membrane certain ions are known to be more concentrated, and on the inside other ions are more concentrated. The semipermeable membrane acts as a barrier to these ions, which brings about a condition like that of a "charged battery"; or, as is said, the membrane is *polarized*, or there is a *potential gradient* across the membrane.

When the fiber is stimulated at any point, a molecular change is thought to take place in the fiber membrane and there is a change in relative permeabilities to K^+, Na^+, and Cl^+. The axon membrane

[16] Herbert S. Gasser, *Analysis of Nervous Action,* 1940.

becomes relatively impermeable to K^+ and temporarily selectively permeable to Na^+. Na^+ moves from the interstitial tissue fluid into the nerve fiber and eventually the Na^+ concentration inside the fiber exceeds that in the interstitial fluid. This causes a reversal of the resting ratio of these ions and reverses the sign of the resting membrane polarization; in other words, it becomes polarized in the reverse direction. The fiber has thus received a stimulus, the nerve impulse has been brought about and moves along the fiber as a self-propagated physicochemical disturbance evidenced by a wave of electrical negativity. An electric current transmits the nerve im-

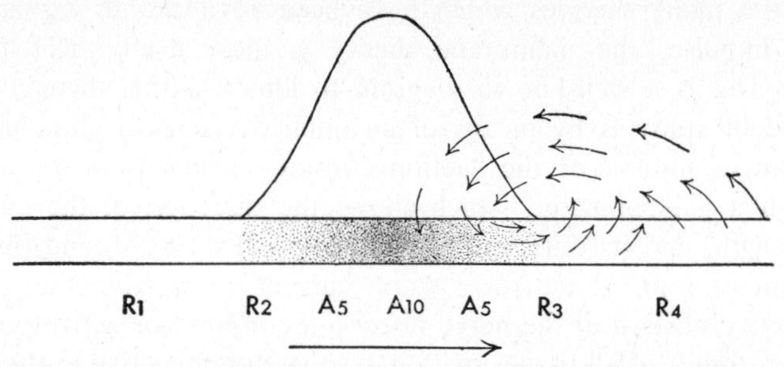

Fig. 162. Diagram of frog's motor nerve axon. Shaded region between R_2 and R_3 is occupied by the excitation wave. Excitation beginning at R_3 has reached its maximum at A_{10} and has subsided at R_2. Small arrows indicate direction of *bioelectric current*. Between R_3 and R_4 its intensity is sufficient to excite the fiber. Excitation is thus always being initiated in advance of the area of excitation. The *curve* represents the microvoltage of the bioelectric current, highest at A_{10}. (R. S. Lillie, *Protoplasmic Action and Nervous Action.* Courtesy of the University of Chicago Press.)

pulse, but this current is generated by acetylcholine, which is rapidly produced (by enzymes) and rapidly removed.[17] As the wave passes along, polarity is again established. Until polarity is established (refractory period), another stimulus cannot be received. Hence, a series of nerve impulses (areas of electrical negativity) rather than a continuous flow of nervous energy moves along a nerve fiber. Nerve impulses move in both directions from a point of stimulation, but normally nerve fibers are *stimulated at their ends*—sensory nerve fibers at their peripheral ends (receptors) and motor fibers from cell bodies which are centrally located.

In the body *the nervous impulse in nerve fibers* is considered to be

[17] C. H. Best and N. B. Taylor, *The Physiological Basis of Medical Practice,* 1955.

a self-propagated disturbance (wave of electrical negativity) which, when initiated at the end organ, travels along the fiber by virtue of local electrical changes. If the traveling nervous impulse and the moving wave of electrical change are not identical, at least the nervous impulse is always accompanied by an electrical change. Therefore variations of electrical change (*action currents*) in nerve fibers during excitation and conduction are used to study the characteristics of nervous impulses.

Studied in this way, *the excitability of nerve fiber shows three periods*—the absolute refractory, the relative refractory, and the normal period.

The *absolute refractory period* is the short period following a stimulus during which the nerve fiber is inexcitable. This period makes it impossible for a fiber continuously to respond to stimuli from a receptor. There is therefore a "rhythmic discharge" of nervous impulses from receptor to nerve fiber rather than a continuous flow, hence the expression "nervous impulses." Small nerve fibers have longer refractory periods than large fibers.

The *relative refractory period* follows the absolute refractory period. During this time the nerve fiber gradually resumes its excitability and finally returns to the normal period.

Threshold value of stimulus. As stated, the intensity of the stimulus at the receptor end of a fiber must reach threshold value before it can excite the fiber. During the *relative refractory period* the stimulus must be stronger than during the *normal phase.*[18] The smaller the fiber, the higher the threshold value of the stimulus must be to produce excitation.

Frequency of nerve impulses usually varies directly with the strength of the stimulus. *Velocity* of conduction of nerve impulses varies with diameter of nerve fiber, the thicker fibers conducting more rapidly than the finer fibers.

Strength of nerve impulses is measured by strength of electrical charges. This seems to be related to the diameter of the fiber. This leads to the *"all-or-none law,"* which states that fibers give a maximum response to a stimulus or none at all; or, in other words, no increase of stimulus above threshold value increases the strength of the electric charge.

[18] A period of hyperexcitability or supernormal phase occurs, according to Adrian, only when the *p*H of perifiber lymph is lower than that of plasma.

Fatigue of nerve fibers is practically impossible if oxygen supply is adequate, which is usually the case in normal physiology. While a fiber is conducting, there is first given off a small amount of "initial" heat, and this is followed by "recovery" heat, which is 10 to 30 times as much as initial heat and is given off slowly. This implies chemical changes which in some respects, but not all, seem to be similar to the metabolic changes in muscle cells.

Physiology of nerves and tracts. Nerves are composed of nerve fibers of various diameters, and most nerves contain both sensory and motor nerve fibers. It has been said that in general the larger the fiber the lower its threshold stimulus, the more rapidly is the excitatory process set up, the shorter is the chronaxie, the greater is the velocity of nerve impulses, the shorter is the refractory period, the greater the strength of the "charges" set up in it, etc. Erlanger and Gasser[19] have suggested grouping nerve fibers into A, B, and C groups on the basis of their size and activities. The velocity in the A group is about 100 m per second. To this group belong large motor fibers (8 to 18 μ diameter), fibers from muscle spindles, fibers from receptors for touch and temperature, etc. The fibers of the B group are postganglionic fibers with conduction velocities of about 10 to 20 m per second, and those of the C group are preganglionic myelinated and afferent amyelinated fibers with conduction rates of 0.3 to 1.6 m per second. The relation of size to activity is not as clear in the B and C groups as in the larger fibers of the A group. The ventral roots of spinal nerves are composed largely of A-group fibers, whereas the dorsal roots contain fibers of all groups.

Synapses and the nerve impulse. Synapses are found in gray matter of brain, spinal cord, and ganglia. Conduction at synapses differs from that along nerve fibers. The chief differences are:

1. Conduction at the synapse is slower than conduction along a nerve fiber. This suggests that there is some sort of obstruction at the synapse.
2. There is greater variability in the ease of transmission.
3. Summation and inhibition seem to be brought about at the synapses.
4. The refractory period is more highly variable.
5. The synapses are more readily fatigued than the nerve fibers

[19] Joseph Erlanger and H. S. Gasser, *American Journal of Phyisology,* 1930.

and are more readily affected by anesthetics and such substances as nicotine.

6. In regions containing many synapses, the blood supply is very rich, which suggests more active metabolism.

7. A nerve fiber is capable of transmitting an impulse either to or from the cell body, but at the synapse the nerve impulse can pass in only one direction, which is from the axon of one neuron to the dendrite of another. In this way a synapse appears to function as a factor in establishing the *polarity* of neurons.

It is at the synapse that conditions determine where the nerve impulse is to go; for instance, in *C*, Fig. 160, the nerve impulse from a receptor may go to the upper muscle, to the lower muscle, to both, or to neither, depending probably on conditions at the synapse and cell bodies. Thus groups of synapses and cell bodies are often called *adjustors* or centers.

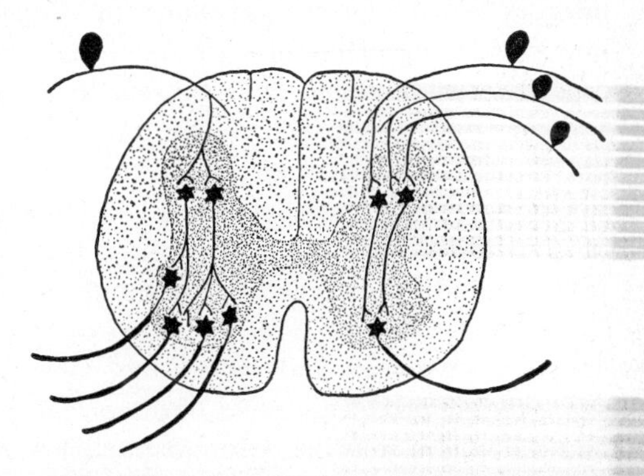

Fig. 163. A cross section of the spinal cord, showing on the right a converging reflex, on the left a spreading reflex. (See also Fig. 160.)

It is possible that at the synapse there are somewhat similar conditions to those at the receptor end organ—that is, a disturbance of equilibrium takes place as the nerve impulses reach the synapse, so that the connecting and motor neurons have a tendency to approach equilibrium in relation to this change in their environment; this is to say they receive and pass on the nerve impulse. This disturbance and tendency to re-establish equilibrium (in ionic or membrane potential, etc.) may be between cells or may be between the parts of the synapse and the body fluids between them. This gives rise to what has been called the central excitatory state (c.e.s.) and the central inhibitory state (c.i.s.), both of which are known to exist and offer many problems for research. Although results of experiments today are conflicting, the conception of c.e.s. and c.i.s. is in line with the neurohumoral idea of integrating activity and is most helpful in extending knowledge of how the nervous system through its unit pattern, the reflex arc, carries out the varied and minutely integrated work of the nervous system, linking together rapidly and accurately the work of distant parts of the body.

Spreading of impulses. All parts of the nervous system are bound together by connecting pathways. Some of these pathways are long, well-defined bundles of fibers connected in a way to facilitate uniform and definite responses to stimulation. Other pathways are diffuse, poorly integrated, and not well adapted to conduct impulses for long distances. Incoming impulses have many pathways open to them. Usually the varying resistance offered at different available synapses prevents their too wide diffusion, and the formation of pathways reinforces this limitation. However, a strong stimulus may produce a great spreading of impulses through the gray matter; and in extreme cases efferent impulses may be sent to all the muscles, causing contractions and resulting in convulsions. In early life spreading of impulses occurs much more readily than in later years, which accounts for the fact that conditions which cause marked muscular contractions or convulsions in a child often cause only mild contractions or a chill in adults. In some cases light stimuli may cause marked spreading. It is thought that (1) certain drugs such as strychnine and caffeine affect the synapses and render the passage of impulses easier; (2) some toxins produce a state of abnormal irritability or sensitiveness; and (3) there may be a summation of mild stimuli at the synapses caused by various irritations.

Convergence of impulses. Nerve impulses from many receptors may converge to one effector mechanism over a nerve fiber which becomes the *final common path.* Each muscle is thus connected with a number of receptors, and it follows that the stimulation of two receptors will produce a more vigorous response than the stimulation of one; three will be more effective than two, etc. The energy of one impulse is added to the energy of other impulses activated at the same time, when all have more or less open pathways to a common effector.

Facilitation. After repetition of a reflex act, its establishment becomes increasingly easy, and the reaction time lessens, within limits. This is known as facilitation and is the basis of habit formation.

Reaction time. This term is applied to the time elapsing between the application of a stimulus and the beginning of a response. Reaction time varies in different individuals and in the same individual under different conditions, depending upon the strength of the stimulus (the stronger the stimulus, the more prompt the response), the nature of the stimulus (e.g., the response to sound is more prompt than that to light), and the number of synapses through which impulses are transmitted. The number of synapses determines to some extent the length of time required for response.

It is sometimes stated that the time required for conduction within the nerve centers may be 12 times as great as that required for conduction along the sensory and motor nerve fibers. The time within the center varies, depending upon (1) the strength of the stimulus, (2) the condition of the nerve centers, the time being lengthened by any condition, such as fatigue, that lessens the irritability of the center.

The *speed* at which a nerve impulse travels along a nerve fiber varies in different varieties of fibers but is estimated to be about 80 to 120 m (393 ft) per second in a medullated nerve fiber in man. In general, the speed of the nerve impulse varies with the diameter (or cross-sectional area) of the fiber. The velocity of the nerve impulse over any nerve refers to the velocity over the largest fibers of that nerve.

There seems to be a close relationship between the *chronaxie* and the diameter of nerve fibers. The finer fibers are less excitable; that is, they have longer chronaxies than the large ones.

Chronaxie is the measure of the excitability of a tissue. The chronaxie of tissue is the shortest *duration of time* over which an electrical current of twice the rheobasic strength must be applied to stimulate the tissue.[20]

Automatic action. Some of the nerve centers are in a state of constant stimulation due to some unknown inherent property and to substances contained in the circulating blood which are constantly acting upon them. In consequence they are constantly discharging impulses to the organs they innervate. The centers controlling respiration and the action of the heart are of this type and are described as automatic.

Inhibition. If every stimulation were followed by its full response, the body would be "on the jump" all the time. This overactivity is

[20] C. H. Best and N. B. Taylor, *The Physiological Basis of Medical Practice*, 1955, p. 916.

checked by the partial or complete *blocking* of impulses or their results. This is known as inhibition. In addition to blocking, inhibition brings about an *actual depression* of the activity of the effector, e.g., reduces the tone of muscle. All effector mechanisms can be inhibited as well as stimulated. Spreading and convergence of impulses, facilitation, and inhibition make a wide range and greater variety of reactions possible. Physiological conditions at the synapses are important in relation to the ease and direction of the traveling of the nerve impulses over these reflex mechanisms.

Reflexes whose adjusting mechanism is in the cord may be inhibited by centers in the cerebrum. Micturition is an example. Micturition is probably brought about as a spinal reflex, the stimulus starting from receptors in the bladder itself, so that when the bladder is filled, it automatically empties itself. By training in early infancy, this simple reflex may be inhibited from the cerebrum, so that micturition takes place only under voluntary control. The same is true of defecation. Such reflexes are called *educated,* or *conditioned, reflexes.* If the conducting paths to the cerebrum are interrupted, the spinal reflex may control, resulting in involuntary micturition and defecation.

Effectors, or **the ends of motor nerve fibers,** are discussed on pp. 225 and 228.

Fatigue of nerve cells. Various authorities state that a normal neuron, when stimulated, first increases in size (due to increased metabolism), but that long-continued activity decreases the size of the cell bodies and reduces the granular material of the cytoplasm and the chromatic material of the nucleus. If the fatigue is not excessive, a period of rest will restore the cells to a normal condition. Nerve fatigue is induced by both mental and muscular work.

The onset of fatigue is favored by poor health and mental conflict of any kind, such as regrets, worry, fear, etc. Conscious effort to keep the attention concentrated induces fatigue. Work done under compulsion, as from a sense of duty, results more readily in fatigue than when interest is the driving motive.

Constant irritation such as eyestrain, abnormal conditions of the feet, prolonged distention of the intestine with feces such as occurs in chronic constipation—in fact anything that causes pain or a continued sense of discomfort brings about conditions in the nervous system similar to those caused by fatigue.

One of the results of fatigue is increased resistance to the passage of impulses at the synapses. The higher cerebral centers are most easily and quickly affected. It is for this reason that young children "go stale" very quickly unless their mental training is properly balanced by rest and play.

It is often stated that change of activity is equivalent to rest. It is questioned whether this is true of mental activity; but it may be true of muscular activity, and the explanation offered is that when nervous stimuli are altered, new pathways and new groups of muscles are called into play. For example, a person who is tired of housework may go out of doors and walk, because the fresh air, changed environment, and different thoughts serve as new stimuli, travel different brain paths, and throw into activity different groups of muscles.

Normally the fatigue of nerve cells resulting from each day's activities is repaired by a night's sleep; but if day after day for an extended period the fatigue is excessive, it accumulates, and a pathological condition results. This is more apt to happen if mental conflicts add their quota. If in addition monotony—which means the same neurons are activated by the same kind of stimuli each day —is added, abnormal states, variously known as brainfag, neurasthenia, and nervous prostration, are likely to develop. The depression of the higher cerebral centers is usually the starting point of such a condition. If the will power is weakened, the power of concentration lessened, and the higher mental capacities reduced, the imagination is likely to substitute morbid ideas and fears (phobias) that under ordinary circumstances would not be thought of, or if they were, would be rejected or reasoned away.

Worry and fear stimulate parts of the autonomic system and bring about conditions in the alimentary canal that favor loss of appetite, indigestion and constipation, and other complaints that are associated with a very wide range of neurasthenic conditions.

Hypersensitiveness, or excessive response to stimuli, often follows the depression of the higher inhibition capacities. Even slight stimuli may be interpreted as pain, and ordinary sounds are regarded as loud noises. It is while in this condition that an individual is susceptible to suggestion which may help him to regain his control or, on the other hand, may be his undoing, for if he hears of abnormal symptoms, he is likely to think he has them, and he may acquire them, for it has been proved that the vascularity and sensitiveness of a part may be increased if the attention is centered upon it. It is to this state that the advertisements of patent medicines, which include long lists of symptoms that can be cured by the particular remedy advertised, make their appeal; and this group of sufferers are the ready victims of many kinds of quacks.

SUMMARY

The nervous system cooperates with the body fluids in coordinating activities of the body

Classification of parts of nervous system. See p. 217.

- **Neurons**
  - **Consist of**
    - **Cell body,** or cyton, source of energy and affords nutrient
    - **Cell processes**
      - *Dendrites*, short, thick, rough outline, branch freely
      - *Axons*, long, smooth, few branches
  - **Function,** to receive impulses and convey them to other cells
  - **Classification based on number of processes**
    - **Bipolar** cells
    - **Multipolar** cells
  - **Classification based on functions**
    - **Afferent** (receptor, sensory) carry impulses from periphery to center
    - **Efferent** (effector, motor) carry impulses from center to periphery
  - **Nerve fibers**
    - **Medullated,** white
      - Consist of
        - Axis cylinder
        - Medulla, or myelin sheath
        - Neurilemma, or sheath of Schwann
      - Function of myelin sheath is insulation, and it possibly plays part in chemical processes involved in the production of nerve impulses
      - Nodes of Ranvier, constriction, myelin sheath absent, branching of fiber
    - **Nonmedullated,** gray or yellow
      - Myelin sheath reduced or absent
  - **Synapse,** probably interlacing of branches of axon of one neuron with dendrites of another neuron; no protoplasmic continuity
  - **End organs**
    - Peripheral structures related to nerve fibers
    - (1) **Sensory,** or receptor, for Herrick's classification—see p. 225
      - a) *Somesthetic* receptors, concerned with general body sensibility
      - b) *Special sense organs*

- **Neurons**
  - **End organs**
    - c) *Kinesthetic* receptors, concerned with muscle tone
    - d) Receptors for *superficial sensibility*
    - e) Receptors for *deep sensibility*
    - f) Exteroceptors
      - Protopathic
      - Epicritic
    - (2) **Motor,** or effector
      - Motor end plates
      - Visceral motor nerve endings
  - **Nucleus,** a group of the cell bodies within the central nervous system, the fibers of which form one anatomical nerve, or tract, within the brain or cord
  - **Ganglion,** a group of the cell bodies of several neurons outside the central nervous system
  - **Center,** a group of neurons regulating a certain function; may be either a nucleus or a ganglion
  - **Nerves**
    - (1) **Funiculus,** a bundle of fibers enclosed in a tubular sheath
    - (2) **Endoneurium,** connective tissue between the individual fibers
    - (3) **Perineurium,** connective tissue surrounding each funiculus
    - (4) **Epineurium,** connective tissue covering several funiculi

Reflex arc, structural unit of nervous system
Reflex act, functional unit of nervous system

- **Reflex Arc and Act**
  - **Classification of activities or reactions**
    - (1) Simple reflexes such as withdrawal of finger from prick
    - (2) Involving neurons extending up into brain stem and cerebellum, such as crying out when finger is pricked
    - (3) Involving neurons in the cerebral cortex, such as dressing of wound after finger is pricked
  - **Classification of reflexes**
    - Simple—single muscle involved
    - Coordinated, or complex—several muscles involved in orderly activity
    - Convulsive or disorderly activity
  - **Receptor end organs**
    - (1) May be simple fiber ends, or these ends may be surrounded by accessory structures
    - (2) *Stimuli* are physical or chemical changes in the immediate environment
    - (3) Changes in environment are converted into a rhythmic succession of nervous impulses

Reflex Arc and Act

- **Receptor stimuli**
  - (1) *Adequate* stimulus, the *kind* of change to which the receptor is sensitive
  - (2) *Minimal* stimulus, the *least* change that can excite a receptor and produce a response in effector
  - (3) *Subminimal* stimulus, a stimulus below threshold level
  - (4) *Intensity* of stimulus, probably related to the degree of change at the receptor and reflected in the frequency of impulses initiated
- **Train of impulses**
  - Adaptation of end organs varies in relation to the speed with which they reach approximate equilibrium with their environment
    - (1) *Slow* adaptation, long trains (sequences) of nerve impulses issue from end organs
    - (2) *Rapid* adaptation, short trains of impulses issue from end organs
- **The nerve impulse**
  - (1) All impulses of similar nature whether carried by sensory or motor nerves
  - (2) In nature probably a self-propagated disturbance
  - (3) Excitability of nerve fiber shows three periods
    - *a*) Normal period or phase, the most excitable period
    - *b*) Absolute refractory period, a short period during which the nerve fiber is inexcitable
    - *c*) Relative refractory period, period of less excitability than normal phase
  - (4) Frequency of nerve impulse usually varies with the strength of the stimulus
  - (5) Velocity of conduction of nerve impulse varies with diameter of fiber
  - (6) Strength of nerve impulse measured by amplitude of electrical changes
  - (7) Spreading of impulses greater in early life
  - (8) Convergence of impulses increases the vigorousness of responses
  - (9) Inhibition of impulses prevents overactivity and reduces the tone of muscle
- **Automatic action** (constant stimulation) of some nerve centers due to inherent property and to substances in blood
- **Reaction time,** time elapsing between application of a stimulus and beginning of response

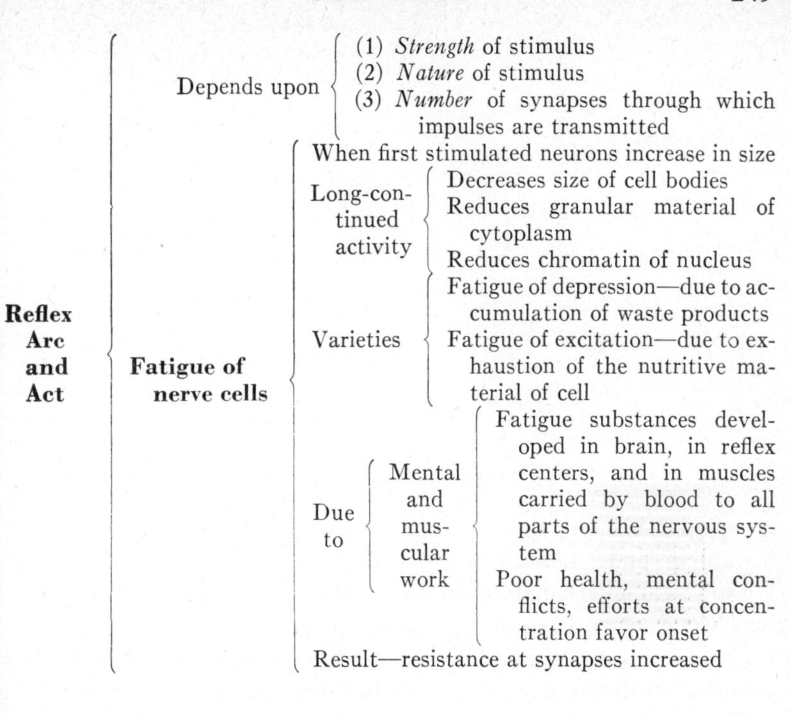

- **Reflex Arc and Act**
  - Depends upon
    - (1) *Strength* of stimulus
    - (2) *Nature* of stimulus
    - (3) *Number* of synapses through which impulses are transmitted
  - **Fatigue of nerve cells**
    - When first stimulated neurons increase in size
    - Long-continued activity
      - Decreases size of cell bodies
      - Reduces granular material of cytoplasm
      - Reduces chromatin of nucleus
    - Varieties
      - Fatigue of depression—due to accumulation of waste products
      - Fatigue of excitation—due to exhaustion of the nutritive material of cell
    - Due to
      - Mental and muscular work
        - Fatigue substances developed in brain, in reflex centers, and in muscles carried by blood to all parts of the nervous system
        - Poor health, mental conflicts, efforts at concentration favor onset
    - Result—resistance at synapses increased

CHAPTER 9

THE SPINAL CORD AND SPINAL NERVES THE BRAIN AND CRANIAL NERVES THE AUTONOMIC SYSTEM—ITS STRUCTURE AND FUNCTION

THE SPINAL CORD

A brief sketch of the lower animals, characterized as segmental, is helpful in understanding the structure and functions of the spinal cord. Segmental animals are made up of a number of smaller units which may be capable of leading an independent existence. This is made possible by the fact that each segment possesses separate circulatory, digestive, excretory, and nervous systems, so that the segments may be separated without endangering or seriously impairing their life processes. As far as the nervous system is concerned, we find that each segment of these animals is equipped with a centrally placed ganglion from which nerve fibers extend in all directions to the different tissues of the segment. A stimulus applied to its surface is soon followed by movement or some other motor response. It appears, therefore, that the nervous elements allotted to each segment are arranged in the form of reflex circuits, their centers being grouped in the shape of a central ganglion. The life of the animal as a whole, however, requires a certain *correlation* between the activities of its different segments and a *subordination* of the latter to the functional necessities of the whole. This end is attained first by intermediary neurons which unite the successive ganglia with one another and, secondly, by a hyperdevelopment of the head ganglion, which thus gains a directing control over the others.

A nervous system of this kind is reflex in its nature and forms the basal stem around which the nervous system as it appears in the highest animals is eventually developed. The head ganglion is comparable to the brain and the segmental ganglia to the spinal cord, and from these parts the afferent and efferent nerve fibers arise.

The spinal cord is lodged within the spinal canal of the vertebral column. It consists of gray and white matter extending from the

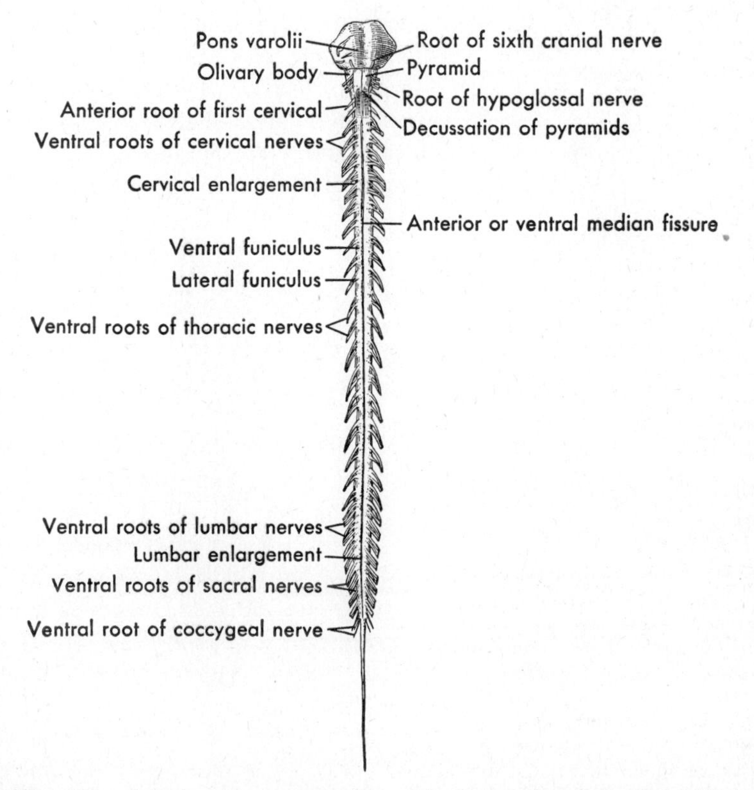

Fig. 164. Spinal cord, ventral view. The actual length is about 18 in. (Toldt.)

foramen magnum of the skull, where it is continuous with the medulla oblongata, to about the second lumbar vertebra.

The spinal cord diminishes slightly in size from above downward and presents two spindlelike enlargements—the cervical enlargement (level of fourth cervical to second thoracic vertebrae) and the lumbar enlargement (level of tenth thoracic, widest at twelfth thoracic), which dwindles as the *conus medullaris* ending at the level of the first or second lumbar vertebra, where it gives rise to the nonnervous threadlike *filum terminale.* The filum terminale punctures the dura mater at the level of the second sacral vertebra and terminates on the first coccygeal vertebra (Fig. 165). At these enlargements the nerves are given off to the arms and legs, respectively. The average length of the cord is about 45 cm (18 in.). It is incompletely divided into lateral halves by a ventral fissure and a dorsal sulcus, the ventral fissure dividing it in the middle line in front and the dorsal sulcus in the middle line behind. Because of these fissures only a narrow bridge of the substance of the cord connects its two halves. This bridge, the *transverse commissure,* is traversed throughout its entire length by a minute *central canal,* which opens into the fourth ventricle at its upper end and, at its lower, terminates blindly in the filum terminale.

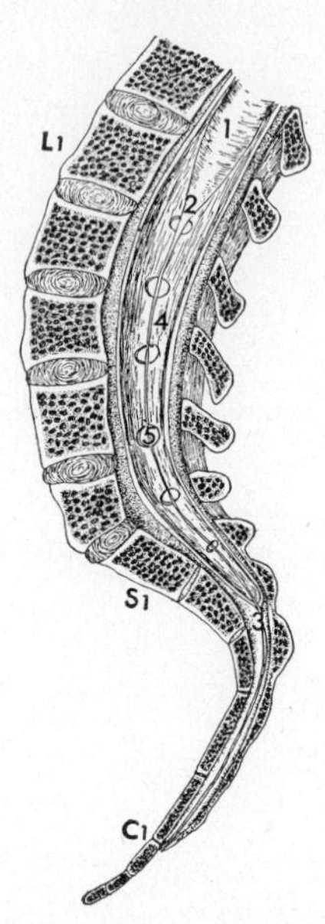

Fig. 165. Longitudinal section of the vertebral column, showing the end of the spinal cord. *1*, beginning of conus; *2*, end of conus and beginning of filum terminale; *3*, filum punctures dura; *4*, spinal canal; *5*, foramen for exit of spinal nerve; C_1, first coccygeal vertebra; L_1, first lumbar vertebra; S_1, first sacral vertebra.

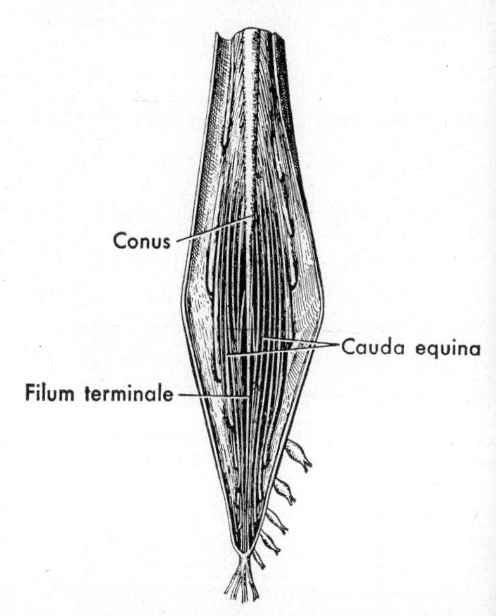

Fig. 166. The conus, filum terminale, and cauda equina. (Toldt.)

Meninges, or membranes. The spinal cord does not fit as closely into the spinal canal as the brain does into the cranial cavity but is suspended within the canal. It is protected and nourished by three membranes and the cerebrospinal fluid which circulates between them,

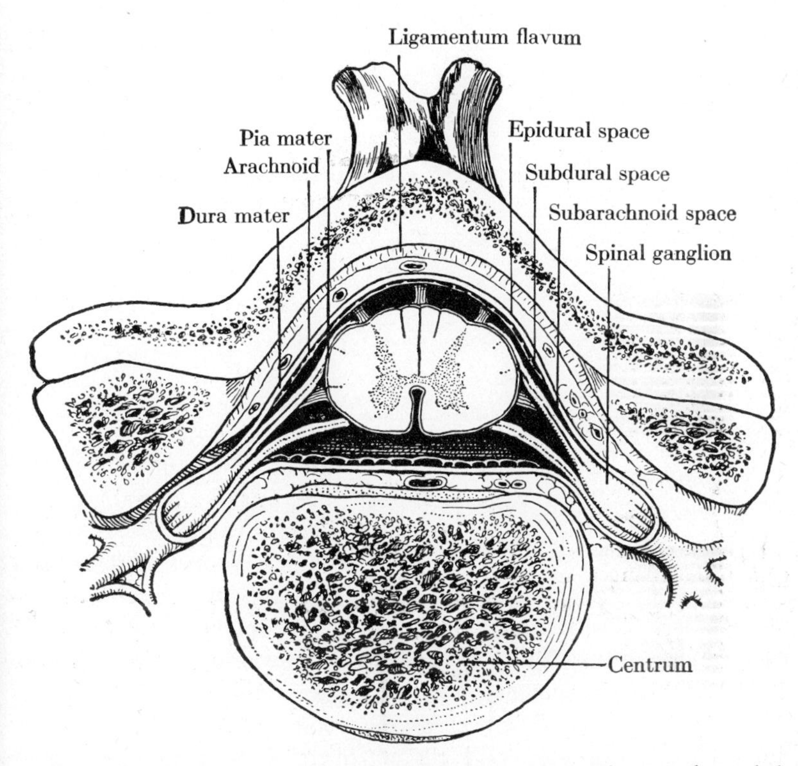

Fig. 167. Transverse section of spinal cord and vertebra to show their relative positions. Note dorsal root ganglia lying in the intervertebral foramina and the dorsal and ventral roots in the spinal canal. The spinal nerve is shown outside the vertebra, with branches to the dorsal body wall, ventral body wall, and viscera.

especially in the subarachnoid space. These membranes are continuous with the membranes covering the brain and are called by the same names, the *pia mater* closely investing the cord, the *arachnoid mater,* and the *dura mater* outside. The spinal fluid serves to moisten and lubricate the cord and protect it against changing pressures.

Structure of the cord. The cord consists of gray and white matter. The gray matter consists of nerve cells and nerve fibers held together by neuroglia. The white matter consists of nerve fibers embedded in a network of neuroglia. The gray and white matter are supported by connective tissue and are well supplied with blood and lymph vessels. The blood supply is from the anterior spinal artery and from a succession of small arteries which enter the spinal canal through the intervertebral foramina; these branches are derived from the vertebrae, the intercostals, and the arteries in the abdomen and pelvis.

The *gray matter* is in the interior surrounding the central canal and on cross section appears to be arranged in the form of the letter H. The transverse bar on the H is called the *gray commissure* and connects the two lateral masses of gray matter. On each side the gray matter presents a ventral, or anterior, and a dorsal, or posterior, column.[1] The former is short and bulky, whereas the latter is long and slender. The ventral column contains the cell bodies from which the efferent (motor) fibers of the spinal nerves arise. These nerves pass out through openings in the meninges and the intervertebral foramina. The lateral aspect of the ventral column contains cell bodies which give rise to the efferent fibers of the white *rami communicantes*[2] (Fig. 168), or preganglionic fibers, which connect with

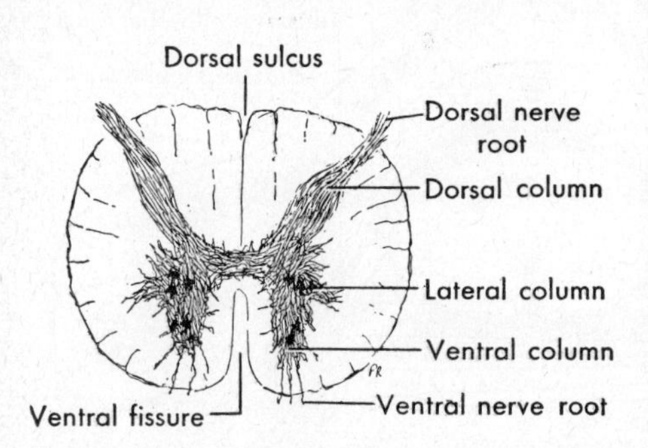

Fig. 168. Transverse section of spinal cord, middle thoracic region.

[1] This text uses *ventral* and *anterior* for the front aspect, *dorsal* and *posterior* for the back and *columns* for what were formerly described as *horns* of the gray matter. For each of the three major divisions of the white matter of each half of the cord, *funiculus* instead of *column* is used.

[2] Preganglionic fibers are efferent, medullated fibers (*white rami*) which pass from the spinal cord by way of the ventral column and end around the cells of the sympathetic ganglia. From the cells of the ganglia, postganglionic fibers go to the various effector organs, i.e., smooth muscle, cardiac muscle, and glands. The latter are nonmedullated. Some pass back to the spinal nerves and form the *gray rami*.

the sympathetic ganglia outside the vertebral column. The dorsal column contains cell bodies from which afferent, ascending fibers go up to higher levels of the spinal cord and to the brain. The fibers of the spinal nerves entering the cord form synapses with the neurons in these columns. The gray matter also contains a great number of central or connecting neurons which serve for the passage of impulses (1) from the dorsal to the ventral roots of the spinal nerves, (2) from one side of the cord to the other, (3) from one level of the cord to another.

The *white matter* is arranged around and between the columns of gray matter, the proportion of gray and white varying in different regions of the cord. On each side the white matter may be said to consist of three portions, or funiculi, namely, a ventral, a lateral, and a dorsal funiculus. Each funiculus is in turn divided into smaller segments, or fasciculi.

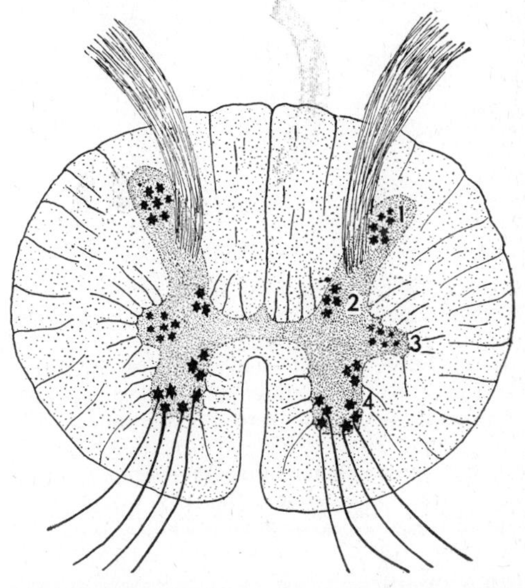

Fig. 169. Cross section of the spinal cord to show some of the groups of nerve cells. *1*, dorsal-column cells; *2*, Clarke's-column cells; *3*, lateral-column cells; *4*, ventral-column cells.

Some of these funiculi consist of fasciculi made up of fibers which are *ascending,* or sensory (ascending tracts). They serve as pathways to the brain for impulses entering the cord over afferent fibers of spinal nerves. Other funiculi consist of fasciculi which are *descending,* or motor (descending tracts). They transfer impulses from the brain to the motor neurons of the spinal nerves. They begin in the gray matter of the brain, descend, and terminate in the gray matter of the cord, e.g., *3* and *6*. Other funiculi (white in Fig. 170) are made up chiefly of short ascending and descending fibers beginning in one region of the spinal cord and ending in another.

The funiculi in the dorsal portion are chiefly ascending. Injury to these funiculi will interfere with the passage of sensory impulses and possibly (depending on location and extent of injury) result in loss of sensation in the parts from which the passage of impulses is blocked.

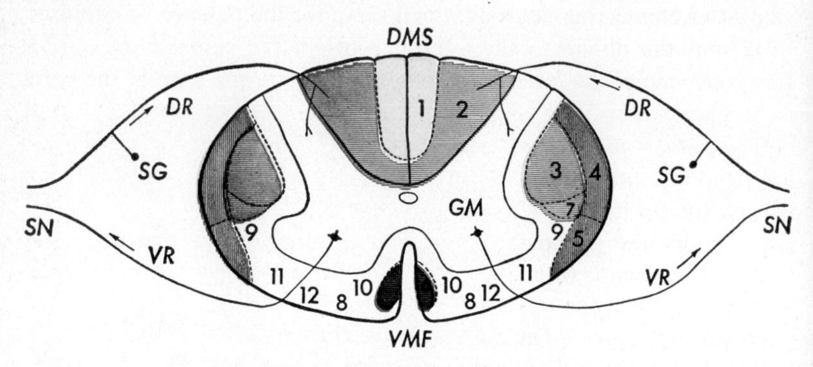

Fig. 170. Diagram to show general location of some of the conduction paths as seen in a transverse section of the spinal cord. *DMS,* dorsal median sulcus; *DR,* dorsal root; *GM,* gray matter; *SG,* spinal ganglion; *SN,* spinal nerve; *VMF,* ventral median fissure; *VR,* ventral root. *1,* fasciculus gracilis (tract of Goll); *2,* fasciculus cuneatus (tract of Burdach); *3,* lateral cerebrospinal fasciculus (crossed pyramidal tract); *4,* dorsal spinocerebellar fasciculus; *5,* ventrolateral spinocerebellar fasciculus (Gowers' tract); *6,* ventral cerebrospinal fasciculus (direct pyramidal tract); *7,* rubrospinal tract; *8,* ventral spinothalamic tracts; *9,* lateral spinothalamic tracts; *10,* vestibulospinal tracts; *11,* tectospinal tracts; *12,* olivospinal tracts.

Locomotor ataxia, or *tabes dorsalis,* is a degeneration of the posterior fasciculi and posterior columns of the cord, resulting in disturbances of sensation, interference with reflexes, and consequently with movements such as walking, which after it is once learned is largely reflex. *Anterior poliomyelitis,* or infantile paralysis, is inflammation of the anterior areas of gray matter and results in paralysis of the parts supplied with motor nerves from the diseased portion of the cord.

Functions of the spinal cord. It is an important center of reflex action for the trunk and limbs, and it consists of the principal conducting paths to and from the higher centers in the cord and brain. From the standpoint of function, the spinal cord may be regarded as consisting of more or less independent segments. Each segment is related by afferent and efferent nerve fibers to its own definite segmental area of the body, as well as to the segments above and below.

The pathways of the cord

Ascending Pathways of the Cord

1. *Of the dorsal portion of the spinal cord* (Fasciculus gracilis and cuneatus)
   a) *Fasciculus gracilis*
   The fibers of fasciculus gracilis are medially placed in the dorsal part of the cord and are made up of long ascending fibers from the sacral, lumbar, and lower thoracic dorsal ganglia. These fibers terminate around cells in the nucleus gracilis in the low medulla. The fibers from the cells in the nucleus gracilis cross to the opposite side (sensory decussation), ascend, and terminate in the thalamus. The cells in the thalamus send fibers to the postcentral gyrus of the cortex (sensory area).
   The peripheral fibers of the dorsal ganglia form the sensory fibers of the peripheral nerves from the lower extremity and lower part of the trunk.
   b) *Fasciculus cuneatus*
   The fibers of fasciculus cuneatus are more laterally placed in the dorsal part of the cord and are made up of long ascending fibers from the upper thoracic and cervical ganglia. These fibers terminate around cells in the nucleus cuneatus in the medulla. The fibers from the cells of the nucleus cuneatus cross to the opposite side, ascend, and terminate in the thalamus. Both of these pathways conduct impulses from proprioceptors (on bones, muscles, tendons and joints) that give rise to sensations of movement and position. Other impulses conducted are for spatial discrimination, for more exact tactile localization, for vibratory sensations, and for two-point discrimination.
   The peripheral fibers of the dorsal ganglia form the sensory fibers of the peripheral nerves from the upper extremity, upper trunk, and neck.
2. *The spinothalamic pathways*
   These pathways arise from large cells in the dorsal horn of the gray of the cord. Most of the fibers cross in the cord and ascend in the white matter of the opposite side as the lateral and ventral

spinothalamic pathways (Fig. 170, areas 8 and 9). These fibers eventually terminate in the thalamus. The lateral spinothalamic pathway conveys impulses of pain and temperature; the ventral fibers convey impulses of touch and pressure. The cells in the cord receive impulses from sensory fibers of the peripheral nerves.

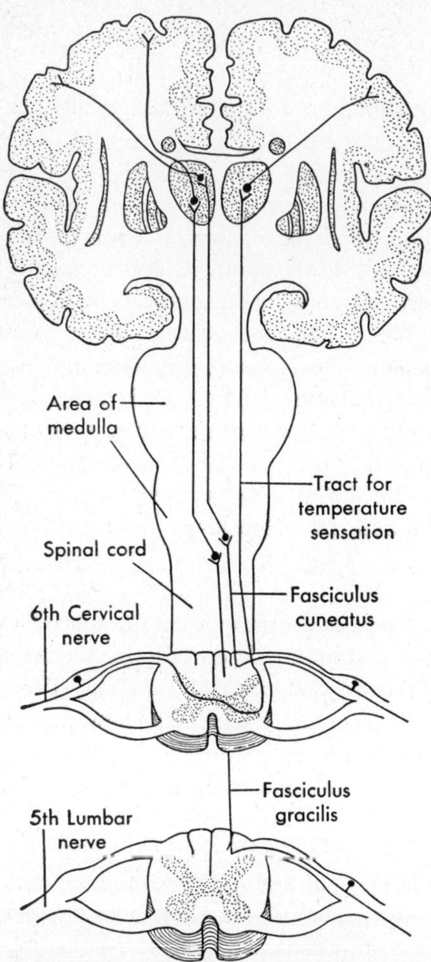

Fig. 171. Ascending tracts, three shown. Fasciculus (tract) gracilis concerned with proprioceptor impulses from lower part of body, fasciculus cuneatus concerned with proprioceptor impulses from upper part of body, and the spinothalamic fasciculus concerned with pain and temperature sensations.

3. *The dorsal spinocerebellar pathways* arise from cells in the medial gray of the cord (Clarke's column) and pass to the white of the cord of the same side (Fig. 170, area 4) and ascend to the medulla. Here they form the inferior cerebellar peduncle and terminate in the cortex of the cerebellum. This pathway conveys impulses from the lower extremity and trunk.

The ventral spinocerebellar pathway arises from cells in the intermediate gray of the cord. Some of the fibers cross in the cord, some do not, and ascend in the cord (Fig. 170, area 5) and reach the cerebellum via the superior cerebellar peduncle. These fibers convey proprioceptive impulses from all parts of the body including the upper extremities and neck.

Both of the spinocerebellar pathways convey impulses from receptors on muscles, tendons, and joints to the cerebellum. This enables the cerebellum to exert its synergizing and regulative tonic influence upon all voluntary muscles. Both of these pathways are composed of two neurons, the spinal ganglion cells and the cell bodies located in the spinal cord.

The Descending Pathways

1. *The pyramidal pathways* (corticospinal). The fibers arise from large pyramidal cells in the precentral gyrus of the cerebral cortex and other adjacent areas. They converge and descend through the internal capsule, pons, and medulla. As they descend, fibers are given off to the motor nuclei of the cranial nerves. In the medulla the majority of the fibers cross to the opposite side (pyramidal decussation) and descend as the crossed pyramidal pathway (Fig. 170, area 3). The remaining fibers do not cross but descend on the same side as the direct pyramidal pathway (Fig. 170, area 6). All of these fibers terminate around the large motor cells in the ventral gray of the cord, and convey impulses which bring about volitional movements, especially fine individual movements that are essential for developing motor skills. The fibers are large and heavily myelinated. Myelinization begins before birth and is completed about the beginning of the third year.
2. *The vestibulospinal pathway* originates from cells in the vestibular nucleus of the medulla and descends uncrossed in the cord and terminates around the large motor cells in the ventral gray of the cord (Fig. 170, area 10). Since the vestibular nucleus receives fibers from the vestibular portion of the eighth nerve and from the cerebellum, the tract conveys impulses from the middle ear and cerebellum which exert a tonic influence on the muscles of the extremities and trunk, which aids in maintaining equilibrium and position.
3. *The rubrospinal pathway* originates from cells in the red nucleus of the midbrain. The fibers cross and descend in the cord and terminate around cells in the dorsal part of the ventral gray of the

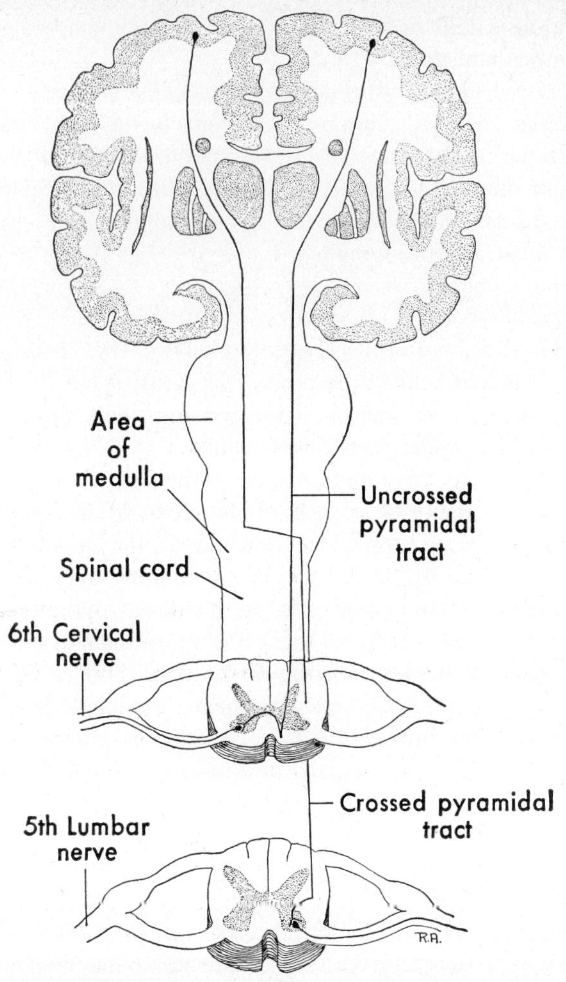

Fig. 172. Descending tracts, two shown. The lateral corticospinal tract (crossed pyramidal) and the ventral corticospinal tract (direct pyramidal). These tracts contain motor fibers to skeletal muscles.

cord of the thoracic region. The red nucleus relays impulses from the cerebellum and vestibular apparatus to the motor nuclei of the brain stem and cord, and in this way reflex postural adjustments are made possible.

4. *The tectospinal pathways* originate from cells in the colliculi of the midbrain. The fibers cross and descend in the cord to termi-

nate around cells in the ventral gray. The fibers convey impulses which mediate reflex activity of the muscles of the head and neck in response to optic stimuli and perhaps auditory stimuli. (Fig. 170, area 11.)

5. *The olivospinal pathway* originates from cells in the olivary nucleus of the medulla and perhaps other higher centers and terminates around cells in the ventral gray of the upper part of the cord (Fig. 170, area 12). The olivary nucleus receives fibers from and sends fibers to the cerebellum.

The impulses from all of these descending pathways eventually reach the voluntary muscles via the large motor cells in the ventral gray of the cord, and exert regulatory control over reflex and other activities.

There are also descending fibers in the cord that innervate smooth muscle, cardiac muscle, and glandular epithelium. The hypothalamus is the coordinating center of the autonomic pathway. The hypothalamus is under control of the thalamus and higher autonomic centers of the cortex. The fibers descend from the hypothalamus in a rather diffuse manner and terminate around autonomic cells in the ventrolateral gray of the cord.

In addition to the long descending and ascending fiber tracts, there are other short fiber tracts that form part of the intrinsic reflex mechanisms of the spinal cord.

THE SPINAL NERVES

There are 31 pairs of spinal nerves, arranged in the following groups, and named for the region of the vertebral column from which they emerge:

Cervical	8 pairs
Thoracic	12 pairs
Lumbar	5 pairs
Sacral	5 pairs
Coccygeal	1 pair

The first cervical nerves arise from the medulla oblongata and leave the spinal canal between the occipital bone and the atlas. The other spinal nerves arise from the spinal cord, and each leaves the spinal canal through an intervertebral foramen behind the vertebra whose number it bears,[3] e.g., the sixth thoracic nerve emerges through the foramen between the sixth and seventh vertebrae. The coccygeal nerve passes from the lower extremity of the canal.

Mixed nerves. The spinal nerves consist mainly of medullated nerve fibers and are called mixed nerves because they contain both motor and sensory fibers. Each spinal nerve has two roots, a ventral root and a dorsal root. The fibers of the ventral root *arise from nerve cells comprising the gray matter* in the ventral column and convey impulses from the spinal cord to the periphery.

The fibers of the dorsal root arise from the *cells composing the enlargement,* or *ganglion,* of the dorsal root situated in the openings between the arches of the vertebrae. These cell bodies give off a single fiber which divides in a T-shaped manner into two processes. One extends to a sensory end organ of the skin or of a muscle, tendon, joint, etc. The other extends into the spinal cord, forming the dorsal root of a spinal nerve.

The fibers that enter the cord directly do not pass into the gray matter immediately; some extend upward and some downward in the white matter before doing so. Sooner or later they all enter the gray matter of the spinal cord or brain, where they form synapses with central or motor neurons.

The ventral roots have their origin within the spinal cord and contain motor fibers. The dorsal roots have their origin outside the cord, i.e., in the spinal ganglia, and contain sensory fibers. The fibers connected with these two roots are collected into one bundle and form a spinal nerve just before leaving the canal through the intervertebral openings.

Distribution of terminal branches of the spinal nerves. After leaving the spinal column, each spinal nerve divides into four main trunks known as the meningeal, or recurrent (distributed to the meninges), the dorsal, ventral, and visceral branches. The dorsal branches supply the muscles and skin of the back of the head, neck, and trunk. The ventral branches supply the extremities and parts of the body wall in front of the spine. The visceral branches connect with the sympathetic ganglia by means of fibers which pass from the nerve to the ganglia and vice versa. Extending from the sympathetic ganglia to their final distribution, these nerves, called autonomic

[3] In the cervical region the eight spinal nerves emerge from the vertebral column above the vertebra whose name they bear.

nerves, form plexuses called the cardiac, the celiac, or solar, the hypogastric, the pelvic, and the enteric. In passing to the viscera, muscles, skin, etc., terminal branches of these nerves follow the same pathways as the blood vessels.

After the nerves emerge from the cord, a plexus is formed in the cervical, brachial, lumbar, and sacral segments from which the peripheral nerves are formed.

The cervical plexus. This plexus is formed by the anterior branches of the first four cervical nerves. Nerves 2, 3, 4 divide into an upper and a lower branch; these in turn unite to form three loops from which peripheral nerves are distributed. There is communication between these nerves and the hypoglossal, vagus, and accessory cranial nerves to the head and neck musculature.

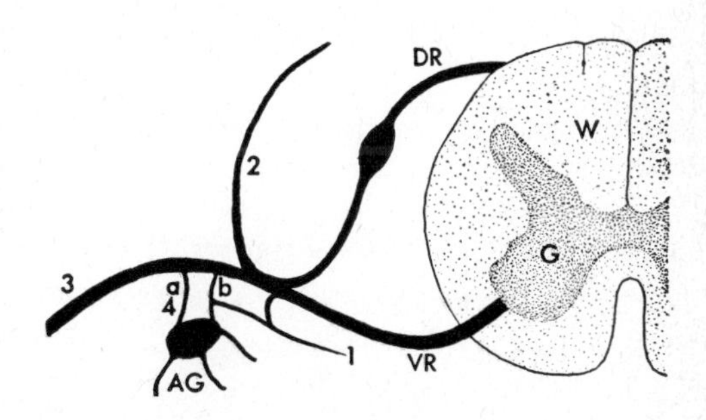

Fig. 173. The four branches of a typical thoracic spinal nerve. *1,* meningeal, or recurrent, branch; *2,* dorsal branch; *3,* ventral branch; *4,* visceral branch; *a,* gray ramus; *b,* white ramus; *AG,* autonomic ganglion; *DR,* dorsal root of spinal nerve; *G,* gray matter; *VR,* ventral root of spinal nerve; *W,* white matter.

The brachial plexus. This plexus is formed by the union of the anterior branch of the last four cervical and the first thoracic nerves. See Fig. 174 for fiber communications and nerves formed. There are three main cords formed, the lateral, medial and posterior. Important nerves formed are the median, ulnar and radial nerves which supply the upper extremity.

The lumbar plexus. This plexus is formed by a few fibers from the twelfth thoracic and the anterior primary divisions of the first four lumbar nerves. Figure 175 illustrates the communications and distribution of the fibers to the muscles. The largest nerves formed are the femoral and obturator nerves.

The sacral plexus. This plexus is formed by a few fibers from lumbar nerve 4, all of 5 and sacral nerves 1, 2, and 3. Fig. 175 illustrates the intercommunication of these fibers and the great nerves

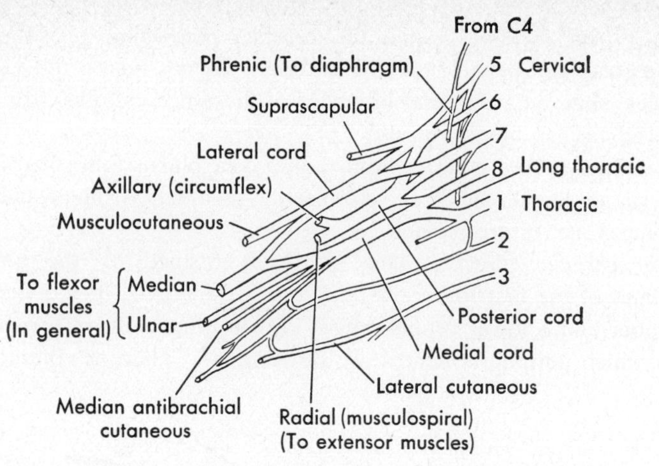

Fig. 174. Diagram of the brachial plexus, showing distribution of some of the larger nerves. See p. 310 for distribution of the nerves.

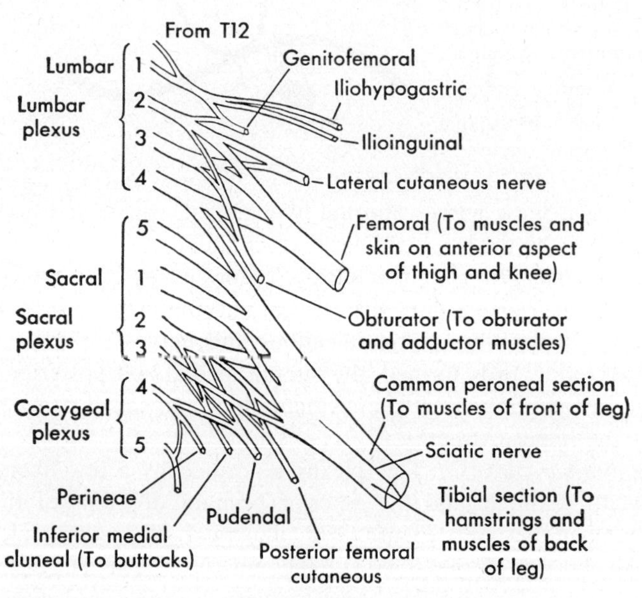

Fig. 175. Diagram of lumbosacral plexus and distribution of some of the larger nerves. See p. 310 for distribution of the nerves.

formed. The largest is the great sciatic, which supplies the muscles of the lower extremity.

In the thoracic region a plexus is not formed, but the fibers pass as intercostal nerves out into the intercosal spaces to supply the intercostal muscles, the upper abdominal muscles, and the skin of the abdomen and chest.

Names of peripheral nerves. Many of the larger branches given off from the spinal nerves bear the same name as the artery which they accompany or the part which they supply. Thus, the radial nerve passes down the radial side of the forearm in company with the radial artery; the intercostal nerves pass between the ribs in company with the intercostal arteries. An exception to this is the two sciatic nerves, which pass down from the sacral plexus, one on each side of the body near the center of each buttock and the back of each thigh, to the popliteal region, where each divides into two large branches which supply the leg and foot. Motor branches from these nerves pass to the muscles of the legs and feet, and they receive sensory branches from the skin of the lower extremities.

Degeneration and regeneration of nerves. Since the cell body is essential for the nutrition of the whole cell, it follows that if the processes of a neuron are cut off, they will suffer from malnutrition and die. If, for instance, a spinal nerve is cut, all the peripheral part will die, since the fibers composing it have been cut off from their cell bodies situated in the cord or in the spinal ganglia. The divided ends of a nerve that has been cut across readily reunite by cicatricial tissue, that is, the connective-tissue framework unites, but the cut ends of the fibers themselves do not unite. On the contrary, the peripheral or severed portion of the nerve fiber begins to degenerate, its medullary sheath breaks up into a mass of fatty molecules and is gradually absorbed, and finally the central fiber also disappears. In regeneration, the new fiber grows from the central

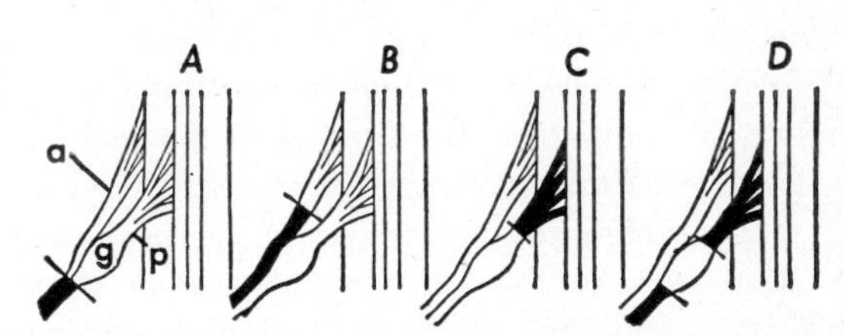

Fig. 176. Degeneration of spinal nerves and nerve roots after section. *A*, section of nerve trunk beyond the ganglion; *B*, section of anterior root; *C*, section of posterior root; *D*, excision of ganglion. *a*, anterior root; *p*, posterior root; *g*, ganglion. Black indicates the portion of the nerve which degenerates after section.

end of the severed nerve fiber and penetrating into the peripheral end of the neurilemma grows along this as the fiber of the new nerve, the fiber after a time becoming surrounded with a medullary sheath. The nuclei of the neurilemma are thought to play an important part in both the degeneration and regeneration of the cut fiber. Restoration of function in the nerve may not occur for several months, during which time it is presumed the new nerve fibers are slowly finding their way along the course of those which have been destroyed. Since in the white matter of the brain and cord and in the cranial nerve fibers the neurilemma is very much reduced or lacking, regeneration of an injured cranial or white-matter nerve fiber rarely, if ever, occurs.

THE BRAIN

The brain is the largest and most complex mass of nervous tissue in the body. It is contained in the cranial cavity and comprises five fairly distinct connected parts: the cerebrum, the midbrain, the cerebellum, the pons Varolii,[4] and the medulla oblongata.

In early embryonic life the brain, or encephalon, consists of three hollow vesicles named the forebrain, or prosencephalon, the midbrain, or mesencephalon, and the hindbrain, or rhombencephalon. During growth the cerebral hemispheres, their commissures, and the first, second, and third ventricles are developed from the forebrain; the corpora quadrigemina, the cerebral peduncles, and the cerebral aqueduct (a tubular connection between the third and fourth ventricles) are developed from the midbrain; the medulla oblongata, the pons Varolii, the cerebellum, and the included fourth ventricle are developed from the hindbrain.

The weight of the brain in the adult male is about 1,380 gm (48.6 oz); in the adult female, about 1,250 gm (44 oz). The weight of the brain is an indication of growth, which in early life depends upon the enlargement of the cells and their processes, the myelination of the nerve fibers, and an increase in the amount of neuroglia. The brain grows rapidly up to the fifth year and in general ceases to grow much beyond about the twentieth year. In advanced age the brain gradually loses weight.

The development of the brain is not only a matter of growth but also a matter of forming new pathways, i.e., new synapses and a permanent modification of the synapses that are functionally active during various forms of mental activity. The nature of the brain proto-

[4] Constanzio Varoli, Italian anatomist (1543–1575).

plasm and the use to which it is put determine to some extent the length of time during which development may continue. Mental exercise keeps the brain active and capable of development, just as exercising a muscle tends to prevent atrophy or loss of function.

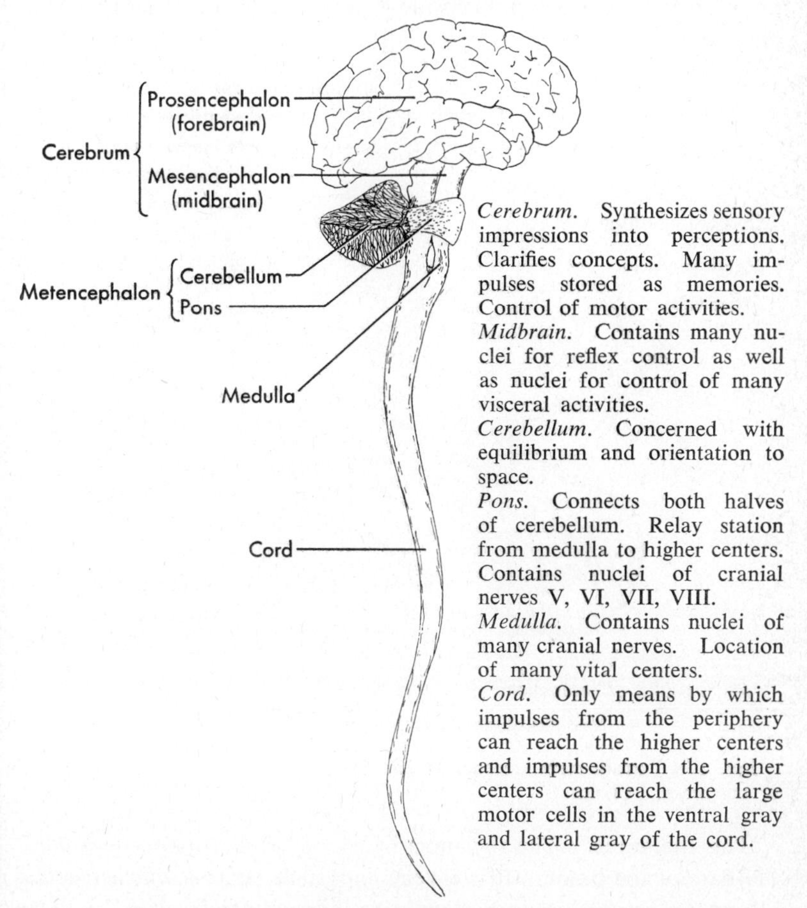

Cerebrum
Prosencephalon (forebrain)
Mesencephalon (midbrain)
Metencephalon
Cerebellum
Pons
Medulla
Cord
Cerebrum. Synthesizes sensory impressions into perceptions. Clarifies concepts. Many impulses stored as memories. Control of motor activities.
Midbrain. Contains many nuclei for reflex control as well as nuclei for control of many visceral activities.
Cerebellum. Concerned with equilibrium and orientation to space.
Pons. Connects both halves of cerebellum. Relay station from medulla to higher centers. Contains nuclei of cranial nerves V, VI, VII, VIII.
Medulla. Contains nuclei of many cranial nerves. Location of many vital centers.
Cord. Only means by which impulses from the periphery can reach the higher centers and impulses from the higher centers can reach the large motor cells in the ventral gray and lateral gray of the cord.

Fig. 177. Diagram illustrating gross divisions of the central nervous system.

The parts of the brain may be grouped as the brain stem (or axis), the cerebrum, and the cerebellum.

The brain stem. In the *brain stem* are found the pons, the medulla, and the midbrain containing the cerebral and cerebellar peduncles, corpora quadrigemina, red nucleus, etc.

The basal nuclei. Deep in each cerebral hemisphere and close to the brain stem is an aggregation of gray matter named the *basal nucleus.* The basal nucleus is composed of the thalamus, the corpus striatum—made up of the lentiform nucleus, the caudate nucleus, and between them the internal capsule—the amygdaloid nucleus, and the claustrum.

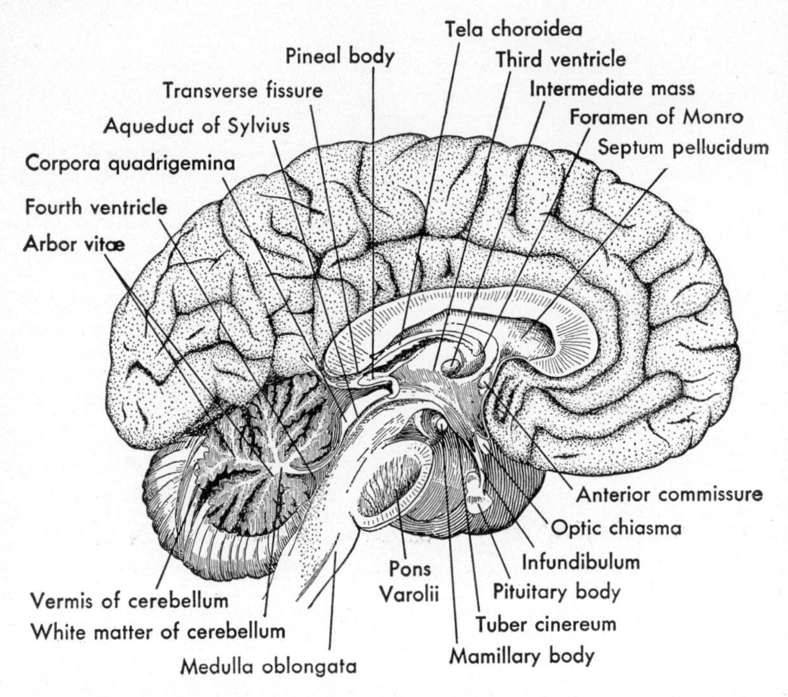

Fig. 178. Brain stem, cerebellum, and cerebrum, seen in a sagittal section of the brain. (Toldt.)

The thalamus. The thalamus is a large oval structure located at the base of the brain. It is a very important nucleus which receives all sensory impulses either directly or indirectly from all parts of the body. It also receives fibers from the cerebral cortex, cerebellum, spinal cord, and other nuclei. It thus forms an important relay station and sorts out afferent impulses from lower levels and sends them on to their special areas of the cerebral cortex.

The hypothalamus. The hypothalamus lies below the thalamus and forms part of the lateral walls and the floor of the third ventricle.

The known functions of the hypothalamus are as follows: (1) Through fiber connections with the posterior lobe of the pituitary gland the hypothalamus regulates water metabolism. (2) Carbohydrate and fat metabolism are also in some way regulated through the hypothalamus. (3) Through centers in the anterior part of the hypothalamus heat-loss functions such as sweating and vasodilatation

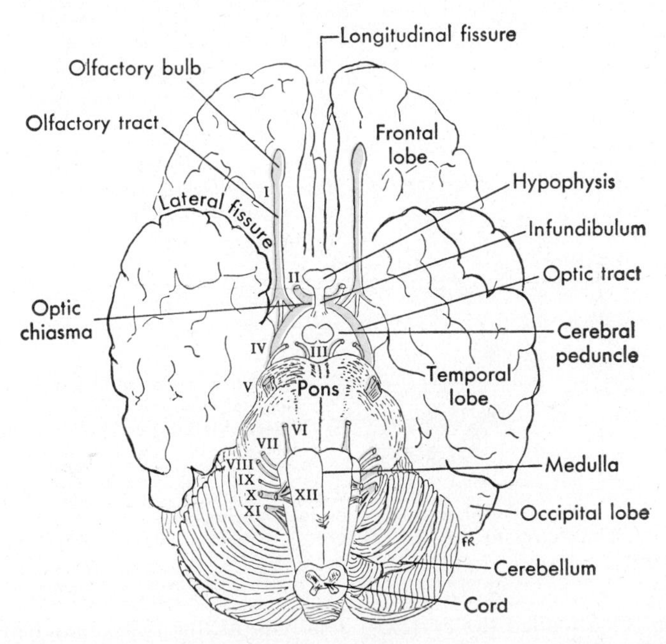

Fig. 179. Under surface of brain showing cerebrum, cerebellum, the pons, and medulla; note the infundibulum to which the pituitary, or hypophysis, is attached. The numerals indicate the cranial nerves. Note the olfactory bulb and tract and the optic chiasma.

are regulated. The posterior hypothalamus controls heat production by shivering and prevention of heat loss by vasoconstriction. (4) When the hypothalamus is not under cortical control, various emotional manifestations become evident. When the posterior hypothalamus is destroyed, emotional lethargy, sleepiness, and a drop in temperature (due to reduction of somatic and visceral activities) results. This gives evidence that there is a center for controlling emotional states and sleep. (5) Gastric movements and genital functions are also in some way regulated through the hypothalamus. (6)

Other groups of cells are believed to regulate parasympathetic activity.

The brain stem and basal nuclei are composed of the cranial nuclei and the tracts of the sensory and motor nerve fibers connecting the spinal cord and brain and also various parts of the brain with each other.

The cerebrum is by far the largest part of the brain. It is egg-shaped and fills the whole of the upper portion of the skull. The entire surface, both upper and under, is composed of layers of gray

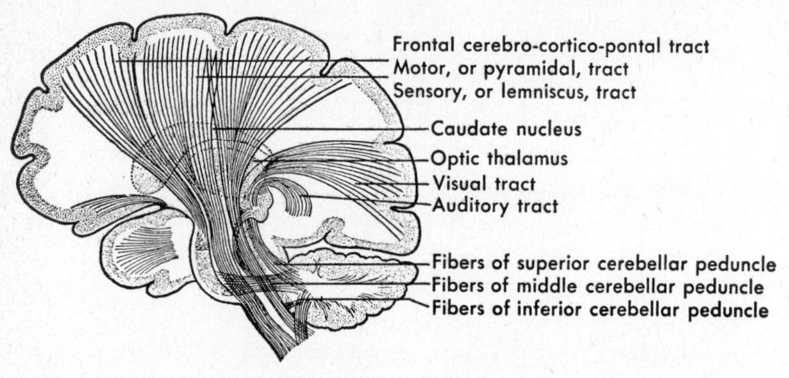

Fig. 180. Dorsoventral section of brain, showing some of the fiber tracts from the spinal cord to the cerebral cortex. Fiber tracts to cerebellum are also shown.

matter and is called the cortex. The bulk of the white matter in the interior of the cerebrum consists of very small fibers running in three principal directions: (1) from above downward—*projection fibers,* connecting the cerebrum with other parts of the brain and spinal cord; (2) from the front backward—*association fibers,* connecting gyri on the same side of the cerebrum; and (3) from side to side—*commissural fibers,* connecting the right and left sides of the cerebrum. The fibers link the different parts of the brain together and connect the brain with the spinal cord (Figs. 180 and 181). There are also autonomic fibers that link the brain with the spinal cord.

Fissures and convolutions. In early life the cortex of the cerebrum is comparatively smooth, but as time passes and the brain develops, the surface becomes covered with furrows, which vary in depth. The deeper furrows are called *fissures,* the shallow ones, *sulci,* and the ridges between the sulci are called *gyri* or *convolutions.* The fissures and sulci are infoldings of gray matter; consequently the more numerous and deeper they are, the greater is the amount of gray matter. The number, length, and depth of these fissures and sulci and the prominence and sizes of the convolutions vary. The cortex on the top of the convolutions is thick but thins out toward the floor of the fissures and sulci.

There are five important fissures which are landmarks.

1. The *longitudinal cerebral fissure* extends from the back to the front of the cerebrum and almost completely divides it into two hemispheres, the two halves, however, being connected in the center by a broad, transverse band of white fibers called the *corpus callosum* (Figs. 179 and 181). A process of the dura mater extends down into this fissure and separates the two cerebral hemispheres. It is called the *falx cerebri,* because it is narrow in front and broader behind, thus resembling a sickle in shape. Blood is returned from the brain in venous channels called sinuses. Two important sinuses are lodged between the layers of the falx cerebri. The *superior sagittal sinus* is contained in the upper border and the *inferior longitudinal* (inferior sagittal) *sinus* in the lower border. Figure 182 shows these two sinuses, also the *straight sinus,* the *transverse,* or *lateral,* and the *superior petrosal* sinuses of one side of the head. *Confluence of sinuses* (*torcular Herophilus*[5]) is the name applied to the dilated extremity of the superior sagittal sinus.

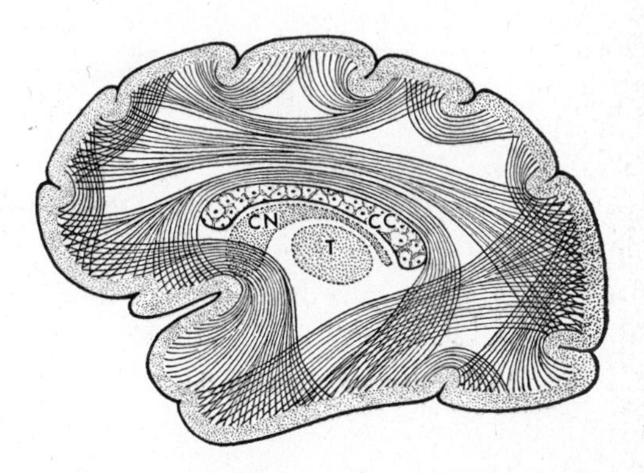

Fig. 181. Section of brain, showing association fibers connecting the gyri, and commissural fibers connecting the two sides of the brain. The commissural fibers are cut across and appear as dots. The myelin sheaths surrounding these fibers appear as white areas. *CC,* corpus callosum; *CN,* caudate nucleus; *T,* thalamus.

2. The *transverse fissure* is between the cerebrum and the cerebellum. A process of the dura also extends into this fissure and

[5] Herophilus, Greek physician (335–280 B.C.).

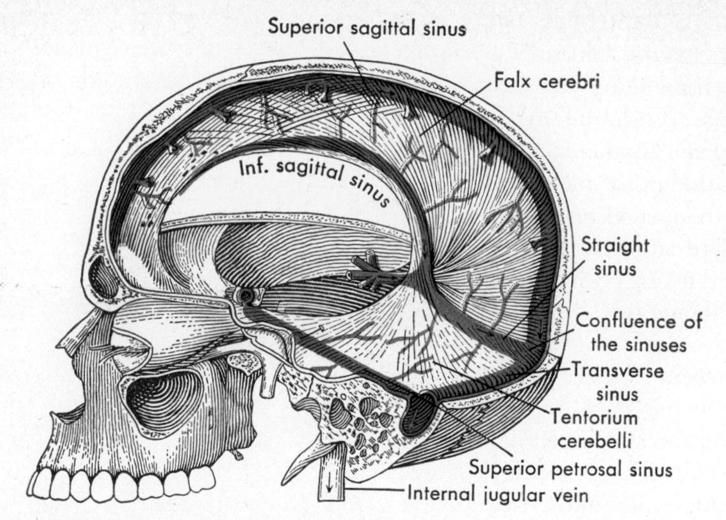

Fig. 182. Diagram showing the great blood sinuses of the head. Other structures are also shown.

covers the upper surface of the cerebellum and the under surface of the cerebrum. It is called the *tentorium cerebelli.*

3. *Central sulcus*, or *fissure of Rolando* 4. *Lateral cerebral fissure*, or *fissure of Sylvius* 5. *Parieto-occipital fissure*	There is one of each in each hemisphere. For location see Fig. 183.

Lobes of the cerebrum. The longitudinal fissure divides the cerebrum into two hemispheres, and the transverse fissure divides the cerebrum from the cerebellum. The three remaining fissures, assisted by certain arbitrary lines, divide each hemisphere in five lobes. With one exception these lobes are named from the bones of the cranium under which they lie; hence they are known as: (1) *frontal lobe,* (2) *parietal lobe,* (3) *temporal lobe,* (4) *occipital lobe,* (5) the *insula* (*island of Reil*).

THE FRONTAL LOBE is that portion of the cerebrum lying in front of the central sulcus and usually consists of four main convolutions.

THE PARIETAL LOBE is bounded in front by the central sulcus and behind by the parieto-occipital fissure.

THE TEMPORAL LOBE lies below the lateral cerebral fissure and in front of the occipital lobe.

THE OCCIPITAL LOBE occupies the posterior extremity of the cerebral hemisphere. When one examines the external surface of the hemisphere, there is no marked separation of the occipital lobe from the parietal and temporal lobes that lie to the front, but when the surface of the longitudinal cleft is examined, the parieto-occipital fissure serves as a boundary anteriorly for the occipital lobe.

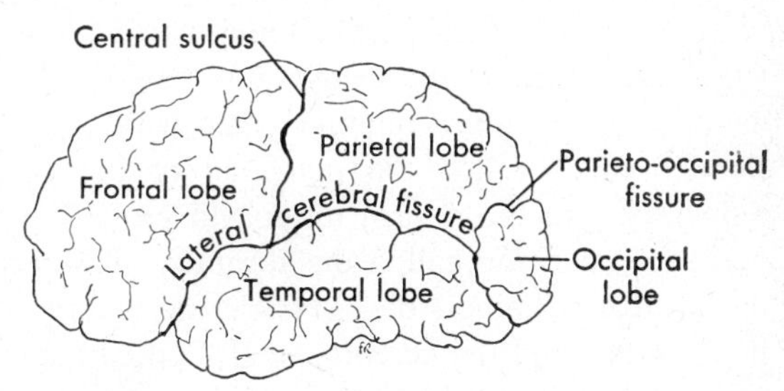

Fig. 183. Left side of brain, showing lobes and fissures.

THE INSULA (island of Reil[6]) is not seen when the surface of the hemisphere is examined, for it lies within the lateral cerebral fissure,

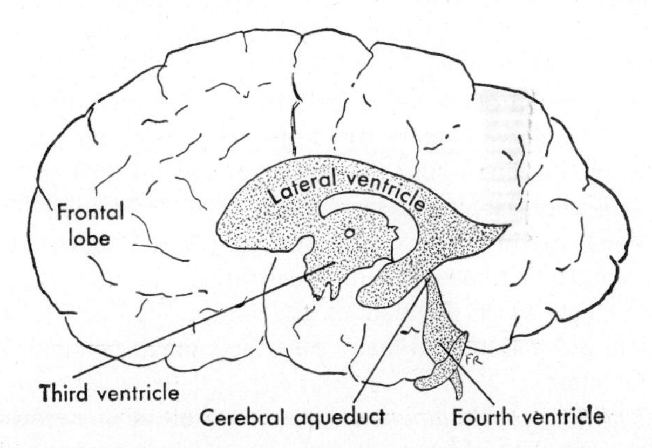

Fig. 184. Diagram showing ventricles of the brain. Side view.

and the overlying convolutions of the parietal and frontal lobes must be lifted up before the insula comes into view.

VENTRICLES OF THE BRAIN. The brain contains cavities called *ventricles*. The *two lateral ventricles* are situated one in each of the

[6] Johann Christian Reil, Dutch anatomist (1759–1813).

cerebral hemispheres under the mass of white fibers called the corpus callosum, which connects the two hemispheres. The *basal nuclei* of the brain are in the floor of the lateral ventricles. The *third ventricle* is behind the lateral ventricles but connected with each one by means of small openings called the foramina of Monro.[7] The *fourth ventricle* is in front of the cerebellum, behind the pons Varolii and the medulla. The third communicates with the fourth by means of a slender canal called the *aqueduct* of the cerebrum (aqueduct of Sylvius[8]). In the roof of the fourth ventricle are three openings, the median one being called the foramen of Magendie.[9] By means of these three openings, the ventricles communicate with the subarachnoid cavity, and the cerebrospinal fluid can circulate from one to the other. The so-called fifth ventricle is not a portion of the general cavity—not a true ventricle. It is a narrow space in front of the third, having no connection with the other ventricles. The cavity of the lateral ventricles is large enough to occupy an appreciable space and may become overdistended with cerebrospinal fluid in certain pathological conditions.

Functions of cerebrum. In the higher vertebrates the cerebrum constitutes a larger proportion of the central nervous system than in the lower forms. It is especially large in animals that are capable of profiting by experience; hence it has come to be regarded as the organ for the recall of memories of former experiences or associative memory. The nerve centers which govern all our mental activities —reason, intelligence, will, and memory—are located in the cerebrum. It is the seat of consciousness, the interpreter of sensations (correlation), the instigator and coordinator of voluntary acts, and it exerts a controlling force (both accelerating and inhibiting) upon many reflex acts which originate as involuntary. Laughing, weeping, micturition, defecation, and many other acts might be cited as examples of the latter.

Localization of brain function. As the result of numerous experiments on animals and close observation of the effects of electrical stimulation of the cerebral cortex on human individuals and clinical results of cerebral disease, physiologists have been able to localize

[7] Alexander Monro, Scottish anatomist (1697–1767).

[8] Jacobus Sylvius, French anatomist (1478–1555). A teacher of Andreas Vesalius.

[9] François Magendie, French physiologist (1783–1855).

certain areas in the brain which control motor, sensory, and other activities. Some knowledge has been gained concerning the areas in the cerebrum which are concerned with the higher mental activities. In no case, however, is the control of a function limited to a single center, for practically all mental processes involve the discharge of nervous energy from one center to another. All parts of the cerebrum are connected. Change in the nervous activity of any part disturbs the equilibrium of the whole. Any activity, therefore, is the result of all the changes throughout the whole of the cortex. No one area acts alone to govern a particular function. As Herrick[10] says, such areas "are merely nodal points in an exceedingly complex system of neurons which must act as a whole in order to perform any function whatsoever."

Names of areas. The portions of the cerebrum which govern muscular movement are known as *motor areas,* those controlling sensation as the *sensory areas,* and those connected with the higher faculties, such as reason and will, as *association areas.*

MOTOR AREA. The surface of the brain involved in the function of motion is the precentral gyrus of the frontal lobe, i.e., the gray matter immediately in front of the central sulcus. The large pyramidal cells whose fibers from the corticospinal pathways are located in the precentral gyrus and arranged so that motor cells for toe movement are located in the upper part and motor cells for face movement are located near the lateral cerebral fissure. A study of Figs. 185 and 186A illustrates the arrangement of both motor and sensory areas.

SENSORY AREA. The sensory area occupies the portions of the cerebrum behind the central sulcus and can be divided into regions like those of the motor area just in front of the sulcus. The *visual area* is situated in the posterior part of the occipital lobe and the *auditory area,* in the superior part of the temporal lobe.

The area for the sense of taste has been located deep in the fissure of Sylvius near the island of Reil. See Fig. 186A for alimentary system areas. The sense of smell is located on the medial aspect of the temporal lobe (uncus).

Research shows that there are two bilateral areas for speech, one in the superior frontal lobe and one just anterior to the precentral gyrus and above the fissure of Sylvius.

There are three other cortical areas concerned with speech in the

[10] C. J. Herrick, *An Introduction to Neurology.*

frontal, parietal, and temporal lobes. See Fig. 186B. These do not develop in both hemispheres. In right-handed persons they become more fully developed in the left hemisphere; and in left-handed persons, in the right hemisphere. The basis of language is a series of memory pictures. The mind must know and recall the names of things in order to mention them; it must have seen or heard things in

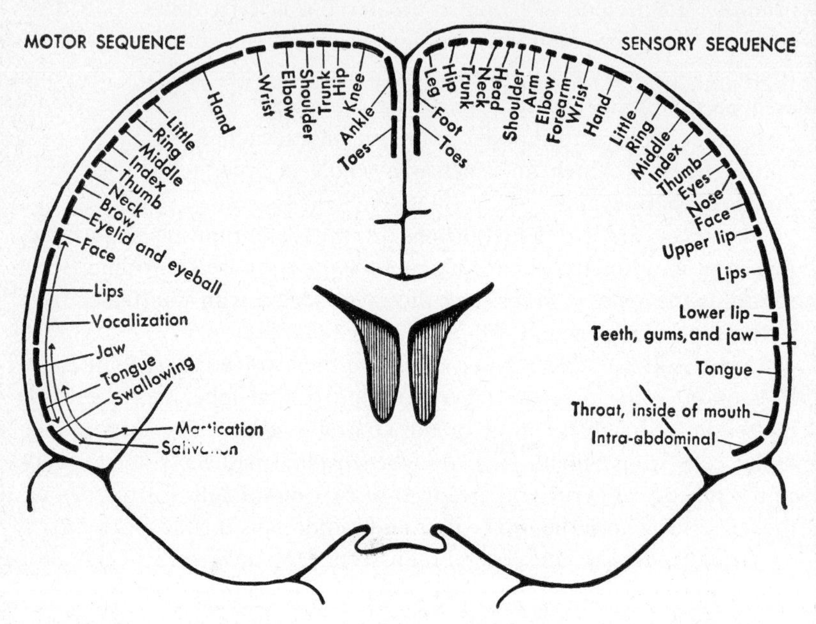

Fig. 185. Cross section of the cerebrum through the sensorimotor region with the motor and sensory sequences indicated. The lengths of the solid bars represent an estimate of the average relative cortical areas from which the corresponding responses were elicited. (W. Penfield and T. Rassmussen, *The Cerebral Cortex of Man.* Courtesy of The Macmillan Company.)

order to describe them and to have learned the words to express these ideas. Even this is not enough. All these factors must work together under the influence of the center for articulate speech, which, as seen in Fig. 186B, is in close connection with those for the larynx, tongue, and the muscles of the face. Injury to these centers results in some form of inability to speak (aphasia), to write (agraphia), or to understand spoken words (word deafness) or written words (word blindness). It is customary, therefore, to distinguish two types of aphasia, i.e., motor and sensory. By *motor aphasia* is meant the condition of those who are unable to speak although there is no paralysis of the muscles of articulation. By *sensory aphasia* is meant the condition of those who are unable to understand written, printed, or spoken symbols of words, although the sense of vision and that of

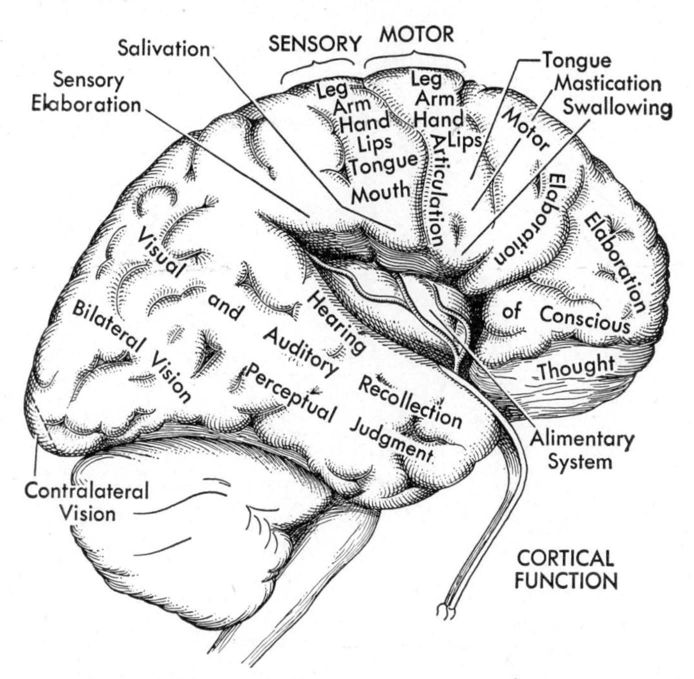

Fig. 186A. Cortical function. This illustration will serve as a summary restatement of conclusions, some hypothetical (e.g., the elaboration zones), others firmly established. The suggestion that the anterior portion of the occipital cortex is related to both fields of vision rather than to one alone is derived from the results of stimulation. (W. Penfield and T. Rasmussen, *The Cerebral Cortex of Man.* Courtesy of The Macmillan Company.)

hearing are unimpaired. These centers are really memory centers, and aphasia is due to loss of memory of words, of the meaning of words seen or heard, or of how to form letters.

ASSOCIATION AREAS. The motor and sensory areas form, so to speak, small islands which are surrounded on all sides by cerebral tissue in which as yet no definite functions have been localized. These regions are designated as association areas and are made up of association fibers which connect motor and sensory areas. Animals that are capable of acquiring habits and conditioned reflexes have a greater development of these areas. It is thought that the association areas are plastic and register the effects of individual experience.

In the absence of the cerebrum, any animal becomes a simple reflex animal. In other words, all its actions are then removed from volition, in fact, from consciousness. All responses that depend

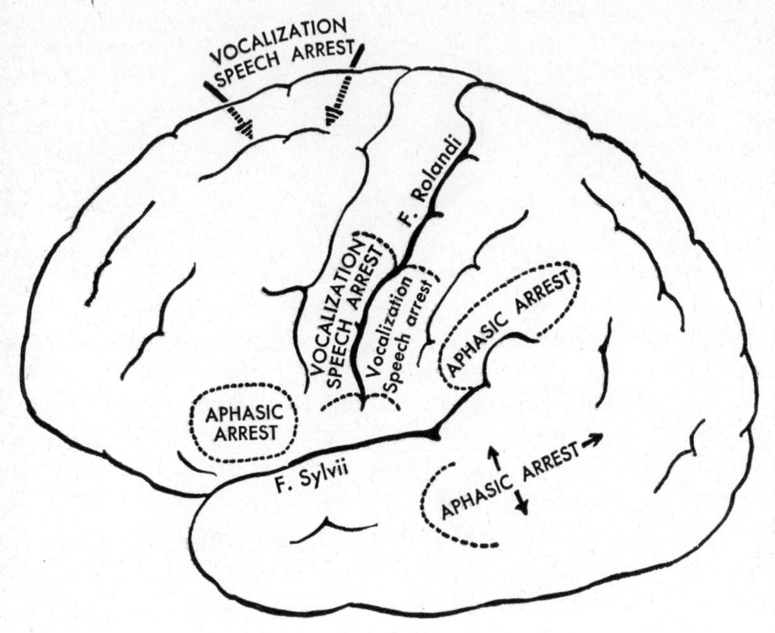

Fig. 186B. Summary of areas in which stimulation may interfere with speech or produce vocalization in the dominant hemisphere. Speech interference produced by stimulation of the superior intermediate frontal area within the longitudinal fissure has in certain cases produced evidence of aphasia rather than simple arrest, an observation that calls for further study. (W. Penfield and T. Rasmussen, *The Cerebral Cortex of Man.* Courtesy of The Macmillan Company.)

upon memory of acquired and inherited experience have been destroyed.

The midbrain is a short, constricted portion which connects the pons and cerebellum with the hemispheres of the cerebrum. It is directed upward and forward and consists of (1) a pair of cylindrical bodies called the cerebral *peduncles,* which are made up largely of the descending and ascending fiber tracts from the cerebrum above, the cerebellum, medulla, and spinal cord below; (2) four rounded eminences, called the *corpora quadrigemina* (Fig. 178), which contain important correlation centers and also nuclei concerned with motor coordination; and (3) an intervening passage or tunnel, the cerebral aqueduct (aqueduct of Sylvius), which serves as a communication between the third and fourth ventricles.

The cerebellum and its functions. The cerebellum occupies the lower and back part of the skull cavity. It is below the posterior portion of the cerebrum, from which it is separated by the tentorium cerebelli, a fold of the dura mater, and behind the pons and the upper part of the medulla. It is oval in form, constricted in the center, and flattened from above downward. The constricted central portion is called the *vermis,* and the lateral expanded portions are called the *hemispheres.*

The surface of the cerebellum consists of gray matter and is not convoluted but is traversed by numerous furrows, or sulci. The gray matter contains cells from which fibers pass to form synapses in other areas of the brain and cells with which fibers entering the cerebellum from other parts of the brain form synapses. The cerebellum is connected with the cerebrum by the *superior peduncles,* with the pons by the *middle peduncles,* and with the medulla oblongata by the *inferior peduncles* (Fig. 180). These peduncles are bundles of fibers. Impulses from the motor centers in the cerebrum, from the semicircular canals of the inner ear, and from the muscles enter the cerebellum by way of these bundles. Outgoing impulses are transmitted to the motor centers in the cerebrum, down the cord, and thence to the muscles.

The cerebellum receives sensory impulses from the semicircular canals, from the sensory centers of the body wall, and from the cerebral cortex. It sends nerve impulses into all the motor centers of the body wall and helps to maintain posture and equilibrium, and the tone of the voluntary muscles. None of its activities come into consciousness. In man, injury to the cerebellum results in muscular weakness, loss of tone, and inability to direct the movements of the skeletal muscles. There may be difficulty in walking due to inability to control the muscles of the legs or difficulty in talking due to lack of coordination of the muscles moving the tongue and jaw. The area of the body affected is determined by the location and extent of the injury to the cerebellum. Only parts of the body on the same side as the injury to the cerebellum are involved, unless both sides of the cerebellum are injured, and then the lack of muscle tone and coordination may be so great that the person is helpless.

The pons (Varolii) is situated in front of the cerebellum between the midbrain and the medulla oblongata. It consists of interlaced transverse and longitudinal white nerve fibers intermixed with gray matter. The transverse fibers are those derived from the middle

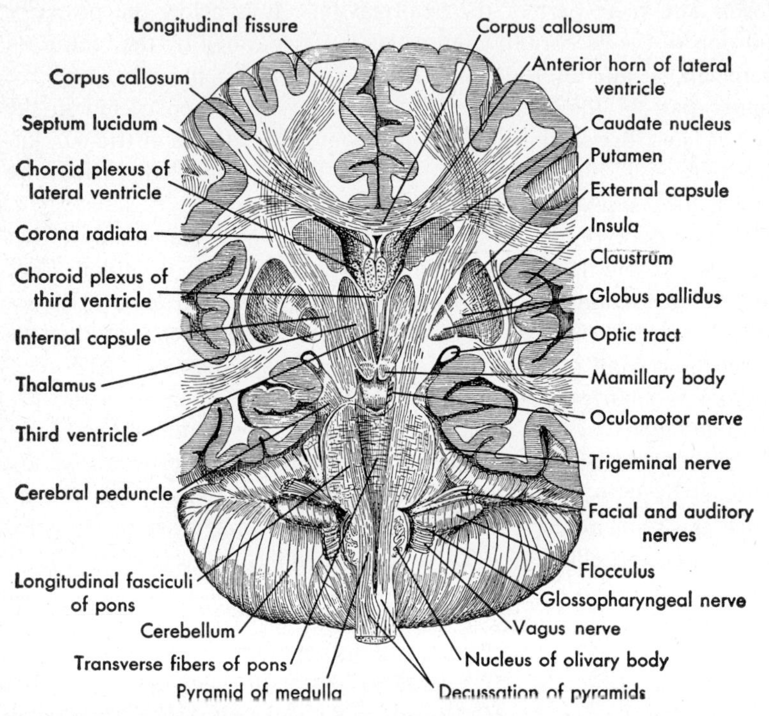

Fig. 187. Section of brain to show basal nuclei, internal capsule, pons, and medulla. (Toldt.)

peduncles of the cerebellum and serve to join its two halves. The longitudinal fibers connect the medulla with the cerebrum. In it also are the nuclei of all or a part of the fibers of the cranial nerves V, VI, VII, VIII.

Function. The pons is a bridge of union between the two halves of the cerebellum and a bridge between the medulla and the cerebrum. The fifth (trigeminal) nerve emerges from the side of the pons near its upper border. The sixth (abducent), seventh (facial), and eighth (acoustic) nerves emerge in the superficial furrow which separates the pons from the medulla in front (Fig. 179).

The medulla oblongata (spinal bulb) is continuous with the spinal cord, which, on passing into the cranial cavity through the foramen magnum, widens into a pyramid-shaped mass which extends to the lower margin of the pons. Externally, the medulla resembles the upper part of the spinal cord, but the internal structure is different. All of the afferent and efferent tracts of the spinal cord are represented in the medulla, and many of them decussate, or cross,[11] from one side to the other, whereas others end in the medulla. The nerve cells of the medulla are grouped to form *nuclei,* some of which are centers in which the cranial nerves arise. The motor fibers of the glossopharyngeal and of the vagus nerves, also the cranial portion of the accessory nerves, arise in the *nucleus ambiguus.* The hypoglossal nerve arises in the *hypoglossal nucleus.* Some of the nuclei are relay stations of sensory tracts to the brain, e.g., the *nucleus gracilis* and *nucleus cuneatus.* Some serve as centers for the control of bodily functions, e.g., the *cardiac, vasoconstrictor,* and *respiratory centers.*

Functions. The medulla serves as an organ of conduction for the passage of impulses between the cord and the brain. It contains (1) the cardiac, (2) the vasoconstrictor, and (3) the respiratory centers and controls many reflex activities.

The cardiac inhibitory center consists of a bilateral group of cells lying in the medulla at the level of the nucleus of the *vagus nerve.* The fibers from this center accompany the vagi to the heart and unite with the cardiac branches from the thoracolumbar nerves to form the *cardiac plexus* (Fig. 210), which envelops the arch and ascending portion of the aorta. From the cardiac plexus the heart receives *inhibitory fibers.* It is believed this center constantly discharges impulses which tend to hold the heart to a slower rate than it would assume if this check did not exist. The activity of the heart is also affected by nerve impulses from the *cardiac sympathetic nerves,* which increase the rate of the heartbeat and are called accelerator nerves. The inhibitory and accelerator fibers are true antagonists acting in opposite ways upon the heart.

[11] In many cases of paralysis or convulsions, it is possible to locate the portion of the brain that is affected by observation of the part of the body involved in the loss of function or in the convulsion.

The vasoconstrictor center consists of a bilateral group of cells in the medulla. Fibers from these cells descend in the cord, and at various levels they form synapses with spinal neurons in the ventral columns of gray matter. The spinal neuron serves as a preganglionic vasoconstrictor fiber, which terminates in a sympathetic ganglion (Fig. 210). The path is further continued by a postganglionic fiber. The center and the fibers are in a state of constant activity and can be stimulated reflexly through sensory nerves. One must conceive of different cells in this center being connected by definite vasoconstrictor paths with different parts of the body, e.g., the intestines or the skin. Further, the different parts of the center may be acted upon separately.

It is thought that the tonicity of this center may be increased (excited) or decreased (inhibited). The fibers which, when stimulated, cause an excitation of the vasoconstrictor center, resulting in peripheral vasoconstriction and rise of arterial pressure, are called *pressor fibers*. The fibers which, when stimulated, cause an opposite effect, i.e., decrease the tone of the center, resulting in peripheral vasodilatation and fall of arterial pressure, are called *depressor fibers*.

Vasodilator fibers are efferent fibers which, when stimulated, cause a dilation of the arteries in the region supplied. Such fibers have been demonstrated in the facial and glossopharyngeal nerves, in the sympathetic nerves, and in the pelvic nerves. There is no experimental evidence that the vasodilators are in a state of tonic activity, or that their activity is controlled from a center in the brain.

The respiratory center consists of a bilateral group of cells located in the medulla. The results of various experiments tend to support the belief that the respiratory center is automatic, that is, it possesses an inherent rhythmic activity similar to that of the heart center; but it is very sensitive to reflex stimulation. It is thought that the respiratory center is in connection with the sensory fibers of all the cranial and spinal nerves and with all the pathways between the cerebrum and the medulla. Stimulation of any of the sensory nerves of the body (e.g., by a dash of cold water, by unusual sights or sounds) or emotional states may affect the respiratory rate. The effect of the sensory nerves upon the activity of the respiratory center may be to increase or to decrease the rate or the amplitude of the respiration. Fibers which have a stimulating or augmenting effect are called respiratory pressor fibers, and those which have an inhibiting effect are called respiratory depressor fibers. It is thought that these fibers may by means of collateral connections produce cffects upon the heart and blood vessels as well as upon respirations. It is also sensitive to the chemical composition of blood.

Inasmuch as normal respiration consists of an active inspiration and a passive expiration, it has been suggested that the respiratory center should be called the inspiratory center. However, we do have active expirations independent of the respirations proper, as in coughing, laughing, or the straining of defecation, micturition, or parturition, and as an integral part of the respirations in dyspnea. In dyspnea the coordinated activity of the expiratory muscles suggests the possibility of an expiratory center. There is no definite knowledge of such a center, but if it exists, it is probably located in the medulla.

In addition to the control of respiration and circulation, many other reflex activities are effected through the medulla by means of the vagus and other cranial nerves, which originate in this region. Such reflex activities are sneezing, coughing, vomiting, winking, and the movements and secretions of the alimentary canal.

Meninges. The brain and spinal cord are enclosed within three membranes. These are named from without inward: the dura mater, arachnoid mater, and pia mater.

The dura mater is a dense membrane of fibrous connective tissue containing a great many blood vessels. The cranial and spinal portions of the dura mater differ and are described separately, but they form one complete membrane. The *cranial dura mater* is arranged in two layers which are closely connected except where they separate to form sinuses for the passage of venous blood. The outer, or endosteal, layer is adherent to the bones of the skull and forms their internal periosteum. The inner, or meningeal, layer covers the brain and sends numerous prolongations inward for the support and protection of the different lobes of the brain. These projections also form sinuses that return the blood from the brain, and sheaths for the nerves that pass out of the skull. The *spinal dura mater* forms a loose sheath around the spinal cord and consists of only the inner layer of the dura mater; the outer layer ceases at the foramen magnum, and its place is taken by the periosteum lining the vertebral canal. Between the spinal dura mater and the arachnoid mater is a potential cavity, the *subdural cavity,* which contains only enough fluid to moisten their contiguous surfaces.

The arachnoid mater is a delicate serous membrane placed between the dura mater and the pia mater. The cranial portion invests the brain loosely and, with the exception of the longitudinal fissure, it passes over the various convolutions and sulci and does not dip down into them. The spinal portion is tubular and surrounds the cord

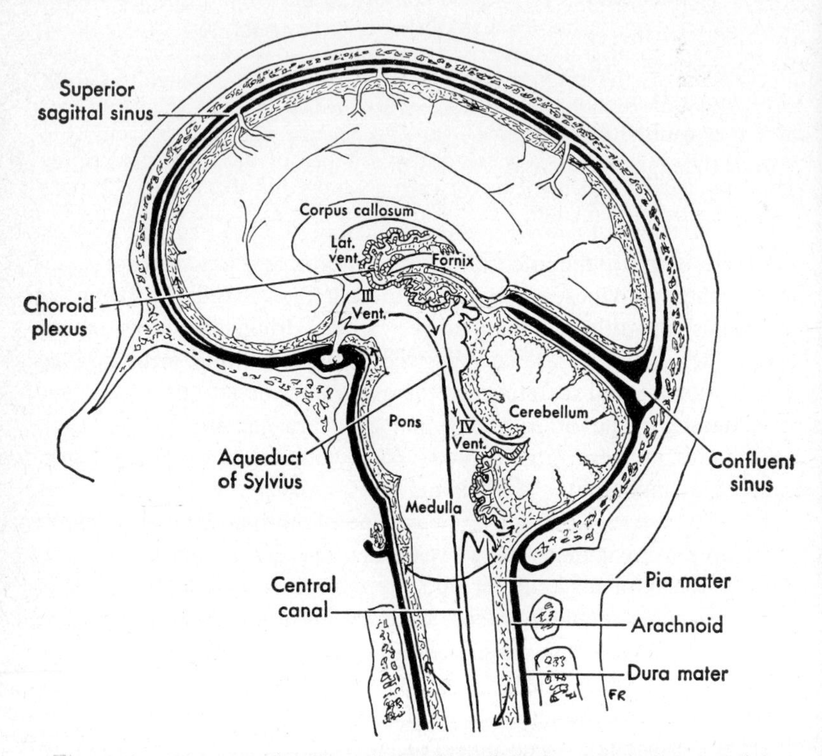

Fig. 188. Diagram showing relationship of brain and cord to meninges. The choroid plexus forms the cerebrospinal fluid. (Modified from *The Principal Nervous Pathways,* 4th ed., by Andrew T. Rasmussen, The Macmillan Company.)

loosely. The *subarachnoid cavity,* between the arachnoid mater and the pia mater, is occupied by a spongy connective tissue and intercommunicating channels in which the subarachnoid fluid is contained.

The pia mater is a vascular membrane consisting of a plexus of blood vessels held together by fine areolar connective tissue. The cranial portion invests the surface of the brain and dips down between the convolutions. The spinal portion is thicker and less vascular than the cranial. It is closely adherent to the entire surface of the spinal cord and sends a process into the ventral fissure.

The cerebrospinal fluid. The meningeal membranes and the spaces filled with fluid form a pad enclosing the brain and cord on all sides. Cerebrospinal fluid is probably secreted and diffused from the blood by the epithelial cells which cover the *choroid plexuses* of the ventricles.[12] After filling the lateral ventricles it escapes by the foramen of Monro into the third ventricle and thence by the aqueduct into the fourth ventricle. From the fourth ventricle the fluid is poured through the medial foramen of Magendie and the two lateral foramina of Luschka[13] into the subarachnoid spaces and reaches the cisterna magna. From the cisterna magna the cerebrospinal fluid may pass down the spinal canal within the arachnoid mater and return upward in the subarachnoid space. From the cisterna magna this fluid also bathes all parts of the brain. From the subarachnoid spaces it is absorbed through the villi of the arachnoid mater, which project into the dural venous sinuses; a small amount passes into the perineural lymphatics of the cranial and spinal nerves. Experimentally, it has been found that dyes added to the cerebrospinal fluid travel along the course of certain cranial nerves, especially the olfactory. This loophole affords an opportunity for the entry of infection from the nasal cavities to the cerebral cavity.

The cerebrospinal fluid is highly variable in quantity, which is usually given as from 80 to 200 cc. It is colorless, alkaline, and has a specific gravity of 1.004 to 1.008. It consists of water with traces of protein,[14] some glucose, some salts as in blood plasma, a few lymphocytes, a relatively large quantity of carbon dioxide, and some pituitary hormones. The cerebrospinal fluid serves to keep the brain and spinal cord moist, lubricates them, and protects them from varying pressures.

Infection and inflammation of the meninges of the brain will quickly spread to those of the cord. Such inflammation results in increased secretion, which, as it collects in a confined bony cavity, gives rise to symptoms of pressure, such as headache, slow pulse, slow respirations, and

[12] The choroid plexuses are highly vascular folds or processes of the pia mater, which are found in the ventricles. The capillary network is intricate.

[13] Herbert von Luschka, German anatomist (1820–1875).

[14] If the membranes of the brain or cord are inflamed, there is an increase in the protein present.

partial or complete unconsciousness. Cerebrospinal fluid may be removed by lumbar puncture. The needle by which the fluid is withdrawn is usually inserted between the third and fourth lumbar vertebrae into the subarachnoid space on the right side, the patient usually lying upon his left side with his knees drawn up, so as to arch the back. The fluid, or exudate, will contain the products of the inflammatory process and the organisms causing it. Lumbar puncture is used for (1) diagnosis of meningitis, syphilis, intracranial pressure, and cerebral hemorrhage; (2) therapeutic effect, (*a*) to relieve pressure in meningitis, hydrocephalus, uremia, and convulsions in children; (*b*) for the introduction of sera, such as antimeningitis serum, antipneumococcus serum, and tetanus antitoxin.

THE CRANIAL NERVES

Twelve pairs of cranial nerves emerge from the under surface of the brain and pass through the foramina in the base of the cranium. They are classified as motor, sensory, and mixed nerves (Fig. 179). For simplicity one nerve of each pair is described.

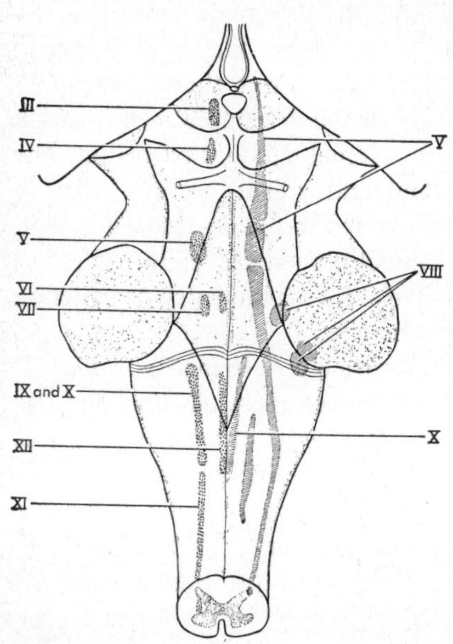

Fig. 189. Dorsal view of brain stem, showing nuclei of origin of the cranial nerves. Sensory nuclei are striped and shown on the right. Motor nuclei are dotted and shown on the left. (Modified from Toldt.)

The origin of the cranial nerves is comparable to that of the spinal nerves. The motor fibers of the spinal nerves arise from cell bodies in the ventral columns of the cord, and the sensory fibers arise from cell bodies in the ganglia outside the cord. The motor cranial nerves arise from cell bodies within the brain, which constitute their *nuclei of origin*. The sensory cranial nerves arise from groups of nerve cells outside the brain. These cells may form ganglia on the trunks of the nerves, or they may be located in peripheral sensory organs, such as the nose and eyes. The central processes of the sensory nerves run into the brain and end by arborizing around

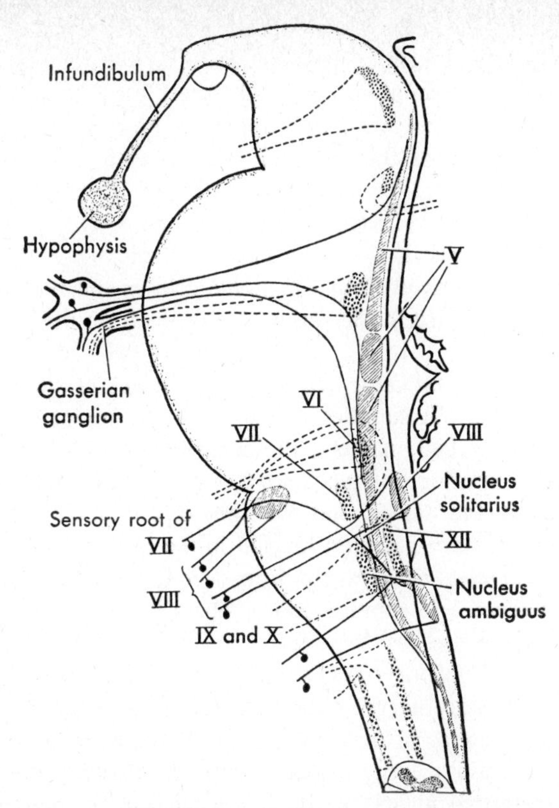

Fig. 190. Brain stem; lateral view showing hypophysis and the nuclei of origin of some of the cranial nerves. (Modified from Toldt.)

nerve cells which form their *nuclei of termination.* The nuclei of origin of the motor nerves and the nuclei of termination of the sensory nerves are connected with the cerebral cortex.

Numbers and names. The cranial nerves are named according to the order in which they arise from the brain, and also by names which describe their nature, function, or distribution.

I.	Olfactory	Sensory	VII.	Facial	Mixed
II.	Optic	Sensory	VIII.	Acoustic	Sensory
III.	Oculomotor	Motor	IX.	Glossopharyngeal	Mixed
IV.	Trochlear	Motor	X.	Vagus	Mixed
V.	Trigeminal	Mixed	XI.	Accessory	Motor
VI.	Abducent	Motor	XII.	Hypoglossal	Motor

I. The olfactory nerve is the special nerve of the sense of smell. It arises from the central or deep processes of the olfactory cells of the nasal mucous membrane, where its fibers form a network (Fig. 346), and are then collected into about 20 branches, which pierce the cribriform plate of the ethmoid bone in two groups and form synapses with the cells of the olfactory bulb. From the olfactory bulb other fibers extend inward to centers in the cerebrum.

II. The optic nerve is the special nerve of the sense of sight. It consists of fibers derived from ganglionic cells in the retina. These cells are probably third in the series of neurons from the receptors to the brain.

III. The oculomotor nerve arises from a nucleus in the floor of the cerebral aqueduct. It supplies motor fibers to four of the extrinsic muscles of the eyeball, namely, the superior rectus, inferior rectus, medial rectus, inferior oblique, and to two intrinsic muscles of the eyeball, namely the ciliaris and the sphincter pupillae. It is associated with the ciliary ganglion located in the back part of the orbit.

IV. The trochlear nerve arises from a nucleus in the floor of the cerebral aqueduct. It supplies motor fibers to the superior oblique muscle of the eye.

V. The trigeminal, or *trifacial, nerve* is the largest cranial nerve and the chief sensory nerve of the face and head. The motor fibers extend to the muscles of mastication. It emerges from the brain by a small motor and a large sensory root. The fibers of the motor root arise from two nuclei, a superior, located in the cerebral aqueduct, and an inferior, located in the upper part of the pons. It is uncertain whether the fibers from the superior nucleus are motor or sensory. The fibers of the sensory root arise from cells in the trigeminal ganglion (semilunar, or Gasserian), which lies in a cavity of the dura mater near the apex of the petrous portion of the temporal bone. The fibers from the two roots coalesce into one trunk and then subdivide into three large branches: (1) the ophthalmic, (2) the maxillary, and (3) the mandibular.

THE OPHTHALMIC BRANCH is the smallest and is a sensory nerve. It divides into three branches—the lacrimal, the frontal, the nasociliary—and communicates with the oculomotor, the trochlear, and the abducent. It supplies branches to the cornea, ciliary body, and iris; to the lacrimal gland and conjunctiva; to part of the mucous

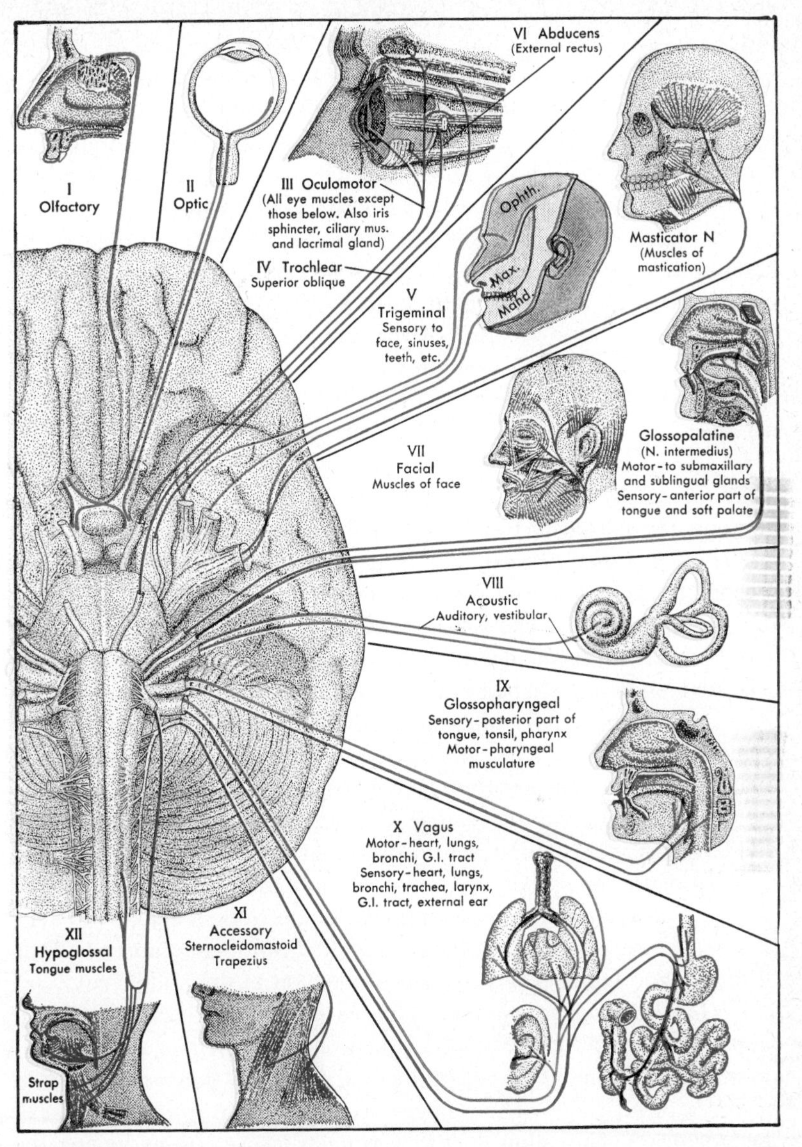

Fig. 191. Diagram showing base of brain, the emergence of the cranial nerves, their distribution to the structures, and the functions with which they are concerned. (Reproduced from full-color illustration in *Ciba Collection of Medical Illustrations,* Vol. 1, "Nervous System." Courtesy of Ciba Pharmaceutical Products, Inc., Summit, N. J.)

membrane of the nasal cavity; to the skin of the eyelid, eyebrow, forehead, and nose.

THE MAXILLARY, the second division of the trigeminal, is also a sensory nerve. It divides into many branches, which are distributed to the dura mater, the forehead, the lower eyelid, the lateral angle of the orbit, the upper lip, the gums and teeth of the upper jaw, and the mucous membrane and skin of the cheek and of the nose.

THE MANDIBULAR is the largest of the three divisions of the trigeminal. It is both a sensory and a motor nerve; it divides into many branches which are distributed to the temple, the pinna of the ear, the lower lip, the lower part of the face, the teeth and gums of the mandible, and the muscles of mastication. It also supplies the mucous membrane of the anterior part of the tongue with the lingual nerve.

VI. The abducent nerve arises in a small nucleus lying beneath the floor of the fourth ventricle. It is a motor nerve and supplies fibers to the lateral rectus muscle of the eye.

VII. The facial nerve is a mixed nerve, consisting of motor and sensory fibers. The motor fibers arise from a nucleus in the lower part of the pons. The sensory fibers arise from the geniculate ganglion on the facial nerve. The single process of the ganglionic cells divides in a T-shaped manner into central and peripheral fibers. The central fibers pass into the medulla oblongata and end in the terminal nucleus of the glossopharyngeal nerve. The peripheral fibers form the sensory root and emerge from the brain with the motor root. Behind the ramus of the mandible the facial nerve divides into many branches. Motor fibers are supplied to the muscles of the face, part of the scalp, the pinna, and muscles of the neck. Vasodilator fibers are supplied to the submaxillary and sublingual glands. Sensory fibers are supplied to the anterior two-thirds of the tongue (taste), and a few to the region of the middle ear.

VIII. The acoustic (*auditory*) *nerve* is a sensory nerve and contains two distinct sets of fibers, which differ in their origin, destination, and function. One set of fibers is known as the *cochlear nerve*, or nerve of hearing. These fibers originate in the spiral ganglion of the cochlea. The other is the *vestibular nerve*, or nerve for the maintenance of equilibrium. The fibers originate in the vestibular ganglion of the internal auditory meatus (Fig. 320).

IX. The glossopharyngeal nerve contains both sensory and motor fibers and is distributed, as its name indicates, to the tongue and pharynx. The sensory fibers arise from the superior and petrous ganglia, which are situated on the trunk of the nerve, the former in the jugular foramen, the latter in the petrous portion of the temporal bone. The motor fibers arise from the nucleus ambiguus, common to this and the tenth nerve, which is situated in the medulla. This nerve supplies sensory fibers to the mucous membrane of the fauces, tonsils, pharynx, and the posterior third of the tongue, giving the sense of taste. It also supplies motor fibers to the muscles of the pharynx and secretory fibers to the parotid gland.

X. The vagus, or *pneumogastric, nerve* has a more extensive distribution than any of the other cranial nerves, since it passes through the neck and thorax to the abdomen. It is a mixed nerve. Its motor fibers arise from the nucleus ambiguus. These fibers supply the muscles of the pharynx, larynx, trachea, heart, mouths of the large arteries and veins, aortic arch, esophagus, stomach, small intestine, pancreas, liver, spleen, ascending colon, kidneys, and visceral blood vessels. The heart is supplied with inhibitory fibers, and the gastric and pancreatic glands with secretory fibers. Its sensory fibers arise from cells of the jugular ganglion and from the ganglion nodosum located on the trunk of the nerve. The sensory fibers are distributed to the mucous membrane of the larynx, trachea, lungs, esophagus, stomach, intestines, and gallbladder.

XI. The accessory nerve is a motor nerve, consisting of two parts, cranial and spinal. The cranial part arises from the nucleus ambiguus in the medulla, and its fibers are distributed to the pharyngeal and superior laryngeal branches of the vagus. The spinal part arises from the spinal cord as low as the fifth cervical nerve, ascends, enters the skull through the foramen magnum, is directed to the jugular foramen, through which it passes, and descends to the sternocleidomastoid and trapezius muscles. Some fibers are distributed in the vagus nerve.

XII. The hypoglossal nerve arises from the hypoglossal nucleus in the medulla. It is a motor nerve supplying the muscles of the tongue and hyoid bone.

The so-called olfactory nerve is described as the olfactory lobe even in the nasal mucosa. The optic is described as the optic tract even in the retina. The auditory is described as a sensory nerve. The other cranial nerves, except perhaps the spinal accessory, are described as containing many fine sensory fibers along with the more prominent myelinated motor fibers.

THE AUTONOMIC SYSTEM

The division of the nervous system into the cerebrospinal system and the visceral system is based on a difference in function and not on an actual anatomical separation. The *visceral* system is both afferent and efferent in function. Many physiologists use the term *autonomic* in the sense of visceral. Others restrict the term *autonomic* to the *efferent* side of the *visceral* system only. Since discussions of the visceral system are almost wholly concerned with its efferent side, there is very little difference in the actual use of the terms *visceral* and *autonomic*. The visceral system possesses a certain independence of the cerebrospinal system. It regulates and controls vital activities. There is no consciousness of these activities, except as they contribute in a general way to a sense of well-being. Most of the centers controlling these processes are located within the central nervous system; but reflex and coordinating centers occur in the walls of the viscera, and these centers are capable of controlling such activities as the contractions of the stomach, the peristalsis of the intestines, glandular activity, etc., without impulses emanating from the central nervous system. Not only is this true, but some of the visceral functions can be performed quite apart from any nervous control whatever. The heart muscle contracts automatically; some of the glands are excited to secretion by chemical substances in the blood, as, for instance, the secretion of pancreatic fluid due to the stimulus of the hormone *secretion*. Even though such activities are not directly excited by the nervous system, they may be brought under the control of the nervous system, because in all the visceral functions the nonnervous and the nervous co-operate in a most intimate way.

In the autonomic (efferent visceral) system two neurons connect the central nervous system and the organ to be stimulated. The fiber of a neuron belonging to the *central nervous system* extends to an *autonomic* ganglion and ends around the dendrites of an autonomic neuron. The fiber of the second neuron passes from the ganglion to the organ to be innervated. The fiber of the first neuron is called *preganglionic* and the fiber of the second neuron is called *postganglionic*. (See Fig. 192.)

Much confusion has arisen because of the various names used to describe this system and the various ways in which it has been classified. The terms *sympathetic*, *parasympathetic*, *visceral*, *vegetative*, *splanchnic*, *involuntary*, and *autonomic* are used in different ways in various texts. Langley's classification follows: According to this classification the autonomic nerves fall into three classes, parasympathetic, sympathetic, and enteric. The parasympathetic system is called the craniosacral system. The sympathetic system is also called the thoracolumbar (thoracicolumbar) system.

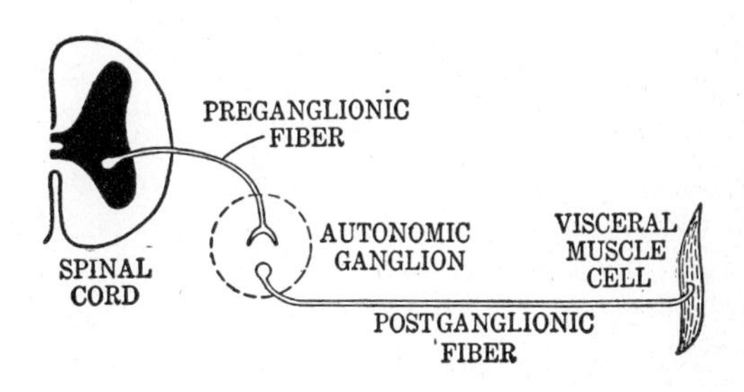

Fig. 192. Diagram to show the relation of the preganglionic and postganglionic fibers of the autonomic system.[15]

Craniosacral, or parasympathetic, system. This includes all the fibers which arise from the midbrain (tectal autonomics), from the medulla (bulbar autonomics), and from the sacral region of the cord (sacral autonomics).

The tectal autonomics arise from roots in the midbrain, send preganglionic fibers with the oculomotor nerve into the orbit, and pass to the ciliary ganglion, where they terminate by forming synapses with motor neurons whose axons (postganglionic fibers) proceed as the short ciliary nerves to the eyeballs. These fibers convey motor impulses to the circular muscle of the iris and the ciliary muscle of the eye.

The bulbar autonomics arise from roots in the medulla which emerge in the seventh, ninth, tenth, and eleventh cranial nerves. Fibers of the facial and glossopharyngeal nerves supply vasodilator fibers to the glands and blood vessels of the nose and mouth. Preganglionic fibers of these nerves terminate in some of the ganglia

[15] In some diagrams the preganglionic fibers are shown as solid lines, the postganglionic as dashed lines. In other diagrams the preganglionic fibers are shown as dashed lines, the postganglionic fibers as solid lines.

which are found in this region, namely, the sphenopalatine, otic, submaxillary, and sublingual. The autonomic fibers that arise with the vagus (and the accessory) convey motor impulses to the plain muscular tissue of the larynx, esophagus, stomach, small intestine, and part of the large intestine; secretory impulses to the stomach, liver, and pancreas; and inhibitory impulses to the heart. Preganglionic fibers of these nerves terminate in the local ganglia in or near the organs which they innervate.

The sacral autonomics include autonomic fibers which emerge from the cord. Neurons of the second, third, and fourth sacral spinal nerves send fibers to the pelvis, where they are collected to form the pelvic nerve, which proceeds to the pelvic plexus, from which postganglionic fibers are distributed to the pelvic viscera. Motor fibers pass to the smooth muscle of the descending colon, rectum, anus, and bladder. Vasodilator fibers are distributed to these organs and to the external genitals (penis, clitoris, vulva), while inhibitory fibers pass to the smooth muscles of the external genitals. The parts supplied by these nerves are indicated in Fig. 194.

Thoracolumbar, or sympathetic, system. This includes: (1) centers in the cervical, thoracic, and lumbar regions of the cord and preganglionic fibers arising in these centers; (2) the sympathetic ganglia and their fibers—the lateral chain, or sympathetic trunk; and (3) the great prevertebral plexuses. Postganglionic fibers may arise either from a ganglion in the lateral chain or from a ganglion in one of the great plexuses.

The sympathetic centers in the spinal cord are composed of groups of cells lying in the lateral columns of the gray matter of the cord. They give rise to preganglionic fibers which make their first termination in one of the sympathetic ganglia.

The sympathetic ganglia (Fig. 193) consist of paired chains of ganglia which lie along the ventrolateral aspects of the vertebral column, extending from the base of the skull to the coccyx. They are grouped as cervical, thoracic, lumbar, and sacral, and, except in the neck, they correspond in number to the vertebrae against which they lie:

Cervical	*Thoracic*	*Lumbar*	*Sacral*
3 pairs	10–12 pairs	4 pairs	4–5 pairs

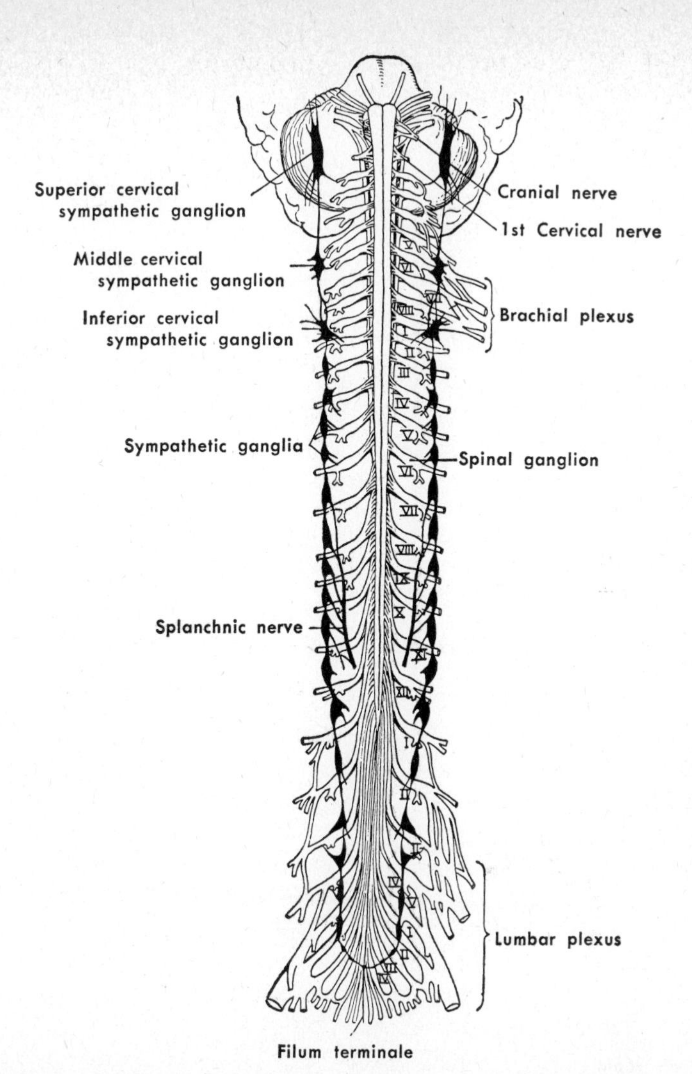

Fig. 193. Diagram of spinal cord, spinal nerves, the right and left chains of autonomic ganglia. At the top the medulla is seen, with some of the cranial nerves. The cerebellum is seen behind the medulla at the sides, and behind the cerebellum the cerebrum is shown at the sides. (From Huxley, after Allen Thomson.)

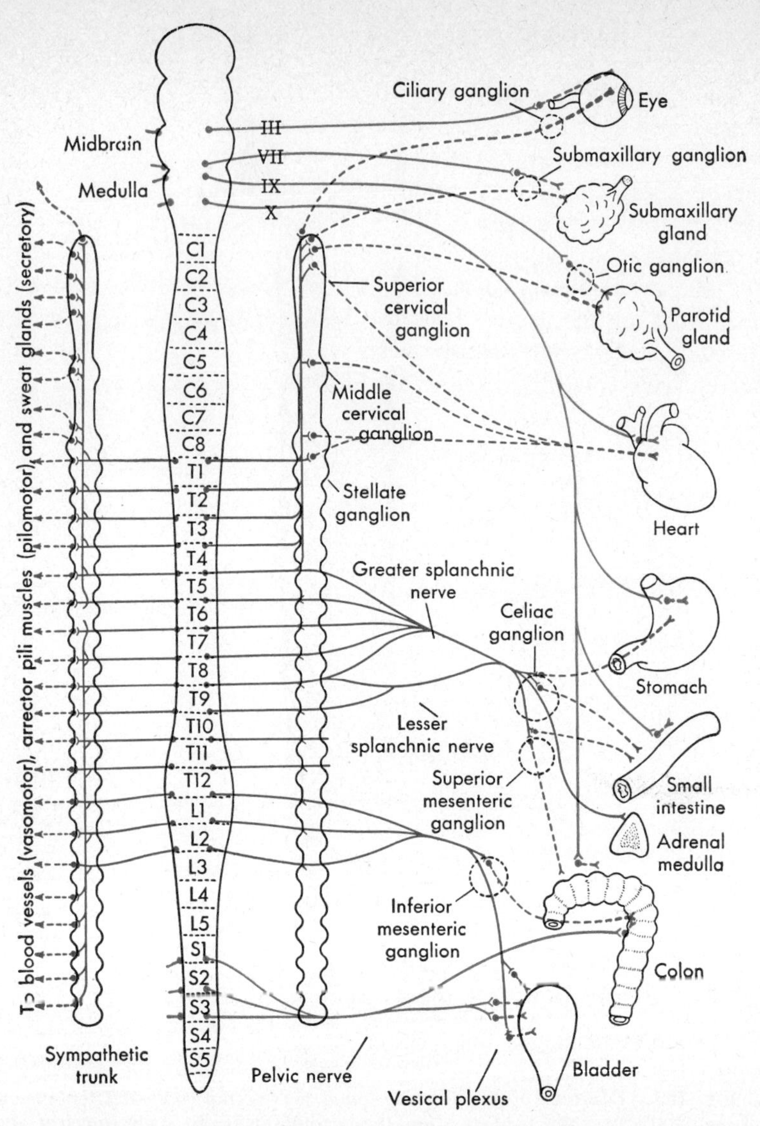

Fig. 194. Diagrammatic representation of some of the chief conduction pathways of the autonomic nervous system. For clearness, the nerves to blood vessels, arrector pili muscles, and sweat glands are shown on the left side of the figure and the pathways to other visceral structures only on the right side. The sympathetic division is shown in red, the parasympathetic in blue. Solid lines represent preganglionic fibers; broken lines represent postganglionic fibers. (Modified from *Bailey's Textbook of Histology*, 13th ed., revised by P. E. Smith and W. M. Copenhaver. Courtesy of The Williams and Wilkins Company.)

The sympathetic ganglia are connected with each other by nerve fibers called gangliated cords, and with the spinal nerves by branches which are called *rami communicantes.* In the thoracic and lumbar regions these communications consist of two rami, a white ramus and a gray ramus (Fig. 195). The white ramus consists of myelinated motor fibers passing between the central nervous system and the sympathetic. The gray ramus consists of nonmyelinated fibers that are the axons of the cells in the sympathetic ganglia distributed chiefly with the peripheral branches of all of the spinal nerves. On entering the ganglion, the fiber may end around a sym-

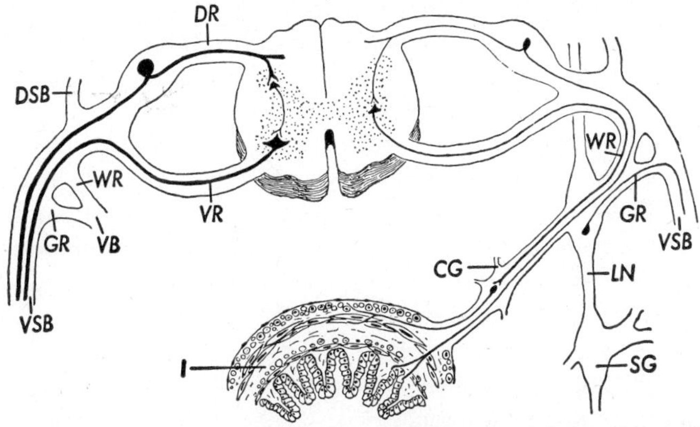

Fig. 195. Diagram of spinal cord, showing its relation to autonomic ganglia and viscera. *CG,* collateral ganglion; *DR,* dorsal root of spinal nerve; *DSB,* dorsal somatic branch of spinal nerve; *GR,* gray ramus of visceral branch of spinal nerve; *I,* intestine; *LN,* longitudinal nerve; *SG,* sympathetic ganglion; *VB,* visceral branch; *VR,* ventral root of spinal nerve; *VSB,* ventral somatic branch; *WR,* white ramus.

pathetic neuron, from which a postganglionic fiber joins the spinal nerve, or it may pass up or down in the chain for some distance before ending around a sympathetic neuron. Just as the connector elements (tracts) in the spinal cord may ascend or descend, so the connector elements in the autonomic system ascend or descend in the sympathetic trunks.

The fibers that are distributed to the skin areas of the body—head, limbs, and trunk—return by way of the gray rami to the spinal nerves and are distributed with these nerves, the final distribution being somewhat different for the different varieties of fibers—vasomotor, sweat, and pilomotor.

The fibers that are distributed to the blood vessels, glands, and walls of the viscera take a different course. For the head region the fibers, after entering the sympathetic chain, pass upward and end in the *superior cervical ganglion;* from this ganglion postganglionic fibers emerge by the various plexuses that arise from this ganglion. Fibers from the fifth to the tenth, and sometimes the eleventh, thoracic ganglia converge to form two main nerve trunks, the *great splanchnic* and the *small splanchnic.* Branches from these nerves form synapses in the celiac plexus.

The great plexuses of the thoracolumbar system consist of ganglia and fibers derived from the lateral chain ganglia and the spinal cord. They are situated in the thoracic, abdominal, and pelvic cavities, and are named the cardiac, celiac, and hypogastric plexuses.

1. THE CARDIAC PLEXUS is situated at the base of the heart, lying on the arch and the ascending portion of the aorta.

2. THE CELIAC PLEXUS (solar plexus) is situated behind the stomach, between the suprarenal glands. It surrounds the celiac artery and the root of the superior mesenteric artery. It consists of two large ganglia and a dense network of nerve fibers uniting them. It receives the greater and lesser splanchnic nerves of both sides and some fibers from the vagi, and gives off numerous secondary plexuses along the neighboring arteries. The names of the secondary plexuses indicate the arteries which they accompany and the organs to which they distribute branches.

Phrenic	Superior gastric	Spermatic
Hepatic	Suprarenal	Superior mesenteric
Splenic	Renal	Abdominal aortic
		Inferior mesenteric

These nerves form intricate networks, and any one organ may receive branches from several nerves. This increases the number of pathways and connections between the organs.

3. THE HYPOGASTRIC PLEXUS is situated in front of the last lumbar vertebra and the promontory of the sacrum. It is formed by the union of numerous filaments which descend on either side from the aortic plexus and from the lumbar ganglia; below, it divides into two pelvic plexuses.

The enteric system. This system includes the myenteric (Auerbach's)[16] and submucous (Meissner's)[17] plexuses of the

[16] Leopold Auerbach, German anatomist (1828–1897).
[17] Georg Meissner, German histologist (1829–1903).

digestive tube. They extend from the upper level of the esophagus to the anal canal. The myenteric plexus is situated between the longitudinal and circular muscular coats. The submucous plexus lies in the submucosa. These plexuses are intimately connected with each other (Fig. 316).

Functions of the autonomic system. The autonomic system innervates (1) all plain muscular tissue in the body, (2) the heart, and (3) the glands. The skeletal muscles may also receive autonomic fibers. The ganglia serve as relay stations for many of the impulses passing from the cerebrum, medulla, or spinal cord, but for the most part they act independently of these influences.

In general every organ has a double autonomic innervation, one from the thoracolumbar system and one from either the cranial or the sacral autonomic system. The functions of these two systems are usually antagonistic. With the exception of nicotine, which paralyzes all autonomic ganglia, most drugs which act on the autonomic system affect principally either the craniosacral system, as do atropine, pilocarpine, and physostigmine, or the thoracolumbar system, as do epinephrine, ergotoxine, and cocaine.

A few examples of the antagonistic action of the craniosacral and thoracolumbar systems are listed, so that the results following stimulus of these two systems may be compared.

Craniosacral	**Thoracolumbar**
(1) Contracts pupils	(1) Dilates pupils
(2) Contracts ciliary muscles, so that eyes are accommodated to see objects near at hand	(2) Lessens tone of ciliary muscles, so that eyes are accommodated to see objects at a distance
(3) Contracts bronchial tubes	(3) Dilates bronchial tubes
(4) Slows and weakens action of heart	(4) Quickens and strengthens action of heart
(5) Dilates blood vessels in nose, salivary glands, and pharynx	(5) Constricts blood vessels supplied with vasoconstrictor nerves, except arterioles of brain and lungs
(6) Increases contractions of gastrointestinal tract. Relaxes sphincters	(6) Lessens contractions of gastrointestinal tract. Constricts sphincters
(7) Increases secretions of glands	(7) Decreases secretions of glands
(8) Increases uterine contractions	(8) Lessens uterine contractions

The thoracolumbar system is strongly stimulated by pain and unpleasant excitement such as anger, fear, or insecurity. The animal responses to anger and fear are fight and flight, and the conditions brought about by stimulating the thoracolumbar system are such as to favor these responses, i.e., the bronchial tubes are relaxed and rapid breathing is rendered easier; the contraction of the blood vessels in the stomach and intestines and the increased heart action force more blood into the skeletal muscles and thus provide them with the extra oxygen and fuel needed for increased muscular activity; the supply of glucose from the liver is also increased, thus furnishing more fuel; and increased perspiration tends to lessen body heat and prevents a rise in temperature. All of these responses are closely connected with the suprarenal glands, which secrete epinephrine. This secretion is increased in amount when the thoracolumbar system is stimulated.

The coordinated activities of the autonomic system and the suprarenal glands are referred to by Cannon[18] as the "sympathetico-adrenal system" which is responsible for the homeostasis (p. 6) of the body. For this reason Cannon refers to the autonomic system as the interofective division of the nervous system and to the voluntary (central and somatic) nervous system as the exterofective division.

The neurohumoral, or *chemical mediatory, theory* of the action of nerve fibers at their effector ends, either on muscle, on gland, or at a synapse, postulates chemical substances, or mediators, as a result of nerve impulses which reach the fluid matrix at the tissue-neural area. These substances[19] are acetylcholine and adrenaline, or closely related substance quickly appearing and as quickly disappearing as a result of enzyme activity or a sequence of such activities. On this basis autonomic fibers have been classed by H. H. Dale[20] as cholinergic and adrenergic fibers.

Cholinergic	**Adrenergic**
All parasympathetic fibers Preganglionic sympathetic fibers	All postganglionic sympathetic fibers (except those to arterioles of muscles, to sweat glands, and to uterus, which are cholinergic)

[18] W. B. Cannon, *The Wisdom of the Body.*

[19] O. Loewi demonstrated that vagus inhibition of the heart is due to acetylcholine or similar substance and that cardiac acceleration is due to adrenaline or like substance.

[20] H. H. Dale and W. Feldberg, *Journal of Physiology,* 1934.

Cholinergic fibers bring about inhibition, while adrenergic fibers bring about stimulation. Dale suggests that it is possible that this classification may include the fibers of the cranial and spinal nerves as well.

Interdependence of the craniosacral and thoracolumbar systems. Marked stimulation of one system, or even part of one system, is likely to stimulate some part of the other system, thus checking excessive stimulation with the bad results that might follow. For example: stimulation of the part of the vagus that supplies the bronchial tubes may cause such marked contraction of the tubes that interference with breathing, pain, and distress may result; this in turn stimulates the thoracolumbar system to lessen the contraction of the tubes. Or another example: the afferent branch of the vagus that is connected with the heart is stimulated by marked contraction of the blood vessels, and this transmits: (1) inhibitory impulses to the heart, thus slowing its action; (2) inhibitory impulses to the vasoconstrictor center, thus lessening the contraction of the vessels.

SUMMARY

- **Spinal Cord**
  - Located in spinal canal
  - Extends from foramen magnum to second lumbar vertebra; filum terminale to first coccygeal vertebra
  - Average length about 45 cm
  - Consists of
    - Gray matter in form of H enclosed within white matter
    - On each side white matter is in funiculi
      - anterior, or ventral
      - lateral
      - posterior, or dorsal
  - **Fissures**
    - Ventral divides front portion into lateral halves
    - Dorsal sulcus divides back portion into lateral halves
  - **Transverse commissure**—connects lateral halves
  - **Central canal**—center of isthmus
  - **Membranes**
    - **Pia mater**—inner membrane, closely invests spinal cord
    - **Arachnoid mater**—middle membrane
    - **Dura mater**—outer membrane
  - **Functions**
    - Important center of reflex action for the trunk and limbs
    - Consists of the principal conducting paths to and from the higher centers in the brain

Spinal Nerves

- **Number**
  - Cervical — 8 pairs
  - Thoracic — 12 pairs
  - Lumbar — 5 pairs
  - Sacral — 5 pairs
  - Coccygeal — 1 pair
  - 31 pairs
- **Variety**
  - Medullated
  - Mixed
    - Sensory
    - Motor
- **Origin**—two roots
  - Ventral, or motor, in gray matter of cord
  - Dorsal, or sensory, in spinal ganglia
- **Distribution**—four main branches
  - Ventral supplies extremities and parts of body in front of spine
  - Dorsal supplies muscles and parts of the body in back of the spine
  - Visceral extends to lateral chain of ganglia and viscera
  - Meningeal, or recurrent, branch returns to meninges

Brain Consists of Three Hollow Vesicles in Early Embryonic Life

- (1) **From the prosencephalon, or forebrain** — The cerebral hemispheres, their commissures, the first, second, and third ventricles are developed
- (2) **From the mesencephalon, or midbrain** — The corpora quadrigemina, the cerebral peduncles, and the cerebral aqueduct are developed
- (3) **From the rhombencephalon, or hindbrain** — The medulla oblongata, the pons Varolii, and the cerebellum are developed

Brain

- Located in cranial cavity
- Covered by meninges—same as spinal cord
- **Divisions**
  - Cerebrum
    - Cortex
    - Basal nuclei
  - Cerebellum
  - Brain stem
    - Midbrain
    - Pons Varolii
    - Medulla oblongata

Average Weight of Human Brain

- Male—about 1,380 gm
- Female—about 1,250 gm

Development of Brain

- Not only a matter of growth but of forming new pathways
- Nature of brain protoplasm and use to which it is put determine length of time during which development continues
- Mental exercise tends to keep brain active

- **Cerebrum**
  - **Description**
    - Ovoid in shape
    - Fills upper portion of skull
    - Cerebral cortex and basal nuclei
    - Gray matter on outside
      - Fissures
      - Sulci
      - Convolutions
    - White matter on inside
    - **Fissures**
      - Longitudinal cerebral fissure
      - Transverse fissure
      - Central sulcus, or Rolandic
      - Lateral cerebral, or Sylvian
      - Parieto-occipital
    - **Lobes**
      - Frontal
      - Parietal
      - Occipital
      - Temporal
      - Insula, or island of Reil
    - **Ventricles**
      - Lateral ventricles (two)
      - Third ventricle
      - Fourth ventricle
      - Fifth ventricle—not a true ventricle
  - **Function**
    - Governs all mental activities
      - Organ of associative memory
      - Reason
      - Intelligence
      - Will
      - Seat of consciousness
      - Interpreter of sensations
      - Instigator of voluntary acts
      - Exerts a controlling force on reflex acts
  - **Names of areas**
    - **Motor area**—in front of central sulcus
    - **Sensory areas**
      - Behind the central sulcus
      - Visual—occipital lobe
      - Auditory—superior part of the temporal lobe
      - Olfactory, Gustatory — anterior part of temporal lobe
    - **Association areas**—cerebral tissue surrounding motor and sensory areas, in which as yet no definite functions have been localized
- **Midbrain**
  - **Description**
    - Short, constricted portion connects pons and cerebellum with the hemispheres of the cerebrum
    - Consists of
      - Pair of cerebral peduncles
      - The corpora quadrigemina
      - The cerebral aqueduct

- **Cerebellum**
  - **Description**
    - Oval in form, constricted in center
    - Central portion called vermis
    - Lateral portions called hemispheres
    - Gray matter on exterior
    - White matter in interior
    - Connected with cerebrum by superior peduncles
    - Connected with pons by middle peduncles
    - Connected with medulla by inferior peduncles
  - **Function**
    - Helps to maintain equilibrium and the tone of voluntary muscle

- **Pons Varolii**
  - **Description**
    - Situated between the midbrain and the medulla oblongata. Consists of interlaced transverse and longitudinal white fibers mixed with gray matter
  - **Function**
    - Connects two halves of cerebellum and also medulla with cerebrum
    - Place of exit for trigeminal, abducent, facial, and acoustic nerves

- **Medulla Oblongata**
  - **Description**
    - Pyramid-shaped mass, upward continuation of cord. Sensory and motor tracts of spinal cord represented. Many of them cross from one side to the other in the medulla; some end in medulla
    - Gray matter forms nuclei
    - Nuclei serve as
      - Centers in which cranial nerves arise, centers for control of bodily functions
      - Relay stations of sensory tracts to brain
  - **Function**
    - Vital centers
      - Cardiac center
      - Vasoconstrictor center
      - Respiratory center
    - Controls such reflex activities as
      - Sneezing
      - Coughing
      - Vomiting
      - Winking
      - Movements and secretions of alimentary canal

- **Meninges, or Membranes, of Brain and Cord**
  - **Cranial dura mater**—arranged in two layers. Outer layer adherent to bones of skull; inner layer covers the brain
  - **Spinal dura mater**—consists of only the inner layers, forms a loose sheath around the cord

Meninges, or Membranes, of Brain and Cord
- **Arachnoid mater**—serous membrane placed between the dura mater and pia mater of both brain and cord
- **Cranial pia mater**—vascular membrane, invests brain and dips down into crevices and depressions
- **Spinal pia mater**—is closely adherent to cord and sends a process into the anterior fissure

Cerebrospinal Fluid
- Found in meningeal spaces of brain and cord and ventricles of the brain
- Formed in choroid plexuses of the ventricles from blood
- Clear, limpid fluid, specific gravity 1.004 to 1.008
- Quantity variable, 80–200 cc
- Contains traces of protein, glucose, salts, lymphocytes, carbon dioxide, and pituitary hormones
- **Function**
  - Nutritive medium for nerve cells
  - Acts as a water bed

Autonomic, or Efferent Visceral, System
- **Cranio-sacral, or parasympathetic**
  - Ocular autonomics
    - Tectal autonomics—neurons that arise from roots in midbrain, pass to ciliary ganglia—terminate by forming synapses with motor neurons, whose axons proceed as ciliary nerves to the eyeballs
  - Oroanal autonomics
    - Bulbar autonomics—neurons arising from roots in medulla which emerge in seventh, ninth, tenth, and eleventh cranial nerves
    - Sacral autonomics—neurons of the second, third, and fourth sacral spinal nerves send preganglionic fibers to pelvis—form pelvic nerve, which proceeds to pelvic plexus
- **Thoraco-lumbar**
  - **Vertebral, or thoraco-lumbar**
    - Chain of ganglia which lie along ventrolateral aspects of vertebral column
    - **Grouped as**
      - Cervical 3 pairs
      - Thoracic 10–12 pairs
      - Lumbar 4 pairs
      - Sacral 4–5 pairs
    - **Connected**
      - (1) With each other by gangliated cords
      - (2) With spinal nerves by rami communicantes — White and gray
    - **Distributed**
      - (1) In spinal nerves to blood vessels, glands, walls of viscera, and skin areas of body

Autonomic, or Efferent Visceral, System

- **Thoracolumbar**
  - **Vertebral, or thoracolumbar**
    - **Distributed**
      - (2) Fibers from fifth to tenth or eleventh thoracic ganglia converge to form
        - Great splanchnic
        - Small splanchnic
    - Consist of masses of gray matter in thoracic and abdominal cavities
  - **Three great plexuses**
    - Form
      - Cardiac plexus
      - Celiac plexus
      - Hypogastric plexus
      - Connect with many others embedded in thoracic and abdominal viscera
- **Enteric**
  - Myenteric plexus—situated between the circular and longitudinal coats of digestive tube
  - Submucous plexus—lies in the submucosa
- **Funtions**—Brings about homeostasis. Innervates all plain muscular tissue, the heart, and the glands. Most important factor is reflex stimulation. Ganglia serve as relay stations

Craniosacral and Thoracolumbar Systems

- Many of the viscera are supplied with nerves from both craniosacral and thoracolumbar systems—functions of these two sets are often antagonistic
- These two systems are interdependent, stimulation of one system or part of one system likely to stimulate some part of the other system
- Thoracolumbar system is stimulated by intense excitement
- Craniosacral system is not
- Nicotine paralyzes all autonomic ganglia. Most drugs affect either the craniosacral of the thoracolumbar, not both

TABLE OF THE CRANIAL NERVES

Name	*Nuclei of Origin and Termination*	*Distribution*	*Function*
I. **Olfactory** (Sensory)	Central or deep process of olfactory bulb	Nasal mucous membranes	Sense of smell
II. **Optic** (Sensory)	Ganglionic cells of retina	Retina of eye	Sense of sight
III. **Oculomotor** (Motor)	Nucleus in floor of cerebral aqueduct	Superior, inferior, and medial recti; inferior oblique, ciliaris, and sphincter pupillae muscles	Motion
IV. **Trochlear** (Motor)	Nucleus in floor of cerebral aqueduct	Superior oblique of eye	Motion
V. **Trigeminal** (Sensory and motor)	Fibers of *sensory* root arise from the semilunar ganglion, which lies in cavity of dura mater near the apex of the petrous portion of the temporal bone Fibers of the *motor* root arise from superior and inferior nuclei in pons Fibers from the *two roots coalesce* into one trunk and then subdivide into (1) the ophthalmic, (2) the maxillary, and (3) the mandibular	(1) *Ophthalmic* distributes nerves to cornea, ciliary body, iris, lacrimal gland, conjunctiva, part of the mucous membrane of the nasal cavity, skin of the forehead, eyelid, eyebrow, and nose (2) *Maxillary* distributes nerves to the dura mater, forehead, lower eyelid, lateral angle of orbit, upper lip, gums and teeth of upper jaw, mucous membrane and skin of cheek and nose (3) *Mandibular* distributes branches to the temple, auricle of ear, lower lip, lower part of face, teeth and gums of mandible, and muscles of mastication. Lingual nerve to mucous membrane of anterior part of tongue	Sensation Sensation Sensation and motion Some fibers of VII reach the anterior tongue via lingual branch of V (taste)

TABLE OF THE CRANIAL NERVES—(*Continued*)

Name	*Nuclei of Origin and Termination*	*Distribution*	*Function*
VI. **Abducent** (Motor)	Nucleus beneath floor of fourth ventricle	Lateral rectus of the eye*	Motion
VII. **Facial** (Sensory and motor)	*Sensory* fibers arise from the geniculate ganglion on the facial nerve	Distributes nerves to the anterior two-thirds of the tongue (taste) and a few to the region of the middle ear	Sense of taste
	Motor fibers arise from a nucleus in the lower part of the pons	Distributes nerves to the muscles of the face, part of the scalp, the auricle, and muscles of the neck	Motion
		Vasodilator fibers are distributed to the submaxillary and sublingual glands	Secretion
VIII. **Acoustic** (Sensory) Two sets of fibers	*Cochlear* from bipolar cells in the spiral ganglion of the cochlea	To the organ of Corti	Sense of hearing
	Vestibular from bipolar cells situated in upper part of the outer end of the internal auditory meatus	To the semicircular canals	Equilibrium
IX. **Glossopharyngeal** (Sensory and motor)	*Sensory* fibers arise from the superior and petrous ganglia, situated on the trunk of the nerve, the former in the jugular foramen, the latter in the petrous portion of the temporal bone	Distributes sensory nerves to mucous membrane of fauces, tonsils, pharynx, and posterior third of tongue	Sense of taste
	Motor fibers arise from nucleus ambiguus in the medulla	Distributes motor fibers to the muscles of the pharynx, and secretory fibers to the parotid gland	Motion Secretion

* Fibers from nucleus of nerve VI communicate with nucleus of nerve IV to coordinate the activity of the rectus lateralis and rectus medialis of the two sides.

TABLE OF THE CRANIAL NERVES—(*Continued*)

Name	*Nuclei of Origin and Termination*	*Distribution*	*Function*
X. **Vagus** (Sensory and motor)	*Sensory* fibers arise from jugular ganglion and ganglion nodosum situated on trunk of nerve after it passes through the jugular foramen	Distributes sensory nerves to the mucous membrane of the larynx, trachea, lungs, esophagus, stomach, intestines, and gallbladder	Sensation
	Motor fibers arise from nucleus ambiguus in the medulla	Distributes motor nerves to larynx, esophagus, stomach, small intestine, and part of the large intestine	Motion
		Distributes inhibitory fibers to heart	
		Distributes secretory fibers to gastric and pancreatic glands	Secretion
XI. **Accessory Nerve, or Spinal Accessory** (Consists of two parts, cranial and spinal) (Motor)	*Cranial* fibers arise from nucleus ambiguus in the medulla	Distributes fibers to the pharyngeal and superior laryngeal branches of the vagus	Motion
	Spinal fibers arise from spinal cord as low as the fifth cervical nerve	Distributes nerves to the sternocleidomastoid and trapezius muscles	Motion
		Some fibers are distributed with X	
XII. **Hypoglossal**	Arises from the hypoglossal nucleus in the medulla	Distributes nerves to the muscles of the tongue	Motion

SPINAL NERVES

Spinal Nerves		Plexuses Formed	Some Main Nerves	Distribution to a Few Muscles	
Cervical	**1**			To muscles of occipital triangle	
	2	Cervical plexus C2–C4	Branches from plexus	Skin and muscles of cervical region and neck; trapezius, etc.	
			Phrenic nerve (chiefly C4)	Motor to diaphragm	
				Deltoid	Supraspinatus
	3		Branches from plexus	Pectoralis	Infraspinatus
				Rhomboides	Biceps
	4			Flexor carpi radialis	Flexor carpi ulnaris
			Median and ulnar nerve	Flexor digitorum sublimis	Flexor digitorum profundus
				Flexor pollicis longus	Flexor pollicis brevis
	5	Brachial plexus		Pronators	
	6	C5–T1		Triceps	Extensor carpi radialis
				Brachialis	Extensor carpi ulnaris
			Radial nerve	Brachioradialis	Extensor pollicis
	7			Supinator	Extensor indicis proprius
				Extensor digitorum	
	8				
	1				
				Levatores costarum	Back muscles
Thoracic	**1–12**		Intercostal nerves	Intercostal muscles	
				Abdominal muscles	
Lumbar	**1**		Femoral nerve	Iliopsoas	Quadriceps femoris
	2			Sartorius	Knee
	3	Lumbosacral T12–S3	Ventral branches of L5–S2	External rotators of thigh	
	4		Obturator nerve	Gracilis, adductor muscles	
	5		Gluteal nerve	Gluteal muscles	

SPINAL NERVES—(*Continued*)

Spinal Nerves		*Plexuses Formed*	*Some Main Nerves*	*Distribution to a Few Muscles*
Sacral	**1**	Lumbosacral T12–S3	Sciatic nerve	Biceps femoris (long head) Semitendinosus Semimembranosus
	2		Medial and posterior popliteal and tibial nerves	Posterior tibial, Flexor digitorum Gastrocnemius, Flexor hallucis longus Soleus, Small muscles of foot Plantaris
	3			
	4		Lateral popliteal and anterior tibial nerves	Biceps femoris (short head) Anterior tibialis Extensor digitorum longus and brevis Extensor hallucis longus Peroneus longus, brevis, tertius
	5			
Coccygeal	**1**			

UNIT III

The Structural and Functional Relationship for Correlation and Coordination of Internal Activities—Metabolism

CHAPTER 10

THE BLOOD
CHARACTERISTICS, COMPOSITION
FUNCTION

Each of the enormous number of living cells which make up the body is supplied with materials to enable it to carry on its activities, and at the same time materials resulting from its activities are removed. Most cells are far from the source of supplies and the organs of elimination; hence the need of a medium to distribute supplies and collect materials not needed by them. This need is met by the liquid tissues—blood, lymph, and tissue fluid—which consist of cells and an intercellular liquid.

CHARACTERISTICS

The most striking external feature of the blood is its well-known color, which is bright red, approaching scarlet in the arteries, but of a dark-red or crimson tint in the veins.

It is a somewhat viscous or sticky liquid. Its viscosity is about $4\frac{1}{2}$ to $5\frac{1}{2}$ times that of water, or it flows approximately $4\frac{1}{2}$ to $5\frac{1}{2}$ times more slowly than water under the same conditions. It is a little heavier than water; its specific gravity varies between 1.041 and 1.067.[1] The blood of men has a somewhat higher specific gravity than that of women. In general 1.058 is taken as a fair average. It has a peculiar odor, a saltish taste, a temperature of about 38° C (100.3° F), and a *p*H value of from approximately 7.38 to 7.4.[2] These ranges cover the values for both arterial and venous blood.

[1] The specific gravity of a liquid is the weight of the liquid (blood, urine, etc.) compared with the weight of an equal volume of distilled water at 15° C (60° F), the weight of the water being considered 1.000.

[2] An alkaline solution is one in which the hydroxyl ions (OH) are in excess

(*Footnote continued on p. 316.*)

Quantity of blood in the body. In the adult, blood volume has been estimated to be about $\frac{1}{12}$ or $\frac{1}{13}$ of body weight, and plasma volume about $\frac{1}{20}$ to $\frac{1}{25}$ of body weight. This means that in an individual weighing 70 kg (154 lb) there would be about 6 qt of blood. Normally there is little variation in quantity of blood. The ratio between blood quantity and tissue-fluid quantity, however, is not constant. Probably many factors cause this ratio to change. An example is body response to changes in environmental temperature. It is thought that as environmental temperature rises the quantity of blood increases in relation to the tissue fluid, whereas with decrease in environmental temperature there is increase in tissue fluid with a decrease in blood quantity.

The quantity of blood varies with age, sex, muscularity, adiposity, activity, state of hydration, condition of the heart and blood vessels, and many other factors. There are also wide and unpredictable individual variations.

In pathological conditions changes may appear; plethora, or an increase in volume, may occur in polycythemia and sometimes in anemia, etc.

Plasma volume may be measured by giving known quantities of the blue dye T1824 intravenously; after thorough mixing, samples of blood are taken to determine how much the dye has been diluted. There are other methods of determining blood volume.

Composition of blood. Seen with the naked eye, the blood appears opaque and homogeneous; but on microscopic examination, it is seen to consist of cells, or *corpuscles,* in an intercellular liquid, the *plasma.* The volume of cells and plasma is approximately equal.

- **Blood**
  - Cells
    - Erythrocytes, or red cells
    - White cells
      - Lymphocytes
      - Monocytes
      - Granulocytes
    - Platelets, or thrombocytes
  - Plasma

---

of the hydrogen ions. An acid solution is one in which the hydrogen ions (H) are in excess. A neutral solution is one in which the hydroxyl and hydrogen ions are in equal concentration. In physiology the concentration of hydrogen ions is expressed for convenience' sake as the hydrogen exponent (symbol *p*H). The potentiometer is an instrument for measuring the relative concentration of hydrogen and hydroxyl ions. Substances having a *p*H of 7 are *neutral;* above, from 7 to 14, are increasingly *alkaline;* and below, from 7 to 1, are increasingly *acid.*

Erythrocytes. Under the microscope erythrocytes are seen to be homogeneous circular disks, without nuclei, and biconcave in profile. The average size is about 7.7 μ[3] (0.0077 mm) in diameter. The area of an average cell is about 0.000128 sq mm. On microscopic examination, when viewed singly by transmitted light, they have a yellowish-red tinge. It is only when great numbers of them are gathered together that a distinctly red color is produced.

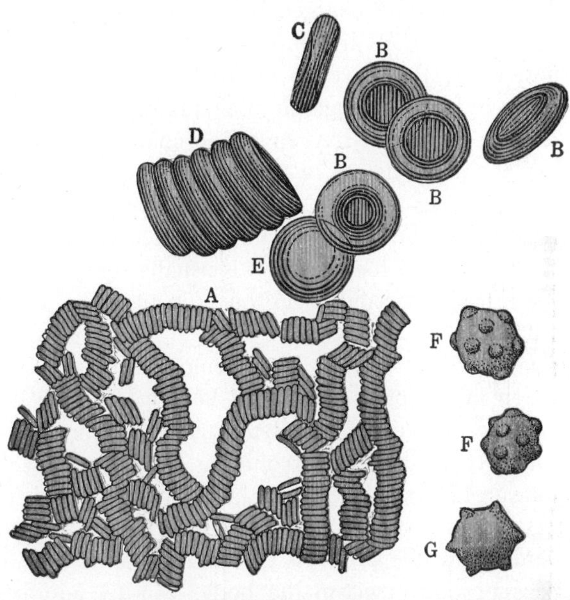

Fig. 196. Erythrocytes of the blood, magnified. *A*, moderately magnified. The erythrocytes are seen lying in rouleau. *B*, erythrocytes much more highly magnified, face view; *C*, in profile; *D*, in rouleau, more highly magnified; *E*, an erythrocyte swollen into a sphere by imbibition of water; *F*, erythrocytes puckered or crenated all over; *G*, same at edge only.

Erythrocytes consist of a colorless, filmy, elastic framework, or stroma, in which hemoglobin is deposited. They are soft, flexible, and elastic, so that they readily squeeze through apertures and passages narrower than their own diameters and immediately resume their normal shape.

Hemoglobin is a conjugated protein consisting of a protein named *globin* and a nonprotein pigment portion named *hematin*, which contains iron. The proportion of iron in hemoglobin is about 0.33 per

[3] The micron (symbol μ) equals 1/1000 of a millimeter (0.001 mm).

cent. Hemoglobin has the power to combine with oxygen to form oxyhemoglobin. In the tissues it gives up this oxygen and is then known as reduced hemoglobin.

Functions of erythrocytes. The erythrocytes have many functions, such as carrying oxygen to the tissues, carrying carbon dioxide from the tissues, and maintenance of normal acid-base balance (*p*H value), viscosity, specific gravity, etc. In the capillaries of the lungs hemoglobin becomes fully charged with oxygen, forming oxyhemoglobin. The erythrocytes carry this oxyhemoglobin to the capillaries of the tissues, where they give up the oxygen. Here the oxyhemoglobin becomes reduced hemoglobin and is ready to be carried to the lungs for a fresh supply of oxygen. The color of the blood is dependent upon the combination of the hemoglobin with oxygen; when the hemoglobin has its full complement of oxygen, the blood has a bright-red hue, and when the amount is decreased, it changes to a dark-crimson hue. The scarlet blood is usually found in the arteries and is called arterial; the dark-crimson is in the veins and is called venous blood.

Hemolysis, or laking. The loss of hemoglobin from erythrocytes and its solution in the plasma is called hemolysis. Substances that cause this action are called hemolytic agents. Hemolysis may be brought about (1) by hypotonic solutions, which diminish the concentration of substances in the plasma, (2) by the action of foreign blood serums, (3) by such agents as snake venom, the products of defective metabolism, the products of bacterial activity, or immunizing substances produced within the body, (4) by adding ether or chloroform, (5) by adding salts or fatty acids, (6) by adding bile salts, (7) by alternate freezing and thawing, (8) by amyl alcohol or saponin, and (9) by ammonia and other alkalis. Erythrocytes which have lost their hemoglobin are colorless and incapable of serving as oxygen carriers.

Number of erythrocytes. The average number of erythrocytes in a cubic millimeter of normal blood is given as 5,000,000 for men and 4,500,000 for women. This would give about 20,000,000,000,000 as the total number in the blood of the body. Since the area of one cell is approximately 0.000128 sq mm, the area of the total number of cells would be about 2,560 sq m.[4] Pathological conditions may cause a marked diminution in number, and differ-

[4] Frequently given as 25 trillion R. B. C. with a total area of 3,200 sq m.

ences have been observed in health. The number varies with altitude, temperature, the constitution, nutrition, and mode of life; with age, being greatest in the fetus and newborn child; with the time of day, showing a diminution after meals.

The instrument used for counting the number of blood cells is called a *hemocytometer.* This consists of a counting slide and diluting pipettes for counting red and white cells. If the red-cell pipette is filled to the 0.5 mark with blood and then to the 101 mark with physiological saline (Hayem's solution, etc.), the bulb will contain a 1:200 mixture of blood and saline.

The slide is ruled in both directions by parallel lines $1/20$ mm apart, and the cover glass is held $1/10$ mm above the slide (Fig. 197). If a drop of the mixture from the bulb of the pipette is introduced under the cover glass, examination through the microscope will show $1/4000$ cu mm of mixture over the area of one of the smallest squares. Of this $1/200$ is blood; therefore, the cells seen will be those contained in 1/800,000 cu mm of blood ($1/20 \times 1/20 \times 1/10 \times 1/200$). If the cells in 80 of these squares are counted and this figure is multiplied by 10,000, the result will be the number of red cells in 1 cu mm of blood; e.g., if 450 cells are counted in 80 squares, this number multiplied by 10,000 gives 4,500,000. Exact technique is necessary in making blood-cell counts. For this technique reference should be made to the laboratory manual and to textbooks of clinical diagnosis.

Polycythemia. The condition in which there is an increase of erythrocytes above the normal is called polycythemia. Conditions associated with cyanosis and residence in high altitudes are usually followed by polycythemia. It is thought that low atmospheric pressure existing in high altitudes decreases the ability of hemoglobin to combine with oxygen, and this reduction of oxygen tends to stimulate the formation of new cells. This result represents the chief benefit anemic people derive from residence in high altitudes.

In shock due to diffusion of plasma to tissues, after profuse perspiration, diarrhea, etc., there is an apparent (not real) increase in erythrocytes.

Anemia. This term is applied to conditions associated with a deficiency of erythrocytes or a deficiency of hemoglobin in the cells. A deficiency of erythrocytes results from (1) hemorrhage, (2) hemolysis, (3) inability to produce new erythrocytes due to lack of nutritious food, diseases of the bone marrow, and various infections.

Sufferers from anemia are greatly benefited by a diet rich in iron. The visceral meats, such as beef liver, gastric mucosa, heart, and brain, are rich sources of iron. Certain vegetables, such as kale, spinach, lentils,

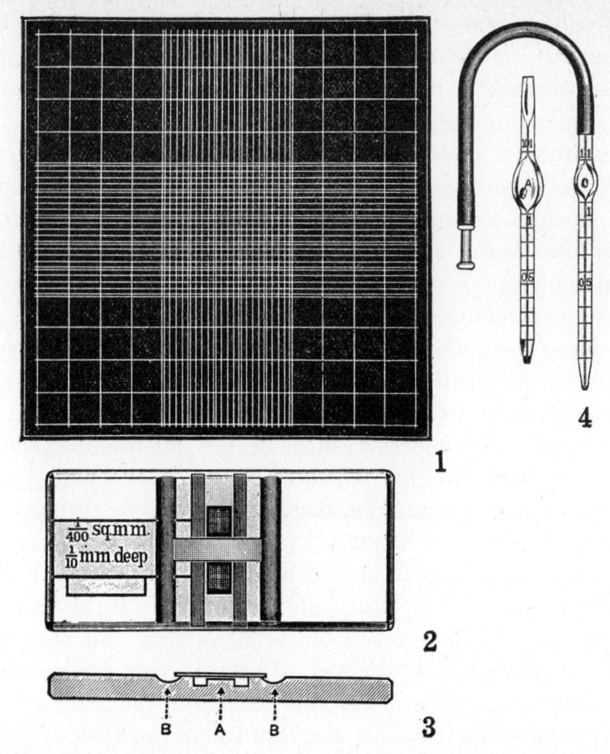

Fig. 197. A hemocytometer, consisting of a counting chamber and pipettes for making counts of blood cells. *1,* counting chamber; *2,* counting slide showing two counting chambers; *3,* edge view of counting slide; *4,* Thoma diluting pipettes. *A,* counting chamber, i.e., space between slide and cover glass—0.1 mm in depth; *B,* groove ensuring resting of cover glass on cover-glass rests. The pipette with 101 above the bulb is for diluting blood for red-cell counts; the pipette with 11 above the bulb is for diluting blood for white-cell counts. (Modified from *Practical Bacteriology, Blood Work and Animal Parasitology,* by E. R. Stitt. Courtesy of The Blakiston Company and A. H. Thomas Company.)[5]

peas, and beans, and cereals, such as oatmeal and whole-grain wheat, contain a high percentage of utilizable iron.

Hemoglobin. In adults 100 cc of normal blood contain on the average between 11.5 gm and 19 gm of hemoglobin—in males the average is 14 to 18 gm; in females, 11.5 to 16 gm. It has been

[5] There is a more complete description of the hemocytometer and directions for its use in *Laboratory Manual in Anatomy and Physiology,* 2nd ed., 1948, by C. E. Stackpole and L. C. Leavell.

suggested that 16.6 gm be taken as 100 per cent. In children (4 to 13 years) the average is 12 gm; at birth it is about 17.2 gm of hemoglobin per 100 cc of blood. The *percentage of hemoglobin* can be obtained by comparing the color of blood with standard color comparators.

Color index is an expression which indicates the amount of hemoglobin in each erythrocyte compared with the amount considered normal for the cell. The number of erythrocytes is obtained by microscopic examination, and the normal number is considered 100 per cent. The percentage of hemoglobin has been indicated just above. From these the *color index* can be derived as follows: use the percentage of hemoglobin as the numerator and the percentage of erythrocytes as the denominator of a fraction, e.g., $^{100}/_{100}$. If the number of cells is normal, it is reckoned as 100 per cent; and if the hemoglobin is reduced to 70 per cent, this gives us $^{70}/_{100}$ or $^{7}/_{10}$ and shows that the cells contain only $^{7}/_{10}$ of their normal amount of hemoglobin. Color index is significant in the anemias, especially in pernicious anemia.

The life cycle of erythrocytes is not definitely known. It is thought that they arise from endothelial cells of the capillaries of the *red marrow* of bone which by cell divisions form *erythroblasts.* An erythroblast loses its nucleus and cytoplasmic granules, becomes smaller, assumes the shape of the erythrocytes, and gradually develops hemoglobin before reaching the blood stream. Immature erythrocytes, reticulocytes, normoblasts (nucleated), etc., are sometimes found in blood. In the embryo erythrocytopoietic tissue is found also in spleen and liver. The life span of these cells is thought to be from 20 to 50 days. These calculations are based on the daily loss of bile pigments in excretions. Sections of bone marrow show many stages of blood-cell formation. When and how the erythrocytes disintegrate is not known. One supposition is that as they age they undergo hemolysis and fragmentation in the blood. Another is that they are destroyed in the spleen, lymph nodes, and liver. About 85 per cent of the iron in erythrocytes is re-used and about 15 per cent must be replaced by the diet.

For complete maturation of the red cells an antianemia principle is necessary. Substances in food form the *extrinsic factor.* The mucosa of the fundus of the stomach forms a substance called the *intrinsic factor.* These two substances are taken to the liver, where the antianemia principle is formed and stored and, as needed, taken to the bone marrow where it functions in the complete maturation of the red blood cells. Primary anemia results if the stomach fails to elaborate the intrinsic factor.

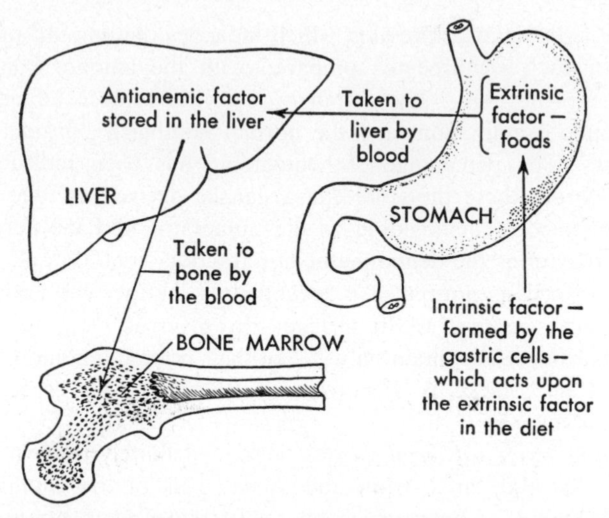

Fig. 198. Diagram showing relationship of factors responsible for normal development of the red blood cells.

White cells are minute ameboid cells, variable in size. White cells may be called *leukocytes,* or the term *leukocyte* may be restricted to granulocytes only.

The number of white cells in a cubic millimeter of blood is from 5,000 to 9,000 (about 1 white to 700 red). An increase in number is designated as *leukocytosis* and occurs in such infections as pneumonia, appendicitis, or an abscess in the body. A decrease in the number of leukocytes is designated as *leukopenia.*[6] It is a characteristic symptom of typhoid fever and tuberculosis. Physiological leukocytosis up to 10,000 occurs under normal conditions, such as digestion, exercise, pregnancy, cold baths, etc. Ten thousand or more per cubic millimeter usually indicate pathological leukocytosis.

[6] This should not be confused with *leukemia,* a disease characterized by an increase in the white cells of the blood. *Temporary* increases to 20,000 or more after exercise, etc., are thought to be due to changes in circulation.

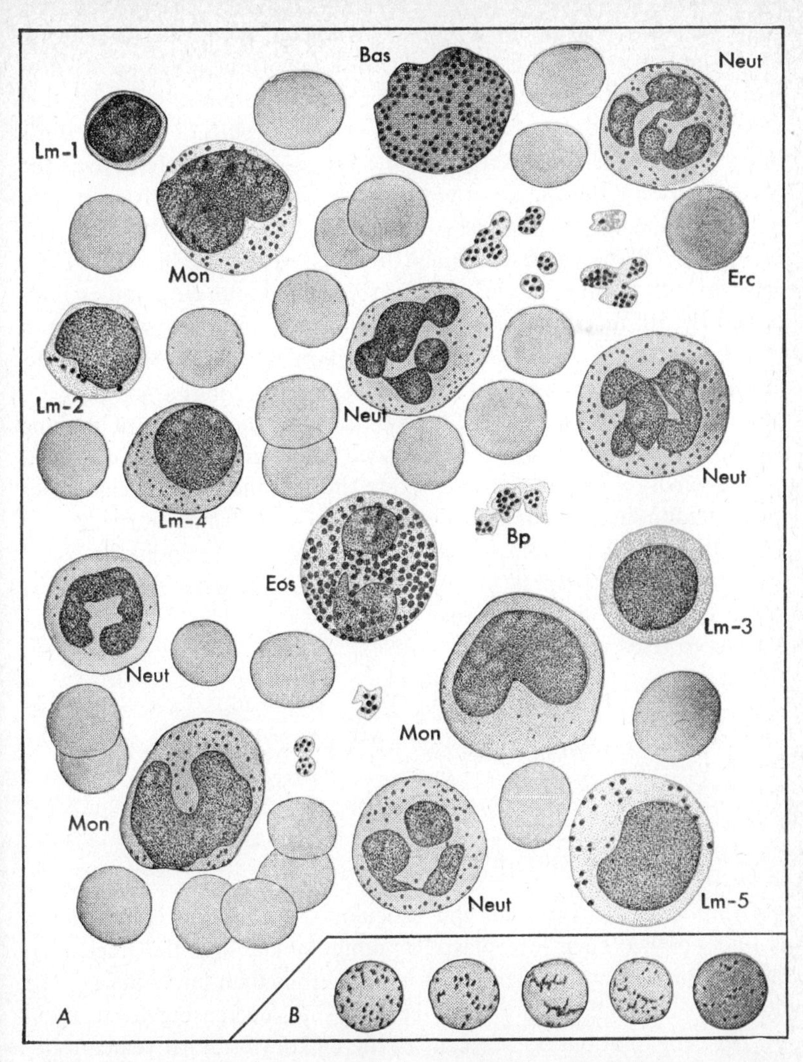

Fig. 199. *A,* cells from normal human blood. (Wright's stain.) *Bas,* basophil leukocyte; *Bp,* aggregations of blood platelets; *Eos;* eosinophil leukocyte; *Erc,* erythrocytes; *Lm 1–5,* lymphocytes (*1–3* are small and medium sizes and *4–5* are the less numerous larger forms); *Mon,* monocytes; *Neut,* neutrophil leukocytes. *B,* reticulocytes from normal human blood stained with dilute cresyl blue. (Modified from *Bailey's Textbook of Histology,* 13th ed., revised by P. E. Smith and W. M. Copenhaver. Courtesy of The Williams and Wilkins Company.)

If the white-cell pipette (Fig. 197) is filled to the 0.5 mark with blood and then to the 11 mark with $1\frac{1}{2}$ per cent acetic acid solution (Toison's solution, etc.), the bulb will contain a 1:20 dilution of blood. If a drop of the solution from the bulb of the pipette is introduced under the cover glass, the microscope will show over the area of one of the large corner squares, each of which has an area of 1 sq mm, $\frac{1}{10}$ cu mm of solution, $\frac{1}{20}$ of which is blood. The cells seen will be those contained in $\frac{1}{200}$ cu mm of blood. If the cells in four of these large squares are counted and the number multiplied by 50, the result will be the number of white cells in 1 cu mm of blood; e.g., if 140 cells are counted in four large squares, this number multiplied by 50 will give 7,000.

Ameboid movement. White cells possess the power of making ameboid movements, which has earned for them the name of "wandering" cells. This power to migrate through the walls of capillaries into surrounding tissues is called *diapedesis;* it occurs normally but is greatly stimulated and increased by pathological conditions.

Varieties of white cells. White cells may be classified in many ways, depending upon structure, cytoplasmic granules, and their reactions to dyes.

White Cells[7]	(1) **Lymphocytes** (20–25%)	*a*) Small	6–10 μ
		b) Large	
	(2) **Monocytes** (3–8%)	*a*) Mononuclear	9–12 μ
		b) Transitional	
	(3) **Granular leukocytes** (60–70%)	*a*) Neutrophils	7–12 μ
		b) Eosinophils	9–14 μ
		c) Basophils	7–9 μ

Lymphocytes arise from the reticular tissue of the lymph nodes of the body. Their cytoplasm is nongranular, and the nucleus is large. Small lymphocytes are more numerous than large ones. The number of lymphocytes is high in early life, decreasing from about 50 per cent to about 35 per cent of the leukocytes at 10 years.

Monocytes include the large mononuclear and transitional types. They are large cells, each with an indented excentric nucleus, and they can act as phagocytes.

Granulocytes, or *granular leukocytes,* show marked pseudopodial movement.

[7] All these figures variable in technical books.

1. NEUTROPHILS, OR POLYMORPHONUCLEAR LEUKOCYTES, have a nucleus that is lobulated, and the granules of the cytoplasm stain with neutral dyes. They form from 60 to 70 per cent of the total number of leukocytes. They ingest bacteria (*phagocytosis*) and in the adult are formed in the marrow tissue.

2. EOSINOPHILS are similar in size and structure to the neutrophils, but the granules of the cytoplasm are larger and stain with acid dyes such as eosin. Normally they are present in small numbers (2 to 4 per cent), but under certain pathological conditions they show a marked increase. It is thought they arise in the bone marrow, as do the neutrophils. After the administration of ACTH or cortisone there is a decrease in the number of circulating eosinophils. Therefore, eosinophil counts are important when giving these hormones.

3. BASOPHILS have a polymorphic nucleus, and the granules of the cytoplasm stain with basic dyes. They are found in small numbers ($\frac{1}{2}$ per cent).

The proportion of the different classes of leukocytes in the blood varies during diseased conditions, especially during infections. Differential counts have a great practical value in diagnosis.

Functions of the white cells. The functions of the white blood cells are many and are incompletely understood. Important functions follow: (1) They help to protect the body from pathogenic organisms. It is assumed that they either ingest bacteria and thus destroy them directly, or that they form certain substances, called *bacteriolysins,* which have the power of dissolving them. Leukocytes which ingest bacteria are called *phagocytes,* and the process is called *phagocytosis.* The neutrophils are thought to be most active in attacking bacteria, and Metchnikoff[8] called them *microphages.* According to some authorities phagocytosis is dependent upon certain substances called *opsonins,* which are found in the blood and which prepare the bacteria for ingestion by the leukocytes. (2) They cooperate in promoting tissue repair and regeneration. It is thought that the cells of connective and epithelial tissue cannot obtain material for growth directly from the blood. The leukocytes, however, can synthesize growth-promoting substances directly from the blood. It is proposed to call these substances *trephones.* (3) The white cells aid in absorption from the intestine. (4) They take part in the

[8] Elie Metchnikoff, Russian biologist (1845–1916).

clotting of blood. (5) They help to maintain the normal supply of blood proteins found in the plasma. The blood proteins are not those found in digested food, and it may be that the leukocytes assist in forming the typical blood proteins and, as a result of their metabolism, aid in keeping up the normal supply of blood proteins.

Inflammation. When tissues become inflamed either from injury or from infection, there is irritation, followed by an increased supply of blood to the part. If the irritation continues or is severe, the flow of blood slackens, and a condition of stasis, or engorgement, results. The leukocytes become active and migrate in large numbers through the walls of the blood vessels (diapedesis) into the infected tissues. Some of the blood plasma exudes, and a few erythrocytes are forced through the capillary walls. This constitutes inflammation; and the symptoms of redness, heat, swelling, pain, and loss of function are due to irritation caused by the toxins of the bacteria, to the increased supply of blood, to the engorgement of the blood vessels, and to the collection of fluid in the tissues (edema), which is spoken of as inflammatory exudate. Under these conditions a death struggle between the leukocytes and bacteria takes place. If the leukocytes win, they kill the bacteria, remove every vestige of the struggle, and find their way back to the blood. This process of recovery is described as *resolution* and is dependent upon the individual's resistance, i.e., the rapid formation of phagocytes, opsonins, etc. If the bacteria are victorious, large numbers of phagocytes and tissue cells will be destroyed, and *suppuration*, i.e., the formation of pus, ensues. *Pus* consists of dead and living bacteria, phagocytes, necrotic tissue, and material that has exuded from the blood vessels.

Also, in the case of a wound, the leukocytes accumulate in the region of the wound and act as barriers against infection. When inflammation is deep and the local symptoms cannot be observed, knowledge of the increase of the white cells is of assistance in determining the severity of the infection and the degree of resistance being offered by the body. This requires not only an absolute count, i.e., the total number of white cells in 1 cu mm of blood, but also the differential count, i.e., the relative number of each type of leukocytes, particularly the number of neutrophils. In making a *differential count,* the number of each kind of white cells in 100 is counted. A high absolute count with a high neutrophil percentage indicates severe infection and good body resistance. A high absolute count with a moderate neutrophil percentage indicates a moderate infection and good resistance. A low absolute count with a high neutrophil percentage indicates severe infection and weak resistance.

Life cycle of white cells. Little is known of the life period of white cells, which is probably short, but they are thought to be destroyed in the spleen, liver, and bone marrow.

Blood platelets, or **thrombocytes,** are disk-shaped bodies about 2 to 4 μ in diameter. In edge view they appear as short rods; in face view they appear as round plates. The average number is 300,000 per cubic millimeter of blood. When removed from the blood, they agglutinate and disintegrate very rapidly unless an anticoagulant is added.

Function. It is thought that the blood platelets bring about the clotting of blood. When exposed to air and rough surfaces (conditions accompanying a wound and hemorrhage), large numbers of blood platelets disintegrate and set free a substance known as tissue extract. This substance is essential to clotting.[9] The platelets also contain a major portion of the histamine of the blood.

The plasma of the blood is a complex fluid of a clear amber color. It contains a great variety of substances, as might be inferred from its double relation to the cells, serving as it does as a source of nutrition and as a means of removing the waste products that result from their metabolism.

Water. About 9/10 of the plasma is water. This proportion is kept fairly constant by water intake and by water output of kidney, coupled with the continual exchanges of fluid which take place between the blood, intercellular tissue fluid, and the cells.

Blood proteins. Three proteins are usually described in the plasma of circulating blood—fibrinogen, serum globulin, and serum albumin—but there are indications of others. The first two of these proteins belong to the group of globulins and hence have many properties in common. Serum albumin belongs to the group of albumins[10] of which white of egg constitutes another member. It is thought that fibrinogen, prothrombin, and serum albumin are formed in the liver. Just where serum globulin is formed is rather uncertain. It has been suggested that in the adult plasma proteins may also be formed by the disintegration of both white and red blood cells, the general tissue cells of the body, and reticuloendothelial cells of bone marrow, spleen, and liver.

The blood proteins serve to maintain osmotic characteristics of blood, give viscosity to it, and aid in regulation of the acid-base

[9] The active principle of tissue extract has been given various names; for instance, thromboplastin, thrombokinase, cytozyme (Bordet), cephalin, etc.

[10] Albumins and globulins give the same general tests; they are both coagulated by heat, and the chief difference is in their solubilities.

balance. Immune substances are associated with serum globulin. Fibrinogen is essential for blood clotting.

Nutrients. These are the end products resulting from the digestion of food—amino acids, glucose, and neutral fats. Under normal conditions amino acids are present in a small proportion; glucose and fat are present in about the same proportion, i.e., 0.08 to 0.18 per cent. Temporary increases in these amounts may follow the ingestion of a large quantity of food.

The salts found in the blood are derived from food and from the chemical reactions going on in the body. The most abundant is sodium chloride. An 0.85 per cent sodium chloride solution is isotonic with blood plasma. See also p. 41.

Other organic substances. Urea, uric acid, creatinine, purine bases, and many substances from the cells are in the blood on their way to be excreted by the kidneys or other organs of elimination.

Gases. Dissolved gases—oxygen, nitrogen, and carbon dioxide—are found in the blood. Carbonic acid is continually entering the blood from the tissues, but the blood contains certain buffer substances, i.e., sodium bicarbonate, sodium phosphate, protein, red blood cells, etc., which enter into combination with the carbon dioxide so that only a small per cent is present in simple solution.

Special substances. The blood serves as a medium to carry internal secretions and enzymes, also antithrombin, antiprothrombin, prothrombin, etc.

Antibodies. This term is applied to substances which are antagonistic to invading organisms. Recovery from many infections is due to an accumulation of these substances in the blood and to the success of the phagocytes in destroying the invading organisms. When bacteria enter the body, they stimulate the production of antibodies. Antibodies may be classified as (1) lysins, which act by dissolving organisms, (2) opsonins, which aid the white cells by sensitizing or preparing the organisms for ingestion, and (3) agglutinins, which clump the organisms in masses. Antitoxins are also classed as antibodies, because they neutralize the toxins formed by pathogenic organisms. The antibodies existing in the blood at any given time depend upon the condition of health, freedom from infection, etc.

Functions of the blood. Blood is the transporting medium of the body. The functions as commonly listed are:

It carries oxygen from the lungs to the tissues.

It carries to the tissues nutritive material absorbed from the intestine.

It carries products formed in one tissue to other tissues where they are used. In other words, it carries hormones and internal secretions.

It carries the waste products of metabolism to the organs of excretion—the lungs, kidneys, intestine, and skin.

It aids in maintaining the temperature of the body at the normal level.

It aids in maintaining the normal acid-base balance of the tissues.

It constitutes a defense mechanism against the invasion of harmful organisms.

It aids in maintaining fluid balance between blood and tissues.

It clots, preventing loss of blood after trauma.

THE CLOTTING OF BLOOD

Blood drawn from a living body is fluid. It soon becomes viscid and, if left undisturbed, forms a soft jelly. As the cells settle out of the plasma, a pale, straw-colored liquid begins to form on the

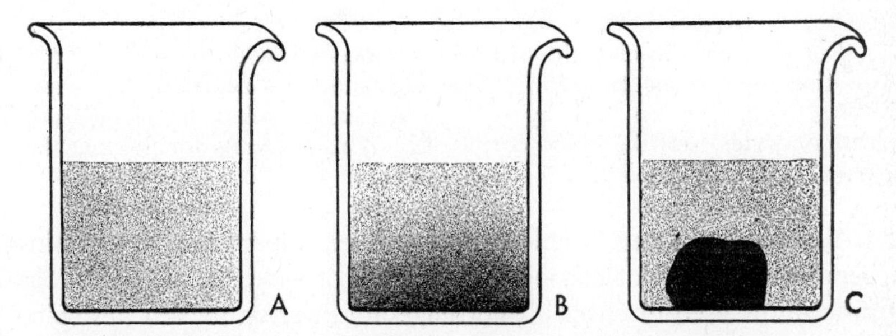

Fig. 200. The clotting of blood. Beaker *A* contains fresh blood. *B* contains recently clotted blood. *C,* after syneresis has occurred, the serum has been squeezed out.

surface, and finally the entire jelly separates into a firm mass, or *clot,* and a liquid called *blood serum.* If a portion of the clot is examined under the microscope, it is seen to consist of a network of fine needle-like fibers, in the meshes of which are entangled the red and some of the white cells. As the clot shrinks, the red cells are held more firmly by this network; but some of the white cells, owing to their power of ameboid movement, escape into the serum. The needle-like fibers are composed of fibrin. Many theories have been advanced to account for the formation of the insoluble fibrin from soluble fibrinogen. The exact process is not known, but it is thought to be comparable to the clotting of milk under the influence of rennin. Blood contains two substances—antithrombin and antiprothrombin (heparin)—concerned with preventing the clotting of blood in the blood vessels, and three substances concerned with the clotting of blood. The latter are (1) fibrinogen, (2) calcium salts, and (3) prothrombin (thrombogen). When blood clots, prothrombin and calcium salts form thrombin, and thrombin changes fibrinogen to fibrin, which is insoluble. The fibrin and the blood cells form the clot. This may be represented in diagrammatic form as follows:

Injured cells of blood and tissues → tissue extract (cephalin)
Tissue extract neutralizes antithrombin and antiprothrombin
Prothrombin + calcium + thromboplastic substance
+ platelet accelerator → thrombin
Thrombin + serum activator → active thrombin
Active thrombin + fibrinogen + platelet factor → insoluble fibrin
Fibrin + cells of blood → clot

For the blood to clot, the two agents concerned with the prevention of clotting must be neutralized. These substances are neutralized by tissue extract which is set free by the crushed tissue cells, the platelets, or thrombocytes, and the blood corpuscles. This accounts for the fact that blood clots only when tissues are wounded.

Value of clotting. This property is of importance in arresting hemorrhage, the clot closing the openings of wounded vessels. The procedures used to check hemorrhage are directed toward hastening the formation of a clot and stimulating the blood vessels to contract so that a smaller-sized clot will be sufficient.

The time it takes for the blood of human beings to clot is usually about 5 minutes.[11] In rare individuals the blood does not clot readily

[11] Estimation of coagulation time is important as a preliminary to operation when there is any reason to expect dangerous capillary oozing, as in tonsillectomies or operations upon jaundiced persons. The normal time depends on type of test used.

or at all, so that any injury or operation involving hemorrhage is dangerous. This condition is called *hemophilia.* Only males suffer from this condition. Adult females are exempt from hemophilia, but they may transmit it to their offspring.

Conditions affecting clotting. Clotting is *hastened* by:

Injury to the walls of the blood vessels.

Contact with a rough surface or any foreign substance. Clotting is hastened when gauze or a like substance is put into a wound.

The venom of certain snakes.

A temperature above 46° C (116° F) (e.g., the use of hot sponges or towels applied to a wound) hastens clotting, probably by accelerating the formation of thrombin and the chemical changes of clotting.

Rest, which tends to prevent the dislodgment of clots forming at the opening of vessels. If blood is contained in a dish, agitation hastens the disintegration of the thrombocytes and thus favors the formation of tissue extract.

Clotting is *hindered* by:

Contact with the smooth lining of the heart and blood vessels.

A deficiency of the normal calcium salts.

The addition of citrates[12] or oxalates to the blood, because they interact with the calcium.

A very low temperature. Cold hinders the formation of a clot but is often used to check hemorrhage because it stimulates the blood vessels to contract.

A deficiency or abnormal condition of thrombocytes.

Concentrated solutions of such salts as magnesium sulfate, sodium sulfate, and sodium fluoride.

Leech extracts and the venom of certain snakes.

Low fibrinogen.

Deficiency of vitamin K. This vitamin is necessary for adequate production of prothrombin. Patients with obstructive jaundice may have prolonged clotting time because bile is necessary for the absorption of vitamin K from the intestine.

Removal of fibrin. If fresh blood, before it has time to clot, is whipped with a bundle of fine rods, fibrin will form on the rods. If

[12] In transfusions the donor's blood is often rendered incoagulable as it is withdrawn by adding sodium citrate. It is then injected into the vein of the patient.

the whipping of the blood is continued until all the fibrin has been removed, the blood will have lost the power of clotting. Such blood is called *defibrinated.*

Withdrawing the blood into a container lined with a coating of oil or paraffin.

Why blood does not clot within the blood vessels. In accordance with the theory of clotting which we have considered, blood does not clot within the blood vessels because of: the absence of tissue extract and the presence of antiprothrombin and antithrombin.

Intravascular clotting. It is well known that clots occasionally form within the blood vessels. The most frequent causes are:

1. Any foreign material, even air, that is introduced into the blood and not absorbed may stimulate the formation of thrombin and a clot.

2. When the internal coat of a blood vessel is injured, as for instance by a ligature or the bruising incidental to operations, the endothelial cells are altered and may act as foreign substance. If in addition there is a stasis of blood at this point, disintegration of the blood platelets and white cells may result in the formation of thrombin and a clot. The products of bacteria and other toxic substances may injure the lining of a blood vessel and produce the same result. Inflammation of the lining of a vein is called *phlebitis.*

Thrombus and embolus. A clot which forms inside a blood vessel is called a thrombus, and the condition is called *thrombosis.* A thrombus may bc broken up and disappear, but the danger is that it may lodge in the heart or certain parts of the brain, where it blocks circulation and causes instant death. A thrombus that becomes dislodged from its place of formation is called an embolus. Such a condition is called *embolism.*

Hemorrhage. During hemorrhage blood pressure falls and the heart rate is accelerated in an effort to maintain cardiac output. The liver and spleen give up all possible blood to increase venous return. If hemorrhage is not controlled, there is further reduction in arterial blood pressure. Vasoconstriction is marked, the pulse is thready and rapid, the skin clammy and cold, the individual is restless, anxious, and air hungry. Blood flow to the tissues is decreased and the cell needs in relation to oxygen are not met. The resulting anoxia initiates processes which cause a rapid formation of a vasoexcitatory material (VEM) which constricts the arterioles, thereby reducing blood movement through the capillaries. The resulting ischemia of the muscles releases a vasodepressor material (VDM) into circulation, but its effect is masked by greater release of VEM. It is at this time or before that blood transfusions must be given if they are to be of value.

If hemorrhage has not been controlled and pressure continues to fall, the arterioles and precapillary sphincters relax, open, and more blood moves into the capillaries. The net effects of VDM and VEM are the same, blood flow through the kidneys and tissues is reduced and finally ceases. The liver releases more VDM, vasodilatation is more marked, blood moves more slowly. Excretion and destruction of VDM fails, blood stagnates in all of the capillaries, and the individual is in what is called irreversible shock due to hemorrhage. Blood transfusions at this time have little or no effect.

Regeneration of the blood after hemorrhage. During hemorrhage it is probable that a healthy individual may recover from the loss of blood amounting to 3 per cent of the body weight. Experiments on animals show that the plasma of blood regains its normal volume within a few hours after a slight hemorrhage and within 24 to 48 hours if much blood has been lost. The number of red cells and hemoglobin are restored slowly, returning to normal after a number of days or even weeks.

When the need for an increased volume of blood is urgent, hypodermoclysis, intravenous infusion, or transfusion may be practiced. *Hypodermoclysis* means the injection of fluids into the subcutaneous tissue. The fluid most frequently used is physiological salt solution, i.e., a 0.85 per cent solution of sodium chloride. The solution is introduced where there is loose tissue to favor absorption of fluid, e.g., in the subcutaneous tissues, in the abdominal wall above the crest of the ileum, and in the front of the thighs midway between the knee and hip.

Intravenous infusion is the injection of a solution directly into a vein. Physiological salt solution and various electrolyte solutions that approximate plasma concentration are used for this purpose. The solution is frequently introduced into the cephalic or the median basilic vein (vena mediana cubiti). These veins are usually the largest, the most prominent, and nearest to the surface of the arm.

The disadvantage of intravenous infusion of normal saline is that the results effected are temporary, water being rapidly lost to the tissues, rendering them edematous. Infusion of blood plasma gives more permanent and satisfying results. Dextran is used as a plasma substitute.

Transfusion is the transfer of blood from one person (the donor) to another (the recipient). This may be accomplished by (1) the direct method, in which the blood flows through tubing from a needle inserted into the donor's vein to a needle inserted into the recipient's vein, or (2) the indirect method, more frequently used, in which donor's blood is withdrawn in a flask and prevented from clotting by the use of sodium citrate, etc. It is then injected intravenously. Before blood is used, several laboratory tests are necessary: a blood-typing test, including Rh, for blood of recipient and donor should be of the same type; a test for isohemolysins in the serum,[13] which would hemolyze red cells; a Wassermann test to exclude the possibility of transmitting syphilis; etc.

[13] Some authorities consider this unnecessary, since if agglutinins are absent, it is thought hemolysins will also be absent.

BLOOD TYPING

Blood typing, or classification into groups, is dependent upon agglutination of blood cells. Figure 201 tabulates laboratory findings when blood of various types is added to serum of a known type. It also shows all combinations of agglutinogens and agglutinins possible in blood. To explain this phenomenon, blood cells are said to contain two substances called agglutinogens, designated by the capital letters A and B. Serum is said to have two agglutinins designated by *a* and *b.* Clumping occurs when an agglutinogen and an agglutinin of the same letter come into contact. These groups, or types, have been variously designated. Systems of nomenclature in current use are shown below with their relationship to the agglutinogens and agglutinins in each type. Landsteiner[14] named the groups in terms of the agglutinogens in the cells. He indicated the existence of other groups or subdivisions of these groups, for instance the MN factor.

AGGLUTININS IN SERUM

AGGLUTINOGENS	CELLS	ab	b	a	o
	O	—	—	—	—
	A	+	—	+	—
	B	+	+	—	—
	AB	+	+	+	—

Fig. 201. Blood typing.

+ denotes agglutination
— denotes absence of agglutination

A donor of group O is called a *universal donor.* Since the blood of group O has no agglutinogens, it cannot be agglutinated by agglutinins of the serum of blood of any type; and if it is introduced into the blood of a recipient very slowly, it will be so greatly diluted as not to agglutinate the recipient's red cells. A recipient whose blood is AB is called a *universal recipient,* since his blood cannot agglutinate the red cells of any group.

[14] Karl Landsteiner, Austrian physician and researcher in America (1868–1943).

The descriptive names O, A, B, AB are in terms of the agglutinogens in the cells.

Classification of Blood Groups
Landsteiner's Findings

Blood Group Name	*Agglutinogens in Cells*	*Agglutinins in Serum*
O	O	a and b
A	A	b
B	B	a
AB	AB	O

Blood types are determined by adding whole blood to serum of a known type. If the cells are agglutinated by serum *a*, agglutinogen A must be present; similarly, agglutination with serum *b* indicates blood of type B. As indicated on the figure, in practice it is necessary only to use serum *a* and serum *b* to test for the four blood groups.

The Rh factor. Landsteiner and other researchers have discovered an agglutinogen in human blood which is also present in the rhesus monkey. For this reason it is called the Rh factor. In the United States, a study of the white population shows that about 85 per cent are Rh positive and 15 per cent are Rh negative. There are several subgroups of Rh-positive blood.

In giving a blood transfusion, if an Rh-negative person receives Rh-positive blood, the recipient will develop an anti-Rh agglutinin which may cause hemolytic reaction.

It is also believed that most cases (about 90 per cent) of *erythroblastosis fetalis* is caused by the production of anti-Rh agglutinins in the mother's blood (if the mother is Rh negative and the father is Rh positive, the child will be Rh positive). It is thought that leakage of agglutinogens through fetal circulation into mother's circulation causes formation of antiagglutinins which in turn destroy the red cells of the fetus. However, not all children born of such parents develop hemolytic reactions.

Rh-negative blood contains a factor designated as the Hr factor to indicate its relationship to the Rh factor. The Hr factors are weakly antigenic. It is possible for an Rh-positive mother to develop an anti-Hr agglutinin in response to an Hr factor if the fetus is Rh negative.

Blood groups are inherited, as Mendelian dominants; therefore group O is recessive to groups A, B, and AB.

In the fetus agglutinogens are found in the red blood cells about the sixth week. At birth the concentration is about one-fifth of the adult level. Normal concentrations are reached during adolescence. Agglutinins as a rule are not present in the blood of the newborn. Specific agglutinins are formed in blood plasma within two weeks and reach the highest concentration at about 10 years of age. Agglutinin concentration is variable in all individuals at all ages. Once established, blood groups do not change—that is, once a group B always a group B.

SUMMARY

- **Blood**
  - **Description**
    - Color
      - Bright red in arteries
      - Dark red in veins
    - Sticky fluid
    - Specific gravity varies between 1.041 and 1.067
    - Reaction varies from *p*H 7.38–7.4
    - Temperature, 38° C
    - Peculiar odor. Salty taste
    - $\frac{1}{13}$ of the body weight
  - **Composition**
    - Cells, about $\frac{1}{2}$ of volume
      - Erythrocytes
      - Leukocytes
        - Lymphocytes
        - Monocytes
        - Granular leukocytes
      - Blood platelets, or thrombocytes
    - Plasma, about $\frac{1}{2}$ of volume
      - Intercellular liquid
- **Erythrocytes, or Red Cells**
  - **Description**
    - Biconcave disks about 7.7 μ in diameter
    - Stroma containing hemoglobin
      - Nonprotein, named hematin
      - Protein, named globin
    - Have no nuclei
    - Soft, flexible, and elastic
  - **Number**
    - Cubic millimeter of blood contains about
      - 5,000,000 for men
      - 4,500,000 for women
      - Varies even in health
    - Polycythemia—increase above normal
    - Anemia—deficiency of erythrocytes or deficiency of hemoglobin in the cells
  - **Functions**
    - Oxygen carriers
    - Maintenance of viscosity, *p*H value, etc.
  - **Hemolysis**—loss of hemoglobin from the erythrocyte is called hemolysis

- **Erythrocytes, or Red Cells**
  - **Life cycle**
    - Before birth—originate in liver, spleen, and red marrow
    - After birth may originate in endothelial cells of blood capillaries of the red marrow of bones
    - Lose their nuclei before being forced into the circulation, which suggests that their term of existence is short
    - Disintegrate
      - (1) Undergo hemolysis in blood stream
      - (2) Destroyed in spleen, lymph nodes, and liver

- **White Cells**
  - **Description**
    - Minute masses of nucleated protoplasm
    - Variable in size, sometimes smaller than red cells; majority are larger
    - Gray in color
  - **Number**
    - Cubic millimeter of blood
      - 5,000 to 9,000
      - Increase = leukocytosis
      - Decrease = leukopenia
  - **Varieties**
    - Lymphocytes
      - *a*) Small
      - *b*) Large
    - Monocytes
      - *a*) Large mononuclear
      - *b*) Transitional
    - Granular leukocytes
      - *a*) Polymorphonuclear, or neutrophils
      - *b*) Eosinophils, or acidophils
      - *c*) Basophils
  - **Functions**
    - Supposed to be different for different forms
    - (1) Protect the body from pathogenic bacteria
    - (2) Promote tissue repair
    - (3) Aid in absorption from intestine
    - (4) Take part in the clotting of blood
    - (5) Probably help to form typical blood proteins
  - **Life cycle**
    - Lymphocytes arise from the reticular tissue of the lymph nodes of the body
    - Granular leukocytes arise from cells of bone marrow
    - Numbers lost in
      - (1) Battles against bacteria
      - (2) Hemorrhage
      - (3) Formation of granulation tissue or tissue regeneration

- **Inflammation**
  - (1) Irritation resulting from injury or infection
  - (2) Increased supply of blood
  - (3) Engorgement of blood vessels
  - (4) Migration of white cells
  - (5) Exudation of plasma
  - (6) Erythrocytes forced through capillary walls

- **Inflammation**
  - **Symptoms**
    - Redness
    - Heat
    - Swelling
    - Pain
    - Loss of function
  - **Result**
    - *a*) Resolution—white cells destroy bacteria, clear up debris, and return to blood
    - *b*) Suppuration—bacteria destroy white cells and form pus
    - *c*) Pus consists of
      - Bacteria
        - Dead
        - Living
      - Phagocytes
      - Disintegrated tissue cells
      - Exudate from blood vessels
- **Blood Platelets, or Thrombocytes**
  - **Description**
    - Disk-shaped bodies. Always smaller than red or white cells
    - Origin, fate, and function are open questions
    - Assist in clotting of blood
- **Plasma**
  - **Water**—about 90%
  - **Blood proteins**
    - Fibrinogen
    - Serum globulin
    - Serum albumin
  - **Nutrients**
    - Amino acids
    - Glucose
    - Neutral fats
  - **Salts**
    - Chlorides, Sulphates, Phoshates, Carbonates of Sodium, Calcium, Magnesium, Potassium, Iron — Nutrients, wastes, or regulatory substances
  - **Other organic substances**
    - Purine bases[15]
    - Urea
    - Uric acid
    - Creatine
    - Creatinine
  - **Gases**
    - Oxygen
    - Carbon dioxide—waste or regulatory
    - Nitrogen
  - **Special substances**
    - Internal secretions—antithrombin—antiprothrombin
    - Enzymes—prothrombin
  - **Antibodies**
    - Lysins—dissolve bacteria
    - Opsonins—prepare bacteria for ingestion
    - Agglutinins—clump bacteria in masses
    - Antitoxins—neutralize toxins

[15] This list is by no means complete.

- **Functions of Blood**
  - Carries oxygen from lungs to tissues
  - Carries food material to tissues
  - Carries hormones and internal secretions
  - Carries waste products to organs of excretion
  - Aids in maintaining normal temperature
  - Aids in maintaining acid-base balance of tissues
  - White cells constitute defense mechanism against infection
  - Aids in maintaining internal fluid pressure
  - Clots, preventing loss of blood after trauma
- **Clotting**
  - **Serum**—blood minus fibrin and cells
  - **Theory**
    - Blood contains two substances concerned with preventing the clotting of blood in the blood vessels, namely, antithrombin and antiprothrombin
    - Blood contains three substances concerned with the clotting of blood, namely, fibrinogen, calcium salts, and prothrombin
  - **Process**
    - Cellular elements of blood and tissues → tissue extract
    - Tissue extract neutralizes antithrombin and antiprothrombin
    - Prothrombin + calcium + thromboplastic substance + platelet accelerator → thrombin
    - Thrombin + serum activator → active thrombin
    - Active thrombin + fibrinogen + platelet factor → insoluble fibrin
    - Fibrin + cells of blood → clot
  - **Value**—checks hemorrhage
  - **Hastened by**
    - Injury to the walls of the vessels
    - Contact with a rough surface or any foreign material
    - The venom of certain snakes
    - A temperature above 116° F
    - Rest
    - Agitation
  - **Hindered by**
    - Contact with smooth lining of vessels
    - A deficiency of the normal calcium salts
    - The addition of citrates or oxalates to the blood
    - A very low temperature
    - A deficiency or abnormal condition of the thrombocytes
    - Concentrated solutions of magnesium sulphate, sodium sulphate, and sodium fluoride
    - Deficiency of vitamin K
    - Leech extracts and the venom of certain snakes
    - Low fibrinogen
    - Removal of fibrin
    - Reception in vessel coated with oil or paraffin

Intravascular Clotting
- **Theory to account for absence of clotting**
  - Absence of tissue extracts
  - Presence of antithrombin and antiprothrombin
- **Causes**
  - Any foreign material introduced into blood and not absorbed will stimulate clotting
  - Injury to internal coat of blood vessels
- **Thrombus**—name given to clot which forms inside vessel
- **Embolus**
  - A thrombus that has become dislodged from place of formation

Regeneration of Blood after Hemorrhage
- Plasma is regenerated rapidly, red cells within a few days or weeks

Treatments to Increase Volume of Blood
- **Hypodermoclysis**—injection of fluids into subcutaneous tissue
- **Intravenous infusion**—injection of solution into vein
- **Transfusion**—transfer of blood of one person to another

Blood Typing
- *Unknown blood* to be typed, after dilution with citrated saline, is added to *known b serum* and *a serum*
- **Unknown belongs to**
  - Group O—no agglutination
  - Group A—agglutination in *a* serum only
  - Group B—agglutination in *b* serum only
  - Group AB—agglutination in *b* serum and *a* serum
- Group O—universal donor
- Group AB—universal recipient

Rh Factor
- Individuals
  - 85% positive
  - 15% negative

Hr Factor—Present in individuals who are Rh negative

CHAPTER 11

THE BLOOD VASCULAR SYSTEM—ANATOMY OF THE HEART, ARTERIES, CAPILLARIES, VEINS

The blood circulates continuously throughout the body in a network of blood vessels. It is driven along these blood vessels by the action of the heart, which is placed in the center of the vascular system. The arteries conduct the blood out from the heart and distribute it to the different parts of the body; the veins bring it back to the heart again. From the arteries the blood flows through a network of minute vessels, the capillaries, into the veins. The whole forms a closed system of tubes.

HEART

The heart is a hollow, muscular organ, situated in the thorax between the lungs and above the central depression of the diaphragm.

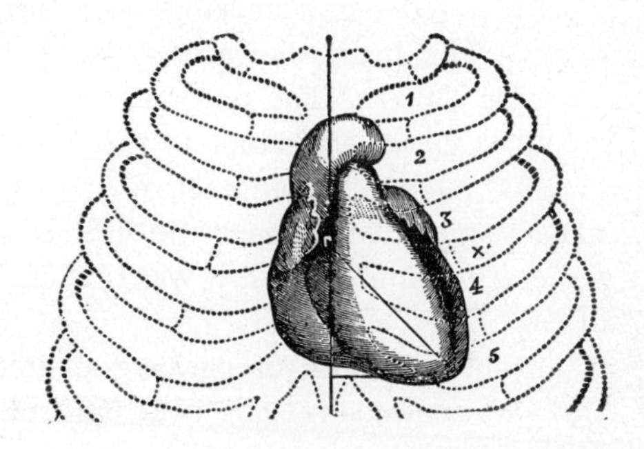

Fig. 202. Heart *in situ; 1, 2, 3, 4, 5,* intercostal spaces; vertical line represents median line. The space outlined by the triangle indicates the superficial cardiac region; *x* shows the location of the nipple on the fourth rib. (Dalton.)

It is about the size of the closed fist, shaped like a blunt cone, and so suspended by the great vessels that the broader end, or base, is directed upward, backward, and to the right. The pointed end, or apex, points downward, forward, and to the left. As placed in the body, it has an oblique position, and the right side is almost in front of the left. The impulse of the heart against the chest wall is felt in the space between the fifth and sixth ribs, a little below the left nipple, and about 8 cm (3 in.) to the left of the median line.

Pericardium. The heart is covered by a membranous sac called the *pericardium*. It consists of two parts: (1) an external fibrous portion and (2) an internal serous portion.

1. *The external fibrous* pericardium is composed of fibrous tissue and is attached by its upper surface to the large blood vessels which emerge from the heart. It covers these vessels for about 3.8 cm ($1\frac{1}{2}$ in.) and blends with their sheaths. The lower border is adherent to the diaphragm, and the front surface is attached to the sternum.

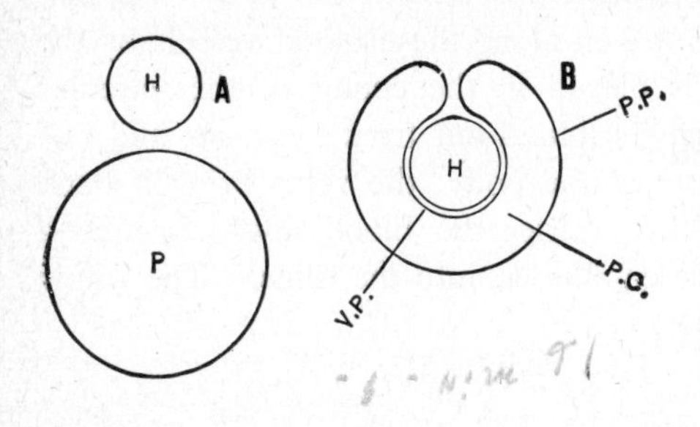

Fig. 203. Diagram of the heart and serous pericardium. *A* shows heart and pericardium lying separately; *B* shows the pericardium invaginated by the heart; *P.C.*, pericardium cavity which actually is a very narrow space filled with pericardial fluid; *P.P.*, parietal layer that lines the fibrous pericardium; *V.P.*, visceral layer that clings close to the heart muscle.

2. *The internal,* or *serous, portion* of the pericardium is a completely closed sac; it envelops the heart and lines the *fibrous* pericardium. The heart, however, is not within the cavity of the closed sac (Fig. 203). The portion of the serous pericardium which lines it and is closely adherent to the heart is called the *visceral* portion (*viscus*, an organ); the remaining part of the serous pericardium, namely, that which lines the fibrous pericardium, is known as the *parietal* portion (*paries*, a wall). The visceral and parietal portions of this serous membrane are everywhere in contact. Between them is a small quantity of pericardial fluid preventing friction as their surfaces continually slide over each other with the constant beating of the heart.

Endocardium. The inner surface of the cavities of the heart is lined by serous membrane called *endocardium*. It covers the valves, surrounds the chordae tendineae, and is continuous with the lining membrane of the large blood vessels. Inflammation of the endocardium is called *endocarditis*.

Myocardium. The main substance of the heart is cardiac muscle, called *myocardium*. This tissue includes the muscle bundles of (1) the atria (auricles), (2) the ventricles, and (3) the atrioventricular bundle (of His).[1] Inflammation of the myocardium is known as myocarditis.

1. The principal muscle bundles of the atria radiate from the area which surrounds the orifice of the superior vena cava. One, the interatrial bundle, connects the anterior surfaces of the two atria. The other atrial muscle bundles are confined to their respective atria, though they merge more or less.

2. The muscle bundles of the ventricles[2] begin in the atrioventricular fibrous rings. They form U-shaped bundles with the apex of the U toward the apex of the heart. There are many of these bundles, but for general description they may be divided into two groups. One group begins at the *left* atrioventricular ring, passes toward the right and the apex, where it forms whorls, and then ends either in the left ventricular wall, the papillary muscles, or the septum and the right ventricular wall.

The other bundles repeat this path except that they start at the *right* ventricular ring, pass to the left in the anterior wall of the heart, and end in the same structures as above.

3. The muscular tissue of the atria is not continuous with that of the ventricles. The walls are connected by fibrous tissue and the atrioventricular bundle of modified muscle cells. This bundle arises in connection with the atrioventricular (A-V) node, which lies near the orifice of the coronary sinus in the right atrium. From this node the atrioventricular bundle passes forward to the membranous septum between the ventricles, where it divides into right and left bundles, one for each ventricle. In the muscular septum between the ventricles each bundle divides into numerous strands, which spread over the internal surface just under the endocardium. The greater part of the atrioventricular bundle consists of spindle-shaped muscle cells. The significance of the atrioventricular bundle is discussed on p. 412.

[1] Wilhelm His, Jr., German anatomist (1864–1934).

[2] If more details are needed see textbooks of anatomy. Toldt, Gray, Morris-Shaeffer, Cunningham, etc.

The cavities of the heart. The heart is divided into a right and a left half, frequently called the right heart and the left heart, by a muscular partition, the ventricular septum, which extends from the base of the ventricles to the apex of the heart. The atrial septum is inconspicuous. The two sides of the heart have no communication

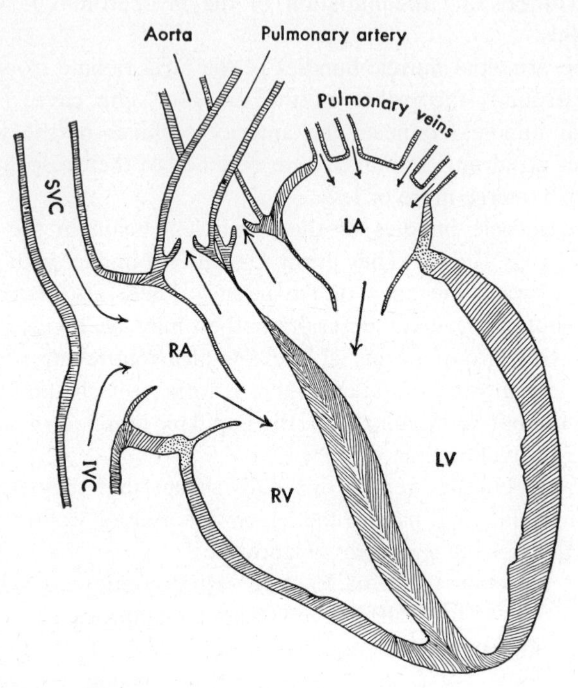

Fig. 204. A diagram to show the four chambers of the heart and valves which guard their openings, seen from the front. The muscular interventricular septum is shown between the right and left ventricles. The upper part of the septum (black line) is called the membranous septum. *RA* and *LA*, right and left atria; *RV* and *LV*, right and left ventricles; *IVC* and *SVC*, inferior and superior venae cavae. Note also atrioventricular openings, the atrioventricular, aortic, and pulmonary valves. The arrows indicate the direction of the blood flow.

with each other after birth. The right side contains *venous* and the left side *arterial* blood. Each half is subdivided into two cavities: the upper, called the *atrium* (auricle); the lower, the *ventricle*.

The walls of the atria are thinner than the walls of the ventricles, and the wall of the right ventricle is thinner than that of the left (the proportion being as 1 to 3). See Fig. 207. This difference is accounted for by the greater amount of work the ventricles, as compared with the atria, have to do.

Muscular columns, called the *trabeculae carneae* (*columnae carneae*), project from the inner surface of the ventricles. They are of three kinds: the first are attached along their entire length and

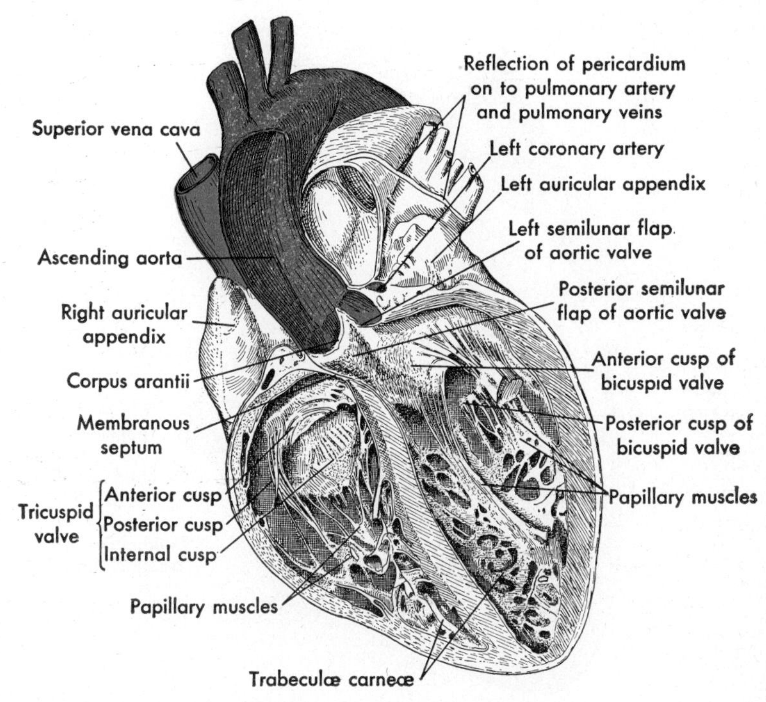

Fig. 205. The heart, seen from the front. (Toldt.) What is the function of the right heart?—the left heart? Why are valves needed?

form ridges, or columns; the second are attached at their extremities but are free in the middle (the *moderator band* of the right ventricle is the best example and is prominent in a sheep's heart); the third are the so-called *papillary muscles*, which are continuous with the wall of each ventricle at its base (Fig. 205). The apices of the papillary muscles give rise to fibrous cords, called the *chordae tendineae*, which are attached to the cusps of the atrioventricular valves.

Orifices of the heart. The orifices comprise the left and right atrioventricular orifices and the orifices of eight large blood vessels connected with the heart.

On the right side of the heart, the superior and inferior venae cavae and coronary sinus empty into the atrium, and the pulmonary artery leaves the ventricle.

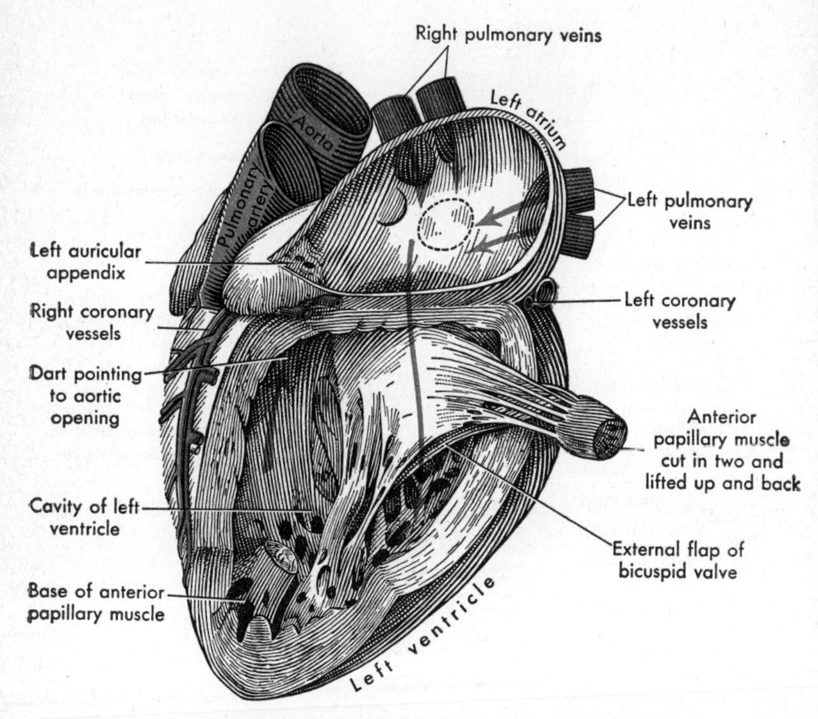

Fig. 206. Left atrium and ventricle, the dorsal wall of each removed. The trabeculae carneae are muscle columns which project from the inner surface of the ventricles. Some of them give origin to the papillary muscles. (Gerrish.)

On the left side of the heart, four pulmonary veins empty into the atrium, and the aorta leaves the ventricle. There are some smaller openings to receive blood directly from the heart substance, and before birth there is an opening between the right and left atria called the *foramen ovale*. Normally this closes soon after birth. Its location is visible as the foramen ovale (Fig. 386).

Valves of the heart. Between each atrium and ventricle there is a somewhat constricted opening, the atrioventricular orifice, which is strengthened by fibrous rings and protected by valves. The openings into the aorta and pulmonary artery are also guarded by valves.

The tricuspid valve. The right atrioventricular valve is composed of three irregular-shaped flaps, or cusps, and hence is named *tricuspid.* The flaps are formed mainly of fibrous tissue covered by endocardium. At their bases they are continuous with one another and form a ring-shaped membrane around the margin of the atrial openings; their pointed ends project into the ventricle and are attached by the chordae tendineae to small muscular pillars, the papillary muscles, in the interior of the ventricles.

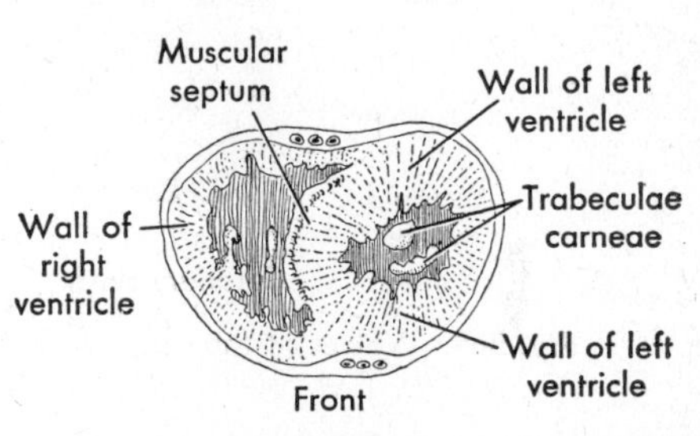

Fig. 207. Cross section through ventricles, showing relative thickness of their walls and shape of cavities.

The bicuspid valve. The left atrioventricular valve consists of two flaps, or cusps, and is named the *bicuspid,* or *mitral* valve. It is attached in the same manner as the tricuspid valve, which it closely resembles in structure except that it is much stronger and thicker in all its parts. Chordae tendineae are attached to the cusps and papillary muscles in the same way as on the right side; they are less numerous but thicker and stronger.

Function. The tricuspid and bicuspid valves freely permit the flow of blood from the atria into the ventricles because the free edges of the flaps are pointed in the direction of the blood current; but any flow forced backward gets between the flaps and the walls of the ventricles and drives the flaps upward until, meeting at their edges, they unite and form a complete transverse partition between the atria and ventricles. Being restrained by the chordae tendineae, the expanded flaps of the valves resist any pressure of the blood which might otherwise force them to open into the atria; at the same time the papillary muscles, to which the chordae tendineae are attached, contract and shorten, thus keeping the chordae tendineae taut.

Semilunar valves. The orifice between the right ventricle and the pulmonary artery is guarded by the *pulmonary valve,* and the orifice between the left ventricle and the aorta is guarded by the *aortic valve.* These two valves are called *semilunar valves* and consist of three half-moon-shaped pockets, each pocket being attached by its

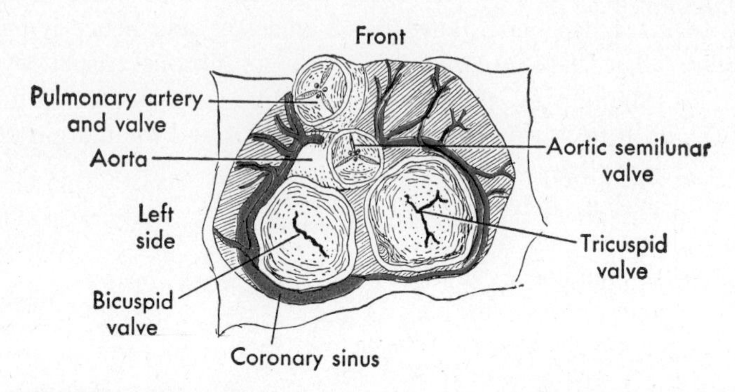

Fig. 208. Valves of the heart as seen from above, atria removed.

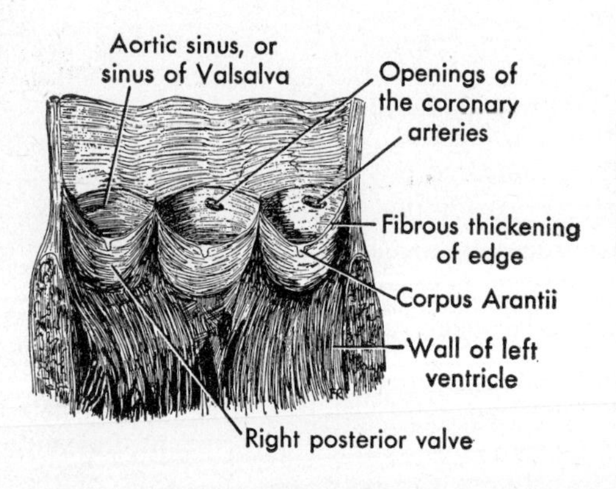

Fig. 209. Aortic valve. The artery has been cut open to show the aortic semilunar valve. The sinuses of Valsalva are three pouches or slight dilatations which are located between the semilunar valves and the wall of the aorta. These sinuses are also found on the pulmonary artery between the semilunar valves and the wall of the artery.

convex margin to the inside of the artery where it joins the ventricle, while its free border projects into the lumen of the vessel. Small nodular bodies, called the *corpora Arantii,*[3] are attached to the center of the free edge of each pocket. Opposite the valves the aorta presents slight dilations called the *aortic sinuses,* or sinuses of Valsalva.[4]

[3] Julius Caesar Arantius, Italian anatomist (1530–1589).
[4] Antoine Marie Valsalva, Italian anatomist (1666–1723).

Function. The semilunar valves offer no resistance to the passage of blood from the heart into the arteries, as the free borders project into the arteries, but they form a complete barrier to the passage of blood in the opposite direction. In this case each pocket becomes filled with blood, and the free borders are floated out and distended so that they meet in the center of the vessel. The corpora Arantii assist in the closure of the valves and help to make the barrier complete.

The orifices between the two caval veins and the right atrium and the orifices between the left atrium and the four pulmonary veins are not guarded by valves. The opening into the inferior vena cava is partly covered by a membrane known as the caval (Eustachian) valve.

Blood supply. Just above the attached margins of the aortic valve, the aorta gives off two branches, called the *right* and *left coronary* arteries. They encircle the heart like a crown, hence their name. They supply the substance of the heart with blood, as the blood contained within the cavities of the heart nourishes only the endocardium. The blood distributed by the coronary arteries after flowing through the capillaries distributed throughout the substance of the wall of the heart is returned by two sets of veins: (1) those which return the blood to the coronary sinus—a wide venous channel (2.25 cm long) emptying into the right atrium; (2) three or four small veins which return the blood directly to the right atrium. A few of the smallest veins[5] empty into the atria and ventricles.

Nerve supply. The heart is supplied with two sets of motor nerve fibers. One set reaches the heart through the vagus nerves of the craniosacral system. Nerve impulses over these nerve fibers have a tendency to slow or stop the heartbeat and are called *inhibitory*. The other set passes to the heart by way of the spinal cord, the superior, middle, and inferior cardiac nerves, and fibers in the visceral branches of the first five thoracic spinal nerves of the thoracolumbar system. They quicken and augment the heartbeat and are called *accelerator*. Both sets of nerve fibers originate in the medulla oblongata, and through this center either set may be stimulated.

In addition, the heart is supplied with afferent nerve fibers: one

[5] These are called veins of Thebesius. Adam Christianus Thebesius, German physician (1686–1732).

set from the aortic arch, called *depressor* fibers; the other set from the right side of the heart, called *pressor* fibers. Both sets of afferent fibers run to the cardiac center in the medulla within the sheath of the vagi. Impulses over the depressor fibers bring about reflex inhibition of the heart—aortic, or Marey's, reflex. Impulses over the

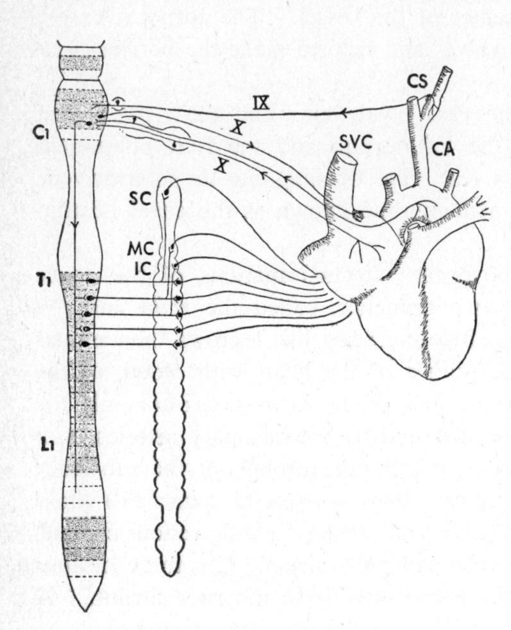

Fig. 210. Nerve supply of the heart. See Fig. 193 also. *IX,* fiber of glossopharyngeal nerve; *X,* vagus nerve; C_1, level of first cervical spinal nerve; *CA,* carotid artery; *CS,* carotid sinus; *IC,* inferior cervical ganglion; L_1, level of first lumbar spinal nerve; *MC,* middle cervical ganglion; *SC,* superior cervical ganglion; *SVC,* superior vena cava; T_1, level of first thoracic spinal nerve. *SC, MC,* and *IC* give rise to the superior, middle, and inferior cardiac nerves.

pressor fibers bring about reflex acceleration of the heart—right heart, or Bainbridge's, reflex.

ARTERIES

The arteries are tubes that carry blood from the heart to the capillaries. They are composed of three coats:

1. An inner coat (*tunica intima*) consisting of three layers—a layer of endothelial cells, a layer of delicate connective tissue which is found only in vessels of considerable size, and an elastic layer consisting of a membrane or network of elastic fibers in which, under the microscope, perforations may be seen. On this account it is sometimes called the *fenestrated membrane.*

2. A middle coat (*tunica media*) of muscular and elastic tissue. The muscular tissue consists of fine bundles of plain muscle fibers arranged in layers and circularly disposed around the vessel. In the large arteries elastic fibers form layers which alternate with the layers of muscle fibers. In the largest arteries white connective-tissue fibers have been found in this coat.

3. An external coat (*tunica externa*, or *adventitia*) of areolar connective tissue in which there are scattered smooth muscle cells or bundles of them arranged longitudinally.[6] In all but the smallest arteries this coat contains some elastic tissue.

By virtue of the structure of the middle coat, the arteries are both extensile and elastic. The proper functioning of the arteries depends upon their extensibility and elasticity.

The great extensibility of the arteries adapts them to receive the additional amount of blood forced into them at each contraction of the heart. Their elasticity adapts them to squeeze the blood outward and to return to their original diameters in time to receive the volume of blood leaving the heart at the next beat.

The strength of an artery depends largely upon the outer coat; it is far less easily cut or torn than the other coats and serves to resist undue expansion of the vessel.

The arteries do not *collapse* when empty; and when an artery is severed, the orifice remains open. The muscular coat, however, contracts somewhat in the neighborhood of the opening; and the elastic fibers cause the artery to retract a little within its sheath, so as to diminish its caliber and permit a blood clot to plug the orifice. This property of a severed artery is an important factor in the arrest of hemorrhage.

Most of the arteries are accompanied by a nerve and one or two veins, all surrounded by a sheath of connective tissue, which helps to support and hold these structures in position.

Size of the arteries. The largest arteries in the body, the aorta and pulmonary artery, measure more than 3 cm ($1\frac{1}{4}$ in.) in diameter at their connection with the heart. These arteries give off branches which divide and subdivide into smaller branches. The smallest arteries are called *arterioles;* and at their distal ends, where only the internal coat remains, the capillaries begin. The arteriole walls con-

[6] J. L. Bremer and H. L. Weatherford, *A Textbook of Histology,* 1944.

tain a great proportion of smooth muscle in relation to elastic tissue, and they are to be thought of as muscular rather than elastic.

The muscular arteries. These include the arteries of medium size, and their middle coat is chiefly muscular. Muscular arteries are also called *distributing* arteries because they distribute the blood to the various organs and by contraction or relaxation they aid in regulating the volume of blood passing to structures to meet varying functional demands.

The elastic arteries. These include the large arteries and are called *conducting* arteries because they conduct blood from the heart to the medium-sized arteries. The middle coat contains a large amount of elastic tissue, and the wall is comparatively thin for the size of the vessel.

Blood supply of the arteries. The blood which flows through the arteries nourishes only the inner coat. The external coat is supplied with arteries, capillaries, and veins, called *vasa vasorum,* or blood vessels of the blood vessels.

CAPILLARIES

The capillaries are exceedingly minute vessels which average about 8 μ (0.008 mm) in diameter. They connect the arterioles (smallest arteries) with the venules (smallest veins).

Structure. The walls of the capillaries consist of one layer of endothelial cells continuous with the layer which lines the arteries, the veins, and the heart. These cells are held together by cell cement. There is a substance called hyaluronic acid that forms a gelatinous material in the cell cement and tissue spaces. It holds cells together and binds water in the tissues.

Distribution. The capillaries communicate freely with one another and form interlacing networks of variable form and size in the different tissues. All the tissues, with the exception of the cartilages, hair, nails, cuticle, and cornea of the eye, are traversed by these networks of capillary vessels. The capillary diameter is so small that the blood cells often must pass through them in single file, and very frequently the cell is larger than the caliber of the vessel and becomes distorted as it passes through. In many parts the capillaries lie so close together that a pin's point cannot be inserted between them. They are most abundant and form the finest networks in those organs where the blood is needed for purposes other than local nutrition, as, for example, secretion or absorption.

Function. In the glandular organs the capillaries supply the substances requisite for secretion; in the ductless glands they also take up the products of secretion; in the alimentary canal they take up some of the elements of digested food; in the lungs they absorb oxygen and give up carbon dioxide; in the kidneys they discharge the waste products collected from other parts; all the time, everywhere in the body, through their walls an interchange is going on which is essential to the life of the body. It is in the capillaries, then, that the chief work of the blood is done; and the object of the vascular mechanism is to cause the blood to flow through these vessels in a steady stream. Krogh estimated that there are about 70,000 sq ft of blood capillaries in the adult body.

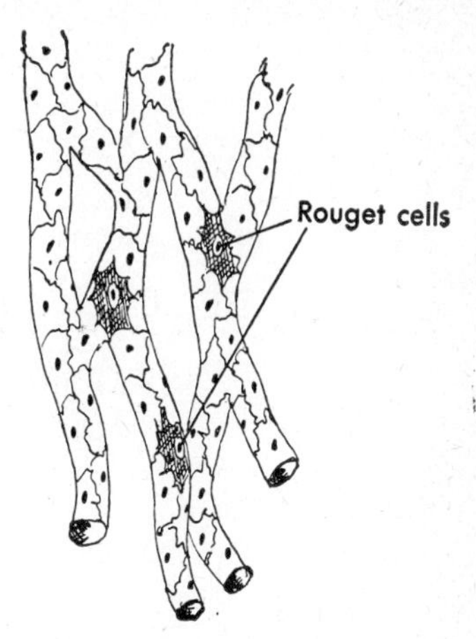

Fig. 211. Capillary networks form the means by which cells may receive oxygen and nutrient materials. There are about 70,000 sq ft of capillaries in the adult.

VEINS

The veins carry blood to the heart and are made by the gathering together of the capillaries. The structure of the veins is similar to that of the arteries. They have three coats: (1) an inner endothelial lining, (2) a middle muscular layer, and (3) an external layer of areolar connective tissue. The main differences between the veins and arteries are: (1) the middle coat is not as well developed and not as elastic in the veins; (2) many of the veins are provided with valves; (3) the walls of veins are much thinner than those of arteries, hence tend to collapse when not filled with blood.

Valves. The valves are semilunar folds of the internal coat of the veins and usually consist of two flaps, rarely one or three.

The convex border is attached to the side of the vein, and the free edge points toward the heart. Their function is to prevent reflux of the blood and keep it flowing in the right direction, i.e., toward the heart.

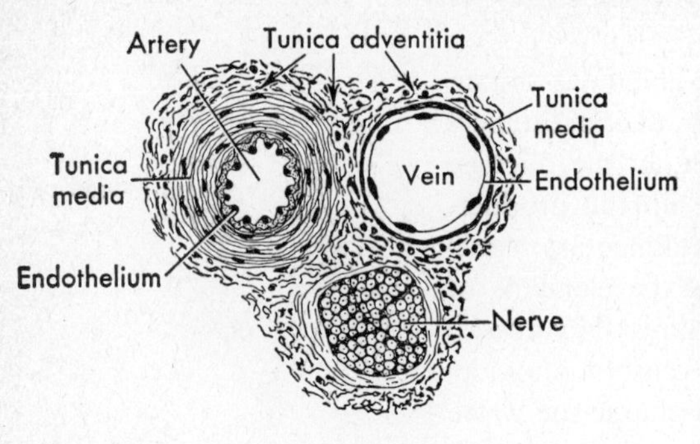

Fig. 212. Transverse section through artery, vein, nerve. See text for structure detail. Arteries, veins, and nerves run along together.

If for any reason the blood on its onward course toward the heart is driven backward, the refluent blood, getting between the wall of the vein and the flaps of the valve, will press them inward until their edges meet in the middle of the channel and close it. The valves are most numerous in the veins where reflux is most likely to occur, i.e., the veins of the extremities. For the same reason a greater number are found in the lower than in the upper limbs. They are absent in many of the small veins, in the large veins of the trunk, and in veins not subjected to muscular pressure. The veins, like the arteries, are supplied with blood vessels.

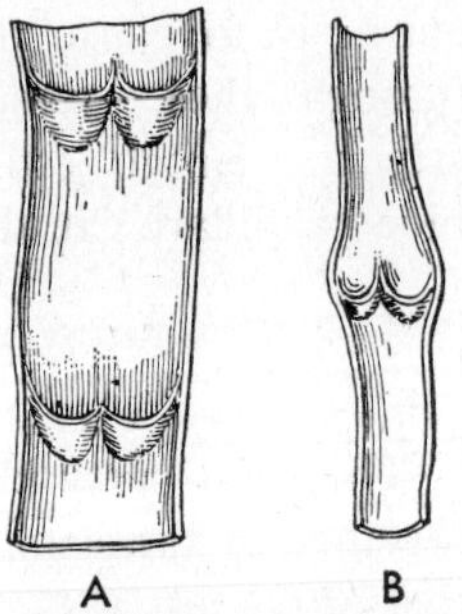

Fig. 213. Diagram showing valves of veins. *A*, part of a vein, laid open, with two pairs of valves; *B*, longitudinal section of vein, showing valves closed.

It must be remembered that, although the arteries, capillaries, and veins have each the distinctive structure described, it is difficult to draw the line between the arteriole and the large capillary and between the large capillary and the venule. The veins on leaving the capillary networks only gradually assume their several coats, and the arteries dispense with their coats in the same imperceptible way as they approach the capillaries.

Vasomotor nerves. The muscular tissue in the walls of the blood vessels is well supplied with nerve fibers, chiefly from the sympathetic portion of the autonomic system. These nerve fibers are called *vasomotor* and are divided into two sets: (1) vasoconstrictor and (2) vasodilator.

They connect with the vasoconstrictor center in the medulla oblongata, which is constantly sending impulses to the vessels, thus keeping them in a state of tone. The vasoconstrictor center is a reflex center and is connected with afferent fibers coming from all parts of the body. Arteries are supplied with vasodilator fibers which, when stimulated, inhibit the action of the vasoconstrictors and cause dilatation of the vessels. The widening and narrowing of the arteries not only affect the local circulation in different parts of the body, but the amount of resistance they oppose to the arterial impulse also influences in some degree the character of the heartbeat.

SUMMARY

- **Blood Vascular System**
  - Heart
  - Arteries—small arteries are named arterioles
  - Capillaries
  - Veins—small veins are named venules
- **Heart**
  - **Location**
    - Between lungs
    - Above diaphragm
  - **Structure**
    - Outside covering—*pericardium*
      - Fibrous portion
      - Serous
        - Visceral
        - Parietal
    - Muscle substance—*myocardium*
      - Muscle bundles of the atria
      - Muscle bundles of the ventricles
      - Atrioventricular bundle—bundle of muscular and nervous tissue located in septum between right and left heart, which connects the musculature of atria and ventricles
    - Smooth lining on inside—*endocardium*
  - **Cavities**
    - **Right heart**
      - **Right atrium**
        - Receives blood
        - Thin walls
      - **Right ventricle**
        - Expels blood into pulmonary artery
        - Thick walls
    - **Left heart**
      - **Left atrium**
        - Receives blood
        - Thin walls
      - **Left ventricle**
        - Expels blood into aorta
        - Very thick walls
  - **Orifices**
    - **Right heart**
      - **Right atrium**
        - Superior vena cava—returns blood from upper portion of body
        - Inferior vena cava—returns blood from lower portion of body

- **Heart**
  - **Orifices**
    - **Right heart**
      - **Right atrium**
        - Coronary sinus—returns blood from heart muscle
      - Atrioventricular orifice between atrium and ventricle
      - **Right ventricle**
        - Pulmonary artery—carries blood from heart to lungs
    - **Left heart**
      - **Left atrium**
        - Two right pulmonary veins; Two left pulmonary veins — Return blood from lungs
      - Atrioventricular orifice between atrium and ventricle
      - **Left ventricle**
        - Aorta—distributes blood to all parts of the body
  - **Valves**
    - **Right atrioventricular or tricuspid valve**—composed of three cusps, situated in the right ventricle
    - **Left atrioventricular or bicuspid, or mitral, valve**—composed of two strong, thick cusps, situated in the left ventricle
    - **Function**—prevent flow of blood from ventricles into atria
    - **Semilunar valves**
      - **Aortic**—composed of three half-moon-shaped pockets between aorta and left ventricle
      - **Pulmonary**—composed of three half-moon-shaped pockets between pulmonary artery and right ventricle
      - **Function**—prevent flow of blood from arteries into ventricles
  - **Blood Supply**
    - **Arteries**
      - Right coronary; Left coronary — branches from aorta
    - **Veins**
      - (1) Cardiac veins empty into coronary sinus
      - (2) Three or four small veins empty into right atrium
      - (3) Veins of Thebesius empty into atrium and ventricles
  - **Nerve Supply**
    - **Craniosacral system**—vagus nerves, *inhibitory fibers*, slow the heart
    - **Thoracolumbar system**—*accelerator fibers* increase rapidity and force of heart
    - **Afferent fibers**
      - Depressor—*reflex inhibitory*, afferent fibers in vagi from aortic arch, left heart, etc.
      - Pressor—*reflex accelerator*, afferent fibers in vagi from right heart, etc.

- **Arteries** (Characterized by Elasticity)
  - **Hollow tubes**—carry blood *from* heart, break up into capillaries
  - **Coats**
    - (1) Inner lining (Intima)
      - Layer of endothelial cells
      - Layer of connective tissue

- **Arteries** (Characterized by Elasticity)
  - **Coats**
    - (1) Inner lining (Intima)
      - Layer of elastic tissue—fenestrated membrane
    - (2) Middle coat (Media)
      - Muscular and elastic tissue
      - A few bundles of white connective tissue
    - (3) External coat (Adventitia)
      - Areolar connective tissue with scattered smooth muscle cells
  - **Size**—aorta more than 1 in. in diameter. Arteries grow smaller as they subdivide. Smallest ones are microscopic and are called *arterioles*
- **Capillaries** (Characterized by Multiplicity)
  - **Tiny tubes**—about 8 μ in diameter. Connect arterioles and venules
  - One layer of endothelial cells
  - Communicate freely—form networks
- **Veins** (Characterized by Valves)
  - **Collapsible tubes**—smallest ones, called *venules*, begin where capillaries end
  - Carry blood to heart
  - Three coats, same as arteries but thinner
  - Valves—semilunar pockets

Vasa vasorum—the term applied to blood vessels that are supplied to coats of other blood vessels

Vasomotor—term applied to *nerve fibers* supplied to blood vessels
- Vasoconstrictor, well known
- Vasodilator, not as well known as fibers

CHAPTER 12

DIVISIONS OF THE VASCULAR SYSTEM ARTERIES, VEINS, PORTAL SYSTEM

The arteries are distributed throughout the body in a systematic manner. The vessels leaving the heart are large but soon divide into branches. This division continues until minute branches are distributed to all parts of the body.

At each division the branches are smaller; but since they are numerous, the total of their diameters is much greater than that of the artery from which they sprang. This means that as the blood flows from the heart toward the capillaries it flows in an "ever widening bed." The diameter of the aorta at the heart is usually given as 1 in.; the sum of the diameters of the systemic capillaries is about 600 to 800 in. The branches of the large arteries leave them at abrupt angles; the branches of the smaller arteries take progressively less abrupt changes of direction.

Division. The way in which the arteries divide varies. (1) An artery may give off several branches in succession and still continue as a main trunk, e.g., the thoracic or abdominal portion of the aorta. (2) A short trunk may subdivide into several branches at the same point, e.g., the celiac artery. (3) An artery may divide into two branches of nearly equal size, e.g., the division of the aorta into the two common iliacs.

Anastomosis. The distal ends of arteries unite at frequent intervals, when they are said to anastomose. Such anastomoses permit free communication between the currents of the blood, tend to obviate the effects of local interruption, and promote equality of distribution and of pressure. Anastomoses occur between the larger as well as the smaller arteries. Where great activity of the circulation is necessary, as in the brain, two branches of equal size unite,

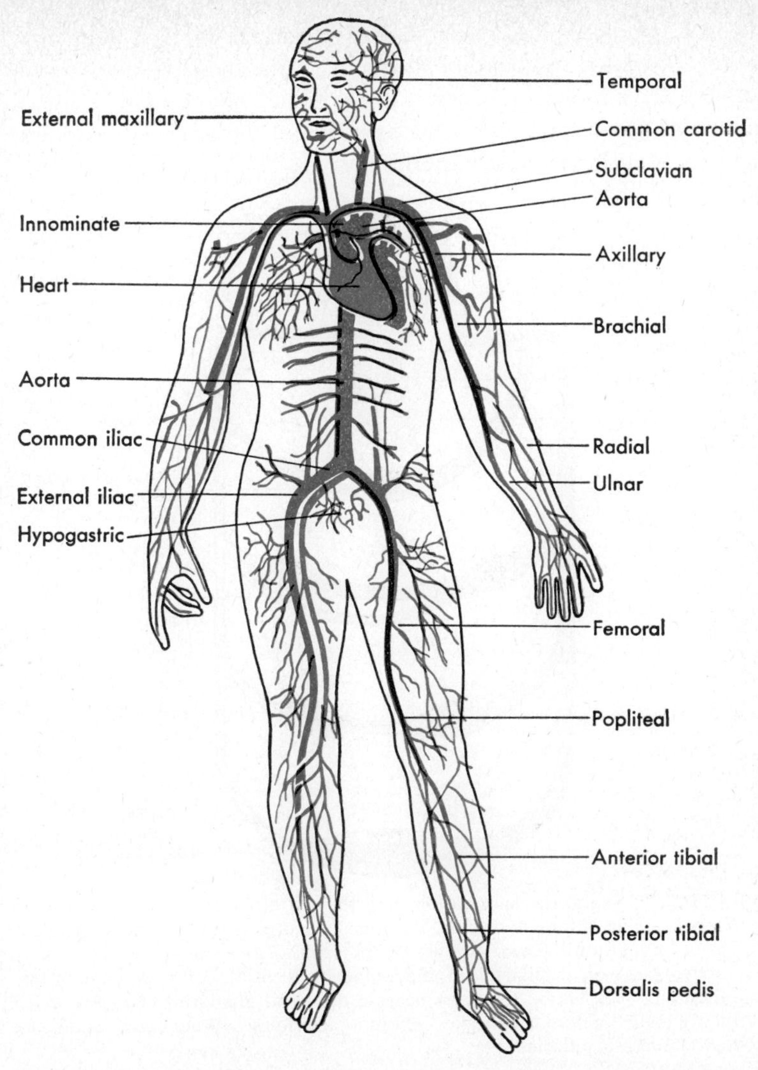

Fig. 214. A general diagram of the circulation. Many arteries are named.

e.g., the two vertebral arteries unite to form the basilar (Fig. 224). In the abdomen, the intestinal arteries have frequent anastomoses between their larger branches. In the limbs, anastomoses are most numerous around the joints, the branches of the arteries above uniting with branches from the arteries below.

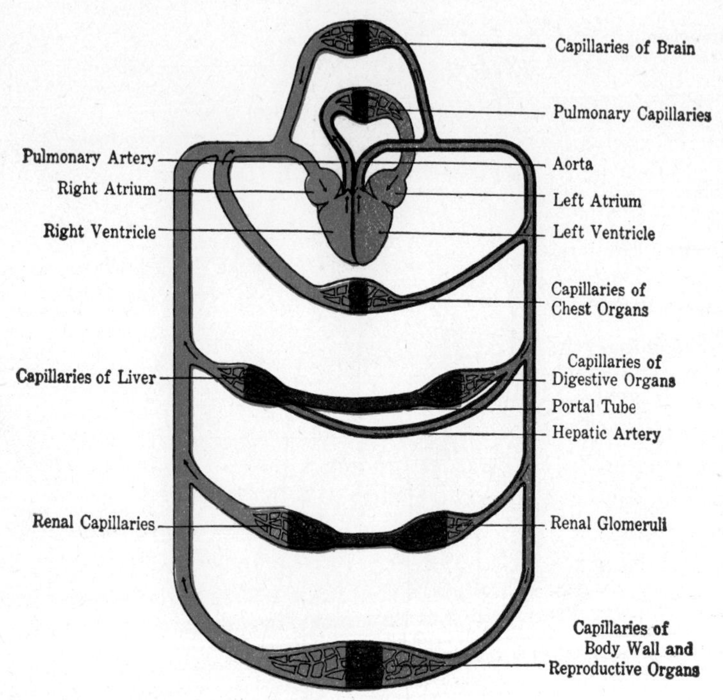

Fig. 215. Diagram of circulation. In the portal circulation the two sets of capillaries are in different organs, the digestive organs and the liver, while in the renal circulation the two sets are in the same organ, the kidney.

The two sets of capillaries in the kidney are visible only on microscopic examination. The blood tubes connecting the renal glomeruli and renal capillaries are called efferent tubes or efferent arterioles. What work is accomplished in these capillaries?

Anastomoses are of importance to the surgeon. By their enlargement, a collateral circulation is established after an artery is ligated.

A *plexus,* or network, is formed by the anastomosis of a number of arteries in a limited area. Arteries usually occupy situations *protected* against accidental injury or the effects of local pressure. Arteries usually pursue a fairly straight course, but in some parts of the body they are tortuous. The external maxillary (facial) artery, both in the neck and on the face, and the arteries of the lips (inferior and superior labial) are extremely tortuous and thereby accommodate themselves to the varied movements occurring in speaking, laughing, turning the head, etc.

DIVISIONS OF THE VASCULAR SYSTEM

The blood vessels of the body are arranged in two main systems: (1) The *pulmonary,* which is the lesser system, provides for the circulation of the blood from the right ventricle to the lungs and back to the left atrium. (2) The *systemic,* which is the larger system, provides for the circulation of the blood from the left ventricle to all parts of the body by means of the aorta and its branches and the return to the right atrium by means of the venae cavae.

Blood vessels of the pulmonary system. The blood vessels of the pulmonary system are (1) the pulmonary artery and all its branches, (2) the capillaries which connect these branches with the veins, and (3) the pulmonary veins.

The pulmonary artery conveys venous blood from the right ventricle to the lungs. The main trunk is a short, wide vessel about 5 cm (2 in.) in length and a little more than 3 cm ($1\frac{1}{4}$ in.) in width. It arises from the right ventricle and passes upward, backward, and to the left. About the level of the intervertebral disk between the fifth and sixth thoracic vertebrae, it divides into two branches, the right and left pulmonary arteries, which pass to the right and left lungs. Before entering the lungs, each artery divides into two branches. After entering, the branches divide and subdivide, grow smaller in size, and finally merge into capillaries which form a network upon the walls of the air cells. These capillaries unite, grow larger in size, and gradually assume the characteristics of veins. The veins unite to form the pulmonary veins.

The pulmonary veins are four short veins, two from each lung, which convey the blood from the lungs to the left atrium. They carry oxygenated blood to be distributed by the systemic arteries. The pulmonary veins have no valves.

Blood vessels carrying blood away from the heart (aorta and coronary and pulmonary arteries) must be elastic to receive blood from the ven-

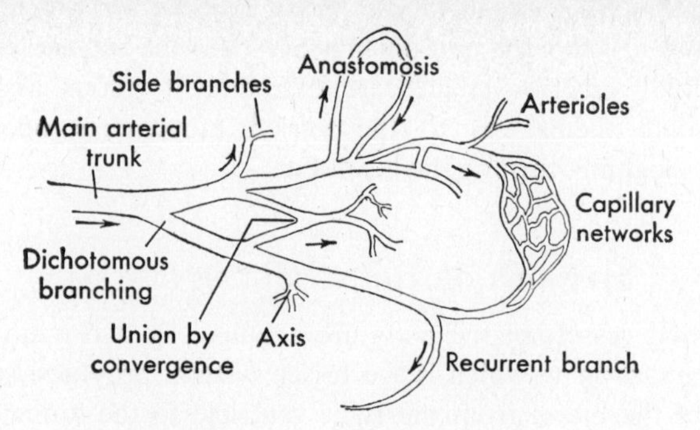

Fig. 216. Diagram showing anastomosis, branching, and confluence of arteries. This arrangement permits free movement of blood.

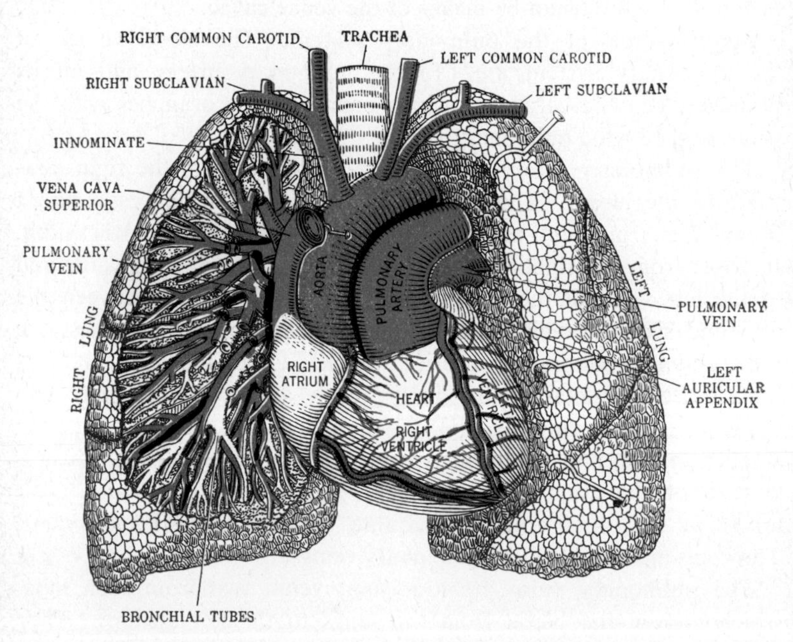

Fig. 217. The pulmonary artery and aorta. The front part of the right lung has been removed, and the pulmonary vessels and the bronchial tubes are thus exposed. (Gerrish.)

tricles during systole. Blood enters the arteries under pressure during ventricular systole and against resistance in the arterial bed; hence blood vessels carrying blood to the lungs are elastic to adjust to the onward movement of blood. The pressure in the arteries of the pulmonary tree (while not as high as in the systemic vessels) keeps blood moving onward through the pulmonary capillaries and finally on into the veins. Blood enters the left atrium during diastole.

Blood vessels of the systemic system consist of (1) the *aorta* and all the arteries that originate from it, including the terminal

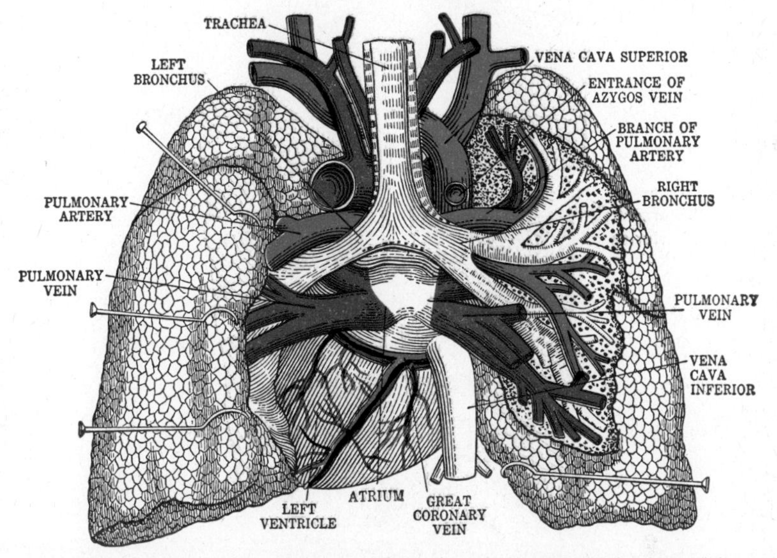

Fig. 218. Pulmonary vessels, seen in a dorsal view of the heart and lungs. The left lung is pulled to the left, and the right lung has been partly cut away to show the ramifications of the air tubes and blood vessels. (Gerrish.)

branches called arterioles; (2) the capillaries which connect the arterioles and venules; and (3) all the venules and veins of the body which empty into the superior and inferior venae cavae and then into the heart, as well as those which empty directly into the heart (coronary veins).

The aorta is the main trunk of the arterial system. Springing from the left ventricle of the heart, it passes toward the right under the pulmonary artery, then arches toward the back over the roof of the left lung, descends along the vertebral column, and, after passing through the diaphragm into the abdominal region, ends opposite the

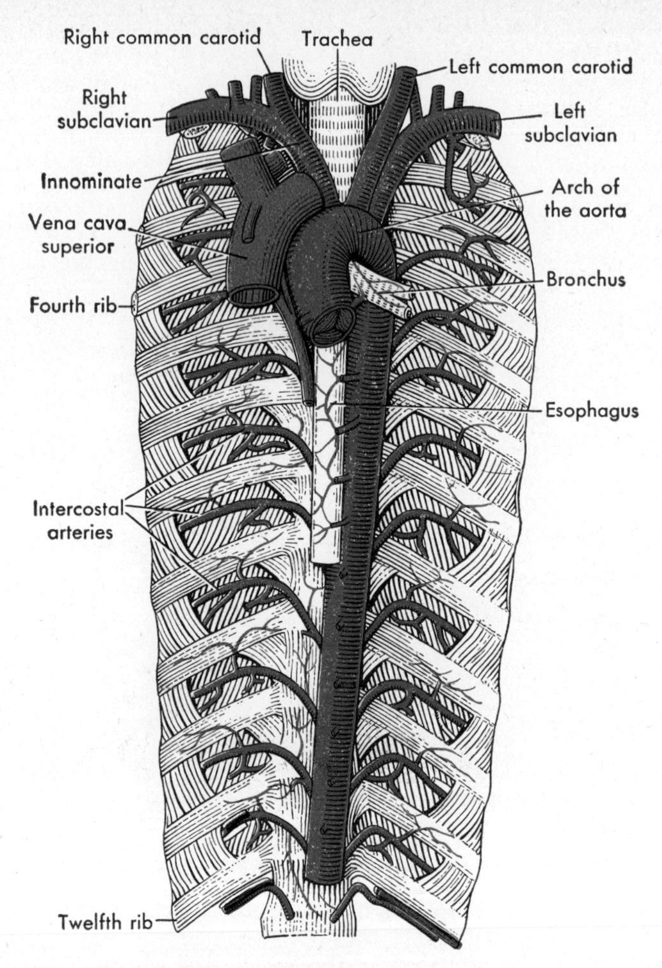

Fig. 219. Thoracic aorta. The thoracic aorta extends from the fourth to the twelfth thoracic vertebra. (Gerrish.)

fourth lumbar vertebra by dividing into the right and left common iliac arteries. In this course the aorta forms a continuous trunk, which gradually diminishes in size from its commencement to its termination. It gives off large and small branches along its course.

The aorta is called by different names throughout its length: (1) the ascending aorta, (2) the arch of the aorta, and (3) the descending aorta, which (*a*) above the diaphragm is referred to as the thoracic aorta and (*b*) below the diaphragm is called the abdominal aorta.

1. *The ascending aorta* is short, about 5 cm (2 in.) in length, and is contained within the pericardium. The only branches of the as-

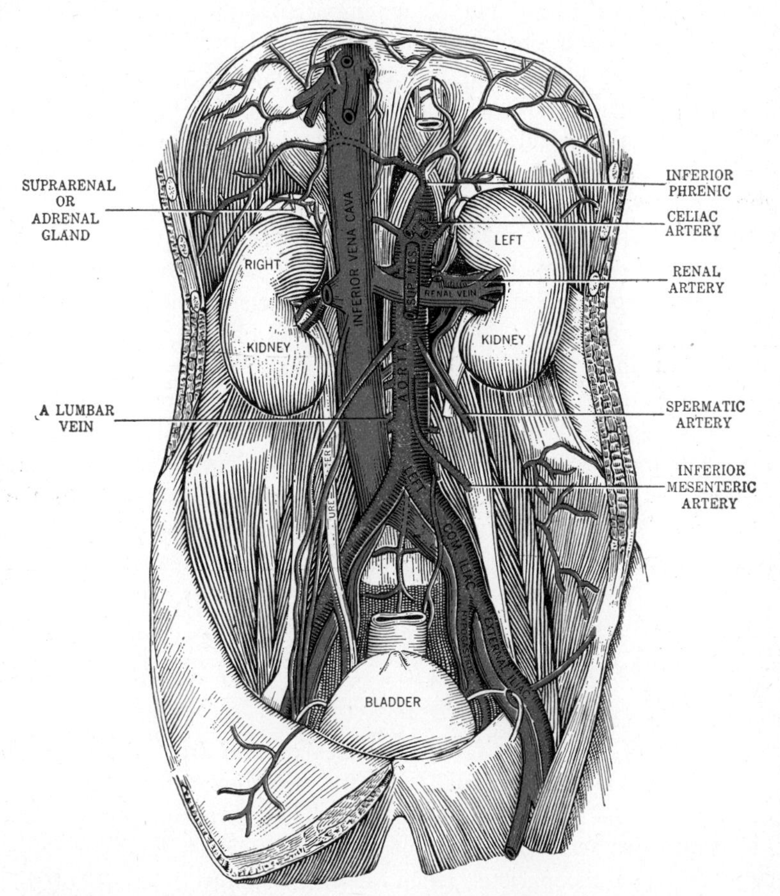

Fig. 220. The abdominal aorta and inferior vena cava. The abdominal aorta bifurcates into the right and left common iliac arteries opposite the fourth lumbar vertebra. (Gerrish.)

cending aorta are the right and left *coronary arteries,* which have been described.

2. *The arch* extends from the ascending aorta upward, backward, and to the left in front of the trachea, then backward and downward

on the left side of the body of the fourth thoracic vertebra, where it becomes continuous with the descending aorta. Three branches are given off from the arch of the aorta—the *innominate,* the *left common carotid,* and the *left subclavian* arteries. Branches of these arteries supply the head and the upper extremities.

The innominate (brachiocephalic) artery arises from the right upper surface of the arch, and ascends obliquely toward the right until, arriving on a level with the upper margin of the clavicle, it divides into the right common carotid and right subclavian arteries.

3. *The descending aorta* extends from the body of the fourth thoracic vertebra to the body of the fourth lumbar vertebra.

a) THE THORACIC AORTA is comparatively straight and extends from the fourth thoracic vertebra on the left side to the aortic opening in the diaphragm in front of the last thoracic vertebra. Branches from the thoracic aorta supply the body wall of the chest cavity and the viscera which it contains.

b) THE ABDOMINAL AORTA commences at the aortic opening of the diaphragm, in front of the lower border of the last thoracic vertebra, and terminates below by dividing into the two common iliac arteries. The bifurcation usually occurs opposite the body of the fourth lumbar vertebra, which corresponds to a spot on the front of the abdomen slightly below and to the left of the umbilicus. Branches from the abdominal aorta supply the body wall of the abdominal cavity and the viscera which it contains.

ARTERIES OF THE TRUNK

Arteries of the chest. The branches derived from the thoracic aorta are numerous but small, and the consequent decrease in the diameter of the aorta is not marked. These branches may be divided into two sets: (*a*) the visceral, or those which supply the viscera, and (*b*) the parietal, or those which supply the walls of the chest cavity.

Visceral Group	Parietal Group
Pericardial arteries	Intercostal arteries
Bronchial arteries	Subcostal arteries
Esophageal arteries	Superior phrenic arteries
Mediastinal arteries	

The pericardial arteries are small and are distributed to the pericardium.

The bronchial arteries extend to the lungs. They vary in number, size, and origin. As a rule, there are two left bronchial arteries, which arise from the thoracic aorta, and one right bronchial artery, which arises from the first aortic intercostal or from the upper left bronchial. Each vessel runs along the back part of the corresponding bronchus, dividing and subdividing along the bronchial tubes, supplying them and the cellular tissue of the lungs.

The esophageal arteries are four or five in number; they arise from the front of the aorta and form a chain of anastomoses along the esophagus. They anastomose with the esophageal branches of the thyroid arteries above and with ascending branches from the left gastric and the left inferior phrenic arteries below.

The mediastinal arteries are numerous small arteries which supply the nodes and areolar tissue in the posterior mediastinum.

The intercostal arteries are usually nine in number on each side; they arise from the back of the aorta and are distributed to the lower nine intercostal spaces. Each intercostal artery is accompanied by a vein and a nerve, and each one gives off numerous branches to the muscles and skin (Fig. 219) and to the vertebral column and its contents.

The subcostal arteries lie below the last ribs and are the lowest pair of branches derived from the thoracic aorta.

The superior phrenic arteries are small. They arise from the lower part of the thoracic aorta and are distributed to the posterior part of the upper surface of the diaphragm.

Arteries of the abdomen. The branches derived from the abdominal aorta may be subdivided into two groups.

Visceral Branches	Parietal Branches
Celiac (celiac axis)	Inferior phrenics
Superior mesenteric	Lumbars
Middle suprarenals	Middle sacral
Renals	
Internal spermatics (male), ovarian (female)	
Inferior mesenteric	

The celiac artery is a short, wide vessel, usually not more than 1.25 cm ($\frac{1}{2}$ in.) in length, which arises from the front of the aorta just below the opening in the diaphragm. It divides into three branches: the *left gastric*, the *hepatic*, and the *splenic*, or *lienal* (Figs. 220 and 221).

THE LEFT GASTRIC courses along the lesser curvature of the stomach from left to right, distributing branches to both surfaces. It anastomoses with the esophageal arteries at one end of its course and with the right gastric artery at the other.

THE HEPATIC ARTERY supplies the liver with blood direct from the aorta via the celiac artery. It gives off three branches: (*a*) the *right gastric*, which courses from right to left along the lesser curvature of the stomach and anastomoses with the left gastric; (*b*) the *gastroduodenal*, which splits into two vessels: one (superior pancreaticoduodenal) supplies the duodenum and the head of the pancreas; the other (right gastroepiploic or gastro-omental) courses from right to left along the greater curvature of the stomach, distributes branches to it, and anastomoses with a branch of the splenic artery; (*c*) the *cystic artery*, which supplies the gallbladder. Before entering the liver, the hepatic artery divides into two branches, right and left, which supply the corresponding lobes of the liver.

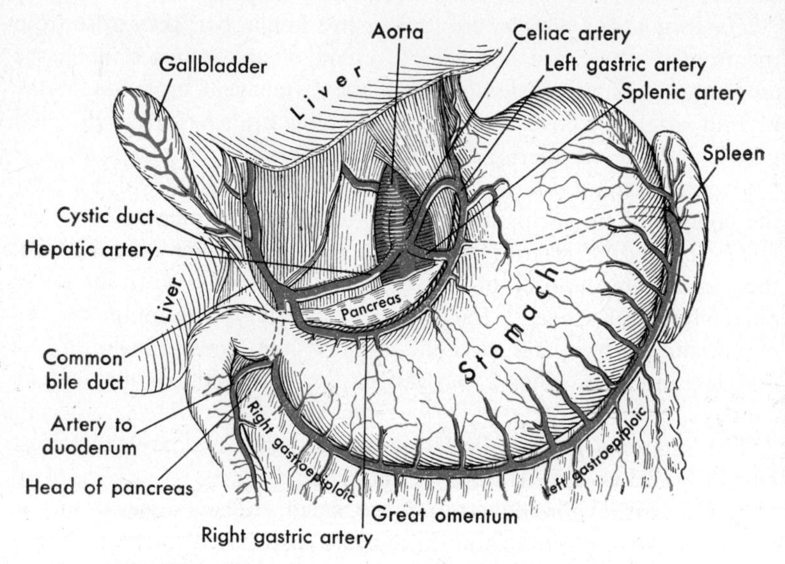

Fig. 221. The celiac artery and its branches. (Morris.)

THE SPLENIC, OR LIENAL, ARTERY is the largest of the three branches of the celiac. It distributes numerous vessels to the pancreas and several small and one large vessel—the left gastroepiploic—to the stomach. The left gastroepiploic (gastro-omental) runs along the

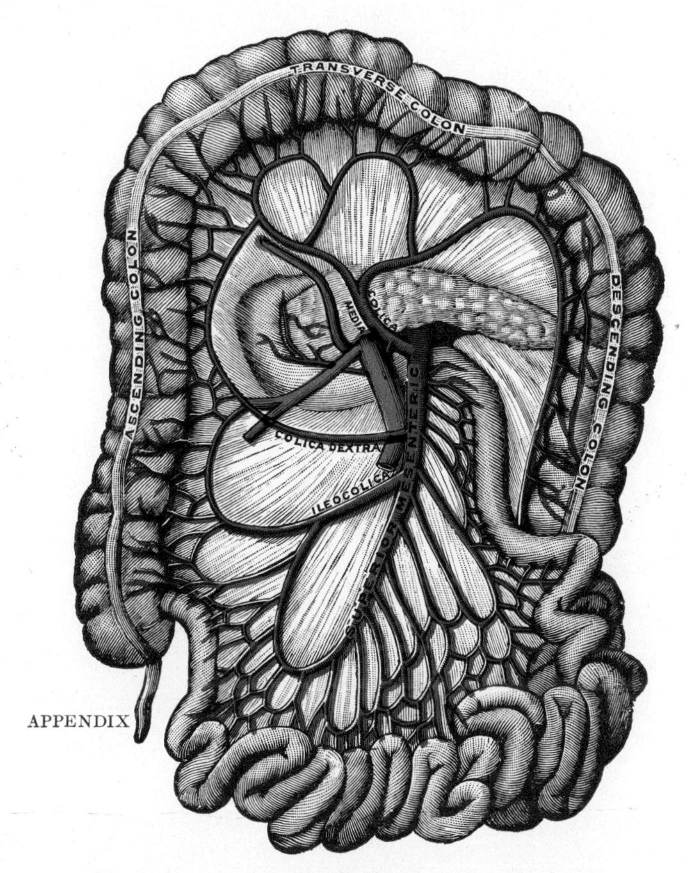

Fig. 222. Superior mesenteric artery. (Gerrish.)

greater curvature of the stomach from left to right and anastomoses with the right gastroepiploic.

The superior mesenteric artery arises from the front part of the aorta, a little below the celiac artery. It supplies all of the small intestine except the duodenum. It also supplies the cecum, the ascending colon, and half of the transverse colon (Fig. 222).

The middle suprarenal arteries are of small size. They arise from the side of the aorta and pass to the suprarenal, or adrenal, glands, where they anastomose with branches of the phrenic and renal arteries.

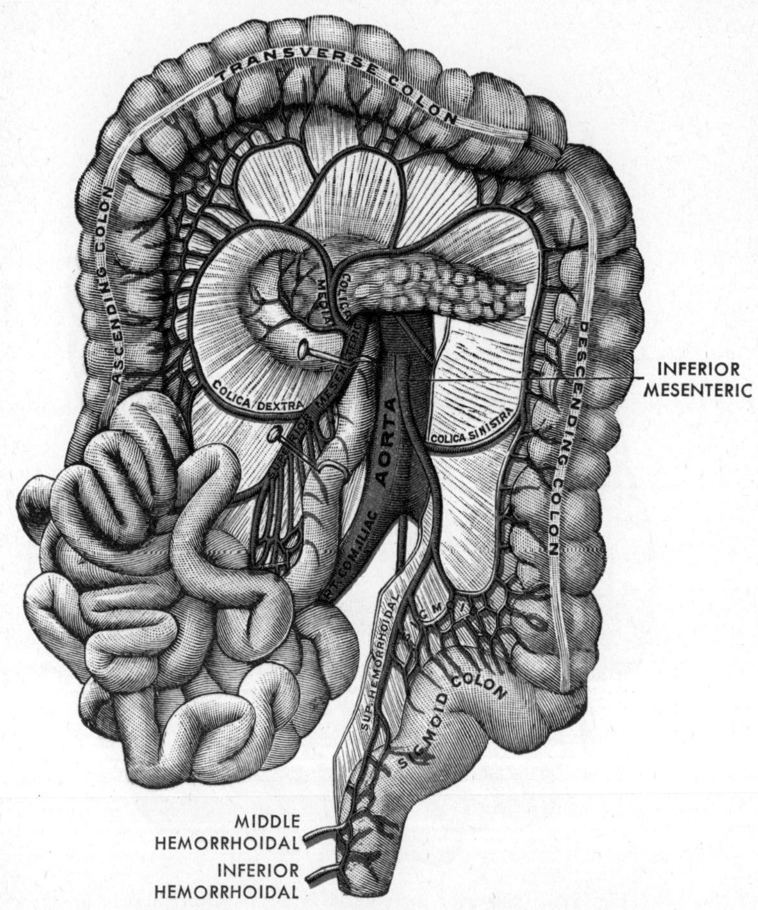

Fig. 223. Inferior mesenteric artery. (Gerrish.)

The renal arteries, right and left, arise from the sides of the aorta, below the superior mesenteric artery. The right is generally a little lower than the left. Each is directed outward, so as to form nearly a right angle with the aorta, and each divides into four or five branches before reaching the hilum of the kidney (Fig. 220).

The internal spermatic arteries arise from the front of the aorta, a little below the renal arteries. They supply the testes (Fig. 220).

The ovarian arteries in the female arise from the same portion of the aorta as the spermatic arteries in the male. They supply the ovaries and send small branches to the ureters and uterine tubes. One branch unites with the uterine artery (a branch of the hypogastric) and assists in supplying the uterus. During pregnancy the ovarian arteries become considerably enlarged.

The inferior mesenteric artery arises from in front of the aorta about 3.8 cm ($1\frac{1}{2}$ in.) above the division of the aorta into the common iliacs. It distributes branches to the left half of the transverse colon and to the descending and sigmoid colon; continued as the superior rectal artery[1] (BNA, superior hemorrhoidal), it also takes part in the blood supply of the rectum. Blood is returned by tributaries to the inferior mesenteric vein, which flows into the portal vein and then into the vena cava.

The inferior phrenic arteries (two) may arise separately or by a common trunk from the aorta or celiac artery. They are distributed to the undersurface of the diaphragm.

The lumbar arteries, usually four in number on each side, are analogous to the intercostals. They arise from the back of the aorta opposite the bodies of the upper four lumbar vertebrae. Occasionally a fifth pair arises from the middle sacral artery. These arteries distribute branches to the muscles and skin of the back; a spinal branch enters the vertebral canal and is distributed to the spinal cord and its membranes, also to the lumbar vertebrae.

The middle sacral artery arises from the back part of the abdominal aorta and passes down in front of the fourth and fifth lumbar vertebrae, the sacrum, and the coccyx to the coccygeal gland.[2]

Arteries of the pelvis. When the descending aorta reaches the body of the fourth lumbar vertebra, it divides into the two *common iliac* arteries. These arteries pass downward and outward for about

[1] The middle and inferior rectal arteries (hemorrhoidal), branches of the internal iliac (hypogastric) artery, also supply the rectum, anal canal, sphincter muscles, and levatores ani. Branches of these arteries form a plexus, and, after circulating in the capillaries of the region, the blood via the venous plexuses and then via the rectal veins joins the hypogastric, either directly or via the internal pudendal vein.

[2] The coccygeal gland consists of irregular masses of cells. It lies in front of, or just below, the coccyx.

5 cm (2 in.), and then each divides into the hypogastric, or internal, and external iliac arteries.

The hypogastric, or *internal iliac, arteries* send branches to the pelvic walls, pelvic viscera, the external genitals, the buttocks, and the medial side of each thigh. The uterine arteries in the female, which supply the tissues of the uterus with blood, are very important branches of the hypogastrics.

The external iliacs are placed within the abdomen and extend from the bifurcation of the common iliacs to a point halfway between the anterior superior spines of the ilia and the symphysis pubis.

The external iliacs send small branches to the psoas major muscles and to the neighboring lymph nodes, and each gives off the *inferior epigastric* and the *deep iliac circumflex.* These arteries are of considerable size and distribute branches to the abdominal muscles and peritoneum, also to the region of the pubes.

ARTERIES OF THE HEAD AND NECK

The principal arteries of the head and neck are the two common carotids (Fig. 214).

The left common carotid arises from the middle of the upper surface of the arch of the aorta, and the *right common carotid* arises at the division of the innominate; consequently the left carotid is an inch or two longer than the right. They ascend obliquely on either side of the neck until, on a level with the upper border of the thyroid cartilage (Adam's apple), they divide into two great branches: (1) the external carotid, (2) the internal carotid. At the root of the neck the common carotids are separated from each other by only a narrow interval, corresponding to the width of the trachea; but at the upper part, the thyroid gland, the larynx, and the pharynx project forward between them.

The external carotid is the more superficial and is placed nearer the middle line than the internal carotid. Each external carotid has nine branches, which in turn break up into smaller branches. These supply the thyroid gland, the tongue, throat, face, and ears; and the meningeal branches pass inside the cranium to the dura mater.

Each internal carotid has many branches, which are distributed to the brain, the eye and its appendages, the forehead, and the nose. Important branches are the cerebral, distributed to the brain, and the ophthalmic, which enters the orbital cavity through the optic foramen and distributes branches to the orbit, the muscles, and the bulb of the eye.

On the internal carotid, at the point where it diverges from the external carotid, is a slight enlargement known as the *carotid sinus* and an epithelial body, the *carotid body*. From receptors in their walls sensory fibers reach the cardiac center in the brain via the IX

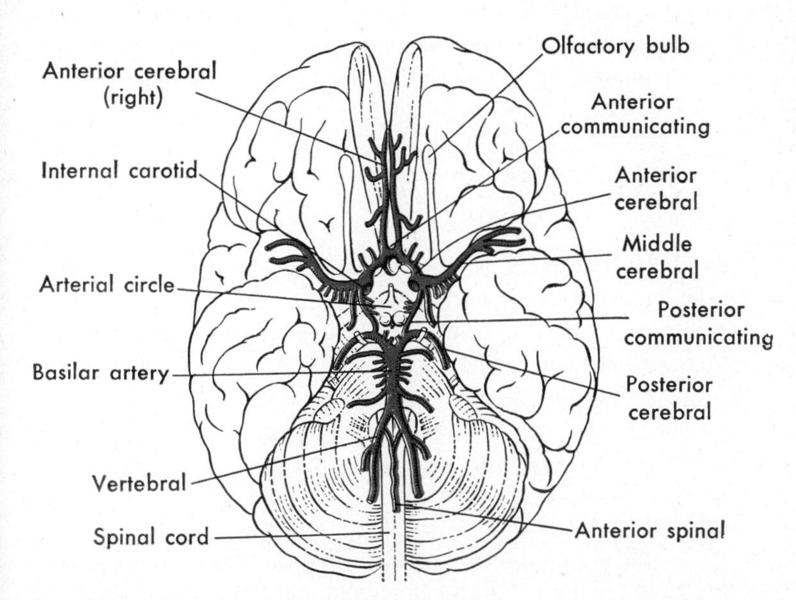

Fig. 224. Diagram of the arterial circulation at the base of the brain, showing the arterial circle of Willis. From this circle the anterior, middle, and posterior cerebral arteries extend to each cerebral hemisphere. The anterior spinal artery supplies the cord.

and X cranial nerves. They are closely related to the maintenance of blood pressure.

Circle of Willis.[3] This is an arterial anastomosis at the base of the brain. It is formed by the union of the *anterior cerebral arteries,* which are branches of the internal carotid, and the *posterior cerebral arteries,* which are branches of the basilar. The *basilar* is formed by the union of the two *vertebrals,* which are branches of the subclavian arteries.

[3] Thomas Willis, English anatomist (1621–1675).

The vertebrals ascend on either side of the vertebral column, pass through the foramen magnum, and unite at the base of the brain to form the basilar artery. The basilar artery extends from the lower to the upper border of the pons, lying in the median groove. It ends by dividing into the two posterior cerebral arteries. These two arteries are connected on either side with the internal carotid by the posterior communicating arteries. In front, the anterior cerebral arteries are connected by the anterior communicating arteries. These arteries form a complete circle (Fig. 224). This arrangement (1) equalizes the circulation of the blood in the brain and (2) in case of destruction of one of the arteries, provides for the blood reaching the brain through other vessels.

ARTERIES OF THE UPPER EXTREMITIES

The subclavian artery is the first portion of a long trunk which forms the main artery of each upper limb. Different portions are given different names, according to the regions through which they pass, viz., subclavian, axillary, brachial. At the elbow the subclavian divides into the radial and ulnar arteries.

The *right subclavian* arises at the division of the innominate, and the *left subclavian* from the arch of the aorta. They pass a short way up into the neck and then turn downward to rest on the first ribs. At the outer border of the first ribs they cease to be called subclavian and are continued as the axillaries. While these arteries continue as the main arteries of the upper extremities, they distribute branches to (1) the base of the brain, (2) the shoulder regions, and (3) the chest.

1. *The vertebrals* have been described above.

2. *The thyrocervical* sends branches to the thyroid, trachea, esophagus, muscles of the neck, and scapula.

3. *The internal mammary artery* extends down just under the costal cartilages to the level of the sixth intercostal space, where it branches into the musculophrenic and superior epigastric arteries. It sends branches to the mammary glands,[4] the diaphragm, the areolar tissue and lymph nodes in the mediastinum, the intercostal muscles, the pericardium, and the abdominal muscles.

[4] The perforating branches of the internal mammary arteries are distributed to the mammary glands and are of large size during lactation.

4. *The costocervical* sends branches to the upper part of the back, the neck, and the spinal cord and its membranes.

The axillary artery (continuation of the subclavian) extends from the outer border of the first rib to the lower border of the tendon of the teres major muscle, where it becomes the brachial. Its direc-

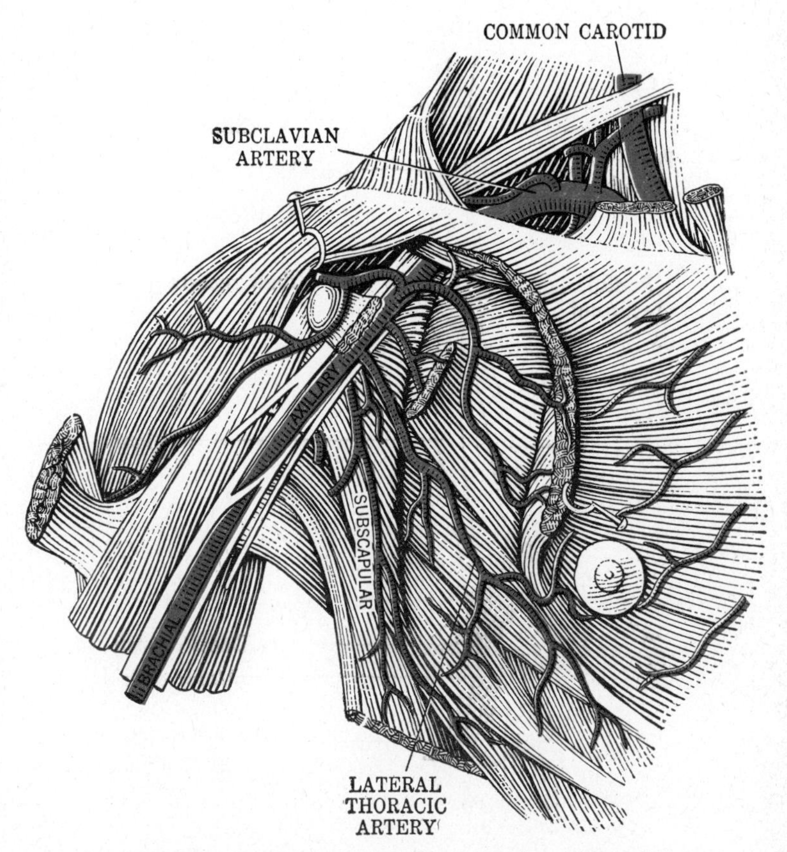

Fig. 225. Subclavian and axillary arteries. (Gerrish.)

tion varies with the position of the upper limb. At the beginning it is deeply situated, but near its termination it is superficial. It gives off branches to the chest, shoulder, and arm.

The brachial artery (continuation of the axillary) extends from the lower margin of the tendon of the teres major muscle to a short

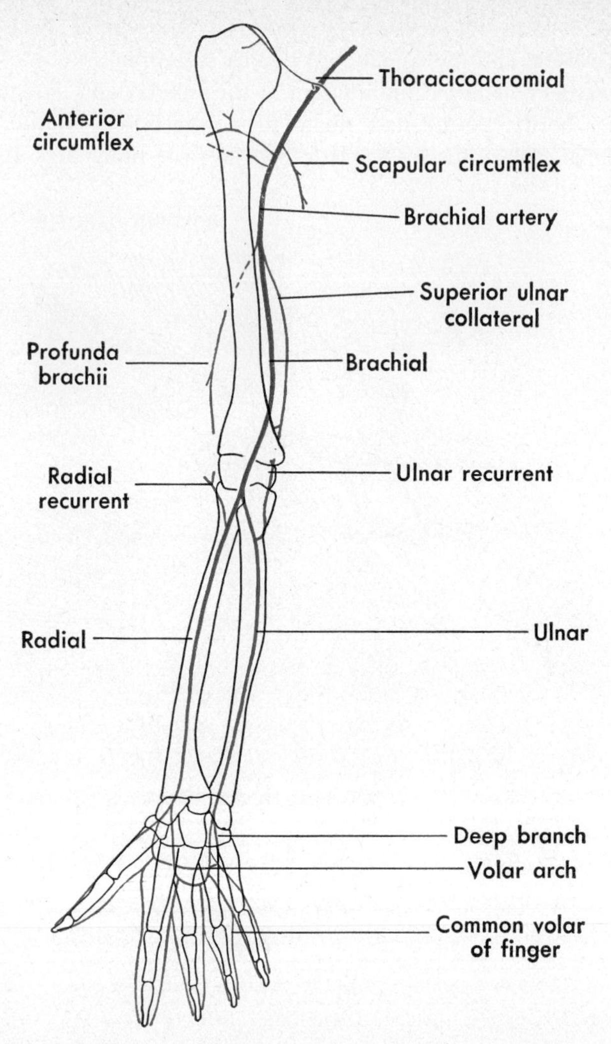

Fig. 226. Anterior view of the arteries of the arm, forearm, and hand.

distance (1 cm) below the elbow, where it divides into the radial and ulnar arteries. The upper part lies medial to the humerus; but as it passes down the arm it gradually gets in front of the bone, and at the bend of the elbow it lies midway between its epicondyles. It lies in the depression along the inner border of the biceps muscle.

Pressure made at this point from within outward against the humerus will control the blood supply to the arm.

The ulnar, the larger of the two vessels into which the brachial divides, extends along the ulnar border of the forearm into the palm of the hand, where it divides into the branches which enter into the formation of the superficial and deep volar arches (palmar arches).

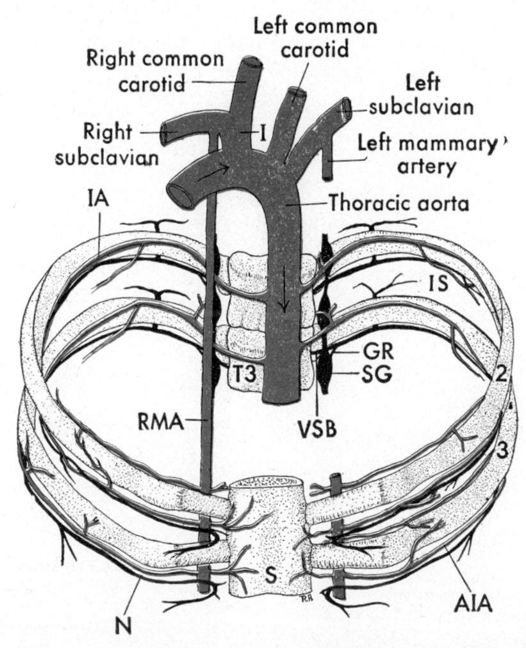

Fig. 227. Internal mammary artery and its branches to the intercostal muscles. *2* and *3,* second and third ribs; *AIA,* anterior intercostal artery; *GR,* gray ramus; *I,* innominate artery; *IA,* intercostal artery; *IS,* intercostal space; *N,* nerve; *RMA,* right mammary artery; *S,* sternum; *SG,* sympathetic ganglion; *T3,* third thoracic vertebra; *VSB,* ventral somatic branch of spinal nerve.

The radial artery appears by its direction to be a continuation of the brachial, although it does not equal the ulnar in size. It extends along the radial (thumb) side of the forearm as far as the lower end of the radius, below which it turns around the lateral side of the wrist and passes forward into the palm of the hand, where it unites with the deep volar branch of the ulnar artery to form the deep volar arch. The superficial and deep volar arches anastomose and supply the hand with blood.

SUMMARY

Blood Circulation in Upper Extremity

Blood leaves the left ventricle, traverses arteries, arterioles, capillaries, venules, and veins and is returned to the right atrium via the superior vena cava.

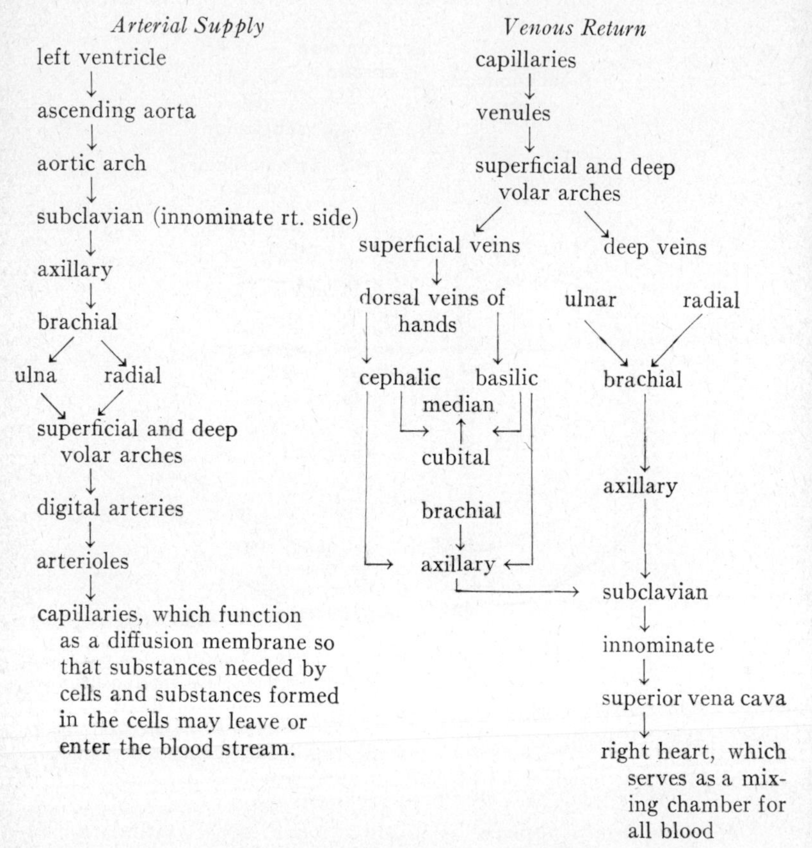

ARTERIES OF THE LOWER EXTREMITIES

The external iliac forms a large, continuous trunk, which extends downward in the lower limb and is named, in successive parts of its course, femoral, popliteal, and posterior tibial.

The femoral artery lies in the upper three-fourths of the thigh, its limits being marked above by the inguinal (Poupart's) ligament and

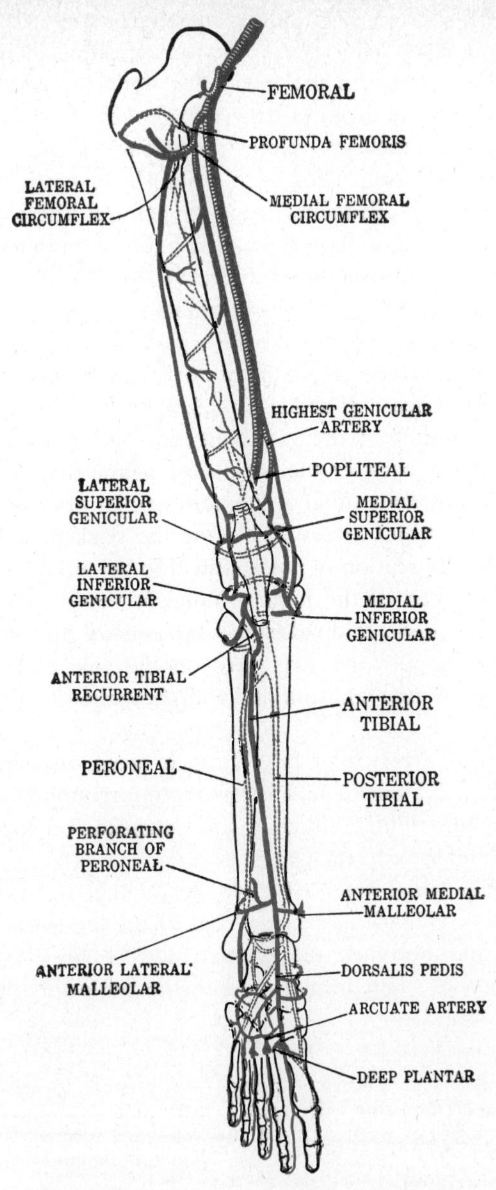

Fig. 228. Diagram of arteries of the leg.

below by the opening in the adductor magnus muscle. After passing through this opening, the artery receives the name of popliteal. In the first part of its course the artery lies along the middle of the depression on the inner aspect of the thigh, known as the femoral triangle (Scarpa's triangle).[5] Here the beating of the artery may be felt, and the circulation through the vessel may be most easily controlled by pressure. Branches from the femoral artery extend to the abdominal walls, the external genitals, and the muscles and fasciae of the thigh; and a descending branch, the *lateral femoral circumflex,* anastomoses with branches of the popliteal to form the *circumpatellar anastomosis,* which surrounds the knee joints.

The popliteal artery, a continuation of the femoral, is placed at the back of the knee. It sends branches to the knee joint, the posterior femoral muscles,[6] the gastrocnemius and soleus muscles, and to the skin of the back of the leg. Just below the knee joint it divides into the posterior tibial and anterior tibial arteries.

The posterior tibial artery lies along the back of the leg and extends from the bifurcation of the popliteal to the ankle. It distributes branches to the calf of the leg and nutrient vessels to the tibia and fibula. At the ankle it divides into the *medial* and *lateral plantar* arteries, which supply the structures on the sole of the foot, form the plantar arch, and anastomose with branches from the dorsalis pedis.

The peroneal artery is a large branch given off by the posterior tibial just about 2.5 cm (1 in.) below the bifurcation of the popliteal. The peroneal distributes blood to the structures on the medial side of the fibula and the calcaneus.

The anterior tibial artery, the smaller of the two divisions of the popliteal trunk, extends along the front of the leg to the front of the ankle joint and becomes the *dorsalis pedis artery.* The dorsalis pedis anastomoses with branches from the posterior tibial and supplies blood to the foot.

[5] The femoral triangle (Scarpa's triangle) corresponds to the depression just below the fold of the groin. Its apex is directed downward. It is bounded above by the inguinal ligament, and the sides are formed laterally by the sartorius muscle and medially by the adductor longus. Antonio Scarpa, Italian anatomist (1747–1832).

[6] These are the biceps femoris, semitendinosus, and semimembranosus.

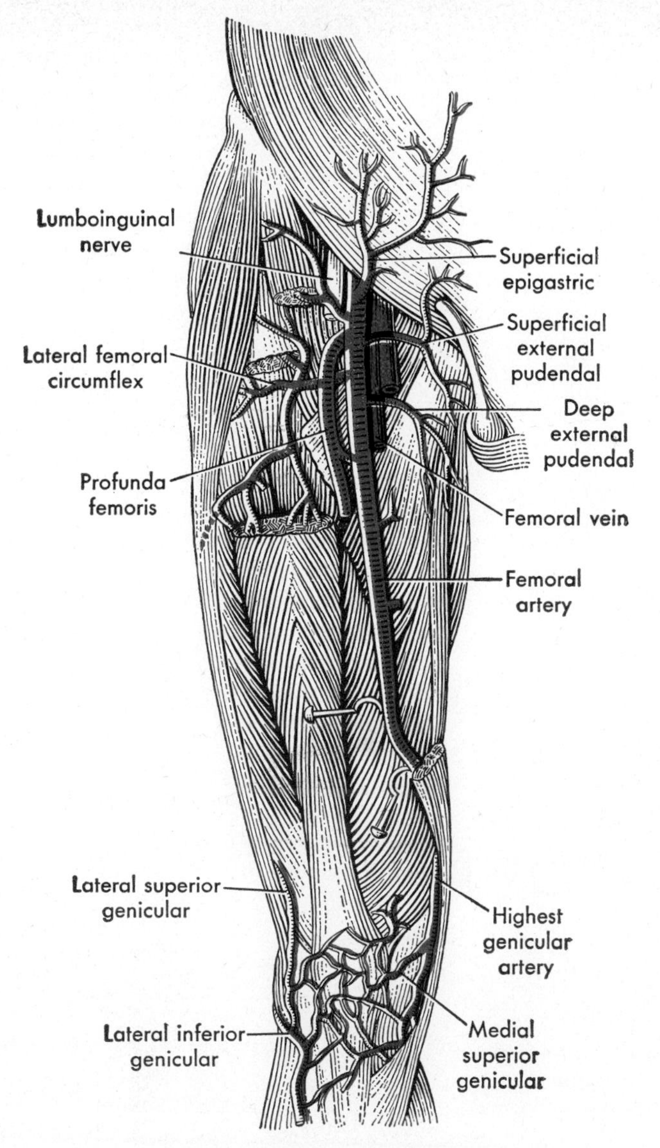

Fig. 229. Femoral artery. (Gerrish.)

SUMMARY

Blood Circulation in Lower Extremity

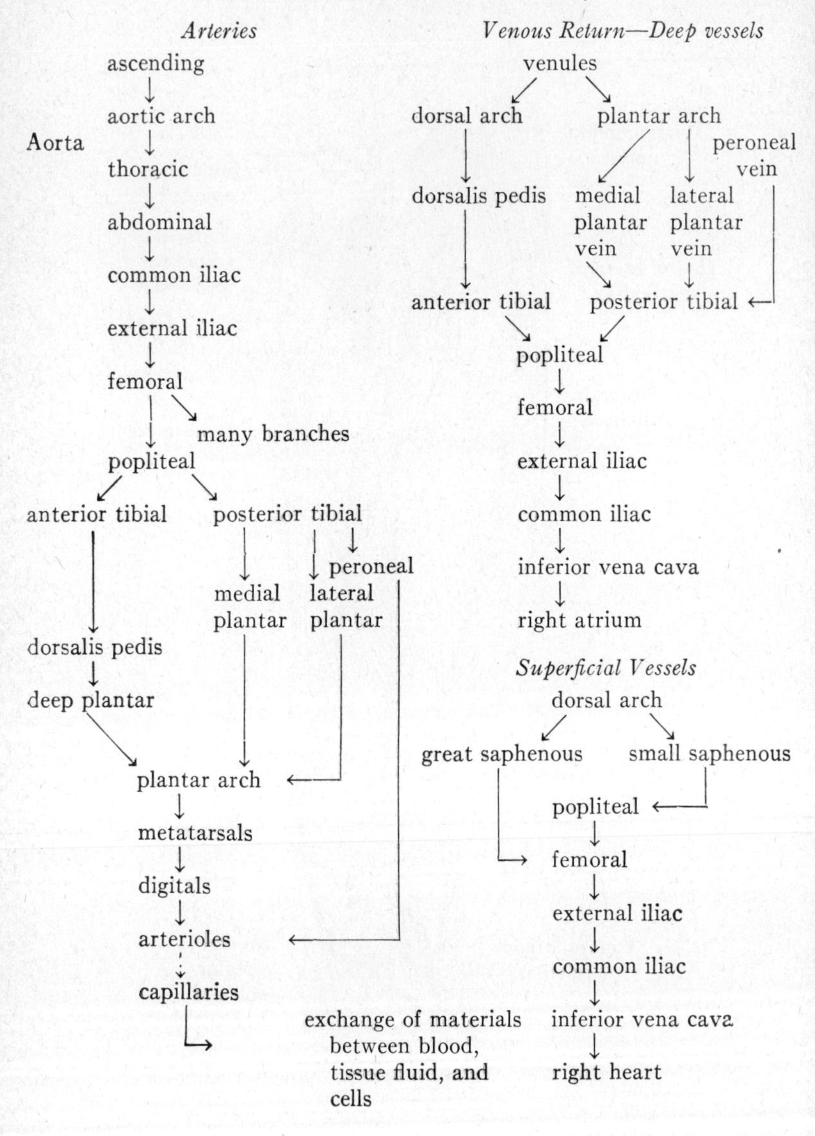

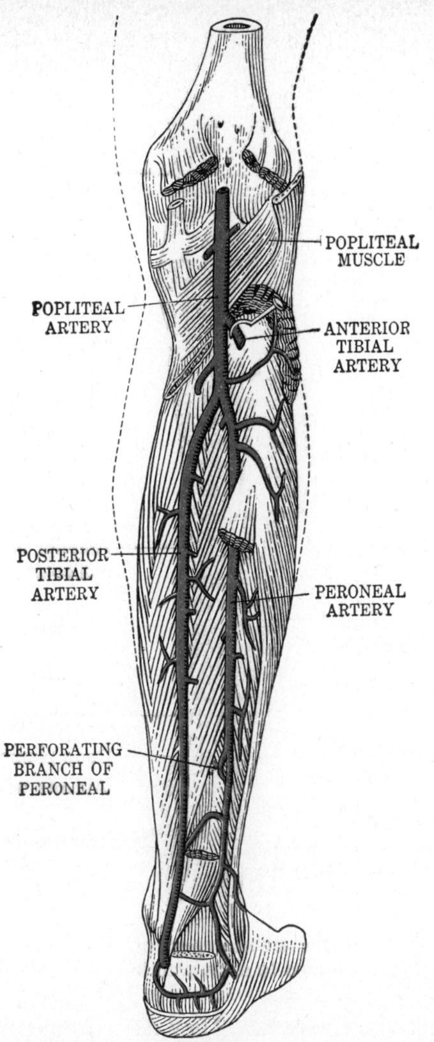

Fig. 230. Arteries in the dorsal part of the leg. **(Gerrish.)**

VEINS

The arteries begin as large trunks, which gradually become smaller and smaller until they end in arterioles, which merge into capillaries, while the veins begin as small branches, called venules, which at first are scarcely distinguishable from the capillaries and which unite to

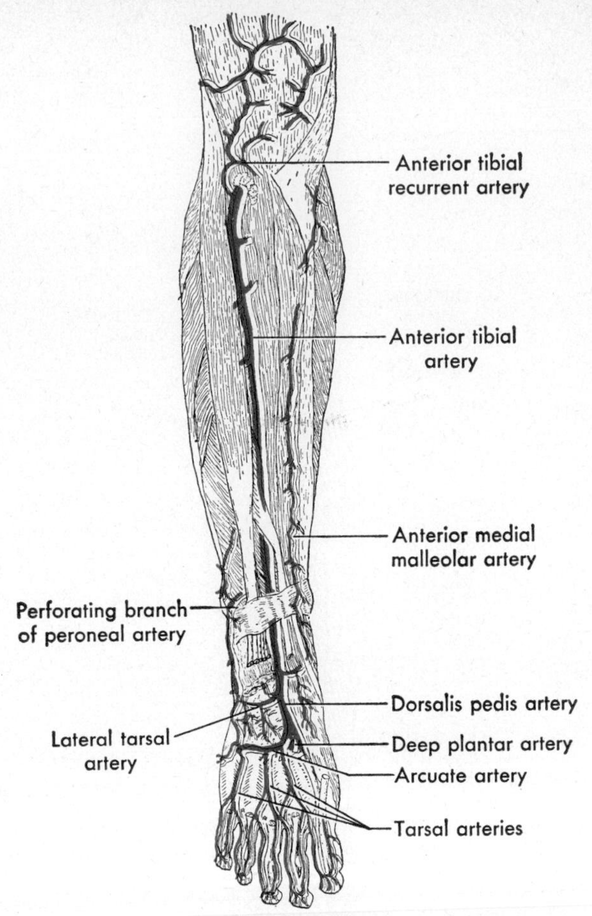

Fig. 231. Arteries on the anterior part of the leg and dorsal part of the foot. Muscles of leg separated to show arteries.

form larger and larger vessels. They differ from the arteries in their larger size, greater number, thinner walls, and the presence of valves in many of them which prevent backward circulation. There are a greater number of veins carrying blood away from an organ than arteries carrying blood to it. Speaking generally, it can be said that the total diameter of the veins returning the blood from any organ is at least twice the diameter of the arteries carrying blood to that organ.

In consequence, the description of the path of venous return cannot be as full as the description of the arteries.

The veins consist of two sets of vessels, the *pulmonary* and the *systemic veins.*

The pulmonary veins convey oxygenated blood from the lungs to the left atrium. These veins commence in the capillary network upon the air cells and unite to form one vein for each lobule. These further unite to form one vein for each lobe, two for the left lung and three for the right. The vein from the middle lobe of the right lung usually unites with that from the upper lobe, and finally two trunks from each lung are formed. They have no valves and open separately into the left atrium.

The systemic veins return the blood from all parts of the body to the right atrium of the heart. In other words, the blood distributed by the systemic arteries is returned by the systemic veins. The systemic veins are divided into three sets—superficial veins, deep veins, and venous sinuses.

The superficial veins are found just beneath the skin in the superficial fascia and return the blood from these structures. They are sometimes called cutaneous veins. The superficial and deep veins very frequently unite. The anastomoses of veins are more numerous than those of arteries.

The deep veins accompany the arteries and are usually enclosed in the same sheath. The deep veins accompanying the smaller arteries, such as the brachial, ulnar, radial, peroneal, and tibial, are found in pairs, one on each side of the vessel, and are called venae comitantes, or companion veins. Usually the larger arteries, such as the femoral, popliteal, axillary, and subclavian, have only one accompanying vein (vena comes).

In certain parts of the body the deep veins do not accompany the arteries. Examples are the veins in the skull and the vertebral canal, the hepatic veins in the liver, and the larger veins which return blood from the bones.

The venous sinuses are canals found only in the interior of the skull. They are formed by a separation of the layers of the dura mater, the fibrous membrane which covers the brain. Their outer wall consists of the dura mater, and their inner lining of endothelium is continuous with the lining membrane of the vessels that communicate with them (Fig. 182). The superior and inferior sagittal sinuses, the straight sinus, and the paired occipital sinuses are the largest of these.

The systemic veins are divided into three groups: (1) veins that empty into the heart, (2) veins that empty into the superior vena cava, and (3) veins that empty into the inferior vena cava.

Five of the veins of the heart empty into the right atrium by way of the coronary sinus. Some smaller veins empty directly into the atria and ventricles (p. 349).

The veins of the head, neck, upper extremities, and thorax, and the azygos veins empty into the *superior vena cava,* which carries the blood to the right atrium.

The veins of the lower extremities and of the abdomen and pelvis empty into the *inferior vena cava,* which carries the blood to the right atrium. The azygos veins, however, deliver part of this blood to the superior vena cava.

VEINS OF THE NECK

The blood returning from the head and face flows on each side into two principal veins, the external and internal jugular.

The external jugular veins are the chief *superficial* veins of the neck. They are formed in the substance of the parotid glands by the union of the posterior facial and the posterior auricular veins of each side of the face. This union takes place on a level with the angle of the mandible, and each vein descends almost vertically down the neck to its termination in the subclavian vein. These two veins receive the blood from the deep parts of the face and the exterior of the cranium.

The internal jugular veins are continuous with the lateral sinuses and begin in the jugular foramen at the base of the skull. They descend on either side of the neck, first with the external carotid, then with the common carotid, and join the subclavian at a right angle to form the innominate (brachiocephalic) vein. They receive the blood from the veins and sinuses of the cranial cavity, from the superficial parts of the face, and from the neck. In a general way the tributaries of the internal jugular veins correspond to the branches of the external carotid arteries (Fig. 234).

VEINS OF THE UPPER EXTREMITIES

The blood from the upper limbs is returned by a deep and a superficial set of veins. The deep veins are the venae comitantes of the forearm and arm and are called by the same names as the arteries, i.e., the deep volar venous arches, metacarpal veins, radial

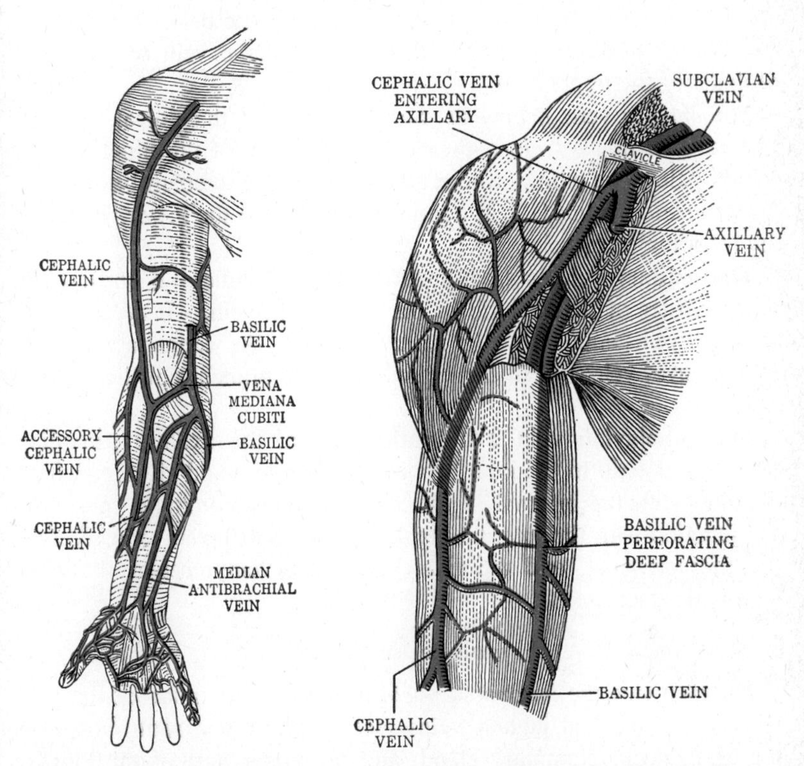

Fig. 232. Superficial veins of the forearm, arm, and hand.

Fig. 233. Superficial veins of front of arm and shoulder. (Gerrish.)

and ulnar veins, brachial veins, axillary veins, and subclavian veins. The deep veins have numerous anastomoses with one another and with the superficial veins.

The superficial veins are much larger than the deep veins and take a greater share in returning the blood, especially from the distal portion of the limb. They commence in two plexuses, one on the back of the hand, formed by the dorsal metacarpal veins—the dorsal venous network—and another plexus situated over the thenar and hypothenar eminences and across the front of the wrist. They include the following:

The cephalic vein begins in the dorsal network and winds upward around the radial border of the forearm to a little below the bend of the elbow, where it joins the accessory cephalic vein to form the cephalic of the upper arm.

The basilic vein begins in the ulnar part of the dorsal network and extends upward along the posterior surface of the ulnar side to a little below the elbow, where it is joined by the median basilic vein (*vena mediana cubiti*). It continues upward to the lower border of the teres major muscle.

The axillary vein is a continuation of the basilic. It ends at the outer border of the first rib in the subclavian vein. It receives the brachial veins and, close to its termination, the cephalic vein. It also receives veins which correspond with the branches of the axillary artery.

The subclavian vein is a continuation of the axillary extending from the first rib to the joint between the sternum and clavicle, where it unites with the internal jugular to form the innominate vein. At the junction with the internal jugular the left subclavian vein receives the thoracic duct, and the right subclavian vein receives the right lymphatic duct.

VEINS OF THE THORAX

On each side the *innominate vein* formed by the union of the subclavian and internal jugular veins receives the blood returning from the head, neck, mammary gland, and the upper part of the thorax. These veins transmit this blood to the superior vena cava.

The right innominate is about 2.5 cm (1 in.) in length, and the left is about 6 cm ($2\frac{1}{2}$ in.) in length.

The superior vena cava is formed by the union of the right and left innominate veins, just behind the junction of the first right costal cartilage with the sternum. It is about 7.5 cm (3 in.) long and opens into the right atrium, opposite the third right costal cartilage.

A supplementary channel between the inferior and superior venae cavae is formed by the azygos veins. In case of obstruction, these

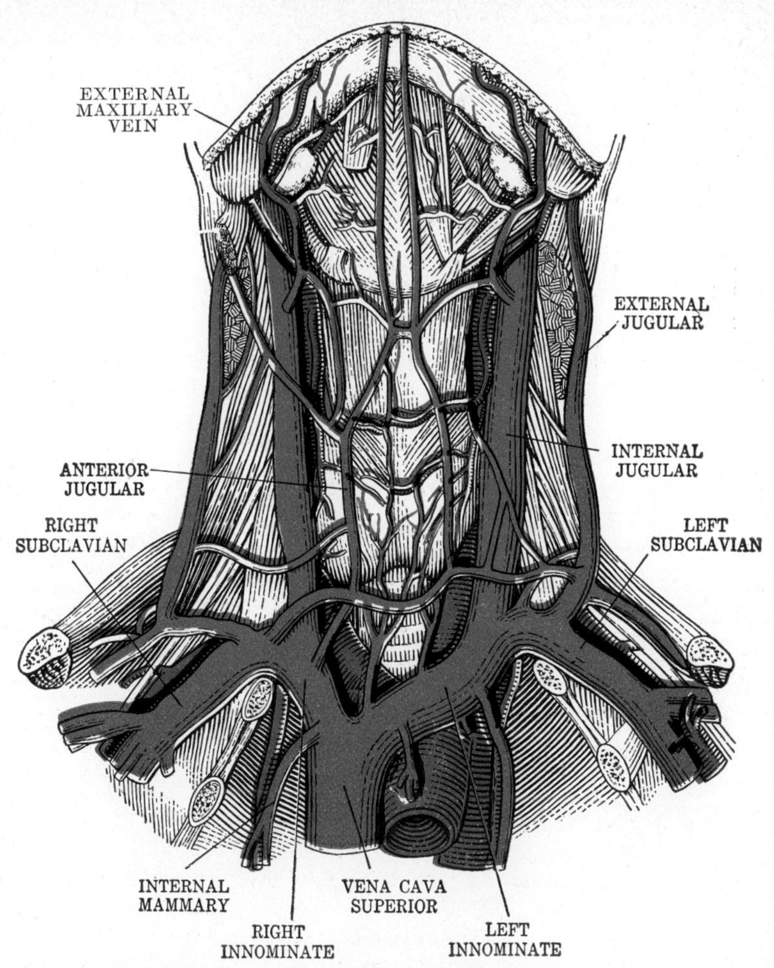

Fig. 234. Veins of the neck and upper part of thorax, front view. (Gerrish.)

veins form a channel by which blood can be conveyed from the lower part of the body to the superior vena cava. They are three in number and lie along the front of the vertebral column.

The azygos vein (right, or major, azygos) begins opposite the first or second lumbar vertebra as the *right ascending lumbar vein*[7] or by a branch of the *right renal vein* or from the *inferior vena cava.*

[7] The lumbar veins empty into the inferior vena cava. They correspond to the lumbar arteries given off by the abdominal aorta and return the blood from the muscles and skin of the loins and walls of the abdomen.

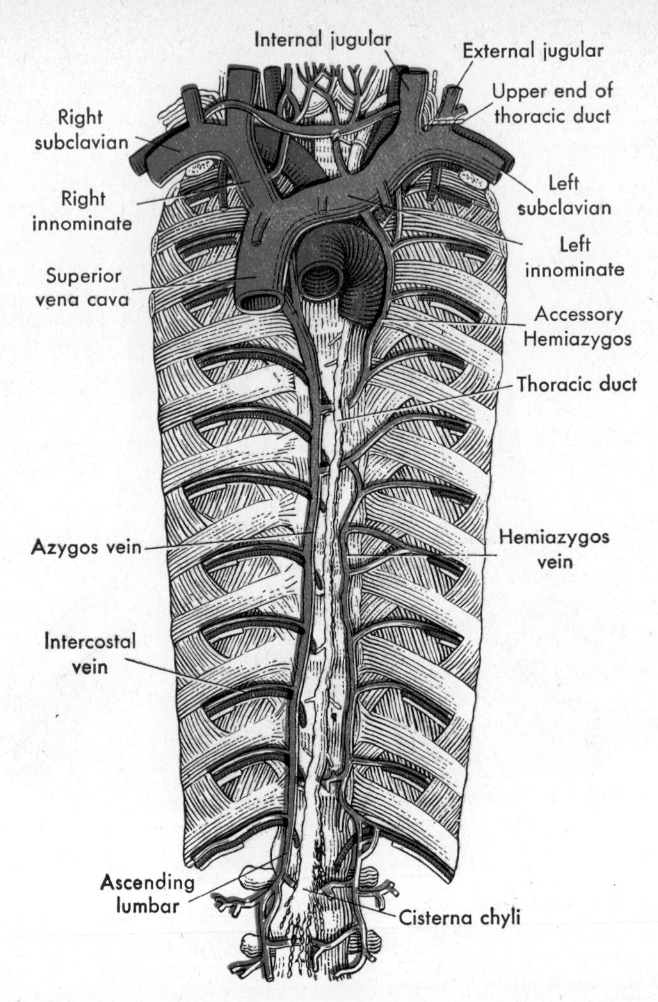

Fig. 235. Azygos and intercostal veins, also thoracic duct. (Gerrish.)

It ascends on the right side of the vertebral column to the level of the fourth thoracic vertebra, where it arches over the root of the right lung and empties into the superior vena cava.

The hemiazygos vein (left lower, or minor, azygos) begins in the left lumbar or renal vein. It ascends on the left side of the vertebral column, and at about the level of the ninth thoracic vertebra it connects with the right azygos vein. It receives the lower four or five intercostal veins of the left side and some esophageal and mediastinal veins.

The accessory hemiazygos vein (left upper azygos) connects above with the highest left intercostal vein and opens below into either the azygos or the hemiazygos. It varies considerably in size, position, and arrangement. It receives veins from the three or four intercostal spaces between the highest left intercostal vein and highest tributary of the hemiazygos; the left bronchial vein sometimes opens into it.

The azygos veins return blood from the intercostal muscles, etc., to the superior vena cava. The internal mammary veins are venae comitantes for the internal mammary arteries and are tributary to the right and the left innominate veins respectively.

The bronchial veins. A bronchial vein is formed at the root of each lung and returns the blood from the larger bronchi and from the structures at the root of the lung; that of the right side opens into the azygos vein near its termination, that of the left side into the highest left intercostal or the accessory hemiazygos vein. A considerable quantity of the blood which is carried to the lungs through the bronchial arteries is returned to the left side of the heart through the pulmonary veins.

VEINS OF THE LOWER EXTREMITIES

The blood from the lower limbs is returned by a superficial and a deep set of veins. The superficial veins are beneath the skin between the layers of superficial fascia. The deep veins accompany the arteries. Both sets are provided with valves, which are more numerous in the deep than in the superficial veins.

The superficial veins of the lower extremities are the great saphenous veins, the small saphenous veins, and their tributaries. The superficial veins of the foot form venous arches on the dorsum and sole of the foot. These arches communicate with each other and receive branches from the deep veins. They drain the blood into a *medial* and *lateral* marginal vein.

The great saphenous vein begins in the medial marginal vein of the dorsum of the foot, extends upward on the medial side of the leg and thigh, and ends in the femoral vein a little more than 3 cm ($1\frac{1}{4}$ in.) below the inguinal ligament. At the ankle it receives branches from the sole of the foot; in the leg it anastomoses with the small saphenous vein and receives many cutaneous veins. In the thigh it receives many branches. Those from the posterior and medial aspects of the thigh frequently unite to form an accessory saphenous vein which joins the great saphenous.

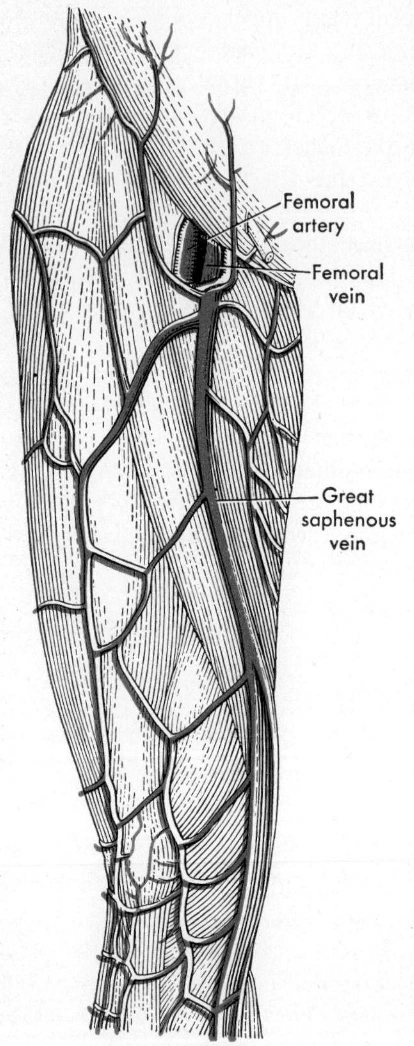

Fig. 236. Superficial veins of the front of the right thigh. (Gerrish.)

The small saphenous begins behind the lateral malleolus, as a continuation of the lateral marginal vein, and passes up the back of the leg to end in the deep popliteal vein. It receives many branches from the deep veins on the dorsum of the foot and from the back of the leg. Before it joins the popliteal, it gives off a branch that runs upward and forward and joins the great saphenous.

The deep veins accompany the arteries below the knee. They are in pairs and are called by the same names as the arteries. The veins from the foot empty into the *anterior tibial* and *posterior tibial veins.* They unite to form the single *popliteal vein,* which is continued as the *femoral* and becomes the *external iliac.*

The femoral veins are continuations of the popliteal veins and extend from the opening in the adductor magnus muscles to the level of the inguinal ligament. Each one receives numerous branches, and near its termination it is joined by the great saphenous vein.

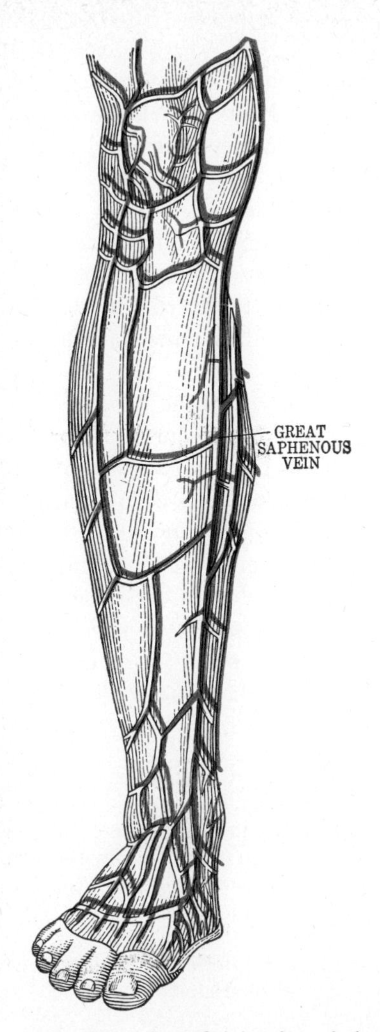

Fig. 237. Superficial veins of the front of the leg and foot. (Gerrish.)

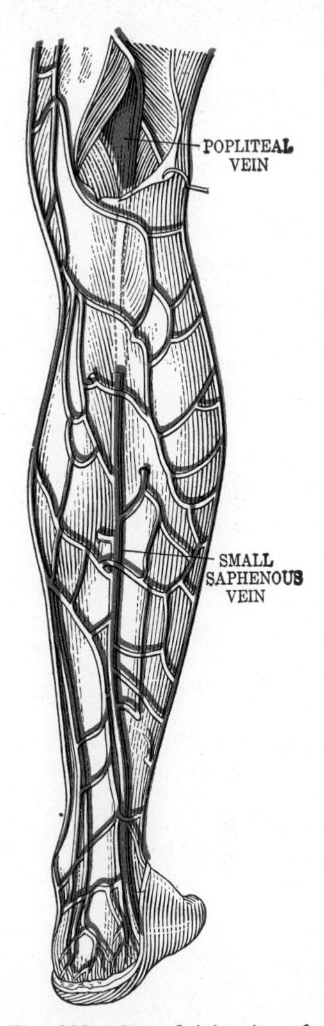

Fig. 238. Superficial veins of the dosum of the leg. (Gerrish.)

Go back and study all of the figures of arteries and veins of the extremities. Then think through the following questions: (1) To control arterial bleeding should pressure or a tourniquet be applied between the heart and injury, or below the injury? Explain. (2) To control venous bleeding where should the pressure or a tourniquet be applied? Explain.

VEINS OF THE ABDOMEN AND PELVIS

The external iliac veins are continuations of the femoral veins and extend from the level of the inguinal (Poupart's) ligaments on either side to the joint between the sacrum and the ilium.

The hypogastric veins are formed by the union of veins corresponding to the branches of the hypogastric arteries. They accompany the hypogastric arteries and unite with the external iliac veins to form the common iliacs (Fig. 220).

The common iliacs extend from the base of the sacrum to the fifth lumbar vertebra, and then the two common iliacs unite to form the inferior vena cava (Fig. 220).

The inferior vena cava begins at the junction of the two common iliacs and thence ascends along the right side of the aorta, perforates the diaphragm, and terminates by entering the right atrium of the heart. The shape and position of the inferior vena cava are comparable to those of the abdominal aorta, and the vein returns blood from the parts below the diaphragm. It receives veins having the same names as the parietal and visceral branches of the abdominal aorta. These veins are (1) lumbar, (2) renal, (3) suprarenal, (4) inferior phrenic, (5) hepatic, (6) right spermatic or ovarian. Most of these veins accompany the arteries of the same names.

There are a few exceptions:

1. The right suprarenal vein empties into the inferior vena cava; the left empties into the left renal or left inferior phrenic.
2. The right inferior phrenic empties into the inferior vena cava; the left often consists of two branches, one of which empties into the left renal or suprarenal vein and the other into the inferior vena cava.
3. The hepatic veins empty into the inferior vena cava, but they commence in the sinusoids of the liver.
4. The right spermatic vein empties into the inferior vena cava, but the left empties into the left renal vein. The ovarian veins end in the same way as the spermatic veins in the male.

The veins that return the blood distributed by the branches of the celiac and the superior and inferior mesenteric are included in the portal system.

The portal system. The veins which bring back the blood from the spleen, stomach, pancreas, and intestines are included in the portal system. Blood is collected from the spleen by veins which unite to form the *splenic vein.* This vein passes back of the pancreas

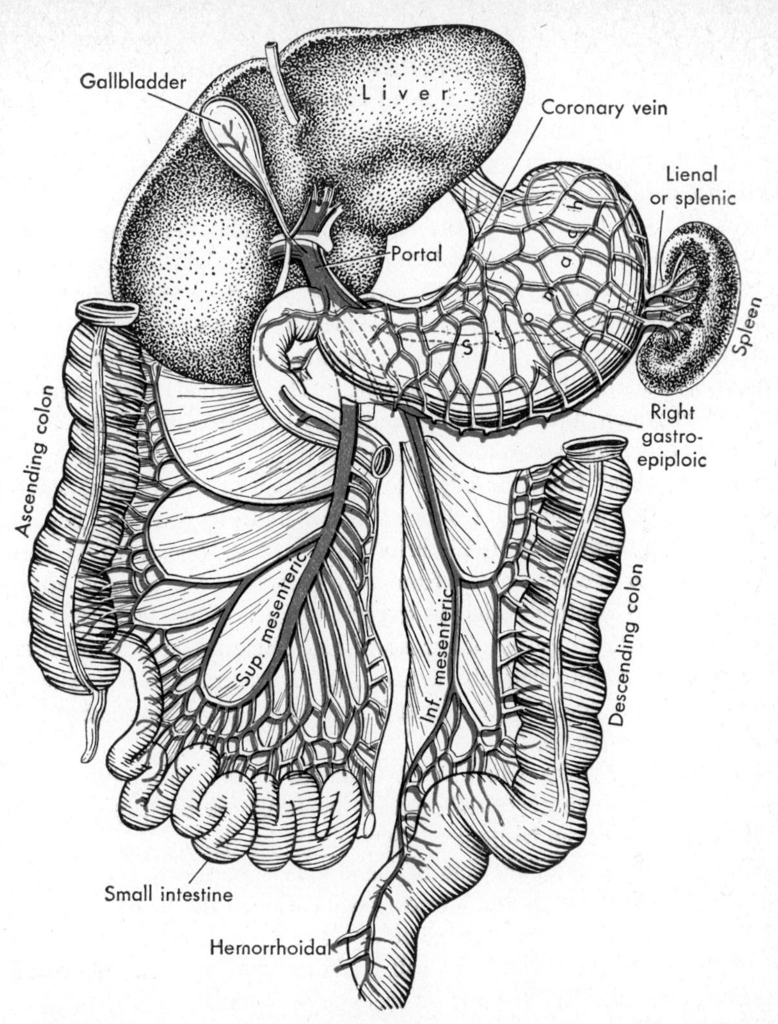

Fig. 239. Portal system of veins. The liver is turned upward and backward, and the transverse colon and most of the small intestine are removed. (Gerrish.)

from left to right and ends by uniting with the *superior mesenteric* to form the *portal tube* (*vein*). Before this union takes place, the splenic receives *gastric veins* (BNA, vena coronaria), *pancreatic veins,* and the *inferior mesenteric vein,* which returns the blood from the rectum, sigmoid, and descending colon. The *superior mesen-*

teric vein returns the blood from the small intestine, the cecum, and the ascending and transverse portions of the colon (Fig. 239).

The portal tube, or *vein,* formed at the level of the second lumbar vertebra by the union of the splenic and the superior mesenteric, passes upward and to the right to the transverse fissure of the liver. Here it divides into a right and a left branch, which accompany the right and left branches of the hepatic artery into the right and left lobes of the liver. Before entering the liver, the right branch usually receives the cystic vein, which returns blood from the gallbladder. The hepatic artery brings blood direct from the aorta, via the celiac artery to the liver. In the liver, blood from both sets of vessels enters into the interlobular vessels. These divide into smaller branches which finally form the capillary-like network of the liver. These are called sinusoids. Many of these capillaries follow the "chain of cells" from the edges to the centers of the lobules, where they form the intralobular veins. These unite to form the hepatic veins which finally enter the inferior vena cava. Thus, it will be seen that the liver receives blood from two sources. The hepatic artery carries blood relatively high in oxygen, since it comes directly from the left ventricle by way of the aorta and celiac artery. The portal tube carries blood relatively high in soluble materials such as simple sugars, amino acids, and digested fats which have been absorbed from the digestive tract.

Some of the veins which are tributaries to the portal tube have small branches whose blood reaches the heart via the superior and inferior venae cavae without going through the liver. For example, branches of the coronary vein of the stomach unite with the esophageal vein which enters the azygos, on its way to the heart, thus bypassing the liver. They are called *accessory portal tubes* (*veins*).

The portal tube (vein) and all its branches constitute the portal system, described as a third, or accessory, system.

SUMMARY

Arteries
- Begin as large trunks, grow smaller
- Usually deep-seated for protection
- Division
  - Trunk gives off several branches
  - One branch that divides into several
  - Two branches of nearly equal size
- Anastomosis—distal ends unite
- Plexus—many anastomoses within limited area

- **Division of the Vascular System**
  - **Pulmonary system**
    - (1) Pulmonary artery conveys venous blood from right ventricle to lungs
      - Right pulmonary artery—right lung
      - Left pulmonary artery—left lung
    - (2) Capillaries connect arterioles and venules
    - (3) Four pulmonary veins—two from each lung—convey oxygenated blood to left atrium
  - **Systemic system**
    - Provides for systemic circulation
    - (1) Aorta and all its branches
    - (2) Capillaries connect arterioles and venules
    - (3) Veins empty into heart either directly or by means of inferior and superior venae cavae

- **Aorta**
  - **I Ascending aorta**
    - About 2 in. long
    - Branches
      - Right coronary
      - Left coronary
  - **II Arch of aorta**
    - Extends from ascending aorta to body of fourth thoracic vertebra
    - Branches
      - Innominate
        - Right common carotid
        - Right subclavian
      - Left common carotid
      - Left subclavian
  - **III Descending aorta**
    - Thoracic aorta
      - Extends from lower border of fourth thoracic vertebra to aortic opening in diaphragm
      - Branches supply body wall and viscera of thorax
    - Abdominal aorta
      - Extends from aortic opening in diaphragm to body of fourth lumbar vertebra. Branches supply body wall and viscera of abdominal cavity

- **Arteries of the Chest**
  - **Visceral group**
    - **Pericardials**—to pericardium
    - **Bronchials**—nutrient vessels of lungs
    - **Esophageals**—to esophagus
    - **Mediastinals**—to nodes and areolar tissue in mediastinum
  - **Parietal group**
    - **Intercostals**—to lower nine intercostal spaces
    - **Subcostals**—anastomose with superior epigastrics, lower intercostals, and lumbar arteries
    - **Superior phrenics**—upper surface of diaphragm

- **Arteries of the Abdomen, Visceral Branches**
  - **Celiac artery**
    - **Left gastric**—lesser curvature of stomach left to right
    - **Hepatic artery** divides into right and left branches before entering liver
      - **Right gastric**—lesser curvature of stomach right to left
      - **Gastroduodenal**
        - *Superior pancreaticoduodenal* to duodenum and head of pancreas
        - *Right gastroepiploic*—greater curvature of stomach right to left
      - **Cystic**—gallbladder
    - **Lienal, or splenic**
      - Branches to pancreas
      - **Left gastroepiploic**—greater curvature of stomach left to right
  - **Superior mesenteric**
    - Supplies small intestine except duodenum, cecum, ascending colon, half of transverse colon
  - **Inferior mesenteric**
    - Supplies left half transverse colon, whole of descending and sigmoid colon, continued as *superior hemorrhoidal artery* to rectum
  - **Middle suprarenals**
    - To suprarenal glands—anastomose with branches of phrenic and renal arteries
  - **Renal arteries**
    - Each divides into four or five branches before entering kidney
  - **Internal spermatics**
    - Supply the testes
  - **Ovarian arteries**
    - Supply ovaries, send small branches to the ureters and uterine tubes; one branch unites with uterine artery and assists in supplying uterus
- **Arteries of the Abdomen, Parietal Branches**
  - **Inferior phrenics**—distributed to under surface of diaphragm
  - **Lumbar arteries**
    - Distribute branches to muscles and skin of back, to the spinal cord and its membranes, and to the lumbar vertebrae
  - **Middle sacral**—passes down to coccygeal gland
- **Arteries of the Pelvis**
  - **Common iliacs** about 5 cm
    - **Hypogastric, or Internal iliacs**
      - Branches to pelvic walls, viscera, genitals, buttocks, medial side of thighs
    - **External iliacs**
      - Branches to psoas major, etc.
      - Inferior epigastric
      - Deep iliac circumflex

Arteries of the Head and Face

- **Left common carotid** arises from arch of aorta
  - **External carotid** — Branches supply thyroid gland, tongue, throat, face, ear, and dura mater
  - **Internal carotid** — Branches same as on right side
- **Right common carotid** arises at division of innominate
  - **External carotid** — Branches same as on left side
  - **Internal carotid** — Branches supply brain, eye and its appendages, forehead, and nose
- **Circle of Willis**
  - Anastomosis at base of brain
  - Formed by union of
    - Anterior cerebral arteries connected by anterior communicating arteries
    - Posterior cerebral arteries connected by posterior communicating arteries

Arteries of the Upper Extremities

- **Subclavian**—forms main artery of each upper limb. Different portions named according to regions through which they pass
- **Right subclavian**
  - Extends from the innominate to the first rib
  - Branches are
    - Vertebral, thyrocervical
    - Internal mammary
    - Costocervical
- **Left subclavian**
  - Extends from arch of aorta to first rib
  - Same branches as right subclavian
- **Axillary arteries**
  - In axillary regions
  - Branches to shoulders, chest, and arms
- **Brachial arteries**
  - Extend from axillary arteries to below bend of elbows, where they divide into ulnar and radial arteries
- **Ulnar arteries**
  - Extend along ulnar border of forearms to palms of hands
- **Radial arteries**
  - Extend along radial side of forearms to palms of hands
- (Ulnar and radial arteries) Form the superficial and deep *volar arches*

Arteries of the Lower Extremities

- **External iliacs** form main arteries of lower limbs. Different portions named according to regions through which they pass
- **Femoral arteries**
  - Extend along inguinal ligaments to openings in adductor magnus muscles. Send branches to abdominal walls, external genitals, muscles and fasciae of the thighs

Arteries of the Lower Extremities

- **Popliteal arteries** — Back of knees, send branches to knee joints, posterior femoral muscles, gastrocnemius, soleus, skin of back of legs. Below knee joints divide into posterior tibials and anterior tibials
- **Posterior tibials** — Back of legs, from bifurcation of popliteal to ankle. Send branches to back of legs and to tibiae and fibulae. Give off peroneal arteries about 1 in. below bifurcation of popliteals. *Peroneals* distribute blood to fibulae and calcaneus bones
- **Anterior tibials** — Front of legs from bifurcation of popliteal to ankle—then become the *dorsalis pedis arteries*
- **Dorsalis pedis arteries** — Dorsum of each foot, anastomose with branches from posterior tibials and supply blood to feet

Veins

- Begin small, grow larger
- **Differ from arteries**
  - Larger size
  - Greater number
  - Thinner walls
  - Valves
  - More frequent anastomoses
- **Sets**
  - Superficial, or cutaneous, beneath the skin
  - Deep
    - Usually accompany the arteries
    - Exceptions are
      - Veins in skull and vertebral canal
      - Hepatic veins
      - Large coronary veins
  - Venous sinuses—canals formed by separation of layers of dura mater
- **Venae comitantes** — Deep veins accompanying smaller arteries, such as brachial, radial, ulnar, peroneal, tibial, are in pairs. A single deep vein accompanying a larger artery, such as femoral, popliteal, axillary, and subclavian artery, is called a *vena comes*
- **Three groups**
  - **Coronary veins** from heart
  - **Superior vena cava** — Veins of head, neck, thorax, and upper extremities empty into this vein
  - **Inferior vena cava** — Veins of abdomen, pelvis, and lower extremities empty into this vein

Veins of the Neck

- **External jugulars** — Formed in parotid glands, terminate in the subclavians. Receive blood from deep parts of the face and the exterior of the cranium

Veins of the Neck	**Internal jugulars**	Continuous with the lateral sinuses, unite with subclavians to form the innominates. Receive blood from the veins and sinuses of the cranial cavity, superficial parts of face and neck	
Veins of the Upper Extremities	**Superficial veins**	Are larger, take a greater share in returning blood	
		Cephalics	Begin in dorsal venous network, join accessory cephalics of arms below elbows, empty into axillaries
		Basilics	Begin in dorsal venous network, are joined by median basilics below elbows, are continued as axillaries
	Deep veins	Accompany arteries, are called by same names, i.e., metacarpals, radials, ulnars, brachials, axillaries, and subclavians	
		Axillaries	Are continuations of the basilics, end at outer border of first ribs, receive brachials, cephalics, and deep veins
		Subclavians	Are continuations of the axillaries, unite with internal jugulars to form innominates
Veins of the Thorax	**Innominates**	Formed by union of internal jugular and subclavians. Receive internal mammary veins One on each side of body	
	Superior vena cava	Formed by union of right and left innominate veins. 7.5 cm long Opens into right atrium	
	Supplementary channel	(1) **Azygos vein** (2) **Hemiazygos vein** (3) **Accessory hemiazygos vein**	Connect with superior vena cava above and inferior vena cava below
	Bronchial veins	Formed at the root of each lung Return blood from larger bronchi and structures at root of lungs Right bronchial vein empties into azygos Left bronchial vein empties into highest left intercostal or the accessory hemiazygos	

Veins of the Lower Extremities

- **Superficial veins**
  - Are between the layers of superficial fasciae
  - Provided with valves
  - **Great saphenous veins**
    - Begin in medial marginal veins, extend upward, and end in femoral veins. Receive branches from soles of feet. Anastomose with small saphenous veins, receive cutaneous veins and accessory saphenous veins
  - **Small saphenous**
    - Continuation of lateral marginal veins, end in deep popliteal veins
    - Receive branches on dorsum of each foot and back of each leg
- **Deep veins**
  - Accompany the arteries and are called by same names. Provided with many valves
  - **Popliteals**
    - Formed by union of anterior tibials and posterior tibials
  - **Femorals**
    - Continuation of the popliteals and extend from opening in adductor magnus muscles to the inguinal ligaments
    - (1) Receive blood from the superficial veins
    - (2) Receive blood from deep veins of feet, legs, and thighs

Veins of the Abdomen and Pelvis

- **External iliacs**
  - Continuation of femoral veins. Extend from inguinal ligaments to the joints between sacral and iliac bones
- **Hypogastrics**
  - Formed by union of veins corresponding to branches of hypogastric arteries
- **Common iliacs**
  - Formed by union of external iliacs and hypogastrics. Extend from base of sacrum to the fifth lumbar vertebra
- **Inferior vena cava**
  - Formed by union of the common iliacs
  - Extends from fifth lumbar vertebra to the right atrium of the heart
  - Receives many tributaries corresponding to arteries given off from the aorta

Portal System

- Blood is collected from spleen by veins that unite to form lienal vein
- **Lienal vein receives**
  - *Gastric veins* (coronary veins) return blood from stomach

Portal System	**Lienal vein** receives	*Pancreatic veins* return blood from pancreas *Inferior mesenteric vein* returns blood from rectum, sigmoid, descending colon
	Superior mesenteric vein	Returns blood from small intestine, ascending and transverse colon and unites with lienal vein to form portal vein or tube
	Portal tube	Carries blood to liver, breaks up into capillary-like vessels termed sinusoids, then unites with capillaries from hepatic artery to form hepatic vein
	Hepatic veins	Formed by union of sinusoids and capillaries from hepatic artery Empty into inferior vena cava

CHAPTER 13

MAINTAINING CIRCULATION

CIRCULATION: PULMONARY, SYSTEMIC, CORONARY
THE HEART: FUNCTION, CARDIAC CYCLE, CONTROLS
FACTORS MAINTAINING CIRCULATION
BLOOD PRESSURE: ARTERIAL, CAPILLARY, VENOUS

The function of the heart is to adjust the circulation in relation to the metabolic rate of body cells. By chemical and nervous control, the needs of these cells are met promptly through the adjustments of pulse rate and pulse volume, which increase and decrease the velocity and volume of blood in the tissue capillaries. Variable cellular needs are thus met by changes in the number, size, and area of the open capillaries and in the temperature and minute volume of the blood in these open capillaries.

The blood is contained in a closed set of tubes, which it completely fills. Interposed in this set of tubes is the heart, which fills with blood from the veins and then contracts, thereby forcing this blood into the capillaries of all parts of the body. The summaries on pp. 405 and 406 give the details of circulation.

This is a description of the general circulation, but the student must understand that both sides of the heart contract almost simultaneously, i.e., the blood fills the atria and ventricles on both sides of the heart at the same time; both atria contract practically together,[1] forcing the blood over the open valves into the ventricles. After a brief pause both ventricles contract, and the blood is forced into the pulmonary

[1] Careful measurements have shown that the contraction of the left atrium lags behind that of the right atrium from 0.01 to 0.03 seconds.

artery and into the aorta. The ventricles pump out equal quantities of blood, but the blood from the left ventricle is sent on a longer journey than the blood from the right.

The pulmonary circulation. The lesser circulation, from the right ventricle to the left atrium, is called the *pulmonary circulation.* The purpose of the pulmonary circulation is to carry the blood which has been through the body, giving up oxygen and collecting carbon dioxide, to the air sacs of the lungs, where the red cells are recharged

Movement of Blood through the Right Heart and Lungs

Superior, inferior venae cavae, coronary sinus—blood enters ↓	During diastole of the atria, via the superior and the inferior venae cavae, coronary sinus, and other small vessels, blood enters and fills the right atrium and ventricle, which for the time may be thought of as a single chamber with the tricuspid valve open.
Right atrium	The atrium contracts (atrial systole) and forces the blood over the open valve into the ventricle, which has been passively filled and now becomes well distended by the extra supply.
↓ Tricuspid valve	After a brief pause (0.1 sec) the ventricle contracts, and the blood gets behind the cusps of the tricuspid valve and closes the valve.
Right ventricle ↓ Semilunar valve	Pressure rises within the ventricle until it exceeds the pressure in the pulmonary artery and the pulmonary semilunar valve opens and blood moves on into the pulmonary artery.
Pulmonary artery ↓ Divides into two branches	The pulmonary artery divides into the right and left branches and takes blood into the lungs.
↓ Lungs	Here blood passes through innumerable capillaries that surround the alveoli of the lungs. Blood gives up carbon dioxide and the red cells are recharged with oxygen.
↓ Capillaries unite to form veins ↓ Pulmonary veins ↓ Left atrium	The venules unite to form larger veins until finally two pulmonary veins from each lung are formed. These return the oxygenated blood to the heart and complete the pulmonary circulation.

Movement of Blood through the Left Heart and to the Somatic Capillaries

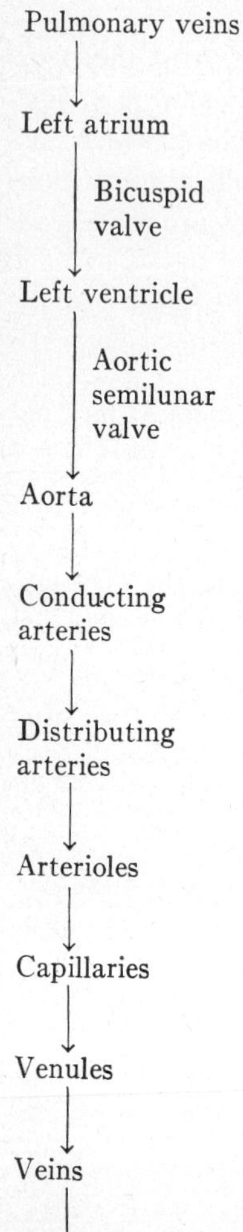

During diastole of the atria, oxygenated blood from the pulmonary veins enters the left atrium and fills it. The left ventricle relaxes and the pressure within it falls, the bicuspid valve opens and blood enters the ventricle and fills it. Systole of the atrium begins and ventricular filling is completed.

Blood gets behind the cusps of the bicuspid valve and closes them. Ventricular systole is initiated. Intraventricular pressure rises, and when it exceeds the pressure in the aorta the aortic semilunar valve opens and blood moves on into the aorta under high pressure.

Blood moves on into the conducting or elastic arteries (innominate, subclavian, common carotids, internal iliac, femoral, etc.).

Blood is forced onward by the recoil of these stretched elastic arteries into the distributing (muscular arteries) such as the axillary, radial, popliteal, tibial, etc., and finally into the arterioles, where blood is moving in a steady stream, and then on into the capillaries, where the main work of the vascular bed is accomplished.

The capillaries unite to form venules and these in turn unite to form veins, then larger veins, until blood finally reaches the right atrium and the circuit begins again. This is known as systemic circulation.

with oxygen and the carbon dioxide is reduced to the standard amount.

The systemic circulation. The more extensive circulation, from the left ventricle to all parts of the body and the return to the right atrium, is known as the *systemic circulation*. The purpose of the systemic circulation is to carry oxygen and nutritive material to the tissues and gather up waste products from the tissues. After leaving the left ventricle, portions of the blood pursue different courses; some portions enter the coronary arteries, some go to the head, some to the upper and lower extremities, and some to the different internal organs.

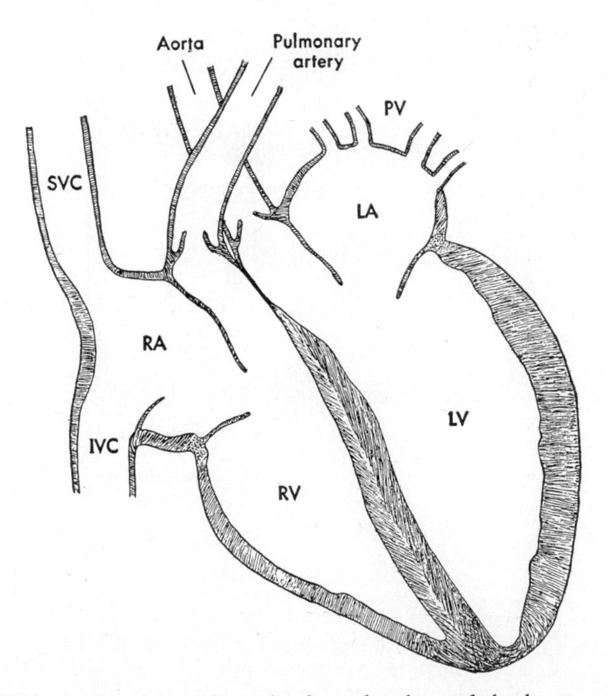

Fig. 240. A diagram to show the four chambers of the heart, seen from the front. The muscular interventricular septum is shown between the right and left ventricles. The upper part of this septum (black line) is called the membranous septum. *RA* and *LA*, right and left atria; *RV* and *LV*, right and left ventricles; *IVC* and *SVC*, inferior and superior venae cavae; *PV*, the four pulmonary veins. Note also atrioventricular openings and the atrioventricular, aortic, and pulmonary valves.

Some portions go on shorter journeys and arrive back at the heart sooner than other portions that travel farther away from the center.

An example of a long journey in the systemic circulation is the circulation from the left heart to the toes and back to the right heart, then to the lungs and to the left heart.

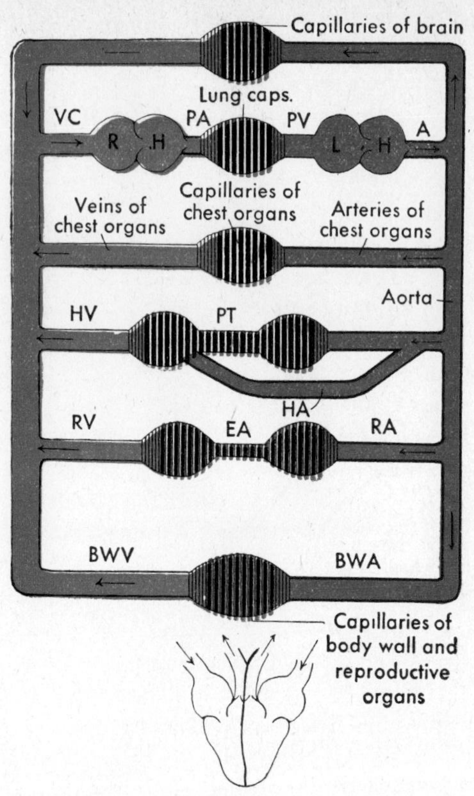

Fig. 241. Diagram of circulation from the viewpoint of physiology. The lung capillaries have been placed between the right and left heart, indicating that the blood in the right heart must go through lung capillaries to reach the left heart. The inset shows the anatomical relations of the chambers of the heart. *A*, aorta; *BWA*, body-wall arteries; *BWV*, body-wall veins; *EA*, efferent arterioles; *HA*, hepatic artery; *HV*, hepatic vein; *LH*, left heart; *PA*, pulmonary artery; *PT*, portal tube; *PV*, pulmonary vein; *RA*, renal artery; *RH*, right heart; *RV*, renal vein; *VC*, vena cava. (See legend, Fig. 215, p. 360.)

An example of a short journey in the systemic circulation is the circulation of blood through the walls of the heart itself.

The coronary circulation. The purpose of the coronary circulation is to distribute blood, containing oxygen, nutrients, etc., to the cardiac muscle cells and return to general circulation the products of metabolism.

The coronary arteries leave the aorta close to the heart. These arteries fill during diastole and empty during systole of the heart. Under normal conditions the rate of blood flow through these vessels is from 50 to 75 cc of blood per 100 gm of heart muscle per minute, depending upon the heart rate and heart volume. In other words, if the heart weighs 300 gm, from 150 to 225 cc of blood will flow through the coronary vessels per minute. Since the output of the heart has been estimated to be about 3 to 4 liters per minute, this means that about 10 per cent of the heart output flows through the coronary arteries. This blood is returned to the heart via the coronary sinus, etc., which opens directly into the right atrium. The coronary vessels carry supplies to, and wastes from, the tissues of the heart. If this circulation is interfered with (occlusion, etc.), normal contractions of the heart are impossible.

The heart muscle receives 10 per cent of cardiac output; the brain 15 per cent; the liver, stomach, and intestines 25 per cent; the kidneys 20 per cent; and the soma 30 per cent of the cardiac output.

In man it is thought to require about 23 seconds[2] to complete a circuit of medium length from the left ventricle to the right atrium (systemic circulation). The blood which enters the right atrium goes through the lungs (pulmonary circulation) before it gets back to the left atrium. This double circulation, pulmonary and systemic, is constantly going on, as each half of the heart is in a literal sense a force pump.

The heart, the cause of circulation. The heart has four chambers, two thin-walled atria above and two thick-walled ventricles. It is divided by a septum into the right and left halves, commonly called the right and left heart. The right atrium has three orifices, the superior and inferior venae cavae, and the atrioventricular orifice. This orifice is guarded by the right atrioventricular valve. The pulmonary artery leaves the right ventricle. It is guarded by the pulmonary valve. The left atrium has five orifices, the four pulmonary veins and the atrioventricular orifice. This orifice is guarded by the left atrioventricular valve. The aorta leaves the left ventricle. It is guarded by the aortic valve.

The heart as a pump. The muscles of the atria and ventricles are so arranged that when they contract they lessen the capacity of the

[2] This shows how rapidly substances introduced into the blood stream can make their way through the body.

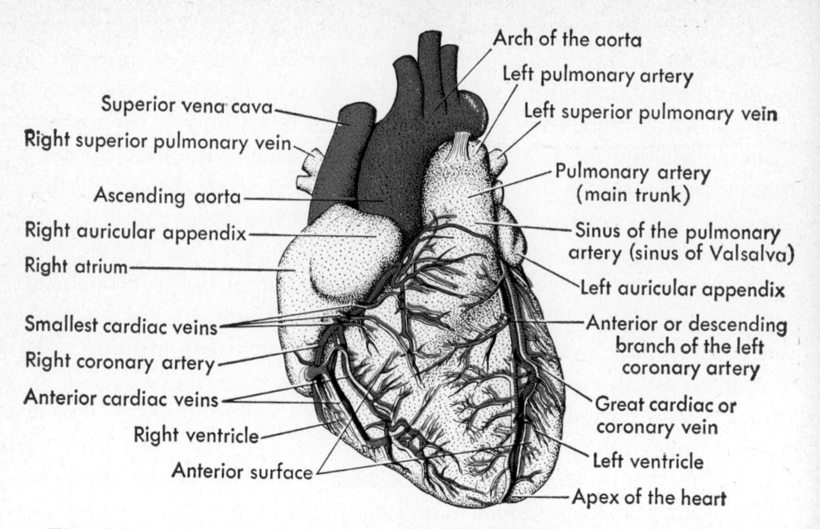

Fig. 242. Diagram of the heart showing coronary arteries and veins. Anterior view. The great cardiac vein and artery indicate location of the septa. (Toldt.)

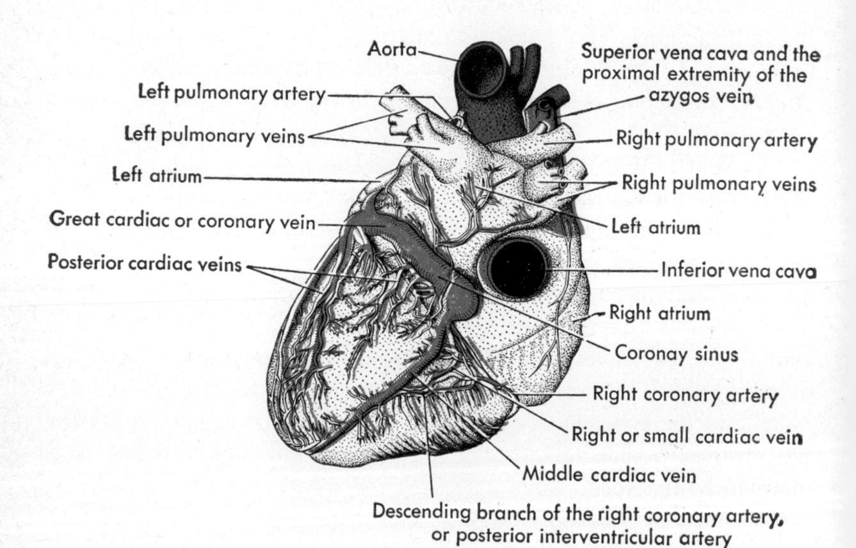

Fig. 243. Diagram of the heart showing the coronary sinus and large veins on the dorsal wall of the heart. Which ventricle is to the left in the diagram? (Toldt.)

chambers which they enclose. The contracting chambers drive the blood through the heart to the arteries (Fig. 240).

The first visible sign of contraction is noted where the superior vena cava empties into the right atrium. The sinoatrial node (S-A node), which lies near the base of the superior vena cava, is the causal factor. It is therefore sometimes called the *pacemaker* of the

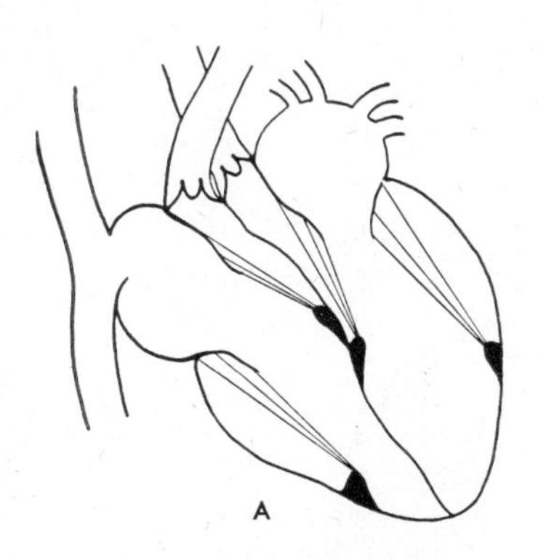

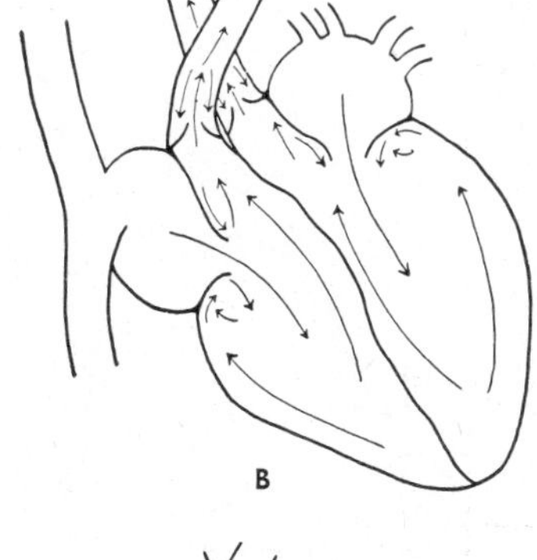

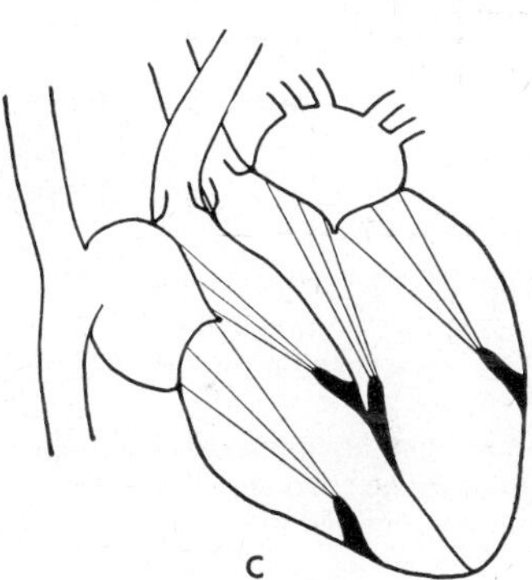

Fig. 244. Diagrams to show position of heart valves in the cardiac cycle. No attempt has been made to show contraction.

In *A*, atrioventricular valves are open, and aortic and pulmonary valves are closed. In *B*, the large arrows show direction of blood flow. The small arrows show "swirling" of blood behind the cusps of valves. In *C*, the atrioventricular valves are closed, and the aortic and pulmonary valves are open.

heart. From this spot the wave of contraction passes over the muscles of both atria. These contract simultaneously, driving the blood into the ventricles. The wave of contraction now spreads over the ventricles, causing them to contract simultaneously, driving the blood into the arteries.

The wave of contraction. If a stimulus is applied to one end of a muscle, a wave of contraction sweeps over the entire tissue. It is therefore easy to conceive how a wave of contraction can sweep over the muscular tissue of the atria, which is practically continuous. The question is—how is this wave transmitted to the muscular tissue of the ventricles, which is *not* continuous with that of the atria? The connecting pathway is furnished by the atrioventricular node (A-V node), which transmits the nerve impulses by means of the A-V bundle and causes the wave of contraction to spread from the atrio-

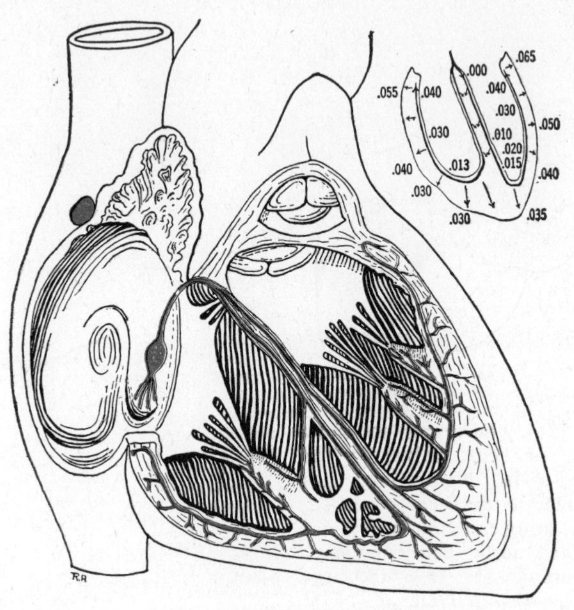

Fig. 245. Diagram of the atrioventricular bundle of His. The atrioventricular (*A-V*) node can be seen near the opening of the coronary sinus in the right atrium. At the upper end of the ventricular septum the bundle divides. The two branches run down in the ventricular septum and give off many smaller branches to the papillary muscles and to the muscular walls of the ventricles. Bundle indicated in red. Part of the sinoatrial (*S-A*) node is seen in red between the base of the superior vena cava and the right auricular appendix. The tip of the left ventricular chamber is not shown. The figures in the small inset indicate the time taken for the nerve impulses to reach the areas indicated.

ventricular openings over the ventricles to the mouths of the pulmonary artery and the aorta.

The electrocardiogram. During contraction, electrical changes constantly take place in heart muscle. Active cardiac muscle fibers are electrically negative to resting fibers. The electrical differences in various parts of the heart which occur constantly can be led off from the surface of the body by electrodes placed on the extremities and connected to a galvanometer. The standard leads used are: (1) from right and left arms; (2) from right arm and left leg; (3) from left arm and left leg.

Each lead shows the potential differences in the heart as recorded at the body surface. There are other leads used. The record made is called the electrocardiogram and shows the events of the cardiac cycle. Each electrical heart cycle begins with a peaked elevation, called the P wave, which is caused by the spread of excitation from the S-A node out over the atria and to the A-V node. The QRS

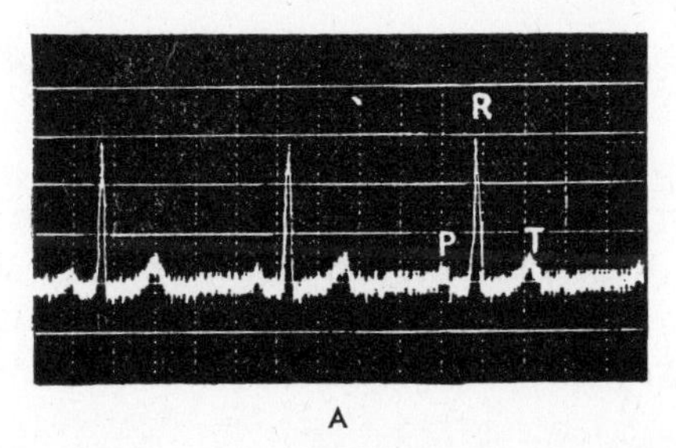

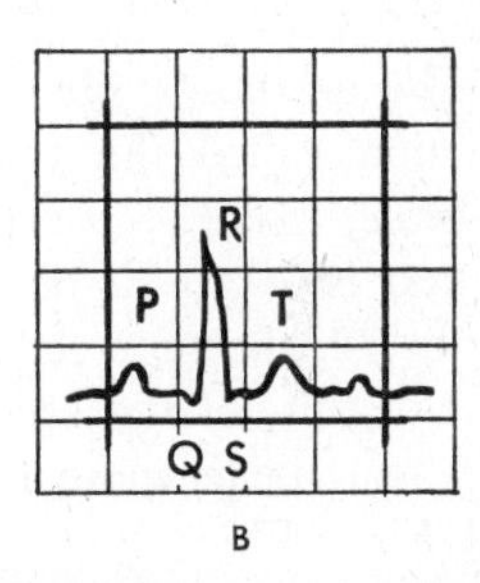

Fig. 246. *A*, electrocardiogram. Lead II, showing contraction of different parts of the heart. *B*, diagram to show excitation during one cycle. Wave *P* occurs during atrial excitation or systole; waves *QRS* occur during ventricular systole; wave *T* occurs as ventricular excitation subsides.

deflections are recorded as the electrical impulse spreads down the A-V bundle and out over the ventricles (ventricular systole). The T wave is recorded as ventricular excitation subsides.

The electrocardiogram is of value in diagnosis.

Heart block. Experimentally, the A-V node and bundle may be damaged, with the result that they lose their power to conduct nerve impulses from the atria to the ventricles. The atria will continue to contract at the rate established by the nerve impulses, but the ventricles adopt a slower rate, usually about 30 or 40 a minute. Since the pulse is caused by ventricular contractions, the pulse drops to 30 or 40 per minute. This condition is known as heart block and is caused by arteriosclerosis, chronic myocarditis, syphilis, etc., or by overdoses of digitalis, etc.

Fibrillation. In fibrillation the rhythmic contraction of the atria is interfered with, and groups of muscles contract independently of other groups. In consequence the atria undergo irregular twitching movements. This condition affects the contractions of the ventricles, which become irregular and rapid. The pulse is also irregular and rapid. The cause of fibrillation is not known, but it is thought to be related to the formation of nerve impulses in many areas of the heart tissue. Some of the ventricular contractions are so feeble as not to be transmitted to the arteries, and although they may be heard at the heart, they are not felt at the pulse. There may be a pulse rate of 100, and yet, on listening, it may be discovered that the heart rate is 120. The difference is called the pulse deficit, which in this case is 20. This condition is associated with mitral stenosis[3] and other diseases of the heart valves. Digitalis reduces fibrillation because its general effect upon the heart is to induce slow, strong, regular, rhythmic contractions, which produce a slow, strong, regular pulse and reduce the pulse deficit. Quinidine also produces a slow, strong, regular pulse, though its action is different.

The cardiac cycle consists of three phases: (1) a period of contraction called the *systole*, (2) a period of dilatation called the *diastole*, and (3) a period of rest. The average heart rate of man at rest is 70 to 72 beats per minute. If we assume a pulse rate of 70 to 72, the time required for a cardiac cycle is 0.8^+ seconds, and half of this, or 0.4 seconds, represents the quiescent phase. When the heart beats more rapidly, it is the rest period that is shortened. See Fig. 247.

	Systole	*Diastole*	*Total*
Atria	0.1 sec	0.7 sec	0.8^+ sec
Ventricles	0.3	0.5	0.8^+ sec
	0.4		

Systole, starting at the mouths of the veins, moves over the atria, then, after a very brief pause at the fibrous rings surrounding the atrioventricular openings, continues over the ventricles in such a way as to lift the blood up into the great arteries.

At each systole a volume of blood variously estimated at around 70 cc (man at rest) is forced from the left ventricle into the aorta.

[3] Mitral stenosis is a disease of the mitral valve in which the atrioventricular opening becomes constricted.

This is known as the *stroke volume.* A similar amount is forced from the right ventricle into the pulmonary artery. The total cardiac output per beat is, therefore, 140 cc.

Taking a pulse rate of 70, 4.9 liters (70 × 70 cc) of blood leave the left ventricle per minute. This is known as the *minute volume.* A similar amount leaves the right ventricle, giving a *total cardiac output per minute* of 9.8 liters of blood. With an increase or decrease in stroke volume, in pulse rate, or in both, the total cardiac output per minute would be increased or decreased.

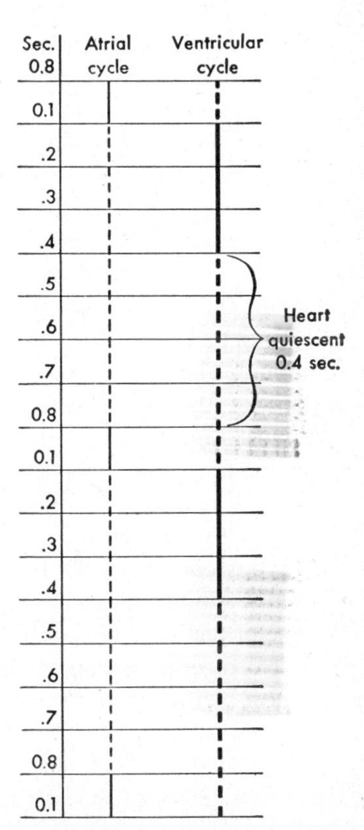

Fig. 247. Atrial cycle and ventricular cycle, showing overlapping of diastole giving 0.4-second quiescent period of whole heart. Solid lines, systole; dotted lines, diastole.

Heart sounds and murmurs. If the ear is applied over the heart, certain sounds are heard, which recur with great regularity. Two chief sounds can be heard during each cardiac cycle. The first sound is a comparatively long, booming sound; the second, a short, sharp one. The sounds resemble the syllables *lubb dup.* The first sound is thought to be due to the contracting muscle and to vibrations caused by the closure of the atrioventricular valves; the second, to the sudden closure of the semilunar valves. In certain diseases of the heart these sounds become changed and are called *murmurs.* These are often due to failure of the valves to close properly, thus allowing regurgitation.

Cause of the heartbeat. The cause of the heartbeat is still unknown. General belief favors the *myogenic theory,* that is, the theory that the function of the nerve tissue in the heart is regulatory, that the contractions are due to the inherent power of contraction possessed by the muscle cells of the heart themselves, and that the stimulus which excites the contractions is a chemical one dependent upon the presence of definite proportions of inorganic salts in the blood. Three are especially important, namely, calcium, sodium, and potassium, which are always present in blood. It is natural to question why the heart is not in a state of continuous contraction. In answer, there is a well-marked antagonism between the effects of calcium and the effects of sodium and potassium; calcium promotes contraction, sodium and potassium promote relaxation. It is possible, therefore, that the contraction and relaxation characteristic of heart muscle are brought about by interactions of these salts and the muscles of the heart.

Automaticity. The most remarkable power of cardiac muscle is its automaticity. By this we mean that the stimuli which excite it to activity arise within the tissue itself. This may be demonstrated by removing the heart of a frog from the body of the animal. The heart will continue to beat for hours provided it is kept moist with Ringer's solution. The degree of automatic power possessed by different regions of the heart varies. Some parts beat faster than others. The most rapidly contracting part is the S-A node. It is from this node that the wave of concentration radiates through the atrial muscle to the A-V node. From here, it is transmitted over the atrioventricular bundle to the ventricles.

Nervous control of the heart. Although the heart contracts automatically and rhythmically, the continuously changing frequency and volume of the heart are controlled by two sets of nerve fibers. These consist of a craniosacral set in inhibitory fibers, extending from the inhibitory center in the medulla via the vagus nerves (afferent and efferent fibers) to the heart, and accelerator nerve fibers from the superior, middle, and inferior cardiac accelerator nerves and the visceral branches of the first five thoracic spinal nerves, belonging to the thoracolumbar system. It is assumed that the cardioaccelerator center is also located in the medulla. The way in which these nerve fibers regulate the heart's action is not known, but it is generally believed that both the inhibitory and accelerator nerve fibers are in a state of constant, though slight, activity. This means that the heartbeat is controlled by two antagonistic influences, one tending to slow the heart action and the other to quicken it. If the inhibitory center is stimulated to greater activity, the heart is slowed still further. If the activity of this center is depressed, the heart rate is increased, because the inhibitory action is removed. Stimulation of the accelerator nerves results in a quickened heartbeat.

Right-heart reflex. There are receptors in the large veins entering the right heart that are sensitive to changes in venous pressure. Increased pressure will accelerate heart action. Impulses are conveyed over afferent vagal fibers to the cardiac center in the medulla.

The aortic reflex. There are pressoreceptors in the arch of the aorta (aortic sinus) that are sensitive to changes in blood pressure. Fluctuation in arterial pressure activates these receptors and impulses are conveyed over afferent nerve fibers of cranial nerves IX and X to the cardiac centers and the heart is slowed. There are also chemoreceptors in the aortic arch that are sensitive to oxygen lack. Impulses are conveyed to the cardiac center in the medulla and the heartbeat is accelerated. Cardiac output is increased and more blood is moved on to the tissue cells.

The carotid sinus reflex. The carotid sinus is a slightly dilated area of the internal carotid artery at the bifurcation of the carotid into the internal and external carotid arteries. Afferent nerve fibers in the cardiac branch of the glossopharyngeal nerve carry nerve impulses from pressure receptors in the carotid sinus to the cardiac center influencing heart action.

There are chemoreceptors in the carotid bodies that are sensitive to oxygen lack. Impulses from these receptors are conveyed to the cardiac center and the heart rate is accelerated, thereby increasing cardiac output. See Fig. 210, p. 350.

Factors affecting the frequency and strength of the heart's action. The frequency and strength of the heartbeat are affected by: blood pressure; emotional excitement or keen interest; reflex influences which are of an unconscious character; the temperature of the blood; such characteristics of heart muscle as tone, irritability, contractility, and conductivity; physical factors such as size, sex, age, posture, and muscular exercise; changes in the condition of the blood vessels; and certain internal secretions.

Under normal conditions the pulse rate is inversely related to the arterial blood pressure (Marey's reflex, or aortic reflex); that is, a rise in the arterial blood pressure causes a decrease in pulse rate, and a decrease in arterial blood pressure causes an increase in pulse rate. On the other hand, the pulse rate is directly related to venous blood pressure (Bainbridge's reflex, or right heart reflex); that is, a rise in the pressure of blood entering the right atrium causes an increase in heart rate.

The pulse rate is very susceptible to changing sensations. Especially is this true in emotional excitement or keen interest. The heart also responds to reflex influences which are of an unconscious character, such as activity of the visceral organs. After meals the heart increases in rate and strength of beat.

Experimentally it has been demonstrated that abnormally high or low temperatures of the blood affect the frequency of the beat. If the heart is perfused with hot liquid, the rate is increased in proportion to the temperature until the maximum point, about 44° C (111° F), is reached. If the temperature is raised above this, the heart soon ceases to beat. In fever the increased rate of the heart action is thought to be due partly to the effect of the higher temperature of the blood on the heart muscle. On the other hand, if cold liquid is perfused through an animal heart, the rate is decreased, and the heart ceases to beat at about 17° C (62° F).

Conditions that affect the *irritability, contractility,* and *conductivity* of the heart muscle or reduce its normal *tone* are likely to change the frequency of the heartbeat, either accelerating or slowing the action. If the tone is below par, the strength of the contractions is diminished.

In almost all warm-blooded animals the frequency of the heartbeat is in inverse proportion to the size of the body. An elephant's heart beats about 25 times per minute, a mouse's heart about 700 times per minute. Generally speaking, the smaller the animal, the more rapid is the consumption of oxygen in its tissues. The increased need for oxygen is met partly by a faster heart rate.

The heartbeat is somewhat more rapid in women than in men. The heart rate of a female fetus is generally 140 to 145 per minute, that of the male, 130 to 135.

Age has a marked influence. At birth the rate is about 140 per minute, at 3 years about 100, in youth about 90, in adult life about 75, in old age about 70, and in extreme old age 75 to 80.

The *posture* of the body influences the rate of the heartbeat. Typical figures are: standing, 80; sitting, 70; and recumbent, 66. If an individual remains in a recumbent position and keeps quiet, the work of the heart may be decreased considerably. This is the reason why patients with certain types of heart disease are kept in a recumbent position.

Muscular exercise increases the heart rate. It is due to (1) the activity of the cardioinhibitory center in the medulla being depressed by the motor impulses from the more anterior portions of the brain to the muscles, probably by means of collateral fibers to the cardiac center; (2) a stimulation of the cardioaccelerator center; (3) an increased secretion of adrenaline and other hormones which accelerate heart action; (4) increased temperature of the blood; (5) the pressure of the contracting muscles on the veins sending more blood to the heart, so that the right side is filled more rapidly. This increase of venous pressure reflexly accelerates the heartbeat.

In order to function properly, the heart requires a certain amount of resistance, and normally this is offered by the blood vessels. The heart will beat more slowly and strongly in response to increased resistance, provided the resistance is not too great. In the latter case the heart is likely to dilate, and its action becomes frequent and weak. The most common causes of abnormally high resistance are lack of elasticity and hardening of the walls of the arteries (arteriosclerosis) and such interference with the venous circulation as occurs in some forms of heart and kidney diseases. When the resistance is below normal, the heartbeats are frequent and weak. Lessened resistance is due either to a relaxed condition of the blood vessels or to the loss of much blood or of much fluid from the blood.

Certain internal secretions affect the frequency and strength of the heartbeat. *Thyroxin* produces a faster pulse. The partial removal of excessively active thyroid glands results in a slower heart rate. *Epinephrine* from the adrenal glands increases the frequency and force of the heartbeat. Other hormones, including those from the pancreas, liver, smooth muscles, etc., also affect the action of the heart, directly and indirectly.

MODIFYING FACTORS OF CIRCULATION

Distribution of blood to different parts of the body. In health the distribution of blood varies, as determined by the needs of the different parts. When the digestive organs are active, they need an extra supply of blood, which may be furnished in one of two ways, possibly both. The blood supply to less active organs may be decreased, or the heart rate may be increased with consequent increase in the output of blood. Other causes may result in an increased supply of blood to an organ. If the skin is exposed to high temperatures, the arterioles which bring blood to it are dilated, and the blood flow near the surface is increased. This aids in the radiation of heat and in the control of body temperature. On the other hand, slight chilling causes contraction of the skin arterioles and resulting paleness. The rate of blood supply to the brain is subject to adjustment in accordance with the mental activity and emotional states. During normal sleep the blood supply to the brain is reduced, and this is accompanied by an increased supply to the muscles and skin regions.

Normally the blood vessels maintain a state of tone about halfway between contraction and dilatation. It is thought that adjustments in the blood supply to various parts are brought about by increasing or decreasing the tone of the local blood vessels. Two factors are important, (1) vasomotor nerve fibers and (2) chemical stimuli.

1. The vasomotor nerve fibers consist of two antagonistic sets. The vasoconstrictors cause the muscular coats of the blood vessels to contract, lessen the diameter of the vessels, and thereby increase resistance to blood flow. The vasodilator fibers increase the diameter of the blood vessels, probably by allowing the muscular coats to relax, and thereby decrease resistance to blood flow.

2. Chemical substances, such as the lactic acid and carbon dioxide produced during muscular activity, may lessen the tone of the blood vessels in the part affected, resulting in local dilatation and an increased supply of blood to the part needing it. At the same time, the acids carried in the blood to the vasoconstrictor center stimulate it and thereby increase the tone of blood vessels in other parts of the body. On the other hand, hormones, such as epinephrine and Pituitrin, cause contraction of the blood vessels.

Epinephrine and digitalis are used medicinally to cause vasoconstriction, and amyl nitrite is inhaled to bring about vasodilatation, particularly when a condition like angina pectoris[4] makes quick relief necessary. It was formerly thought that changes in the size of the blood vessels were limited to the arteries. It is now thought that not only the arteries, but the capillaries and veins, are capable of

[4] Angina pectoris is a disease characterized by attacks of severe constricting pains in the chest, which radiate into the left arm. It is accompanied by a great sense of cardiac oppression and usually occurs in arteriocapillary fibrosis with myocarditis (Fig. 341).

dilatation or constriction under the influence of nerve fibers or chemical stimuli.

In surgical shock there is marked interference with the circulation of the blood. Owing to dilatation of the capillary bed and consequent increase in the volume of blood in the capillaries, there is marked decrease in arterial pressure, which may fall below the level essential to the welfare of the tissues. The pulse becomes rapid and weak, and respiration increases. It is thought that the dilatation of the capillaries may be brought about by substances such as histamine formed in injured tissues.

Factors maintaining arterial circulation. The most important factors maintaining arterial circulation are the pumping action of the heart, the extensibility and elasticity of the arterial walls, the peripheral resistance in the region of the small arteries, especially arterioles in the splanchnic areas, and the quantity of blood in the body.

The extensibility and elasticity of the arterial walls. During each systole the ventricles force blood into arteries that are already full.[5] The extensibility of the arteries enables them to distend and receive this extra supply of blood. This period of distention corresponds to the systole of the heart. Just as soon as the force is removed, the elasticity of the arteries causes them to recoil to their former diameter, and this exerts such a pressure on the contained blood that this blood is forced into the capillaries just rapidly enough to allow the arteries time to reach their usual size during diastole of the heart. The arteries thus not only serve as conducting tubes but exert a force that assists the heart in driving the blood into the capillaries.

The extensibility and elasticity of the arteries change with the health and age of the individual. Sometimes as the result of disease, and usually with age, the arterial walls grow less elastic and become less well adapted for the unceasing work they are called upon to perform. This condition is known as arteriosclerosis.

Peripheral resistance. It is the function of the vasomotor fibers, etc., to "set" the diameters of the muscular arterioles in relation to constantly varying local needs for blood. The *elastic* arteries compensate for heart systole and diastole, thus maintaining a *steady flow* of blood in the capillaries, the arteries accommodating the extra blood forced into them during heart systole and by their recoil forcing this blood toward the capillaries during heart diastole. Inasmuch as local needs for blood vary constantly and through constantly varying

[5] Estimated from 60 to 90 ml (2 to 3 oz) each.

limits, it is the function of the autonomic nervous system (and locally produced chemical substances such as CO_2), reflecting these needs, to set the diameters of the arterioles so that the peripheral resistance meets these local needs (much blood needed, wide arterioles; less blood needed, narrower arterioles). *On the basis of peripheral*

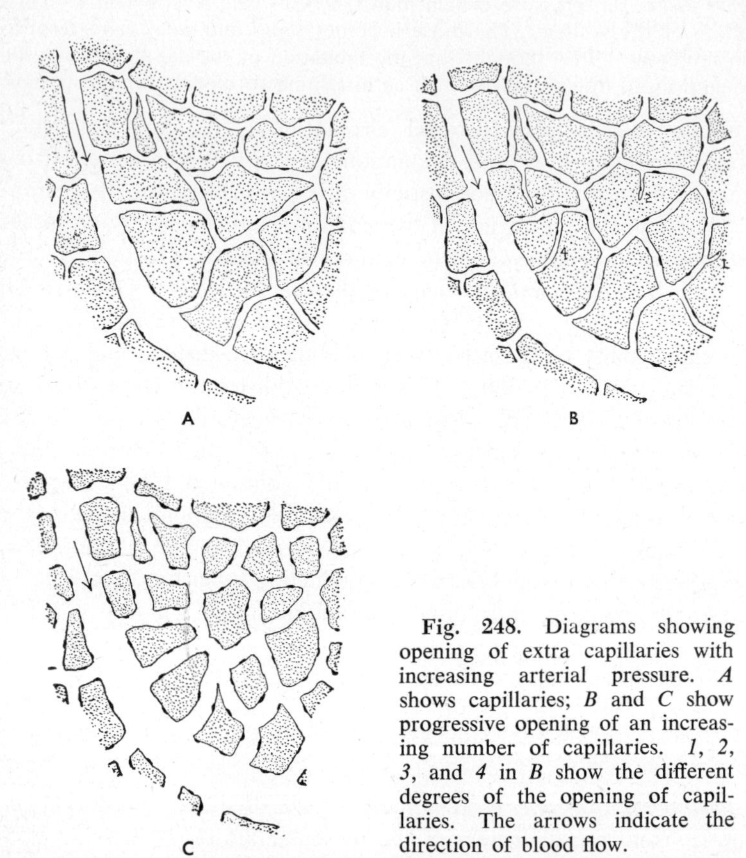

Fig. 248. Diagrams showing opening of extra capillaries with increasing arterial pressure. *A* shows capillaries; *B* and *C* show progressive opening of an increasing number of capillaries. *1*, *2*, *3*, and *4* in *B* show the different degrees of the opening of capillaries. The arrows indicate the direction of blood flow.

resistance thus established, it is the function of the arterioles, etc., to expand and recoil, changing an intermittent flow in the arteries to a steady flow in capillaries. It is easily seen that this is a *fine* adjustment, the elastic arteries giving a *steady* flow in capillaries on *many bases of diameter* of arterioles set by local needs. This fine adjustment (coupled with optimum activity of the heart) is the mechanism by means of which homeostasis, or state of constancy, of body fluids is maintained and is the aim of all hygiene, external and internal.

Quantity of blood. It is evident that, other things being equal, the quantity of blood to be moved is an important factor. Except in cases of severe hemorrhage loss of blood is compensated for by a flow of liquid from the tissues into the blood vessels.

Factors maintaining venous circulation. The effect of the pumping action of the heart is not entirely spent in forcing the blood through the arteries and capillaries. A little force still remains to

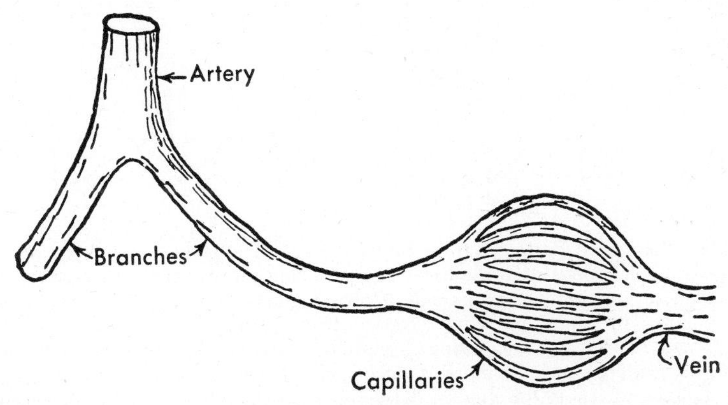

Fig. 249. Diagram to illustrate variations in velocity of blood flow. A vessel divided into two branches; these are individually of smaller cross section than the main trunk, but united they exceed it. Linear velocity will be lower in the branches than in the parent artery. The sum of the cross-sectional areas of the capillaries is greater than that of the artery or vein.

propel the blood back to the heart again, and the presence of valves keeps it flowing in the right direction, i.e., toward the heart. The return flow is also favored by (1) the suction action of the heart caused by the emptying of the atria, (2) the heart and respiratory movements, which cause continual changes of pressure against the thin-walled veins in the thorax and abdomen, (3) the contractions of the skeletal and visceral muscles, which exercise a massaging action upon the veins and, aided by the valves, propel the blood toward the heart.

The velocity of the blood flow. In all the large arteries the blood moves rapidly; in the capillaries, very slowly; in the veins the velocity is augmented as they increase in size, but it never equals that in the aorta. The underlying principle is that in any stream the velocity is greatest where the cross section of the channel is least, and lowest where the cross section is greatest. When a vessel divides, the sum of the cross sections of the two branches is greater than that of the main trunk. Consequently the velocity will be reduced when arteries divide and increased when veins unite. One reason why the velocity in the veins never equals that in the aorta is that the cross section of the venae cavae is greater than the cross section of the aorta. The actual interchange of materials between the blood and the tissues takes place in the capillaries (since the walls of the arteries and veins

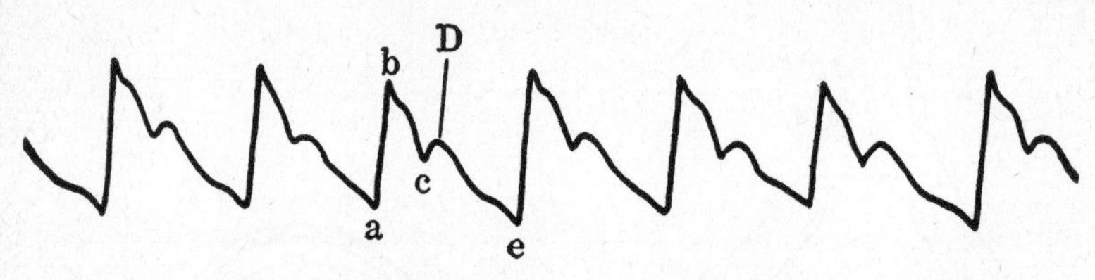

Fig. 250. Sphygmogram from radial artery. The lever of a sphygmograph can be made to record the pulse wave as a line on a moving surface. Each pulse wave consists of an ascending portion, or *anacrotic* limb (*ab*), and a descending, or *catacrotic,* limb (*be*). The ascending limb is smooth and steep and records the increasing distention or systolic pressure of the artery. The descending limb is more slanted and shows smaller waves, the most constant of which is the *dicrotic* wave (*D*) which is preceded by the dicrotic notch (*c*). The artery dilates rapidly and steadily, but its diameter decreases slowly and irregularly. The dicrotic wave is thought to be due to the closure of the semilunar valve of the aorta.

are too thick to permit of diffusion); hence the value of the slow, steady flow of blood in this region.

The pulse. The alternate dilatation and contraction of an artery constitute the pulse. When the finger is placed on an artery which approaches the surface of the body and is located over a bone, a sense of resistance is felt, which seems to be increased at intervals corresponding to the heartbeat. In certain arteries the pulse may be seen with the eye. When the finger is placed on a vein, very little resistance is felt; and under ordinary circumstances no pulse can be perceived by the touch or by the eye.

As each expansion of an artery is produced by a contraction of the heart, the pulse as felt in any superficial artery is a convenient guide for ascertaining the character of the heart's action.

All arteries have a pulse, but it is more readily counted wherever an artery approaches the surface of the body. These locations are as follows: The *radial* artery, at the wrist—the radial artery is usually employed for this purpose on account of its accessible situation; the *temporal* artery, above and to the outer side of the eye; the *external maxillary* (*facial*) artery, where it passes over the lower jawbone, which is about on a line with the corners of the mouth; the *carotid* artery, on the side of the neck; the *brachial* artery, along the inner side of the biceps; the *femoral* artery, where it passes over the pelvic bone; the *popliteal* artery, under the knee; the *dorsalis pedis,* over the instep of the foot.

Points to note in feeling a pulse. In feeling a pulse, the following points should be noted:

1. The *frequency*, or *number of beats per minute*, should be normal for the individual concerned. The intervals between the beats should be of equal length. A pulse may be irregular in frequency and rhythm. When a pulsation is missed at regular or irregular intervals, the pulse is described as *intermittent.*
2. The *force*, or *strength*, of the heartbeat. Each beat should be of equal strength. Irregularity of strength is due to lack of tone of the cardiac muscle or of the arteries. Occasionally the heartbeat appears to be divided, and two pulsations are felt, the second being weaker than the first. This is known as a *dicrotic* pulse. The pulse is studied by the aid of a sphygmograph, which is an instrument that makes graphic tracings of the rise and fall of an artery. It consists of a tension spring to which a button is attached. The button is placed over the artery, and the pulsations are communicated to a lever, which records the tracings on paper (Fig. 250). Such a tracing shows that a pulse wave consists of two phases: (1) an upstroke, called the *anacrotic limb*, which is caused by the distention of the vessel and indicates the force of the heartbeat; and (2) a downstroke, called the *catacrotic limb*, which is caused by the recoil of the vessel. Normally the upstroke is smooth, but the downstroke shows several waves. One in the middle is called the *dicrotic wave.* In certain diseases this is so exaggerated that it gives the sensation of a double beat and is called a dicrotic pulse.
3. The *tension*, or *resistance* offered by the artery to the finger, is an indication of the pressure of the blood within the vessels and the elasticity or inelasticity of the arterial walls. A pulse is described as *soft* when the tension is low and the wall of the artery elastic. A pulse is described as *hard* when the tension is high and the wall of the artery is stiff, thick, and unyielding.

Average frequency of the pulse. The average frequency of the pulse in men is 65 to 70; in women, 70 to 80. A person in perfect health may have a much higher or a much lower rate. The relative frequency of the pulse and respirations is about four heartbeats to one respiration.

As a rule, the rapidity of the heart's action is in inverse ratio to its force. An infrequent pulse, within physiological limits, is usually

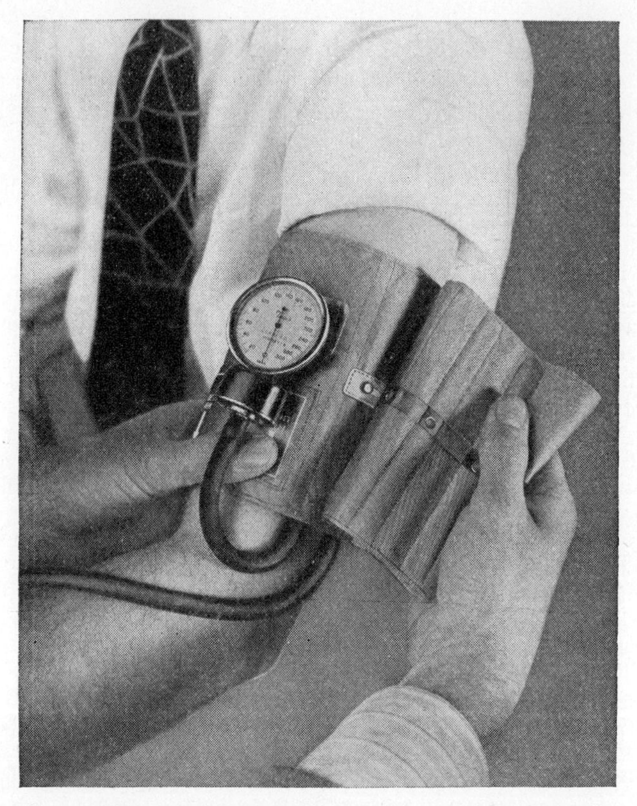

Fig. 251. Method of using aneroid sphygmomanometer for measuring arterial blood pressure. (Courtesy of Taylor Instrument Company.)

a strong one, and a frequent pulse comparatively feeble, the pulse in fever or debilitating affections becoming weaker as it grows more rapid. As the pulse is an indication of the frequency of the heartbeat, it follows that the factors which influence the heartbeat will also influence the pulse.

Blood pressure. Blood pressure is defined as the pressure the blood exerts against the walls of the vessels in which it is contained. The term includes arterial, capillary, and venous pressure; but it is commonly applied to pressure existing in the large arteries, usually the left brachial artery just above the elbow. A vein is easily flattened under the finger; an artery offers a stronger resistance. This is an indication of a great difference between arterial and venous pressure. This difference is also shown when an artery and a vein are cut; the blood springs from the artery in a pulsating spurt, indicating a high pressure, whereas the flow from the vein is continuous, and even when copious "wells up" rather than "spurts out," indicating a low pressure.

Blood pressure is highest in the arteries during the period of ventricular systole. This is systolic pressure. During ventricular diastole blood pressure tends to fall and reaches a minimum just before the beginning of the next systole. The minimum is called diastolic pressure. Pressure in the arteries is high and fluctuating, slightly higher in the large trunks than in their branches. When the blood reaches the capillaries, the surface is multiplied and the friction increased. This offers resistance to the flow, and the result is a *decided drop in the pressure.* Pressure in the veins is low and relatively constant. It must be higher in the small veins than in the large ones they unite to form, as the direction of the blood flow is from the smaller to the larger veins. Their chief effect on blood flow is their great relative ability to hold large volumes of blood under low pressure.

Method of determining blood pressure. A rough estimate of systolic blood pressure may be determined at the radial artery. If the radial artery is hard and incompressible, it may indicate either that some change has occurred in the vessel or that the pressure is high. If, however, the pulse is easy to obliterate with the fingers, it is usual to find a low pressure.

Many forms of apparatus have been devised by which a more accurate knowledge of this condition can be obtained (Fig. 251). The apparatus is called a sphygmomanometer and consists of a scaled column of mercury (mercurial manometer) marked in millimeters, which is connected by rubber tubing with an elastic air bag contained in a fabric cuff. The air bag is in turn connected with a small hand pump. Some instruments are constructed with a spring scale (aneroid manometer), but the principle is the same. The air bag contained in the sleeve is wrapped snugly about the arm just above the elbow over the brachial artery. By placing the finger upon the pulse at the wrist (as the bag is inflated), a point is finally reached where the pulse disappears; then the bag is very slowly deflated until the pulse can just be felt. The pressure in the bag, therefore, against the artery from the outside, as indicated by the reading on the instru-

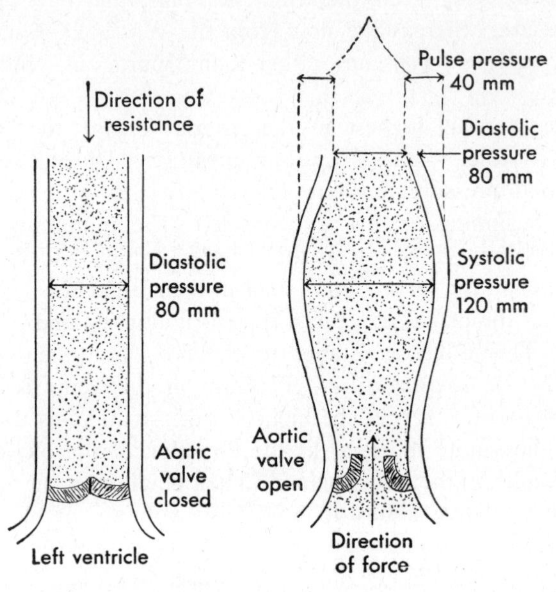

Fig. 252. Diagram showing relationship between direction of resistance and direction of force in measuring arterial blood pressure.

ment, is approximately equal to the pressure which the blood exerts against the wall of the artery from the inside. This is known as the *systolic pressure* and is the greatest pressure which cardiac systole causes in the brachial artery. In the auscultation method of reading blood pressure, a stethoscope is placed over the brachial artery in the bend of the elbow. Blood pressure is then indicated by sounds heard through the stethoscope. The bag is inflated as before until all sounds cease. It is then slowly deflated until the pulse can just be heard. The reading on the manometer at this time indicates systolic pressure. The deflation of the bag is then continued, and the reading on the manometer just before the last sound of the disappearing pulse indicates *diastolic pressure,* which is the lowest pressure which cardiac diastole causes in the brachial artery.

As pressure falls in the sphygmomanometer, the sounds heard change. First a clear, sharp, tapping sound is heard that corresponds to systolic pressure. The next sounds become softer, and as pressure continues to fall the sound gets louder again, then becomes muffled—this corresponds to diastolic pressure. The sound lasts during the next 4 or 6 mm of Hg fall, then all sounds cease to be heard. Diastolic pressure is usually recorded when the muffled sound is heard and when the sound is completely lost—$\frac{120}{80\text{–}75}$.

Pulse pressure is the difference between the systolic and diastolic pressure; viz., if the systolic is 115 mm and the diastolic 75 mm, the pulse pressure is 40 mm. The pulse pressure is an indication of (1) how well the heart is overcoming the resistance offered it, (2) how successfully it is driving the blood to the periphery, and (3) the condition of the arteries, for in arteriosclerosis the pulse pressure is high. Pulse pressure varies and is dependent upon (1) the energy of the heart, (2) the elasticity of the blood vessels, (3) the peripheral resistance, and (4) the quantity of circulating blood.

Capillary pressure is the pressure of the blood within the capillaries. Capillary pressure in man when in a sitting position is on the average about 12 to 32 mm of mercury. It is somewhat higher when standing and lower when lying down.

Venous pressure is the pressure blood exerts within the veins. Normal venous pressure is on the average 40 to 60 mm of water in a recumbent position. It is somewhat higher when in an erect position. Increasing attention is being given to venous pressure, as it is a valuable index in determining the efficiency of heart muscle.

Normal degree of blood pressure. The average blood pressure of an adult male as recorded by the sphygmomanometer over the brachial artery is about 110 to 120 mm systolic and 65 to 80 mm diastolic. Some observers report that the systolic pressure is higher in men than in women. Individual variations are not uncommon, but 140 mm for men and 130 mm for women are considered the normal upper limits. A systolic pressure of 150 mm suggests hypertension. It varies during the mental and muscular work and shows a tendency to fall during fatigue. Cold, drugs, etc., which constrict the arterial pulse may raise the blood pressure. Heat and the drugs of the vasodilator group, like nitroglycerin, may lower it.

Blood pressure is dependent upon the force of the contraction of the ventricles, the elasticity of the arteries and the tone of the muscu-

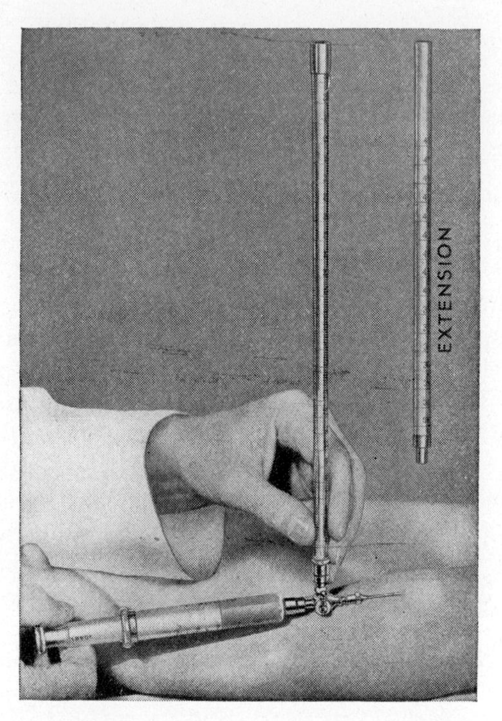

Fig. 253. Use of water manometer for measuring venous blood pressure. (Courtesy of Taylor Instrument Company.)

lar tissue in their walls, and the resistance offered to the flow of blood through the vessels. Minor factors are respiration and the accompanying pressure changes in the chest cavity, the amount of blood in the body, and gravity. Gravity tends to increase pressure in arteries below the level of the heart and to decrease pressure in arteries at levels above the heart.

Venous pressure is low; but when standing, the pressure in the veins of the legs and feet is high, due to gravity—hence the frequency of varicose veins in the lower limbs. Walking relieves this pressure because the contraction of the muscles forces the blood upward in the veins and the valves of the veins favor this movement.

Variations in blood pressure under normal conditions. Variation

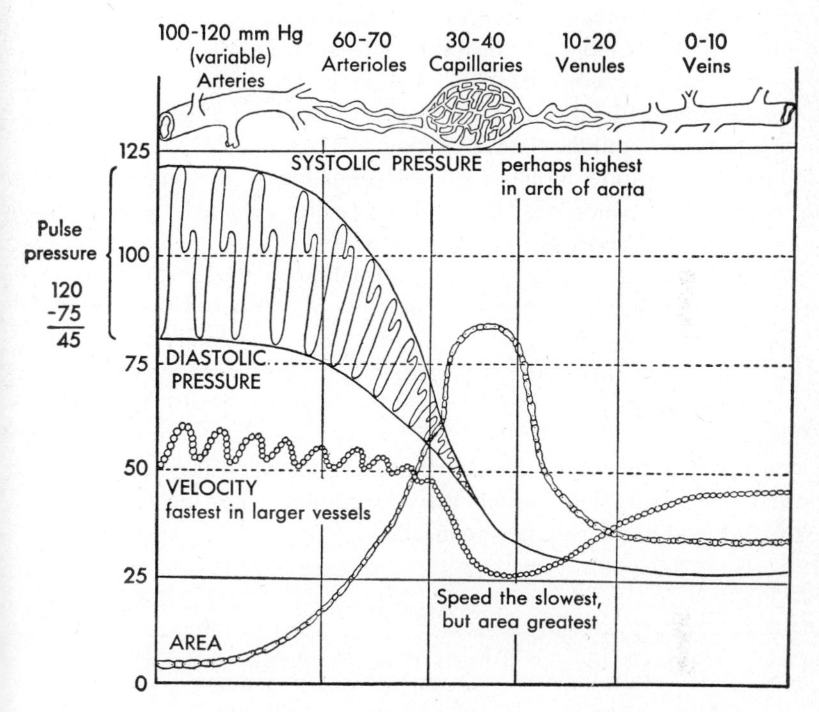

Fig. 254. Diagram showing relationships between arterial, capillary, and venous blood pressures, relationship of area in various blood vessels, and speed with which blood moves through them. (Caroline E. Stackpole and Lutie C. Leavell, *Textbook of Physiology.* Courtesy of the Macmillan Company.)

in arterial blood pressure is compatible with health and is affected by age, sex, muscular activity, digestion, emotions, position, and sleep.

At birth the average systolic pressure is 40 mm of mercury. It increases rapidly during the first month to 80 mm. Then it increases slowly, and at the age of 12 years the average reaches 105 mm. At puberty a somewhat sudden increase occurs; the average is 120 mm. A steady, slow, but not marked increase in blood pressure occurs normally from adolescence throughout life.

Blood pressure is increased by muscular activity. The amount of increase depends upon the amount of energy required for the activity and upon individual differences. Systolic pressure is raised slightly after meals. Emotional factors, such as fear, worry, etc., raise systolic pressure considerably. During quiet, restful sleep systolic pressure falls; the lowest point is reached during the first few hours. It rises slowly until the time of waking.

Systolic pressure is about 8 to 10 mm lower in women than in men. After menopause there is an increase, and the pressure remains a little above the male average.

Blood pressure is raised above normal when the elasticity of the arteries is reduced, as in arteriosclerosis; by various diseases of the heart, liver, and kidneys which interfere with the venous circulation; by stimulation from the vasoconstrictor center in the medulla; usually by fever and increased intracranial pressure, as in fracture of the skull.

Blood pressure is decreased below normal when the heartbeat is weak, when the blood vessels are relaxed, and when the total quantity of blood in the vessels is reduced.

SUMMARY

- **Blood Vascular System**
  - **Pulmonary circulation**
    - Right ventricle, then pulmonary arteries to lungs. Capillary system. Return by pulmonary veins to left atrium
    - **Purpose**—to increase oxygen and decrease carbon dioxide to standard amount, etc.
  - **General, or systemic, circulation**
    - Left ventricle, then by means of aorta and its branches to all parts of the body. Capillary system. Return by veins which empty into superior and inferior venae cavae. These empty into right atrium
    - Requires about 23 sec for medium circuit
    - **Purpose**
      - Carry, and give up to the cells
        - Oxygen
        - Nutritive materials
      - Take from the cells
        - Excess carbon dioxide
        - Other waste products

- **Heart**
  - **Pumping action**
    - By rhythmic contractions blood is moved from veins through heart into arteries
  - **Wave of contraction**
    - Starts at sinoatrial node, transmitted through the atrial muscle to the A-V node, which in turn transmits it via the A-V bundle to the ventricles
    - *Heart block*—condition resulting from damage to atrioventricular bundle and consequent failure to transmit impulses from atria to ventricles. Rate of contraction of ventricles slower than that of atria
    - *Fibrillation*—rhythmic contractions interfered with. Atria undergo irregular twitching movements. Contractions of ventricles irregular and rapid. Difference between heart rate and pulse called *pulse deficit*
  - **Heartbeat**
    - Coordinated contraction of cardiac muscle
    - (1) **Systole**—contraction (2) **Diastole**—dilatation (3) **Rest**—quiescent
      - Cardiac cycle, 70–72 per minute
      - Occupies about 0.8 sec
      - Systolic and rest period each about 0.4 sec
    - **Cause**
      - (1) Unknown. Myogenic theory makes automatic contractility of muscle cells responsible
        - Stimulated by adrenaline-like substance
        - Inhibited by acetylcholine
      - (2) S-A and A-V nodes control sweep of the beat over the heart from veins through heart to arteries
      - (3) Rate and strength of beat fit needs of body under chemical and nervous control
    - **Reflexes**
      - Inhibition (Marey's reflex) over vagi
      - Acceleration (Bainbridge's reflex) over the thoracolumbar system
  - **Heart sounds**
    - Lubb ($\overline{oo}$) — Vibrations caused by closure of atrioventricular valves and contraction of the ventricles
    - Dup (ŭ) — Vibrations caused by closure of the semilunar valves
    - Other sounds have been described

Factors Affecting the Frequency and Strength of the Heart's Action

- **Temperature of Blood**
  - Elevated temperature increases rate, and low temperature decreases rate of heartbeat
- **Characteristics of Heart Muscle**
  - **Tone**, **Irritability**, **Conductivity**, **Contractility** — Anything affecting these factors likely to affect frequency and force of heart's action. Cardiac muscle specially dependent on property of conductivity
- **Physical**
  - **Size**—frequency of heartbeat is in inverse proportion to size of animal. Elephant 25 per minute, mouse 700 per minute
  - **Sex**—frequency of heartbeat higher in women than in men
  - **Age**
    - At birth about 140 per minute
    - At 3 years about 100 per minute
    - In youth about 90 per minute
    - Adult life about 75
    - Old age 75–80 per minute
  - **Posture**
    - (1) Erect posture—about 80 per minute
    - (2) Sitting position—about 70
    - (3) Recumbent—about 66
  - **Muscular exercise**
    - Increases frequency of heartbeat
    - (1) Activity of the cardiac inhibitory center in the medulla depressed by motor impulses from brain to muscles
    - (2) Stimulation of cardiac accelerator center
    - (3) Heart action accelerated by adrenaline and other hormones
    - (4) Increased temperature of the blood
    - (5) Pressure of contracting muscles sends more blood to the heart
  - **Resistance**
    - Normally heart requires certain amount of resistance—offered by blood vessels
    - Normal amount—heart action slow and strong
    - Increased amount—heart action frequent and weak
  - **Condition of blood vessels**
    - Arteriosclerosis—loss of elasticity of arteries
    - Relaxation due to loss of blood volume
- **Internal Secretions**
  - The thyroid—thyroxin stimulates
  - The adrenals—epinephrine stimulates
  - The pituitary—Pituitrin depresses

Distribution of Blood to Different Parts of the Body
- Quantity of blood in body usually about the same, but quantity in any given part adjusted to needs
- **Adjustments dependent on**
  - (1) Vasomotor nerves
  - (2) Chemical stimuli

Factors Maintaining Arterial Circulation
- (1) Pumping action of the heart
- (2) The extensibility and elasticity of the arterial walls
- (3) Peripheral resistance
- (4) The quantity of blood in the body

Factors Maintaining Venous Circulation
- (1) Some force due to pumping action of the heart
- (2) Suction action of the heart
- (3) Changes of pressure in thorax and abdomen due to heart and respiratory movements
- (4) Contractions of the skeletal muscles

Velocity of Blood Flow
- Arteries—blood moves rapidly in large arteries, more slowly in smaller ones
- Capillaries—blood moves very slowly
- Veins—blood moves slowly in small veins, more rapidly in larger veins, but never as rapidly as in arteries

Pulse
- Alternate dilatation and contraction of artery, corresponding to heartbeat
- **Locations where pulse may be counted**
  - Radial artery, temporal artery, external maxillary artery, carotid artery, brachial artery, femoral artery, popliteal artery, dorsalis pedis
- **Points to note**
  - Frequency
  - Force, or strength
  - Tension, or resistance
    - Hard
    - Soft
- **Pulse rate**
  - Factors which influence heartbeat also influence pulse
  - Average
    - 65–70 in men
    - 70–80 in women
  - Ratio of pulse to respiration is about 4 to 1

Blood Pressure
- Pressure blood exerts against walls of vessels
- **Arterial**
  - High and fluctuating
  - Not uniform
    - (1) Highest during ventricular contraction = systolic pressure
    - (2) Lowest before beginning of next systole = diastolic pressure
    - (3) Increases with age
    - (4) Decreases if heart or arteries lose their tone
- **Venous**—low and constant

Blood Pressure

- **Systolic**—greatest pressure which contractions of heart cause. Average systolic pressure in brachial artery of adult 110–120 mm
- **Diastolic**—lowest point to which blood pressure drops between beats. Average diastolic pressure in brachial artery of adult 65–80 mm
- **Varies**
  - During mental and muscular work and shows a tendency to fall during fatigue
  - Cold, drugs, etc., which constrict arterial pulse may raise it
  - Heat, drugs of the vasodilator group may lower it
- **Dependent upon**
  - Strength of the heartbeat
  - Elasticity of the arteries and tone of muscular tissue in walls
  - Resistance offered
  - Minor factors
    - Respiration and resulting changes in chest cavity
    - Amount of blood in body
    - Gravity
- **Increased by**
  - Arteriosclerosis
  - Heart, liver, and kidney diseases which interfere with venous circulation
  - Stimulus from vasoconstrictor center in the medulla which constricts arteries and veins
  - Fever
- **Reduced**
  - When the heartbeat is weak
  - When the blood vessels are relaxed
  - When the total quantity of blood in the vessels is reduced
- **Determined by use of sphygmomanometer**
  - (1) Two types
    - mercurial
    - aneroid
  - (2) Types similar in principle; each consists of an air bag for attachment to arm, a hand pump for inflating the bag, and a scaled device for measurement of pressure in bag, which is equal to pressure of blood against wall of artery

Pulse Pressure

- Difference between systolic and diastolic pressure
- **Indicates**
  - How well the heart is overcoming resistance offered
  - How successfully it is driving blood to the periphery
  - Condition of arteries
- **Dependent upon**
  - Energy of heart
  - Elasticity of blood vessels
  - Peripheral resistance
  - Quantity of blood circulating

CHAPTER 14

LYMPH, LYMPH VASCULAR SYSTEM—PHYSIOLOGY

LYMPH

Composition of lymph. Lymph, tissue fluid, and plasma are similar in composition.[1] Lymph consists of a fluid plasma containing a variable number of lymphocytes, a few granulocytes, no blood platelets (hence clots slowly), carbon dioxide, and *very* small quantities of oxygen. Other contained substances vary in kinds and amounts in relation to the location of lymphatic vessels. In the intestine fat content is high during digestion. Water, glucose, and salts are in about the same concentration as in blood plasma. Protein concentration is lower. Enzymes and antibodies are also present. Lymph has a specific gravity between 1.015 and 1.023.

Sources of lymph. Lymph is formed from tissue fluid by the physical process of filtration. (See Chap. 3.) Colloidal substances from tissue fluid are returned to lymph capillaries rather than to the blood. Water, crystalloids, and other substances also enter the lymph capillaries. Since the process of tissue-fluid formation is continuous lymph formation is also continuous. The lymph system supplements the capillaries and veins in the return of the tissue fluid to the blood. This drainage system is called the lymph vascular system. Even with this system, fluid may accumulate in the tissue spaces.

Filtering pressure will under normal conditions be highest in blood capillaries (as compared with tissue-fluid pressure and lymph pressure) because of beating heart and elastic arteries. Substances like the colloids, therefore, which are filtered out of the capillaries cannot enter the capillaries again but can enter the lymph capillaries.

[1] C. K. Drinker and J. M. Yoffey, *Lymphatics, Lymph, and Lymphoid Tissue*, 1941, speak of the "approximate identity of lymph and tissue fluid."

Hence, it is frequently said that one function of the lymph capillaries is to return blood proteins from tissue fluids to the blood stream.

Another function of the lymph capillaries is to maintain volume and pressure conditions in the spaces occupied by the tissue fluids. Hydrostatic pressure maintained by the heart and elastic arteries is sufficient to supply fluids via the tissue spaces to the cells; but lacking this hydrostatic pressure to remove the fluids, the extra lymph capillaries are added. Some authors speak of the great ability of the endothelial cells of capillary walls to make cell cement, which is constantly destroyed by the hydrostatic pressure of the blood escaping into the tissue spaces.

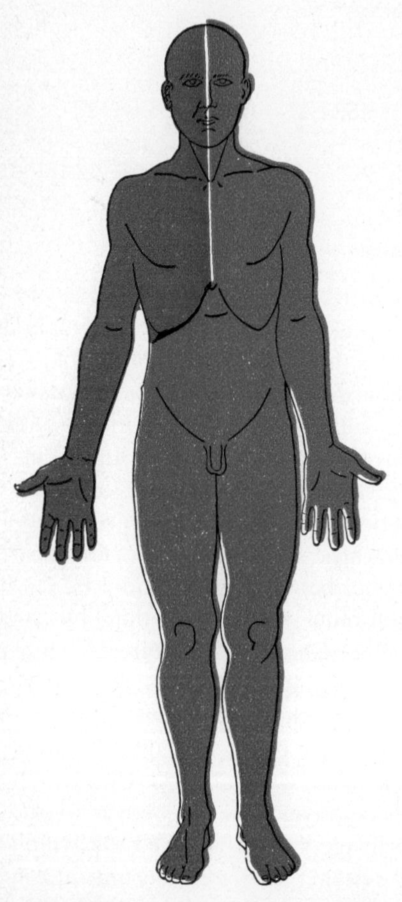

Fig. 255. The regions from which lymph flows into the right lymphatic duct are suggested by the red area, those which are tributary to the thoracic duct by the blue area. (Gerrish.)

The amount of blood plasma filtering from capillaries into tissue spaces will be directly related to the *difference* in hydrostatic pressure against capillary walls from inside and from outside and will be *selective* only in relation to the size of particles passing through the meshwork of the filter.

Some of the substances pass through the tissues by diffusion rather than by filtering out.

LYMPH VASCULAR SYSTEM

Lymph vessels. The plan upon which the lymphatic system is constructed is similar to that of the blood vascular system, if we omit the heart and the arteries. In the tissues we find the *closed* *ends of minute microscopic vessels, called lymph capillaries*, which are comparable to, and often larger than, the blood capillaries. The lymph capillaries are distributed in the same manner as the blood capillaries. Just as the blood capillaries unite to form veins, the lymph capillaries unite to form larger vessels called *lymphatics*. The lymphatics continue to unite and form larger and larger vessels until finally they converge into two main channels, (1) the thoracic duct, and (2) the right lymphatic duct.

- **Lymph Vascular System**
  - **Lymph vessels**
    - Lymph capillaries
    - Lacteals
    - Lymphatics
    - Thoracic duct
    - Right lymphatic duct
  - **Expanded lymph spaces**
    - Pleural cavity
    - Pericardial cavity
    - Peritoneal cavity
    - Meningeal spaces
    - Lymph spaces of eye and ear
    - Synovial bursae
  - **Lymph nodes**

The thoracic duct, or *left lymphatic*, begins in the dilatation called the *cisterna chyli* (chyle cistern), located on the front of the body of the second lumbar vertebra. It ascends upward in front of the bodies of the vertebrae and enters the innominate vein at the angle of junction of the left internal jugular and left subclavian veins. It is from 38 to 45 cm (15 in.) long, about 4 to 6 mm in diameter, and has several valves. At its termination a pair of valves prevent the passage of venous blood into the duct. It receives the lymph from the left side of the head, neck, and chest, all of the abdomen, and both lower limbs, and also the chyle from the lacteals. Its dilatation, the cisterna chyli, receives the lymph from the lower extremities and from the walls and viscera of the pelvis and abdomen (Figs. 255, 256, and 257).

The right lymphatic duct is a short vessel, usually about 1.25 cm ($\frac{1}{2}$ in.) in length. It pours its contents into the innominate vein at the junction of the right internal jugular and subclavian veins. Its orifice is guarded by two semilunar valves.

The lymphatics from the right side of the head, neck, the right arm, and the upper part of the trunk enter the right lymphatic duct. The parts drained by each are suggested by Fig. 255.

Structure of the lymph vessels. The lymphatics resemble the veins in their structure as well as in their arrangement. The smallest consists of a single layer of endothelial cells which have a peculiar dentated outline. The larger vessels have three coats similar to those of the veins, except that they are thinner and more transparent. Their valves are like those of the veins but are so close together that when distended they give the vessel a beaded or jointed appearance. They are usually absent in the smaller networks. The valves allow the passage of material from the smaller to the larger lymphatics and from these into the veins.

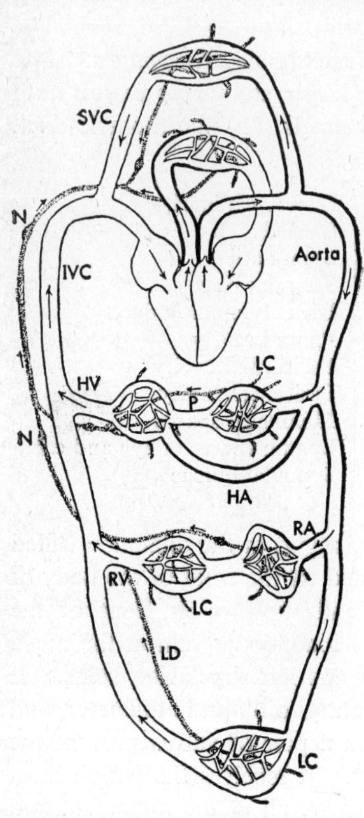

Fig. 256. Diagram to show lymph capillaries, lymph tubes, and lymph nodes in relation to the blood circulatory system. *HA*, hepatic artery; *HV*, hepatic vein; *IVC*, inferior vena cava; *LC*, lymph capillaries; *LD*, lymph duct; *N*, lymph node; *P*, portal tube; *RA*, renal artery; *RV*, renal vein; *SVC*, superior vena cava.

Distribution and classification of lymph vessels. In general the lymph vessels accompany and are closely parallel to the veins. Lymph vessels have been found in nearly every tissue and organ which contains blood vessels. The cartilage, nails, cuticle, and hair are without them, but they permeate all other organs. The lymph, like the blood in the veins, is returned from the limbs and viscera by a superficial and a deep set of vessels. The superficial lymph vessels are placed immediately beneath the skin and accompany the superficial veins. In certain regions they join the deep lymphatics by penetrating the deep fasciae. In the interior of the body they lie in the submucous tissue throughout the whole length of the gastropulmonary and genitourinary tracts and in the subserous tissue of the thoracic and abdominal walls.

The deep lymphatics accompany the deep veins. They are fewer in number and are larger than the superficial lymphatics.

The lymphatics that have their origin in the villi of the small intestine are called *lacteals*. During the process of digestion they are filled with chyle, white in color from the fat particles suspended in it. A close relationship exists between the lymphatics and the serous cavities of the body, i.e., pleural, pericardial, and peritoneal cavities, the meningeal spaces, lymph spaces of eye and ear, and the synovial bursae. These cavities may be considered *expanded lymph spaces*.

Fig. 257. Lacteals and lymphatics, during digestion.

The function of the lymphatics is to carry tissue fluid from the tissues to the veins. Functionally, they may be considered supplementary to the capillaries and the veins, as they gather up a part of the fluid which exudes through the thin capillary walls and return it to the innominate veins. Here it becomes mixed with the blood, enters the superior vena cava, and then the right atrium of the heart. The function of the lacteals is to help in the absorption of digested food, especially fats.

Lymph nodes are small, oval or bean-shaped bodies, varying in size from that of a pinhead to that of an almond, which are placed in the course of the lymphatics. They generally present a slight depression, called the *hilus,* on one side. The blood vessels enter and leave through the hilus. The outer covering is a capsule of connective tissue containing a few smooth muscle fibers. The capsule sends fibrous bands called *trabeculae* into the substance of the node, dividing it into irregular spaces, which communicate freely with each other. The irregular spaces are occupied by a mass of lymphoid tissue, which, however, does not quite fill them as it never touches the capsule or trabeculae but leaves a narrow interval between itself and them. The spaces thus left form channels for the passage of the lymph, which enters by several afferent vessels. After circulating through the node, the lymph is carried out by efferent vessels which emerge from the hilus. The trabeculae support a free supply of blood vessels. It is said that no lymph on its way from the lymph capillaries ever reaches the blood stream without passing through at least one node.

Fig. 258. Diagram illustrating valves of lymphatics. (Gerrish.)

Location of nodes. There are a superficial and a deep set of nodes just as there are a superficial and a deep set of lymphatics and veins. Occasionally, a node exists alone, but they are usually in groups or chains at the sides of the great blood vessels. Lymph nodes are found on the back of the head and neck, draining the scalp; around the sternomastoid muscle, draining the back of the tongue, the pharynx, nasal cavities, roof of the mouth, and face; and under the rami of the mandible, draining the floor of the mouth.

In the upper extremities there are three groups—a small one at the bend of the elbow, which drains the hand and forearm; a larger group in the axillary space, into which the first group drains; and a still larger group under the pectoral muscles. The last-named drains the mammary gland and the skin and muscles of the chest.

In the lower extremities there is usually a small node at the upper part of the anterior tibial vessels, and in the popliteal space back of the knee there are several; but the greater number are massed in the groin. These nodes drain the lower extremities and the lower part of the abdominal wall. The lymph nodes of the abdomen and pelvis are divided into a parietal and a visceral group. The parietal nodes are behind the peritoneum and in close association with the larger blood vessels. The visceral nodes are associated with the visceral arteries. The lymph nodes of the thorax are similarly divided into a parietal set, situated in the thoracic wall, and a visceral set associated with the heart, pericardium, trachea, lungs, pleura, thymus, and esophagus.

Functions of the lymph nodes. The lymph nodes are credited with two important functions.

1. In its passage through the node the lymph takes up fresh lymphocytes, which are continually multiplying by cell division in the substance of the node, which is considered the birthplace of these cells. Serum globulin and antibodies are also added to lymph in the nodes.

2. The nodes are placed in the course of the lymph vessels, and the lymph takes a tortuous course among the cells of the node. This suggests that they serve as filters and are a defense against the spread of infection. The lymph draining from an infected area carries the products of suppuration, and perhaps the infecting organisms themselves, to the first nodes in its pathway. Unless the infection is severe the chances are against the organisms, and the lymph is more or less disinfected before it passes on. Nodes engaged in such a struggle are usually enlarged and tender, and if they are overpowered, they themselves may become the foci of infection.

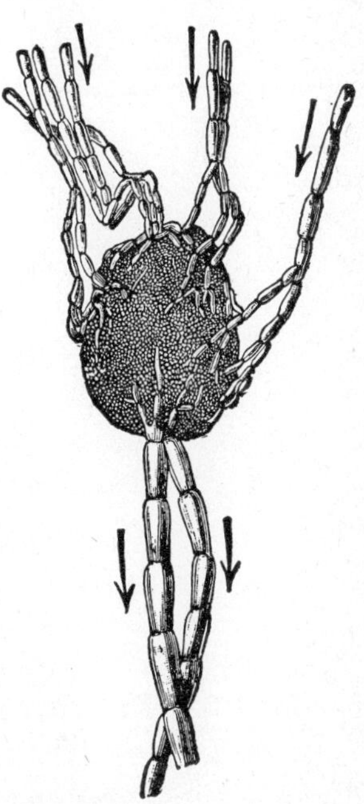

Fig. 259. A lymph node with its afferent and efferent vessels. (Gerrish.)

Factors controlling the flow of lymph. The flow of tissue fluid from the tissue spaces to the lymph capillaries and on to the veins is maintained chiefly by three factors.

1. *Differences in pressure.* The tissue fluid is under greater pressure than the lymph in the lymph capillaries, and the pressure in the larger lymphatics near the ducts is much less than in the smaller

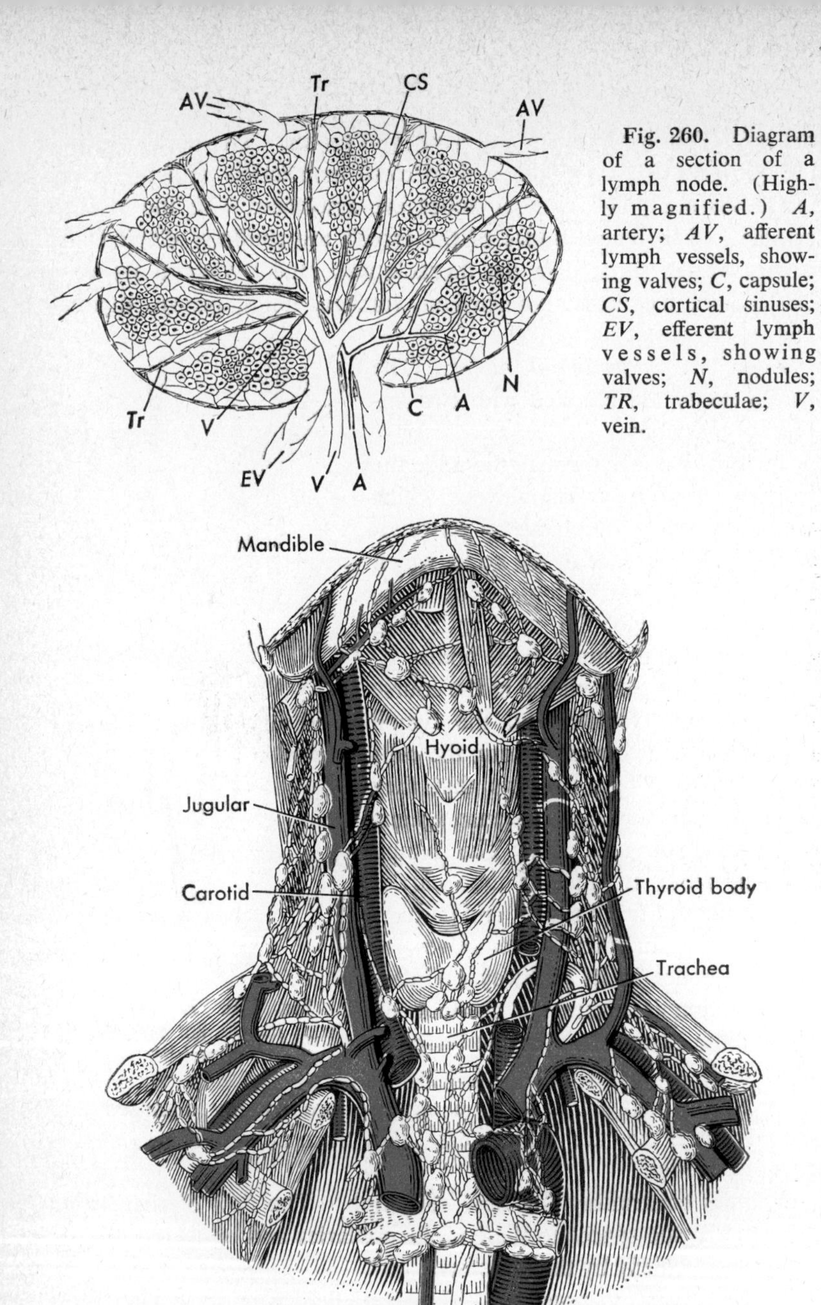

Fig. 260. Diagram of a section of a lymph node. (Highly magnified.) *A*, artery; *AV*, afferent lymph vessels, showing valves; *C*, capsule; *CS*, cortical sinuses; *EV*, efferent lymph vessels, showing valves; *N*, nodules; *TR*, trabeculae; *V*, vein.

Fig. 261. The lymph nodes of the neck and upper part of the thorax. (Gerrish.)

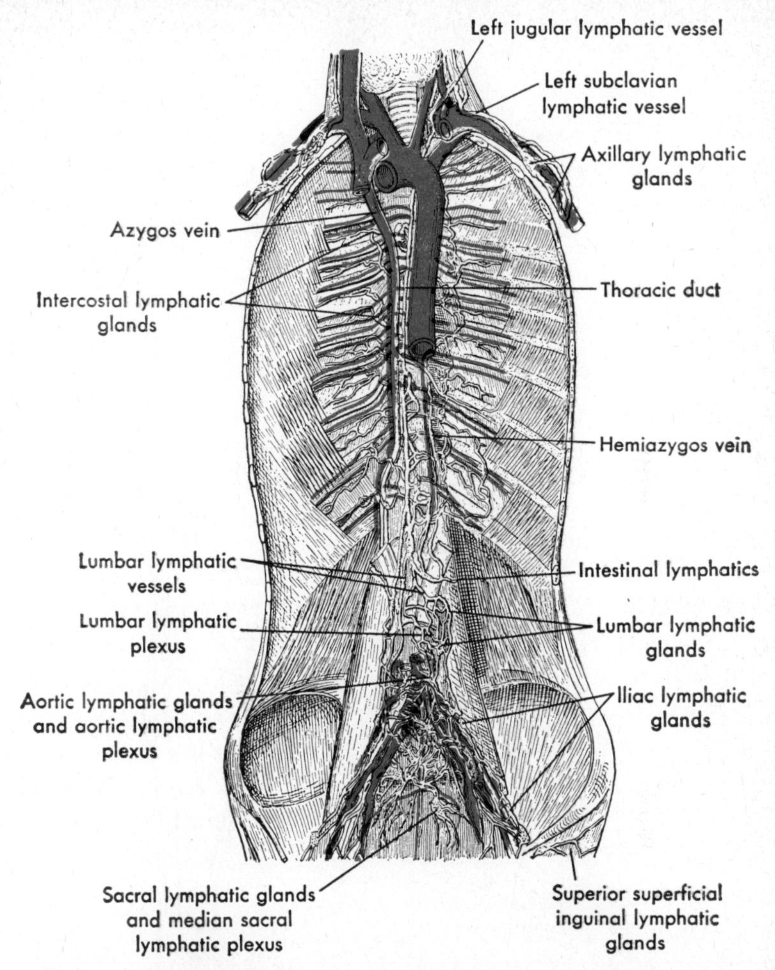

Fig. 262. The lymph nodes and vessels on the dorsal body wall. (Toldt.)

vessels. Consequently we may consider that the lymphatics form a system of vessels leading from a region of high pressure, the tissues, to a region of low pressure, the interior of the large veins of the neck.

2. *Muscular movements and valves.* Contractions of the skeletal muscles compress the lymph vessels and force the lymph on toward the larger ducts. The numerous valves prevent a return flow in the

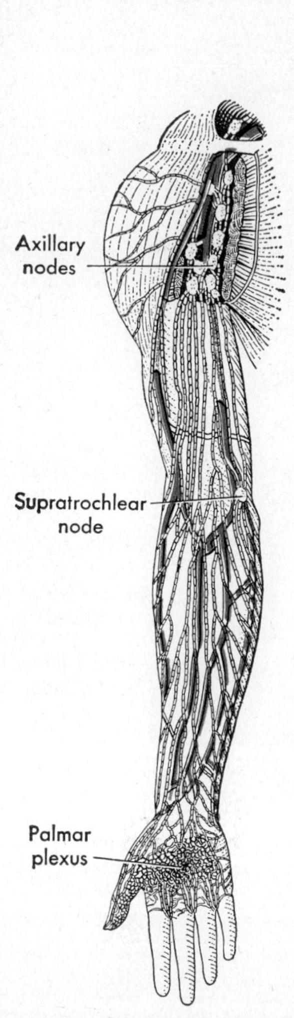

Fig. 263. The lymph nodes and vessels of the upper limb. (Gerrish.)

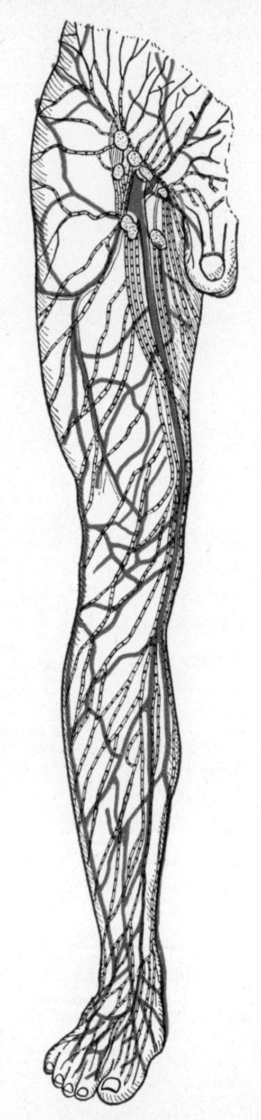

Fig. 264. The lymph nodes and vessels of the lower limb. (Gerrish.)

wrong direction. The flow of lymph from resting muscles is small in quantity, but during muscular exercise and massage it is increased. The flow of chyle is greatly assisted by the peristaltic and rhythmic contractions of the muscular coats of the intestines. Pulsation waves moving over the enormous number of minute arteries existing every-

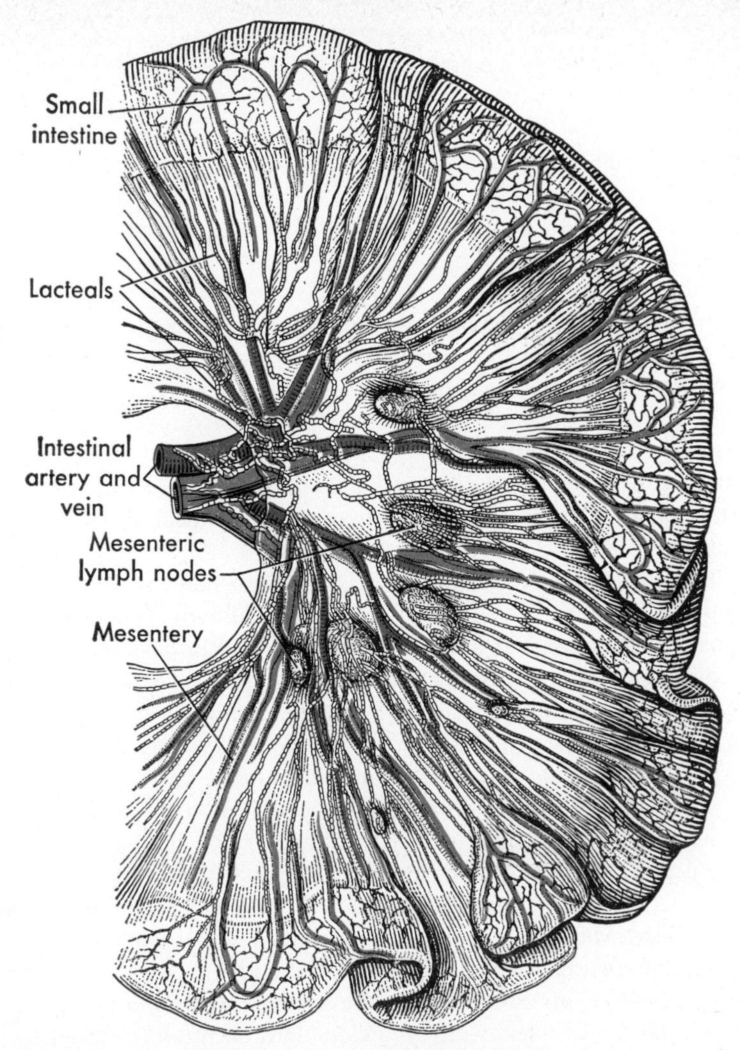

Fig. 265. The lacteals and mesenteric lymph nodes, demonstrated in a loop of small intestine by injection with metallic mercury.

where act to "push ahead" the lymph from valve to valve, on to the larger lymphatics.

3. *Respiratory movements.* During each inspiration the pressure on the thoracic duct is less than on the lymphatics outside the thorax, and lymph is accordingly sucked into the duct. During the succeeding expiration the pressure on the thoracic duct is increased, and

some of its contents, prevented by the valve from escaping below, are pressed out into the innominate veins.

Edema. The lymph in the various tissues of the body varies in amount from time to time, but under normal circumstances remains fairly constant. Under abnormal conditions, these limits may be exceeded, and the result is known as edema. Similar excessive accumulations may also occur in the larger lymph spaces, the serous cavities.

Among the possible causes of edema are:

Any obstruction to the flow of lymph from the tissues.

An excessive formation, the lymph gathering in the tissues faster than it can be carried away by a normal flow.

General or local changes in capillary blood pressure.

Increased permeability of capillary membrane.

It may be a symptom of some other primary conditions, such as certain types of cardiac and renal diseases, mechanical obstruction of veins, etc.

The spleen (lien) is a highly vascular, bean-shaped lymph gland, situated directly beneath the diaphragm, behind and to the left of the stomach. It is covered by peritoneum and held in position by folds of this membrane. Beneath the serous coat is a connective-tissue capsule from which trabeculae run inward, forming a framework, in the interstices of which is found the *splenic pulp*, made up of a network of fibrillae and blood cells—red corpuscles, the various forms of white cells, and large, rounded phagocytic cells (the macrophages of the reticuloendothelial system), which engulf fragmentary and enfeebled red corpuscles and invading organisms such as the Bacillus typhosus. Scattered throughout the pulp are masses of lymphoid tissue called Malpighian follicles.[2] Smooth muscle fibers are found in both the outer capsule and the trabeculae.

The *blood supply* is brought by the splenic artery, a branch of the celiac artery; the splenic artery divides into six or more branches, which enter the concave side of the spleen at a depression called the hilum. The arrangement of the blood vessels is peculiar to this organ. After entering, the arteries divide into many branches and terminate in tufts of arterioles, which open freely into the splenic pulp. The blood is collected by thin-walled veins, which unite to form the splenic vein. The splenic vein unites with the superior

[2] Marcello Malpighi (1628–1694), a physician and professor of comparative anatomy at Bologna.

mesenteric to form the portal tube, which carries the blood to the liver.

Functions. The spleen serves as a possible place of destruction of aged red blood cells or a place of preparation for their destruction by the liver and as a reservoir of blood cells to be liberated under such conditions as exercise or emotional stress. The Malpighian follicles are a place of origin for lymphocytes. It is thought that the spleen produces both erythrocytes and leukocytes during fetal life and also in the adult after certain types of anemia. The spleen undergoes rhythmic variations in size; and by means of this activity, which may be increased under certain physiological demands, it controls mechanically both the quality and volume of the blood by undergoing greater periodic contractions during severe exercise, decrease in barometric pressure, carbon monoxide poisoning, asphyxia, hemorrhage, etc., in which an "emergency call" is made by the tissues for oxygen. Its structure is well adapted to this function as the circulation is, in part, an open one and is sluggish. Also, the distribution of smooth muscle fibers within the trabeculae makes possible these rhythmic contractions about once a minute and slower ones during digestion as well as the more vigorous ones during emergency demands. It is estimated that the spleen can accommodate from one-fifth to one-third of the total volume of blood, but at postmortem its size is greatly reduced.

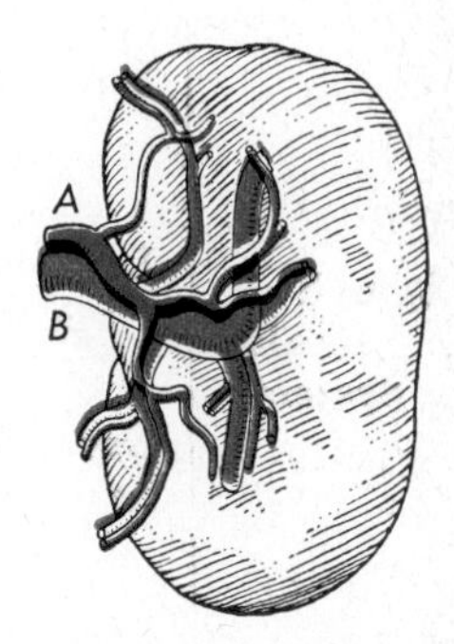

Fig. 266. The spleen, showing the gastric and renal surfaces and the blood vessels. *A*, lienal, or splenic, artery. *B*, lienal, or splenic, vein. (Gerrish.)

Enlargement of the spleen occurs in certain pathological conditions (Banti's disease, Gaucher's disease, etc.), and splenectomy gives favorable results. Enlargement also accompanies malaria, leukemia, Hodgkin's disease, etc.; but removal in these cases is detrimental.

SUMMARY

Lymph	**Description**	Colorless or yellowish liquid Alkaline reaction Salty taste. No odor Consists of blood plasma plus lymphocytes Specific gravity varies between 1.015 and 1.023

- **Lymph**
  - **Description**
    - Contains a relatively low content of blood proteins
    - Contains a relatively low content of nutrients
    - Contains a relatively high content of waste products
    - Clots relatively slowly, does not form a firm clot
  - **Formation**
    - Formed from tissue fluid by the physical process of filtration
  - **Functions**
    - Carries nourishment from blood to tissues
    - Carries waste from tissues to blood
    - Dependent upon diffusion
- **Lymph Vascular System**
  - **Lymph vessels**
    - Lymph capillaries
    - Lymphatics
    - Thoracic duct
    - Right lymphatic duct
    - Lacteals
  - **Expanded lymph spaces**
  - **Lymph nodes**
- **Lymph Vessels**
  - **Lymph capillaries**
    - Origin in tissues
    - One coat of endothelium
    - Start as blind ends of microscopic lymph capillaries, unite to form lymphatics
    - Distribution comparable to that of blood capillaries
  - **Lymphatics**—three coats—numerous valves
  - **Thoracic duct**
    - Begins in cisterna chyli, located on front of body of second lumbar vertebra
    - 38–45 cm long. 4–6 mm in diameter
    - Has three coats—numerous valves
    - Receives lymph from left side of head, neck, and chest, left arm, all of abdomen, and both lower limbs. Receives chyle from lacteals
    - Pours lymph and chyle into left innominate vein
  - **Right lymphatic duct**
    - About 1.25 cm long
    - Receives lymph from right side of head, neck, and chest, also right arm
    - Pours lymph into right innominate vein
  - **Classification**
    - **Superficial**—beneath skin, accompany superficial veins
    - **Deep**—accompany deep blood vessels
  - **Lacteals**
    - Lymphatics of the intestines
    - Many originate in villi of small intestine
    - Contain
      - During digestion—chyle
      - During period of fasting—lymph
    - Absorb fatty substances

Lymph Vessels

- **Expanded lymph spaces**
  - Pleural cavity
  - Pericardial cavity
  - Peritoneal cavity
  - Meningeal spaces
  - Lymph spaces of eye and ear
  - Synovial bursae
- **Functions**—drain off lymph from all parts of the body and return it to the innominate veins

Lymph Nodes

- **Description**
  - Shape
    - Oval
    - Bean-shaped
  - Size varies from that of pinhead to that of almond
  - Outer capsule—connective tissue with some muscle fibers
  - Interior divided into irregular spaces like sponge
  - Spaces partially filled with lymphoid tissue. Communicating channels for lymph, which enters by afferent, leaves by efferent vessels
  - Are well supplied with blood
- **Location**
  - Superficial and deep-set
  - Usually arranged in groups or chains at sides of great blood vessels
  - **Head**
    - (1) Back of head and neck draining scalp
    - (2) Around sternocleidomastoid muscle, draining
      - Back of tongue, pharynx, nasal cavities, roof of mouth, face
    - (3) Under rami of mandible—draining floor of mouth
  - **Upper extremities**
    - (1) Small group at bend of elbow drains
      - Hand
      - Forearm
    - (2) Larger axillary group drains
      - First group
      - Axillary space
    - (3) Larger group under pectoral muscles drains
      - Mammary gland
      - Skin and muscles of chest
  - **Lower extremities**
    - (1) Node at upper part anterior tibial vessels
    - (2) Several in popliteal space
    - (3) Great number massed in groin
    - Drain lower extremities and lower abdominal wall

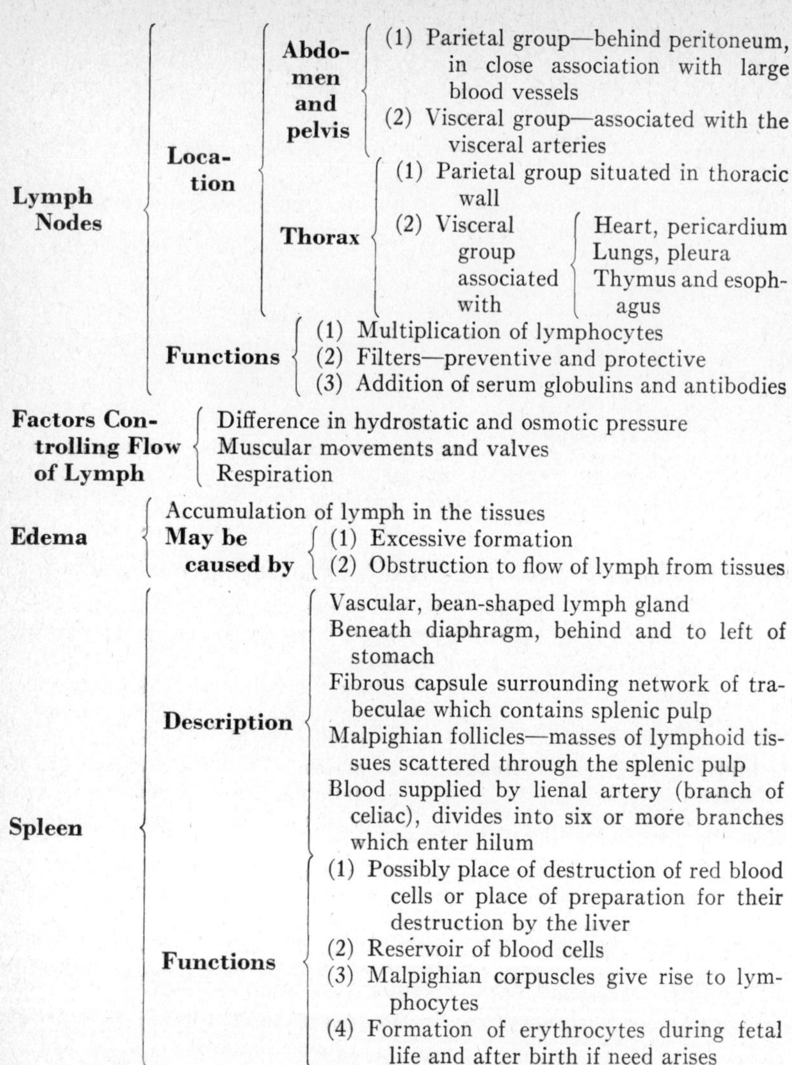

- **Lymph Nodes**
  - **Location**
    - **Abdomen and pelvis**
      - (1) Parietal group—behind peritoneum, in close association with large blood vessels
      - (2) Visceral group—associated with the visceral arteries
    - **Thorax**
      - (1) Parietal group situated in thoracic wall
      - (2) Visceral group associated with
        - Heart, pericardium
        - Lungs, pleura
        - Thymus and esophagus
  - **Functions**
    - (1) Multiplication of lymphocytes
    - (2) Filters—preventive and protective
    - (3) Addition of serum globulins and antibodies
- **Factors Controlling Flow of Lymph**
  - Difference in hydrostatic and osmotic pressure
  - Muscular movements and valves
  - Respiration
- **Edema**
  - Accumulation of lymph in the tissues
  - **May be caused by**
    - (1) Excessive formation
    - (2) Obstruction to flow of lymph from tissues
- **Spleen**
  - **Description**
    - Vascular, bean-shaped lymph gland
    - Beneath diaphragm, behind and to left of stomach
    - Fibrous capsule surrounding network of trabeculae which contains splenic pulp
    - Malpighian follicles—masses of lymphoid tissues scattered through the splenic pulp
    - Blood supplied by lienal artery (branch of celiac), divides into six or more branches which enter hilum
  - **Functions**
    - (1) Possibly place of destruction of red blood cells or place of preparation for their destruction by the liver
    - (2) Reservoir of blood cells
    - (3) Malpighian corpuscles give rise to lymphocytes
    - (4) Formation of erythrocytes during fetal life and after birth if need arises

CHAPTER 15

GLANDS, SECRETIONS, ENZYMES, HORMONES

Physiological organization is brought about by the nervous system and by chemical substances in the circulatory fluids which are carried everywhere in the body, bringing about local changes in equilibrium of physical and chemical conditions and effecting correlations of these changes in a bodywide way.

A gland is a cell or a group of cells which abstracts certain materials from the blood (via tissue fluid) and makes new substances of them. All living cells have this capacity to make new substances and pour them outside themselves. Certain aggregations of cells are known to secrete specific substances either into organs or on the surface of the body, or back into the circulatory fluids. These processes are all subjects of active study today. Students wish to know how secretions are made and what effects they have in relation to organized equilibrium changes or to integration.

Classification. Glands may be classified into groups: (1) *exocrine*, or *duct*, *glands*, which secrete into a cavity or on the body surface, (2) *endocrine*,[1] *incretory*, or *ductless*, *glands*, which secrete into the tissue fluid and blood. Many glands have both exocrine and endocrine secretions, as for instance the pancreas, and may therefore be called *heterocrine glands*.[2] From another point of view glands may be classed as *mucous* if their secretions contain *mucin*, or as *serous* if their secretions contain *serum*. Some glands are of a mixed type, containing both serous cells and mucous cells, as in the submaxillary salivary gland. Again, glands may be grouped according to their structure into *lymphoid* glands and

[1] Endocrine (Greek, *to separate within*).

[2] On this basis all exocrine glands may be classed at heterocrine, since they all also "secrete within." To put it another way, all glands are endocrine, and some (sweat, gastric, etc.) are also exocrine.

epithelial glands. Also every cell may be thought of as a gland, its excretions (as far as the cell itself is concerned) considered as secretions or excretions to the body as a whole, depending on whether they have further value to the body or are useless.

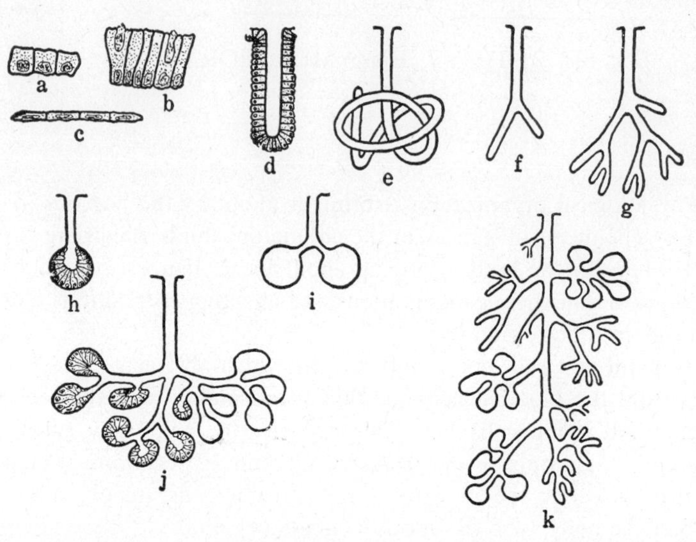

Fig. 267. Diagram showing types of glands. *a*, plain cuboidal secreting cells; *b*, plain columnar secreting cells, two of which are "goblet glands"; *c*, plain flat secreting cells; *d, e, f, g*, tubular glands, simple, twisted, branched, and several times branched. *h, i, j*, saccular or alveolar glands: *h*, simple; *i*, branched; *j*, much-branched. *k*, compound tubuloalveolar gland.

GLANDS OF EXTERNAL SECRETIONS

These glands may consist of a single cell, or may be a simple pocketlike depression of a membrane, or may consist of a vast number of such secreting membranous depressions. The pancreas and liver are examples of the last. The glandular epithelium is supported by areolar connective tissue carrying a dense network of capillaries close to the secreting cells. Nerve fibers are also abundant. The stimuli for varying the amount of the secretory product may be of chemical or nervous origin. The work of the secretory cells consists of two phases: (1) active secretion, including a considerable flow of water, and (2) a period of recovery, during which special substances are produced in the cells. During the second period the protoplasm of the gland cells becomes filled and in some cases distended with granules. During active secretion the granules are lost, the protoplasm clears, and the cells shrink in size. These glands may be classed according to structure:

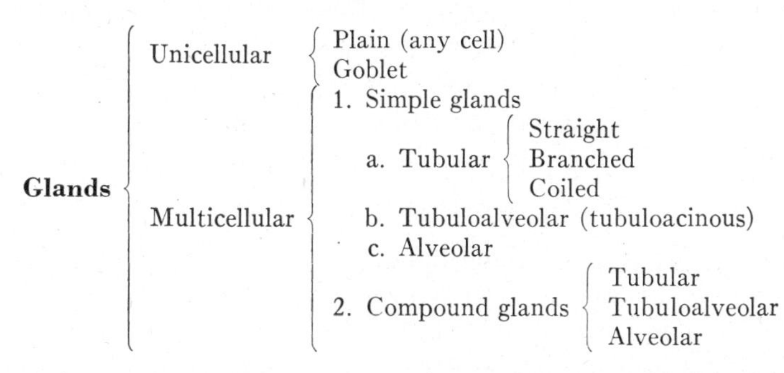

1. The simple glands. The simple tubular glands are divided in relation to the structure of the fundus into straight, branched, and coiled. The *straight* tubular glands are found in the intestine; the branched glands may have several branches and are found in the gastric and uterine mucosa; the coiled tubular glands are found in the skin (the sweat glands).

The simple tubuloalveolar glands are of the branched form. This group includes the smaller glands of the respiratory tract. The mucous glands of the esophagus and Brunner's glands of the duodenum are included in this group by some histologists, since they are frequently enlarged toward their ends.

The simple branched alveolar glands which have a common duct that gives rise to a number of saccules are seen in the Meibomian[3] glands and the large sebaceous glands.

2. The compound glands. The compound glands may be tubular, tubuloalveolar, or alveolar. The *compound tubular* glands have a large number of distinct duct systems, which eventually open into a main or common excretory duct. The liver, kidneys, and testes are good examples of these glands.

The compound tubuloalveolar glands. These glands are numerous, and while the general principle of structure is about the same in all of them, there is considerable variation in their minute structure. These glands also have many distinct duct systems which eventually

[3] Heinrich Meibom, German anatomist (1638–1700).

open into a common duct. All of the salivary glands, the pancreas and some of the larger mucous glands of the esophagus, the seromucous gland of the respiratory pathway, and many of the duodenal glands belong in this group.

The compound alveolar glands. These glands are very much like the other compound glands in general structure; however, the terminal ducts end in alveoli with a dilated saclike form. The mammary glands are good examples of this kind of gland.

THE ACTIVITIES OF GLANDS

A secretion is a new substance made by cells from substances brought to them by the blood and tissue fluid. Examples of external secretions are the digestive fluids (secreted by salivary, gastric, and intestinal glands, the pancreas, and the liver) and the secretions of the lacrimal, Meibomian, ceruminous, sebaceous, sweat, mammary glands, etc. The enzymes of many of these secretions are known and have been named.

Secretion formation. Tissue fluid forms the only source of materials for the manufacture of secretions. The basic materials for the secretions are brought to the tissue fluid by the blood. In Fig. 268 two alveoli of a gland are shown. Assume that these cells are manufacturing a secretion at all times, but that the rate of production varies. The basic materials are filtered and diffused through the walls of the capillaries into the tissue fluid, diffused into the cells of the gland, and leave the gland through its duct as its secretion—pancreatic juice, etc. It is the plasma of the blood that filters and diffuses out from the capillaries into the tissue fluid which bathes these gland cells. The comparison of blood and lymph on p. 46 shows that the serum albumin and the serum fibrinogen do not filter out into the tissue fluid as rapidly as the water and its solutes and that serum globulin practically does not filter out at all. At the same time the by-products of the activity of the glandular cells enter the tissue fluid.

These substances (the extra blood plasma and the products of cell activity) are carried away partly by the blood capillaries and partly by the lymph capillaries. In general, those substances which diffuse readily through capillary walls are carried away by the blood; those, like proteins, which do not diffuse readily are carried away by the lymph capillaries and finally are poured into the blood stream near the heart.

It is easy to see that there is a relationship between the cells of the gland, the tissue fluid which bathes these cells, and the blood, the lymph playing but a subsidiary part. Omitting discussion of the selective permeability of the cell membranes, one item of selectivity

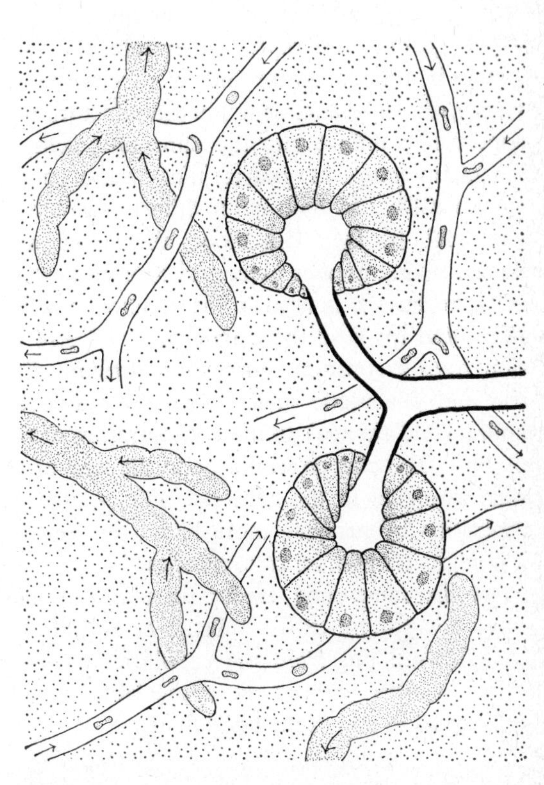

Fig. 268. Diagram of a thin section of two alveoli of a gland lying in areolar connective tissue (stippled). In the near neighborhood are blood capillaries (containing red blood cells), which have been cut in making the section; and lymph capillaries, wider and thinner-walled than the blood capillaries and closed (blind) at the end. The areolar tissue is filled with tissue fluid.

would be related to the specific enzymes within the cell. Any substance (say A or B) which meets an "adjusted enzyme" in the cell would disappear into the manufactured product, e.g., pancreatic juice or gastric juice; hence the concentration of the diffused substance inside the cell will be *kept low*. Then since the blood continues to flow through the capillaries, despite the fact that it is losing substance A or B constantly to the pancreatic cells or gastric cells, the concentration will always be relatively high. And there will be a constant diffusion of substance A into the pancreatic cells and of substance B into the gastric cells.

Then under the conditions when more pancreatic fluid and gastric fluid are being made, the blood flow through the capillaries would be a little greater (pulse rate, pulse volume, velocity of blood a little higher), the capillaries would be a little larger, more capillaries (closed at other times) would be opened up, filtration and diffusion would go on more rapidly; therefore, the tissue fluid would be kept higher in total quantity and in concentration of selected materials. Hence, the substances for the manufacture of pancreatic juice would continue to diffuse into the pancreatic cells, and the substances for the manufacture of gastric juice would continue to diffuse into the gastric cells. In both cases diffusion would continue from a source kept high by the extra blood, the larger and more numerous capillaries giving greater area, and a place kept low for A in pancreatic cells, because it is used in the pancreatic cells for making pancreatic juice, and for B in the gastric cells. There will be a constant diffusion from a constant source to a constant disappearance. The gastric and pancreatic juices would be produced with no change (no appreciable change) in the concentration of the basic materials in the blood. The change would be in the way of making more raw materials available to the cells per unit of time with no change in concentration by bringing more blood per unit of time and distributing it in a greater number of capillaries.

At the same time, under these conditions the lymph capillaries carry away more excess substances and also the blood proteins that have filtered out from the capillaries and so share in the variable work of cells without an appreciable change in concentration of substances in any of the body fluids.

ENZYMES

Enzymes are complex proteins which act as catalyzers; that is, they vary the rate of chemical reactions but do not appear in the final product. Extracellular enzymes, or exoenzymes, are secreted from cells and exert their activities outside these cells, as does ptyalin of saliva. Intracellular enzymes, or endoenzymes, do their work within the cells in which they are made, as does glycogenase. Some characteristics of enzymes may be noted: Body enzymes act best at body temperature (optimum). Each enzyme requires a medium of definite reaction, either acid, alkaline, or neutral—pepsin acts only in an acid medium, whereas trypsin digests protein in either an alkaline or a neutral solution, but not in the presence of free acid. The action of enzymes is *specific*, i.e., enzymes that act upon fats do not act upon carbohydrates. In fact, each one of the sugars seems to require its own special enzyme. Under proper conditions many chemical reactions are *reversible*. For instance, during digestion an enzyme called lipase acts upon neutral fats, changing these fats to fatty acids and glycerin. After absorption, fatty acids and glycerin combine to form body fats, and it is probable that lipase furthers this synthesis.

Nature of enzymes. Most of the studies concerning the nature of enzymes have been carried on in relation to the enzymes having to do with digestion in the alimentary canal. These enzymes always give a protein reaction, and the activity of the enzyme is greatly reduced or lost if the material containing the enzyme has been coagulated or otherwise chemically changed. Willstätter has suggested the following structure in relation to these enzymes.

Enzymes	Colloidal protein	related to the *quantitative* or *catalytic* aspects of enzyme activity
	Specific chemical radicals adsorbed on the protein	related to the *qualitative* or *specific aspects* of enzyme activity

In some instances the vitamin molecule forms part of the enzyme molecule. At the present time there is active research in the field of hormonal influence on enzyme systems and enzyme action. There is evidence that the adrenal-cortex hormones influence both protein and carbohydrate metabolism and that this action is through enzyme systems.

Active and inactive forms. It has been demonstrated that an enzyme may exist in an *inactive* form within the cell producing it and when secreted may still be inactive. This inactive form is designated as a *proferment* or *zymogen*. The zymogen may be stored in the cell in the form of granules which are converted into an active enzyme at the moment of secretion, or it may be secreted in inactive form and require the cooperation of some other substance before it is capable of effecting its normal reaction. In such cases the second substance is said to activate the enzyme. Inorganic substances causing activation are designated *activators*, and organic substances serving the same purpose are called *kinases*. An example of the latter is the enterokinase which activates trypsinogen.

Coenzymes. There are some cases where the action of an enzyme is helped by, or perhaps is dependent upon, the presence of some other substance. An example of this activity is furnished by the influence of bile salts upon lipase. These cases of *coactivity* are to be distinguished from activation by the fact that the combination may be made or unmade. For example, in a mixture of bile salts and lipase, the bile salts may be removed by dialysis. In activation, on the contrary, the active enzyme cannot be changed back to the inactive zymogen.

The endoenzymes. There are many kinds of enzymes within cells, such as the enzymes concerned with hydrolysis, with oxidative processes, with carbohydrate metabolism and the forming of carbohydrates from noncarbohydrate sources; the enzymes concerned with phosphorylations and those concerned with muscle activity. The phosphatases are found in most tissues of the body and are becoming important as diagnostic measures. There are many other kinds of enzymes within cells.

Since enzymes are mostly protein in composition, amino acids are required for their synthesis.

THE ENDOCRINE GLANDS

Endocrine, or ductless, glands pour their secretions directly into tissue fluid and blood and are often called glands of internal secretion. Many and probably all glands with ducts also secrete into the circulatory fluids, as, for instance, the pancreas (insulin), the gastric glands (gastrine), duodenal glands (secretin), liver (glucose), etc. The chemical products of these glands and other glandular tissues are called hormones. Hormones play an important role in general nutrition. They share with the nervous system and with the general constituents of body fluids in correlating and coordinating the activities of the body as a whole.

Hormones differ from enzymes in many ways. Enzymes are rendered inert by boiling, hormones are not; enzymes are not dialyzable, hormones are; enzymes have not been synthesized, while some of the hormones (for example, epinephrine) have been synthesized.

The methods of study of hormones are observation of conditions caused by disease or removal of the glands; feeding of glands, extracts, or synthetic preparations or active principles; and subcutaneous or intravenous injection of glandular extracts to normal animals and to animals in which the glands have been removed, or to subjects thought to present glandular disturbances.

Hypophysis. The hypophyseal gland or pituitary body is a mass of tissue about 1 cm in diameter. It weighs about 0.5 to 0.7 gm and is situated in the sella turcica of the sphenoid bone. (Figs. 56 and 339.) A short stalk connects the hypophysis with the base of the brain, known as the infundibulum. The pituitary gland consists of two lobes—the larger *anterior* lobe of glandular structure and the smaller *posterior* lobe derived from the third ventricle of the brain.

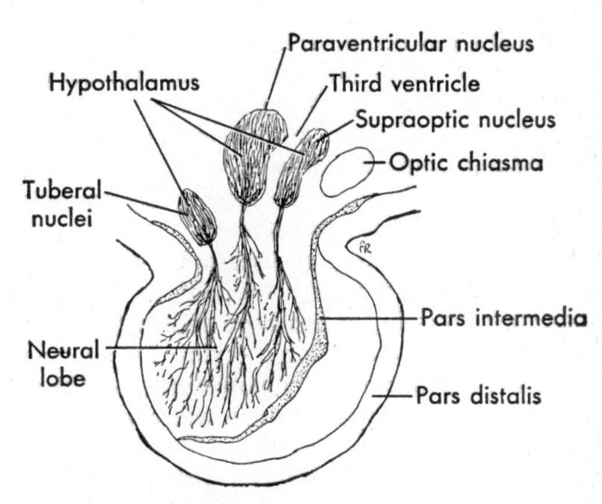

Fig. 269. Diagram to show fibers from hypothalamus to the neurohypophysis.

The posterior lobe consists of two parts: (1) the pars nervosa, or neural lobe, composed chiefly of neuroglia cells and fibers, and (2) the pars intermedia, composed of epithelial cells.

The pituitary is highly vascular and its blood supply is independent of the brain proper. Nerve fibers from the hypothalamus pass through the infundibular stalk, where most of them terminate in the pars nervosa.

The best-known *functions* of the posterior lobe include:

1. Secretion of an *antidiuretic principle* (ADH), which controls water reabsorption in the kidney tubules and in this way helps to regulate the water and electrolyte balance of body fluids.

2. Secretion of an *oxytocic principle* (oxytocin), which causes uterine muscle to contract, especially during parturition. Oxytocin

also has galactagogue action, and hence is given to nursing mothers to increase the flow of milk.

3. Secretion of a *pressor principle.* Whole postpituitary extracts also have stimulating action on smooth muscle in many other parts of the body. It also raises blood pressure and reflexly slows heart rate. Systemic arterioles and capillaries are constricted. Pituitrin constricts the coronary and pulmonary vessels, but dilates the renal and cerebral vessels.

Zondek has demonstrated another hormone of the posterior lobe called *intermedin.* This hormone causes expansion of the melanophores of the frog's skin, etc. In the human it is said that the intermedin may check the excretion of water in diabetes insipidus without increasing the amount of sodium chloride excreted. Its effect, then, is different from that of Pitressin, which while decreasing the excretion of water also decreases the amount of chloride excreted.

Diabetes insipidus, a disease in which the urinary output is greatly increased, was formerly thought to result from hyposecretion of the posterior lobe; but it has been shown that a similar polyuria results from injury to the hypothalamic region (tuber cinereum) of the brain. It is therefore probable that involvement of either the posterior lobe of the pituitary or the hypothalamus will cause the disease.

Functions of the anterior lobe. The anterior lobe is sometimes referred to as a "dominating gland" because of its widespread influence on the other endocrine glands; but it is itself probably in turn influenced by other glands, although the mechanism of the resulting coordination is not entirely clear. It produces a number of hormones and also influences processes for which hormonal action has not yet been demonstrated but is known to exist. By means of the somatotropic hormone it has a specific effect on the growth of bones, muscles, and viscera. This hormone is also apparently related to protein metabolism. The adrenocorticotropic hormone (ACTH) influences the function of the cortex of the adrenal gland. Its greatest influence is on that part of the adrenal cortex that controls electrolyte, water, protein, and carbohydrate metabolism. ACTH also influences the hormones that affect antibody formation and lymphoid-tissue activity. Through these activities the body can more effectively cope with tissue injury and stress situations from any cause.

The thyrotropic hormone (TSH) influences both the structure and the secretory activity of the thyroid gland. There is a hormone

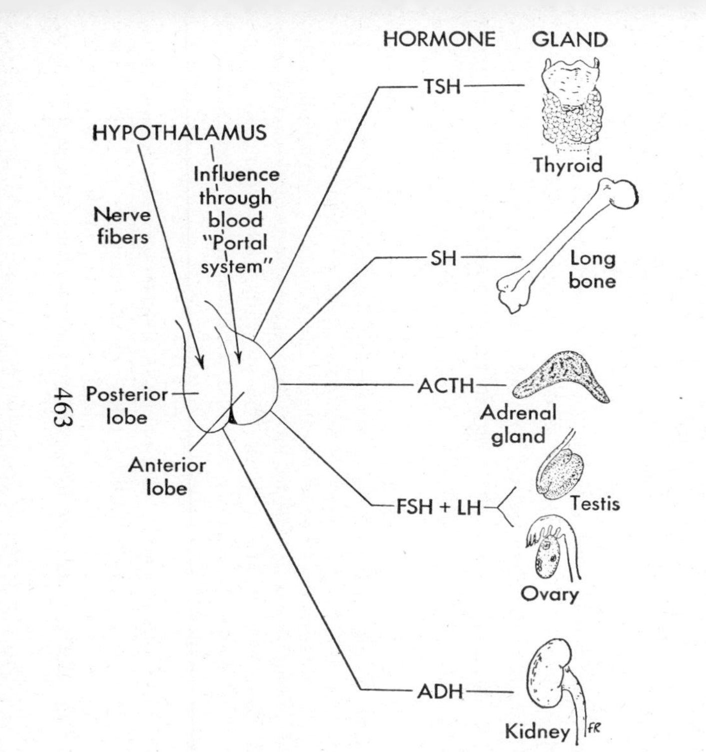

FUNCTIONS (briefly)

TSH (thyroid-stimulating hormone) influences structure and secretory activity of the thyroid. The thyroid controls metabolic rate.

SH (somatotropic hormone) influences growth of bones, muscles, and viscera. In some way it is related to protein metabolism.

The adrenal cortex controls Na^+, Cl^-, K^+ reabsorption in the kidney. Regulates carbohydrate, protein, and fat metabolism. The adrenal medulla functions in relation to "fight and flight" and other reactions.

FSH (follicle-stimulating hormone) is concerned with the ripening of follicles and activity of the seminiferous tubules.
LH (luteinizing hormone) influences secreting cells of the ovaries and testes and maintains their normal activity.
Lactogenic hormone maintains secretion of corpus luteum and initiates lactation.

ADH (antidiuretic hormone) controls water reabsorption in the kidney tubules and in this way helps to regulate water and electrolyte balance of body fluids.

Fig. 270. Diagram showing some of the functions of the pituitary hormones on physiological activities.

that directly influences carbohydrate metabolism and has an opposing physiological effect to the action of insulin, and hence functions in raising blood sugar (diabetogenic or insulin-antagonizing factor).

The anterior pituitary also influences the metabolism of fat (ketogenic factor) and the activity of the parathyroids (parathyrotropic factor) and the activity of the islets of Langerhans (pancreatropic factor).

There are also hormones that control the activity of the gonads, the development of the mammary gland during pregnancy, and the subsequent secretion of milk. Of these the follicle-stimulating hormone (FSH), the luteinizing hormone (LH), and the lactogenic hormone are the most important. FSH is concerned with the ripening of the follicles, the production of estrin, and the maintenance of epithelium in the seminiferous tubules of the testes. LH is concerned with the production of corpus luteum, inhibition of estrin production, suppression of estrus, and stimulation of the interstitial tissues of the testes.

The functions of the anterior pituitary are now believed to be controlled by the hypothalamus through blood supply. The anterior pituitary receives blood from the arterial system and from what is called the *portal circulation* from the hypothalamus. The capillaries in the hypothalamus form a set of veins, and these go by way of the infundibulum to the anterior pituitary lobe, where again they form a capillary network. The blood in these vessels contains the "products of hypothalamic activity," and influence is exerted on the anterior pituitary. The hypothalamus in this way exerts hormonal control of the anterior pituitary.

1. Excess of the growth hormone in early life results in *giantism* (*gigantism*); the result of hyperactivity of the gland in later life is *acromegaly*, in which the jaws, bones, hands, and feet show overdevelopment and the features become enlarged and coarse. Hypoproduction of the growth hormone results in *dwarfism*. Growth of the body and sexual development are arrested. In the adult, the result is called *Simmonds' disease*, in which emaciation, muscular debility, loss of sexual function, and general apathy occur.

2. Research indicates that the thyrotropic hormone of the pituitary controls normal functioning of the thyroid and that cases of hypothyroidism may be primarily disturbances of the pituitary.

3. The general influence of the pituitary on the gonads has long been known through experiments in which removal in young animals inhibited

sexual development and in adults brought about retrogression of sexual activity.

Two gonadotropic hormones are described: (1) the follicle-stimulating principle, which controls general sexual activity, and (2) the luteinizing principle, which stimulates the formation of the testicular hormones in males, and in females stimulates ovulation, development of corpus luteum, and the production of *progesterone.* In the male the follicle-stimulating hormone stimulates the testes to develop sex cells, while the luteinizing principle brings about production of the testicular hormone.

Laboratory tests for human pregnancy make practical use of the follicle-stimulating hormone of the anterior pituitary, which is present in large quantities in the blood and urine in early pregnancy. Injection of extracts of blood serum or urine from a pregnant woman into immature female laboratory animals (mice, rabbits, etc.) brings about changes in the ovaries and genital tracts which provide a basis for determination of pregnancy in the early stages. The test is not infallible, since in conditions such as carcinoma of the genital organs or the menopause the reaction may simulate that of pregnancy.

4. *The lactogenic hormone* is *prolactin,* which controls the production of milk in mammary glands and in the crop of birds. Injections of prolactin into virgin rats stimulates the nesting activities of motherhood, and in chickens brooding habits are stimulated. Removal of the hypophysis during normal periods of lactation inhibits milk secretion.

5. Experiments demonstrate the regulation of the activities of the adrenal glands by the adrenotropic hormone of the pituitary and suggest the possibility that some cases of Addison's disease (p. 461) are primarily of pituitary origin.

Frölich's syndrome (*adiposogenital dystrophy*) has been shown to involve both the hypothalamus of the brain and a deficiency in anterior pituitary lobe secretion. The disease is characterized by obesity and arrested development of the sexual organs.

Clinical disorders of the pituitary are usually shown to be accompanied by tumors of the anterior lobe. These tumors may be the cause of sexual overdevelopment and regressive disorders of growth and sex activities, such disorders as gigantism, acromegaly, and Simmond's disease.

None of the various substances or factors of the anterior lobe have been isolated. It is therefore doubtful that each of the various actions is carried out by a separate hormone.

The pineal body, or **epiphysis,** is a small reddish-gray body about 8 mm in length that develops as an outgrowth of the third ventricle of the brain and remains attached to the roof of the ventricle. In early life it is glandular, and it attains its maximum growth about the seventh year. After this period, and particularly after puberty, it decreases in size, and the glandular tissue is replaced by fibrous tissue. Pathological destruction of the gland is sometimes followed by precocious sexual and skeletal, and possibly also mental, development. The function of the gland is still obscure.

The thyroid gland consists of two lobes, at the sides of the trachea below the thyroid cartilage. These lobes are connected by strands

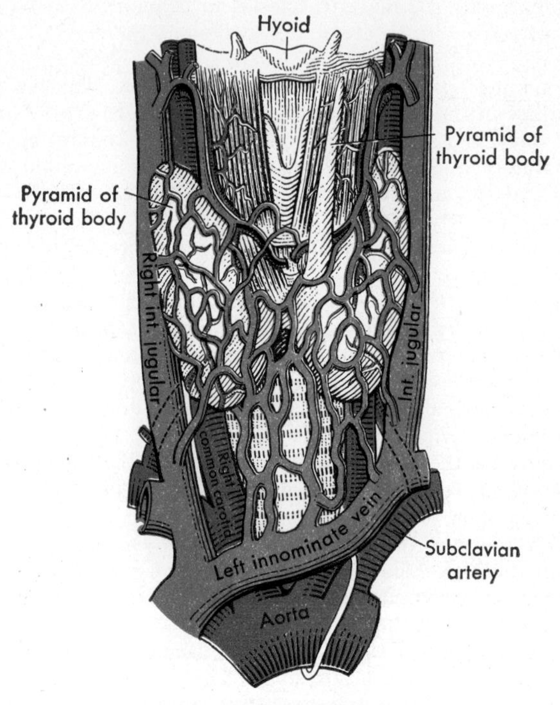

Fig. 271. The thyroid body and the related blood vessels. (Gerrish.)

of thyroid tissue called the *isthmus,* ventral to the trachea. The external layer of the thyroid is connective tissue which extends inward as trabeculae and divides the gland into closed vesicles of irregular size. Centrally, each of these vesicles contains a colloid or jellylike substance which is secreted by the columnar epithelial cells which line the vesicle. This colloid substance is the source of the thyroid hormone. The thyroid gland contains numerous lymphatics.

An abundant blood supply is derived from the external carotids and the subclavian arteries and is returned via the superior, middle, and inferior thyroid veins to the jugular and left innominate veins. It has been estimated that about 4 to 5 qt of blood pour through the gland per hour.

The nerves are derived from the second to fifth thoracic spinal nerves through the superior and middle cervical ganglia of the thoracolumbar system and from the vagus and glossopharyngeal nerves of the craniosacral system.

The functions are to control the metabolic rate and normal growth and development in the young. It supplies thyroxin (C. R. Harington's formula, $C_{15}H_{11}O_4NI_4$) and stores iodine, containing at times as much as 10 to 15 mg.

It is possible that the thyroid hormone functions as a catalyst which increases oxidation in the cells. This action may be through an oxidative enzyme system. One milligram of thyroxin increases the metabolic rate about 2.5 per cent. Carbohydrates, proteins, and fats are all oxidized in increased amounts. When thyroxin is given experimentally to animals or human beings, nitrogen loss exceeds intake. Glucose tolerance is decreased and stored glycogen is also decreased, while blood glucose may be normal or below. The thyroid gland is believed to be related in some way to the metabolism of calcium and phosphorus, since when the hormone is given experimentally, an increased amount of calcium and phosphates is eliminated in the feces and urine. Blood levels remain constant, which indicates that calcium is being lost from bone. Cases of hypoactivity of the gland may show as much as 50 per cent decrease in the metabolic rate; cases of hyperactivity may show an increase up to 40 per cent and more. Injection or feeding of thyroid tissue results in increased basal metabolism, loss of weight, increase in elimination of nitrogen, increased heartbeat—which results at times in an irregularity called *tachycardia*—and nervous excitability.

The size of the thyroid varies with age, sex, and general nutrition. It is relatively larger in youth, in women, and in the well-nourished. Removal of the gland does not cause death but brings about marked changes such as lowered basal metabolism and general malnutrition. Disturbances in the secretion of thyroxin are classed under two headings: (1) hypothyroidism, or lack of secretion, and (2) hyperthyroidism, or excess of secretion.

Goiter is an enlargement of the gland. It may result from increased functional activity due to a decrease in the iodine content of the gland.

This in turn is thought to be due to a decrease or lack of iodine in water and food. Goiter occurs frequently in adolescent girls, but its incidence is greatly reduced if iodine is given. There are other kinds of goiter.

Hypothyroidism. In man certain pathological conditions are due to hypothyroidism, i.e., cretinism and myxedema.

Cretinism is a type of mental deficiency due to congenital defects of the thyroid or to atrophy in early life. The growth of the skeleton ceases, although the bones may become thicker than normal, and there is marked arrest of mental development. Children so afflicted are called cretins. They are not only dwarfed but ill-proportioned, having heavy heads, protruding abdomens and weak muscles, slow speech, etc.

Myxedema is a condition which results from atrophy or removal of the thyroid in adult life. The most marked symptoms of this condition are slowness both of body and mind, usually associated with tremors and twitchings. The skin becomes rough and dry, owing to lack of cutaneous secretions, and assumes a yellow, waxlike appearance. There is an overgrowth of the subcutaneous tissues, which in time is replaced by fat; the hair grows coarse and falls out; the face and hands are swollen and puffy; the metabolic rate is low and the mind apathetic. Cretinism and myxedema are both due to lack of the internal secretion of the thyroid, which may be supplied by feeding the thyroid of other animals. The treatment must be kept up during the patient's life.

Hyperthyroidism. Overactivity of the thyroid gland, i.e., increase in the amount of the internal secretion, produces a condition called *Graves' disease* or *exophthalmic goiter.* It is characterized by protruding eyeballs, quickened and sometimes irregular heart action, elevated temperature, nervousness, and insomnia. The appetite may be, often is, excessive, but this is accompanied by loss of weight due to increased metabolism and digestive disturbances. This condition is sometimes remedied by limiting the blood supply to the gland, by ligating one of the carotid branches, or by removing part of the gland.

The parathyroids, usually four in number and arranged in pairs, are independent of the thyroid both in origin and in function but are usually located on its dorsal surface. These small reddish glands are about 6 to 7 mm long and 2 to 3 mm thick. Accessory nodules are sometimes found surrounding the glands or embedded in the thymus. The glands consist of closely packed epithelial cells richly supplied with capillaries from branches of the inferior and superior thyroid arteries. The nerve supply is from the vagus and glossopharyngeal nerves of the central nervous system and from the cervical autonomics of the thoracolumbar system.

Function. The parathyroids secrete a hormone, *parathyroid hormone*, or *parathormone*, which plays a part in the maintenance of the normal calcium level of the blood and the irritability of the nervous system and muscles. It also regulates phosphorus metabolism. Recent investigation suggests that the primary action of the hormone might be to regulate the excretion of inorganic phosphate in the urine. Vitamin D influences the absorption and intestinal excretion of both phosphorus and calcium. Parathormone may also exert action on bone and stimulate osteoclastic activity. There is a functional relationship between the parathyroids and the anterior lobe of the hypophysis.

Acute symptoms of *hypoparathyroidism*, known clinically as *tetany*, may result from removal of the parathyroids or may possibly occur spontaneously; the concentration of blood calcium is reduced and may result in death if untreated. Symptoms are relieved by giving solutions of calcium; parathyroid hormone is also used in treatment.

In *hyperparathyroidism* there is muscular weakness, pain in the bones, and an increase in the calcium in blood and urine. The bones show decalcification and deformity, and spontaneous fractures may occur. These symptoms are sometimes accompanied by a parathyroid tumor, the removal of which brings about a reduction in the blood calcium; and considerable resolidification of the bones occurs. Deposits of calcium may occur, especially in the kidney, and nitrogenous wastes may become excessive in blood and lymph.

The suprarenal glands, or adrenal bodies, are two small bodies, one above the upper end of each kidney. The right adrenal gland is somewhat triangular in shape and the left one more semilunar. They vary in size, and the average weight of each is about 5–9 gm. Each gland is surrounded by a thin capsule and consists of two parts known as the *cortex* or external tissue, and the *medulla*, or chromophil tissue. These parts differ in origin and function. The arteries supplying this highly vascular gland are derived from the aorta, the inferior phrenics, and the renal arteries. Blood is returned via the suprarenal veins. It has been estimated that a quantity of blood equal to about six times each gland's weight passes through it per minute. The nerve fibers are derived from the celiac and renal plexuses (splanchnic nerves). Removal of the gland has long been known to be followed by prostration, muscular weakness, and lowered vascular tone, with subsequent death in a few days. Recent investigation has shown these symptoms to be caused by the removal of the cortex. Removal of the medulla causes no serious disturbance.

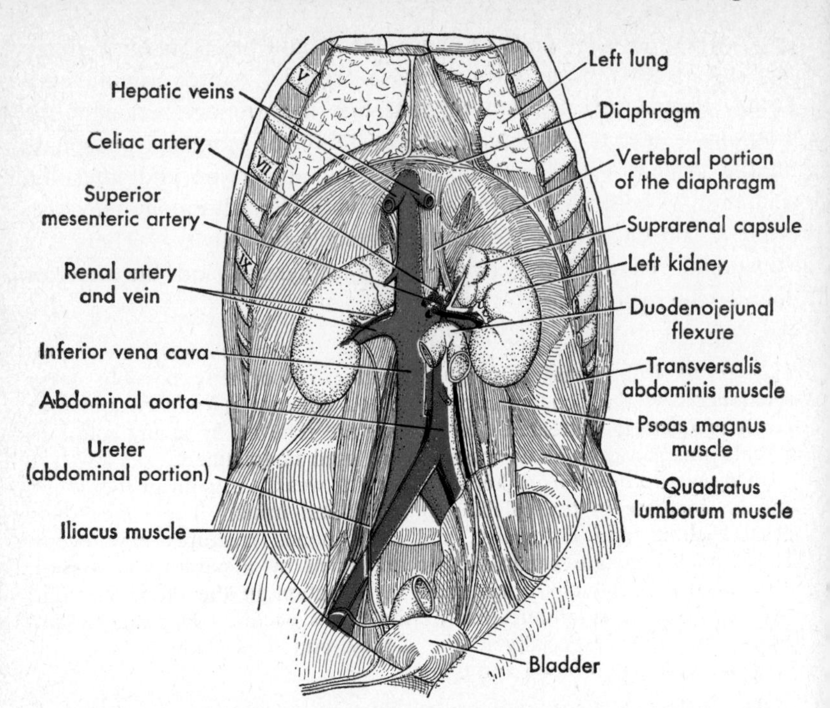

Fig. 272. Diagram showing position of adrenal glands and kidneys. Note blood-vessel arrangement and relationship to other structures.

The adrenal cortex. The *cortex* is derived from that portion of the mesoderm which gives rise to the kidneys and consists of masses of epithelical cells arranged in columns. The adrenal cortex is essential to life, and if the glands are removed from an animal, death results, chiefly from circulatory and renal failure. The chief known functions of the adrenal cortex include the following:

1. REGULATION OF ELECTROLYTE AND WATER BALANCE. The salt or mineral hormone functions at the renal tubule and stimulates the reabsorption of Na^+ (and perhaps Cl^-) into the blood and in this way the Na^+ (and Cl^-) content of the extracellular fluids is maintained. Water is reabsorbed with the salt. The reabsorption of K^+ is depressed and hence the Na^+–K^+ ratio is controlled in these fluids. Fluid balance is also controlled through regulation of these electrolytes. Desoxycorticosterone acetate (DOCA) is the most commonly known in this group.

2. INFLUENCE ON METABOLISM. These are hormones that regulate the metabolism of carbohydrate, protein, and fat. These hormones stimulate the formation of glycogen from proteins in the liver and inhibit glucose utilization in the tissues by preventing glucose conversion into hexosephosphate; the breakdown of tissue protein to amino acids is promoted and fat mobilization from fat depots is increased. Cortisone is widely known in this group.

3. EFFECT ON LYMPH AND BLOOD. The adrenal cortex influences the activity of lymphoid tissue and the number of eosinophils in circulating blood.

4. The adrenal cortex in some way aids the body to cope more effectively with adverse environmental situations or what is commonly known as *stress* situations.

5. Several hormones have been isolated from the adrenal cortex that have influence on the sexual organs. Cortin is probably the most commonly known. The exact relationship is not completely understood. Research shows that when the adrenal cortex, ovaries, or testes are removed certain physiological changes take place. Removal of the ovaries or testes results in an enlargement of the adrenal cortex.

The adrenal cortex activity is regulated by the anterior pituitary hormone, adrenocorticotropin, ACTH.

Cortical hypofunction is recognized as the cause of Addison's disease, which, if untreated, is fatal within from 1 to 3 years.

Symptoms of Addison's disease are muscular weakness and general apathy, gastrointestinal disturbances, pigmentation of the skin and mucous membrane, loss of weight, and depressed sexual function. The pigmentation is the outstanding symptom and is due to excessive deposition of the normal cutaneous pigment, *melanin*. Treatment with cortical extract gives favorable results.

Cortical hyperfunction appears to be associated with cortical tumors. It results in the young in precocious sexual development with profuse growth of hair, and in the case of adult females, in the development of secondary male characteristics.

The adrenal medulla. The medulla is derived from the same portion of the ectoderm that gives rise to the sympathetic ganglia, and it consists of large, granular cells arranged in a network. It secretes a hormone called *epinephrine* (*adrenaline*). The probable formula is $C_9H_{13}NO_3$. The cells which compose the larger portion of the medulla give a characteristic reaction when treated with chromic acid,[4] and application of this treatment has shown that similar cells are located along the abdominal aorta. All cells showing this reaction are assumed to have a similar function.

A number of theories exist with regard to the function of epinephrine under normal conditions, and these are conflicting and difficult to summarize satisfactorily. Among them is the "emergency theory" of Cannon and his associates, in which the supply of epinephrine to the blood by the adrenal medulla is increased under emotional stress. This increase results in a more rapid heartbeat; a greater flow of blood to the muscles, central nervous system, and heart; an increased output of glucose from the liver; inhibition of the intestinal muscle coat and closure of the sphincters. The muscle coat of the bronchi is relaxed and there is contraction of the splenic capsule, which gives more blood to circulation. By means of these reactions the stable environment of cells is maintained during periods of great functional demands. Whether these reactions are in direct response to the increased secretion of the medulla or are brought about by the action of the sympathetic nervous system alone is undetermined. Nevertheless, it may be said that the medulla in normal physiological activities carries on certain emergency functions such as bringing about a rise in general blood pressure and increased output of the heart, the liberation of carbohydrate supplies from the liver, the contraction of both smooth and striated muscles, decrease in the coagulation time of the blood, discharge of red blood cells from the spleen, and an increase in the depth and rate of respiration. These and other accompanying reactions, such as emotional responses, result in equipping the individual to meet the emergency at hand. The adrenal medulla also secretes another hormone called *nor-epinephrine* (*nor-adrenaline*), which is adrenaline minus its terminal CH_3 group. There are differences in the action of these two hormones, but at the present time the action of nor-epinephrine has not been clarified. Removal of the medulla causes no serious physiological disturbance.

Examples of the use of epinephrine in medicine are: to raise blood pressure in surgical emergencies or by injection into cardiac tissue to resuscitate a heart which has ceased to beat; to bring about vasoconstriction, which prolongs the action of local anesthetics and reduces the absorption of the substance producing the anesthesia; to relax the bronchioles in asthma; to contract mucous membranes and the arterioles of the skin, thus reducing the loss of blood in minor operations.

[4] Tissues which give the chromophil reaction are grouped as parts of the chromophil system.

Internal secretions of the gonads. The *ovary* produces at least two internal secretions whose actions are understood. (1) Follicular hormone or *estrin,* or theelin, $C_{18}H_{22}O_2$, is secreted by the follicles of the ovary and controls the uterus and Fallopian tubes, the vagina, and the mammary glands in their changes throughout the phases of the menstrual cycle. Theelin is present in the blood of females from puberty to the menopause, reaching its highest concentration just before menstruation. It is present in large amounts in the urine during pregnancy. It is also present in the urine of males and has been found in the tissues of growing plants and animals. (2) The corpus luteum hormone or progesterone (progestin, $C_{21}H_{30}O_2$) supplements the action of theelin, carrying on further development of the uterine mucosa in preparation for implantation of the developing ovum. It suppresses estrus and ovulation and is essential for the growth of the mammary glands. It is essential to the complete development of the maternal portion of the placenta and to the development of the mammary glands during pregnancy. This effect may be due partly to a stimulation of the anterior pituitary.

Recent texts refer to the possibility of there being another hormone, *relaxin,* of the corpus luteum. It is believed to bring about relaxation of the pelvic ligaments in preparation for parturition. It suppresses the estrus cycle.

Two male hormones have been isolated—the androgens: *androsterone* (probable formula $C_{19}H_{30}O_2$) from male and also female urine, and *testosterone* (probable formula, $C_{19}H_{28}O_2$) from the testicle. Androsterone is also found in the testis and the blood. In experimental animals, injections of these male hormones stimulate the development of the comb, wattles, and spurs of capons, and in hens ovulation is inhibited. In guinea pigs from which the testes (except the epididymis) have been removed, the period of motility of spermatozoa is greatly prolonged by injection of testicular extracts.

Attention is called to the fact that both the male and female sex hormones are produced in both sexes. The sex of the animal determines whether male or female hormones will be secreted in preponderance.

The adrenal cortex is also a source of androgens and estrogens found in normal urine and is the only source of the substances in urine after ovariectomy or castration.

The placenta. The placenta is a source of estrone or theelin and estriol or theelol and of chorionic gonadotropin or the anterior-pituitary-like substance which has similar action to the gonadotropic principles of the anterior pituitary hormones. The placenta is also a source of a rich supply of immune bodies and of a blood coagulant.

The thymus usually consists of two lobes, but they may unite to form a single lobe or may have an intermediate lobe between them. It is situated in the upper chest cavity along the trachea, overlapping the great blood vessels as they leave the heart. Each lobe is composed of several lobules, each of which is composed of an outer

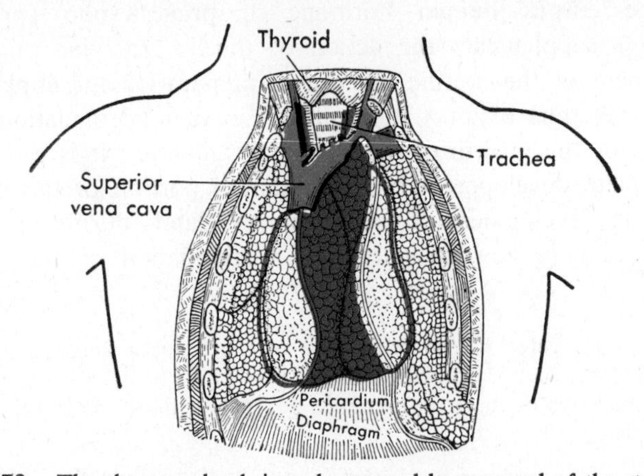

Fig. 273. The thymus gland, in red, exposed by removal of the sternal and costal cartilages. In children the thymus is large, its growth continuing until puberty. At this time atrophic changes begin to take place, and in an adult of advanced years the gland may be scarcely recognizable.

cortex, or lymphoid portion, in which a few reticular cells are scattered, and a central *medulla*, in which the reticulum is coarser and the lymphoid cells fewer in number. In the medulla are rounded nests of cells, 30 to 100 μ in diameter, called the corpuscles of Hassall.[5] The arteries are derived from an internal mammary and the superior and inferior thyroids. The nerves are derived from the vagi of the craniosacral system and from the thoracolumbar system. The thymus gradually decreases in size after puberty, the corpuscles of Hassall disappearing more slowly than other portions. Experiments on rats indicate that an internal secretion of the thymus plays a part in the early development of maturity with all its characteristics. It does not cause increase in size, but rats attain adult size earlier.

[5] Arthur Hill Hassall, English physician (1817–1894).

It is credited with inhibiting the growth of the gonads, and in birds the absence of it inhibits the formation of the eggshell. It is suggested that the Hassall corpuscles may be concerned with antibody formation.

The liver forms an internal secretion which was the first of these substances to be discovered. From glycogen (a nondiffusible colloid) stored within its cells, the liver forms *dextrose* and secretes it into the blood. Dextrose is, therefore, an internal secretion. The liver also forms *urea* from amino acids and liberates it into the blood to be eliminated by the kidneys. Urea is, therefore, also an internal secretion. There are other internal secretions formed by the liver.

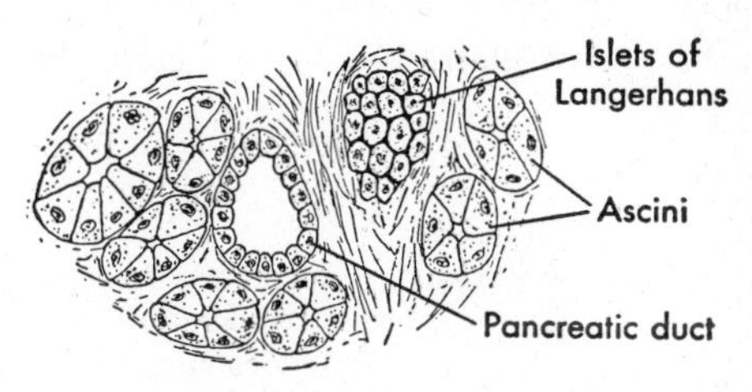

Fig. 274. Diagram of thin section of a bit of the pancreas. (Highly magnified.)

The pancreas, in special groups of cells, the islets of Langerhans, forms an internal secretion called *insulin,* which is essential for normal sugar metabolism. There are several types of cells in the islet group. The beta cells secrete insulin, and it is possible that the alpha cells produce a hormone which influences the normal metabolism of fats.

Internal secretions of the gastric and intestinal mucous membranes. *Gastrin* is a hormone secreted into the blood by the mucous membrane lining the pyloric end of the stomach. This hormone is carried by the blood to the peptic and pyloric glands, which it stimulates. Some investigators believe that the mucous membrane of the intestine, particularly the duodenum, contains cells which secrete a substance known as *prosecretin*, which is inactive until the medium is acid. When the acid chyme from the stomach enters the duodenum, prosecretin is liberated, changed to *secretin*, absorbed by the blood, and carried to the pancreas, liver, and intestines, stimulating each organ to secretory activity. There are other hormones produced in the intestinal mucosa. See Chap. 18.

Tissue hormones. All tissues and perhaps all various local aggregates of tissues secrete substances characteristic of themselves. These are called *tissue hormones.*

It should be recognized that hormones are but one factor in the complex chemical system of the body fluids controlling physiological integration, health, development, and behavior. It is evident that if experimental studies were made bringing about disturbances in the kinds and concentrations and physical states of the other constituents of these heterogeneous body fluids, the results would be as striking as those described in relation to experiments with hormones.

SUMMARY

Physiological integration of different systems of body brought about in part by secretions of glands

- **Glands**
  - **A gland** is a cell or a group of cells which forms new substances
  - **Classification**
    - According to place of secretion
      - Exocrine, or duct glands
      - Endocrine, or ductless glands
      - Heterocrine, both exocrine and endocrine
    - According to kind of secretion
      - Mucous (mucin)
      - Serous (serum)
    - According to structure
      - Lymphoid
      - Epithelial
  - **Glands of external secretion**—see p. 455 for classification
  - **Secretion,** a new substance made by cells from substances obtained from blood or tissue fluid
  - **Enzymes**
    - Digestive fluids owe their power to promote chemical reactions to **enzymes**
    - **Definition**—substances produced by living cells which act by catalysis
    - **Characteristics**
      - Act best at body temperature
      - Require medium of definite *p*H
      - Action is specific and may be reversible
    - **Zymogen**
      - Antecedent or inactive form activated by
        - Activators—inorganic
        - Kinases—organic
    - **Coenzymes**—substances which help or act with an enzyme
    - **Endoenzymes**—substances found within cells which promote chemical reactions
  - **Endocrine glands**
    - Hormones stimulate activities
    - Methods of study
      - (1) Observation of pathological conditions or removal
      - (2) Feeding of glands or extracts or synthetic principles of glands
      - (3) Observation of results of injections of extracts
    - Hypophysis, pineal body, thyroid, parathyroids, suprarenals, gonads, placenta, thymus, liver, pancreas, intestinal mucosa, tissue hormones

Name of Gland	Location	Secretion	Probable Function	Diseases Associated with It
Pituitary, or Hypophysis Cerebri Mass of reddish-gray tissue about 1 cm in diameter. Consists of an anterior and posterior lobe and a tuberal part containing colloid	Lodged in the sella turcica of the sphenoid bone	Posterior lobe—antidiuretic	Controls water reabsorption in the kidney tubule	**Hypofunction** (or hypothalamic lesion) results in *diabetes insipidus*
		Oxytocin	*Oxytocic effect*—muscles of uterus contract	
		Pituitrin *Pitressin*	*Pressor effect*—(1) blood pressure raised and heartbeat slowed, (2) plain muscle stimulated, (3) arterioles and capillaries constricted	
		Anterior lobe—	"Dominating gland" controlling (1) skeletal growth, (2) thyroid secretion, (3) activity of gonads, (4) lactation, (5) activity of adrenal cortex, and (6) other metabolic processes	**Hypofunction** may result in Simmonds' disease, dwarfism, polyuria, depressed sexual function, and scanty growth of hair **Hyperfunction** may result in gigantism (early life), acromegaly (adult life), and glycosuria
Pineal Body, or Epiphysis Reddish-gray body about 8 mm in length, develops as an outgrowth of third ventricle of brain	Attached to roof of third ventricle of brain		Not known	Tumors of the pineal gland sometimes develop in children. In boys the main symptoms are associated with precocious development of sex organs and precocious physical and mental development
Thyroid Weighs about 1 oz Consists of two lobes connected by an isthmus	In front of trachea below thyroid cartilage	*Thyroxin* formula—($C_{15}H_{11}O_4NI_4$), contains 65% iodine	Influences the general rate of oxidation in the body, also growth and development in the young	**Hypothyroidism** may result in cretinism (early life) and myxedema (adult life) **Hyperthyroidism** may result in Graves' disease (exophthalmic goiter)

Name of Gland	*Location*	*Secretion*	*Probable Function*	*Diseases Associated with It*
Parathyroids Four small glands	Between the posterior borders of the lobes of the thyroid gland and its capsule	*Parathyroid hormone* (*parathormone*)	Regulates the blood-calcium level and the irritability of the nervous system and muscles	When the parathyroids are removed, **tetany** develops **Hyperparathyroidism** { Muscular weakness; High blood calcium
Suprarenal bodies, or adrenals Two small glands, each surrounded by a fibrous capsule and consisting of two parts { Cortex; Medulla	Placed above and in front of the upper end of each kidney			
Cortex consists of epithelial cells arranged in columns. These cells are derived from the part of the mesoderm that gives rise to the kidneys		*Cortin*	Regulates electrolyte and water balance. Influences fat, carbohydrate, and protein metabolism; activity of lymphoid tissue and sexual organs. Aids body to cope more effectively with stress situations	Removal causes death **Hypofunction** results in Addison's disease **Hyperfunction** appears to be associated with precocious sexual development
Medulla consists of a network of large granular cells which when treated with chromic acid give a yellow or brown reaction. Derived from neural crest of ectoderm		Medulla—*epinephrine*	Constitutes a reserve mechanism that comes into action at times of stress. Epinephrine augments the response of sympathetic nerves, increases the heartbeat, increases blood supply to muscles, nervous system, and heart, and increases output of glucose from liver	Removal of the medulla causes no serious physiological disturbance

Name of Gland	*Location*	*Secretion*	*Probable Function*	*Diseases Associated with It*
Gonads—*ovaries* Two almond-shaped bodies which weigh 2–3.5 gm	One on each side of the uterus, attached to the back of the broad ligament behind and below the uterine tubes	*Estrone*, or *theelin* ($C_{18}H_{22}O_2$), is formed by vesicular follicles *Progesterone* ($C_{21}H_{30}O_2$) is formed by cells of corpus luteum	Seems to act by maintaining nutrition and mature size of female reproductive organs Sensitizes the mucous membrane of the uterus so that it responds to the contact of the developing ovum and assists in implantation	Excessive ovarian function produces precocious puberty Diminished ovarian function characterized by late onset of menstruation, faulty development of genital organs, delayed menstruation, etc.
Testes Two glandular organs which weigh 10.5–14 gm	In the scrotum	*Androsterone* ($C_{19}H_{30}O_2$) and *testosterone* ($C_{19}H_{28}O_2$)	Experimental evidence suggests that these hormones control the development of secondary sex characteristics in the male	The tendency to become obese after castration
Placenta	Pregnant uterus	Estrone chorionic gonadotropin	Similar action to the anterior pituitary hormones	
Thymus Consists of two lateral lobes which occasionally unite to form a single mass Temporary organ Reaches greatest size at puberty	Upper part of chest cavity, along the trachea, and overlapping the great blood vessels as they leave the heart	Thymus secretion	Experimental results suggest varied functions	
Liver Largest gland in body, weighs 42–56 oz	Under the diaphragm on right side	*Dextrose* and *urea*	(1) Changes glucose to glycogen and glycogen to glucose (2) Changes ammonium carbonate to urea	

Name of Gland	*Location*	*Secretion*	*Probable Function*	*Diseases Associated with It*
Pancreas A compound gland which weighs between 2 and 3 oz	In front of the first and second lumbar vertebrae behind the stomach	Islets of Langerhans furnish an internal secretion containing *insulin*	(1) Restores the power to utilize the glucose of the blood (2) Accelerates the synthesis of sugar to glycogen and the storage of glycogen (3) Restricts production of sugar in liver from protein and fat (4) Counteracts the tendency to acidosis	**Diabetes Mellitus** See Chap. 18
Gastric Mucosa	Stomach	*Gastrin* is hormone of internal secretion	Carried by blood to fundic and pyloric glands and stimulates secretion	
Intestinal Mucosa	Intestine-duodenum	Cells of duodenum contain *prosecretin*, changed by acid to *secretin*	Secretin stimulates the pancreas, liver, and intestines to activity	

CHAPTER 16

RESPIRATORY SYSTEM: ANATOMY, HISTOLOGY, PHYSIOLOGY OF RESPIRATION

The normal course of the chemical changes in tissue cells is dependent upon oxygen; hence the need of a continuous supply. One of the chief end products of these chemical changes is carbon dioxide; hence the need for continuous elimination of carbon dioxide. In unicellular animals the intake of oxygen and the output of carbon dioxide occur at the surface of the one-celled animal. As organisms increase in size and complexity, some form of apparatus is developed the function of which is to bring oxygen to the cells of the organism. In man the circulating blood in the alveolar capillaries takes up oxygen and gives up carbon dioxide, and later in the capillaries of the tissues it gives up oxygen and takes up carbon dioxide.

This exchange of gases is known as respiration and is dependent upon the proper functioning of certain organs, the *respiratory system.* The essentials of a respiratory system consist of a moist and permeable membrane, with a moving stream of blood containing a relatively high percentage of carbon dioxide on one side and air or fluid containing a relatively high percentage of oxygen on the other. In most aquatic animals the respiratory organs are external in the form of gills; in mammals the respiratory organs are situated internally in the form of lungs. The lungs are placed in communication with the nose and mouth by means of the bronchi, trachea, and larynx.

NOSE

The nose is the special organ of the sense of smell, but it also serves as a passageway for air going to and from the lungs. It filters, warms, and moistens the entering air and also helps in phonation.

It consists of two parts—the external feature, the nose, and the internal cavities, the nasal fossae.

The external nose is composed of a triangular framework of bone and cartilage, covered by skin and lined by mucous membrane. On its under surface are two oval openings, the nostrils (*anterior nares*), which are the external openings of the nasal cavities. The margins of the nostrils are usually provided with a number of hairs.

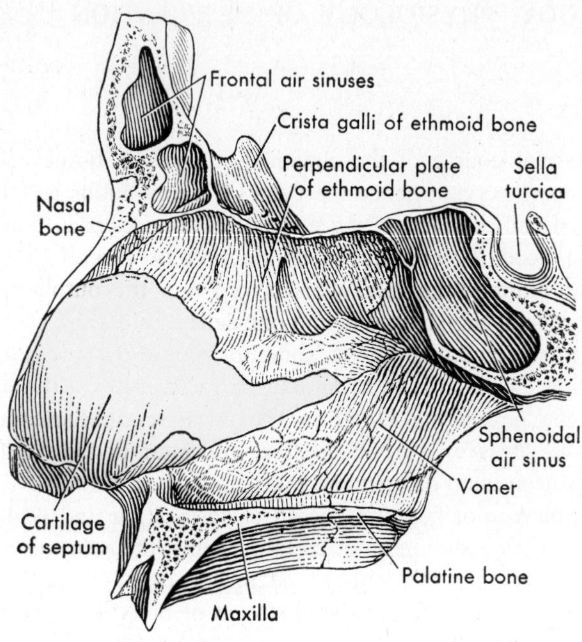

Fig. 275. Bones and cartilage of septum of nose, left side. (After Gray's *Anatomy*.)

The nasal cavities are two wedge-shaped cavities, separated from each other by a partition, or septum. The septum is formed in front by the crest of the nasal bones and the frontal spine; in the middle, by the perpendicular plate of the ethmoid; behind, by the vomer and sphenoid; below, by the crest of the maxillae and palatine bones. The septum is usually bent more to one side than the other, a condition to be remembered in giving nasal treatments.

The conchae and processes of the ethmoid, which are exceedingly light and spongy, project into the nasal cavities and divide them into three incomplete passages from before backwards—the superior, middle, and inferior meatus. The palate and maxillae separate the nasal cavities from the mouth, and the horizontal plate of the ethmoid forms the partition between the cranial and nasal cavities.

The nasal cavities[1] communicate with the air in front by the anterior nares, and behind they open into the nasopharynx by the two

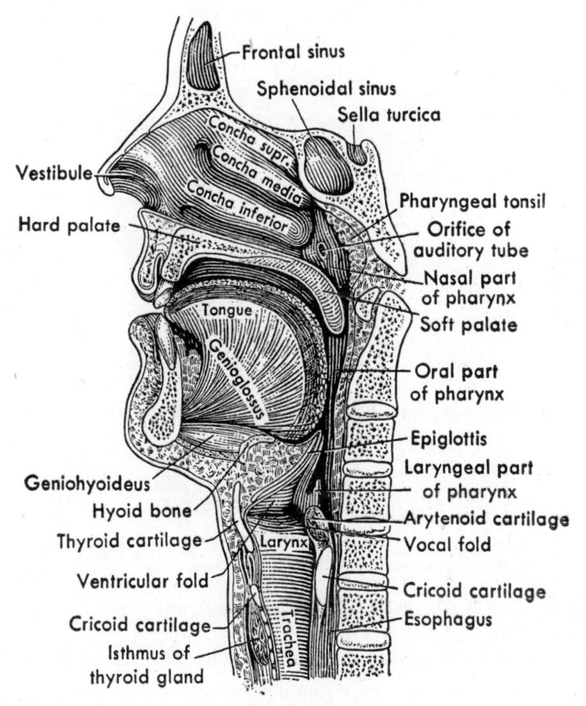

Fig. 276. Sagittal section of the nose, mouth, pharynx, and larynx. (After Gray's *Anatomy*.)

posterior nares (choanae). They are lined with mucous membrane. At the entrance each cavity or vestibule is lined with thick, stratified, squamous epithelium containing sebaceous glands and numerous

[1] Eleven bones enter into the formation of the nasal cavities: the floor is formed by the palatine (2) and part of the maxillae bones (2); the roof is formed chiefly by the horizontal plate of the ethmoid bone (1), the sphenoid (1), and the small nasal bones (2); in the outer walls we find, in addition to processes from other bones, the two conchae (2). The vomer (1) forms part of the septum.

coarse hairs. The middle, or respiratory, portion of the cavity is lined with pseudostratified epithelium with many ciliated and goblet cells. The upper, or olfactory, portion is lined with neuroepithelium which contains olfactory cells which are the receptors for smell. This membrane, which is highly vascular, is continuous externally with the skin and internally with the mucous membrane lining the

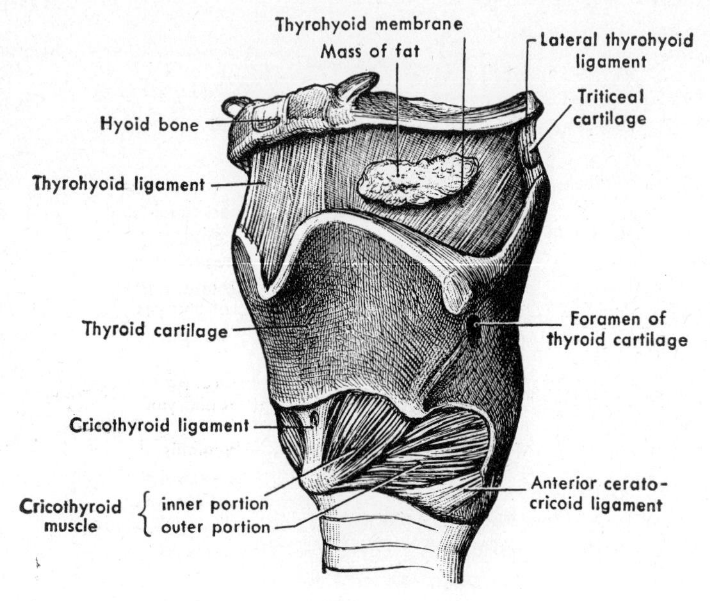

Fig. 277. Larynx, seen from the left side and front. (Toldt).

sinuses and other structures connected with the nasal passages. Inflammatory conditions of the nasal mucous membrane may extend into the sinuses.[2]

Advantage of nasal breathing. Under normal conditions breathing should take place through the nose. The arrangement of the conchae makes the upper part of the nasal passages very narrow;

[2] One reason for the emphasis on the prevention or early cure of head colds is due to this possibility. The infection may spread upward into the nasolacrimal duct, lacrimal sac, and conjunctiva; into the head sinuses, such as the frontal, ethmoidal, sphenoidal, or the antrum of Highmore; through the pharynx into the larynx, trachea, and bronchi; or through the Eustachian tube to the middle ear and the mastoid portion of the temporal bones. Perhaps the most serious possibility is the extension of the infection to the meninges by way of the olfactory nerve.

these passages are thickly lined and freely supplied with blood, which keeps the temperature relatively high and makes it possible to moisten and warm the air before it reaches the lungs. The hairs at the entrance to the nostrils and the cilia of the epithelium serve as filters to remove particles which may be carried in with the inspired air.

Nerves and blood vessels. The mucous membrane of the septum contains the endings of the olfactory nerve fibers. The nerve fibers for the muscles of the nose are fibers of the facial (seventh cranial), and the skin receives fibers from the ophthalmic and maxillary nerves, which are branches of the trigeminal (fifth cranial). Blood is supplied to the external nose by branches from the external and internal maxillary arteries, which are derived from the external carotid. The lateral walls and the septum of the nasal cavities are supplied with nasal branches of the ethmoidal arteries, which are derived from the internal carotid.

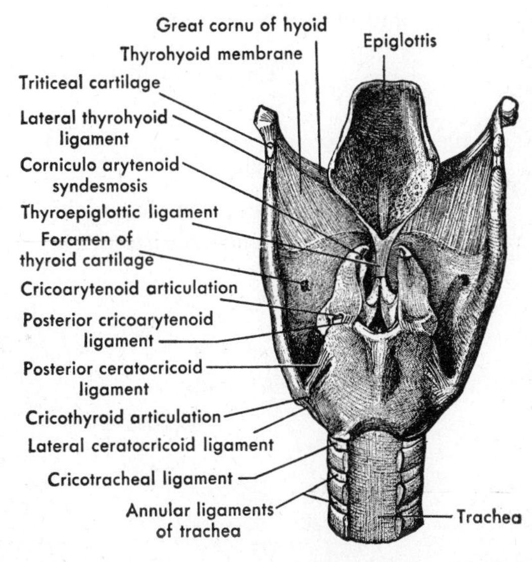

Fig. 278. Larynx, seen from behind. (Toldt.)

The mouth serves as a passageway for the entrance of air, and the pharynx transmits the air from the nose or mouth to the larynx, but both are closely associated with digestion and will be described with the digestive organs.

RESPIRATORY TRACT

The respiratory tract is composed of the following organs in addition to the nose and nasopharynx already described: (1) larynx, (2) trachea, (3) bronchi, (4) lungs.

The larynx, or organ of voice, is placed in the upper and front part of the neck between the root of the tongue and the trachea. Above and behind it lies the pharynx, which opens into the esophagus, or gullet; and on either side of it lie the great vessels of the neck. The larynx is broad above and shaped somewhat like a triangular box, with flat sides and prominent ridge in front. Below, it is narrow and rounded where it blends with the trachea. It is made up of nine fibrocartilages, united by extrinsic and intrinsic ligaments and moved by numerous muscles.

Cartilages of the Larynx

Single Cartilages	Paired Cartilages
Thyroid	Arytenoid
Cricoid	Corniculate
Epiglottis	Cuneiform

The thyroid cartilage resembles a shield and is the largest. It rests upon the cricoid and consists of two square plates, or laminae

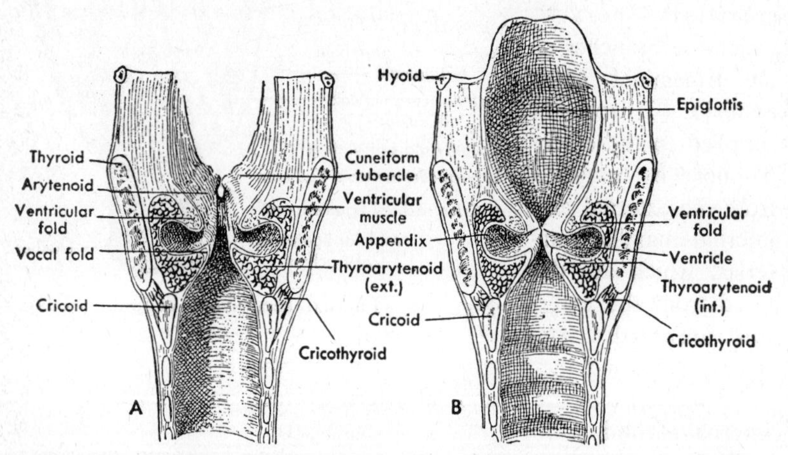

Fig. 279. Larynx, frontal section. *A,* posterior segment; *B,* anterior segment. (J. P. Schaeffer, *Morris' Human Anatomy.* Courtesy of The Blakiston Company.)

(right and left), which are joined at an acute angle in the middle line in front and form by their union the laryngeal prominence (Adam's apple).

The cricoid cartilage resembles a seal ring with the hoop part in front and the signet part in the back.

The epiglottis is shaped like a leaf. The stem is inserted in the notch between the two plates of the thyroid.

The arytenoid cartilages resemble a pyramid in form and rest on either side of the upper border of the lamina of the cricoid cartilage.

The corniculate cartilages are small, conical nodules of elastic cartilage articulating with the upper inner surface of the arytenoid cartilage. They prolong the arytenoid cartilages backward and medialward.

The cuneiform cartilages are small, elongated pieces of elastic cartilage placed on each side in the aryepiglottic fold.

The larynx is lined throughout with mucous membrane, which is continuous above with that lining the pharynx and below with that lining the trachea.

The cavity of the larynx is divided into two parts by two folds of mucous membrane stretching from front to back but not quite meeting in the middle line. They thus leave an elongated fissure called the *glottis,* which is the narrowest segment of the air passages. The glottis is protected by a lid of fibrocartilage called the *epiglottis.*

The vocal folds. Embedded in the mucous membrane at the edges of the slit are fibrous and elastic ligaments, which strengthen the edges of the glottis and give them elasticity. These ligaments, covered with mucous membrane, are firmly attached at both ends to the cartilages of the larynx and are called the inferior or *true vocal folds,* because they function in the production of the voice. Above the vocal folds are two ventricular folds, which do not function in the production of the voice but serve to keep the true vocal folds moist, in holding the breath, and in protecting the larynx during the swallowing of food.

The glottis varies in shape and size according to the action of muscles upon the laryngeal walls. When the larynx is at rest during quiet breathing, the glottis is V-shaped; during a deep inspiration it becomes almost round, while during the production of a high note the edges of the folds approximate so closely as to leave scarcely any opening at all.

Muscles of the larynx. Many of the muscles of the neck, face, lips, and tongue are concerned with speech. The muscles of the larynx are extrinsic and intrinsic. The extrinsic muscles include the:

1. Infrahyoid Muscles	**2. Suprahyoid Muscles** (some of)
omohyoid	stylopharyngeus
sternohyoid	palatopharyngeus
thyrohyoid	inferior and middle constrictors of the pharynx
sternothyroid	

Functions. (1) In prolonged inspiratory efforts, such as in singing, they produce tension on the lower part of the cervical fascia and hence prevent the apices of the lungs and the large blood vessels from being compressed. (2) In the act of swallowing, the larynx and hyoid bone are drawn up with the pharynx—these muscles depress them. (3) They elevate and depress the thyroid cartilage.

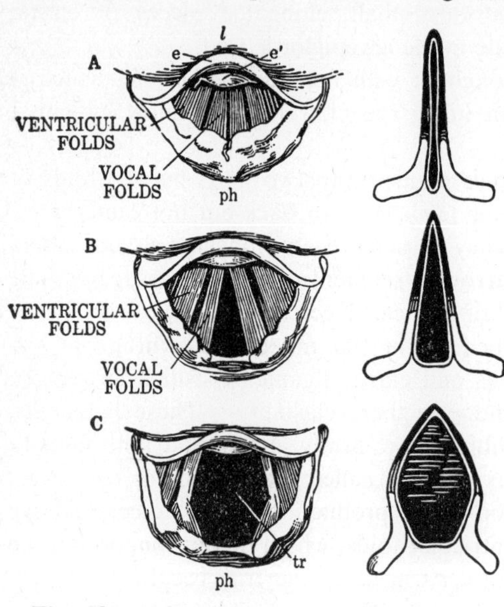

Fig. 280. The larynx as seen by means of the laryngoscope in different conditions of the glottis. *A*, while singing a high note; *B*, in quiet breathing; *C*, during a deep inspiration; *l*, base of tongue; *e*, upper free edge of epiglottis; *e'*, cushion of epiglottis; *ph*, part of anterior wall of pharynx; *tr*, trachea.

The intrinsic muscles are confined entirely to the larynx and are shown in Fig. 277.

Functions. The posterior cricoarytenoid muscles rotate the arytenoid cartilages outward, thereby separating the vocal cords. The lateral cricoarytenoid muscles rotate the arytenoid cartilages inward, thereby approximating the vocal cords. The arytenoid muscles approximate the arytenoid cartilages, especially at their back parts. These muscles also open and close the glottis.

The cricothyroid muscles elevate the arch of the cricoid cartilage in front, causing the lamina to be depressed and thereby increasing the distance between the vocal processes and thyroid cartilages.

The thyroarytenoid muscles draw the arytenoid cartilages forward toward the thyroid, and in this way they shorten and relax the vocal cords. Working together, these muscles regulate the degree of tension on the vocal cords.

Nerves and blood vessels. The laryngeal nerves are derived from the internal and external branches of the superior laryngeal, branches of the vagi. Blood is supplied to the larynx by branches of the superior thyroid artery which arises from the external carotid, and from the inferior thyroid, a branch of the thyroid axis, which arises from the subclavian artery.

Phonation. This term is applied to the production of vocal sounds. All of the respiratory organs function in the production of vocal sounds, but the vocal folds, the larynx, and the parts above are specially concerned. The speech centers and parts of the brain which control the movements of the tongue and jaw, also the tongue itself, are of special importance. The organs of phonation in man are similar to those of many animals much lower in the scale of life; the association areas of the brain account for the greater variety of sounds that man can produce.

Voice. The vocal folds produce the voice. Air driven by an expiratory movement out of the lungs throws the two elastic folds into vibrations. These impart their vibrations to the column of air above them and so give rise to the sound which we call the voice. The pharynx, mouth, and nasal cavities above the glottis act as resonating cavities. The volume and force of the expired air and the amplitude of the vibrations of the vocal folds determine the loudness or intensity of the voice. The pitch of the voice depends upon the number of vibrations occurring in a given unit of time. This in turn is dependent on the length, thickness, and degree of elasticity of the vocal folds and the tension by which they are held. When the folds are tightly stretched and the glottis almost closed, the highest sounds are emitted.

Differences between the male and female voice. The size of the larynx varies in different individuals, and this is one reason for differences in pitch. At the time of puberty, the growth of the larynx and the vocal folds is much more rapid and accentuated in the male than in the female. The increase in the size of the larynx causes an increase in the length of the vocal folds and also gives rise to the laryngeal prominence. These changes in structure are accompanied by changes in the voice, which becomes deeper and lower.

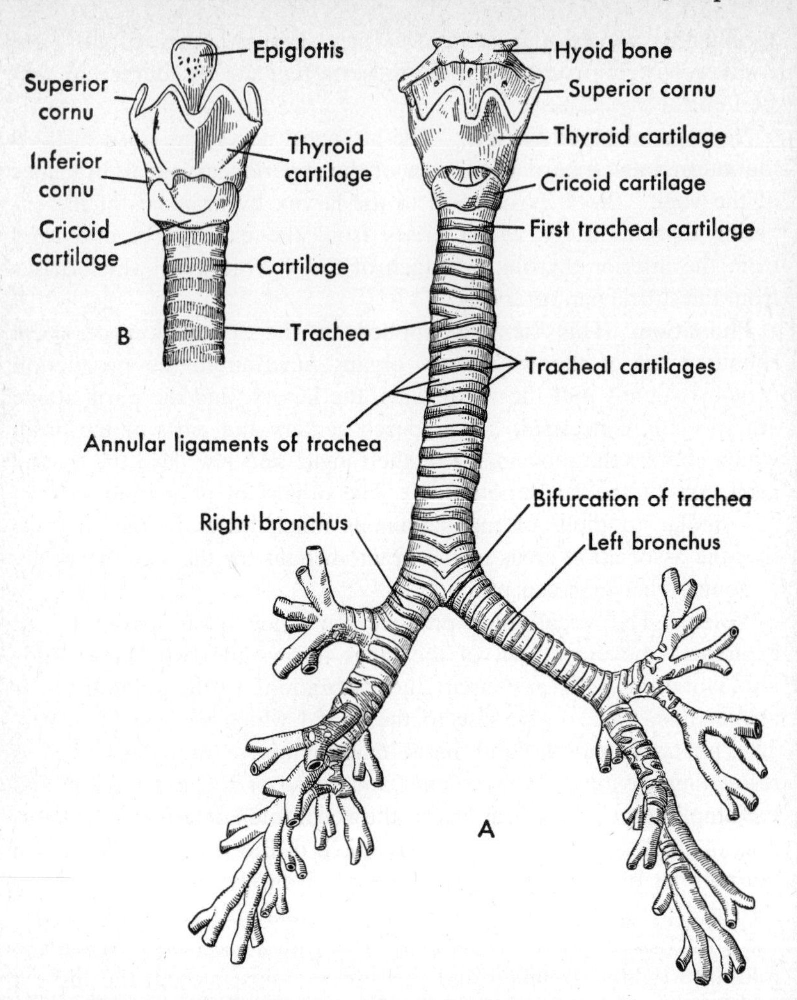

Fig. 281. The trachea and bronchial ramification, front view. *A*, showing ramifications; *B*, showing details of glottis and epiglottis. (Toldt.)

Before the characteristic adult voice is attained, there occurs what is described as a break in the voice, due to the inability of the individual to control the longer vocal folds.

The trachea, or **windpipe,** is a membranous and cartilaginous tube, cylindrical in shape, about 11.2 cm ($4\frac{1}{2}$ in.) in length and about 2 to 2.5 cm (1 in.) from side to side. It lies in front of the esophagus and extends from the larynx on the level of the sixth cervical vertebra to the level of the upper border of the fifth thoracic vertebra, where it divides into the two bronchi, one for each lung.

The walls are strengthened and rendered more rigid by rings of cartilage embedded in the fibrous tissue. These rings are C-shaped and incomplete behind, the cartilaginous rings being completed by

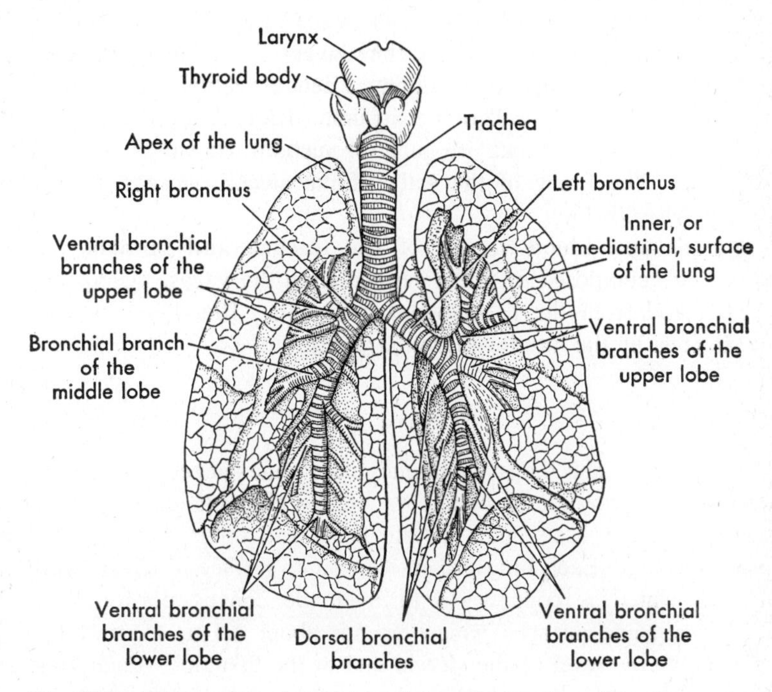

Fig. 282. The trachea, bronchi, and bronchial tubes. The tissues have been removed to show the air tubes.

bands of plain muscular tissue where the trachea is flattened and comes in contact with the esophagus. (See Fig. 281.) Like the larynx, it is lined with mucous membrane and has a ciliated epithelium upon its inner surface. The mucous membrane, which extends into the bronchial tubes, keeps the internal surface of the air passages free from dust particles, etc., the mucus entangles particles inhaled, and the movements of the cilia continually sweep this dustladen mucus upward into the pharynx.

The bronchi. The two bronchi into which the trachea divides differ slightly, the right bronchus being shorter, wider, and more vertical in direction than the left. They enter the right and left lung, respectively, and then break up into a great number of smaller branches, which are called the bronchial tubes and bronchioles. The two bronchi resemble the trachea in structure; but as the bronchial tubes divide and subdivide, their walls become thinner, the small plates of cartilage cease, the fibrous tissue disappears, and the finer tubes are composed of only a thin layer of muscular and elastic tissue lined by ciliated epithelium. Each bronchiole terminates in an elongated saccule called the atrium. Each atrium bears on all parts of its surface small, irregular projections known as alveoli, or air cells. The lining of the walls of the alveoli consists of simple squamous epithelium.

Nerves and blood vessels of the trachea and bronchi. The nerves are composed of fibers derived from the vagus, the recurrent nerves, and from the thoracolumbar system. Blood is supplied to the trachea by the inferior thyroid arteries.

LUNGS

Anatomy. The lungs (pulmones) are cone-shaped organs which occupy the two lateral chambers of the thoracic cavity and are separated from each other by the heart and other contents of the mediastinum. Each lung presents an outer surface which is convex, a base which is concave to fit over the convex portion of the diaphragm, and an apex which extends about 2.5 to 4 cm (1 to 1½ in.) above the level of the sternal end of the first rib. Each lung is connected to the heart and trachea by the pulmonary artery, pulmonary vein, bronchial arteries and veins, the bronchus, plexuses of nerves, lymphatics, lymph nodes, and areolar tissue, which are covered by the pleura and constitute the *root* of the lung. On the inner surface is a vertical notch called the *hilum*, which gives passage to the structures which form the root of the lung. Below and in front of the hilum there is a deep concavity, called the cardiac impression, where the heart lies; it is larger and deeper on the left than on the right lung, because the heart projects farther to the left side.

The right lung is larger and broader than the left, owing to the inclination of the heart to the left side; it is also shorter by 1 in., in consequence of the diaphragm rising higher on the right side to accommodate the liver. It is divided by fissures into three lobes—superior, middle, and inferior.

The left lung is smaller, narrower, and longer than the right. It is divided into two lobes—superior and inferior.

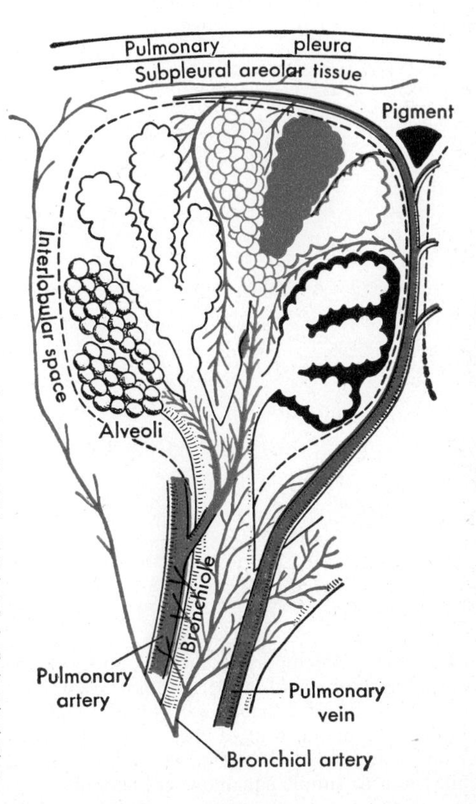

Fig. 283. Diagram of a lobule of the lung. A bronchiole is seen dividing into two branches, one of which runs upward and ends in the lobule. At the left, two atria are seen from the outside. Next are three atria in vertical section, the alveoli of each opening into the common passageway. In the next group the first atrium shows a pulmonary arteriole surrounding the opening of each alveolus, and the second gives the same with the addition of the close capillary network in the wall of each alveolus. Around the fourth group is a deep deposit of pigment, such as occurs in old age and in the lungs of those who inhale coal dust, etc. On the bronchiole lies a branch of the pulmonary artery (blue), bringing blood to the atria for aeration. Beginning between the atria are the radicles of the pulmonary vein (red). The bronchial artery is shown as a small vessel bringing nutrient blood to the bronchiole. (Gerrish.)

The substance of the lungs is porous and spongy; owing to the presence of air it crepitates when handled and floats in water. It consists of bronchial tubes and their terminal dilatations, numerous blood vessels, lymphatics, and nerves, and an abundance of elastic connective tissue. Each *lobe* of the lung is composed of many *lobules,* and into each lobule a *bronchiole* enters and terminates in an *atrium.* Each atrium presents a series of air cells, or *alveoli,* 700,000,000 or more in number. Each alveolus is somewhat globular in form with a diameter of about 100 μ. The amount of surface exposed to the air and covered by the capillaries is enormous. It is estimated that the entire inner surface of the lungs amounts to about 90 sq m, more than 100 times the skin surface of the adult body. Of this lung area about 70 sq m are respiratory.

Nerves of the lungs. The craniosacral nerve supply is made up of fibers in the vagus nerves.

The thoracolumbar nerve supply is made up of fibers in the visceral branches of the first four thoracic spinal nerves.

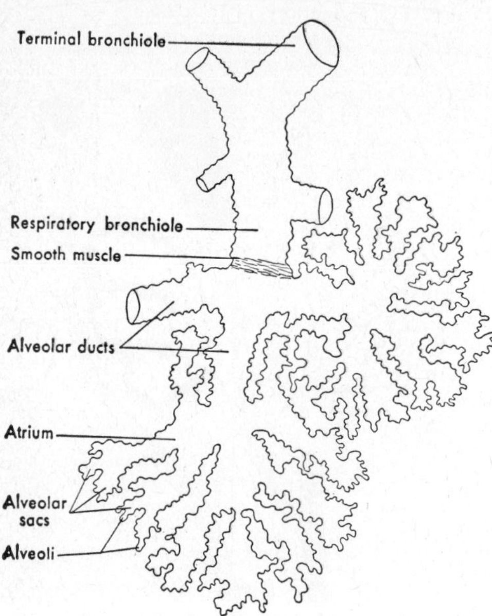

Fig. 284. A pulmonary, or lung, unit. There is cartilage in the wall of the terminal bronchiole, and the lining is ciliated epithelium. There is no cartilage in the wall of the respiratory bronchiole, and the lining changes from ciliated epithelium to simple squamous epithelium.

Blood vessels of the lungs. Two sets of vessels are distributed to the lungs: (1) branches of the pulmonary artery which bring blood from the right ventricle, and (2) branches of the bronchial arteries, which bring blood from the aorta.

1. The branches of the pulmonary artery accompany the bronchial tubes and form a plexus of capillaries around the alveoli. The walls of the alveoli are thin; they consist of a single layer of simple squamous epithelium surrounded by a fine, elastic connective tissue. The capillary plexus lies in the elastic connective tissue, and the air in the alveoli is separated from the blood in the capillaries only by the thin membranes forming their respective walls. The air sacs are surrounded by capillaries and form a surface area of about 150 sq m for the exchange of oxygen and carbon dioxide. Blood pressure in the pulmonary arteries is about 20 to

22 mm Hg. In the soma the capillaries are surrounded by tissue fluid, which exerts pressure against the capillaries; but the capillaries in the lungs are not opposed by such pressures. This means that pressures must be lower to prevent disturbances of hydrostatic and osmotic forces of the blood in the lungs. The pulmonary veins begin in the pulmonary capillaries, which coalesce to form larger branches. These run through the substance of the lung, communicate with other branches, and form larger vessels, which accompany the arteries and bronchial tubes to the hilum. Finally the pulmonary veins open into the left atrium.

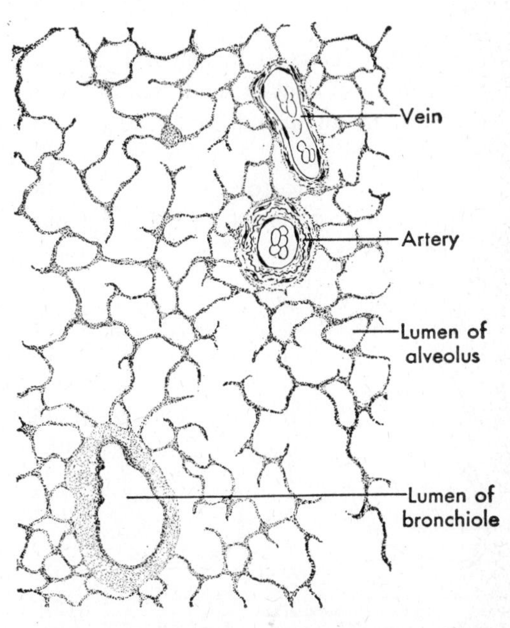

Fig. 285. Diagram of lung section. There are about 700 to 800 millions of alveoli in the lungs. These are surrounded by a network of capillaries for the exchange of oxygen and carbon dioxide.

2. The branches of the bronchial arteries supply blood to the lung substance—the bronchial tubes, coats of the blood vessels, the lymph nodes, and the pleura. The bronchial veins formed at the root of each lung receive veins which correspond to the branches of the bronchial arteries. Some of the blood supplied by the bronchial arteries passes into the pulmonary veins, but the greater amount is returned to the bronchial veins. The right bronchial vein ends in the azygos vein, the left in the highest intercostal or hemiazygos vein.

Pleura. Each lung is enclosed in a serous sac, the pleura, one layer of which is closely adherent to the walls of the chest and diaphragm (parietal); the other closely covers the lung (visceral, or pulmonary).

Named areas of the pleura	Visceral—pulmonary (outer layer of wall of lungs)	
	Parietal—(inner layer of chest wall)	Cervical—neck region of chest wall
		Costal—rib region of chest wall
		Diaphragmatic—upper layer of diaphragm
		Mediastinal—outer layer of mediastinum

The two layers of the pleural sacs, moistened by serum, are normally in close contact, and the so-called pleural cavity is a potential rather than an actual cavity; they move easily upon one another with each respiration. If the surface of the pleura becomes inflamed (pleurisy), friction results; and the sounds produced by this friction can be heard if the ear is applied to the chest. In health only a small amount of fluid is secreted, and its absorption by the lymphatics keeps pace with its secretion, so that normally the amount of serum is very small. In pleurisy the amount may be considerably increased, owing to the extra activity of the irritated secretory cells and excessive transudation from the congested blood vessels. The amount may be sufficient to separate the two layers of the pleura, thus changing the potential pleural cavity into an actual one.[3] This is known as pleurisy with effusion. If the effusion becomes purulent, the condition is called empyema.[4]

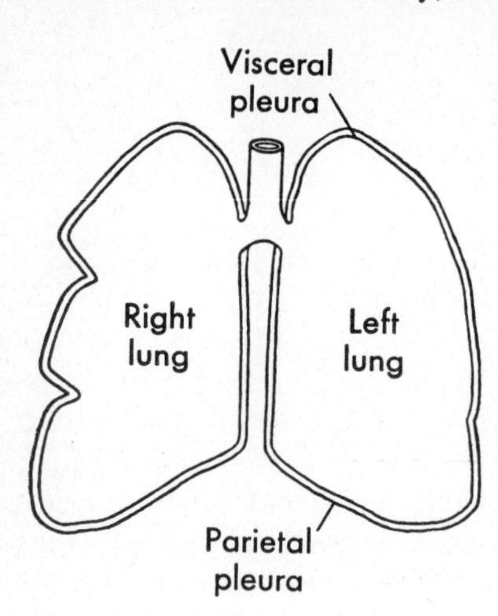

Fig. 286. Diagram showing parietal and visceral pleura. Parietal pleura is the inner layer of the chest wall. Visceral pleura is the outer layer of wall of lung. (See also Fig. 5.)

[3] If a puncture is made through the chest walls so that air enters between the two layers of the pleura (pneumothorax), the lung will collapse. If the puncture is closed, the air will be gradually absorbed, and the lung will resume its normal position. Certain types of tuberculosis are treated by artificial pneumothorax. This is accomplished by surgical removal of part of the chest wall or by injections of nitrogen into the chest cavity.

[4] The treatment of empyema is surgical, to provide for free drainage.

The mediastinum, or interpleural space, lies between the right and left pleurae in the median plane of the chest. It extends from the sternum to the spinal column and is entirely filled with the thoracic viscera, namely, the thymus, heart, aorta and its branches, pulmonary artery and veins, venae cavae, azygos vein, various veins, trachea, esophagus, thoracic duct, lymph nodes, and lymph vessels, all lying in connective tissue.

RESPIRATION

The main purposes of respiration is to supply the cells of the body with oxygen and rid them of the excess carbon dioxide which results from oxidation. Respiration also helps to equalize the temperature of the body and get rid of excess water. It is common to discuss respiration under three headings:

1. The process of *breathing* may be subdivided into inspiration, or breathing in, and expiration, or breathing out.

2. *External respiration* includes external oxygen supply, or the passage of oxygen from the alveoli of the lungs to the blood, and external carbon dioxide elimination, or the passage of carbon dioxide from the blood to the lungs.

3. *Internal respiration* includes internal oxygen supply, or the passage of oxygen from the blood to the tissue cells, and internal carbon dioxide elimination, or the passage of carbon dioxide from the tissue cells to the blood.

It is evident that external respiration is a process which takes place in the lungs and internal respiration is a process which takes place in the cells that make up the tissues.

1. Breathing. The thorax is a closed cavity which contains the lungs. The lungs may be thought of as elastic sacs, the interiors of which remain permanently open to the outside air by way of the bronchi, trachea, glottis, etc. Intrapulmonic pressure, or the pressure against the lungs from inside, is therefore practically outside pressure and is usually given as 760 mm of mercury. The outside of the lungs is protected from atmospheric pressure by the walls of the chest. Intrathoracic (intrapleural) pressure, or pressure against the lungs from the thoracic cavity, varies with thoracic conditions. It is given as about 751 mm at the end of inspiration and as 754 mm at the end of expiration, as taken by a mercurial manometer connected with the intrapleural cavity.

During life the size of the thoracic cavity is constantly changing with the respiratory movements. When all the muscles of respiration are at rest, that is, at the end of a normal expiration, the size and position of the chest may be regarded as normal. Any enlargement constitutes active inspiration, the result of which is to bring more air into the lungs. In quiet breathing, following this active inspiration the thoracic cavity returns to its normal position, giving an expiration. Normal quiet respiratory movements are of this type, an *active inspiration* followed by a *passive expiration.* In deeper or more rapid breathing expiration may also be active.

Mechanism of inspiration and expiration. Active inspiration is the result of the contraction of the muscles of inspiration; passive expiration is due mainly to the elastic recoil of the parts previously stretched. In inspiration the thoracic cavity is enlarged in all directions—vertical, dorsoventral, and lateral. The increase in the vertical diameter is brought about by the contraction of the diaphragmatic muscle, which draws the central tendon downward. The dorsoventral and lateral diameters are increased by the contraction of the intercostal and other muscles which cause the sternum and ribs to move upward and outward. The lungs are expanded in proportion to the increase in the size of the thorax. As in the heart the atrial systole, the ventricular systole, and then a pause follow in regular order; so in the lungs the inspiration, the expiration, and then a pause succeed one another. At the end of inspiration, when the lungs are momentarily quiet, and at the end of an expiration, when they are again quiet, the forces tending to stretch the elastic lungs and the forces tending to collapse them are equal. At the end of inspiration 760 mm intrapulmonic pressure is balanced by 751 mm intrathoracic pressure plus the tendency of the elastic lungs to recoil (estimated at 9 mm). At the end of expiration 760 mm intrapulmonic pressure is balanced by 754 mm plus the tendency of the elastic lungs to recoil (estimated at 6 mm).

Muscles of inspiration. The number of muscles used in inspiration varies greatly. The diaphragm and all the muscles that contract simultaneously with it are classed as inspiratory. Those that contract alternately are classed as expiratory. The following are the inspiratory muscles: muscles of diaphragm, external[5] intercostals, the

[5] To some, the front portions of the internal intercostals may be inspiratory.

scaleni,[6] the sternocleidomastoid, the pectoralis minor, and the serratus posticus superior.[7] In forced inspirations the action of these muscles is supplemented by additional muscles of the trunk, larynx, pharynx, and face.

Muscles of expiration. Normal expiration is considered a passive act due to gravity and the elastic recoil of the lungs. But in forced expirations diminution in the size of the thorax may be accomplished in two ways: (1) by forcing the diaphragm farther up into the thoracic cavity, a result obtained *not* by direct action of the diaphragm but by contracting the muscular walls of the abdomen, the external and internal oblique, the rectus, and the transversalis; and (2) by depressing the ribs. The muscles which depress the ribs are the internal intercostals and the triangularis sterni.[8] Some authorities add the iliocostalis,[9] serratus posticus inferior,[10] and the quadratus lumborum, but it has not been definitely determined whether they act simultaneously with the diaphragm or alternately.

Types of respiration. Two types of respiration are noted. The sequence of movements is the distinguishing factor. In the costal type the upper ribs move first and the abdomen second. The elevation of the ribs is the more noticeable movement. In the diaphragmatic type, the abdomen bulges outward first, and this is followed by a movement of the thorax. Diaphragmatic respirations are deeper. Restriction of the action of the diaphragm by tight clothing is thought to be the cause of costal respiration.

The respiratory center is located in the medulla oblongata, and this center has connections in the pons. Both centers share in the control of respiration. The medulla is the center for the nervous control of the depth and frequency of respiration, that is, the quantity

[6] Singular, *scalenus*. There are three: scalenus anterior, scalenus medius, and scalenus posterior. They arise from the transverse processes of the cervical vertebrae and are inserted in the first and second ribs.

[7] The serratus posticus superior extends from the spinous processes of the seventh cervical and upper two or three thoracic vertebrae to the upper borders of the second, third, fourth, and fifth ribs.

[8] Transversus thoracis, or triangularis sterni, is found on the front and inner side of the thoracic wall. Its fibers pass from the sternum running upward and outward to be inserted in the costal cartilages from the second to the sixth rib.

[9] The iliocostalis is one of the divisions of the sacrospinalis and is inserted into the inferior borders of the lower six or seven ribs.

[10] The serratus posticus inferior arises from the spinous processes of the lower two thoracic and upper second or third lumbar vertebra, and the insertion is in the inferior borders of the lower four ribs.

of air moved through the lungs per minute (lung ventilation). This center has been found to be sensitive to changes in the acidity, per cent of carbon dioxide, per cent of oxygen, the temperature of the blood, and also to the blood pressure in the blood vessels around it. Nerve impulses from all the sense organs reach this center, as do also nerve impulses from the cerebral cortex and from the hypothalamus. From the respiratory center, nerve impulses pass via the spinal cord and spinal nerve to the intercostal muscles, the muscles of the diaphragm, and the abdominal muscles, adjusting respiratory depth and rhythm to body needs. The respiratory rhythm is regulated also reflexly from the lungs themselves. This reflex is called the lung reflex and is perhaps the most powerful in regulating respiration. Inspiratory movements start nerve impulses which reach the respiratory center over afferent fibers of the vagus nerves, inhibiting the center and hence bringing about expiratory movements. Chemoreceptors in the carotid and aortic bodies are sensitive to carbon dioxide excess and oxygen lack. The carotid sinus and aortic arch also respond to changes in pressure; as arterial pressure rises, the pressoreceptors are stimulated and respiration is inhibited. Stimulation of the chemoreceptors increases the depth and rate of respiration.

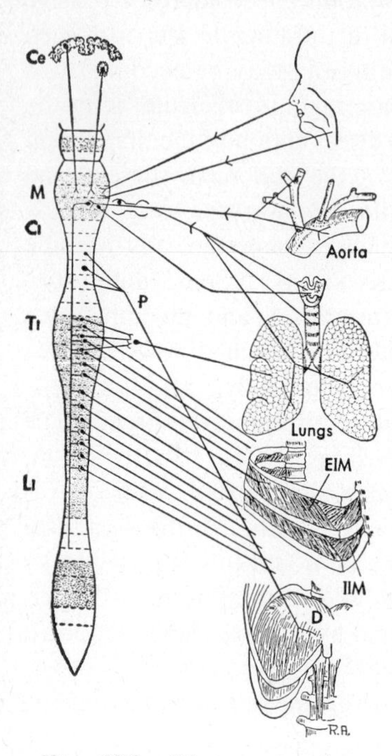

Fig. 287. Diagram to show nervous control of respiration. *Ce,* cerebral cortex; C_1, spinal cord at first cervical level; *D,* diaphragm; *EIM* and *IIM,* external and internal intercostal muscles; L_1, spinal cord at first lumbar level; *M,* medulla; *P,* phrenic nerve; T_1, spinal cord at first thoracic level. Nerves to abdominal muscles are not shown.

Control of respiratory rate. It is possible to increase or decrease the respiratory rate within certain limits, by voluntary effort, for a short time. If we arrest the respirations or diminish their frequency, the carbon dioxide concentration in the blood increases, and eventually the stimulus becomes too strong to be controlled. According to some observers, the "breaking point" is reached in 23 to 27 sec. If, before holding the breath, several breaths of pure oxygen are taken, the breaking point may be postponed; or, if the lungs are thoroughly aerated by forced breathing, so that the carbon dioxide is forced out and pure oxygen is breathed in, the breaking point may be postponed as long as 8 minutes.

Cause of the first respiration. The human fetus makes respiratory movements while in the uterus, possibly moving amniotic fluid in and out of the lungs. This may play a part in dilation of the future air passages. After birth and the interruption of the placental circulation, the first breath is taken. The immediate cause of this activity on the part of the respiratory center must be connected, if not identical, with the cause of the automatic activity of the center during life. Three views are held regarding the immediate cause: that it is due to the increased amount of carbon dioxide in the blood, brought about by cutting the cord; that it is due to stimulation of the sensory nerves of the skin, due to cooler air, handling, drying, etc.; and that it is due to a combination of these causes.

If stimulation through the blood and stimulation through the nerves normally cooperate, it may be that the essential cause is the increased tension of the carbon dioxide, and therefore the increased concentration in acidity, following the cutting of the cord.

During intrauterine life, the fetal blood is aerated so well by exchange with the maternal blood that it does not act as a stimulus to the fetal respiratory center.

Depth of respiration. This is usually given as about 500 cc for quiet breathing. Necessity for increased oxygen supply and need of eliminating carbon dioxide are met by increased depth of respiration before increased rate is brought into play.

In the table on p. 502 it will be noted that the minute ventilation of lungs is multiplied more than three and a half times, and mainly by increasing the completeness of contraction of the respiratory muscles rather than by causing them to contract a greater number of times.

Frequency of respiration. The average rate of respiration for an adult is about 14 to 20 per minute. In health this rate may be increased by muscular exercise, emotion, etc. Anything that affects the heartbeat will have a similar effect on the respirations. Age has

INCREASE IN RESPIRATORY DEPTH AND RATE ON INCREASING CO_2 IN INSPIRED AIR

% CO_2 in Inspired Air[11]	*Average Depth of Respiration, Cc*	*Average Respiratory Rate*	*Minute Lung Volume, Liters*
Normal (0.04)	673	14	9.4
3.07	1216	15	18.2
5.14	1771	19	33.6

a marked influence. The average rate during the first year of life is about 44 per minute, and at the age of 5 yrs, 26 per minute. It is reduced between the ages of 15 and 25 to the normal standard.

2. External respiration. This term is applied to the interchange of gases that takes place in the lungs. There is a continuous flow of blood through the capillaries, so that at least once or twice each minute all the blood in the body passes through the capillaries of the lungs. This means that the time during which any portion of blood is in a position for respiratory exchange is only a second or two. Yet during this time, the following changes take place: the blood loses carbon dioxide and moisture; it gains oxygen, which combines with the reduced hemoglobin of the red cells, or erythrocytes, forming oxyhemoglobin, and as a result of this the crimson color shifts to scarlet; and its temperature is slightly reduced.

It is helpful to compare the average amounts of oxygen and carbon dioxide found in venous and in arterial blood. The actual amounts of oxygen and carbon dioxide in venous blood vary with the metabolic activity of the tissues and differ, therefore, in the various organs according to the state of activity of each organ and the volume of its blood supply per unit of time. The main result of the respiratory exchange is to keep the gas content of the arterial blood nearly constant at the figures given. Under normal conditions it is not possible to increase appreciably the amount of oxygen absorbed by the blood flowing through the lungs.[12]

[11] Figures in first three columns taken from J. S. Haldane, "The Regulation of Lung Ventilation," *Journal of Physiology,* May, 1905.

[12] It is sometimes difficult to reconcile this fact with the practice of using pure oxygen in critical cases of pulmonary disease. The relief of the pneumonia patient who inhales pure oxygen is usually marked because the blood

	Oxygen %	*Carbon Dioxide* %	*Nitrogen* %
Venous blood contains	12	56	1.7
Arterial blood contains	20	50	1.7

Diffusion of oxygen and carbon dioxide. The physical theory of respiration assumes that the passage of oxygen to, and of carbon dioxide from, the blood in the lungs and the passage of oxygen from, and of carbon dioxide to, the blood in the tissues take place by diffusion. If two solutions of any gas at different concentrations are separated by a permeable membrane, the molecules will pass through the membrane in both directions until the concentrations are equal on both sides. The concentration of oxygen in the alveolar air is higher than that in the venous blood, and the concentration of carbon dioxide is higher in venous blood than that in the alveolar air. The oxygen and carbon dioxide in the alveoli and in the blood vessels are separated by the permeable membranes forming their respective walls. The diffusion of oxygen is from the alveolar air to the blood, and the diffusion of carbon dioxide is from the blood to the alveolar air. In the tissues the concentration conditions are reversed; and in consequence the diffusion of oxygen is from the blood to the tissues, and the diffusion of carbon dioxide is from the tissues to the blood.

Oxygen, nitrogen, and carbon dioxide. The amount of oxygen carried in arterial blood is about 19 or 20 volumes per cent; about 0.3 per cent is held in solution and the remainder as oxyhemoglobin in erythrocytes. The amount of oxygen carried in venous blood is about 12 to 14 volumes per cent; about 0.3 per cent is held in solution and the remainder as oxyhemoglobin in the erythrocytes. It is evident that in general about half the oxyhemoglobin is reduced, giving its oxygen to tissue cells while blood is in the tissue capillaries.

The amount of carbon dioxide carried in venous blood is about 56 volumes per cent; part is in the physical solution and the remainder held mainly as sodium bicarbonate, etc., in plasma and as potassium

absorbs an increased amount of oxygen. The reason for this seeming contradiction is that normal blood cannot absorb an increased amount of oxygen, but in the case of the pneumonia patient the composition of the blood as regards oxygen is below normal, and the inhalation of pure oxygen brings it up to the standard; hence the marked relief.

bicarbonate, carbamino compounds, etc., in erythrocytes. The amount of carbon dioxide carried in arterial blood is about 50 volumes per cent. Roughton has demonstrated an enzyme in erythrocytes catalyzing formation of carbonic acid from water and carbon dioxide. This may account for the rapidity with which carbon dioxide is received from the tissues and given off to the lungs. If the concentration of carbon dioxide in blood is low, respiration is interfered with, and the condition is known as *acapnia*. The nitrogen in the blood is held in physical solution and circulates without exerting any known immediate effect upon the tissues.

Capacity of the lungs. After the lungs are once filled with air they are never completely emptied. In other words, no expiration ever completely empties the alveoli; neither are they completely filled. The quantity of air which a person can expel by a forcible expiration, after the deepest inspiration possible, is called the vital capacity and averages about 3,500 to 4,100 cc (7–8 pt) for an adult man. It is the sum of tidal, complemental, and supplemental air.

	Inspiratory reserve (complemental) air	1,800 cc	Vital capacity 4,100 cc
	Tidal air	500 cc	
Reserve, or retained, air 3,000 cc	Expiratory reserve (supplemental) air	1,800 cc	
	Residual air	1,200 cc	
		5,300 cc	Total capacity of lungs

Tidal air designates the amount of air that flows in and out of the lungs with each quiet respiratory movement. The average figure for an adult male is 500 cc.

Inspiratory reserve volume designates the amount of air that can be breathed in over and above the tidal air by the deepest possible inspiration. It is estimated at about 1,800 cc.

Expiratory reserve volume is the amount of air that can be breathed out after a quiet expiration by the most forcible expiration. It is equal to about 1,800 cc.

Residual air is the amount of air remaining in the lungs after the most powerful expiration. This has been estimated on the cadaver and is about 1,200 cc.

Reserve air is the residual air plus the supplemental air in the lungs under conditions of normal breathing, that is, about 3,000 cc.

Minimal air. When the thorax is opened, the lungs collapse, driving out the supplemental and residual air; but before the alveoli are entirely emptied, the small bronchi leading to them collapse and entrap a little air in the alveoli. The small amount of air caught in this way is designated as minimal air.[13]

The inspired and expired air. As the expirations never completely empty the lungs of air, it follows that the air entering with each fresh breath becomes mixed with that in the alveoli. Consequently the air in the alveoli is never quite the same as inspired air, but normally the difference is not great. From a physiological standpoint the essential constituents of atmospheric air are oxygen, nitrogen, and carbon dioxide. The average composition of inspired and expired air in volumes per cent is shown below.

	Oxygen %	*Carbon Dioxide* %	*Nitrogen* %
Inspired air	20.96	0.04	79
Expired air	16.02	4.38	79+
	——	——	—
Change	4.94 loss	4.34 gain	0+
Alveolar air	14	5.5	79+

This table shows that in passing through the lungs the air gains 4.34 volumes of carbon dioxide to each 100 and loses 4.94 volumes of oxygen. This is the main fact of external respiration. Other changes occur.

However dry the external air may be, the expired air is nearly, or quite, saturated with moisture. An average of about 1 pt of water is eliminated daily in the breath. Whatever the temperature of the external air, the expired air is nearly as warm as the blood, having a temperature between 36.7 and 37.8° C (98 and 100° F). In man, breathing is one of the subsidiary means by which the temperature and the water content of the body are regulated. The heat

[13] Before birth the lungs contain no air. If after birth respirations are made, the lungs do not collapse completely on account of the capture of minimal air. Whether or not the lungs will float has constituted one of the facts used in medicolegal cases to determine if a child was stillborn. J. H. Fulton, *Howell's Textbook of Physiology,* 1946.

required to warm the expired air and vaporize the moisture is taken from the body and represents a daily loss of heat. It requires about 0.5 Cal to vaporize 1 gm water.

Taking the respiratory rate as 18 per minute and respiratory depth at 500 cc, one breathes in and out 12,000 liters of air per day. Since inspired air contains about 20 per cent oxygen and expired air about 16 per cent, this difference of 4 per cent represents the oxygen retained by the body—some 480 liters per day—and used by the tissues. This amount is often stated as about 350 cc per minute.

3. Internal respiration. The exchange of oxygen and carbon dioxide in the tissues constitutes internal respiration and consists of the passage of oxygen from the blood into the tissue fluid and from the tissue fluid into the tissue cells and the passage of carbon dioxide from the cells into the tissue fluid and from the tissue fluid into the blood.

After the exchange of oxygen and carbon dioxide in the lungs, the aerated blood is returned to the heart and distributed to all parts of the body. In passing through the tissue capillaries, the blood is brought into exchange with tissue fluid in which the oxygen pressure is low. Consequently the blood in passing through the capillaries gives up much of its oxygen, which passes to the tissue fluid and from the latter to the tissue cells. On the contrary, the concentration of carbon dioxide is higher in the cells than in the blood, and this facilitates the passage of carbon dioxide from the cells to the tissue fluid and from the latter to the blood.

It is important to remember that the blood does not give up all its oxygen to the tissues. On the whole the concentration of oxygen in the blood going to the tissues is about 20 per cent or 20 volumes of oxygen per 100 volumes of blood (19 per cent as oxyhemoglobin and less than 1 per cent in solution). Neither does the blood give up all of its carbon dioxide in the lungs. On the whole the concentration of carbon dioxide in the blood going to the lungs is about 56 per cent and returning to the tissues from the lungs is about 50 per cent. Excessive amounts of carbon dioxide will cause death by asphyxia, but in normal amounts it is as essential to life as oxygen.

As shown in Fig. 288, oxygen reaches the cells in three steps: (1) environment to lungs, (2) lungs to blood, (3) blood to cells. In each step the method of oxygen transfer is diffusion, and each step is continuous inasmuch as there is a constant source (*S*) of oxygen on one side of the diffusion plane (or membrane), which keeps the concentration of diffusible (in solution) oxygen relatively high, and a place of disappearance (*D*) of diffusible oxygen on the other side of the diffusion plane, which keeps the concentration of oxygen relatively low. Then, too, the place of disappearance of oxygen in step (1) becomes the source of oxygen in step (2), and the place of disappearance of oxygen in step (2) becomes the source for step (3). In this way the periodic intake of oxygen (respiratory rate 14

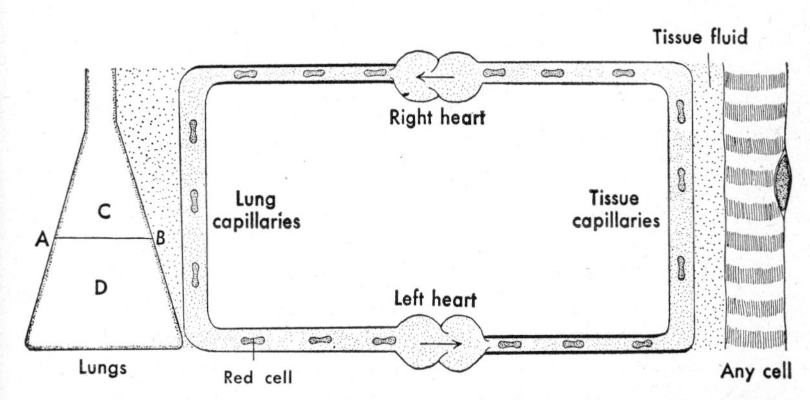

Fig. 288. Diagram for physiology of respiration. Line *AB* represents a plane where *renewed* air (oxygen, 20 per cent; CO_2, 0.04 per cent) and *retained* air (oxygen, 14 per cent; CO_2, 4 per cent) come into contact for diffusion. *C* represents renewed air, 500 cc in quiet respiration; *D* represents retained air, around 3,000 cc.

to 18) is changed to a steady flow to the cells for normally variable activities. To illustrate: In quiet breathing there would be diffusion of oxygen at the plane *AB* between about 500 cc of renewed air (above *AB*) at 20 volumes per cent oxygen and about 2,600 cc of retained air (below *AB*) at 14 volumes per cent oxygen. In muscular activity the renewed air is increased in volume a little or much depending upon the activity—as this is increased, the plane *AB* is increased in area, and the volume of air below *AB* is diminished; hence, more oxygen diffuses through area *AB*, and since this oxygen is distributed to less air here, its concentration is increased to a still greater extent. The figures on p. 504 give maximum renewed air at 4,100 cc and retained air at 1,200 cc. Increased frequency of respiration also keeps the concentration of oxygen above *AB* up.

As the blood flows through the lung capillaries, oxygen diffuses from the alveoli into the plasma and then into the red blood corpuscles and combines with the hemoglobin to form oxyhemoglobin. On leaving the lungs, practically all the hemoglobin exists as oxyhemoglobin, and the plasma is saturated with oxygen in solution. When the blood reaches the tissue capillaries, there is a continuous diffusion of oxygen from erythrocytes to plasma, plasma to tissue fluid, tissue fluid to cells. The rate of this diffusion of oxygen depends upon the rate of use of oxygen by the cells. These diffusions are started by the use of oxygen in the cell.

DIFFUSION OF OXYGEN FROM THE LUNGS TO THE TISSUE CELLS

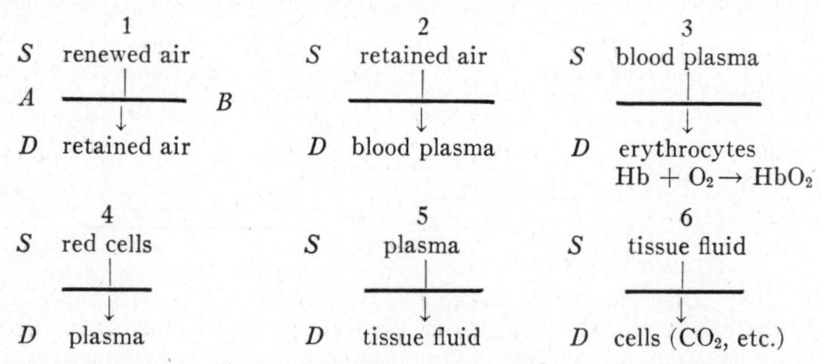

A, diffusion area or membrane; *D*, place into which oxygen diffuses; *S*, source of oxygen.

The blood carries from the lungs about 20 volumes of oxygen per 100 volumes of blood, more than 19 per cent as oxyhemoglobin (stored as oxyhemoglobin) in the erythrocytes and less than 1 per cent (diffusible) in solution in the water of the blood, hence mainly in plasma. While the blood is flowing through the alveolar capillaries, oxygen diffuses from a place kept relatively high in diffusible oxygen (by breathing) to a place kept relatively low in diffusible oxygen (circulating blood).

The blood leaves in the tissues about 10 volumes of oxygen per 100 volumes of blood. While the blood is flowing through the tissue capillaries, oxygen diffuses from a place kept relatively high in diffusible oxygen (flowing blood) to a place kept relatively low in diffusible oxygen (used in cells).

DIFFUSION OF CARBON DIOXIDE FROM CELLS TO LUNGS

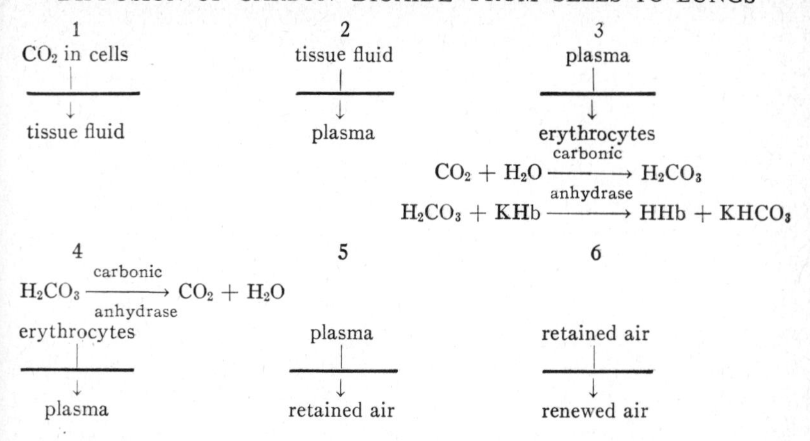

Carbon dioxide is produced in living cells at varying rates, depending upon activity, etc. It is much more soluble in water than is oxygen (2.7 volumes per cent). It also forms carbonic acid with water. This reaction is reversible but is known to proceed slowly at body *p*H (approximately 7.4). There are many theories advanced to explain the way in which carbon dioxide is carried by the blood, since the blood holds more than can be accounted for by solution.[14]

Carbon dioxide is described as diffusing into the plasma and then from the plasma into the erythrocytes, where an enzyme *carbonic anhydrase* catalyzes the formation of carbonic acid ($H_2O + CO_2 \rightarrow H_2CO_3$), which then interacts in many ways with substances in the erythrocytes. One of the ways considered important in the theory is as follows: $H_2CO_3 + KHb \rightleftarrows HHb + KHCO_3$. As oxyhemoglobin gives off oxygen to the tissues, the hemoglobin (a weaker acid than oxyhemoglobin) offers a greater number of basic ions (potassium, etc.) to form compounds with the H_2CO_3, and hence more $KHCO_3$, etc., are formed in the cells. This gives a relatively high concentration of HCO_3^- ions in the erythrocytes. Most of these HCO_3^- ions diffuse back into the plasma and form carbonates there, chiefly $NaHCO_3$ (necessitating the so-called "chloride shift" to erythrocytes where bases combine with the chloride ions leaving

[14] See also books on biochemistry.

bases in plasma to neutralize the HCO_3 ions). It is estimated that from 85 to more than 90 per cent of the carbon dioxide is carried as carbonates in the plasma. Carbon dioxide also forms carbamino compounds (carbhemoglobin) with hemoglobin. Varying quantities of carbon dioxide reaching the blood (variations in rate of metabolism) cause its *p*H to vary and in consequence limit the amount of carbhemoglobin that can be carried; hence, the main factor in the ability of the blood to *take up carbon dioxide in the tissues and give it off in the lungs* is that the hemoglobin *gives off* oxygen in the tissues and, becoming less acid in consequence, takes in another *available* acid (carbonic acid) to take its place. Also, when in the lungs, oxygen is *available* and the carbon dioxide can *escape,* so there is again an exchange; this time oxygen takes the place of the carbon dioxide in the erythrocytes. The *concentration gradients* of the two substances concerned—oxygen and carbon dioxide—change places in the blood in the tissue capillaries and in the blood in the alveolar capillaries.

Respiratory Phenomena

Each intake of air is accompanied by a low rustling sound, which can be heard if the ear is applied to the chest wall. It is thought that the dilatation of the alveoli produces this sound, and absence of it indicates that the air is not entering the alveoli over which no sound is heard or that the lung is separated from the chest wall by effused fluid. The air passing in and out of the larynx, trachea, and bronchial tubes produces a louder sound, which is called a bronchial murmur. Normally this murmur is heard directly above or behind the tubes; but when the lung is consolidated as in pneumonia, it conducts sound more readily than usual, and the murmur is heard in other parts of the chest. In diseased conditions the normal sounds are modified in various ways and are then spoken of under the name of *rales.*

Eupnea. This term is applied to ordinary quiet respiration made without obvious effort.

Dyspnea. In its widest sense, *dyspnea* means any increase in the force or rate of the respiratory movements. These movements show many degrees of intensity, corresponding with the strength of the stimulus. Usually the term *dyspnea* is reserved for the more labored breathing, in which the expirations are active and forced. Dyspnea may be caused by (1) stimulation of the sensory nerves, particularly the pain nerves, (2) an increase in the hydrogen ion concentration of the blood, and (3) any condition that interferes with the normal rate of the respirations or of the heart action or prevents the passage of air in or out of the lungs.

Hyperpnea. The word *hyperpnea* is applied to the initial stages of dyspnea, when the respirations are simply increased.

Apnea. The word means a lack of breathing. In physiological literature, it is used to describe the cessation of breathing movements due to lack of stimulation of the respiratory center, brought about by rapid and prolonged ventilation of the lungs. In medical literature, the term is sometimes used as a synonym for asphyxia, or suffocation.

Cheyne-Stokes respirations. This is a type of respiration which was first described by the two physicians[15] whose names it bears. It is an exaggeration of the type of respiration which is often seen during sleep in normal people. The respirations increase in force and frequency up to a certain point and then gradually decrease until they cease altogether; there is a short period of apnea, then the respirations recommence, and

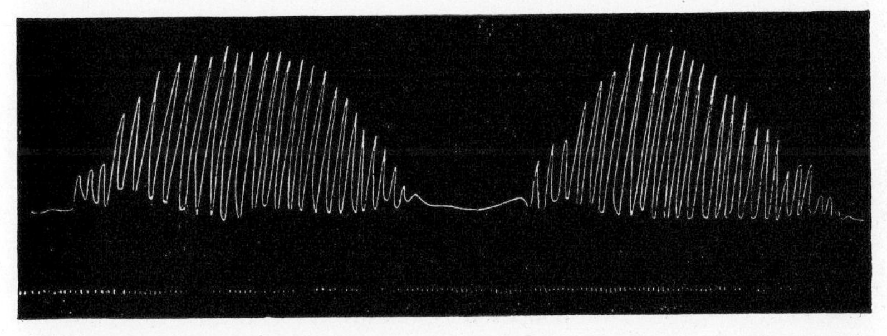

Fig. 289. Stethograph tracing of Cheyne-Stokes respirations. The time is marked in seconds. (Halliburton.)

the cycle is repeated. Cheyne-Stokes respirations are associated with conditions that depress the respiratory center, especially in brain, heart, and kidney diseases.

Edematous respiration. When the air cells become infiltrated with fluid from the blood, the breathing becomes edematous and is recognized by the moist, rattling sounds, or rales, caused by the passage of the air through the fluid. It is a serious condition because it interferes with aeration of the blood and often results in asphyxia.

Asphyxia. Asphyxia is produced by any condition that causes prolonged interference with the aeration of the blood, viz., obstruction to the entrance of air to the lungs, depression of the respiratory center, an insufficient supply of oxygen, or a lack of hemoglobin in the blood. The first stages are associated with dyspnea and convulsive movements; then the respirations become slow and shallow and are finally reduced to mere twitches. The skin is cyanosed, the pupils of the eyes dilate, the reflexes are abolished, and respirations cease. If the heart continues to beat,

[15] John Cheyne, Scotch physician (1777–1836). William Stokes, Irish physician (1804–1878).

resuscitation is often accomplished by artificial respiration, even after breathing has ceased.

Ventilation. Since at every breath the external air gains carbon dioxide and loses oxygen, it was formerly taught that the general discomfort, headache, and languor that result from sitting in a badly ventilated room were due to the increase in carbon dioxide and the loss of oxygen. The results of many experiments seem to prove that the air in badly ventilated rooms does not vary in oxygen and carbon dioxide content as much as had been supposed, though odors given off from the body and its clothing when present in any amount may affect the nervous system disagreeably. It is now thought that the injurious effects of poor ventilation are due to interference with the heat-regulating mechanism of the body. Under favorable conditions the surface of the human body is kept comfortably cool by air currents and by the evaporation of perspiration. In a confined space there is a lack of movement of the air, and it tends to become warm and humid. Moisture is not taken from the skin promptly, and the temperature rises. This results in a dilatation of the blood vessels of the skin, and an increased amount of blood is sent to the surface of the body, thereby increasing the unpleasant warmth. There is likely to be some reduction of the general blood pressure, leading to drowsiness or at least a feeling of inertia. Under such conditions the ability to do muscular work is diminished, mental efficiency is decreased, and fatigue comes on more quickly. In accordance with these views, the most effective precautions that can be taken to secure comfort in a room are to keep it cool and have some moisture and some movement of the air. It has been shown that starting an electric fan in a close room may relieve an almost intolerable condition, as it favors the removal of heat from the bodies of the individuals present. One writer suggests that the real difficulty with a stuffy room is that there is a lack of stimulation for the nervous system. One becomes relaxed and indolent because the nerve endings in the skin are not being played upon as they would be by a constant change in environmental conditions. Because of these facts proper ventilation is based on many things. (1) There must be continuous movement of the air. (2) The temperature and degree of humidity must favor the evaporation of perspiration from the skin. Experiments indicate that the temperature should approximate 18 to 20° C (65 to 70° F). It must, however, be borne in mind that it is the *temperature of the skin* rather than the temperature of the environment which determines physiological response. The temperature of the skin will depend on the rate of heat production (metabolism) on the one hand, and the temperature of the environment, the rate of movement of the air, the humidity, and the heat-insulating quality of clothing, etc., on the other hand. Taking a skin temperature of 33° C as indicating "thermal comfort," a heat-production level determined by the type of work the individuals must do, and a room temperature of 20° C with air flow at about 20 ft per minute and a humidity of less than 50 per cent, there is the clothing that may be varied by different individuals working in the same room to keep the skin temperature at 33° C. Generally speaking, above 20° C the higher the humidity the lower the temperature should be. (3) Odors from skin, clothing, light, and other sources must be eliminated. These requirements can be met only if the size of occupied rooms is in proportion (*a*) to the number of people occupying them, (*b*) to the facilities for ventilation, and (*c*) to the degree of air contamination likely to occur.

OXYGEN AND AVIATION

Several changes that take place in the atmosphere as altitude increases affect man's reactions in high-altitude flying. These may be briefly stated as follows:

Oxygen concentration in air at 40,000 feet is about the same as at sea level, but barometric pressure is lower.

Oxygen begins to decrease at 80,000 feet, and above this height oxygen and nitrogen amounts decrease and the amounts of helium and hydrogen increase. Barometric pressure is also an important factor in high altitudes. At sea level the pressure is 760 mm of mercury, while at 18,000 feet pressure is about half, or 380 mm of mercury; and as altitude increases, barometric pressure decreases. Another factor is temperature; the higher the altitude the lower the temperature. With every 500 feet of rise in altitude there is about 1° C drop in temperature.

These atmospheric changes affect man considerably as ascent in an airplane is made. Oxygen lack causes anoxia, the barometric pressure changes and extremes of cold also directly affect and modify man's reactions. Above 15,000 feet symptoms of oxygen lack are present. The individual becomes quiet, and the lips, ear lobes, and nail beds become slightly bluish. As a higher altitude is reached the blue color deepens, and weakness and dizziness occur. Since the brain cells are dependent upon oxygen, deprivation causes specific symptoms to become evident. The power of attention is diminished, muscular coordination is lessened, and mental confusion is present. Breathing is embarrassed. At still higher altitudes these symptoms become more apparent; the individual may become irritable, the mental confusion increases, and speech is interfered with. Muscular coordination may be so lost that writing is impossible. There is difficulty in understanding written words. There may be drowsiness, headache, apathy, and loss of self-control. Vision and memory are impaired, judgments are unsound, pain sensations are dulled, and appreciation of passage of time is altered. Cyanosis becomes more apparent, dyspnea and vomiting may occur. Each sense is finally lost.

In airplane travel oxygen tension, pressures, and temperature are regulated by artificial means, thus preventing any physiological changes that occur at high altitudes.

SUMMARY

Respiration	All living organisms require continual supply of oxygen Chemical changes in tissue cells dependent upon it Carbon dioxide is one end product of chemical changes in cells, hence need for elimination of excess Exchange of these gases in lungs and cells constitutes respiration
Essentials of Human Respiratory System	(1) Air containing a high percentage of oxygen on one side (2) Moist and permeable membrane (3) Moving stream of blood with a high percentage of carbon dioxide on other side
Respiratory System	Air passes through nose or mouth to (1) Larynx (3) Bronchi (2) Trachea (4) Lungs

Nose	**Function**	Special organ of the sense of smell Passageway for entrance of air to the respiratory organs Helps in phonation
	External nose	Framework of bone (nasal) and cartilage Covered with skin, lined with mucous membrane Nostrils are oval openings on under surface, separated by a partition

- **Nose**
  - **Internal cavities, or nasal fossae**
    - Two wedge-shaped cavities
    - Extend from nostrils to pharynx
    - Lined by mucous membrane, vascular and ciliated
    - Formed by
      - 2 palatine
      - 2 maxillae
      - 1 ethmoid
      - 1 sphenoid
      - 2 nasal
      - 2 conchae, and processes of the ethmoid
        - Superior meatus
        - Middle meatus
        - Inferior meatus
      - 1 vomer
      - —
      - 11 bones
  - **Advantages of nasal breathing**
    - Air is
      - Warmed
      - Moistened
      - Filtered
  - **Communicating sinuses**
    - (1) Frontal
    - (2) Ethmoid
    - (3) Maxillary, or antra of Highmore
    - (4) Sphenoid
  - **Nerves**
    - (1) Olfactory nerve—sense of smell
    - (2) Facial nerve
    - (3) Ophthalmic and maxillary
  - **Arteries**
    - External maxillary, Internal maxillary — derived from the external carotid
    - Ethmoidal arteries derived from internal carotid
- **Larynx**
  - Special organ of voice
  - Triangular box made up of nine pieces of cartilage
  - Situated between the tongue and trachea
  - Contains vocal folds
  - Slit or opening between cords called *glottis*, which is protected by leaf-shaped lid called *epiglottis*
  - Connected with external air by
    - Mouth
    - Nose
  - **Nerves**—derived from
    - Internal branches of superior laryngeal
    - External branches of superior laryngeal
  - **Arteries**
    - Superior thyroid, branch of external carotid
    - Inferior thyroid, branch of thyroid axis
- **Phonation**
  - **Phonation**—production of vocal sounds
  - **Organs of phonation**
    - Respiratory organs
    - Vocal folds
      - Lower pitch of male voice is due to greater length of vocal folds
    - Larynx, pharynx, mouth, nose, and tongue
    - Speech centers and parts of brain which control movements of the tongue and jaw, also association centers

Trachea
- Membranous and cartilaginous tube, $4\frac{1}{2}$ in. long
- Strengthened by C-shaped rings of cartilage
  - Complete in front
  - Incomplete behind
- In front of esophagus—extends from larynx to upper border of fifth thoracic vertebra, where it divides into two bronchi
- **Nerves**
  - Branches of vagus
  - Recurrent nerves
  - Autonomics
- **Arteries**—inferior thyroid

Bronchi and Bronchioles
- Right and left—structure similar to trachea
- Right—shorter, wider, more vertical than left
- Divide into innumerable bronchial tubes, or bronchioles
- As tubes divide, their walls become thinner. Finer tubes consist of thin layer of muscular and elastic tissue lined by ciliated epithelium
- Each bronchiole terminates in elongated saccule called *atrium* (*infundibulum*)
- Each atrium bears on its surface small projections known as *alveoli*, or air cells

Lungs
- **Location**—lateral chambers of thoracic cavity, separated by structures contained in mediastinum
- **Cone-shaped organs**
  - Outer surface convex to fit in concave cavity
  - Base concave to fit over convex diaphragm
  - Apex about 1 or $1\frac{1}{2}$ in. above the level of sternal end of first rib
  - Hilum, or depression on inner surface, gives passage to bronchi, blood vessels, lymphatics, and nerves
- **Right**—larger, broader, shorter—three lobes
- **Left**—smaller, narrower, longer—two lobes
- **Anatomy**
  - Porous, spongy organs. Consist of bronchial tubes, atria, alveoli, also blood vessels, lymphatics, and nerves held together by connective tissues
- **Blood vessels**
  - **Pulmonary artery**
    - Blood for aeration
    - Accompanies bronchial tubes
    - Plexus of capillaries around alveoli
    - Returned by pulmonary veins
  - **Bronchial arteries**—supply lung substance

Pleura
- Closed sac—envelops lungs, but they are not in it
- Two layers
  - Pulmonary, or visceral—next to lung
  - Parietal—outside of visceral
    - Normally in close contact—potential cavity
    - Moistened by serum
- **Function**—to lessen friction

- **Mediastinum**
  - Space between pleural sacs. Extends from sternum to spinal column. Contains the heart, large blood vessels connected with heart, trachea, esophagus, thoracic duct, various veins, lymph nodes, and nerves
- **Respiration**
  - **Function**
    - Increase the amount of oxygen
    - Decrease the amount of carbon dioxide
    - Help to maintain temperature
    - Help to eliminate waste
  - **Processes**
    - **Breathing**
      - **Inspiration**—process of taking air into lungs
      - **Expiration**—process of expelling air from lungs
    - **External respiration**
      - External oxygen supply
      - External carbon dioxide elimination
      - Takes place in the lungs
    - **Internal respiration**
      - Internal oxygen supply
      - Internal carbon dioxide elimination
      - Takes place in the cells
- **Lungs**
  - May be regarded as membranous sacs
  - Interior communicates with outside air by bronchi, trachea, glottis
  - Outside protected by walls of chest
- **Normal Respiratory Movements**
  - Normal size and position of chest are at end of normal respiration
  - Active inspiration
    - Any enlargement which forces more air into lungs
  - Passive expiration—chest returns to normal, no effort involved
- **Mechanism of Inspiration and Expiration**
  - **Enlargement of cavity**
    - Vertical
    - Dorsoventral
    - Lateral
  - **Inspiration**
    - Chest cavity enlarged
      - Elevation of ribs, dependent upon contraction of muscles of inspiration
      - Descent of diaphragm by contraction of diaphragmatic muscles
    - Enlargement of lungs—in proportion to enlargement of cavity—lungs in contact with chest walls
    - Air rushes in through trachea and bronchi
  - **Expiration**
    - Chest cavity made smaller
      - Inspiratory muscles relax
      - Recoil of elastic thorax
      - Recoil of elastic lungs
    - Air forced out through trachea

Muscles of Inspiration
- The muscles that contract simultaneously with diaphragm
- The diaphragm
- paired:
  - The levatores costarum
  - The external intercostals
  - The scaleni
  - The sternocleidomastoid
  - The pectoralis minor
  - The serratus posticus superior

Forced Expiration
- The muscles that contract alternately with diaphragm. Size of chest cavity lessened. Accomplished in two ways:
  - (1) Force diaphragm farther up into thoracic cavity
    - Contraction of muscular walls of the abdomen, i.e., the external and internal oblique, the rectus, and transversalis of both sides
  - (2) Depress the ribs — paired:
    - Internal intercostals
    - Triangularis sterni
    - Some authorities add
      - Iliocostalis
      - Serratus posticus inferior
      - Quadratus lumborum

Types of Respiration
- Sequence of movements is distinguishing factor
- Costal—upper ribs move first, abdomen second
- Abdominal—abdomen bulges outward first, followed by movement of thorax
- Abdominal respirations are deeper

Respiratory Center
- Located in medulla oblongata and pons
  - Center for inspiration is respiratory center
  - Center for expiration—assumed
- Efferent fibers from respiratory center travel down spinal cord and connect with fibers of vagi and sympathetic nerves distributed in the lung tissue
- Afferent fibers lead to the respiratory center
- Connection with the sensory fibers of all the cranial and spinal nerves assumed
- **Action**
  - Automatic, i.e., it is constantly sending impulses over afferent fibers
  - **Rate and rhythm dependent on**
    - Vagus nerves
      - Two kinds of fibers:
        - (1) Inhibits inspiration
        - (2) Inhibits expiration
    - Chemical condition of blood, i.e., hydrogen ion concentration of the blood

Control of Respiration
- Voluntary control for a short time
- Breaking point reached in 23–77 sec
- If lungs are thoroughly aerated by forced breathing, breaking point may be postponed as long as 8 min

Cause of First Respiration
- (1) Increased amount of carbon dioxide in blood
- (2) Stimulation of sensory nerves of skin
- (3) Combination of these two causes

Respiratory Rate
- 16 to 18 times per minute — influenced by
  - Muscular exercise
  - Emotion
  - Heartbeat
  - Age

External Respiration
- Takes place in lungs
- Blood
  - Loses about 6% of carbon dioxide
  - Gains about 8% of oxygen
    - Oxyhemoglobin
    - Scarlet color
  - Temperature is slightly reduced

Capacity of Lungs
- Exchange dependent on diffusion of gases
- After lungs are once filled, they are never emptied during life
- Vital capacity—quantity of air person can expel by forcible expiration after deepest inspiration possible—averages from 3,500–4,100 cc
- **Terms in use**
  - Tidal
  - Inspiratory reserve (complemental)
  - Expiratory reserve (supplemental)
  - Residual
  - Reserve
  - Minimal

Inspired and Expired Air
- **Changes effected**
  - (1) Moisture increased. Expired air is saturated with moisture
  - (2) Temperature increased. Expired air is as warm as blood
  - (3) Heat to warm air and vaporize moisture taken from body
  - (4) Oxygen decreased by 4.94%
  - (5) Carbon dioxide increased by 4.34%

Internal Respiration
- Exchange of gases in the tissues
- Consists of
  - Passage of oxygen from blood into tissue fluid and from tissue fluid into cells
  - Passage of carbon dioxide from tissue cells into tissue fluid and from tissue fluid into blood
- Important to remember blood does not give up all its oxygen to the tissues nor all of its carbon dioxide in the lungs

Respiratory Phenomena
- Air passing into alveoli produces a fine, rustling sound
- Air passing in and out of larynx, trachea, and bronchial tubes produces louder sound called bronchial murmur
- In diseased conditions modified sounds are called *rales*

Respiratory Phenomena

- Eupnea—ordinary quiet respiration
- Dyspnea—difficult breathing
- Hyperpnea—excessive breathing
- Apnea—lack of breathing
- Cheyne-Stokes — Respirations increase in force and frequency, then gradually decrease and stop. Cycle repeated
- Edematous—air cells filled with fluid, hence moist, rattling sounds
- Asphyxia—oxygen starvation

Ventilation Requirements

(1) Continuous movement of the air
(2) The temperature and degree of humidity must favor the evaporation of perspiration from the skin
The temperature should approximate 65–70° F (18–20° C). Generally speaking, above 20° C the higher the humidity, the lower the temperature should be
(3) Disagreeable odors must be eliminated

CHAPTER 17

ANATOMY AND HISTOLOGY OF THE DIGESTIVE SYSTEM

Within the digestive tract is carried on the necessary transformation of ingested complex food substances into the simpler, diffusible substances which may pass into the blood stream and be distributed to the cells of the body. The changes are both physical and chemical and constitute the digestive processes; the organs which take part in them form the digestive system.

These changes can be classified into two groups. First are those concerned with the moving of the foods along through the alimentary tract with optimum speed. This means slowly enough for all the necessary changes in each organ to be accomplished in preparation for those in the next organ and yet fast enough so that proper absorption shall take place and bacterial decomposition or deleterious changes shall not occur.

The other group is concerned with the comminution of the food to particles small enough to diffuse through the wall of the alimentary tract into the body fluids. This is a physical comminution brought about by mastication and by the various types of muscular activity that are described, and a subsequent chemical comminution of large molecules to molecules sufficiently small to diffuse into the blood. It can easily be seen that the normal motility of the alimentary tract and the proper functioning of the neuromuscular mechanisms by which it is carried on are essential to health.

The digestive system consists of the alimentary canal and the accessory organs: tongue, teeth, salivary glands, pancreas, and liver.

The alimentary canal, or digestive tube, is a continuous tube about 9 m (30 ft) long, which extends from the mouth to the anus.

The greater part is coiled up in the cavity of the abdomen. From the esophagus to the rectum it is composed of four coats. From within outward they are as follows:

1. The *mucous coat* is a soft lining membrane containing glands which secrete digestive fluids.

2. The *submucous coat* is composed of areolar connective tissue and serves to connect the mucous membrane to the parts beneath. Most of the large blood and lymph vessels which give rise to branches for the other coats are found here.

3. The *muscular coat* consists of nonstriated muscular tissue arranged in two layers. The cells of the internal layer are arranged circularly around the tube, and the cells of the external layer are longitudinal to the tube. The combined contractions of muscle cells arranged in this fashion produce a movement called a *peristaltic wave.* A dilated portion of the tube, brought about by contraction of the longitudinal muscles, is followed by a portion that is constricted by contraction of the circular muscles. This contraction forces the contents onward into the dilated portion.

4. Below the diaphragm the fourth coat is the *serous coat,* the outer layer of which is a part of the visceral peritoneum.

The esophagus is above the diaphragm, lying in the mediastinum, and the outer coat is an *adventitia* of areolar connective tissue without a limiting serous membrane. The muscles in the upper part of the esophagus are striated. These are gradually replaced by nonstriated muscle tissue until in the lower part of the esophagus all of the muscle tissue is nonstriated.

The peritoneum, the largest serous membrane in the body, in the male consists of a closed sac,[1] the parietal layer of which lines the walls of the abdominal cavity, the visceral layer being reflected over the abdominal organs and the upper surface of some of the pelvic organs. The space between the layers, named the peritoneal cavity, is under normal conditions a potential cavity only, since the parietal and visceral layers are in lubricated contact. The arrangement of the peritoneum is very complex, for elongated sacs and double folds extend from it, to pass in between and either wholly or partially surround the viscera of the abdomen and pelvis. One important fold is the *omentum,* which hangs in front of the stomach and the intestines; another is the *mesentery*, which is a continuation of serous

[1] Figure 32; note that in the female the uterine tubes open into the peritoneal cavity.

coat and attaches the small and much of the large intestine to the posterior abdominal wall.

When the abdominal cavity is opened, the intestines appear to lie loosely coiled within it. If a coil is lifted, a clear, glistening sheet of tissue is found attached to it. This is the mesentery, the dorsal portion of which is gathered into folds which are attached to the dorsal abdominal wall along a short line of insertion, giving the mesentery the appearance of a ruffle or flounce.

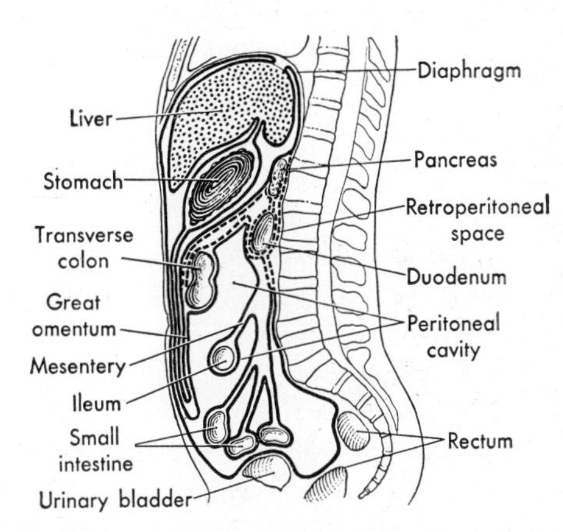

Fig. 290. Diagram of a side view of the body, showing abdominal cavity, peritoneum, mesentery, and omentum. The continuous lines indicate the free surfaces of the peritoneum; the dotted lines indicate those parts of the peritoneum in which the free surfaces have disappeared.

Functions of the peritoneum. The peritoneum serves to prevent friction between contiguous organs by secreting serum, which acts as a lubricant. It aids in holding the abdominal and pelvic organs in position. The omentum usually contains fat, sometimes in considerable amounts.

Divisions of the alimentary canal. The alimentary canal has been given different names in different parts of its course. These names are:

Mouth cavity, containing tongue, orifices of ducts of salivary glands, and teeth
Pharynx
Esophagus
Stomach

Small intestine
- Duodenum
- Jejunum
- Ileum

Large intestine
- Cecum
- Colon ——→
  - Ascending
  - Transverse, or mesial
  - Descending
  - Sigmoid
- Rectum
- Anal canal

MOUTH CAVITY, PHARYNX, ESOPHAGUS

The mouth, oral, or **buccal cavity,** is a cavity bounded laterally and in front by the cheeks and lips; behind, it communicates with the pharynx. The roof is formed by the hard and soft palate, and the greater part of the floor is formed by the tongue and sublingual region and lower jaw. The space bounded externally by the lips and cheeks and internally by the gums and teeth is called the *vestibule.* The cavity behind this is the *mouth cavity proper. The lips,* two musculomembranous folds, surround the orifice of the mouth.

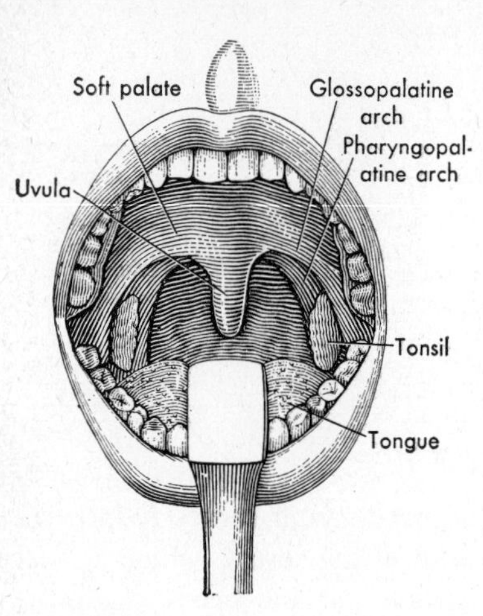

Fig. 291. The soft palate, uvula, and tonsils, as seen from the front. (Gerrish.)

The palate consists of a hard portion in front, formed by processes of the maxillae and palatine bones, which are covered by mucous membrane. Suspended from the posterior border is the soft palate, a movable fold of mucous membrane, enclosing muscle fibers, blood vessels, nerves, adenoid tissue, and mucous glands. Hanging from the middle of its lower border is a conical process called the palatine *uvula.*

The fauces is the name given to the aperture leading from the mouth into the pharynx, or throat cavity. At the base of the uvula on either side is a curved fold of muscular tissue covered by mucous membrane, which shortly after leaving the uvula is split into two pillars; the one runs downward, lateralward, and forward to the side of the base of the tongue; the other downward, lateralward, and backward to the side of the pharynx. These arches are known respectively as the *glossopalatine arch* (anterior pillars of the fauces) and the *pharyngopalatine arch* (posterior pillars of the fauces).

The palatine tonsils[2] are two masses of lymphoid tissue situated, one on either side, in the triangular space between the glossopalatine and the pharyngopalatine arches. The surface of the tonsils is

[2] As commonly used, the name *tonsil* refers to the palatine tonsils.

marked by openings called crypts, which communicate with channels that course through the substance of the tissue. They are supplied with blood from the lingual and internal maxillary arteries, which are derived from the external carotid arteries. They receive nerve fibers from both divisions of the autonomic nervous system. Situated below the tongue are masses of lymphoid tissue called the *lingual tonsils.*

The function of the tonsils is similar to that of other lymph nodes. They aid in the formation of white blood cells and help to protect the body from infection by acting as filters and preventing the entrance of microorganisms. If they are abnormal, their protective function is reduced, and they may serve as foci of infection, which passes directly into the lymph and so into the blood. If they are much enlarged, they tend to fill the throat cavity and interfere with the passage of air to the lungs. Inflammation of the palatine tonsils is called tonsillitis.

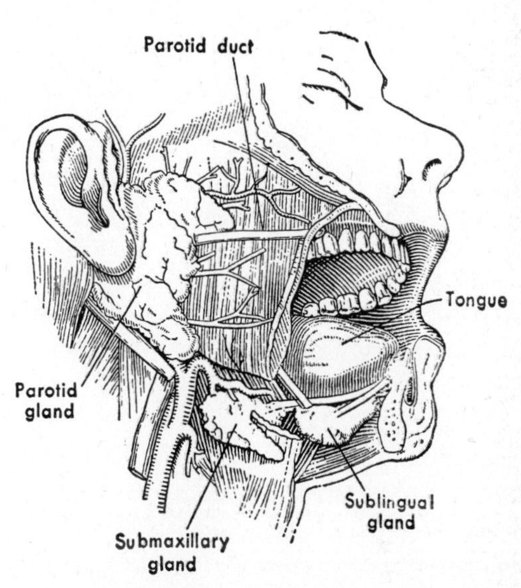

Fig. 292. The salivary glands and their ducts. These glands manufacture about 1,500 cc of saliva in 24 hours.

The palate, uvula, palatine arches, and tonsils are plainly seen if the mouth is widely opened and the tongue depressed.

The tongue is the special organ of the sense of taste. It assists in mastication, deglutition, and digestion by movements which help to move the food and keep it between the teeth; the glands of the tongue secrete mucus, which lubricates the food and makes swallowing easier; and stimulation of the end organs (taste buds) of the nerves of the sense of taste increases the secretion of saliva and starts the first flow of gastric fluid.

The salivary glands. The mucous membrane lining the mouth contains many minute glands called *buccal glands,* which pour their secretion into the mouth. The chief secretion, however, is supplied by three pairs of compound saccular glands, the salivary glands, named parotid, submaxillary, and sublingual glands. Each *parotid* gland is placed just under and in front of the ear; its duct, the parotid (Stensen's),[3] opens upon the inner surface of the cheek opposite the second molar of the upper jaw. The *submaxillary* and *sublingual* glands lie below the jaw and under the tongue, the submaxillary being placed farther back than the sublingual. One duct (Wharton's) from each submaxillary and a number of small ducts from each sublingual open in the floor of the mouth beneath the tongue. The secretion of the salivary glands, mixed with that of the small glands of the mouth, the buccal secretion, is called *saliva.*

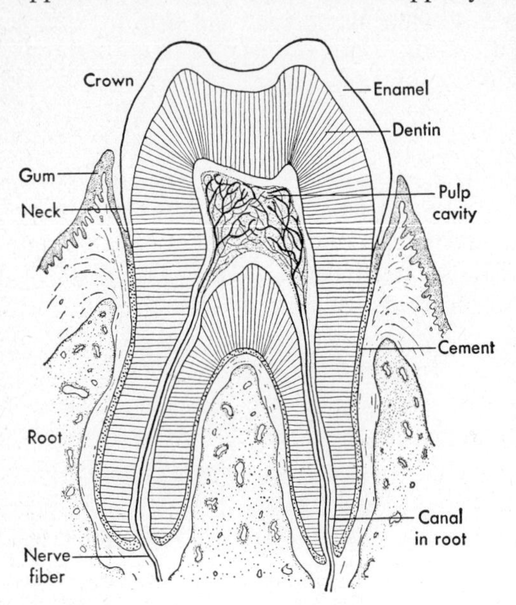

Fig. 293. Section of human molar tooth. In the pulp cavity are located blood vessels and nerves.

Nerves and blood vessels. The facial (VII) and glossopharyngeal (IX) nerves supply these glands. The fibers are both secretory and vasomotor and are derived from the craniosacral and thoracolumbar systems. Blood is supplied to the salivary glands by branches of the external carotid artery and is returned, after traveling through many branch arteries and capillaries, via the jugular veins.

The teeth (dentes). The alveolar processes of the maxillae and mandible contain *alveoli,* or sockets, for the teeth. Dense connective tissue covered by smooth mucous membrane—the gums, or

[3] Nicolaus Stensen, Danish anatomist, 1638–1686.

gingivae—covers these processes and extends a little way into each socket. The sockets are lined with periosteum, which connects with the gums and serves to attach the teeth to their sockets and as a source of nourishment.

Each tooth consists of three portions: the *root,* consisting of one to three fangs contained in the socket; the *crown,* which projects beyond the level of the gums; and the *neck,* or constricted portion between the root and the crown.

Each tooth is composed principally of ivory, or *dentin,* which gives it shape and encloses a cavity, the pulp cavity. The dentin of the crown is capped by a dense layer of *enamel.* The dentin of the root is covered by *cement.* These three substances—enamel, dentin, and cement—are all harder than bone, enamel being the hardest substance found in the body. They are developed from epithelial tissue. The pulp cavity is just under the crown and is continuous with a canal that traverses the center of each root and opens by a small aperture at its extremity. It is filled with dental pulp, which consists of connective tissue holding a number of blood vessels and nerves, which enter by means of the canal from the root.

There are two sets of teeth developed during life: the first, deciduous, or milk, teeth; and the second, permanent.

Deciduous teeth. In the first set are 20 teeth, 10 in each jaw: 4 incisors, 2 canines, and 4 molars. The cutting of these teeth usually begins at 6 months and ends at about the age of 2 years. In nearly all cases the teeth of the lower jaw appear before the corresponding ones of the upper jaw.

DECIDUOUS TEETH

	Molars	*Canine*	*Incisors*	*Canine*	*Molars*
Upper	2	1	4	1	2
Lower	2	1	4	1	2

The deciduous teeth are usually cut in the following order:

Lower central incisors	6–9 months
Upper incisors	8–10 months
Lower lateral incisors and first molars	15–21 months
Canines	16–20 months
Second molars	20–24 months

Permanent teeth. During childhood the temporary teeth are replaced by the permanent. In the second set are 32 permanent teeth, 16 in each jaw. The first molar usually appears between 5 and 7 years of age.

PERMANENT TEETH

	Molars	*Premolars*	*Canine*	*Incisors*	*Canine*	*Premolars*	*Molars*
Upper	3	2	1	4	1	2	3
Lower	3	2	1	4	1	2	3

The permanent teeth appear at about the following periods:

First molars	6 years
Two central incisors	7 years
Two lateral incisors	8 years
First premolars	9 years
Second premolars	10 years
Canine	11–12 years
Second molars	12–13 years
Third molars	17–25 years

According to their shape and use the teeth are divided into incisors, canines, premolars, or bicuspids, and molars. *Incisors,* eight in number, form the four front teeth of each jaw. They have a sharp cutting edge and are especially adapted for biting food. *Canines* are four in number, two in each jaw. They have sharp, pointed edges, are longer than the incisors, and serve the same purpose in biting and tearing. *Premolars,* or *bicuspids,* are eight in number in the permanent set (none in the temporary set). There are four in each jaw, two placed just behind each of the canine teeth. They are broad, with two points or cusps on each crown, and have only one root, which is more or less completely divided into two. Their function is to grind food. *Molars* are 12 in number in the permanent set (eight in the deciduous set). They have broad crowns with small, pointed projections, which makes them well fitted for crushing food. Each upper molar has three roots, and each lower molar has two roots, which are grooved and indicate a tendency to division. The 12 molars do not all replace temporary teeth but are gradually added with the growth of the jaws. The hindmost molars are the last teeth to be added. They may not appear until 25 years of age, hence called *late teeth* or "wisdom teeth."

Long before the teeth appear through the gums their formation and growth are in progress. The deciduous set begins to develop about the sixth week of intrauterine life; and the permanent set, with the exception of the second and third molars, begins to develop about the sixteenth week. About the third month after birth, the second molars begin to grow, and about the third year the third molars, or wisdom teeth, do likewise. Diseases such as rickets and marasmus retard the eruption of the temporary teeth, and severe illness during childhood may interfere with the normal development of the permanent teeth so that they are marked with notches and ridges. Moreover, cavities form in them readily. The diet of the mother during pregnancy and the diet of the child during the first years of life are important factors in determining the quality of the teeth and the development of caries. When the central incisors are notched along their cutting edges and the lateral incisors are pegged, they are named Hutchinson's[4] teeth and are a diagnostic sign of congenital syphilis.

Function. The teeth assist in the process of mastication by comminuting food. It might be thought that the vigorous employment of the teeth for this purpose would only hasten their wear and tear. This may be true at a time when their life is nearly extinct, but at an earlier period mastication of the more solid foods is good for the teeth because they are made to sink and rise in their sockets with a massaging effect upon the gums, which tends to promote circulation in the pulp.

The pharynx, or throat cavity, is a musculomembranous tube shaped somewhat like a cone, with its broad end turned upward and its constricted end downward to end in the esophagus. It may be divided from above downward into three parts, nasal, oral, and laryngeal. The upper, or *nasopharynx,* lies behind the posterior nares and above the soft palate. The middle, or *oral,* part of the pharynx reaches from the soft palate to the level of the hyoid bone. The *laryngeal* part reaches from the hyoid bone to the esophagus. The pharynx communicates with the nose, ears, mouth, and larynx by seven apertures: two in front above, leading into the back of the nose, the *posterior nares;* two on the lateral walls of the nasopharynx, leading into the auditory tubes, which communicate with the ears; one midway in front, the *fauces,* connecting with the mouth in front; two below—one, the well-defined glottis, opening into the larynx, and the other, the poorly defined opening into the esophagus.

The mucous membrane lining the pharynx is continuous with that lining the nasal cavities, the mouth, the auditory tubes, and the larynx. It is well supplied with mucous glands. About the center of the posterior wall of the nasopharynx is a mass of lymphoid tissue, the pharyngeal tonsil. When abnormally large it is called *adenoids.*

Usually lymphoid tissue is larger in children than in adults and tends to grow smaller with age. Owing to their position, adenoids may become

[4] Sir Jonathan Hutchinson, English surgeon (1828–1913).

infected or enlarged, block the auditory tubes, and interfere with the passage of air through the nose.

Nerves and blood vessels. Both divisions of the autonomic system supply nerve fibers to the pharynx. There are both sensory and motor fibers within the glossopharyngeal and vagus nerves. Blood is supplied by branches from the external carotid artery.

Functions. The pharynx transmits the air from the nose or mouth to the larynx and serves as a resonating cavity in the production of the voice. It also serves as a channel to transmit food from the mouth to the esophagus. When the act of swallowing is about to be performed, the muscles draw the pharynx upward and dilate it to receive the food; they then relax, the pharynx sinks, and, the other muscles contracting upon the food, it is pressed downward and onward into the esophagus.

The esophagus, or **gullet,** is a muscular tube, about 23 to 25 cm (9 to 10 in.) long and 25 to 30 mm wide, which begins at the lower end of the pharynx, behind the trachea. It descends in the mediastinum in front of the vertebral column, passes through the diaphragm at the level of the tenth thoracic vertebra, and terminates in the upper, or cardiac, end of the stomach, about the level of the xiphoid process.

Structure. The walls of the esophagus are composed of four coats: (1) an external, or fibrous, (2) a muscular, (3) a submucous, or areolar; and (4) an internal, or mucous, coat. The muscular coat consists of an external longitudinal and an internal circular layer. Contractions of these layers produce peristaltic waves which propel food to the stomach. The areolar coat serves to connect the muscular and mucous coats and to carry the larger blood and lymph tubes. The mucous membrane is disposed in longitudinal folds which disappear when the tube is distended by the passage of food. It is studded with minute papillae and small glands, which secrete mucous to lubricate the canal.

Nerves and blood vessels. The nerve fibers are from the vagus and the thoracolumbar nervous system. They form a plexus between the layers of the muscular coat and another in the submucous coat. Blood is supplied to the esophagus by arteries from the inferior thyroid branch of the thyrocervical trunk, which arises from the subclavian; from the thoracic aorta; from the left gastric branch of the celiac artery; and from the left inferior phrenic of the abdominal aorta. Blood is returned via the azygos, thyroid, and left gastric veins of the stomach.

Functions. The esophagus receives food from the pharynx and by a series of peristaltic contractions passes it on to the stomach.

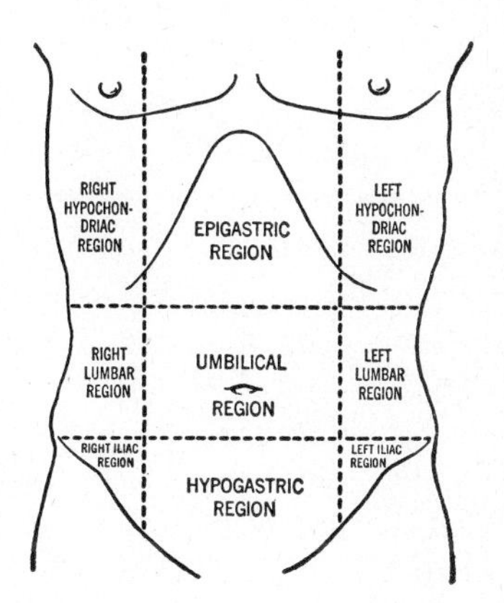

Fig. 294. The nine regions of the abdomen. The abdomen may be artifically divided into nine regions by drawing the following arbitrary lines:[5]

1. Draw a circular line around the body at the level of the lowest point of the tenth costal cartilages.
2. Draw another circular line at the level of the tubercles on the crests of the ilia.
3. Draw a verticle line on each side from the center of Poupart's ligament upward.

These lines are to be considered as edges of planes which divide the abdomen into the nine regions. (Cunningham.)

THE STOMACH

In the abdominal cavity the esophagus ends in the stomach (gaster), which is a collapsible, saclike dilatation of the alimentary canal serving as a temporary receptacle for food. It lies obliquely in the epigastric, umbilical, and left hypochondriac regions of the abdomen, directly under the diaphragm. The shape and position of the stomach are modified by changes within itself and in the surrounding organs. These modifications are determined by the amount of the stomach contents, the stage of digestion which has been reached, the degree of development and power of the muscular walls, and the condition of the adjacent intestines. It is never entirely empty, but always contains a little gastric fluid and mucin. When contracted, the shape of the stomach as seen from the front is comparable to that of a sickle. At an early stage of gastric digestion, the stomach usually consists of two segments, a large globular portion on the left and a narrow tubular portion on the right. When distended with food, it has the shape shown in Fig. 296. The stomach presents two openings and two borders, or curvatures.

[5] C. E. Stackpole and L. C. Leavell, in *Laboratory Manual in Anatomy and Physiology,* 2nd ed., 1948, give a diagram of the four regions of the abdomen.

Openings. The opening by which the esophagus communicates with the stomach is known as the *cardiac*, or esophageal, orifice; the orifice which communicates with the duodenum is known as the *pyloric*. Both the cardiac and pyloric apertures are guarded by ringlike muscles, or sphincters, which when contracted keep the orifices closed. By this arrangement the food is kept in the stomach until it is ready for intestinal digestion, when the circular fibers guarding the pyloric aperture relax. The relaxation of this aperture may be connected with the consistency of the stomach contents and with the regular peristaltic waves moving over the stomach onto the duodenum. Cannon's[6] idea that hydrochloric acid in the stomach seems to favor or produce a relaxation of the pyloric sphincter is questioned today.

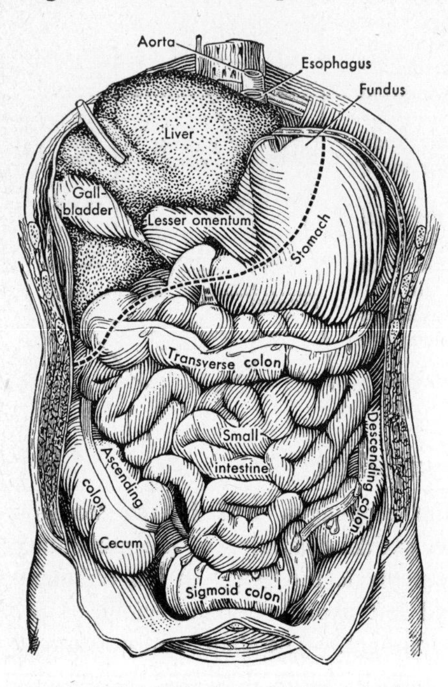

Fig. 295. The stomach and intestines, front view, the great omentum having been removed and the liver turned up and to the right. The dotted line shows the normal position of the anterior border of the liver. (Gerrish.)

Curvatures. In all positions the stomach is more or less curved upon itself. A line drawn from the cardiac orifice along the concave border to the pyloric orifice follows the lesser curvature. A line connecting the same points, but following the convex border, follows the greater curvature.

Component parts. The *fundus* is the rounded end of the stomach, above the entrance of the esophagus. The opposite, or smaller, end is the *pyloric portion*. The central portion, between the fundus and the pyloric portion, is called the *body*.

Structure. The wall of the stomach consists of four coats: serous, muscular, submucous, or areolar, and mucous.

[6] Walter B. Cannon, American physiologist (1871–1946).

1. The *serous* coat is part of the peritoneum and covers the organ. At the lesser curvature the two layers come together and are continued upward to the liver as the *lesser omentum.* At the greater curvature the two layers are continued downward as the apronlike *greater omentum,* which is suspended in front of the intestines.

2. The *muscular* coat of the stomach is beneath the serous coat and closely connected with it. It consists of three layers of unstriped muscular tissue: an outer, longitudinal layer; a middle, or circular layer; and an inner, less well-developed, oblique layer limited chiefly to the cardiac end of the stomach. This arrangement facilitates the

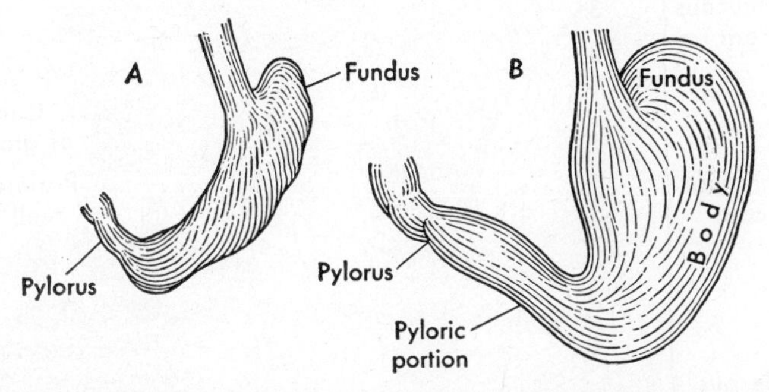

Fig. 296. Form and outline of the stomach at different stages of digestion when seen from the front. *A,* contracted; *B,* early stage of digestion.

muscular actions of the stomach by which it presses upon food and moves it back and forth.

3. The *submucous* coat consists of loose areolar connective tissue connecting the muscular and mucous coats.

4. The *mucous* coat is very soft and thick, the thickness being mainly due to the fact that it is densely packed with small glands embedded in areolar connective tissue. It is covered with columnar epithelium and in its undistended condition is thrown into folds, or *rugae.* The surface is honeycombed by tiny, shallow pits, into which the ducts or mouths of the glands open. Figure 297 and its legend describe the coats and the tissues which compose them.

The gastric glands are of three varieties: cardiac, fundic, or oxyntic,[7] and pyloric.

[7] Also called true gastric or peptic.

Cardiac glands occur close to the cardiac orifice. They are of two kinds—simple tubular glands with short ducts, and compound racemose glands. *Fundic glands* are simple tubular glands which are found in the body and fundus of the stomach. These glands are lined with epithelial cells, of which there are two varieties. (*a*) One

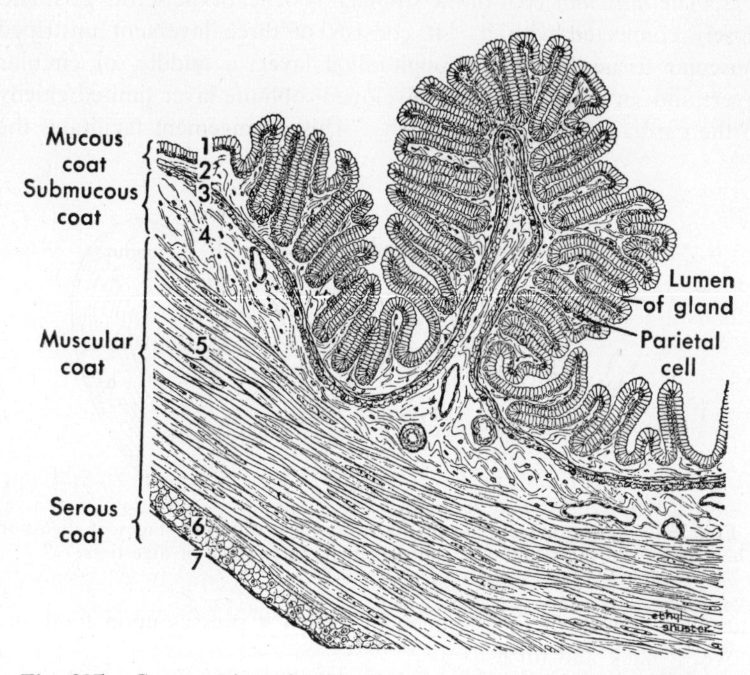

Fig. 297. Cross section of a bit of the wall of the stomach, highly magnified to show coats. One ruga covered with glands is shown. Parietal cells are shown communicating with the lumen of the glands by clefts between the chief cells which line the lumen. *1,* columnar epithelium; *2,* areolar connective tissues; *3,* muscularis mucosae; *4,* areolar connective tissue; *5,* circular layer of smooth muscle; *6,* longitudinal layer of smooth muscle; *7,* areolar connective tissue and mesothelium.

variety of cells is found lining the lumen of the tube. These are called chief or central cells and secrete pepsinogen. (*b*) A second variety, called parietal or oxyntic cells, is found behind the chief cells and does not come in contact with the lumen. These cells secrete hydrochloric acid; and pepsinogen, in the presence of acid, is converted into pepsin. *Pyloric glands* are branched tubular glands found most plentifully about the pylorus. They secrete pepsinogen and mucin.

The combined secretion of these glands forms the gastric fluid.

Some of the cells of the mucous membrane secrete a hormone known as *gastrin*, which is carried by the blood to the fundic and pyloric glands and stimulates them to secretory activity.

Nerves and blood vessels. The stomach is supplied with thoracolumbar nerve fibers from the celiac plexus. Terminal branches of the right vagus are distributed to the back of the organ; branches from the left vagus are distributed to the front. Stimulation of the vagus fibers increases secretion and peristalsis. Stimulation of the thoracolumbar autonomic fibers has just the opposite effect, i.e., inhibits secretion and peristalsis. The blood vessels are derived from the three divisions of the celiac artery, i.e., the left gastric, hepatic, and splenic. Blood is returned via the right gastroepiploic, which joins the superior mesenteric, the left gastroepiploic and several short gastric veins which join the splenic, and the left gastric. All of these eventually join the portal tube. A small quantity of blood is returned to the azygos vein instead of entering the portal tube.

The functions of the stomach are to hold the food while it undergoes certain mechanical and chemical changes which reduce it to a semiliquid condition (chyme), to secrete gastric juice, and at frequent intervals to pass small amounts of chyme into the intestine.

THE SMALL INTESTINE

The small, or thin, intestine extends from the pylorus to the colic valve. It is a convoluted tube about 7 m (23 ft) in length and is contained in the central and lower part of the abdominal cavity.

At the beginning the diameter is about 3.8 cm ($1\frac{1}{2}$ in.), but it gradually diminishes and is hardly 1 in. at its lower end. For descriptive purposes the small intestine is divided into three portions: the duodenum, jejunum, and ileum are continuous and show only slight variations.

The duodenum is 25 cm (10 in.) long, and is the shortest and broadest part of the small intestine. It extends from the pyloric end of the stomach to the jejunum. Beginning at the pylorus, the duodenum at first passes upward, backward, and to the right, beneath the liver. It then makes a sharp bend and passes downward in front of the right kidney; it makes a second bend, toward the left, and passes horizontally across the front of the vertebral column. On the left side, it ascends for about 2.5 cm (1 in.) and then ends opposite the second lumbar vertebra in the jejunum.

The jejunum, or empty intestine, so called because it is always found empty after death, constitutes about two-fifths of the remainder, or 2.2 m ($7\frac{1}{2}$ ft), of the small intestine and extends from the duodenum to the ileum.

The ileum, or twisted intestine, so called from its numerous coils, constitutes the remainder of the small intestine and extends from the jejunum to the large intestine, which it joins at a right angle. The orifice is guarded by a sphincter muscle, which acts as a valve and prevents the return of material that has been discharged into the large intestine. This is known as the colic, or ileocecal, valve. There is no definite point at which the jejunum ceases and the ileum begins, although the mucous membranes of the two divisions differ somewhat.

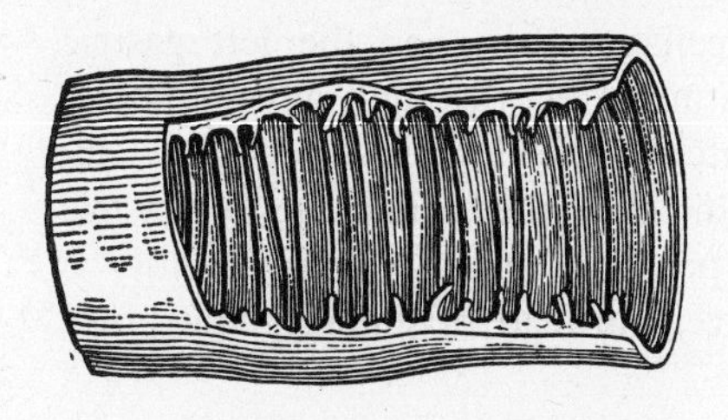

Fig. 298. Portion of small intestine laid open to show circular folds (valvulae conniventes). Not highly enough magnified to show villi.

The coats of the small intestine are four in number and correspond in character and arrangement to those of the stomach. (1) The *serous* coat furnished by the peritoneum forms an almost complete covering for the whole tube except for part of the duodenum. (2) The *muscular* coat of the small intestine has two layers: an outer, thinner and longitudinal; and an inner, thicker and circular. This arrangement aids the peristaltic action of the intestine. (3) The *submucous*, or areolar connective tissue coat, connects the muscular and mucous coats. (4) The *mucous* coat is thick and very vascular.

Circular folds. About 3 or 4 cm (1 or 2 in.) beyond the pylorus the mucous and submucous coats of the small intestine are arranged in circular folds (valvulae conniventes, or plicae circulares) which project into the lumen of the tube (Fig. 299). Some of these folds extend all the way around the circumference of the intestine; others extend part of the way. Unlike the rugae of the stomach, the circular folds do not disappear when the intestine is distended. About the middle of the jejunum they begin to decrease in size, and in the

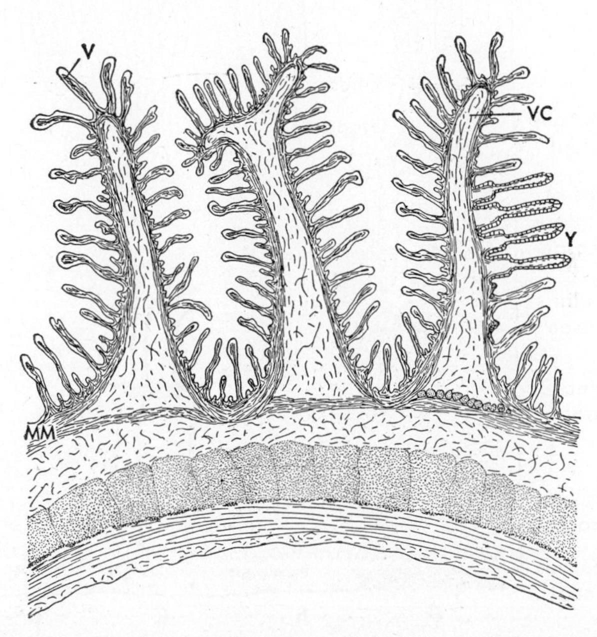

Fig. 299. Longitudinal section of small intestine. Three valvulae conniventes (*VC*) are shown. Many villi (*V*) are shown on the valvulae and between them. At *Y* four villi with glands between them have been diagramed; *MM,* muscularis mucosae.

lower part of the ileum they almost entirely disappear. The purpose of these folds is to delay the food in the intestine slightly and to present a greater surface for secretion of digestive juices and absorption of digested food.

Villi. Throughout the whole length of the small intestine the mucous membrane presents a velvety appearance due to minute, fingerlike projections called *villi,* which are said to number between 4,000,000 and 5,000,000 in man. Each villus consists of a central lymph channel called a *lacteal*, surrounded by a network of blood capillaries held together by lymphoid tissue. This in turn is surrounded by a layer of columnar cells. After the food has been digested, it passes into the capillaries and lacteals of the villi.

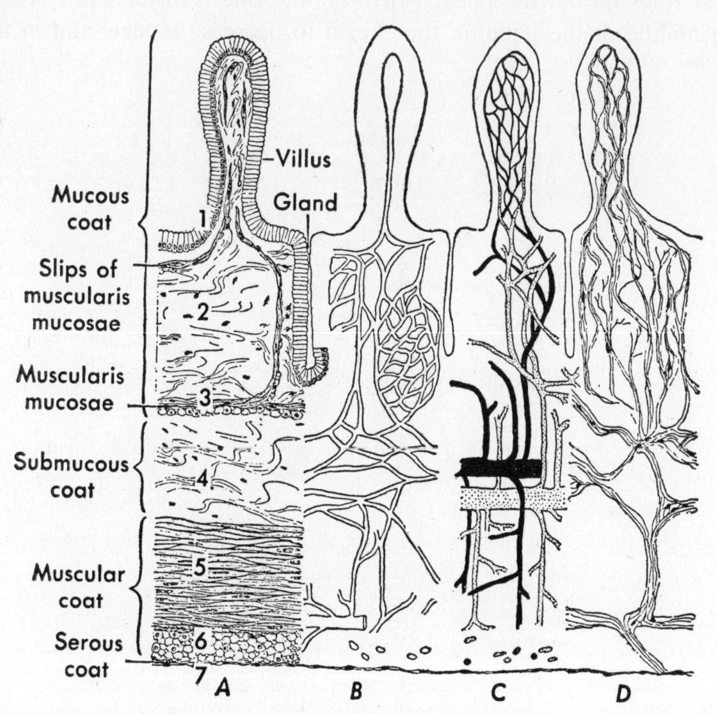

Fig. 300. Diagram of a cross section of small intestine. *A* shows coats of intestinal wall and tissues of coats; *1,* columnar epithelium; *2,* areolar connective tissue; *3,* muscularis mucosae; *4,* areolar connective tissue; *5,* circular layer of smooth muscle; *6,* longitudinal layer of smooth muscle; *7,* areolar connective tissue and endothelium. *B* shows arrangement of central lacteal, lymph nodes, and lymph tubes. *C* shows blood supply; arteries and capillaries black, veins stippled. *D* shows nerve fibers; the submucous plexus lying in the submucosa, the myenteric plexus lying between the circular and longitudinal layers of the muscular coat. (*A* and *C* drawn from microscopic slide of injected specimen; *B* and *D* adapted after Mall.)

Glands and nodes of the small intestine. In addition to these projections the mucous membrane covering the circular folds and between them is thickly studded with secretory glands and nodes. These are known as:

Intestinal glands or crypts of Lieberkuehn[8]
Duodenal or Brunner's glands[9]
Lymph nodules—(1) solitary lymph nodules, (2) aggregated lymph nodules

Intestinal glands are found over every part of the surface of the small intestine. They are simple tubular depressions in the mucous membrane, lined with columnar epithelium and opening upon the surface by circular apertures.

Duodenal glands are compound glands found in the submucous tissue of the duodenum. The intestinal glands and the duodenal glands secrete the intestinal digestive fluid, which is named the *succus entericus*.

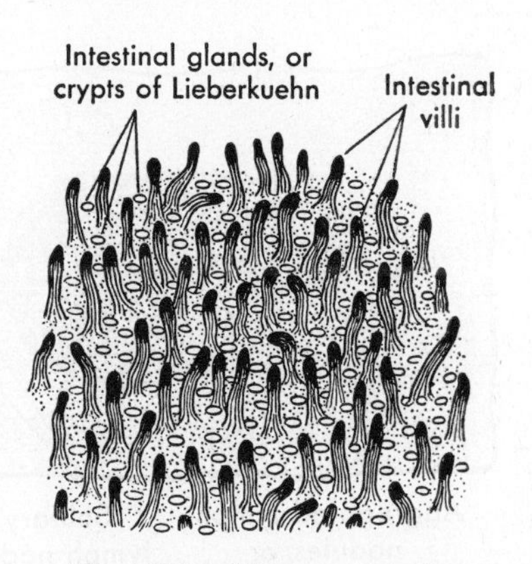

Fig. 301. Mucous membrane of the ileum, showing villi and the mouths of the intestinal glands.

1. *Lymph nodules.* Closely connected with the lymphatic vessels in the walls of the intestines are small, rounded bodies of the size of a small pinhead, called *solitary lymph nodules.* They are most numerous in the lower part of the ileum and consist of a rounded mass of fine lymphoid tissue, the meshes of which are crowded with leukocytes. Into this mass of tissue one or more small arteries enter and form a capillary network, from which the blood is carried away by one or more small veins. Surrounding the mass are lymph channels which are continuous with the lymphatic vessels in the tissue below.

2. *Aggregated lymph nodules* are collections of lymph nodules, commonly called Peyer's[10] patches. These patches are circular or oval in shape, from 10 to 30 in number, and vary in length from about 2.5 to 10 cm (1 to 4 in.). They are largest and most numerous in the ileum. In the lower part of the jejunum they are small and few in number. They are occasionally seen in the duodenum. Peyer's patches may be the seat of local inflammation and ulceration

[8] Johann Nathaniel Lieberkuehn, German anatomist (1711–1756).
[9] Johann Conrad Brunner, Swiss anatomist (1653–1727).
[10] Johann Conrad Peyer, Swiss anatomist (1653–1712).

in typhoid fever and intestinal infections, particularly tuberculosis of the intestines.

Figure 300 describes the coats of the small intestines and the tissues which compose them and the relationship of blood vessels, lymph tubes, and nerve fibers.

Nerves and blood vessels. The vagus nerves supply sensory and motor fibers to the small intestine. Thoracolumbar nerve fibers are derived from the plexuses around the superior mesenteric artery. From this source they run to the myenteric plexus (Auerbach's plexus) of nerves and ganglia situated between the circular and longitudinal muscular fibers. Branches from this plexus are distributed to the muscular coats; and from these branches another plexus, the submucous (Meissner's) plexus, is derived (Fig. 316). It sends fibers to the mucous membrane. The arteries supplying the small intestine are branches of the superior mesenteric. These vessels distribute branches, which lie between the serous and muscular coats and form frequent anastomoses. Blood is returned by the superior mesenteric vein, which unites with the splenic to form the portal tube.

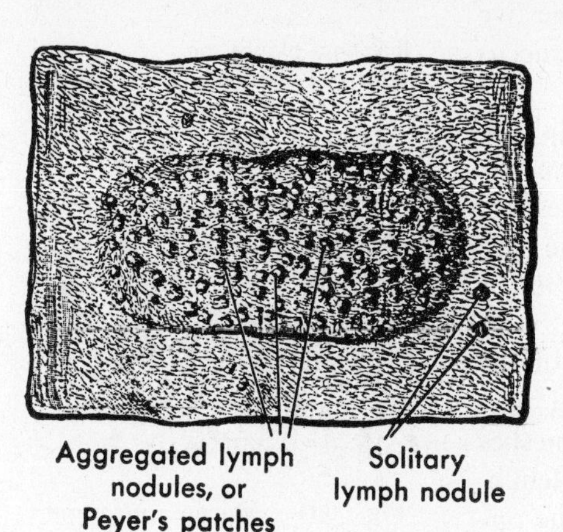

Fig. 302. Aggregated lymph nodules in wall of ileum (Peyer's patches). (Toldt.)

Functions. It is in the small intestine that the greatest amount of digestion and absorption takes place. It receives bile and pancreatic fluid from the liver and the pancreas. The glands of the small intestine secrete the succus entericus. The circular folds delay food so that it is more thoroughly subjected to the action of digestive fluids; and being covered with villi, they increase the surface for absorption. Some of the cells of the mucous membrane (particularly in the duodenum) secrete *prosecretin.* When the acid chyme enters the intestine, prosecretin is changed to *secretin* and carried by the blood to the liver, pancreas, and all parts of the intestine, stimulating them to secretory activity.

THE LARGE INTESTINE

The large, or thick, intestine is about 1.5 m (5 ft) long but is wider than the small intestine, being about 6.3 m ($2\frac{1}{2}$ in.) at the cecum. It extends from the ileum to the anus. It is divided into four parts: the cecum with the vermiform appendix, colon, rectum, and anal canal.

The cecum. The small intestine opens into the side wall of the large intestine about 6 cm ($2\frac{1}{2}$ in.) above the commencement of the

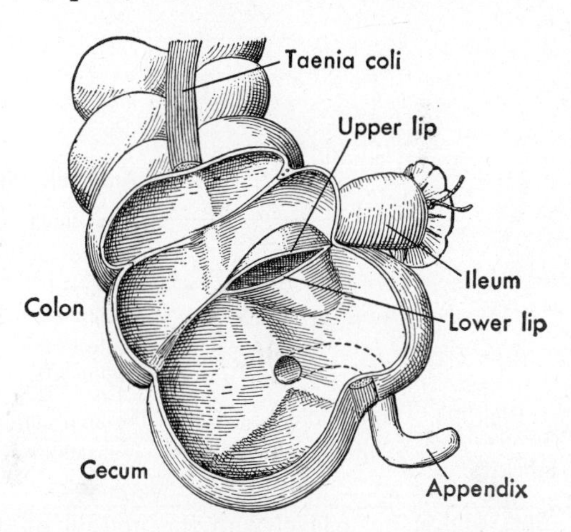

Fig. 303. Cavity of the cecum, its front wall having been cut away. The valve of the colon (colic) and the opening of the appendix are shown. One of the three muscular bands (taenia coli) shows on the outside, and the location of another is shown on the inside.

large intestine. This 6 cm of large intestine forms a blind pouch called the cecum. The opening from the ileum into the large intestine is provided with two large projecting lips of mucous membrane forming the colic, or ileocecal, valve, which allows the passage of material into the large intestine but effectually prevents the passage of material in the opposite direction.

The vermiform appendix is a narrow tube attached to the end of the cecum. The length, diameter, direction, and relations of the appendix are very variable. The average length is about 7.5 cm (3 in.).

The functions of the appendix are not known. It is most fully developed in the young adult and at this time is subject to inflammatory and gangrenous conditions commonly called appendicitis.

The reasons for this are that its structure does not allow for ready drainage, its blood supply is limited, and its circulation is easily interfered with because the vessels anastomose to a very limited extent.

The colon, although one continuous tube, is subdivided into the *ascending, transverse* or *mesial, descending,* and *sigmoid colon.* The ascending portion ascends on the right side of the abdomen until it reaches the under surface of the liver, where it bends abruptly to the left (right colic or hepatic flexure) and is continued across the abdomen as the transverse colon until, reaching the left side, it curves beneath the lower end of the spleen (left colic or splenic flexure) and passes downward as the descending colon. Reaching the left iliac region on a level with the margin of the crest of the ileum, it makes a curve like the letter S—hence its name of sigmoid—and finally ends in the rectum (Fig. 295).

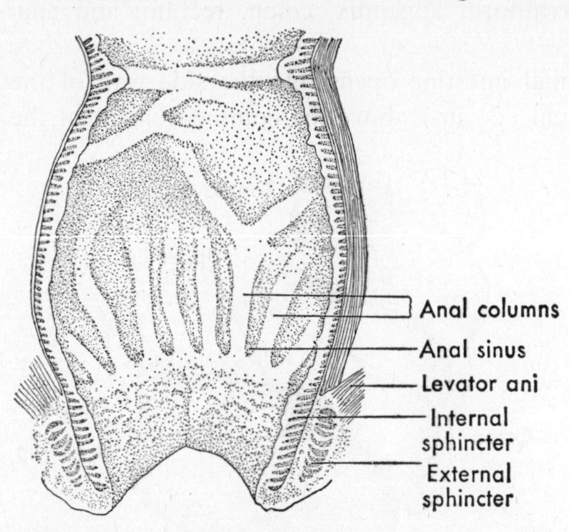

Fig. 304. Longitudinal section of the anal canal. Shows anal columns, anal sinuses, and sphincter muscles.

The rectum is about 12 cm (5 in.) long and is continuous with the sigmoid colon and anal canal. From its origin at the third sacral vertebra it descends downward and forward along the curve of the sacrum and coccyx and finally bends sharply backward into the anal canal. In small children the rectum is much straighter than in adults.

The anal canal is the terminal portion of the large intestine and is about 2.5 to 3.8 cm (1 to $1\frac{1}{2}$ in.) in length. The external aperture, called the *anus,* is guarded by an internal and external sphincter. It is kept closed except during defecation.

The condition known as *piles* or *hemorrhoids* is brought about by enlargement of the veins of the anal canal. Piles may be *external,* wherein enlargement is of the veins just outside the anal orifice, or *internal,* wherein the enlargement is of veins within the canal.

The coats of the large intestine are the usual four, except in some parts where the *serous* coat only partially covers it and in the anal canal, where the serous coat is lacking. The *muscular* coat consists of two layers of fibers, the external arranged longitudinally and the internal circularly. The longitudinal fibers form a thicker layer in some regions than in others. The thick areas form three separate bands, the *taeniae coli,* which extend from the cecum to the beginning of the rectum, where they spread out and form a longitudinal layer which encircles this portion. Because these bands (about 5 to 7 mm wide) are about one-sixth shorter than the rest of the tube, the walls of the tube are puckered into numerous *sacculations.* The third coat consists of *submucous areolar tissue* and the fourth, or inner, coat consists of *mucous membrane.* The mucous coat possesses no villi and no circular folds. It contains intestinal glands and solitary lymph nodules which closely resemble those of the small intestine.

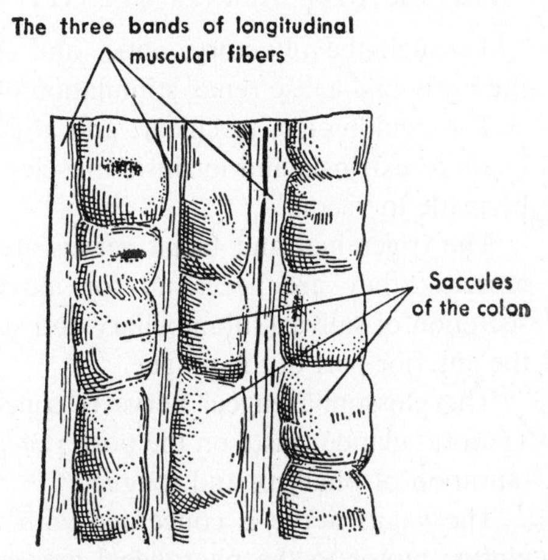

Fig. 305. The muscular coat of the opened large intestine, showing the longitudinal fibers.

Nerves and blood vessels. Fibers from both divisions of the autonomic system reach the large intestine, nerves from the mesenteric and hypogastric plexuses being distributed in a way similar to that found in the small intestine. The arteries are derived mainly from the superior and inferior mesenteric arteries. Branches of the superior mesenteric artery supply the cecum, appendix, and ascending and transverse colon. Branches of the inferior mesenteric artery supply the descending colon and the rectum. The rectum also receives branches from the hypogastric arteries. Blood from the large intestine is returned via the superior and inferior mesenteric veins; and blood from the rectum is returned via the superior rectal, which joins the left colic vein, and the middle and inferior rectal, which join the internal iliac vein.

Functions. The process of digestion is continued by the digestive fluids with which the food becomes mixed in the small intestine; the process of absorption is continued, and the waste products are removed from the body.

THE CRANIAL NERVES RELATED TO DIGESTIVE PROCESSES

Through the olfactory, optic, and acoustic nerves impulses reach the brain and cause reflex stimulation of digestive juices.

The oculomotor, trochlear, and abducens nerves supply the motor fibers to extrinsic and intrinsic muscles of eye so that adjustment may be made to vision.

The trigeminal and facial nerves are sensory to teeth, mouth, and muscles; they are necessary for movement of jaw in mastication, secretion of saliva (submaxillary and sublingual glands), and taste on the anterior part of the tongue.

The glossopharyngeal nerve is concerned with secretion of saliva (parotid gland), taste on the posterior part of the tongue, and general sensation of pharynx and tongue.

The vagus nerve is concerned with taste in the region of the epiglottis; motor to the pharyngeal muscles, sensory of pharynx; motor to esophagus, stomach, small intestine, and part of large intestine; secretory to glands of the stomach, small and large intestine, liver and pancreas.

The accessory nerve is motor to muscles of pharynx and is concerned with the act of swallowing.

The hypoglossal nerve is motor to muscles of the tongue for mastication and swallowing.

ACCESSORY ORGANS OF DIGESTION

The accessory organs of digestion are: (1) the tongue, (2) the teeth, (3) the salivary glands, (4) the pancreas, (5) the liver, and (6) the gallbladder. The first three have been described.

The pancreas is a soft, reddish- or yellowish-gray gland which lies in front of the first and second lumbar vertebrae and behind the stomach. In shape it somewhat resembles a hammer and is divided into head, body, and tail. The right end, or head, is thicker and fills the curve of the duodenum, to which it is firmly attached. The left, free end, is the tail and reaches to the spleen. The intervening portion is the body. Its average weight is between 60 and 90 gm (2 to 3 oz); it is about 12.5 cm (5 in.) long and about 5 cm (2 in.) wide.

Structure. The pancreas is a racemose gland composed of lobules. Each lobule consists of one of the branches of the main duct, which terminates in a cluster of pouches, or alveoli. The lobules are joined together by areolar tissue to form lobes; and the lobes, united in the same manner, form the gland. The small ducts from each lobule open into one main duct about 3 mm in diameter, which runs transversely from the tail to the head through the substance of the gland. This is known as the pancreatic duct or duct of Wirsung.[11] The pancreatic and common bile duct usually unite and enter the duodenum about 7.5 cm (3 in.) below the pylorus. The short tube formed by the union of the two ducts is dilated into an ampulla, called the *ampulla of Vater*.[12] Sometimes the pancreatic duct and the common bile duct open separately into the duodenum, and there is frequently an accessory duct (duct of Santorini)[13] which opens into the duodenum about 1 in. above the orifice of the main duct.

Islets of Langerhans. Between the alveoli small groups of cells are found, which are termed the *islets* of Langerhans[14] (interalveolar cell islets). They are surrounded by a rich capillary network and furnish the internal secretion of the pancreas.

Function. Two secretions are formed in the pancreas. (1) The pancreatic fluid is an external secretion and is poured into the duodenum during intestinal digestion. (2) The secretion formed by the islets of Langerhans is an internal secretion, from which insulin has been extracted. It is absorbed by the blood, carried to the tissues, and aids in regulating glucose metabolism.

The liver (hepar) is the largest gland in the body, weighing ordinarily from 1.2 to 1.6 kg (42 to 56 oz). It is located in the right hypochondriac and epigastric regions and frequently extends into the left hypochondriac region. The upper convex surface fits

[11] Johann Georg Wirsung, Bavarian anatomist (1643).
[12] Abraham Vater, German anatomist (1684–1751).
[13] Giovanni Domenico Santorini, Italian anatomist (1681–1737).
[14] Ernest Robert Langerhans, German histologist (1847–1888).

closely into the under surface of the diaphragm. The under concave surface of the organ fits over the right kidney, the upper portion of the ascending colon, and the pyloric end of the stomach.

Ligaments. The liver is connected to the under surface of the diaphragm and the anterior walls of the abdomen by five ligaments,

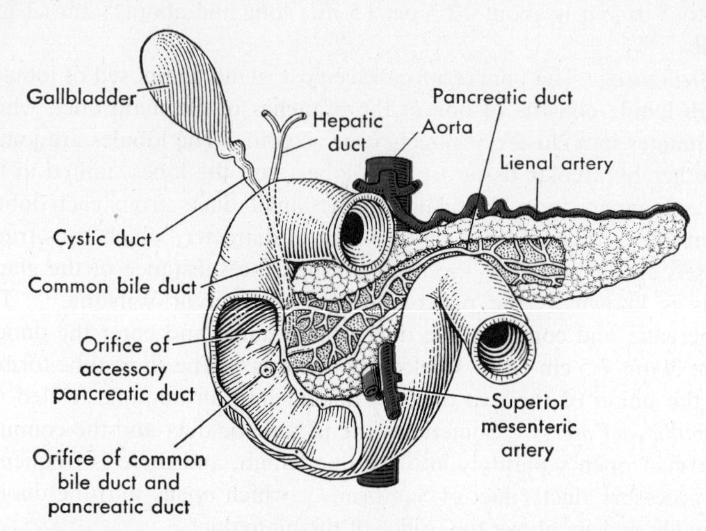

Fig. 306. Ducts of the pancreas. Part of the front of the duodenum is cut away. (Gerrish.)

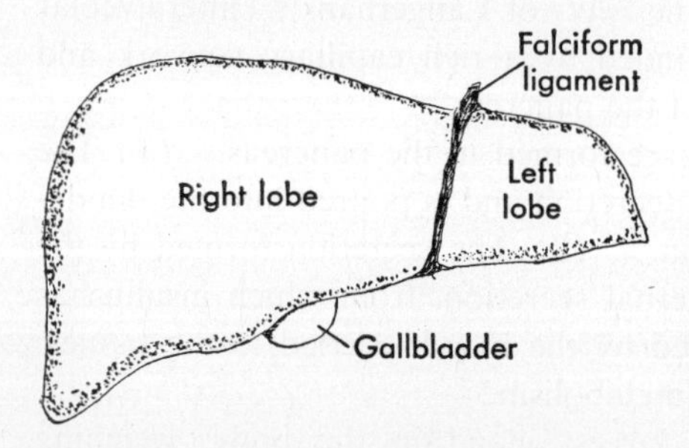

Fig. 307. Superior surface of liver. The liver measures 20 to 22.5 cm (8 to 9 in.) from side to side, 10 to 12.5 cm (4 to 5 in.) from front to back, and 15 to 17.5 cm (6 to 7 in.) from above downward in its thickest part. It has many diverse functions.

four of which—the falciform, the coronary, and the two lateral—are formed by folds of peritoneum. The fifth, or round ligament, is a fibrous cord resulting from the atrophy of the umbilical vein of intrauterine life.

Fossae. The liver is divided by four fossae, or fissures, into four lobes.

The important fossae are the left sagittal; the portal, or transverse, which transmits the portal tube, hepatic artery, nerves, hepatic duct, and lymphatics; the fossa for the gallbladder; and the fossa for the inferior vena cava.

Lobes. The liver is divided into four lobes:

1. Right (largest lobe)
2. Left (smaller and wedge-shaped)
3. Quadrate (square)
4. Caudate (tail-like)

Vessels. The liver has five sets of vessels:

1. Branches of portal tube
2. Bile ducts
3. Branches of hepatic artery
4. Hepatic veins
5. Lymphatics

Nerves and blood vessels. The nerve fibers are derived from the left vagus and the thoracolumbar system. They enter at the transverse fossa and accompany the vessels and ducts to the interlobular spaces. From here fibers are distributed to the coats of the blood vessels and ramify between and within the cells. The blood vessels connected with the liver are the hepatic artery, the portal tube, and the hepatic veins.

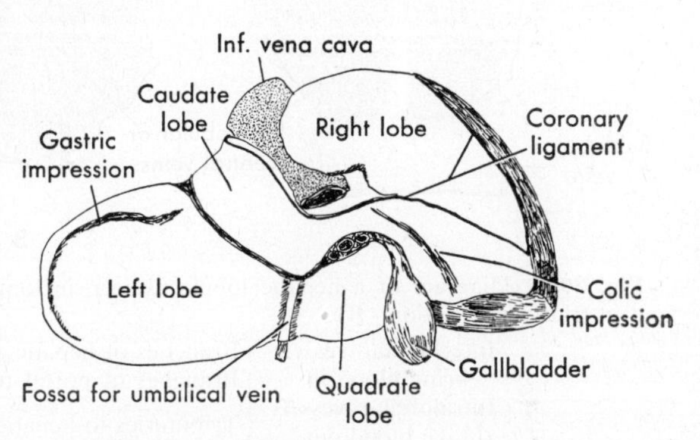

Fig. 308. Diagram of under surface of liver.

Histology of liver. The liver is made up of many minute units called lobules. The lobule is the *unit of gross structure.* Each *lobule* is an irregular body about 1 to 2.5 mm ($\frac{1}{20}$ to $\frac{1}{10}$ in.) in diameter, composed of *chains* or *cords of hepatic cells* held together by areolar connective tissue, in which ramify capillaries derived from the portal tube, the hepatic artery, nerves, and the hepatic ducts. The cords of hepatic cells extending from the edges to the centers of the lobules with their blood and lymph supply are the *units of minute structure.* Together they give an enormous area of contact between liver cells and capillaries for the volume of tissue concerned. Thus each lobule has all the essentials of a gland: (1) blood vessels in close connection with secretory cells, (2) cells which are capable of forming a secretion, and (3) ducts by which the secretion is carried away.

The portal tube brings to the liver blood from the stomach, spleen, pancreas, and intestine. After entering the liver, it divides into a vast number of branches which form a plexus, the interlobular plexus in the spaces between the lobules. From this plexus the blood is

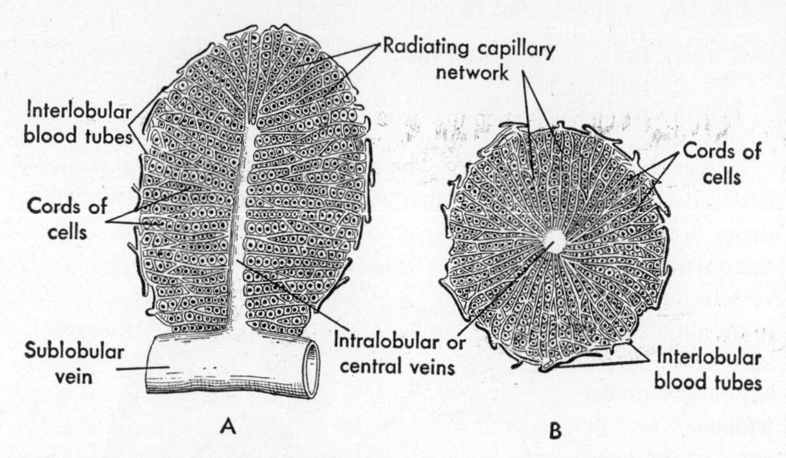

Fig. 309. Diagram of a hepatic lobule as seen in longitudinal section (*A*) and in transverse section (*B*).

Interlobular vessels bring blood in	Branches of hepatic artery Branches of portal tube
Intralobular vessels take blood out	Tributaries to hepatic vein

carried into the lobule by fine branches which converge toward the center. The walls of these small vessels are incomplete, so that the blood is brought in direct contact with each cell. These channels are termed *sinusoids,* and at the center of the lobule they empty the blood into the intralobular vein. The intralobular veins from a number of lobules empty into a much larger vein, upon whose surface a vast number of lobules rest; and therefore the name *sublobular* (under the lobule) is given to these veins. They empty into still larger veins, the *hepatic,* which converge to form three large trunks and empty into the *inferior vena cava,* which is embedded in the posterior surface of the gland.

The hepatic artery. The blood brought to the liver by the portal tube is venous blood; arterial blood is brought by the *hepatic artery.* It enters the liver with the portal tube, divides and subdivides in the same manner as the portal tube, thus forming another network between the lobules, and in the lobules between the cells. The capillaries from the portal tube and the hepatic artery empty into the intralobular vein near the center of each lobule. From here the blood from the hepatic artery and from the portal tube is returned by the hepatic veins, which empty into the inferior vena cava.

Lymphatics. There are a superficial and a deep set of lymphatic vessels. They begin in irregular spaces in the lobules, form networks around the lobules, and run always from the center outward.

The bile ducts. The surfaces of the hepatic cells are grooved, and the grooves on two adjacent cells fit together and form a passage into which the bile is poured as soon as it is formed by the cells. These passages form a network between and around the cells as intricate as the network of blood vessels. They are called *intercellular biliary passages* and radiate to the circumference of the lobule, where they empty into the interlobular bile ducts. These unite and form larger and larger ducts until two main ducts, one from the right and one from the left side of the liver, unite in the portal fossa and form the *hepatic duct.*

The hepatic duct passes downward and to the right for about 5 cm (2 in.) and then joins (at an acute angle) the duct from the gall-bladder, termed the *cystic duct.* The hepatic and cystic ducts together form the *common bile duct* (*ductus choledochus*), which passes downward for about 7.5 cm (3 in.) and enters the duodenum about 7.5 cm below the pylorus. This orifice usually serves as a common opening for both the common bile duct and the pancreatic duct. It is very small and is guarded by a sphincter muscle, which keeps it closed except during digestion.

The liver is invested in an outer capsule of fibrous tissue called *Glisson's*[15] *capsule.* This capsule is reflected inward at the transverse fossa and envelops the vessels and ducts which pass into the liver. With the exception of a few small areas, the liver is enclosed in a serous tunic derived from the peritoneum.

Functions. The liver has many functions. Among these are the formation and secretion of bile, and the transformation of glucose,

[15] Francis Glisson, English anatomist (1597–1677).

which is derived from the carbohydrates of our food, into glycogen, which is stored in the cells of the liver. When the body requires more glucose than food furnishes, enzymes of the liver cells reconvert the glycogen stored in the liver into glucose and admit it into the circulation. Iron and copper are stored in the liver, and the regulation of the concentration of each kind of amino acid in the blood is perhaps the most important function of this organ. This latter function is brought about by the conversion of amino acids into glucose and urea, urea being eliminated by the kidney and glucose disposed of as suggested previously. The liver also plays a part, along with the spleen, in disposing of the products resulting from disintegration of worn-out erythrocytes, and it gives rise to *heparin,* an anticoagulant of the blood. Heparin is found in other tissues (spleen, heart, thymus, muscle) but is present in large quantities in the liver.

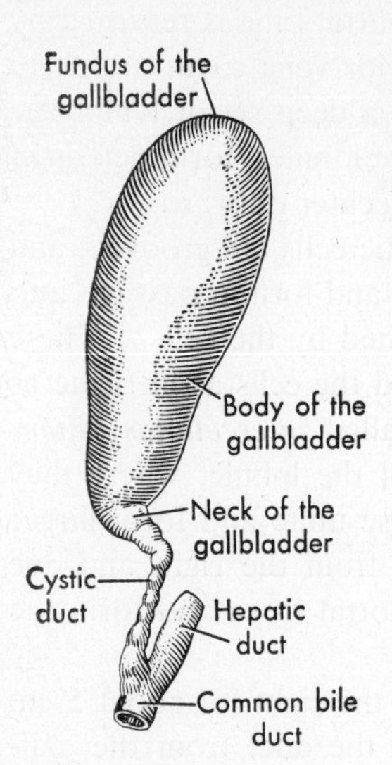

Fig. 310. The gallbladder, moderately distended, with the cystic duct, and the junction of the latter with the hepatic duct to form the common bile duct. (Toldt.)

Bile may be thought of as both an excretion and a secretion. Bile pigments as they are eliminated in the feces may be considered excretory products. As a secretion, bile aids in the digestion and absorption of fats.

The importance of the "protective" role of the liver cannot be overemphasized. It controls the concentration of the various kinds of amino acids in the blood and changes the waste products of protein metabolism into substances which may be eliminated by the kidneys. A large proportion of the amino acids ingested is subjected to this conversion into urea with subsequent removal by the kidneys. Similarly, in cooperation with the pancreas and the duodenum, it controls the concentration of carbohydrates in the blood. It secretes into the bile various poisonous substances derived from putrefaction in the intestine or from foods ingested.

The liver helps to regulate blood volume. It is believed to manufacture serum albumin and serum globulin and fibrinogen. It is the site of blood formation in the embryo. It produces heat and forms vitamin A from carotene. It stores the erythrocyte-maturing factor which is necessary for the complete development of the red blood cell. Water, iron, copper, and vitamins A, D, and the B complex are stored in the liver.

The gallbladder is a pear-shaped sac lodged in the gallbladder fossa on the under surface of the liver, where it is held in place by connective tissue. It is about 7 to 10 cm (3 to 4 in.) long, 2.5 cm (1 in.) wide, holds about 36 cu cm (9 dr), and is composed of three coats: (1) the inner one is mucous membrane; (2) the middle one is muscular and fibrous tissue; and (3) the outer one is serous membrane derived from the peritoneum. It is only occasionally that the peritoneum covers more than the under surface of the organ.

Function. The gallbladder serves as a reservoir for the bile. In the intervals between digestion, i.e., when the duodenum is empty, the sphincter guarding the bile duct is contracted, and the bile is held in the gallbladder. It is thought that the acid chyme entering the duodenum relaxes the sphincter, and the gallbladder contracts and forces out its contents, which pour into the duodenum.

Study of Figs. 290 to 310 and their legends makes a good review of the anatomy and histology of the organs concerned with digestion in preparation for the physiology to follow.

SUMMARY

Digestion. Digestion is dependent on the proper functioning of certain organs that are grouped together and called the digestive system

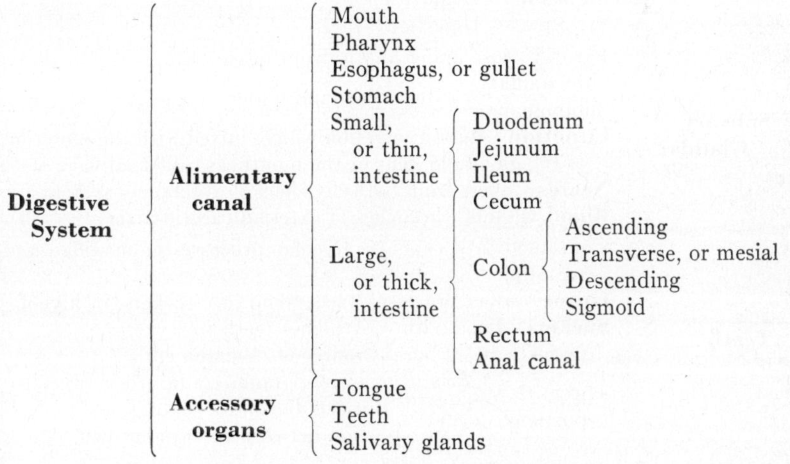

- **Digestive System**
  - **Alimentary canal**
    - Mouth
    - Pharynx
    - Esophagus, or gullet
    - Stomach
    - Small, or thin, intestine
      - Duodenum
      - Jejunum
      - Ileum
    - Large, or thick, intestine
      - Cecum
      - Colon
        - Ascending
        - Transverse, or mesial
        - Descending
        - Sigmoid
      - Rectum
      - Anal canal
  - **Accessory organs**
    - Tongue
    - Teeth
    - Salivary glands

- **Digestive System**
  - **Accessory organs**
    - Pancreas
    - Liver
    - Gallbladder

- **Alimentary Canal**
  - Continuous tube from mouth to anus
  - About 9 m (30 ft) long
  - Esophagus—four coats
    - Internal, or mucous
    - Submucous, or areolar
    - Muscular
    - Fibrous
  - From diaphragm to rectum—four coats
    - Mucous
    - Submucous, or areolar
    - Muscular
    - Peritoneum

- **Mouth, or Buccal, Cavity**
  - Roof—palate
    - (1) Hard palate { Maxillae, Palatine } processes
    - (2) Soft palate—uvula, palatine arches, and tonsils
  - Floor—tongue
  - Bounded laterally and in front by cheeks and lips
  - Behind it communicates with pharynx
  - Contains
    - Tonsils. Orifices of ducts of salivary glands
    - Tongue. Teeth

- **Tonsils**
  - Masses of lymphoid tissue occupy triangular space between palatine arches on either side of throat
  - **Function**
    - Similar to that of other lymph nodes
    - (1) Aid in formation of white cells
    - (2) Act as filters and protect body from infection

- **Tongue**
  - Special organ of sense of taste
  - Assists in
    - Mastication
    - Deglutition
    - Digestion

- **Salivary Glands**
  - Parotid—just under and in front of ear
  - Submaxillary, Sublingual } Below jaw and under tongue
  - **Function**—form a secretion which, mixed with the secretion of the glandular cells of the mouth, is called saliva
  - **Nerves**—fibers from both divisions of autonomic system
  - **Blood vessels**—branches of external carotid artery

- **Teeth**
  - Contained in sockets of alveolar processes of maxillae and mandible
  - **Gums**—cover processes and extend into sockets, or alveoli
  - **Sockets**—lined with periosteum
    - Attaches teeth to sockets
    - Source of nourishment
  - Three portions
    - *Root*—one or more fangs contained in alveolus
    - *Crown*—projects beyond level of gums
    - *Neck*—portion between root and crown

- **Teeth**
  - Composed of three substances developed from epithelium
    - *Dentin*
      - Gives shape. Encloses pulp cavity, which contains nerves and blood vessels that enter by canal from root
    - *Enamel*—caps crown
    - *Cement*—covers root
  - Two sets
    - (1) Deciduous—6 months–2 years: Incisors 8, Canines 4, Molars 8 = 20
    - Begin to develop about the sixth week of intrauterine life
    - (2) Permanent—$6\frac{1}{2}$ years–25 years of age: Incisors 8, Canines 4, Premolars 8, Molars 12 = 32
    - With the exception of the second and third molars the permanent teeth begin to develop about the sixteenth week of intrauterine life
  - **Function**—to assist in the process of mastication

- **Pharynx**
  - Muscular, membranous, cone-shaped tube between mouth and esophagus
  - Three parts
    - Nasal or nasopharynx—behind posterior nares above soft palate
    - Oral—extends from soft palate to hyoid bone
    - Laryngeal—extends from hyoid bone to esophagus
  - **Seven apertures**
    - 2 posterior nares
    - 2 auditory tubes
    - 1 fauces
    - 1 larynx
    - 1 esophagus
  - **Nerves**
    - Receives fibers from both divisions of autonomic system
  - **Blood vessels**—branches from external carotid artery
  - **Function**
    - Transmits air to larynx
    - Serves as a resonating cavity
    - Receives food and passes it to esophagus

- **Esophagus, or Gullet**
  - Tube—23–25 cm (9–10 in.) long. Extends from pharynx to cardiac end of stomach
  - **Four coats**
    - (1) Internal, or mucous
    - (2) Submucous, or areolar
    - (3) Muscular
      - Internal circular layer
      - External longitudinal layer
    - (4) External, or fibrous

Esophagus, or Gullet
- **Nerves**
  - Vagus
  - Thoracolumbar autonomic system
- **Blood vessels**
  - (1) Inferior thyroid branch of thyrocervical trunk
  - (2) Branches from thoracic aorta
  - (3) Left gastric branch of celiac artery
  - (4) Left phrenic branch of abdominal aorta
- **Function**—receives food and passes it on to stomach

Stomach or Gaster
- Dilated portion of canal, size and shape vary
- Oblique position in epigastric, umbilical, and left hypochondriac regions, under diaphragm
- **Openings**
  - Cardiac orifice—connects with esophagus
  - Pyloric orifice—connects with duodenum
  - Guarded by ringlike muscles known as sphincters
- **Curvatures**
  - Lesser curvature—concave border
  - Greater curvature—convex border
- **Parts**
  - Fundus—blind end above entrance of esophagus
  - Body—between fundic and pyloric portions
  - Pyloric portion—smaller end
- **Four coats**
  - (1) Outer—serous—peritoneum
  - (2) Muscular
    - (1) Longitudinal layer
    - (2) Circular layer
    - (3) Oblique layer chiefly at cardiac end
  - (3) Submucous—vascular
  - (4) Mucous-rugae
- **Glands**
  - Some cells of mucous membrane secrete secretin
  - Cardiac—secrete mucin
  - Fundus
    - Chief, or central, cells secrete pepsinogen; parietal, or oxyntic, cells secrete acid
  - Pyloric—secrete pepsinogen and mucin
- **Nerves**
  - Thoracolumbar autonomic nerves from celiac plexus
  - Vagus nerves
- **Blood vessels** from three divisions of celiac
  - Left gastric
  - Hepatic
  - Splenic
- **Function**
  - (1) Receives food in relatively large quantities about three times a day, holds it while it undergoes mechanical and chemical changes, then passes it on in small portions at frequent intervals
  - (2) Secretes mucin and gastric fluid

- **Small, or Thin, Intestine**
  - Convoluted tube extends from stomach to colic valve. About 7 m (23 ft) long, coiled up in abdominal cavity
  - **Three divisions**
    - Duodenum—about 25 cm (10 in.)
    - Jejunum—about 2.2 m ($7\frac{1}{2}$ ft)
    - Ileum—about 4 m ($14\frac{1}{2}$ ft)
  - **Four coats**
    - (1) Serous from peritoneum
    - (2) Muscular
      - Longitudinal layer
      - Circular layer
    - (3) Submucous—connects muscular and mucous coats
    - (4) Mucous
      - Circular folds
      - Villi—contain lacteals
  - **Glands and nodes**
    - Intestinal glands; Duodenal, or Brunner's — Secrete intestinal fluid
    - Lymph nodules
      - Solitary
      - Aggregated lymph nodules called Peyer's patches
  - **Nerves**
    - Vagi supply sensory and motor fibers
    - Thoracolumbar autonomic nerves derived from plexuses around superior mesenteric artery
  - **Blood vessels**
    - Branches of hepatic
    - Branches of superior mesenteric — Distribute arched branches which lie between serous and muscular coats
  - **Function**
    - Digestion
    - Receives bile from liver, pancreatic fluid from pancreas
    - Secretion of succus entericus
    - Absorption
- **Large, or Thick, Intestine**
  - Largeness in width, not in length
  - Length, 1.5 m (5 ft); width, 6.3 cm ($2\frac{1}{2}$ in.) to 3.5 cm ($1\frac{1}{2}$ in.)
  - Extends from ileum to anus
  - **Four parts**
    - Cecum, with vermiform appendix
    - Colon
      - Ascending
      - Transverse, or mesial
      - Descending
      - Sigmoid
    - Rectum—about 12 cm (5 in.)
    - Anal canal—3.5 cm ($1–1\frac{1}{2}$ in.); Anus
      - Internal sphincter
      - External sphincter
  - **Four coats**
    - (1) Serous, except that in some parts it is only a partial covering, and at rectum it is lacking
    - (2) Muscular
      - Longitudinal layer — Arranged in three ribbonlike bands that begin at appendix and extend to rectum
      - Circular layer

Large, or Thick, Intestine
- **Four coats**
  - (3) Submucous
  - (4) Mucous
    - No villi
    - No circular folds
    - Numerous
      - Intestinal glands
      - Solitary lymph nodules
- **Nerves**—fibers from both divisions of autonomic nervous system
- **Blood vessels**
  - Superior mesenteric supplies cecum, ascending and transverse colon
  - Inferior mesenteric supplies descending colon and rectum. Rectum also receives branches from hypogastric arteries
- **Function**
  - Continuance of digestion and absorption
  - Elimination of waste

Pancreas
- In front of first and second lumbar vertebrae, behind stomach
- **Hammer shape**
  - Head attached to duodenum
  - Body in front of vertebra
  - Tail reaches to spleen
- **Size**
  - About 12.5 cm (5 in.) long
  - About 5 cm (2 in.) wide
- **Average weight**—60–90 gm (2–3 oz)
- **Structure**
  - Compound gland—each lobule consists of one of the branches of main duct, which terminates in cluster of pouches, or alveoli
  - Lobules held together by connective tissue form lobes
  - Lobes form gland
  - Duct from each lobule empties into pancreatic duct, also called duct of Wirsung
  - Scattered throughout pancreas are islets of Langerhans
- **Function**
  - (1) Secretes pancreatic fluid—digestive fluid
  - (2) Forms an internal secretion—aids in oxidation of glucose

Liver
- Largest gland in body
- Location
  - Right hypochondriac
  - Epigastric
  - Left hypochondriac
- Convex above—fits under diaphragm
- Concave below—fits over right kidney, ascending colon, and pyloric end of stomach
- **Five ligaments**
  - (1) Falciform
  - (2) Coronary
  - (3) Right lateral
  - (4) Left lateral
  - (1)–(4): Formed by folds of peritoneum

Liver

- **Five ligaments**
  - (5) Round ligament — Results from atrophy of umbilical vein
- **Four fossae**
  - (1) Left sagittal fossa
  - (2) Portal, or transverse, fossa transmits
    - Portal tube
    - Hepatic artery
    - Hepatic duct
    - Lymphatics
    - Nerves
  - (3) Gallbladder fossa
  - (4) Fossa for inferior vena cava
- **Four lobes**
  - (1) Right (largest lobe)
  - (2) Left (smaller and wedge-shaped)
  - (3) Quadrate (square)
  - (4) Caudate (tail-like)
- **Five sets of vessels**
  - (1) Branches of portal tube
  - (2) Branches of hepatic artery
  - (3) Hepatic veins
  - (4) Lymphatics
  - (5) Bile ducts
- **Nerves**—derived from left vagus and thoracolumbar autonomic system
- **Blood vessels**
  - Hepatic artery
  - Portal tube
  - Hepatic veins
- **Anatomy of liver**
  - Cords of hepatic cells are grouped in lobules
  - Lobules 1.0 to 2.5 mm in diameter
  - Branches of portal tube
    - Interlobular veins (between lobules)
    - Intralobular capillaries (within lobules)
    - Sublobular veins (under lobules)
    - Hepatic veins—exit at portal fossa, empty into inferior vena cava
  - Bile ducts
    - Channels between cells form intercellular biliary passages
    - Interlobular ducts
    - Hepatic duct—exit at portal fossa
  - Branches of hepatic artery
    - Interlobular arteries (between lobules)
    - Intralobular capillaries (within lobules)
    - Course beyond intralobular capillaries same as that pursued by blood from portal tube
  - Lymphatics
    - Start in lobules, form network, and run from center to periphery
  - Glisson's capsule invests liver

- **Liver**
  - **Anatomy of liver**
    - Serous membrane from peritoneum almost completely covers it
  - **Function**
    - Bile production
    - Forms an internal secretion which regulates the changing of glucose to glycogen and the reverse of this reaction
    - Gives rise to heparin, or antithrombase
    - Proteins changed into substance that can be eliminated, urea, etc.
    - Finishes disintegration of erythrocytes
    - Secretes into the bile various poisonous substances
    - Regulation of blood volume
    - Manufacture of serum albumin
    - Manufacture of serum globulin
    - Manufacture of fibrinogen
    - Forms blood in embryo
    - Produces heat
    - Forms vitamin A from carotene
    - Storage of iron and copper
    - Agent of disease resistance
    - Doubtless many other functions in relation to the constancy of the internal environment
- **Gallbladder**
  - Pear-shaped sac lodged in gallbladder fossa on under surface of liver
  - **Size**
    - 3–4 in. long
    - 1 in. wide
    - Capacity about 36 cu cm
  - **Three coats**
    - (1) Mucous membrane
    - (2) Fibrous and muscular tissue
    - (3) Serous membrane from peritoneum
  - **Function**
    - Serves as reservoir for bile in intervals between digestion
    - During digestion—pours bile into duodenum

CHAPTER 18

FOOD, VITAMINS, THE PHYSIOLOGY OF DIGESTION

Natural nutrients are in general nondiffusible and held in cells. The cells of plants and animals constitute a natural food supply for man. These foods are taken periodically, digested, absorbed and stored, in general, in nondiffusible form in cells and redigested and delivered to the cells by the circulatory fluids for use. The many processed foods (sugar, flour, tomato juice, meat extracts, salad dressings, etc.) bring about changed dietary habits necessitating varied and detailed studies of food essentials as a basis for healthful dietary habits under artificial conditions.

FOOD

Food is any substance taken into the body to yield energy, to build tissue, and to regulate body processes.

All the body activities require a certain amount of energy; this energy is supplied by food. The energy released in cells during the interaction of oxygen and food is present in the form of potential, or latent, energy, binding the atoms into molecules and the molecules into larger masses. The splitting of these complex molecules into smaller and simpler ones releases this energy as kinetic energy. Food material, over and above what is needed for this purpose, is stored in the body either in the form of glycogen or as fat. This may be regarded as reserve fuel which, when needed, is oxidized to release energy.

Food supplies material for the manufacture of protoplasm, either for growth (increase in the bulk of protoplasm) or repair (replacing the protoplasm incidentally oxidized day by day).

Nutrition and growth are dependent upon certain essential substances called vitamins. Water and inorganic salts are necessary to maintain the normal composition of the tissues.

Classification of food. Chemical analysis shows that the chemical elements found in the body are also found in food. Various combinations of these elements give a great variety of substances which are grouped as shown on p. 35.

Nutrients, or Food Principles		
	Water	Mineral salts
	Carbohydrates	Vitamins
	Fats	
	Proteins	

Water constitutes more than two-thirds of the material ingested daily.

The water content of the body comes from three sources: beverages or other liquids; foods, especially vegetables and fruits; and the water formed in the tissues as the result of metabolic activities. See Chap. 20.

Carbohydrates, or **glucides,** are the most abundant and most economical sources of energy. They contain carbon, hydrogen, and oxygen, in the proportion of 2 atoms of hydrogen to 1 of oxygen.[1] All simple sugars and all substances which can be converted into simple sugars by hydrolysis are carbohydrates. The names of these compounds suggest the number of simple sugar groups they will yield on hydrolysis (p. 577): monosaccharides, disaccharides, and polysaccharides.

1. *Monosaccharides,* or simple sugars, contain one sugar group, $C_6H_{12}O_6$. They are soluble and can be absorbed into the body fluids without further change. They are the units from which all the other carbohydrates are formed.

Monosaccharides		
	Glucose, or dextrose, found in fruits, especially the grape, and in body fluids	$C_6H_{12}O_6$
	Fructose, or levulose, found with glucose in fruits	$C_6H_{12}O_6$
	Galactose, obtained by hydrolysis of lactose and certain gums	$C_6H_{12}O_6$

[1] In the carbohydrate rhamnose this proportion does not occur. There are also a number of compounds which are not carbohydrates, although they do contain carbon, hydrogen, and oxygen, the two latter in the proportion of 2 to 1.

2. *Disaccharides.* The formula, $C_{12}H_{22}O_{11}$, shows that disaccharides consist of two monosaccharide groups. During the process of digestion, they are changed to monosaccharides, either glucose or invert sugar, which consists of a molecule each of glucose and fructose. Only one splitting is necessary, as one molecule of a disaccharide plus one molecule of water will form one molecule of glucose and one of fructose.

Disaccharides	Sucrose, or cane sugar, found in vegetables, fruits, and juices of many plants	$C_{12}H_{22}O_{11}$
	Lactose, or milk sugar, found in the milk of all mammals	$C_{12}H_{22}O_{11}$
	Maltose is an intermediate product in the digestion of starch, found in the body, in germinating cereals, malts, and malt products	$C_{12}H_{22}O_{11}$

$$\underset{\text{Sucrose}}{C_{12}H_{22}O_{11}} + \underset{\text{Water}}{H_2O} \rightarrow \underbrace{\underset{\text{Glucose}}{C_6H_{12}O_6} + \underset{\text{Fructose}}{C_6H_{12}O_6}}_{\text{Invert Sugar}}$$

3. *Polysaccharides* are represented by the molecular formula $(C_6H_{10}O_5)_n$. The elements are present in the same proportion, but the value of *n* may be large and is probably different for the different polysaccharides. For instance, the value of *n* for the starch molecule is said to be between 24 and 30, representing 24 to 30 sugar groups, whereas for the dextrin molecule it is smaller, so that a single molecule of starch when hydrolyzed produces several molecules of dextrin of the same relative composition. Since the polysaccharides are complex, they must pass through several hydrolyses before they are changed to simple sugars. Each splitting of the molecule gives substances with simpler composition, though with the same relative proportion of the constituents, and to each is given a special name. The number of molecules of simple sugar resulting from the hydrolysis of any polysaccharide would depend upon the value of the *n*.[2]

A summary of the hydrolysis of starch to glucose may be expressed as follows.

$$\underset{\text{Starch}}{(C_6H_{10}O_5)_n} + \frac{n}{2}H_2O \rightarrow \underset{\text{Maltose}}{\frac{n}{2}C_{12}H_{22}O_{11}}$$

[2] Dr. Albert P. Mathews, following Haworth's work, assigns to *n* the following values. In starches, $n = 30–24$; in erythrodextrins, 24–18; in achroödextrins, 12–6.

The maltose molecules are, however, formed *successively* as represented below:

$$
\begin{array}{c}
\text{starch} + H_2O \rightarrow \text{maltose} \\
\downarrow \\
\text{soluble starch} + H_2O \rightarrow \text{maltose} \\
\downarrow \\
\text{erythrodextrin} + H_2O \rightarrow \text{maltose} \\
\downarrow \\
\text{achroödextrin } \alpha + H_2O \rightarrow \text{maltose} \\
\downarrow \\
\text{achroödextrin } \beta + H_2O \rightarrow \text{maltose} \\
\downarrow \\
\text{maltose}
\end{array}
$$

Polysaccharides

Polysaccharides	Starch—found in grain, tubers, roots, etc.	$(C_6H_{10}O_5)_n$
	Cellulose—outside covering of starch grains and basis of all woody fibers	$(C_6H_{10}O_5)_n$
	Glycogen—form in which sugar is stored in the liver and muscles, etc.	$(C_6H_{10}O_5)_n$
	Dextrin—formed from starch by partial hydrolysis	$(C_6H_{10}O_5)_n$

Starch is the principal form in which carbohydrate is stored in plants. Three-fourths of the solids of ripe potatoes and one-half to three-fourths of the solid matter of the cereal grains are starch. During the ripening process in some plants (e.g., apple and banana) starch is changed to glucose; in other plants (e.g., corn and peas) the opposite process occurs.

Cellulose constitutes the supporting tissue of plant cells. When derived from mature plants, cellulose is quite resistant to the action of dilute acids or digestive enzymes and passes through the digestive tract unchanged. The chief value of cellulose in human nutrition is to give bulk to the intestinal contents and thereby facilitate peristalsis.

Glycogen is the form in which reserve carbohydrate is stored in the animal body, in greatest quantity in the liver and muscles.

Dextrins are formed from starch by the action of enzymes, acids, or heat.

Fats, or lipids. The word *fat* is sometimes used in an anatomic sense and sometimes in a chemical sense. In an anatomic sense, fat denotes adipose tissue. In a chemical sense, fats are glyceryl esters of fatty acids. In other words, fats are derived from the reaction of three molecules of fatty acid and one molecule of glycerin. The ordinary fats of animal and vegetable food are not simple substances but are mixtures of simple fats named palmitin, stearin, olein, etc., which are derived from the fatty acids—palmitic, stearic, and oleic acid.

Under the influence of certain enzymes found in the body, fats are split by hydrolysis into the substances of which they are built, i.e., glycerin and fatty acids.

$$\underset{\text{(Stearin)}}{C_3H_5(C_{18}H_{35}O_2)_3} + \underset{\text{(Water)}}{3H_2O} \rightarrow \underset{\text{(Glycerin)}}{C_3H_5(OH)_3} + \underset{\text{(Stearic Acid)}}{3H \cdot C_{18}H_{35}O_2}$$

Fats are compounds of carbon, hydrogen, and oxygen, the same elements found in carbohydrates; but these elements are present in different combinations and proportions. The fats contain proportionately less oxygen and more carbon and hydrogen than the carbohydrates; consequently they make a more concentrated form of fuel.

Lipoids, or compound fats, are esters of fatty acids containing groups in addition to an alcohol and a fatty acid. Examples of compound fats are:

1. *Phospholipids,* or *phosphatides,* which contain both phosphorus and nitrogen. The best known are the lecithins, which are abundant in egg yolk and occur in brain and nerve tissue and in all the cells of the body. Cephalins and sphingomyelin are other examples.

2. *Glycolipids,* or *cerebrosides,* are compounds of fatty acids with a carbohydrate and contain nitrogen but no phosphoric acid. Cerebrosides are found in the myelin sheaths of nerve fibers in connection with, possibly in combination with, lecithin.

3. *Sterols* are solid alcohols, which are found in nature combined with fatty acids. They contain carbon, hydrogen, and oxygen. The best known is cholesterol, or cholesterin ($C_{27}H_{45}OH$), which is very widely distributed in the body, being found in the medullary coverings of nerve fibers, in the blood, in all the cells and liquids of the body, in the sebum secreted by the sebaceous glands of the skin, and in the bile. In the blood cholesterol protects the erythrocytes against the action of hemolytic substances, and in the sebum it protects the skin. Under the influence of ultraviolet rays (direct sunlight or from a mercury-vapor quartz lamp), cholesterol may be so changed as to acquire the property of an antirachitic vitamin. It is thought that the cholesterol of the bile is a waste product.

Proteins are more complex than either carbohydrates or fats and differ from them in containing *nitrogen.* Proteins always contain carbon, hydrogen, oxygen, and nitrogen; sometimes sulphur, phosphorus, or iron is present. Proteins are built up of simpler substances called *amino acids.* Amino acids are acids that contain an amino group (NH_2) instead of a replaceable hydrogen atom. In acetic acid, which is one of the simplest organic acids, the formula is

CH_3COOH. If one of the three hydrogen atoms in the CH_3 group is replaced by NH_2, a substance results which has the formula $CH_2(NH_2)COOH$ and is called aminoacetic acid or *glycine* or glycocoll. Another organic acid is propionic acid, which has the formula C_2H_5COOH; if an atom of hydrogen is replaced by the amino group, $C_2H_4(NH_2)COOH$, which is aminopropionic acid, or *alanine,* results.

Some 40 or more amino acids have been described as occurring in nature and many more have been synthesized. However, only 23 amino acids have been unequivocally accepted as common "building stones" of the proteins. No one protein contains all of them, but caseinogen of milk yields 17 or more. The proteins of one animal differ from those of another; even the proteins of different tissues are not identical. This is also true of the proteins of plants. The proteins of milk, fish, egg, cereal, and vegetables represent different combinations of amino acids and are represented by different formulas. Among vegetables the legumes—peas, beans, lentils, and peanuts—are especially rich in proteins.

Examples of protein constituents of food:

- *Albumin*, the white substance seen when egg is heated, the scum that forms on the top of milk when its temperature is raised above 76° C (170° F), the white coating that forms on meat when it has been in a hot oven for a short time
- *Casein*, the substance that is formed into a curd when acid or rennin is added to milk or when milk sours
- *Glutenin*, the gummy substance in wheat
- *Legumin*, a protein substance contained in the legumes
- *Gelatin*, from intercellular substance of connective tissues, including bones and tendons
- *Organic extracts*, protein substances formed in animals and plants as a result of their metabolism. The flavor of meats and some plant foods is due to extractives

AMINO ACIDS THAT HAVE BEEN FOUND IN THE FOOD PROTEINS OF MAN*

Glycine	Tyrosine	Hydroxyglutamic acid
Alanine	Serine	*Arginine*
Valine	*Threonine*	*Lysine*
Leucine	Cystine	*Histidine*
Isoleucine	*Methionine*	Proline
Norleucine	Aspartic acid	Hydroxyproline
Phenylalanine	Glutamic acid	*Tryptophane*

* Those which man cannot make and which, therefore, must be taken in as food are italicized.

This table shows the relative quantities of amino acids found in casein of milk, in gelatin, and in gliadin of wheat.[3]

Figs. in %	G	Al	*V*	*Le*	*Ph*	Ty	S	C	As	Gl	H.g	*Ar*	*Ly*	*Hi*	Pr	H.p	*Tr*
Casein	0.5 %	1.85	6.7	9.7	3.9	6.7	0.50	0.34	4.1	21.8	10.5	3.8	6.3	1.83	8.0	0.23	2.2 %
Gelatin	25.5	8.7	1.0	7.1	1.4	0.0	0.4	0.31	3.4	5.8	0.0	8.2	5.9	0.9	9.5	14.1	0.0
Gliadin	0.4 %	2.0	3.3	6.6	2.4	3.5	0.13	2.1	0.80	43.7	2.4	3.1	0.92	3.4	13.2	?	1.14 %

The following shows a few of the many possible combinations of carbon, hydrogen, oxygen, and nitrogen to form amino acids.

Glycine $CH_2(NH_2)\cdot COOH$
Lysine $CH_2(NH_2)\cdot CH_2\cdot CH_2\cdot CH_2\cdot CH(NH_2)\cdot COOH$
Aspartic acid $COOH\cdot CH_2\cdot CH(NH_2)\cdot COOH$
Tyrosine $CH_2\cdot CH(NH_2)\cdot COOH$

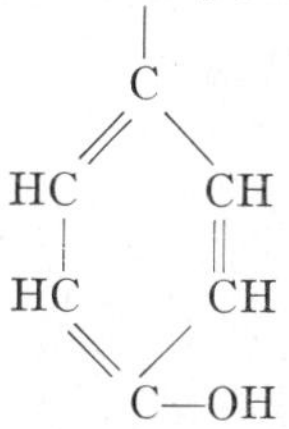

```
             |
             C                          Proline  H2C———CH2
           //  \                                  |       |
        HC      CH                               H2C     CH·COOH
        |       ||                                  \   /
        HC      CH                                   NH
          \\   /
            C—OH
```

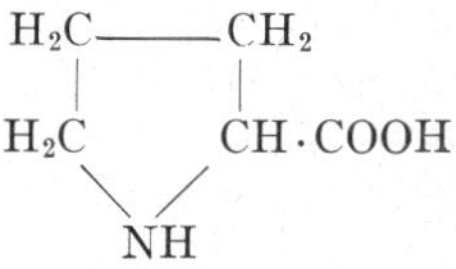

```
Tryptophane CH
          //   \
        HC      C———C·CH2·CH(NH2)·COOH
        |       ||    ||
        HC      C     CH
          \\   / \   /
           CH     NH
```

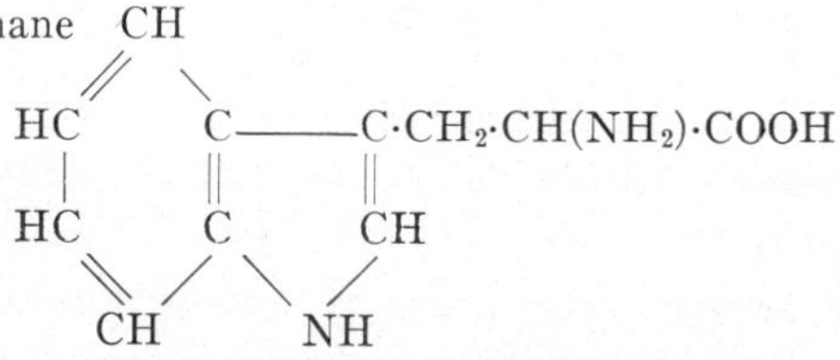

Methionine $CH_3\cdot S\cdot CH_2\cdot CH_2\cdot CH(NH_2)\cdot COOH$

Classification. Proteins are classified in three main groups:

1. Simple proteins when hydrolyzed yield only amino acids or their derivatives. Examples are serum albumin, globulin, albuminoids, etc.

2. Conjugated proteins are substances which contain a simple protein molecule united to some other nonprotein molecule which is usually acid in nature. On hydrolysis they yield amino acids and the other molecule. This molecule is nuclein in the nucleoproteins, a carbohydrate in the glycoproteins, a phospho-body in the phosphoproteins, pigment in the chromoproteins and hemoglobins, and a fatty substance in lecithoproteins.

[3] Figures from H. C. Sherman.

3. Derived proteins are classified into two groups: (*a*) primary protein derivatives, and (*b*) secondary protein derivatives.

a) Primary protein derivatives are derivatives of the protein molecule, apparently formed through hydrolytic changes, which involve only slight alterations of the protein molecule. Examples are fibrin and casein. Each splitting of a protein molecule gives substances with similar composition, and each one of these is given a special name, such as acid albumins, albuminates, etc.

b) Secondary protein derivatives are products of further hydrolytic change. Good examples are proteoses, peptones, and peptides.

Hydrolysis of proteins

proteins + water → proteoses
proteoses + water → peptones
peptones + water → peptides
peptides + water → amino acids

Specific catalysts are pepsin, trypsin, and peptidases

The mineral elements which enter into the composition of the body are listed on p. 35. Since each element enters into the metabolism of body cells, a constant supply of each is necessary to meet the daily loss. These elements are supplied in food. On analysis many of them are classified as ash constituents, since they remain after incineration of the food. These "ash constituents" may function in the body in several ways: as constituents of bone, giving rigidity to the skeleton; as essential elements of all protoplasm; as soluble constituents of the fluids of the body influencing the elasticity and irritability of the muscles and nerves, supplying material for the acidity or alkalinity of the digestive fluids, helping to maintain the acid-base equilibrium of the body fluids as well as their osmotic characteristics and solvent power; and probably in many other ways.

The importance of the optimum concentration of each of these mineral salts in the tissues and fluids of the body is so great that any considerable change from the normal endangers life.

Calcium is a constituent of all protoplasm and of body fluids, and is present in large proportions in bone and teeth. Ninety-nine per cent of body calcium is in bones. Calcium is essential for all cellular activities. It is related to the membrane action of cells, to excitability of muscle, etc., and to normal heart action, and must be present in ionic form for normal blood clotting. Children and pregnant and lactating animals require large amounts. Milk is the best source of calcium, but calcium is also present in leafy vegetables. The body cannot readily adapt itself to calcium shortage; therefore a liberal amount is needed daily. It is estimated that an intake of 0.8 gm of calcium would maintain the body fluids at optimum concentration. Calcium deficiency is a definite problem in the diet of Americans.

A growing child requires at least 1 gm each of calcium and phosphorus per day. A pregnant woman requires at least 1.5 gm of calcium per day during the latter half of pregnancy. This amount should be increased to 2 gm during lactation.

Phosphorus is essential to the normal development of bones and teeth. It is also concerned with maintaining acid-base balance and enzyme systems and their functioning. Good sources are milk, cheese, meat, and eggs.

Iron is found not only in the erythrocytes of the blood but in all cells of the body. It serves to carry oxygen and to activate cell functions. Storage of iron in the body is limited; therefore foods containing it should be included in the daily diet. These foods are whole grains, egg yolk, beef (especially beef liver), fruits, and green vegetables.

When born, a baby has a special store of iron in its body, which serves during the period of lactation. During the latter half of the first year, egg yolk and iron-bearing vegetables should be added gradually to the diet, so that as the reserve iron is used up, fresh supplies will be available. In premature infants this special store of iron is absent; preparations of iron and copper are added to the milk.

Iodine is utilized by the thyroid gland in the preparation of thyroxin. To maintain the body store of iodine and meet the loss in metabolism, it is estimated that a normal adult requires about 0.000014 gm of iodine daily. Regions in which the supply of iodine is insufficient in water report good results in reducing the incidence of goiter by administration of iodine. Milk, leafy vegetables, and fruits grown in nongoitrous regions, fresh and canned salmon, cod, halibut, haddock, lobsters, and oysters are sources of iodine.

Copper is a factor in hemoglobin formation, though it is not a part of the molecule. It is considered an essential element in nutrition, though its role in physiology is not known. It has a wide distribution in foods; good sources are liver, nuts, legumes, fruits, and leafy vegetables.

A diet which furnishes sufficient energy and protein may be lacking in calcium, phosphorus, iron, iodine, and copper unless vegetables and fruits are added in sufficient quantities to prevent this deficiency. The amount of calcium and phosphorus needed is relatively large, and definite provision must be made for it. The amount of iron needed is minute; but since the quantities in food materials are also minute, the sources of supply must be considered in planning a dietary. If the requirements for calcium and iron are met, it is probable that all other minerals will also be supplied in the same foods. Sherman[4] suggests as an average minimum requirement for equilibrium 0.45 gm of calcium, 0.88 gm of phosphorus, and 0.008 gm of iron per person per day. Setting the standard allowance 50 per cent above this indicated average minimum, the allowance is calcium 0.8 gm, phosphorus 1.5 gm, and iron 0.020 gm.

See Chap. 20 for the minerals concerned with water and electrolyte balance.

Vitamins are organic food substances essential for growth and normal metabolism. There is no generally accepted theory as to the way in which they influence nutrition. They are considered to be essential to all cells as factors in oxidative reactions, probably as the active groups of tissue enzymes. Several vitamins (B, C, G) are known to be factors in tissue respiration, and it has been suggested that all of them may serve as respiratory catalysts. Recent experiments[5] emphasize the determination of optimum (as distinguished from merely adequate) amounts of the vitamins and their physiological effects.

THE FAT-SOLUBLE VITAMINS

Vitamin A (antixerophthalmic) and its provitamins, or precursory substances, called alpha-, beta-, and gamma-carotenes and cryptoxanthin, constitute a nutritional factor[6] essential to growth and to

[4] H. C. Sherman, *Chemistry of Food and Nutrition.*

[5] The laboratory rat resembles man in his environment, in feeding habits, and in the chemical nature of his nutritive processes. The cycle of development takes place about 30 times as rapidly in the rat as in man, which makes it possible to observe the entire life of the animal within a comparatively brief period. For these and other reasons, the rat can be used for nutrition experiments, the findings of which are to be applied to human problems.

[6] A *nutritional factor* is a single substance or any group of substances performing a specific function in nutrition.

the efficiency of the general nutritional processes at all ages. The precursory substances occur in yellow pigments of plants such as paprika, carrots, pumpkins, and sweet potatoes, etc., and in the green parts of plants. These plants contain no vitamin A but do contain the precursory substances which the body can make into vitamin A, chiefly in the liver. Hence, as Sherman says, they add to the vitamin-A value of the food but not to its vitamin content. Fish-liver oils contain vitamin A as well as the precursory substances; hence, they add to the vitamin content of the foods as well as to its vitamin value. The probable formula of vitamin A is $C_{20}H_{30}O$. Deficiency results in disturbances associated with nervous and epithelial tissues: (1) failure to gain weight, (2) susceptibility to xerophthalmia, an eye condition conducive to subsequent infection and resulting blindness if unchecked, (3) a dermatosis or dry skin, (4) generally impaired epithelial tissues and resulting increased susceptibility to infections of the lungs, skin, bladder, sinuses, ears, and alimentary tract, and (5) night blindness, which results from cornification of the retina. In night blindness, it is thought, when light acts on the rods of the retina, visual purple, which contains vitamin A, is broken down. In vitamin-A deficiency the blood brings an inadequate new supply of vitamin A for reconstruction of the visual purple, and night blindness occurs. Eating liver has long been known as a cure. It is believed kidney stones are associated with absence or deficiency of vitamin A. Important sources of the vitamin are milk, butter, eggs, cream cheese, green and yellow vegetables, liver oils of halibut, cod, and other fish, and liver and glandular organs in general. Carotenes can be changed to vitamin A in the liver and stored as vitamin A. This is of importance in adult nutrition. Recent clinical observations indicate a need for selection of foods to ensure a more liberal intake of vitamin A because of its imperative need by the body in health and in resistance to disease.

Vitamin D (antirachitic) is essential for the maintenance of an adequate concentration of calcium and phosphorus in the body fluids and their deposition in bone. It is considered a factor in maintaining the normal health of the respiratory system, in the formation of normal teeth, and in protection against dental caries. The crystalline form which has been isolated is *calciferol*, $C_{28}H_{44}O$. It is recognized as probably being of multiple nature. Deficiency in vitamin D has long been known to result in rickets, which may be cured by administration of vitamin D, by direct sunlight, by ultraviolet irradiation of the body, or by administering ergosterol or similar substances produced by irradiation. The effect of ultraviolet irradiation of the skin is through its transformation of provitamin D of skin-gland secretions into vitamin D, which is absorbed by the skin. The effective rays are those that cause tanning. The redness which precedes sunburn indicates too great exposure. Excellent sources of vitamin D are cod- (and other fish) liver oil, egg yolk, whole milk, and butter fats. It is relatively stable in ordinary cooking and also at autoclave temperatures. Since an ordinary diet is commonly somewhat deficient in this vitamin, children should be given it in concentrated form.

Vitamin E (antisterility) prevents sterility in both male and female rats. The probable formula is $C_{29}H_{50}O_2$. It is fat soluble; good sources of it are oil of wheat germ, cereals, green vegetables, muscle, milk, and butter. Its wide distribution in foods leads students to believe that it has no significant relationship to human sterility.

Vitamin K possesses antihemorrhagic properties. It is considered a factor essential to normal clotting of blood. The blood of animals having a deficiency of this vitamin shows a lowered content of prothrombin and a delayed clotting time. Studies also show that bile salts are necessary for the absorption of vitamin K from the intestine. The blood of patients with certain types of diseases of the biliary tract is low in prothrombin and has a delayed clotting time. Vitamin K is fat soluble and appears to be found in a great variety of foods, but knowledge regarding quantitative distribution of this vitamin is still limited. The newborn infant may have an alimentary deficiency, since the vitamin is not readily passed from mother to fetus; hence the need for giving vitamin K to the newborn and to mothers before delivery.

THE WATER-SOLUBLE VITAMINS

Vitamin B is of multiple nature and is usually referred to as "the vitamin-B complex." Three factors (B, or B_1; G, or B_2; and B_6) demand attention; the possibility of several additional components is recognized. Vitamin B_1 or thiamin is present in large quantities in the wheat germ and in bran; but milling, polishing, refining, etc., usually eliminate it from such foods as rice, white flour, hominy, and corn meal. Other important sources of it are tomatoes, eggs, green vegetables, and yeast. The probable formula is $C_{12}H_{18}N_4SCl_2O$. Deficiency in vitamin B is known to cause symptoms of beriberi, or polyneuritis, and imperfect growth. It is essential to normal cell respiration, and an adequate supply of it is necessary for maintenance or normal appetite, normal metabolism of carbohydrates, and the preservation of normal motility of the digestive tract. Sherman believes an adequate supply of vitamin B—and of many other important nutritional factors as well—will be provided if half of the needed food calories are taken as milk, eggs, fruit, and vegetables and half of the breadstuffs and cereals used are whole-grain or "dark" forms.

Beriberi occurs chiefly among Oriental nations that make great use of rice as food. The disease takes a variety of forms, but the symptoms are gastrointestinal disturbances, paralysis, and atrophy of the limbs. This condition is caused by limiting the diet to polished rice. If the polishings are restored to the diet, the condition disappears; or if meat or barley is used with the polished rice, the condition is avoided.

Vitamin G (B_2), or *riboflavin*, is somewhat more heat-stable than is B_1. The probable formula is $C_{17}H_{20}N_4O_6$. It was first isolated from milk and named *lactoflavin* and is frequently referred to as the "flavin factor." It has a wide distribution; the best sources of it are milk, yeast, liver, eggs, fish, and green vegetables. It is a factor in tissue respiration and is considered essential to normal growth and nutrition at all ages. Pellagra was formerly considered a simple vitamin-G deficiency disease but is now thought to be caused by deficiency of more than one vitamin and possibly of protein also. *Niacin* or nicotinic acid, which has been called the pellagra-preventive (P-P) factor, is also essential in the diet for the prevention of pellagra. Since 1945 pellagra has rapidly declined in the United States. Its conquest has been made possible through preventive measures, and in 1952 it was considered practically extinct.[7] Sprue is another disease in which vitamin-G deficiency may be a factor of at least secondary nature.

Pyridoxine (B_6), the chemical formula of which is $C_8H_{11}O_3N$, appears to be essential to normal growth and nutrition. Deficiency produces a dermatitis. It shows a distribution in foods somewhat similar to that of vitamin G, with a few exceptions such as egg

[7] Anna May Wilson, "The Conquest of Pellagra," *Today's Health*, **30**:40, 1952.

white, which supplies G. Meat, cereals, fish, and milk are potent sources of it.

Ascorbic acid (vitamin C) has long been known to be essential in the prevention of scurvy. Early records of sea voyages reveal many epidemics of scurvy, and it was reported from Austria and Russia during World War I. The cause is lack of fresh meat and vegetables; the prevention is in the use of these. Early cures were through the use of citrus fruit juices. Laboratory experiments on animals and men prove conclusively that scurvy is due to lack of vitamin C in the diet. More recently vitamin C has been shown to be of importance in tissue respiration, which is decreased in scurvy. The probable formula is $C_6H_8O_6$; the commercial preparation is *cevitamic acid.* Shortage of vitamin C is shown to reduce general nutrition, to prevent healing of experimental bone wounds, and to be a contributory factor in capillary fragility and the general resistance of the body. Therefore, a liberal daily intake of vitamin C throughout life is recommended. The relationship of deficiency to dental caries is undetermined. However, Sherman says "the soundness both of the teeth themselves and their supporting bones and gums is importantly dependent upon the amount of vitamin C supplied by the food." The citrus fruits (oranges, lemons, grapefruit), tomatoes (raw or canned), and raw cabbage are rich sources of it. All fresh vegetables are sources. It is the least resistant of the vitamins; heating and drying usually destroy it. Destruction is inhibited in the absence of oxygen or in the presence of acid.

Symptoms of scurvy are loss of weight, pallor, weakness, breathlessness, palpitation of the heart, swelling of the gums, loosening of the teeth, pains in the bones and joints, edema, nervousness, and slight hemorrhages appearing as red spots under the skin and forming hidden bleeding places in the muscles and internal organs. The heart hypertrophies and shows degenerative changes, which often cause sudden death.

Vitamin P is thought to control capillary permeability. Its distribution in foods is much like that of vitamin C. Another substance, *rutin,* also helps to prevent capillary fragility.

As the result of experiments, other water-soluble vitamins have been suggested, such as pantothenic acid, green-leaf factor, biotin, inositol, folic acid, choline, etc. At present experimental literature does not disclose enough about them to warrant including them in the table beginning on p. 573.

Vitamins		*Sources*	*Effects of Cooking*	*General Effects of Optimum Intake*	*Evidences of Deficiency*
Fat-soluble	A	Milk, butter, eggs, fish-liver oils, green vegetables (provitamins)	Resists heat in absence of air; readily destroyed by oxidation	A factor in— Decreasing susceptibility to skin infections Preserving general health and vigor Effecting chemistry necessary for vision Promoting growth	Failure to gain weight, susceptibility to xerophthalmia, night blindness, dry skin, impaired epithelial tissues, increased incidence of respiratory diseases and of skin (toad skin), ear and sinus infections, inflammations and infections of alimentary and urinary tracts, degenerative changes in nervous tissues
	D	Egg yolk, whole milk, butter, fish-liver oils	Slight; relatively stable	A factor in well-developed bone and teeth, calcium and phosphorus metabolism	Rickets (in children)
	E	Seeds of plants, eggs, lettuce, spinach, meat, wide distribution	Unusually heat resistant	A factor in normal gestation in rats and probably also in humans	Sterility in rats
	K	Wide distribution, especially green leaves		A factor in normal functioning of liver and normal clotting time	Liver damage Delayed clotting time
Water-soluble	B_1 Thiamin	Whole-grain cereals, legumes, eggs, oranges, bananas, apples, cooking water of vegetables, lean meat	Destroyed by prolonged heating, by temperatures higher than boiling	A factor in— Normal carbohydrate metabolism Maintenance of normal appetite, digestion, absorption Promotion of growth	Beriberi, polyneuritis Stunted growth of children, lowered appetite, reduced intestinal motility, arrested growth

Vitamins		Sources	Effects of Cooking	General Effects of Optimum Intake	Evidences of Deficiency
Water-soluble	G (B_2) Riboflavin	Milk, eggs, green vegetables, fruits, liver, lean meat	Relatively heat-stable	A factor in— Normal growth and in nutrition and vitality at all ages	Dermatitis, pellagra (in part) Well-defined eye lesions
	B_6 Pyridine	Whole-grain cereals, yeast, milk, eggs, meat, fish, liver	Relatively heat-stable	A factor in— Normal activity of the nervous system	Florid type of dermatitis (experimental pellagra in rats) Nervousness, irritability, and insomnia
	C Ascorbic acid	Citrus fruits (raw or canned), tomatoes (raw or canned), green vegetables, potatoes, etc.	Readily destroyed by heat, especially slow cooking	A factor in— Normal development of teeth and maintenance of health of gums Healing of wounds and protection against infections Maintenance of high level of positive health Normal cellular chemistry of all tissues	Low nutritional level Scurvy and possibly predisposition to dental caries and systemic type of pyorrhea

Research on vitamins is widespread and energetic. It is impossible to summarize the conclusions briefly. Experiments on vitamin-deficient diets represent the disturbance of a balance normally maintained. It is therefore difficult to determine to what extent laboratory experiments represent primarily the effect of such disturbance rather than true vitamin deficiency. Experiments indicate that excess of vitamins causes disturbances about which conclusions cannot yet be summarized. All vitamins are needed in minute amounts and are indispensable to health.

ACCESSORY ARTICLES OF DIET

In addition to the foodstuffs proper, foods contain numerous substances which add to their attractiveness, stimulate appetite, and increase secretion of digestive fluids. These may be classified as flavors and condiments, and stimulants. The first group includes the various oils and esters that give odor and taste to food, and the condiments such as salt, pepper, mustard, etc. Stimulants include tea, coffee, cocoa, meat extracts, alcohol, etc. Tea and coffee owe their stimulating effect to caffeine, which has a diuretic effect on the kidneys and a stimulating effect on the nerve centers. It prevents sleepiness, probably because of its action in raising blood pressure. Tea and coffee increase muscular energy and diminish fatigue. Cocoa and the chocolate made from it by the addition of sugar contain nourishment in the form of carbohydrate, fat, and protein. Their stimulating effects are due to theobromine. Meat extracts contain secretagogues which stimulate the gastric glands. Alcohol is rapidly absorbed and quickly oxidized; it yields energy and gives rise to carbon dioxide and water. It is not transformed into glycogen or fat, and hence is not stored. Indirectly it may cause obesity in two ways: moderate drinking creates a keen appetite and so favors overeating; the oxidation of alcohol lessens the need for the oxidation of fats or carbohydrates.

DIGESTIVE PROCESSES

In a broad sense all the processes by which foods are rendered available to an organism are digestive processes. Cooking and many of the industrial processes often initiate the task which the digestive organs complete. In cooking, various changes are brought to pass, such, for example, as changing starch to dextrin, partially splitting fats into glycerin and fatty acids, and changing some proteins to the first stages of their dehydrolyzed products. A second reason for classifying cooking as a digestive process is that the appearance, odor, and taste of food are frequently improved and these changes stimulate the end organs of the optic and olfactory nerves, and the taste buds, causing a reflex stimulation of the digestive mechanisms. In a third way cooking may aid digestion by killing parasites or organisms which otherwise would gain a foothold in the alimentary canal or other parts of the body and thus modify or change digestive processes. Digestive process within the body may be described as mechanical and chemical.

The digestive processes are controlled by the nervous system. Any severe strain or strong emotion which affects the nervous system unpleasantly inhibits the secretion of the digestive fluids and interferes with digestion, often checking the appetite and even preventing the taking of food. On the other hand, pleasurable sensations aid digestion; hence the value of attractively served food, pleasant surroundings, and cheerful conversation.

Mechanical digestion includes the various physical processes that occur in the alimentary canal. It is to be considered as preliminary to chemical digestion. It serves the following purposes: taking food in and moving it along through the alimentary canal just rapidly enough to allow the required chemical changes to take place in each part; lubricating the food by adding the mucin and water secreted by the glands of the alimentary canal; liquefying the food by mixing it with the various digestive fluids; and grinding the food into small particles, thereby increasing the amount of surface to come in contact with the digestive fluids.

The mechanical processes consist of:

1. Mastication—comminution and mixing with saliva
2. Deglutition, or swallowing
3. Peristaltic action of the esophagus
4. Movements of the stomach
5. Movements of the intestines
6. Defecation

Chemical digestion is essentially a process of hydrolysis which is dependent upon the presence of enzymes. Hydrolysis is a double decomposition in which water is one of the interacting substances.

An example of hydrolysis (hydrolytic cleavage) is the splitting of maltose into simple sugars.

$$C_{12}H_{22}O_{11} + H_2O = C_6H_{12}O_6 + C_6H_{12}O_6$$

Necessity for chemical digestion. Chemical digestion is necessary because foods in general cannot diffuse through animal membranes, and the tissues cannot use them; hence, they must be reduced to smaller molecules and to such standard substances as the tissues can use, i.e., (1) simple sugars, resulting from the hydrolysis of all carbohydrate foods; (2) glycerin and fatty acids, resulting from the hydrolysis of fats; and (3) amino acids, resulting from the hydrolysis of proteins.

Agents of chemical digestion. Hydrolytic cleavages similar to those of digestion can be brought about in several ways. Boiling foodstuffs with acids, treatment with alkalies, or the application of superheated steam will accomplish these changes. The remarkable fact is that violent reagents, high temperatures, or both are necessary to produce these changes in the laboratory, whereas in the digestive tract they take place at body temperature and are due to the enzymes present in the digestive fluids. Enzymes have been described (p. 458) as organic substances that act as catalysts.

Classification of enzymes. It has been suggested that each hydrolytic enzyme be designated by the name of the substance (substrate) on which it acts, together with the suffix, *ase.* According to this, the starch-splitting enzymes are called *amylases;* fat-splitting enzymes, *lipases;* protein-splitting enzymes, *proteinases.* This suggestion has been followed in part only, as some of the older names continue to be used, e.g., pepsin, trypsin, etc.

1. The sugar-splitting enzymes. *a.* The inverting enzymes, which hydrolyze disaccharides to monosaccharides. Examples: maltase splits maltose to dextrose; invertase splits cane sugar to dextrose and levulose; and lactase splits milk sugar (lactose) to dextrose and galactose. *b.* The enzymes which split the monosaccharides. Example: enzymes capable of splitting glucose of the tissues into lactic acid.

2. The amylolytic, or starch-splitting, enzymes. Examples: ptyalin, or salivary diastase, and amylase, or pancreatic diastase, cause hydrolysis of starch.

3. The lipolytic, or fat-splitting, enzymes. Example: lipase found in the pancreatic secretion, etc., causes hydrolysis of fat.

4. The proteolytic, or protein-splitting, enzymes. Examples: pepsin of gastric fluid and trypsin of pancreatic fluid, which cause hydrolysis of the proteins.

5. The deaminizing enzymes. Amino acids contain an NH_2 group which is split off by hydrolysis.

6. The phosphatases, enzymes found present in practically all animal tissue.

7. The clotting enzymes, which convert soluble proteins to insoluble forms. Examples: rennin, which causes the clotting of the casein of milk, and thrombin, which causes coagulation of the fibrinogen of the blood.

8. The oxidizing enzymes, or oxidases. Enzymes which cause oxidation.

9. The dehydrogenating enzymes, or dehydrogenases. These bring about the removal of hydrogen before oxidation—in some cases.

10. The mutation enzymes. H. C. Sherman lists enzymes called mutases, which bring about "chemical rearrangement without breaking-down of larger into smaller molecules."[8]

The enzymes that bring about chemical digestion in the alimentary tract belong to the sugar-splitting group, the amylolytic, the lipolytic, and the proteolytic groups.

CHANGES THE FOOD UNDERGOES IN THE MOUTH

Mastication. When solid food is taken into the mouth, its comminution is immediately begun. It is cut and ground by the teeth, being pushed between them again and again by the muscular contractions of the cheeks and the movements of the tongue, until the whole is thoroughly crushed.

Insalivation. During the process of mastication saliva is poured in large quantities into the mouth and, mixing with the food, lubricates, moistens, and reduces it to a softened mass known as a bolus, which can be readily swallowed.

Secretion of saliva. The nerve supply of the salivary glands is derived in part from the craniosacral and in part from the thoracolumbar divisions of the autonomic system. Both sets of nerves carry secretory and vasomotor fibers. The cranial nerves carry vasodilator fibers and, when stimulated by the sight or smell of food, (1) dilate the blood vessels, increasing the volume and temperature of the gland, and (2) cause the glands to produce a secretion that is copious in amount and watery in consistency. This is called a psy-

[8] H. C. Sherman, *Chemistry of Foods and Nutrition,* 1952.

chical secretion. The thoracolumbar nerves carry vasoconstrictor fibers and, when stimulated by the presence of food in the mouth, (1) constrict the blood vessels and (2) produce a smaller amount of thicker secretion. Under normal conditions, the flow of saliva is the result of reflex stimulation of the secretory nerves. Obviously, the taste buds of the tongue, fauces, and cheeks are the sense organs which are stimulated by the presence of food in the mouth.

Saliva. Saliva is secreted by the salivary glands—parotid, submaxillary, and sublingual—and by the numerous minute buccal glands of the mucosa of the mouth.

It consists of a large amount of water (some 99.5 per cent) containing some protein material, mucin, inorganic salts, *salivary amylase*, or ptyalin, etc. It has a specific gravity of about 1.005 and is nearly neutral in reaction (*p*H about 6.4 to 7.0). Although the amount of saliva secreted per day varies considerably, an average amount is from 1 to 1.5 liters. About half of the substances contained in the water of saliva are inorganic salts in solution, mainly chlorides, carbonates, and phosphates of sodium, calcium, and potassium. The other substances are organic, mainly mucin, salivary amylases, serum albumin and globulin, urea, etc. The calcium carbonate and phosphate in combination with organic material may be deposited on the teeth as tartar, especially if the saliva is alkaline and contains considerable mucin. Occasionally these salts may be also deposited in the ducts of the salivary glands.

The functions of saliva are to soften and moisten the food, assisting in mastication and deglutition; to coat the food with mucin, lubricating it and ensuring a smooth passage along the esophagus; to dissolve dry and solid food, providing a necessary step in the process of stimulating the taste buds as taste sensations play a part in the secretion of gastric juice; to digest starch by means of salivary amylase; and to help maintain optimum water content of the body. If the water content of the body is below optimum, the mucous membranes of the mouth and throat become dry; this brings about a sensation of thirst, inducing the drinking of more water.

Salivary amylase. Salivary amylase changes starch to dextrin and maltose. The process of reducing starch to maltose is a gradual one, consisting of a series of hydrolytic changes which take place in successive stages and result in a number of intermediate compounds, such as soluble starch, erythrodextrin, achroödextrin, etc. The change is best effected at the temperature of the body,[9] in a neutral solution, saliva that is distinctly acid hindering or arresting the process. Boiled starch is changed more rapidly and completely than raw, but food is rarely retained in the mouth long enough for the saliva to do more than begin the digestion of starch.

Deglutition, or **swallowing,** is divided into three stages which correspond to the three regions—mouth, pharynx, and esophagus—through which the food passes. The *first stage* consists of the passage of the bolus of food through the fauces. The *second stage* consists of the passage of the bolus through the pharynx. During this stage, the respiratory opening into the larynx is closed by the approximation of the vocal folds which close the glottis, by the elevation of the larynx, and by contraction of the muscles of deglutition. The parts are crowded together by the descent of the base of the tongue, the lifting of the larynx, and the coming together of the vocal folds.

The *third stage* consists in the passage of the bolus through the esophagus. Apparently the consistency of the food affects this stage of the process. Solid or semisolid food is forced down the esophagus by a peristaltic movement and requires from 4 to 8 seconds for passage from mouth to stomach. About half of this time is taken up in the passage through the esophagus, and the remainder is spent in transit through the cardiac orifice of the stomach. Liquid or very soft food is shot through the esophagus and arrives at the lower end in about 0.1 seconds. It may pass into the stomach at once or may be held in the esophagus for moments, depending on the condition of the cardiac sphincter. When the stomach is empty, the cardiac sphincter is probably relaxed. On ingestion of food it becomes tonically contracted, and, as digestion continues, the sphincter becomes more tense. The tension develops in response to the rise of acidity in the liquid just within the cardiac portion of the stomach.

Summary. During the process of mastication, insalivation, and deglutition the food is reduced to a soft, pulpy condition, any starch it contains may begin to be changed into sugar, and the food acquires a more or less alkaline reaction.

[9] A temperature of 100° F (37.7° C) in the alimentary canal is necessary for digestion; hence, iced drinks or iced foods that lower this temperature delay digestion.

Vomiting is controlled by a nerve center in the medulla which can be stimulated by chemical and physical qualities of the tissue fluid in the center and by nerve impulses which reach it. Under ordinary circumstances the contractions of the cardiac sphincter prevent the regurgitation of food, but spasmodic contractions of the abdominal muscles may, if the diaphragm is fixed, force the contents of the stomach through the esophagus and mouth to the exterior. This is called vomiting. It is usually preceded by a sensation of nausea and a reflex of saliva and is itself a reflex act brought about by mechanical irritation of the throat or by irritating substances in the stomach and duodenum, by pain, heart conditions, and certain emotions as fear, worry, repulsion, etc.

CHANGES THE FOOD UNDERGOES IN THE STOMACH

The food which enters the stomach is delayed there by the contraction of the sphincter muscles at the cardiac and pyloric openings. The cavity of the stomach is always the size of its contents, which means that when there is no food in it, it is contracted, but when food enters, it expands just enough to hold it. An investigator fed rats with foods of different colors and found that the portions which had been eaten successively were arranged in definite layers. The food first taken lay next to the wall of the stomach and filled the pyloric region, while the succeeding portions were arranged regularly in the interior in concentric layers. This was interpreted as evidence that the cavity of the stomach is only as large as its contents. The first portion of food entirely filled it, and successive portions were received into the interior because the wall layer was occupied. Within a few minutes after the entrance of food small contractions start in the middle region of the stomach and run toward the pylorus. These contractions are regular and in the pyloric region become more forcible as digestion progresses. As a result of these movements the food in the prepyloric and pyloric portions is macerated, mixed with the acid gastric fluid, and reduced to a thin liquid mass called *chyme*. At certain intervals the pyloric sphincter relaxes, and the wave of contraction forces some of the chyme into the duodenum. The fundic end of the stomach is less actively concerned with these movements but serves as a reservoir for food which is under slight pressure, as the muscles are in a state of continual contraction or tone. Because of lack of movement and muscular tone in this part of the stomach, the food at the fundic end may remain undisturbed for an hour or more and thus escape rapid mixture with the gastric fluid, which, therefore, penetrates slowly to the interior of the mass; hence salivary digestion may continue for a time. As the chyme is gradually forced into the duodenum, the pressure of the fundus forces the food into the pyloric end.

The time required for stomach digestion depends upon the nature of the food eaten. Liquids taken on an empty stomach pass through the pylorus promptly. Small test meals may remain from 1 to 2 hours, but average meals probably stay in the stomach from 3 to 4½ hours. The ejection of chyme through the pylorus occurs at regular intervals and is supposed to be dependent mainly upon the strength of the peristaltic wave, and possibly also the chyme. The muscular activity of the stomach is highly varied both in strength and in type. Some peristaltic waves are weak, whereas others are strong and move over the pylorus, forcing chyme into the intestine. Solid particles forced against the pylorus tend to keep it closed, but a finely divided condition of the chyme, and possibly its acidity, produce relaxation of the pyloric sphincter. In the intestine the presence of chyme causes a closing of the sphincter, which remains closed until the next peristaltic wave opens it again. Either the presence of the chyme or possibly its acidity in the duodenum seems to bring about the closure.

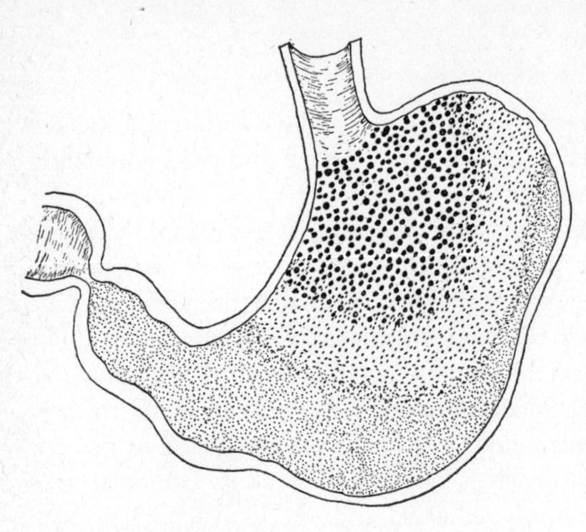

Fig. 311. Diagram of the stomach, showing in fine stippling the position of food first entering it; in coarser and black stippling, the food entering the stomach later.

The secretion of gastric juice is constant. Even in the period of fasting there is a small continuous secretion, but during the act of eating and throughout the period of digestion the rate of secretion is greatly increased.

In an ordinary meal the secretion first started is due to the sensations of eating and the taste and odor of food, which stimulate the sensory end organs in the mouth and nose. This so-called "psychical secretion" ensures the beginning of gastric digestion and is supplemented by chemical action arising in the stomach. Some foods, such as meat juices and extracts, contain substances called *secretagogues* or hormones which are supposed to act directly upon the nerves of the pyloric mucous membrane and form a substance called *gastrin* or *gastric secretin*, which is absorbed into the blood

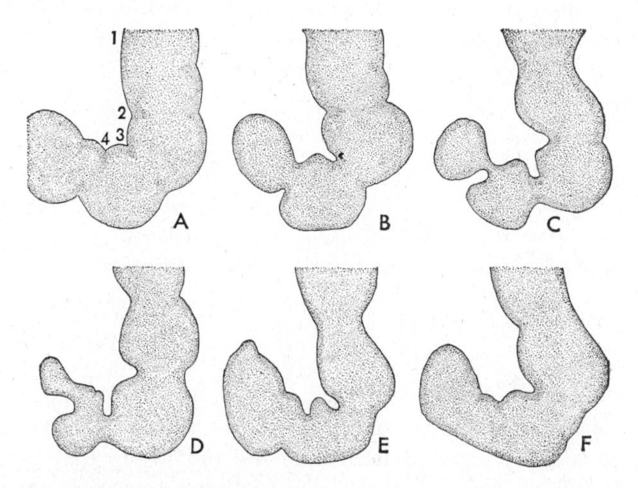

Fig. 312. Diagrams to show peristalsis in the stomach. Locate *1*, *2*, *3*, *4* in succeeding figures to trace a peristaltic wave over the stomach.

and carried to the gastric glands. This substance stimulates the glands to secretion. Other foods, such as milk, bread, white of egg, etc., do not appear to contain these secretagogues. When foods are eaten, a psychical secretion is started; and when this has acted, some products of digestion in turn become capable of stimulating a further secretion of gastric fluid. These steps are (1) cephalic, *psychical* or appetite *secretion*, (2) the secretion due to secretagogues in food, and (3) the secretion due to secretagogues in the products of digestion, probably proteoses and peptones. Recent work emphasizes the effect of increased blood flow in stomach vessels and opens the question of the active or passive cause of this varied blood flow in relation to gastric secretion. Intestinal hormones may also affect gastric secretion. Fats inhibit gastric secretion, possibly by the production of intestinal hormones, e.g., *enterogastrone.*

Gastric fluid is secreted by the gastric glands lining the mucous membrane of the stomach. It is a thin, colorless, or nearly colorless liquid with an acid reaction (pH is about 0.9 to 1.7) and a specific gravity of about 1.003 to 1.008. The quantity secreted depends upon the amount and kind of food to be digested, possibly an average of 1.5 to 2.5 liters per day. Upon analysis it is found to be a watery secretion containing some protein, some mucin, and inorganic salts; but the essential constituents are hydrochloric acid and two or possibly three enzymes—pepsin, rennin, and gastric lipase.

Hydrochloric acid. It is believed that the parietal (acid, or oxyntic) cells of the gastric glands secrete the hydrochloric acid from chlorides found in the blood. The chief chloride is sodium chloride, and by some means this is decomposed; the chlorine combines with hydrogen and is then secreted upon the free surface of the stomach as hydrochloric acid. In normal gastric fluid it is found in the proportion of about 0.5 per cent. It serves to activate pepsinogen and convert it to pepsin; to provide an acid medium, which is necessary for the pepsin to carry on its work; to swell the protein fibers, thus giving easier access to pepsin; to help in the inversion of cane sugar, which is the easiest of the disaccharides to hydrolyze; as an antiseptic and perhaps to destroy many organisms that enter the stomach.

If the secretion of hydrochloric acid is excessive, it causes hyperacidity of the gastric fluid, which is one of the underlying causes of gastritis and gastric ulcer. If the secretion of hydrochloric acid is below normal, it causes hypoacidity of the gastric fluid. This condition is frequently present in pernicious anemia and carcinoma.

Pepsin is supposed to be formed in the pyloric glands and the chief cells of the gastric glands. It is present in these cells in the form of a zymogen, an antecedent inactive substance called *propepsin* or *pepsinogen,* which is quickly changed to active pepsin by the action of hydrochloric acid.

Pepsin is a weak proteolytic enzyme requiring an acid medium in which to work. It has the property of hydrolyzing proteins through several stages into proteoses and peptones. This action is preparatory to the more complete hydrolysis that takes place in the intestine under the influence of trypsin and erepsin, for peptones are not absorbed but undergo a further hydrolysis to amino acids.

Rennin is also supposed to be formed in the chief cells of the gastric glands in a zymogen form, the *prorennin,* which after secretion is converted to the active enzyme. Probably this enzyme acts *only* upon the soluble protein of milk, *casein.* It converts this into paracasein (another terminology says caseinogen to casein), which reacts with calcium to form the curd, the digestion of which is carried on by the pepsin and later by the trypsin. Some researchers question the presence of this enzyme in the human stomach.

Various observers have described other enzymes in addition to the pepsin and rennin, but the evidence regarding these is inconclusive. It is probable that the ptyalin swallowed with the food continues the digestion of starchy material in the fundus for some time. Regarding the fats, it is believed that they undergo no digestion in the stomach. They are set free from their mixture with other foods by the digestive action of the gastric fluid; they are liquefied by the heat of the body and are scattered through the chyme as a coarse emulsion by the movements of the stomach, all of which prepares them for digestion. Emulsified fats such as cream may be acted upon to a limited extent by a third enzyme called *gastric lipase,* but the acid condition of the stomach contents prevents any considerable change of this sort.

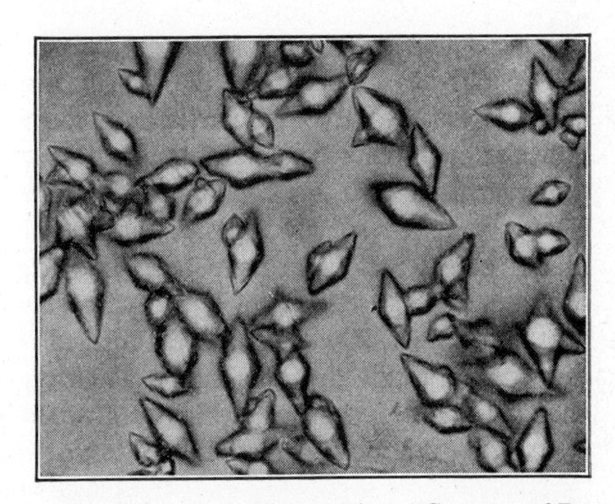

Fig. 313. Crystalline pepsin. (Courtesy of Dr. J. H. Northrop and the *Journal of General Physiology.*)

Summary. The stomach serves as a place for temporary storage and maintains a gradual delivery to the intestine; a place for the continuation of the salivary digestion of starch, the beginning of the digestion of proteins and perhaps fats, and germicidal activity. While the food is in the pyloric region it is subjected to the acidity of the gastric fluid. Doubtless gastric secretion has many other functions.

Inhibition of gastric digestion. The secretion of gastric fluid is inhibited by stimulation of the thoracolumbar system, so that anger, pain, worry, and also a distaste for food may delay digestion. Secretion is also inhibited by active exercise soon after a meal, because active exercise increases the amount of blood in the skeletal muscles and decreases the supply to the stomach. When gastric digestion is much delayed, organisms are likely to cause fermentation of the sugars, producing gas which may cause distress.

CHANGES THE FOOD UNDERGOES IN THE SMALL INTESTINE

The chyme entering the duodenum after an ordinary meal is normally free from coarse particles of food and is acid in reaction, both the hydrochloric acid and the lactic acid produced by fermentation contributing to this condition. Much of the food is undigested. The proteins are partly digested; some progress has been made in hydrolyzing starch; fats have been liquefied and mixed with other food but probably have not been hydrolyzed themselves. If milk is part of the diet, it will have been curdled and redissolved. It is in the small intestine that this mixture undergoes the greatest digestive changes. These changes, which constitute intestinal digestion, are effected by the movements of the intestines, the pancreatic fluid, the succus entericus, or secretion of the intestinal glands, and the bile.

It is convenient to describe the secretion and digestive action of these three fluids separately, but it must be remembered that they act simultaneously. The pancreatic fluid and the bile enter the intestine about 7 to 10 cm beyond the pylorus, and, therefore, the foods in the small intestine throughout its length are subjected to a mixture of pancreatic fluid, bile, and small intestinal fluid.

Movements of the small intestine are described as peristaltic, rhythmical, and pendular.

Peristalsis may be defined as a wave of dilatation brought about by the contraction of longitudinal muscles, followed by a wave of constriction caused by the contraction of circular muscles. The purpose is to pass the food slowly forward.[10] *Peristaltic waves* pass very slowly along short distances of the small intestine, with an

[10] T. Railford and M. J. Mulinos, "The Myenteric Reflex," *American Journal of Physiology*, 1934.

occasional rapid wave known as the *peristaltic rush,* which moves the food along greater distances.

The *rhythmical movements* consist of a series of local constrictions of the intestinal wall which occur rhythmically at points where masses of food lie. These constrictions divide the food into segments. Within a few seconds each of these segments is halved, and the corresponding halves of adjoining segments unite. Again constrictions occur, and these newly formed segments are divided, and the halves re-form. In this way every particle of food is brought into intimate contact with the intestinal mucosa and is thoroughly mixed with the digestive fluids.

Pendular movements are constrictions which move onward or backward for short distances, gradually moving the chyme forward and backward over short distances in the small intestine.

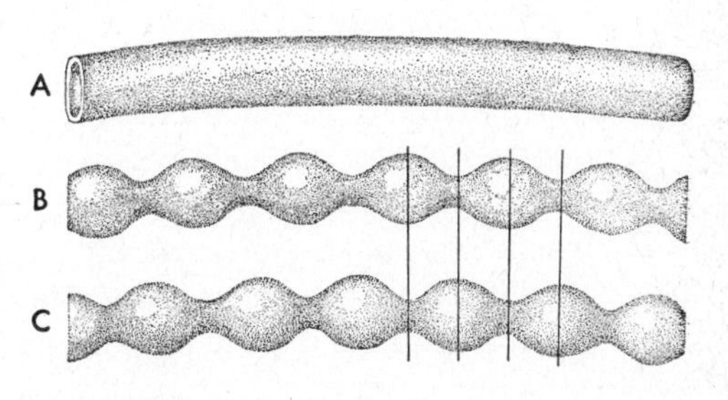

Fig. 314. Three diagrams of a portion of the small intestine, to show rhythmical movements. The small straight lines indicate the same area of the intestine at different intervals.

The varied muscular movements of the small intestine increase the blood supply, bringing materials for secretion and removing absorbed materials faster. They assist the minute glands in emptying their secretion, mix the digestive fluids and food intimately, and bring fresh absorbable material constantly to the mucosa, thereby increasing absorption.

Secretion of pancreatic fluid. Pancreatic secretion, like gastric secretion, consists of two parts: a nervous secretion, caused by the secretory fibers in the vagus and splanchnic nerves, and a chemical secretion, due to the action of a hormone, *secretin.* Two views are held regarding secretin. Acid of the gastric fluid, upon reaching the duodenum, may produce or liberate secretin, which, absorbed by the blood and carried to the pancreas, stimulates the pancreas to secrete. Another view is that the secretin exists in the mucous membrane of the duodenum (or jejunum) and is liberated by the bile which pours into the intestine during gastric digestion. The bile salts, as they are absorbed, carry the secretin into the blood, and in this way it reaches the pancreas. It is thought that the nervous secretion provides pancreatic fluid in the early stages of intestinal digestion and that the chemical secretion maintains the flow until all the stomach contents reach the duodenum.

Pancreatic fluid. The *nervous secretion* of pancreatic fluid is thick, and rich in enzymes and proteins. The trypsin in it may be active. The *chemical secretion* is thin, watery, contains little enzyme or proteins, and is alkaline. Pancreatic fluid contains three groups of enzymes, which act on proteins, carbohydrates, and fats. They are trypsin, amylase, and lipase.

The proteolytic enzyme, trypsin, under favorable conditions may hydrolyze the protein molecule to its constituent amino acids. Trypsin is secreted in an inactive form called *trypsinogen* and is activated by *enterokinase*, a coenzyme which is contained in the mucous membrane of the small intestine.

Trypsin hydrolyzes proteins into proteoses, peptones, and polypeptides. It is probably a complex of many enzymes.

The conditions under which enzymes act and the length of time they are allowed to act determine the actual products that result. Other proteolytic enzymes have been postulated.

The amylolytic enzyme (*amylase*) is similar to ptyalin in action. It causes hydrolysis of starch with the production of maltose. The starchy food that escapes digestion in the mouth and stomach becomes mixed with this enzyme and continues under its action until the colic valve is reached. Maltose is further acted upon by the maltase of the intestinal secretion and is hydrolyzed to glucose.

The lipolytic enzyme (*lipase*) is capable of hydrolyzing fats to glycerin and fatty acids. The process of hydrolysis is preceded by emulsification,[11] in which bile salts play a leading role. The lipase splits some of the fats to fatty acids and glycerin; emulsification increases the surface of fat exposed to the chemical action of the lipase and is a mechanical preparation for the further action of lipase. The glycerin and fatty acids produced by the action of the lipase are

[11] An oil emulsion contains minute globules of oil that do not coalesce. Milk is a natural emulsion. On standing, it slowly separates as the fat rises to the top and forms cream.

absorbed by the epithelium of the intestine. It is thought that the fatty acids form soluble and diffusible compounds with the bile salts and are absorbed in this form. After absorption the fatty acids and glycerin again combine to form fat, but it is probable that they combine in such proportions as to make fat which is characteristic of man. The action of lipase is said to be reversible, i.e., it causes both the splitting of the fats and the synthesis of the split products, not only in the intestine but in the various tissues, during the metab-

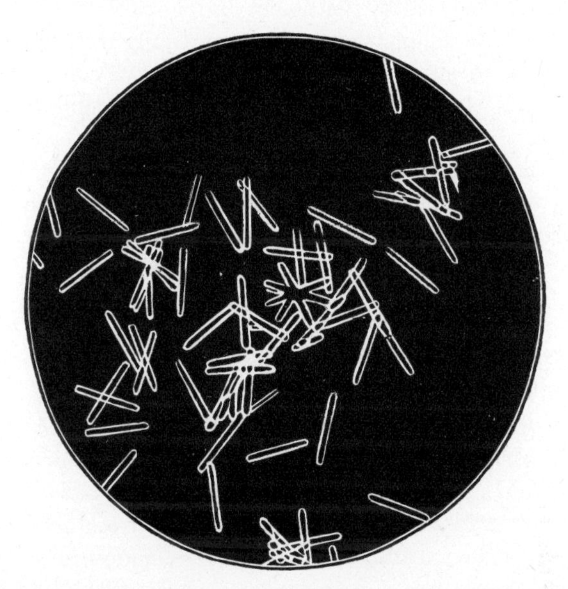

Fig. 315. Crystalline trypsin. (Courtesy of Dr. J. H. Northrop and the *Journal of General Physiology*.)

olism or the storage of fat. Lipase is found in blood and in many of the tissues.

The intestinal secretion (succus entericus) is a clear, yellowish fluid which has a marked alkaline reaction. Extracts of the walls of the small intestine have been found to contain four or five enzymes which influence intestinal digestion to a marked extent. The enzymes are to be found in the secretion, and their actions are as follows.

Enterokinase, an enzyme or coferment which activates the trypsin of the pancreatic fluid; *erepsin*, an enzyme which hydrolyzes peptides to amino acids, thus completing the work begun by pepsin and trypsin.

Hydrolyzing enzymes hydrolyze disaccharides into monosaccharides. *Maltase* acts upon the products formed in the digestion of starches, i.e., maltose and dextrin, and hydrolyzes them to glucose. *Invertase*, or *sucrase*, acts upon sucrose and hydrolyzes it to glucose and fructose. *Lactase* acts upon lactose and hydrolyzes it to glucose and galactose. This inverting action is necessary because disaccharides cannot be used by the tissues and would escape in the urine, but in the form of simple sugars they are readily used by the tissues.

Nuclease is an enzyme which is said to be a constituent of the intestinal fluid. It acts upon the nucleic acid component of nucleoproteins.

Secretin is not an enzyme but a hormone which is secreted or formed in the intestinal mucosa. Under the influence of acids or of acids and bile, it is absorbed and carried to the pancreas and causes a flow of pancreatic fluid.

Bile is formed in the liver and is an alkaline fluid, *p*H about 6.8 to 7.7, the specific gravity of which varies from about 1.010 to 1.050. Approximately 500 to 800 cc are secreted daily. It is usually yellow, brownish yellow, or olive green in color.[12] It consists of water, bile pigments, bile acids, bile salts, cholesterol, lecithin,[13] and neutral fats. The bile acids are glycocholic and taurocholic, occurring as sodium glycocholate and sodium taurocholate. These salts are alternately poured into the duodenum, then reabsorbed, and reappear in the bile. Thus, by continued circulation, the bile repeats its function many times. The mucous membranes of the bile ducts and gallbladder add a mucinlike protein called nucleoalbumin, which, together with some mucin, gives bile its mucilaginous consistency.

SECRETION OF BILE is continuous, but the amount varies, increasing when the blood flow is increased and vice versa. It is thought that the presence of bile in the intestines stimulates secretion in the liver and that this is due to the bile salts, which act as cholagogues. Bile enters the duodenum only during the period of digestion. Between these periods it cannot enter the duodenum because of the sphincter which closes the common bile duct; consequently the bile

[12] The color of bile is determined by the respective amounts of the bile pigments, (1) biliverdin and (2) bilirubin, that are present.

[13] The cholesterol and lecithin are compound fats.

backs up into the gallbladder. Apparently the ejection of chyme into the duodenum excites a contraction of the gallbladder and an inhibition of the sphincter, which results in an ejection of bile. The physiological effects of bile may be grouped as follows:

1. DIGESTIVE SECRETION. Bile salts are essential for the action of lipase. Mixtures of bile and pancreatic fluid split the fats more rapidly than pancreatic fluid alone. The bile acids may activate the lipase or may act as a coferment, and they help in the absorption of the fatty acids.

Bile is essential for the absorption of vitamin K and other fat-soluble vitamins. It also stimulates intestinal motility and neutralizes the acid chyme, creating a favorable hydrogen ion concentration for pancreatic and intestinal enzyme activity; bile salts help to keep cholesterol in solution.

2. EXCRETION. The bile is an excretory channel for toxins, metals, cholesterol, etc. It is thought that the liver cells excrete the bile pigments that are brought to them by the blood, just as the kidney cells remove the urea from the blood. The cholesterol of bile is probably a waste product of cellular disintegration.

3. ANTISEPTIC. It was formerly believed to have an antiputrefactive action, but it is now thought that the greater amount of putrefaction in the absence of bile is brought about by the action of bacteria on proteins and carbohydrates which have remained undigested because of the protective covering of insoluble fat which is found on them in the absence of bile.

Gallstones. Cholesterol may become so concentrated in the gallbladder that it tends to crystallize out, and these crystals form gallstones. Catarrhal conditions, which are often due to the typhoid and colon bacilli or to a change in the character of the bile, may cause this crystallization. Gallstones are usually formed in the gallbladder. Their passage through the cystic and common bile ducts often causes severe pain, called gallbladder colic. They may plug the duct and cause obstructive jaundice.

Jaundice. When the flow of bile through the bile duct is interfered with, bile gets into the blood and is carried to all parts of the body, producing a condition of jaundice, which is characterized by a yellow discoloration of the skin and of the whites of the eyes. The urine is of a greenish hue because of the extra quantity of pigment eliminated by the kidneys, and the stools are grayish in hue, owing to the lack of bile. Gallstones in the gallbladder or ducts, plugging of the ducts with mucus, and constipation may interfere with the excretion of bile. Hemolytic jaundice is due to too rapid destruction of erythrocytes. Infective jaundice is due to the incapacity of the liver cells to eliminate pigments.

Action of organisms in the small intestine hydrolyzes carbohydrates and proteins constantly. Fermentation of the carbohydrates gives rise to organic acids, such as lactic and acetic, but none of the products of this fermentation is considered toxic. On the other hand, the putrefaction of proteins gives rise to a number of end products that are distinctly toxic. Under normal conditions and on a mixed diet, carbohydrate fermentation is the characteristic action of the organisms in the small intestine, whereas protein putrefaction occurs in the large intestine. The reason for this seems to be that carbohydrates serve to protect proteins because some of the organisms of the small intestine, i.e., bacillus coli, will not attack proteins as long as carbohydrate material is present. In addition, the organic acids produced by the fermentation of carbohydrates tend to neutralize the alkalinity of the intestinal secretion and may even give an acid reaction. An acid reaction is unfavorable to the action of the organisms that hydrolyze proteins, and in this way putrefaction in the small intestine is prevented. It seems that the nature of organismal activity in the small intestine depends partly upon the character of the diet, which, therefore, may be chosen so as to favor one or the other kind.

The *Bacillus acidophilus* is the species best adapted to maintaining good intestinal conditions. The ingestion of cultures of this organism and the liberal consumption of lactose, which is favorable to their development, often assist in establishing good intestinal hygiene.

The time required for digestion in the small intestine is influenced by many factors. It depends largely on the varying proportions of the different foods included in a meal. Twenty to thirty-six hours are required for the passage of ingested food material through the gastrointestinal tract of adults who are on a mixed diet. There is considerable variation among individuals, and usually not all of the residue from a single meal is evacuated at the same time. In diarrheal conditions the time is much shortened.

According to observations made upon a patient with a fistula at the end of the small intestine, food begins to pass into the large intestine from 2 to 5 hours after eating, and it requires 9 hours or more after eating before the last of a meal has passed the colic valve.

HORMONES OF THE INTESTINAL PATHWAY

Gastrin is secreted by the pyloric mucosa and excites the fundic glands of the stomach to active secretion.

Enterogastrone is secreted by the pyloric mucosa and inhibits gastric secretion and gastric motility.

Secretin is secreted by the upper intestinal mucosa and excites the pancreas to activity.

Pancreozymin is secreted by the duodenal mucosa and is enzyme-stimulating in the pancreas.

Cholecystokinin is secreted by cells in the upper small intestine and causes contraction of the smooth muscle of the gallbladder, causing it to empty.

Villikinin is secreted by the upper small intestinal mucosa and stimulates movements of the villi, thereby increasing mixing of food and absorption of the end products of digestion.

Enterocrinin is secreted by the duodenal mucosa and excites to activity the glands that secrete succus entericus.

Insulin is secreted by the islet cells of the pancreas and helps to control the level of blood sugar.

Duodenin and *incretin* are believed to be secreted by the small intestinal mucosa and concerned with absorption of glucose. However, these are questioned by some researchers.

All of these hormones are secreted into the tissue fluid or blood stream and, for the most part, pass through the portal vein to the liver, hepatic veins, inferior vena cava, right heart, lungs, left heart, and aorta, out through the arteries and capillaries, and finally to the tissue fluid and cells, where they have action.

CHANGES THE FOOD UNDERGOES IN THE LARGE INTESTINE

Movements of the large intestine. The opening from the small intestine into the large is controlled by the colic valve and the colic sphincter, which is normally in a state of tone. As food passes the colic valve, the cecum becomes filled, and gradually the accumulation reaches higher and higher levels in the ascending colon. The contents of the ascending colon are soft and semisolid, but in the distal end of the transverse colon they attain the consistency of feces.

A type of movement characteristic of the large intestine is called *haustral churning.* The pouches, or sacculations, which are found in the large intestine become distended and from time to time contract and empty themselves. Another type of movement is designated as *mass peristalsis.* It consists of the vigorous contraction of the entire ascending colon, which transfers its contents to the transverse colon. Such movements occur only three or four times a day and last only a short time. Some observers connect them with eating and assume

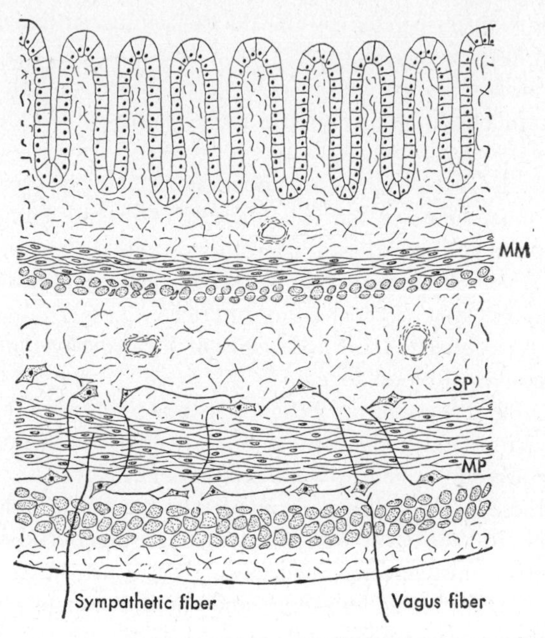

Fig. 316. Cross section of large intestine, showing coats. *MM,* muscularis mucosae; *MP,* myenteric plexus; *SP,* submucous plexus.

that the entrance of food into the stomach sets up a reflex movement in the colon.

The secretion of the large intestine contains much mucin, shows an alkaline reaction, and is said not to contain enzymes. When the contents of the small intestine pass the colic valve, they still contain a certain amount of unabsorbed food material. This remains a long time in the intestine; and since it contains the digestive enzymes received in the duodenum, the process of digestion and absorption continues.

Action of organisms in the large intestine brings about, in an alkaline reaction, constant putrefaction of whatever proteins are present as the result of not having been digested and absorbed in the small intestine. The splitting of the protein molecules by this process is very complete; not only are they hydrolyzed to amino acids, but these amino acids are deaminized and oxidized to simpler groups. The list of simple substances resulting from putrefaction is long and includes peptones, proteoses, ammonia, and amino acids, and also indol, skatol, phenol, fatty acids, carbon dioxide, hydrogen sulphide, etc. Some of these are given off in the feces; others are absorbed and carried to the liver, where they are changed to less toxic compounds, e.g., ethereal sulphates, such as indoxyl sulphate, skatoxyl sulphate, etc., and ultimately excreted in the urine. Therefore the amount of these sulphates in the urine is indicative of the degree of intestinal putrefaction. The ethereal sulphates are said to be produced during putrefaction when oxidation follows deamination. When oxidation precedes deamination, more toxic substances, such as tyramine, tryptamine, and histamine, are said to be formed. Even though these sulphates and allied compounds are less toxic than those from which they are derived, they are more or less poisonous, and the feeling of malaise associated with constipation is thought to be due to the presence of these substances in the blood. Some investigators hold that pressures developed by the retention of material in the lower colon are responsible for much of the malaise, etc., attributed to the absorption of toxic products. Some investigators believe that bacterial putrefaction is harmful. A conservative view is that intestinal organisms are not beneficial, but under normal conditions the body is able to neutralize their effects.

The feces. Two classes of material may be mingled in the contents of the colon, (1) the residues of the diet with microorganisms and their products, and (2) the excretions of the digestive tube and its glands. The proportion existing between these two is variable. The feces consist of water; the undigested and indigestible parts of the food; pigment due to undigested food or to metallic elements contained in it and to the bile pigments; great quantities of microorganisms of different kinds; the products of bacterial decomposition, i.e., indol, skatol, etc.; the products of the secretions; mucous and epithelial cells from the walls of the alimentary tract; cholesterol or a derivative, which is probably derived from the bile; some of the purin bases; inorganic salts of sodium, potassium, calcium, magnesium, and iron.

Defecation. The anal canal is guarded by an internal sphincter and an external sphincter muscle, which are normally in a state of tonic contraction and protect the anal opening. Normally the rectum is empty until just before defecation. Various stimuli (depending on one's habits) will produce peristaltic action of the colon, so that a small quantity of feces enters the rectum. This irritates the sensory nerve endings and causes a desire to defecate. The voluntary contraction of the abdominal muscles, the descent of the diaphragm, and powerful peristalsis of the colon all combine to empty the colon and rectum.

One of the commonest causes of constipation is the retention of feces in the rectum because of failure to act on the desire for defecation. After feces once enter the rectum there is no retroperistalsis to carry them back to the colon, and the sense of irritation becomes blunted. The desire may not recur for 24 hours, during which time the feces continue to lose water and become harder and more difficult to expel. The best means to prevent and overcome constipation are: Act upon the desire for defecation, and have a regular time for doing so. Use a liberal amount of fruit and vegetables and a liberal intake of water. Some authorities teach that a certain amount of indigestible materials in the diet is wholesome. It stimulates the lining of the intestines, promotes peristalsis, and as it is pushed along the tube takes with it the less bulky but more toxic wastes. Form the habit of daily exercise which uses all the muscles, especially the abdominal muscles.

SUMMARY

Food
- Any substance taken into the body
- (1) To yield energy
- (2) To provide material for growth of tissues
- (3) To regulate body processes

Classification
- Chemical analysis shows that elements found in body are found in food
- Nutrients, or Food Principles
  - Water
  - Carbohydrates
  - Fats
  - Proteins
  - Mineral salts
  - Vitamins

Water
- Enters into composition of all tissues; most tissues contain from 75 to 90%

Water
- Greater proportion in young animals and active tissues
- Constitutes about two-thirds of daily intake
- Sources of water content of body
  - Beverages
  - Water contained in food
  - Water formed in the tissues
- Supplies fluid for
  - Secretions
  - Chemical reactions
  - Transfer of food material
  - Elimination of waste
- Important in heat regulation
- Under normal conditions amount in body remains about the same

Carbohydrates
- Most abundant and most economical source of energy. Consist of C, H, and O, the two latter in the proportion to form water
- Include sugars and starches
- **Monosaccharides**
  - Contain one sugar group
  - Glucose, or dextrose, $C_6H_{12}O_6$ } Invert sugar
  - Fructose, or levulose, $C_6H_{12}O_6$ } Invert sugar
- **Disaccharides**
  - Contain two sugar groups
  - Sucrose, or cane sugar, $C_{12}H_{22}O_{11}$
  - Lactose, or milk sugar, $C_{12}H_{22}O_{11}$
  - Maltose, or malt sugar, $C_{12}H_{22}O_{11}$
- **Polysaccharides**
  - Contain many sugar groups
  - Starch $(C_6H_{10}O_5)_n$
  - Cellulose $(C_6H_{10}O_5)_n$
  - Glycogen $(C_6H_{10}O_5)_n$
  - Dextrin $(C_6H_{10}O_5)_n$

Fats, or Lipids
- Used in anatomic sense = adipose tissue
- Used in chemical sense = compound of C, H, and O, but the C and H content is relatively high
- Made from one molecule of glycerin and three molecules of fatty acid. Reaction comparable to neutralization of an acid by a base
- Under influence of body enzymes, split into substances out of which they are built

Stearin + Water → Glycerin + Stearic acid

$$C_3H_5(C_{18}H_{35}O_2)_3 + 3H_2O \rightarrow C_3H_5(OH)_3 + 3H \cdot C_{18}H_{35}O_2$$

Compound Fats, or Lipoids
- Esters of fatty acids containing groups in addition
- *Phospholipids*—contain phosphorus and nitrogen, e.g., lecithin
- *Glycolipids*—compounds of fatty acids with a carbohydrate
- *Sterols*—solid alcohols combined with fatty acids—e.g., cholesterol

Proteins
- Contain C, H, O, N; usually S, sometimes P and Fe may be present

Proteins

- Differ from carbohydrates and fats in having nitrogen
- Built up of simpler substances called amino acids
- Amino acids contain an amino radical (NH_2) instead of a replaceable hydrogen
- Acetic acid—CH_3COOH—for replaceable H substitute (NH_2) $\rightarrow$ $CH_2(NH_2)COOH$ = Aminoacetic acid
- Propionic acid—C_2H_5COOH—for replaceable H substitute (NH_2) $\rightarrow$ $C_2H_4(NH_2)COOH$ = Aminopropionic acid
- About 40 amino acids have been described, and various combinations result in many different kinds of proteins, i.e., milk, meats, fish, egg, peas, beans, lentils, and peanuts
- **Examples of protein constituents of food**
  - Albumin—found in egg, milk, meat
  - Caseinogen—found in milk
  - Glutenin—gummy substance in wheat
  - Legumin—found in legumes
  - Gelatin—derived from connective tissues, including bone and tendon
  - Organic extracts—flavor of meats and some plant foods due to these
- **Classification**
  - Simple
    - Consist only of amino acids
    - Yield only amino acids or derivatives
  - Conjugated
    - Contain protein molecule united to some other molecule otherwise than as a salt—yield amino acids and some other molecule
    - Nucleoproteins—yield amino acids and nuclein
    - Glycoproteins—yield amino acids and a carbohydrate
    - Phosphoproteins—yield amino acids and a phospho-body
    - Hemoglobins—yield amino acids and hematin
    - Lecithoproteins—yield amino acids and a fatty substance
  - Derived
    - Primary—involve only slight alterations of the protein molecule
    - Secondary—products of further hydrolytic cleavage, such as proteoses, peptones, and peptides

Chemical Elements

- Fifteen or more elements enter into composition of body
- Five may be furnished by carbohydrates, fats, proteins, and water
- Ten others to be provided are:
  - Iron, calcium, sodium, potassium, magnesium, phosphorus, chlorine, iodine, fluorine, silicon

- **Chemical Elements**
  - **Function**
    - As constituents of bone
    - As essential elements of soft tissues
    - As soluble salts held in solution in fluids of body
  - **Standard daily allowance** of elements which should be stressed in daily diet
    - Calcium 0.8 gm
    - Phosphorus 1.5 gm
    - Iron 0.020 gm
    - Iodine 0.000014 gm
- **Vitamins**
  - Essential for growth and nutrition
  - Influence the metabolism of foodstuffs
  - Current research stresses determination of optimal amounts
  - Vitamin A
    - Vitamin A and the related carotenes essential to growth, and to nutrition and health at all ages
    - Recent clinical observation indicates desirability of securing optimal quantity
  - Vitamin D
    - Gives protection in childhood against rickets
    - Ergosterol transformed into vitamin D by ultraviolet light
  - Vitamin E
    - Deficiency results in sterility in both male and female rats
    - Its value in treatment of human sterility not yet proved
  - Vitamin K
    - Essential to normal clotting of blood
  - Vitamin B
    - Of multiple nature (B, G, H)
    - Essential to positive health
    - Liberal use of leafy green vegetables, unprocessed flours, and whole-grain cereals provides optimal amount
  - Vitamin C
    - *Ascorbic acid* (commercial product, *cevitamic acid*)
    - Essential to normal development of bones and teeth
    - Subclinical shortage frequent; attention should be given to optimal intake
  - Vitamin P
    - Probably of multiple nature
    - Essential for normal capillary permeability
- **Accessory Articles of Food**
  - **Flavors and condiments**
    - Have no nutritive value, but increase the secretion of gastric fluid
  - **Stimulants**
    - Tea and coffee—stimulating action due to caffeine
    - Cocoa
      - Contains carbohydrate, fat, and protein
      - Stimulating effects due to theobromine

- **Accessory Articles of Food**
  - **Stimulants**
    - Meat extracts
      - Contain secretagogues which stimulate the gastric glands to secretion
  - **Alcohol**
    - Oxidized rapidly, yields energy
    - Waste products
      - Carbon dioxide
      - Water
    - Favors obesity
      - Moderate drinking creates keen appetite and thus favors overeating
      - Lessens need for oxidation of fat or carbohydrates; hence these are spared to accumulate

- **Digestive Processes**
  - **Mechanical**
    - Include various physical processes that are preliminary to the more important chemical digestion
    - Mastication—comminution and mixing
    - Deglutition, or swallowing
    - Peristaltic action of esophagus
    - Movements of stomach
    - Movements of intestines
    - Defecation
  - **Chemical**
    - Splitting of complex substances into simpler ones
    - Process of hydrolysis that is dependent on enzymes
    - Rendered necessary by variety and complexity of foods, which must be reduced to standard and simple substances that the tissues can use, i.e.,
    - End products
      - Simple sugars
      - Glycerin and fatty acids
      - Amino acids

- **Enzymes**
  - Substances produced by living cells which act by catalysis, i.e., vary speed of reactions
  - It is suggested that each hydrolytic enzyme be designated by the name of the substance on which it acts, together with the suffix, *ase*
  - **Classification according to action**
    - (1) Sugar-splitting
      - *a*) Inverting
      - *b*) Enzymes which act on simple sugars
    - (2) Amylolytic, or starch-splitting
    - (3) Lipolytic, or fat-splitting
    - (4) Proteolytic, or protein-splitting
    - (5) Clotting enzymes
    - (6) Oxidizing enzymes, or oxidases
    - (7) Deaminizing enzymes

LIST OF DIGESTIVE FLUIDS AND CHIEF ENZYMES

Digestive Fluids	*Enzymes*	*Functions*
Saliva	Ptyalin, or salivary amylase	Hydrolyzes starch to dextrin and sugar (maltose)
	Maltase	Aids the change of maltose to glucose
Gastric	Pepsin plus hydrochloric acid	Hydrolyzes proteins into proteoses and peptones
	Rennin	Curdles the caseinogen of milk
	Gastric lipase	May initiate the digestion of emulsified fats, such as cream
Pancreatic	Trypsin, trypsinkinase (actually a group of enzymes)	In a slightly acid, neutral, or strongly alkaline medium, these enzymes act at different stages of protein decomposition and reduce proteins to amino acids
Pancreatic	Amylase, or diastase	Hydrolyzes starch to dextrin and sugar (maltose)
	Lipase, or steapsin	Splits fats to glycerin and fatty acids. This action is reversible
Succus Entericus	Enterokinase	Activates the trypsin of the pancreatic fluid or acts as a coenzyme
	Peptidases (erepsin)	Hydrolyze peptides to amino acids
	Inverting: Maltase	Hydrolyzes dextrin and maltose to glucose
	Inverting: Invertase	Hydrolyzes sucrose to dextrose and fructose
	Inverting: Lactase	Hydrolyzes lactose to glucose and galactose
	Nuclease	Acts upon the nucleic acid of nucleoproteins
Bile	No enzyme	Serves as coenzyme and activates the lipase of the pancreatic fluid

Changes Food Undergoes in the Mouth

- Mastication (chewing)—comminution and mixing
- Insalivation (mixing with saliva)
- **Saliva**
  - Secreted by salivary glands
    - Parotid, Submaxillary, Sublingual — and mucous glands of mouth
  - Craniosacral autonomic fibers
    - Carry secretory and vasodilator fibers
    - Stimulated by sight or smell of food
    - Causes a production of a copious amount of watery secretion

Changes Food Undergoes in the Mouth

- **Saliva**
  - Thoraco-lumbar autonomic fibers
    - Carry secretory and vasoconstrictor fibers
    - Stimulated by food in mouth
    - Produce a smaller amount of thicker secretion
  - Consists of water, some protein material, mucin, inorganic salts, and enzymes—**ptyalin** and **maltase**
  - Specific gravity 1.004–1.008. Neutral in reaction—*p*H 6.6–7.1
  - Amount—1 to 1.5 liters per day
  - **Functions**
    - (1) Assists in mastication and deglutition
    - (2) Serves as lubricant
    - (3) Dissolves or liquefies the food, thus stimulating the taste buds and indirectly the secretion of gastric fluid
    - (4) Ptyalin hydrolyzes starch to dextrin and maltose; maltase changes maltose to glucose
- **Deglutition** (swallowing). Passage of food through (1) fauces, (2) pharynx, and (3) esophagus. Consistency of food affects third stage

Changes Food Undergoes in the Stomach

- Time required—depends on nature of food eaten; average meal of mixed food requires 3–4½ hr
- Food held in stomach by contraction of cardiac and pyloric sphincters
- Cavity size of contents—never empty—always few cubic centimeters of gastric fluid in stomach
- When food enters, expands just enough to receive it; contractions start in middle region and run toward pylorus; food in prepyloric and pyloric regions macerated, mixed with gastric fluid, and reduced to **chyme**
- Salivary digestion continues until gastric fluid penetrates bolus of food
- **Dependent on gastric fluid**
  - Periods of fasting—secreted in small amount
  - While eating and during period of digestion—amount increased
  - Secretion
    - Psychical, or appetite, secretion
      - Sensations of eating
      - Taste and odor of food
    - Chemical
      - Secretagogues contained (1) in food and (2) in products of digestion
      - Gastric secretin

- **Changes Food Undergoes in the Stomach**
  - **Dependent on gastric fluid**
    - Secreted by glands of stomach
      - Cardiac
      - Fundus, or oxyntic
      - Pyloric
    - Acid reaction due to free hydrochloric acid
    - Enzymes
      - Pepsin
      - Rennin
      - Gastric lipase
    - Inhibited by
      - Stimulation of sympathetic system, anger, pain, fear, worry, distaste for food
      - Secretion dependent on blood, hence checked if blood supply is diverted

- **Hydrochloric Acid**
  - Secreted by parietal cells of gastric glands from chlorides found in blood
  - NaCl most abundant—chlorine combines with hydrogen to form hydrochloric acid
  - Normal proportion about 0.5%
  - **Functions**
    - Activates pepsinogen and converts it to pepsin
    - Provides acid medium for pepsin to carry on its work
    - Swells protein fibers
    - Helps in inversion of cane sugar, easiest disaccharide to hydrolyze
    - Acts as a disinfectant and may kill bacteria
    - Helps to regulate opening and closing of pyloric valve

- **Pepsin**
  - Formed in pyloric glands and chief cells of gastric glands
  - Pepsinogen—zymogen, changed by HCl to active pepsin
  - Weak proteolytic enzyme—requires acid medium
  - Hydrolyzes proteins through several stages to proteoses and peptones, which action is preparatory to more complete hydrolysis in intestine

- **Rennin**
  - Formed in chief cells of gastric glands
  - Prorennin—zymogen—secretin converts to active enzyme
  - Acts only upon casein, converts it to paracasein, which reacts with calcium to form the curd, preparatory to further action by pepsin and trypsin. This enzyme is questioned in human beings

Gastric Lipase—limited action on emulsified fats like cream

- **Functions of Stomach**
  - Serves as temporary storage reservoir
  - Contractions promote mechanical reduction of food
  - Salivary digestion continues until acidity is established
  - **Gastric digestion**
    - Pepsin hydrolyzes proteins
    - Rennin hydrolyzes caseinogen
    - Gastric lipase may hydrolyze emulsified fats
  - HCl has germicidal action

- **Digestion in the Intestine**
  - **Small intestine**
    - **Movements**
      - Peristaltic—pushes food forward slowly
      - Rhythmical—facilitates mixing with secretions
      - Pendular—back and forth
    - **Secretions**
      - Pancreatic fluid
      - Succus entericus
      - Bile
    - **Bacteria**
      - Decompose carbohydrates
      - Little or no effect on protein
    - **Time required**
      - Depends on proportions of different foodstuffs
      - Food begins to pass into large intestine $2\text{–}5\frac{1}{4}$ hr after eating, requires 9 or more hours before last of meal has passed
  - **Large intestine**
    - **Movements**
      - (1) Antiperistalsis—press mass backward toward colic valve
      - (2) Haustral churning
      - (3) Mass peristalsis moves food from one division to another
    - Secretion—contains mucin, no enzymes, alkaline reaction
    - Digestive enzymes from duodenum continue to act
    - **Bacteria**
      - Mainly putrefaction of proteins with formation of relatively toxic amines or less toxic substances as indol

- **Pancreatic Fluid**
  - Secretion controlled by hormone secretin. Discharged into small intestine during digestion
  - Clear, viscid fluid, alkaline reaction
  - **Secretion**
    - Nervous secretion caused by secretory fibers in vagus and splanchnic. It is thick, rich in enzymes
    - Chemical secretion due to secretin. It is thin, watery, poor in enzymes, alkaline
  - **Enzymes**
    - Proteolytic
      - Formerly thought one enzyme, trypsin, reduced protein to amino acids. Now thought that *trypsin* and *trypsinkinase* are necessary and act at different stages

- **Pancreatic Fluid**
  - **Enzymes**
    - Amylolytic, or amylase
      - Action similar to that of ptyalin
      - Hydrolyzes starch to achroodextrin and maltose
    - Lipolytic, or lipase
      - Hydrolyzes fats to glycerin and fatty acids
      - Emulsification aided by bile salts occurs as soon as small amount of fat is split to fatty acids and glycerin
      - Fatty acid combines with alkaline salts. Emulsification regarded as preparatory process
      - Action is reversible—synthesizes fat, characteristic of human animal, from fatty acids and glycerin
- **Succus Entericus**
  - Secretion thought to be promoted by secretin
  - Clear yellowish fluid, alkaline reaction due to sodium carbonate
  - **Enzymes**
    - Enterokinase—acts as coenzyme
    - Peptidases—hydrolyze peptides to amino acids
    - Maltase—hydrolyzes dextrin and maltose to glucose
    - Invertase—hydrolyzes sucrose to glucose and fructose
    - Lactase—hydrolyzes lactose to glucose and galactose
    - Nuclease—acts upon nucleic acid portion of nucleoproteins
  - **Secretin**
    - (1) Exists in form of prosecretin and is changed by acid of gastric fluid to secretin
    - (2) Is preformed in duodenum and is liberated by the bile
- **Bile**
  - Alkaline liquid, color may be yellow, brownish yellow, or olive green
  - Amount secreted varies with amount of food eaten, estimated at about 500–800 cc daily
  - Consists of water, bile pigments, bile acids, bile salts, cholesterol, lecithin, neutral fats, and nucleoalbumin
  - Bile salts, i.e., sodium taurocholate and sodium glycocholate, thought to stimulate activity of liver
  - Secreted continuously, enters duodenum during period of digestion
  - Digestive secretion aids action of lipase

Bile
- Excretion—eliminates toxins, metals, and cholesterol
- Antiseptic—thought to limit putrefaction

Abnormal Conditions
- Gallstones—concentrated cholesterol which crystallizes out and forms gallstones
- Jaundice—due to absorption of bile by blood; bile carried throughout body; pigments deposited in skin and whites of eyes

Feces
- **Consist of**
  - Residues of diet, microorganisms and their products
  - Excretions of digestive tube and its glands
- Contain (1) water, (2) the residues of food, (3) pigment, (4) microorganisms, (5) products of bacterial decomposition, indol, skatol. etc., (6) products of secretions, (7) mucous and epithelial cells (8) cholesterol, (9) purin bases, and (10) inorganic salts

Defecation—term applied to the act of expelling feces from rectum

CHAPTER 19

ABSORPTION, METABOLISM OF CARBOHYDRATES, FATS, AND PROTEINS BASAL METABOLISM, FOOD REQUIREMENT

By absorption is meant the passage of food in soluble and diffusible form from the cavity of the alimentary canal into the blood.

ABSORPTION

Conditions which determine the amount of absorption which takes place from any part of the alimentary canal are the area of surface for absorption, the length of time food remains in contact with the absorbing surface, the concentration of fully digested material present, and the rapidity with which absorbed food is carried away by the blood.

Absorption in the small intestine. It is in the small intestine that these conditions are most favorable for absorption; therefore it is here that the greatest amount of absorption takes place.

The circular folds and villi of the small intestine increase the internal surface enormously. It is estimated to be more than 10 sq m.

Food remains in the small intestine for several hours; during this time the most complete digestive changes occur.

The blood flows steadily within the wall of the small intestine. The blood in the capillaries is separated from the digested nutrients in the small intestine by the walls of the capillaries and the intestinal mucosa. On the intestinal side of the wall are the products of digestion and the digestive fluids. Sugars, glycerin, fatty acids, and amino acids are relatively abundant and diffuse into the blood, which contains little of these substances. The continuous digestion of foods,

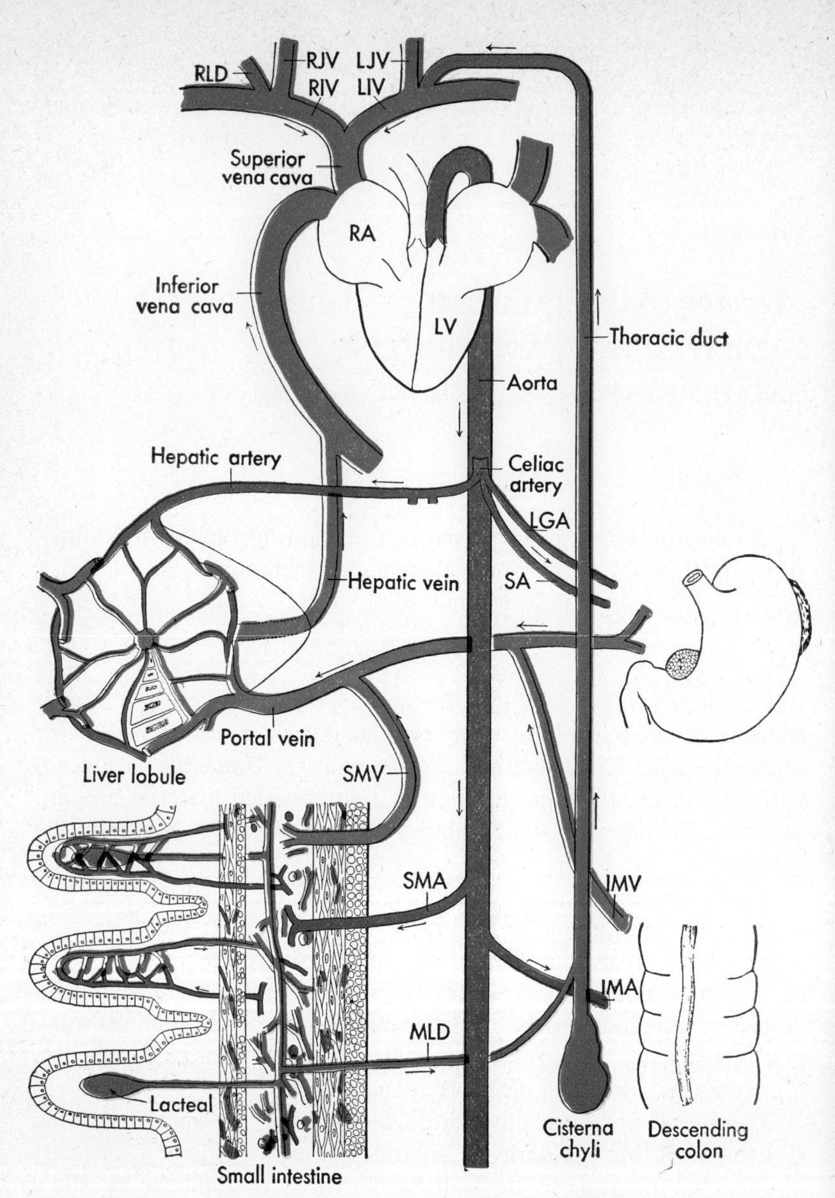

Fig. 317. Diagram to show absorption. *IMA,* inferior mesenteric artery; *IMV,* inferior mesenteric vein; *LGA,* left gastric artery; *LIV,* left innominate vein; *LJV,* left jugular vein; *LV,* left ventricle; *MLD,* mesenteric lymph duct; *RA,* right atrium; *RIV,* right innominate vein; *RJV,* right jugular vein; *RLD,* right lymph duct; *SA,* splenic artery; *SMA,* superior mesenteric artery; *SMV,* superior mesenteric vein.

the muscular activity of the intestinal wall, and the lashing and pumping activities of the villi stir up the intestinal contents and keep relatively high the concentration of absorbable materials in contact with the absorbing membrane. These motions also increase the circulation in the villi, and therefore the absorbed materials are moved on, keeping the concentration in the blood relatively low. Absorption takes place through the membrane from a constantly higher concentration of absorbable particles to a constantly lower concentration until all digested material is absorbed. The more nearly normal the muscular activity of the gastrointestinal tract, the greater will be the

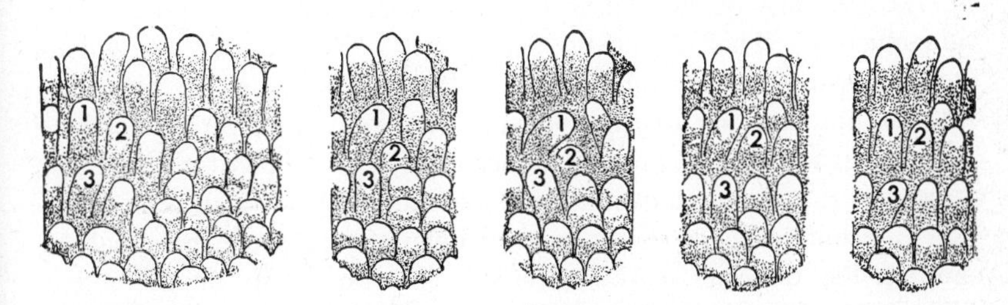

Fig. 318. Diagram of surface view of the lining of the small intestine as it would be seen if the intestine were opened and stretched under a microscope. Several villi "stand" with tips toward observer. Villus 1 bends to right and then straightens up. Villus 2 telescopes itself and then stretches out. Villus 3, bent to right, swings to left and over to right again.

blood flow through its walls, and hence the more prompt will be the absorption of the digested nutrients. It is probably in relation to gastrointestinal motility that thiamin, or vitamin B_1, plays its part in absorption. (See p. 570.)

The paths of absorption by which the products of digestion find their way to the blood are the capillaries of the villi and intestinal mucosa, and the lacteals of the villi and lymph ducts.

Fats are absorbed into the central lymph channel of each villus and forced into the larger lymphatics. After reaching the large lymph vessels, the absorbed material flows to the thoracic duct and enters the innominate vein at the junction of the left internal jugular and left subclavian veins.

The products of carbohydrate and protein digestion and probably some of the glycerin and fatty acids are absorbed by the capillaries of the villi and carried to the portal tube, which in turn carries them to the liver.

Absorption in the stomach is comparatively negligible. In the first place, the area of the gastric mucosa as compared with that of the small intestine is small; also the area of contact at any one time is small, due to comparative shapes of the two organs. The time the food remains in the stomach is short as compared with that in the intestine. The motility of the small intestine is much greater than that of the stomach. In addition to the activity of the musclar wall itself, bringing about peristalsis, rhythmical segmentations, and pendular action, there is the activity of the muscularis mucosa, giving marked motility of the villi as well as of the other superficial parts of the mucosa. This all means not only that the area of the food in contact with the absorbing surface is vastly greater in the small intestine, but that this area is constantly renewed, thus keeping concentration high. Also attention has been called to the fact that chemical digestion is not abundant in the stomach and in general the end products are not reached to any great extent until the small-intestinal digestion has done its work; moreover, as soon as any bit of food is digested and becomes soluble, it is moved on into the small intestine. The temperature of the small intestine is probably higher than that of the stomach. These facts alone would account for the lesser absorption in the stomach, whatever theory of absorption is held.

Absorption in the large intestine. When the contents of the small intestine pass the colic valve, they still contain a certain amount of unabsorbed food material. Enzymes are present, and digestion and absorption continue. The consistency is about that of chyme, because the absorption of water from the small intestine is counterbalanced by diffusion or secretion of water into it. In the large intestine the absorption of water is great, sufficiently so under usual conditions to cause the formation of semisolid or hard feces.

Absorption of water is not appreciable in the stomach. Large quantities are absorbed in the small intestine, but the most conspicuous change in the fluidity of the intestinal contents takes place in the large intestine, mainly in the ascending colon and proximal end of the transverse colon.

Absorption of mineral salts from any part of the intestines depends upon the nature of the salt and the concentration of the solution. Certain salts are readily absorbed, e.g., acetates, chlorides, and most of the ammonium salts; on the other hand, tartrates, citrates, and some of the sulphates are very slowly absorbed. To be absorbed, salts must be in higher concentration in the intestine than in the blood.

Cathartic action of salts in solution in water. The cathartic salts are very slowly absorbed from the gastrointestinal tract. For example, the cation magnesium and the anions citrate, tartrate, sulphate, and phosphate are slowly absorbed. When taken as salts they remain in the intestine for a comparatively long period of time. The tissues between the salts in solution and the blood stream form a semipermeable membrane which is readily permeable to water. Consequently fluid moves from the blood stream to the intestinal cavity, thereby increasing the bulk of the intestinal content.

The end products of digestion are simple sugars, derived from the various carbohydrates; fatty acids and glycerin, derived from the various fats; and amino acids, derived from the various proteins. (See pp. 561 to 564.)

These substances are diffusible, although the foodstuffs from which they are derived are not. Starch, even when boiled for a long time, does not make its way through ordinary membranes; the simple sugars do. Fats are not diffusible in water; fatty acids and glycerin are diffusible. Amino acids pass freely through membranes.

The simple sugars pass into the capillaries and thence by way of the portal tube to the liver.

The fatty acids and glycerin are absorbed as such in the small intestine. The bile furnishes bile salts (sodium glycocholate and sodium taurocholate), which aid in the absorption of the split fats. During their passage through the intestinal walls most of the fatty acids and glycerin recombine to form simple fats. It is believed that this synthesis is due to the action of lipase. The greater part of the fat is absorbed by the lacteals in the villi and carried to the thoracic duct, which empties into the left innominate vein. In addition it is considered probable that some of the fat is absorbed directly as fatty acid and glycerin by the capillaries of the villi and carried by way of the portal tube to the liver, before reaching the general circulation.

The amino acids pass into the capillaries of the villi, although there is experimental evidence that, after excessive feeding of protein, a portion may enter the lymphatics. Amino acids are found in the blood, which distributes them to the tissues.

METABOLISM

General metabolism includes all the processes involved from the time food enters the body until it is excreted, but ordinarily the use of the term is limited to include only the changes that occur in digested foodstuffs from the time of their absorption until their elimination in the excretions. In a more limited sense *metabolism refers to the sum total of the chemical changes which take place within cells.* It is in this sense the term is used here.

Metabolic changes may be classified under two heads: *anabolism,* or constructive processes, and *catabolism,* or destructive processes.

The changes classified as anabolic include the processes by which cells take food substances from the blood and make them a part of their own protoplasm. This involves the conversion of nonliving material into living material and is a building-up, or synthetic, process. The synthesis of glycogen and of fats within the cells is also anabolism. Comparatively little is known of the exact nature of anabolic changes.

The changes classified as catabolic consist of the processes by which cells resolve into simpler substances (1) part of their own protoplasm or (2) substances which have been stored in them. This disintegration yields simpler substances, some of which may be used by other cells, though most of them are excreted. The catabolic processes consist mainly of the simple splitting of complex molecules into smaller ones; hydrolysis, i.e., the splitting of complex molecules into simpler ones with the absorption of water; and oxidation,[1] or the union of oxygen with the constituents of the cells. In the tissues, the participation of oxygen in the chemical changes of the body forms an integral part of the processes of nutrition.

The functions of metabolism as generally stated are the manufacture of protoplasm in growth and repair of tissue and the release of energy.

Factors which promote metabolic changes are oxygen absorbed from the lungs, enzymes formed by the tissue cells, hormones formed by the ductless glands, vitamins furnished by food, and the nervous

[1] The addition of oxygen to a substance or the withdrawal of hydrogen from it. Oxidation and reduction both occur at the same time. An oxidation of one substance means reduction of another substance. Oxidation is also defined as the removal of electrons and reduction as the adding of electrons. There are other definitions.

system. The metabolism of each of the foodstuffs will be considered separately.

METABOLISM OF CARBOHYDRATES

Metabolism of carbohydrates may be considered under three headings—supply, storage, and consumption.

Supply is regulated by the diet.

Storage is temporarily provided for by the liver, the muscles, and the cells of the tissues. During the process of digestion the carbohydrates are changed to simple sugars. Absorption of glucose takes place mainly into the capillaries of the small intestine. These capillaries pour their contents into the portal tube, which carries the blood, rich with glucose, to the liver. The liver cells take this glucose from the blood, and by dehydrolysis soluble glucose is condensed to insoluble glycogen, which is stored in the liver cells. In thus storing up glycogen and doling it out as needed, the liver helps to maintain the normal quantity of glucose—0.08 to 0.12 per cent—in the blood. From the blood stream glucose is taken up by the muscles and other tissues and stored as glycogen until needed. Hence the liver serves as a central storehouse from which the tissues receive a supply whenever their content of glycogen is depleted. The percentage of glycogen in a muscle is small, though the total content of all the muscle cells may equal that of the liver. The maximum storage of glycogen in the body is about 400 gm, or nearly 1 lb. The conversion of glycogen into glucose as it is required by the tissues is thought to be effected by glycogenase, an enzyme contained in the liver cells. The need of the blood for glucose is constant, because it is constantly giving up glucose to the tissues.

The amount of sugar oxidized is controlled by the energy needs of the tissues, particularly muscle tissue, for their activity is the principal factor determining the rate of oxidation; and naturally the amount of fuel required will be in proportion to the rate at which it is used.

Regulation of blood sugar. The regulation of the amount of sugar in the blood is important. Several processes are concerned: the production of glycogen in the liver (glycogenesis), the production of glucose in cells from proteins and from fats (gluconeogenesis), the conversion of glycogen to sugar according to body needs (glycogenolysis), the consumption of sugar in the tissues (glycolysis), and the possible loss of sugar through the kidneys (glycosuria).

These processes must be regulated and adapted one to another. Just how this regulation is effected is not definitely known.

Insulin, the hormone of the pancreas, is essential for the normal course of sugar metabolism. In derangements of carbohydrate metabolism insulin restores the power to utilize the glucose of the blood and accelerates the synthesis of sugar to glycogen and the storage of glycogen in the liver and muscles. It restricts the production of sugar in the liver from protein and fat, a process which probably occurs under normal conditions but is greatly increased in diabetes. It also counteracts the tendency to acidosis.

Insulin is ineffective when taken by mouth, because it is decomposed in the alimentary canal; consequently, it is injected under the skin. The reduction of blood sugar by insulin does not necessarily stop at the normal level; if it proceeds further, prostration may occur. Such reaction is avoided by taking orange juice or sugar when the sensation of weakness is first felt. Best and Taylor (1955) state that the incidence of hypoglycemic reaction with protamine zinc insulin is much less than with regular insulin, and that from 30 to 40 per cent of the total insulin used in the United States and Canada is now in this form. The improvement of patients receiving insulin is marked, but insulin does not cure diabetes. It is palliative in character, and usually dependence on the insulin continues. Diabetes affects chiefly individuals who have allowed themselves to become overweight. In later life, the person who is of normal or less than normal weight is practically insured against diabetes.

Other hormones which influence glucose metabolism include: thyroxin, which increases the rate of oxidation in tissue cells and increases the rate of glycogen formation from noncarbohydrate sources; an anterior pituitary hormone (APH) which inhibits glucose utilization in the tissues and increases glucose formation from glycogen and noncarbohydrate sources; cortin, which decreases glucose oxidation in tissues and increases formation of glycogen from noncarbohydrate sources.

It may be that there is a *sugar-regulating center* in the medulla which controls the conversion of glycogen to sugar; some think that control is exercised indirectly through the medulla of the *adrenal glands.* An increase of sugar concentration in the blood is followed by elimination of sugar in the urine (glycosuria). An instance of such glycosuria is that following strong emotion. One result of emotional excitement is an increased secretion of epinephrine into the blood, and one of the effects of epinephrine is to convert a large amount of glycogen to glucose. It is thought that this release of sugar may be a provision to supply fuel to the muscles under conditions which usually call for strenuous action.

Functions of carbohydrates. The oxidation of glucose serves the following purposes: It furnishes the main source of energy for muscular work. The glycogen of a muscle disappears in proportion to the work done by the muscle, and it is thought the oxidation of the sugar furnishes the energy which is utilized by the muscles. It furnishes an important part of the heat needed to maintain the body temperature. The oxidation of each gram of sugar yields 4 calories of heat; and since the carbohydrates form the largest part of our diet and are easily oxidized, they must be regarded as specially available material for keeping up body heat. It prevents oxidation of the body tissues, because it constitutes a reserve fund that is the first to be drawn upon in time of need. As carbohydrate food is increased, protein food may be diminished to a certain irreducible minimum, which is probably the amount necessary for the reconstruction of new tissue. Carbohydrates, in excess of the amount that can be stored as glycogen in the liver and muscles, are converted into adipose tissue. Nutritional experiments show that the fat of the body may be formed from carbohydrate food. It permits complete oxidation of fatty acids.

Waste products of carbohydrate metabolism. Eventually the glucose derived from the glucose of the blood or from the glycogen of the cell is oxidized by the cell to *carbon dioxide* and *water*.

Derangements of carbohydrate metabolism. The sugar-regulating mechanism of the body may prove inadequate.

The mechanism of conversion of sugar to glycogen in the liver (glycogenesis) breaks down, giving rise to alimentary glycosuria, i.e., sugar in the urine. This may follow the ingestion of a larger amount of sugar than the liver and muscles can store, resulting in an increased amount in the blood (hyperglycemia). A higher percentage of sugar than normal in the blood is irritating to the tissues, and the sugar is excreted in the urine. This is designated as temporary glycosuria.

The mechanism of conversion of glycogen to sugar in the liver (glycogenolysis) breaks down in injuries to the central nervous system, excessive internal secretion by the suprarenal glands, etc.

The mechanism of consumption of sugar in the tissues for energy purposes (glycolysis) breaks down in diabetes mellitus. Removal of the pancreas of an animal is followed by the appearance of sugar in the urine in large amounts. If the extirpation is complete, glycosuria is followed by emaciation and muscular weakness, which finally end in death in 2 or 3 weeks. On the other hand, if a portion of the pancreas is left, even though its connection with the duodenum is interrupted, it may prevent glycosuria partly or completely. This indicates that the internal secretion is the important factor in the metabolism of sugar.

The normal impermeability of the kidney breaks down in phloridzin diabetes. When injected into an animal, this drug causes a temporary glycosuria, which is very complete as long as the action of the drug lasts. Examination of the blood reveals the fact that the percentage of sugar is not increased, so that the immediate cause of the glycosuria is different from that responsible for diabetes of man or of animals without the pancreas. It is thought that phloridzin acts on the kidney itself.

Diabetes mellitus. In mankind derangements of carbohydrate metabolism manifest themselves chiefly in the disease known as diabetes mellitus, the early symptoms of which are excessive secretion of urine containing abnormal amounts of glucose and urea, and abnormal thirst and hunger. In this disease the daily loss of sugar in the urine may be very large. In severe cases all the carbohydrate of the food may be excreted in the form of sugar; and even when no carbohydrate food is eaten, sugar continues to be excreted in considerable amounts. In the latter case the sugar is supposed to have its source in the proteins of the food or of the tissues. The opinion of experts in this field is that a lesion in the islets of Langerhans in the pancreas results in a reduction of the supply of insulin to the body, and in consquence the tissues cannot use the sugar brought to them by the blood. In addition to the sugar found in the urine in diabetes, this secretion may contain considerable amounts of the acetone bodies. These acetone bodies are intermediary products in the metabolism of fats, and their presence is due to incompleteness in the normal process of the oxidation of fat. The accumulation of acetone bodies in the blood and tissues of the diabetic is responsible for the condition called *acidosis.*

METABOLISM OF FATS

The results of experimental work confirm the view that after fat is split into glycerin and fatty acids by the lipase of the pancreatic fluid, it is absorbed by the epithelial cells and in the act of passing through them combines to form fat. This combination is brought about by lipase. The greater portion of the fat passes into the central lymph channel of each villus. From these small lacteals it finds its way through the larger lymphatics in the mesentery to the thoracic duct and then through the thoracic duct to the blood. It seems probable that some of the fat is absorbed by the capillaries in the villi, enters the portal tube, and passes through the liver before reaching the general circulation. Fat is carried by the blood to all parts of the body, and the tissues slowly take it out as they need it in their metabolic processes. Within the tissues it serves as fuel and is oxidized to supply the energy needs of the cells.

Functions of fat. The fat absorbed as food may serve several purposes. It may be oxidized with the liberation of energy. From a chemical standpoint, fats make available more energy, weight for weight, than the proteins or the carbohydrates. If fat is eaten and absorbed in excess of the actual needs of the body, the excess is stored in adipose tissue and represents reserve nourishment to be drawn upon in time of need. Fat may be combined with other substances and form compound fats such as lecithin. The fat of the active tissues of the body, as distinguished from deposits of fat called adipose tissue, consists largely of compound fats and fatlike substances. The glyceryl radical of fat may be converted to glucose.

Oxidation of fat in the body. It is thought that the first step in the splitting of fat is brought about by the lipase found in the tissues. The fat stored in adipose tissue in various parts of the body (e.g., under the skin, peritoneum, etc.) does not undergo oxidation in these places. In time of need it is reabsorbed by the blood and redistributed to the more active tissues. It is thought that lipase controls the output of fat to the blood, just as the liver enzymes control the supply of sugar in the blood. After the action of the lipase, oxidation takes place in a series of steps which reduces the higher fatty acids to simpler ones, and these in turn are oxidized to carbon dioxide and water.

Origin of body fat. The modern view is that the fat of the body is formed from the fats, carbohydrates, and proteins of the food. Proteins are usually a small part of the daily diet, and it is thought that body fat is formed from fat and carbohydrate foodstuffs first. If the amino acids resulting from the digestion of protein food are not built into body protein, they are deaminized, and the organic acid radical left may be converted to sugar, glycogen, and fat.

Derangements of fat metabolism. When fats are completely oxidized, the waste products are carbon dioxide and water. It is thought that such complete oxidation cannot occur unless sugar is being oxidized at the same time; hence the saying, "Fats burn in the flame of carbohydrates." When carbohydrate is removed from the diet and the glycogen of the body has been depleted, during fasting or underfeeding and in severe diabetes, the body lives on its fats and proteins. Under these conditions the oxidation of fats is incomplete, and acetone bodies are found in the urine. Their presence indicates acidosis, which may lead to coma and death.

Because of their tendency to form acetone bodies under certain conditions, fatty acids are said to have ketogenic properties. Certain amino acids and the proteins of which they are constituents are also ketogenic. On the other hand, carbohydrates, the sugar-forming amino acids, and glycerol are antiketogenic.

Obesity is said to be of two kinds. One is caused by eating more food than the body needs, lack of exercise, or both. A diet that is rather bulky but not highly nutritious, including fruit and the coarser vegetables, is recommended for this type of obesity. The second kind is associated with endocrine disturbances. Castration, the menopause, disease of the hypophysis, myxedema, and other physiological and pathological disturbances are usually, though not always, accompanied by deposits of abnormal amounts of fat.

METABOLISM OF PROTEINS

As a result of digestion, proteins are hydrolyzed to amino acids, which are absorbed by the blood capillaries of the villi, pass into the portal tube, are carried through the liver into the blood of the general circulation, and are distributed to the tissues. The tissues select and store certain of these substances, and in each organ subsequent use is made of them, either to build up new tissue or to repair the wastes of metabolism. Thus amino acids constitute the form in which protein food is presented to the tissues, just as glucose constitutes the form in which carbohydrate food is presented. Amino acids not used in building up protoplasm are broken down or deaminized in the liver. In deaminization (NH_2) groups are removed from amino acid molecules. This nitrogenous portion is split off as ammonia and is converted to urea[2] or ammonium salts and eliminated by the kidney.

The nonnitrogenous portion (organic acid group) which is left after deaminization is oxidized to furnish energy or is built up into glycogen and fat and is oxidized later. Therefore, this portion of the amino acid may be regarded as a source of energy equivalent to that furnished by the carbohydrates and fats. According to this, it is obvious that some of the amino acids are selected to construct tissue protein, and the balance not needed for this purpose serves to supply energy.

The blood contains amino acids at all times. Fasting does not

[2] NH_2 is split off with enough hydrogen to form ammonia, NH_3. Ammonia combines with carbonic acid to form ammonium carbonate.

free the tissues from them; neither does a high-protein diet result in any great increase in the blood or tissues. Amino acids are considered intermediary products in the building up and breaking down of body protein. Both the building up and breaking down are thought to occur in all the tissues.

Endogenous and exogenous metabolism. Exogenous metabolism is the metabolism of all the protein ingested in excess of that required for tissue maintenance and growth. Endogenous metabolism is the use of proteins in tissues for repair and growth.

Nucleoproteins are conjugated proteins (p. 565). They are found in the nuclei of cells and are abundant in the nucleated cells of the glandular organs, such as the liver, pancreas, thymus, etc. Foods rich in nucleoproteins are sweetbreads, kidney, roe, liver, and sardines. Foods with a fairly high nucleoprotein content are beef, veal, mutton, pork, chicken, turkey, goose and other game, fish (cod excepted), spinach, asparagus, and beans. In the course of digestion the protein is separated from the nucleic acid and is eventually reduced to amino acids. The nucleic acid gives rise to substances known as purine bodies. Uric acid is the end product of the metabolism of purine bodies, from which it arises as a result of oxidation. There is some difference in the products of the hydrolysis of tissue nucleoproteins and food nucleoproteins, but the process is the same, and uric acid is the waste product of both. Uric acid resulting from the oxidation of the nucleoproteins of the tissues is classed as endogenous. Uric acid resulting from the oxidation of the nucleoproteins of food is classed as exogenous.

Nutritive value of different proteins. Proteins vary in their constituents and in their nutritive value. Because of this they are classed as *adequate* proteins, or those containing all the constituents for the growth and maintenance of the body, and *inadequate* proteins, which furnish material for energy needs but not for growth and the repair of tissue waste. The difference between the two kinds seems to lie in the character of the amino acids of which they are composed. Gelatin is an example of an inadequate protein. It is easily digested and absorbed, undergoes oxidation, which results in the liberation of energy and the production of urea, carbon dioxide, and water, but it does not supply all the material needed for the repair of tissue waste. On the other hand, the casein of milk and the glutenin of wheat contain all the essential amino acids and can furnish energy and build tissue. Another group of food proteins includes those which, while giving adequate quantities of some amino acids, give much too large quantities of others.

BASAL METABOLISM

Sources of the body's energy. After absorption, carbohydrates may, under normal conditions, be oxidized or stored as glycogen or changed into fat; fat may be oxidized or stored, and its glyceryl radical may be converted into carbohydrate; protein absorbed as amino acids may be built up into tissue protein or deaminized and oxidized, or it may yield carbohydrate or share in the production of fat. The body is able to use any or all of the foods as fuel to produce energy. Consequently, the most convenient way to compare food values is in terms of their energy value.

The energy of food is not used directly as heat, but it is customary to measure it in heat units. To determine the amount of energy produced by the oxidation of food, the amount of heat produced on oxidation is measured by a calorimeter in terms of calories. A large calorie is the amount of heat required to raise the temperature of 1 kg (2.2 lb) of water 1° C or that of 1 lb of water 4° F. The large calorie (Cal) is the one referred to in physiology. When undergoing complete oxidation in the bomb calorimeter, the foodstuffs yield the following:

Carbohydrate	1 gm—4.10 Cal
Fat	1 gm—9.45 Cal
Protein	1 gm—5.65 Cal

The oxidation of protein in the body is never quite complete, for the urea, creatinine, uric acid, etc., eliminated in urine still contain about 1.3 Cal per gram of protein catabolized by the cells. Hence, protein yields to body cells only 4.35 Cal (5.65 to 1.3 calories) per gram of protein received by them. The cells oxidize the carbon and the hydrogen completely but only partially oxidize the nitrogen of the proteins which they receive. It is said that on the average about 98 per cent of the carbohydrates, about 95 per cent of the fats, and about 92 per cent of the proteins ingested are absorbed and reach the body cells. Some deductions also are made for what are called *losses in digestion.* The practical figures used in estimating the fuel value of food are therefore:

Carbohydrate	1 gm—4 cal (98% of 4.10 Cal)
Fat	1 gm—9 cal (95% of 9.45 Cal)
Protein	1 gm—4 cal (92% of 4.35 Cal)

Basal metabolism is the rate of energy metabolism required to keep the body alive. The term *basal metabolism* is used clinically to indicate the rate of energy metabolism of the body when the subject is lying quiet and relaxed in a room of comfortable temperature in what is called "postabsorptive" state, i.e., 12 to 18 hours after the

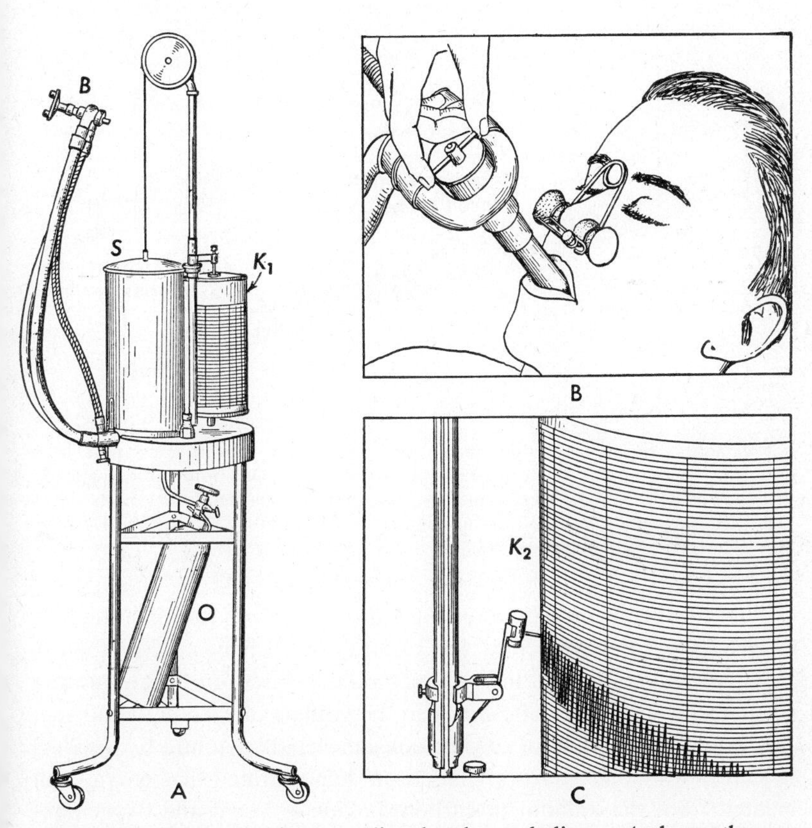

Fig. 319. Apparatus for recording basal metabolism. *A* shows the apparatus complete; *B*, mouthpiece enlarged; *C*, recording being made on a kymograph. K_1, kymograph in position; K_2, kymograph enlarged, showing writing pen in position and a part of its written record. The pen is attached to the spirometer bell and moves up and down with respirations. *O*, oxygen tank; *S*, spirometer bell. (Courtesy of Warren E. Collins, Inc.)

last meal. The digestion and absorption of the meal should be completed, and only such expenditure of energy should occur during the test as is required to maintain body warmth, cellular tone, respiration, and circulation; or, briefly, conditions should represent *functional activity at a minimum.* The basal rate of energy metabolism is used as a starting point for the calculation of total energy requirements in food under varying conditions and as a basis for diagnosis.

Basal metabolism may be determined by *the direct method,* in which the subject is placed in a respiratory chamber and the amount

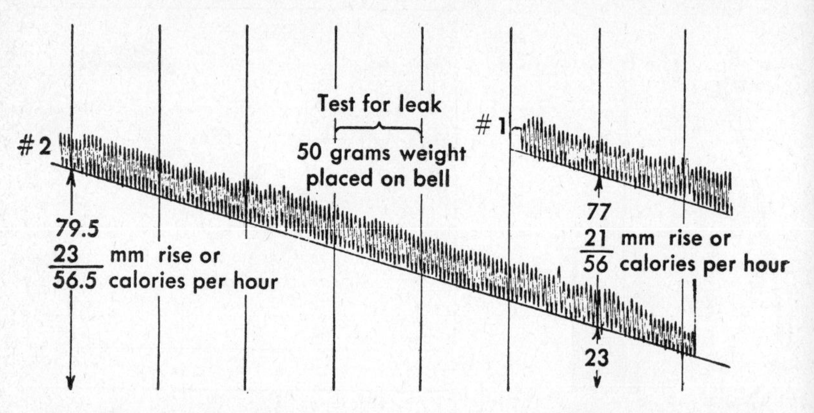

Fig. 320. Basal metabolic rate. Rise in oxygen line is 56.5 mm in 6 minutes. A woman's body surface is 1.53 sq m (Du Bois table), age 27 years. Normal calories per hour, 56. Actual calories per hour, 52.2. This gives a basal metabolic rate of —6.7 per cent (within normal range). Respiration 20, pulse 67, temperature 98.6°F.

of heat evolved is measured, or *the indirect method,* in which the heat given off is computed from the respiratory exchange. Metabolism rates determined by the indirect method are based on the respiratory quotient, which is the ratio between the volume of carbon dioxide excreted and the volume of oxygen consumed. It is found by dividing the former by the latter. It has been demonstrated that energy calculated from the amount of carbon dioxide excreted and oxygen absorbed by a subject is equivalent to the heat given off by the body (pp. 681, 682).

The amount of oxygen required to burn a given amount of carbohydrate, fat, or protein is not the same. In the combustion of carbohydrate, the volume of carbon dioxide produced is equal to the volume of oxygen absorbed. The respiratory quotient is, therefore,

$$\frac{\text{Volume } CO_2 \text{ produced}}{\text{Volume } O_2 \text{ consumed}} \text{ or R.Q.} = 1$$

There are slight variations in the respiratory quotients for different fats, owing to differences in molecular weight. For human fat, the quotient is 0.703. For protein, the quotient is 0.8 to 0.82. These figures show that the carbon dioxide produced is generally less than the oxygen which has disappeared in the exchange.

If the combustion of carbohydrate alone were possible, the respiratory quotient would be 1; if only protein were burned, it would be 0.80 to 0.82; if fat, about 0.7.

Under ordinary conditions, the respiratory quotient is about 0.85, but it may vary within rather wide limits.

Basal metabolism in terms of body weight is not often determined, but it is sometimes convenient to use this method of estimation. If the basal rate is 1 Cal per kilogram per hour, the adult standing 5 ft 8 in. tall and weighing 70 kg would have a basal metabolism of 1,680 Cal for 24 hr ($1 \times 70 \times 24$).

Basal metabolism calculated on the basis of surface area of the body is more nearly accurate than that determined in terms of body weight, because heat loss increases in proportion to surface rather than weight. A table for this purpose has been worked out.[3] According to this table, a woman standing 5 ft 4 in. tall and weighing 56 kg with 1.6 sq m of body surface will have a basal metabolism of about 36.9 Cal per square meter per hour with a total basal metabolism of 1,400 Cal per day ($1.6 \times 24 \times 36.9$). Likewise, a man of 50 yr of age, weighing 70 kg and measuring 5 ft 8 in. in height, will have a body surface of 1.83 sq m, and his basal metabolism will be about 1,700 Cal per day ($1.83 \times 24 \times 39.7$).

Variations in basal metabolism. It has been found that a number of factors, such as age, sex, sleep, and internal secretions influence the basal metabolic rate. In women the rate is lower than in men; it gradually decreases with age; it may be increased by systematic exercise over a long period; prolonged undernourishment reduces it; certain races appear to have a slightly higher rate than others; emotional tension increases it; high barometric pressures reduce it.

[3] E. F. Du Bois, *Basal Metabolism in Health and Disease.* H. C. Sherman, *Chemistry of Foods and Nutrition,* 1952.

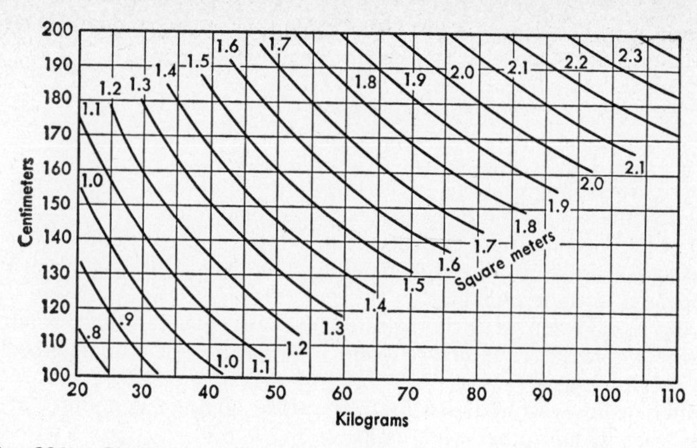

Fig. 321. Chart for determining surface area of adults from weight and height. Height in inches divided by 0.393 gives height in centimeters. Weight in pounds divided by 2.2 gives weight in kilograms. (Courtesy of Dr. Eugene Du Bois and the *Archives of Internal Medicine.*)

DAILY CALORIE REQUIREMENT

If constant weight is to be maintained showing neither gain nor loss, the daily output of calories (as determined by calorimeter tests) must be balanced by calories taken in. On such data the daily food requirement in terms of calories is based. Since basal metabolism represents the heat given off when physiological work is at a minimum (in the morning after a comfortable night's rest, relaxed in bed, serene, before breakfast), it is obvious that any increase in physiological activity, even the slightest (sitting up under the same conditions), increases the metabolic rate over the minimum. Comparison of the basal metabolic rate with the metabolic rates during various types of work shows that work results in an increment in the amount of heat eliminated. Muscular activity, even the slightest, increases the metabolic rate. Such increases have been estimated and graded for the average individual according to the calories required as follows:

	Calories per Hour
Very light work, or sitting at rest	100
Light work	120
Moderate work	175
Severe work	350

Total calorie requirement. The metabolic rate is reduced about 10 per cent beyond the basal level during sleep. Allowance is made for this in estimating the total daily calorie requirement of an individual. On the other hand, the process of digestion of food itself brings about a need for an increase of 6 to 10 per cent of the calorie intake. This is called the *specific dynamic action of food* (S.D.A.), and allowance must be made for it. After consideration of all these factors the total calorie requirement of the average man will be found to be about as follows:

	Calories
8 hr of sleep at 65 Cal	520
2 hr of light exercise at 120 Cal	240
8 hr of moderate work at 175 Cal	1,400
6 hr sitting at rest at 100 Cal	600
	2,760
6–10 per cent for S.D.A.	250
Total requirement for 24 hr	3,010

It will be seen that increase in weight or loss of weight is to be determined largely by control of the caloric intake of food. The three types of food should be considered from the standpoint of their relative nutritional values and of the amount of energy the oxidation of each releases.

The protein requirement of food should be considered from the standpoint of both quantity and quality. Quantitatively, it is estimated that it should be about 1 gm of protein per kilogram of body weight per day, or approximately 50 to 100 gm. Children require, for growth, a higher intake per kilogram of body weight per day. With 75 gm at 4 Cal per gram, this would represent 300 Cal daily. Qualitatively, it is considered desirable to secure amino acids similar to those of human tissues, both as to kinds and relative quantities of the various kinds. It has been suggested that protein foods be grouped into two classes: proteins containing our tissue proteins in approximately the relative quantities to be found in our bodies—these include animal proteins in general and some plant proteins, such as wheat proteins; and proteins in which one or more of our amino acids are lacking (e.g., gelatin lacks tryptophane) or in which there is a preponderance of one or more amino acids needed by the body in only a small quantity (e.g., the chief protein of flour contains about 35 per cent of glutamic acid, whereas milk has only about 22 per cent, which is considered adequate). Certain amino acids such as tryptophane, histidine, tyrosine and sulfur-containing amino acids such as cystine and methionine must be included in the diet since they cannot be made by the body.

Glutamic acid is one of the few amino acids which determine the specific dynamic action of proteins. Since the specific dynamic action of food is greatest in proteins, a diet high in proteins is not suitable for severe muscular activity or in hot weather. On the other hand, through its high specific dynamic action, protein contributes much toward the regulation of body temperature and tends to offset increased heat production due to low environmental temperatures.

Nitrogen equilibrium. Authorities in the field are not in agreement on the question of the desirable protein standard for the diet of an adult. Nitrogen continues to be excreted in the urine even though the diet is devoid of nitrogen. This represents a condition in which the body is oxidizing its own tissues to supply its needs. The nitrogenous portion of the protein molecule of ingested proteins is not stored in the body but is eliminated chiefly in the urine and to a limited extent in the feces. It is therefore important that the body receive daily an amount of protein nitrogen equal to the amount eliminated in the excreta. When this condition exists, the body is said to be in nitrogen equilibrium. If there is a plus balance in favor of the food, it means that protein is being made into body protoplasm; and this is an ideal condition during the period of growth or convalescence from wasting illness. If the balance is minus, it means that the body is oxidizing its own protein. Minimum nitrogen equilibrium can be maintained on about 40 gm of protein or less per day, but it is thought that higher protein intake results in greater resistance to disease and a higher state of physiological efficiency. It is customary to add 50 per cent to the average indicated as the actual requirement based on laboratory experiments, thus bringing the amount up to 60 to 100 gm, which is somewhat more than 1 gm per kilogram of body weight.

Fat requirement. It is recommended that 25 to 35 per cent of the daily calories be in the form of fat. This should be not more than 20 per cent of the carbohydrate grams, or about 65 to 100 gm of fat at 9 Cal per gram (585 to 1,050 Cal) daily. Qualitatively, fats should be similar to those of human tissue both as to kinds and relative quantities of the kinds. Recent evidence indicates the necessity of including some of the unsaturated fatty acids such as linoleic acid and linolenic acid, which are found in most fatty foods.

The carbohydrate requirement is estimated to be about 350 gm, or 1,400 Cal. The amount taken advantageously depends upon the form. The assimilation for sugar varies with the kind but is lower than for starch. In the ordinary diet of a healthy individual, carbohydrates tend to predominate, and there is seldom necessity for estimating fat and carbohydrate separately.

The water requirement should balance the daily outgo, which is about 1,300 cc in urine, 500 cc from skin, 500 cc from lungs, and 100 cc in feces, or a total of 2,400 cc. From this should be deducted 500 cc, representing the water derived from the catabolic processes in cells. The remaining 2,100 cc, about 2½ qt, represent the desirable daily intake of water in beverages and food to balance that lost.

In a diet of 3,000-Cal energy value, the proportions of the main constituents should be approximately as shown below.

	Calories	*Approximate % of Total Calories*	*Grams*
Carbohydrate	1,440	48	380
Fat	1,200	40	133
Protein	360	12	90

Nutritional need considers the means by which the lymph may be kept at optimum constant with regard to the kinds and concentrations of nutrients for the individual cells. The first requisite is an adequate caloric intake to meet the needs of basal metabolism, muscular activity, environmental temperature, etc. With a given caloric intake the food must be of sufficient variety to supply all the nutrients essential to the cells of the body in the proper physiological concentrations; i.e., essential amino acids, carbohydrates, glycerin, essential fatty acids, minerals, vitamins, and water. In general, natural foods or cells are the best sources, especially from the viewpoint of concentrations. The water balance will be maintained with an intake of about 2,100 cc daily, of which about 800 cc will be obtained from food.

Bulk must be considered not only from the viewpoint of total bulk but also in relation to its distribution along the alimentary canal. A diet containing many plant cells contains more indigestible cellulose, and a given caloric intake with a proportionately larger quantity of proteins and carbohydrates than fats will increase bulk since these types of food yield only half as many calories per gram as do fats. Bulk will also be determined by the total quantity of the food and water intake. Obviously, too, the number of meals, the "rate" of eating, and the perfection of digestion and absorption will influence the distribution of bulk along the digestive tract.

Regular spacing of meals tends to prevent overhunger or lack of appetite. Recent investigations indicate the desirability of more frequent meals. Emotional states and fatigue must also be considered as factors in digestion; undue fatigue is often prevented or allayed by an extra intake of readily digested food. Frequent meals of relatively small amounts of food and leisurely enjoyment of meals will tend to bring about a proper distribution of food along the alimentary canal and therefore less variability in concentration of nutrients in body fluids.

SUMMARY

- **Absorption**
  - Passage of digested food material from the cavity of the alimentary canal to the blood
  - **Determining conditions**
    - (1) Area of surface for absorption
    - (2) Length of time food is in contact with absorbing surface
    - (3) Concentration of digested material present
  - **Small intestine**
    - Above conditions are realized in small intestine
    - (1) Circular folds and villi increase internal surface
    - (2) Food remains for several hours
    - (3) Products of digestion higher in intestine, lower in blood
  - **Paths of absorption**
    - (1) Capillaries of villi absorb sugars, amino acids, and some of the glycerin and fatty acids, carry them to portal tube, then to liver

- **Absorption**
  - **Paths of absorption**
    - (2) Central lymph channel of villus absorbs glycerin and fat, empties into larger lymph vessels, then into thoracic duct, superior vena cava, and right atrium of heart
  - **Stomach**
    - Alcohol and alcoholic solutions absorbed
    - Small amounts of sugar, amino acids may be absorbed
    - Water not absorbed
  - **Large intestine**
    - Limited absorption of digested foodstuffs, marked absorption of water
- **Place of Absorption of Digested Foodstuffs**
  - Water—absorbed in small intestine but loss made good by secretion, marked absorption in large intestine
  - Salts—absorption may take place from any part of intestines, depends upon concentration of solution and nature of salt
  - Simple sugars—pass into capillaries and then by way of portal tube to liver
  - Fatty acids and glycerin—absorbed by lacteals in villi, synthesized to form fat. Fat carried to thoracic duct, which empties into left subclavian vein
  - Amino acids—pass into capillaries of villi
- **Metabolism**
  - May include all processes involved from time food enters the body until waste is excreted
  - In this chapter metabolism is limited to include only changes that occur in cells
  - **Consists of**
    - **Anabolism**—processes by which living cells take food substances from the blood and make them into protoplasm and stored products
    - **Catabolism**—processes by which living cells change into simpler substances (1) part of their own protoplasm, or (2) stored products
    - Catabolic processes
      - (1) Simple splitting of complex molecules into simpler ones
      - (2) Hydrolysis, or the splitting of complex molecules into simpler ones with the absorption of water
      - (3) Oxidation, or the union of oxygen with the constituents of the cells. Oxidation and reduction occur at the same time

- **Metabolism**
  - **Functions**
    - Growth and repair of tissue
    - Release of chemical energy in the form of heat, nervous activity, muscular activity, etc.
  - **Factors**
    - (1) Oxygen absorbed from lungs
    - (2) Enzymes secreted by tissue cells
    - (3) Hormones secreted by ductless glands
    - (4) Vitamins furnished by food
    - (5) The nervous system

- **Metabolism of Carbohydrates**
  - Supply regulated by diet
  - Storage provided for, temporarily, by liver, muscles, and cells of tissues. Simple sugars stored as glycogen
  - Consumption controlled by energy needs of tissues
  - **Processes**
    - Glycogenesis, or the production of glycogen in the liver
    - Glycogenolysis, or the conversion of glycogen to sugar according to body needs
    - Glycolysis, or the consumption of sugar in the tissues
    - Glycosuria, or loss of sugar in the urine
  - **Regulation of blood sugar**
    - **Insulin,** the hormone of the internal secretion of the pancreas, is essential. It regulates the metabolism of glucose. In derangements of carbohydrate metabolism it produces four marked effects
      - (1) Restores power to utilize glucose of blood
      - (2) Accelerates synthesis of sugar to glycogen
      - (3) Restricts production of sugar from protein and fat
      - (4) Counteracts tendency to acidosis
      - **Administration**
        - Ineffective by mouth
        - Injected under skin
    - There may be a sugar-regulating center in medulla
    - Control may be exercised indirectly through adrenal glands
    - Influenced by emotional excitement
  - **Functions**
    - Furnish main source of energy for muscular work and all the nutritive processes
    - Help to maintain body temperature
    - Protect body tissues by forming reserve fund for time of need (glycogen)
    - Excess carbohydrates are converted into adipose tissue
    - May be used in constructive processes
    - Assist in oxidation of fats

Metabolism of Carbohydrates

- **Waste products**
  - When completely oxidized, the waste products are carbon dioxide and water
- **Derangements of**
  - (1) Glycogenesis breaks down, giving rise to alimentary glycosuria
  - (2) Glycogenolysis breaks down
  - (3) Glycolysis breaks down
  - (4) Normal impermeability of kidneys breaks down

Metabolism of Fats

- **Reconstruction**—in act of passing through epithelial cells of villi, glycerin and fatty acids combine to form fat
- **Dependent upon lipase**
- **Functions**
  - Serve as fuel, yield heat and other forms of energy
  - Stored as adipose tissue
  - Synthesized to form compound fats and fat-like substances
  - Glyceryl radical may be converted to glucose
- **Oxidation**
  - First step brought about by lipase
  - Lipase controls output of fat to blood
  - Oxidation takes place after lipase has acted
  - Waste products are carbon dioxide and water
  - When oxidation is incomplete, acetone bodies are formed
- **Body fat**
  - Formed from fats, carbohydrates, and proteins of food in order named
- **Obesity**
  - May be caused by eating more food than body needs, by lack of exercise, or both
  - May be due to endocrine disturbances

Metabolism of Proteins

- Absorbed as amino acids. From villi pass to portal tube, thence to liver and general circulation
- Tissues select amino acids
  - (1) To build new tissue
  - (2) To repair wastes of metabolism
- Amino acids not used in building up protoplasm are deaminized and split into nonnitrogenous and nitrogenous portions
- Nonnitrogenous portion oxidized to CO_2 and H_2O, or converted into glycogen
- Nitrogenous portion passes through a series of changes—final waste product urea
- **Classification**
  - Endogenous, includes building up of amino acids to tissue protoplasm and final disintegration to creatinine and urea
  - Exogenous, includes reactions affecting uncombined amino acids, formation of urea from nitrogenous portion, and glucose from nonnitrogenous portion, also secondary production of glycogen

- **Metabolism of Proteins**
  - **Nucleoproteins**
    - Nucleoproteins—process of digestion, protein separated from nucleic acid and hydrolyzed to amino acids, history same as above
    - Nucleic acid gives rise to purine bodies of which uric acid is waste product
    - Endogenous uric acid from nucleoproteins of tissue cells
    - Exogenous uric acid from nucleoproteins of food
  - **Nutritive value**
    - **Adequate** proteins contain all the materials for maintenance and growth of tissue
    - Proteins may yield energy

- **Basal Metabolism**
  - Food source of energy for tissue-building or work
  - **Calorie**
    - Unit of measurement for heat production
    - Large calorie = quantity of heat necessary to raise temperature of 1 kg (2.2 lb) of water 1° C or that of 1 lb of water 4° F
  - **Carbohydrates**—1 gm—4 Cal
  - **Fat**— 1 gm—9 Cal
  - **Protein**— 1 gm—4 Cal
  - **Basal metabolism**—energy necessary to keep body alive
  - **Respiratory quotient**
    - Ratio between carbon dioxide excreted and oxygen absorbed. Ratio between the two is figured by dividing the volume of carbon dioxide by the volume of oxygen
  - **Variations** in basal metabolism influenced by age, sex, internal secretions, etc.

- **Daily Calorie Requirement**
  - **Factors to consider**—body surface, work, sleep, etc.
  - **Distribution**
    - Proteins—10–15%
      - Nitrogen equilibrium maintained on about 40 gm of protein per day
    - Fats—about 25–35%
    - Carbohydrates—depends upon form in which taken, also on amount of fat

- **Nutritional Need**
  - Considers supplying kinds and concentration of all essential nutrients, adequate amounts of all essential chemical elements, adequate caloric intake, selection of proteins from standpoint of suitability for construction of body tissues, etc.

CHAPTER 20

WATER AND ELECTROLYTE BALANCE PHYSIOLOGY; BUFFERS

Water constitutes more than two-thirds of the material ingested daily.

The water content of the body comes from three sources: beverages or other liquids; foods, especially vegetables and fruits; and the water formed in the tissues as the result of metabolic activities.

Water enters into the composition of all the tissues, most tissues containing between 75 and 90 per cent. In actively growing tissues and in young animals there is a greater proportion than in less active tissues or in the bodies of older animals. Water supplies fluid for the secretions and serves as a medium for the chemical changes occurring in digestion. Its most obvious services are in connection with the absorption of food in solution and the removal of dissolved products of metabolism. By its evaporation from the skin and the respiratory passages, it helps to keep the body temperature from rising above normal.

Under normal conditions the amount of water in the body remains about the same, even though the intake may vary considerably. If the intake of water is increased, the blood pressure in the renal vessels is raised, and the secretion of urine is increased. If the intake of water is decreased, the resulting sensation of thirst is such that steps are taken to relieve it.

Distribution of water in the body. Water is distributed in the body in three main compartments:

1. Blood plasma — 5% of body weight
2. Interstitial fluid — 15% of body weight
3. Intracellular fluid — 50% of body weight

This means that if an individual weighs about 70 kg, there are about 3.5 liters of water in the plasma compartment, 10.5 liters of water in the interstitial compartment, and 35 liters of water within the cells. This makes a total of 49 liters of water. The relative volumes are

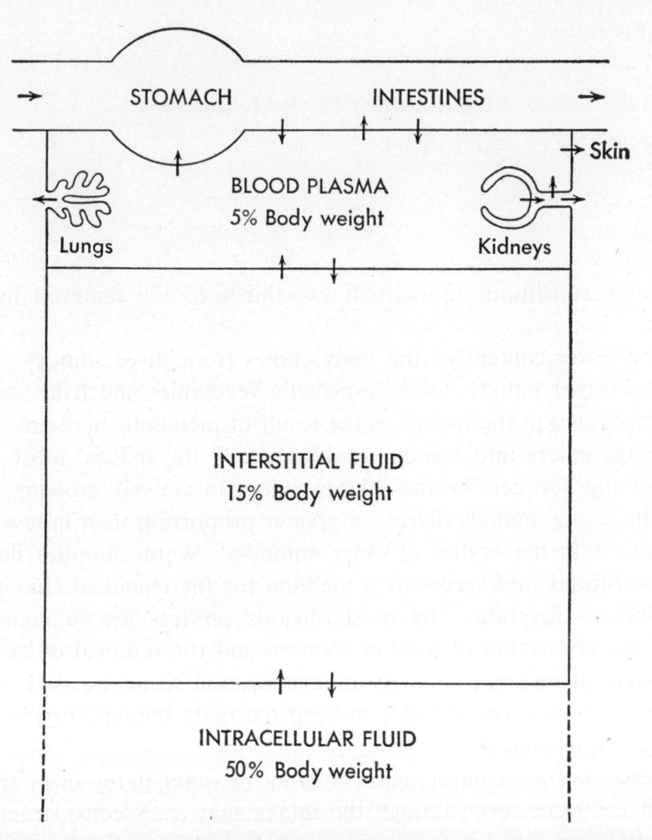

Fig. 322. Diagram illustrating distribution of water in the body and exchange of water between compartments. Normal sources of body fluid and fluid loss. (Reprinted by permission of the publishers from James Lawder Gamble, *Chemical Anatomy, Physiology and Pathology of Extracellular Fluid: A Lecture Syllabus.* Cambridge, Mass.: Harvard University Press, Copyright, 1942, 1947, 1954, by The President and Fellows of Harvard College.)

affected by age, fluid and salt intake, and environment. The amount of fluid in each compartment may be measured with relative accuracy, and the measurement of plasma volume is done frequently.

Plasma volume is measured by giving a known quantity of the blue dye 1824 or other substance intravenously and, after the dye has had opportunity to be mixed well in the blood stream, taking samples of blood to determine how much the dye has been diluted. From these figures plasma volume is determined.

Movement of water between the compartments takes place freely, but the movement of solutes between the compartments does not.

A study of Fig. 322 illustrates the distribution of water; by means of arrows, movement of water from one compartment to the other is also shown. Under normal conditions fluid enters the body via the mouth. In the stomach, fluid moves from the blood plasma to the cells of the gastric glands for the manufacture of about 1,500 cc of gastric juice in 24 hours. This fluid moves, along with ingested foods and fluids, to the intestine. Here again large quantities of water move from the blood plasma to gland cells for the manufacture of large quantities of digestive fluids, and finally the fluids and end products of digestion are returned to the blood stream. Small amounts of fluid (about 150–200 cc) are lost in feces daily. This illustrates the tremendous exchange of fluids between plasma and secreting cells and back to the plasma again.

The fluid in the plasma compartment also loses water via the lungs, the skin, and the kidneys. During respiration, as blood circulates through the lungs, plasma loses about 350 cc of fluid in 24 hours. Water loss via the skin in the form of perspiration helps to control fluid balance as well as the elimination of urea, sodium, chloride, and other salts. The skin also prevents water loss from tissues and capillaries by the protective covering formed by stratified squamous epithelium. Excessive loss of water through perspiration from any cause may deplete the body fluids.

Through the kidney, plasma normally loses about 1,500 cc of water in 24 hours. Since water loss through the kidney is rigorously controlled, urine output varies with fluid intake. The kidney also rigorously regulates the kinds and amounts of substances in solution in the water lost in the form of urine.

In reality interstitial fluid functions as the "middle man" between plasma and cells. Plasma volume and cell volume are kept relatively constant by temporary reduction or increase in the amount of fluid in the interstitial compartment. Movement of fluid in the *extra*-cellular compartments takes place mainly through the cardiovascular system and the lymph channels.

The total forces that hold fluid in the various compartments are not completely understood. However, in the plasma compartment the blood proteins, especially serum albumin, play a large part in the control; and the large percentage of protein within the cell forms an osmotic relationship which holds water within the cells. The electrolytes also function in regulating water movement across membranes.

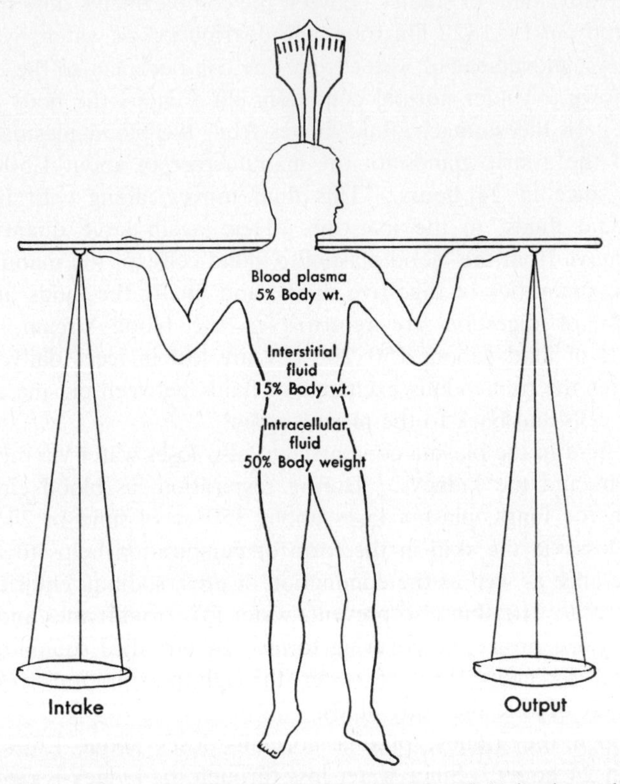

Fig. 323. Diagram illustrating a balance between fluid intake and output. During periods of rapid growth intake must exceed output. If output exceeds intake, dehydration results.

Fluid intake. In Fig. 323 fluid balance is illustrated; however, during periods of active growth more water is needed, and hence intake must exceed output, as cells require a relatively large percentage of water for growth.

Fluid intake is in the form of water, other liquids or beverages and foods. Normally water furnishes about a third or so of the required need, and the rest is furnished by other liquids and foods. Some foods have a much higher water content than do others. Another source of water to the body fluids is the water of metabolism. An ordinary mixed diet yields about 350 cc in 24 hours. On the whole, 100 gm of starch yields 55 gm of water, 100 gm of fat yields 107 gm of water, and 100 gm of protein yields 41 gm of water.

The amount of fluid needed is related to size, weight, and activities of an individual. This means that fluid intake may vary from 1,800 cc to 3,000 cc in 24 hours.

Fluid output. Normally fluid output is directly related to fluid intake. Water loss via the kidneys is rigorously controlled and varies from 1,000 cc to 1,500 cc in 24 hours. As a rule water loss through feces is about 100 cc to 150 cc. Through insensible perspiration about 450 to 800 or so cc of water is lost in 24 hours and via the lungs about 250 to 350 cc. There is a relationship between water loss through kidneys and through the skin. Usually when large quantities of water are lost via the skin, urine volume is decreased.

It is readily understandable that excessive loss of fluid by diarrhea, vomiting, or perspiration, excess loss by the kidneys, loss due to burns, or loss of fluid through hemorrhage seriously disturbs fluid balance. Failure to ingest sufficient quantities of fluid may also deplete the fluid in the interstitial compartment, and eventually the plasma compartment may be disturbed. Concentrations of sodium and chloride are increased, and hemoconcentration may also result. This means that the ratio of blood cells to plasma is increased.

When excess amounts of fluids are taken by mouth, water is rapidly absorbed into the plasma compartment. If large quantities are taken with no food, absorption is complete in about 30 to 40 minutes. Blood volume may be temporarily increased, as well as an increase in cardiac output. However, there is rapid adjustment to the increased water load by the opening of capillary networks to increase the vascular bed; the sinusoids of the liver and spleen open to hold more blood, and fluid is rapidly transferred to the interstitial compartment. The kidneys eliminate excess water and balance is restored within two or three hours.

The antidiuretic hormone (ADH) of the posterior pituitary gland regulates water reabsorption in the kidney tubule, and the salt hormone of the adrenal cortex regulates the sodium-potassium ratio in the plasma and their excretion by the kidney.

In severe stress situations, such as in burns, shock, hemorrhage, etc., electrolytes, especially sodium chloride and the body fluids, must be conserved. In some way ACTH of the pituitary and the adrenal-cortex hormones respond to the stress situation and play a large role in maintaining homeostasis of the body fluids.

The electrolytes. An electrolyte may be an acid, base, or salt which has the power to conduct an electric current. Some are more chemically active than others and ionize readily; hence they are called strong electrolytes. Those less active are called weak electrolytes.

A study of Fig. 324 illustrates the distribution of the electrolytes within the blood plasma, interstitial fluid, and cells. It will be noted that concentrations are expressed in milliequivalents per liter, which show relative magnitudes and interrelationships of the various substances and thus make measurement more meaningful.

To transpose milligrams per 100 cc of blood plasma to milliequivalents, it is necessary to know the number of milligrams per 100 cc and the atomic weight and valence of the substance.

Equivalent weights of ions are one-valent quantities. Equivalent weight is obtained by dividing the atomic or formula weight of the ion by its valence—thus, for Na^+, with an atomic weight of 23, the equivalent weight $= \frac{23}{1} = 23$. To transpose to milligrams per liter:

Na^+ 330 mg per 100 cc blood plasma
3,300 mg per 1,000 cc blood plasma

Therefore $\frac{3{,}300}{23} = 143$ milliequivalents of Na^+ per liter, usually expressed

143 mEq/L

From the illustration it will be noted that the concentration of sodium and chloride are relatively high in blood plasma and interstitial fluid as compared with intracellular fluid. To a large degree the sodium ion controls the movement of water between the cell and interstitial fluid. Potassium is high within cells and relatively low in extracellular fluids and remains in a state of dynamic equilibrium between these compartments. Ca^+, Mg^{++}, SO_4^{--}, HPO_4^{--}, and HCO_3^{--} have about the same concentration in the extracellular fluids. All of these electrolytes function in various ways to hold fluid within compartments, regulate water balance, and help to maintain acid-base balance.

In regulating acid-base balance Na^+, K^+, Ca^{++}, and Mg^{++} function in the body as bases, while HCO_3^{--}, Cl^-, HPO_4^{--}, SO_4^{--}, organic acids, and protein function as acid components. In the elimination of these

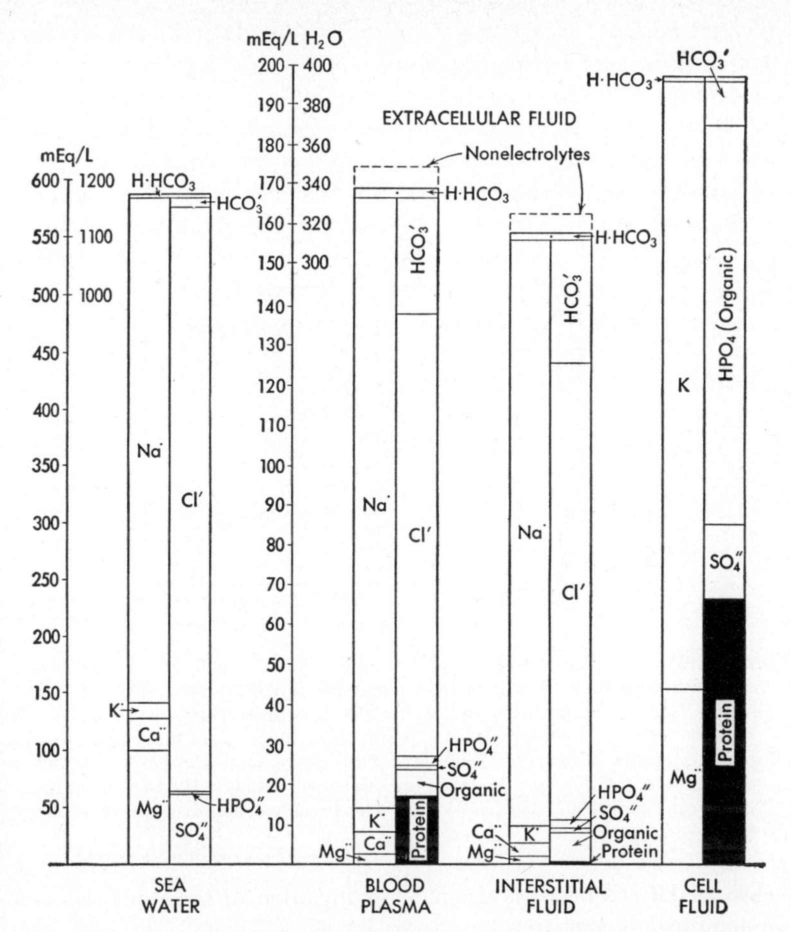

Fig. 324. Chart illustrating the distribution of electrolytes and nonelectrolytes in the body. It will be noted that blood plasma and interstitial fluid are almost identical. Comparison with sea water is very interesting. (Reprinted by permission of the publishers from James Lawder Gamble, *Chemical Anatomy, Physiology and Pathology of Extracellular Fluid: A Lecture Syllabus.* Cambridge, Mass.: Harvard University Press, Copyright, 1942, 1947, 1954, by The President and Fellows of Harvard College.)

substances by the kidney the acid-base balance is regulated by the saving of base by secretion of acid urine, or by the manufacture of ammonia to neutralize fixed acids and by selective reabsorption of large amounts of sodium chloride. In this way the kidneys play a large part in maintaining the normal *p*H of body fluids. The acidity of urine varies from *p*H 5 to 7. Urine with a *p*H of 5.5 is more than 100 times as acid as blood plasma.

Acidosis is a reduction in body base and may result from insufficient base intake or excess acid intake; diarrhea (loss of base); base loss and acid retention by kidney failure; or by certain disturbances of carbohydrate metabolism, such as in diabetes mellitus.

Alkalosis is a reduction of body acid and may result from vomiting

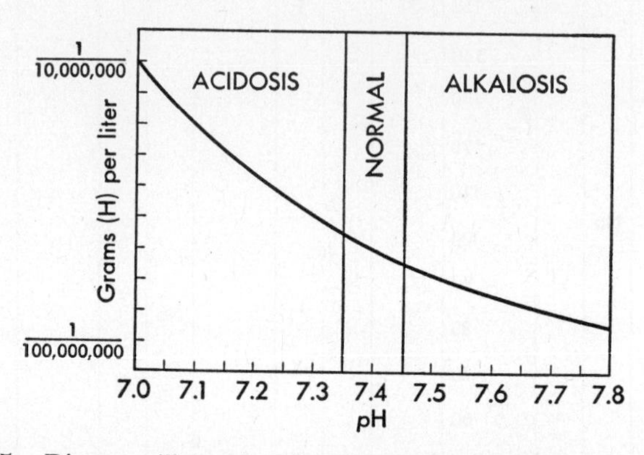

Fig. 325. Diagram illustrating the range of hydrogen ion concentration compatible with life. It will be noted that the physiological range lies on the alkaline side of neutrality. In health the hydrogen range is between *p*H 7.35 and *p*H 7.45. (Reprinted by permission of the publishers from James Lawder Gamble, *Chemical Anatomy, Physiology and Pathology of Extracellular Fluid: A Lecture Syllabus* Cambridge, Mass.: Harvard University Press, Copyright, 1942, 1947, 1954, by The President and Fellows of Harvard College.)

(loss of HCl) or gastric drainage or ingestion of large quantities of sodium bicarbonate.

Acid-base balance is also regulated by what is called *buffering*, accomplished through the exchange of ions and the elimination of certain ions. The sources of substances to be buffered are:

1. THE CELLS. From the metabolism of carbohydrates, fats, and proteins, carbon dioxide, water, and lactic acid; and also from protein, small amounts of phosphoric acid and sulfuric acid are formed.

2. FOODS. From foods absorbed in the intestine various acids and bases are constantly entering the blood stream that must be buffered to maintain the body fluids within a comparatively narrow range of *p*H. For example, from average diets citric, lactic, acetic, and fatty

acids. From fruits and certain foods potassium tartrate and citrate and sodium bicarbonate. These substances react with water and enter the blood stream as tartaric, citric, and carbonic acid along with very small amounts of potassium and sodium hydroxide and hydrochloric acid.

All of these substances must be buffered. Since potassium is

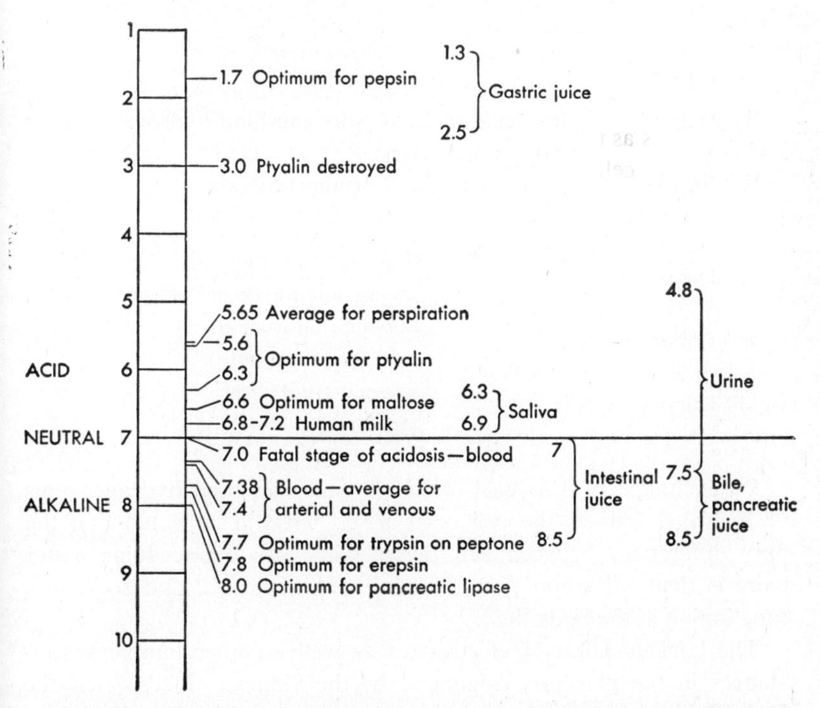

Fig. 326. The *p*H scale runs from 1 to 14. In physiology the *p*H range is comparatively narrow. The *p*H is important in enzyme activity. Body fluids are maintained in optimum range by the buffer systems. The kidneys, the lungs, the hemoglobin also play important roles.

found in relatively large quantities within the cell, it functions effectively in the buffer system, while sodium functions in the extracellular fluids. Buffering is accomplished by exchange of ions, thus binding acids or bases, or forming less acid or more acid salts, or freeing basic or acid ions. All proteins, especially hemoglobin, are effective buffers.

For example, in the extracellular fluids the following buffer pairs function:

1. The phosphate pair	$\frac{\text{sodium monohydrogen phosphate}^{1}}{\text{sodium dihydrogen phosphate}}$
2. The carbonate pair	$\frac{\text{sodium bicarbonate}}{\text{hydrogen carbonate}}$
3. The protein pair	$\frac{\text{sodium proteinate}}{\text{hydrogen proteinate}}$

By exchange of ions each of these pairs can bind hydrogen ions or hydroxyl ions as the body need arises.

Within the cell the following buffer pairs function:

1. The phosphate pair	$\frac{\text{potassium monohydrogen phosphate}}{\text{potassium dihydrogen phosphate}}$
2. The carbonate pair	$\frac{\text{potassium bicarbonate}}{\text{hydrogen carbonate}}$
3. The protein pair	$\frac{\text{potassium proteinate}}{\text{hydrogen proteinate}}$

By exchange of ions each of these pairs can bind hydrogen ions or hydroxyl ions as the cell need arises. It will be noted that the only differences between cell buffer pairs and extracellular buffer pairs is that potassium is functioning within the cell and sodium is functioning in the body fluids.

The concentration of electrolytes as well as other inorganic substances in the plasma is regulated by the kidney. In this way the body fluids are maintained within a comparatively narrow range of *p*H that is necessary for the normal functioning of the various fluids and the maintenance of health.

SUMMARY

Water Balance	**Distribution of water in the body**	1. Blood plasma 5% of body weight 2. Interstitial fluid 15% of body weight 3. Intracellular fluid 50% of body weight

[1] Adapted from C. A. Francis, and Edna Morse, *Fundamentals of Chemistry and Applications.* The Macmillan Company, New York, 1950.

- **Water Balance**
  - **Movement of water**
    - Water moves freely between compartments
    - Solutes do not
  - **Water intake 1,800 to 3,000 cc in 24 hours**
    - Water
    - Other fluids
    - Foods
    - Water metabolism 300–350 cc in 24 hours
  - **Water output**
    - Feces 100 to 150 cc
    - Skin 450 to 800 cc
    - Lungs 250 to 350 cc
    - Urine 1,000 to 1,500 cc
  - **Adjustment to excess fluid intake**
    - Fluid rapidly absorbed into blood stream
    - Blood volume may be temporarily increased
    - Cardiac output may be increased
    - Capillary networks open, vascular bed increased
    - Sinusoids of liver and spleen hold more blood
    - Fluid rapidly transferred to interstitial compartment
    - Kidney eliminates excess water and balance is restored in several hours
  - **Regulation by hormones**
    - ADH regulates water reabsorption in the kidney tubule
    - Salt hormones of the adrenal cortex regulate Na^+–K^+ ratio in plasma and their elimination by the kidney, thus aiding in the regulation of water balance
  - **Stress situations**
    - ACTH and the adrenal cortex hormones respond to stress situations and play a large role in maintaining homeostasis of the body fluids
- **Electrolyte Balance**
  - An electrolyte may be an acid, base, or salt which has the power to conduct an electric current. Some are chemically more active than others.
  - **Distribution**
    - In extracellular fluids
      - Chiefly Na^+, Cl^-, HCO_3^-; small quantities of K^+, Ca^{++}, HPO_4^{--}, SO_4^{--}, Mg^+
    - Intracellular fluids
      - Chiefly K^+, Mg^{++}, HPO_4^{--}; small quantities of SO_4^{--}, HCO_3^-, Na^+
  - **Milliequivalents**
    - Concentrations expressed in terms of milliequivalents
      - Shows relative magnitudes and interrelationships

Electrolyte Balance

- **To transpose milligrams per 100 cc to milliequivalents**
  - Must know:
    - Number of milligrams per 100 cc
    - Atomic weight and valence of the substance
  - Plasma concentrations
    - Na^+ 330 mg per 100 cc $\times 10 \div 23 \times 1$
    - Cl^- 365 mg per 100 cc $\times 10 \div 35 \times 1$
    - Ca^{++} 10 mg per 100 cc $\times 10 \div 40 \times 2$
    - Mg^{++} 2.7 mg per 100 cc $\times 10 \div 24 \times 2$
    - K^+ 20 mg per 100 cc $\times 10 \div 39 \times 1$
  - Whole blood
    - HPO_4 $(mgP)_3$ mg per 100 cc $\times 10 \div 31 \times 1.8$
    - SO_4 $(mgS)_2$ mg per 100 cc $\times 10 \div 32 \times 2$
- **Function of electrolytes**
  - Help to:
    - Control movement of water between cell and interstitial fluid, hold fluid within compartments, regulate water balance, and help to maintain acid-base balance.

Acid-Base Balance

- Electrolytes functioning as bases:
  - Na^+ 142 mEq/L
  - Ca^{++} 5 mEq/L
  - Mg^{++} 3 mEq/L
  - K^+ 5 mEq/L
- Electrolytes functioning as acids:
  - Cl^- 103 mEq/L
  - HCO_3^- 27 mEq/L
  - HPO_4^{--} 2 mEq/L
  - SO_4^{--} 1 mEq/L
  - Organic acids 6 mEq/L
  - Protein 16 mEq/L

Regulation

- Acid-base regulated by:
  1. Saving of base, by secretion of acid urine
  2. Manufacture of ammonia to neutralize fixed acids
  3. Selective reabsorption of large amounts of sodium chloride

Buffering

- Accomplished through the exchange of ions and the elimination of certain ions
- Acids or bases are bound or less acid or more acid salts are formed or basic or acid ions are freed
- Cell proteins, especially hemoglobin, are effective buffers

Sources of Substances to Be Buffered	1. The cell. From the metabolism of:	Carbohydrates Fats Proteins	Carbon dioxide, water, lactic acid; from protein phosphoric and sulfuric acids are formed
	2. From foods. Various acids and bases from average diets:	citric, lactic, acetic, fatty acids sodium bicarbonate	
Substances to Be Buffered	These substances react with water, enter the blood streams as tartaric, citric, and carbonic acids along with very small amounts of potassium, and sodium hydroxide and HCl Potassium pairs of buffers function most effectively within cells Sodium pairs of buffers function most effectively in the extracellular fluids		
Electrolyte Regulation	Most effectively controlled by the kidneys		

CHAPTER 21

THE URINARY SYSTEM: ANATOMY, PHYSIOLOGY, HISTOLOGY URINE FORMATION AND ELIMINATION THE ROLE OF THE KIDNEYS IN MAINTAINING HOMEOSTASIS OF BODY FLUIDS

It has been shown that the blood is constantly supplied by the respiratory and digestive systems with the chemical substances it requires. These substances are carried by the blood to the cells and on entering them become a part of a complex system of substances interacting chemically and tending constantly toward a static equilibrium but failing constantly to reach it since fresh substances are continually brought by the blood and waste substances are continually carried away—thus establishing a *dynamic,* or *active, equilibrium* rather than a static equilibrium. Health depends on the smooth flowing of this dynamic equilibrium. One of the results of these chemical changes is the formation of waste products, which must be removed from the cells and the body.

WASTE PRODUCTS

The waste products of the body are carbon dioxide, water, organic and inorganic salts, hair, nails, dead skin, and the indigestible and undigested portions of foods eaten. To these may be added the dead and living microorganisms constantly excreted in large numbers in the feces.

The wastes of *cell metabolism* may be listed as follows:

1. Liquid—water
2. Gas—carbon dioxide

3. Soluble salts { nitrogenous salts, e.g., urea; inorganic salts, e.g., sodium chloride }
4. Others, e.g., heat

All of these wastes are classed as excreta and the process by which they are removed from the body as excretion, or elimination.

EXCRETORY ORGANS

The organs that function as excretory organs and the products which they eliminate may be tabulated as follows:

	Essential	*Incidental*
Lungs	Carbon dioxide	Water, heat
Kidneys	Water and soluble salts, resulting from metabolism of proteins, neutralization of acids, etc.	Carbon dioxide, heat
Alimentary canal	Solids, secretions, etc.	Water, carbon dioxide, salts, heat
Skin	Heat	Water, carbon dioxide, salts, hair, nails, and dead skin

Since the function of the lungs, the alimentary system, and the skin in the elimination of waste products has been discussed in previous chapters, attention is now turned to the urinary system.

Urinary System {
- 2 kidneys — form urine from materials taken from the blood
- 2 ureters — ducts which convey urine from kidney to bladder
- 1 bladder — reservoir for the reception of urine
- 1 urethra — tube which conveys urine from bladder
}

ANATOMY OF KIDNEYS

The kidneys are compound tubular glands, placed at the back of the abdominal cavity, one on each side of the spinal colum and behind the peritoneal cavity. They correspond in position to the space included between the upper border of the twelfth thoracic and the third lumbar vertebrae. The right kidney is a little lower than the left, because of the large space occupied by the liver.

Each kidney with its vessels is embedded in a mass of fatty tissue termed an *adipose capsule*. The kidney and the adipose capsule are surrounded by a sheath of fibrous tissue called the *renal fascia*. The renal fascia is connected to the fibrous tunic of the kidney by many trabeculae, which are strongest at the lower end. The kidney is held in place partly by the renal fascia, which blends with the fasciae on the quadratus lumborum and psoas major muscles and also with the fascia of the diaphragm, and partly by the pressure and counterpressure of neighboring organs.

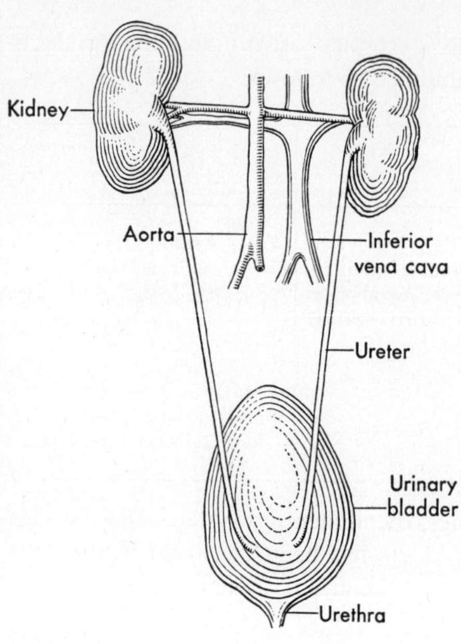

Fig. 327. The urinary system, viewed from behind.

The kidneys are bean-shaped, with the medial or concave border directed toward the medial line of the body. Near the center of the concave border is a fissure called the *hilum* (hilus), which serves as a passageway for the ureter, and for the blood vessels, lymph vessels, and nerves going to and from the kidney. Each kidney is covered by a thin but rather tough envelope of fibrous tissue. At the hilum of the kidney the capsule becomes continuous with the outer coat of the ureter. If a kidney is cut in two lengthwise, it is seen that the upper end of the ureter expands into a basinlike cavity, called the *pelvis* of the kidney. The substance of the kidney consists of an outer portion called the cortical substance (cortex) and an inner portion called the medullary substance (medulla). Between the cortical and medullary substances are the arterial and venous arches. (See Figs. 329, 330, 333.)

The medullary substance is red in color. It consists of from 8 to 18 radially striated cones, the renal pyramids, which have their bases toward the circumference of the kidney, while their apices converge into projections called papillae which are received by the cuplike cavities, or calyces, of the pelvis of the kidney.

The cortical substance is reddish brown, and it contains the renal corpuscles, the convoluted tubules, and blood tubes. It penetrates

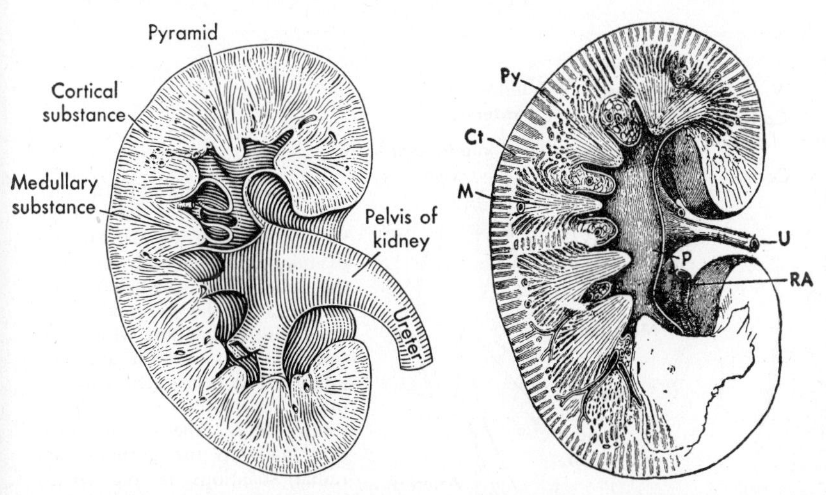

Fig. 328. Diagrammatic longitudinal section of the human kidney. Each kidney is about 11.25 cm long, 5.0 to 7.5 cm broad, and 2.5 cm thick, and weighs about 135 gm. (Henle.)

Fig. 329. Longitudinal section of the kidney. *Ct,* the cortical substance; *M,* the medullary substance; *Py,* the pyramids; *P,* the pelvis of the kidney; *U,* the ureter; *RA,* the renal artery. (T. H. Huxley.)

for a variable distance between the pyramids, separating and supporting them. These interpyramidal extensions are called the *renal columns* (Bertin)[1] and support the blood vessels. A glance across the shiny surface of a freshly cut kidney discloses both a granular structure and also areas showing radial striations. In general these alternate with each other. The granular areas contain the renal capsules, glomeruli, and convoluted areas of the tubule. The radially striated areas contain other parts of the tubule. (See Figs. 330 and 333.)

[1] Exupère Joseph Bertin, French anatomist (1712–1781).

The bulk of the kidney substance, in both cortex and medulla, is composed of minute tubes, or tubules, closely packed together, having only enough connective tissue to carry a large supply of blood vessels and a number of lymphatics and nerve fibers. The appearance of the cortex and medulla is due to the arrangement of these tubules, the *nephrons,* or functional units.

The nephron, or unit pattern, of the kidney consists of a renal tubule with its blood supply. The renal tubules vary in length, and

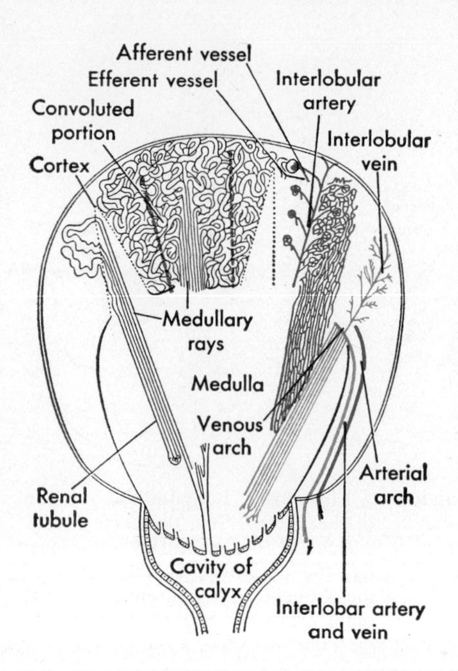

Fig. 330. Diagram of a longitudinal section of a lobe of the kidney, showing the arrangement of tubules and blood vessels in the lobe. The calyx embraces the apex of the pyramid. It is lined with epithelium, which continues from it over the apex, the latter being perforated with the many apertures of excretory tubes. Note the arrangement that leads to the granular and radial striations in the cortex. (Adapted from Gerrish.)

the glomeruli vary in size. The largest ones are found nearest the medulla.

The *renal tubule* begins as a closed, invaginated layer of epithelium, the *renal capsule,* or capsule of Bowman.[2] The inner layer of this globelike expansion closely invests a capillary tuft called a renal *glomerulus.* The glomerulus consists of a few capillary loops which do not anastomose and which are completely encapsulated by the expansion of the tubule except at the point where an *afferent vessel* enters and an *efferent vessel* leaves the capillary tuft. The glomeru-

[2] Sir William Bowman, English anatomist and ophthalmologist (1816–1892).

lus and its enveloping capsule make up a renal corpuscle, or Malpighian[3] body. There are said to be more than 1,000,000 of these corpuscles in the cortex of each kidney. The renal capsule joins the rest of the tubule by a constricted *neck;* and the tubules, after running a very irregular course, open into collecting tubes, which pour their contents through their openings on the pointed ends, or papillae, of the pyramids into the calyces of the kidney. About 20 of these collecting tubes empty into the calyces from each papilla.

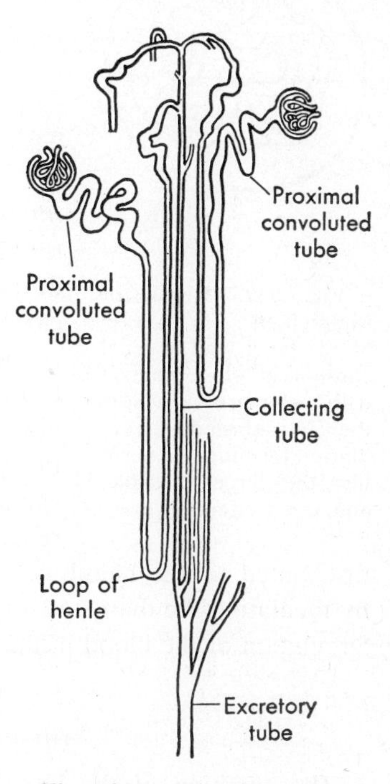

Fig. 331. Diagram of the course of two renal tubules.

The epithelial lining of the renal tubule varies in different parts of the tubule. In the convoluted tubules and ascending limb of Henle's[4] loop the cells are columnar, while in the glomerular capsule and descending limb of Henle's loop they are thin, squamous cells. The collecting tubules have well-defined columnar cells, definitely resembling those of excretory ducts. These and other differences in tubular structure are important in consideration of the physiology of urine formation and conduction (Fig. 333).

The blood supply of the kidney. The kidney is abundantly supplied with blood by the *renal artery,* which is a branch of the abdominal aorta. Before or immediately after entering the kidney at the hilum, each artery divides into several branches, which follow the wall of the ureter into the kidney. They are sometimes called *interlobar arteries.*

When these arteries reach the boundary zone between the cortex and medulla, they divide laterally and form the *arch,* or *arcuate arteries.* From the convexity of these arches, the *interlobular arteries* (cortical) enter the cortex, giving off at intervals minute *afferent*

[3] Marcello Malpighi, Italian anatomist (1628–1694).
[4] Friedrich Gustav Jakob Henle, German anatomist (1809–1885).

arteries, each of which branches out as the *capillaries of a glomerulus.* These capillaries reunite to form an *efferent tube* (artery) much smaller than the afferent artery. The efferent tube breaks up into a close meshwork, or *plexus, of capillaries,* which are in close approximation with both the convoluted tubule in the cortex and the loop of Henle in the medulla. These capillaries unite to form the *interlobular veins* (cortical) and *medullary veins,* which pour their contents into the *arcuate veins* lying between the cortex and the medulla. The arcuate veins converge to form the *interlobar veins.* These merge into the *renal vein,* which emerges from the kidney at the hilum and opens into the inferior vena cava.

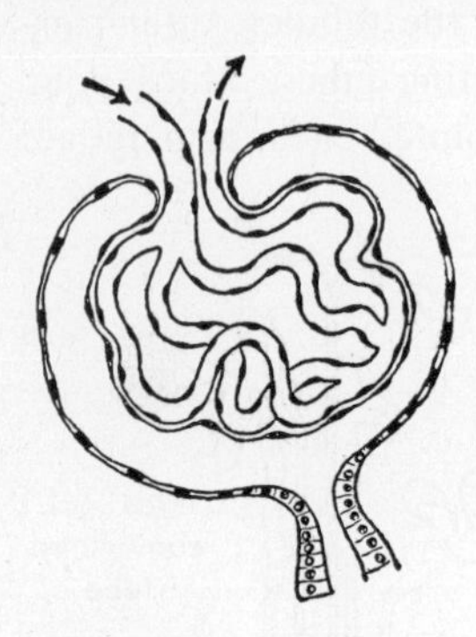

Fig. 332. Renal, or Malpighian, corpuscle. Shows inner wall of renal capsule in close contact with glomerulus. Note the few unbranched capillaries forming glomerulus, the large arteriole, and small efferent vessel.

The nerves of the kidneys are derived from the *renal plexus,* which is formed by branches from the celiac plexus, the aortic plexus, and from the lesser and lowest splanchnic nerves. They accompany the renal arteries and their branches and are distributed to the blood vessels. They are vasomotor nerves, and by regulating the diameters of the small blood vessels they control the circulation of the blood in the kidney.

PHYSIOLOGY OF KIDNEYS

The function of the kidneys is to keep the body fluids at their normal constancy in composition, volume, and *p*H by removing variable amounts of water and organic and inorganic substances from the blood stream as it passes through them. The kidneys extract almost all the nitrogenous waste, the greater part of the salts not needed by the blood, and about one-half the excess water. They also remove foreign substances such as toxins, whether formed within the body or taken in from outside. The concentration of substances in the urine, and not the quantity, is a criterion for judging the amount of work done by the kidneys. It is probable that they are most severely taxed when they have to remove from the blood a maximum of dissolved solids in a minimum of water. In shock the kidneys produce a vasoexcitor material (VEM) which raises blood pressure.

Urine is a complex aqueous solution of organic and inorganic substances, most of which are waste products from the metabolism of body cells. The amounts of these substances in the urine vary greatly; in fact, it is this variability of urine composition that helps

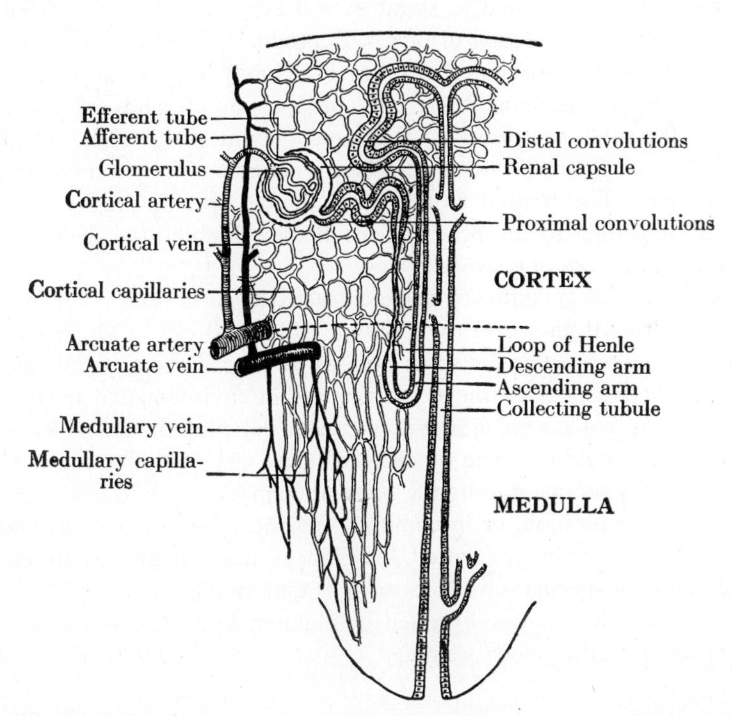

Fig. 333. Nephron and its blood supply. Length of tubule approximately 35 to 40 mm, diameter about 0.02 mm; diameter of capsule about 0.2 mm.

keep the rest of the body fluids in that steady state known as *homeostasis.*

Physical characteristics of urine. Normal urine is usually a yellow, transparent liquid with an aromatic odor. Cloudy urine is not necessarily pathological, for turbidity may be caused by mucin secreted by the lining membrane of the urinary tract; this however, if present in excess does denote abnormal conditions. The color of urine varies with the changing ratios of water and substances in solution and may, of course, be affected by the presence of abnormal materials such as those produced by disease or certain drugs.

Urine is usually slightly acid, though its *p*H may vary between 5 and 7. Diet affects this reaction; a high-protein diet increases acidity, while a vegetable diet increases alkalinity. This variation is due to the difference in the end products of metabolism in each case. If human urine is allowed to stand, it will eventually become alkaline due to the decomposition of urea with production of ammonia.

In health the specific gravity of urine may vary from 1.010 to 1.030, depending upon the relative proportions of solids and water. When the solids are dissolved in a large amount of water, the specific gravity will naturally be lower than when the urine is more concentrated. The normal kidney regulates the specific gravity of the urine according to the needs of the body. Inability to accomplish this is a symptom of disease.

The average quantity of urine extracted by a normal adult in 24 hours varies from 1,200 to 1,500 cc (40 to 50 oz). Much wider variations may occur for short periods of time without pathological significance, as, for example, when a rise in environmental temperature or unusual muscular activity increases perspiration and so lessens the urinary output. The quantity of urine may be affected by the amount of fluid taken in by the body; the amount of fluid lost in perspiration, respiration, or in vomiting, diarrhea, hemorrhage, etc.; the health of the organs concerned, the kidneys, heart, blood vessels, etc.; the action of specific substances such as diuretics.

The amount of urine excreted by children in 24 hours is great in proportion to their body weight.

6 months–2 years	18–20 oz
2–5 years	16–26 oz
5–8 years	20–40 oz
8–14 years	32–48 oz

Chemical composition of urine. Water forms about 95 per cent of urine. The solutes (on chemical examination, precipitated as solids—some 60 gm in 1,500 cc of urine) are organic and inorganic waste products.

Urine 1,500 cc daily	Organic wastes 35 gm	Urea	30 gm
		Creatinine	1–2 gm
		Ammonia	1–2 gm
		Uric acid	1 gm
		etc.	1 gm

Solutes (solids) 60 gm	Inorganic salts 25 gm	Chlorides Sulphates Phosphates of	Sodium Potassium Magnesium Calcium

Sodium chloride is the chief inorganic salt, about 15 gm being excreted daily by the kidneys.

Origin of constituents of urine. The wastes are formed in cells and are both exogenous and endogenous; that is, they are derived both from the metabolism of the excess proteins of the diet and from the catabolism of protoplasm. The table below shows these types. Urea, it is seen, is an example of exogenous waste, varying greatly with high and low protein intake. Creatinine, on the other hand, is probably endogenous, varying little with high and low protein intake.

NITROGEN OUTPUT AS INFLUENCED BY LEVEL OF PROTEIN INTAKE[5]

	High-Protein Diet (Free from Meat)	*Low-Protein Diet (Starch and Cream)*
Total nitrogen	16.8 gm	3.6 gm
Urea nitrogen	14.7 gm, or 87.5%	2.2 gm, or 61.7%
Ammonia nitrogen	0.49 gm, or 2.9%	0.42 gm, or 11.3%
Uric acid nitrogen	0.18 gm, or 1.1%	0.09 gm, or 2.5%
Creatinine nitrogen	0.58 gm, or 3.6%	0.60 gm, or 17.2%
Undetermined nitrogen	0.85 gm, or 4.9%	0.27 gm, or 7.3%
Water output	normal	diminished

Note that urea nitrogen decreased 12.5 gm when the low-protein diet was used. The creatinine nitrogen varied only 0.02 gm. These diets were similar in all respects, except for protein intake.

Urine is formed from the noncolloidal constituents (water, salts, etc.) of blood plasma which filter out from the glomerular capillaries under 50 mm Hg net pressure into the renal capsules and are concentrated into urine in the other parts of the renal tubules. The table on p. 656 shows the amount of concentration that takes place.

It is believed that in the renal tubule enough water may diffuse back into the capillaries to account for the concentration of the majority of substances in urine. However, to account in this way

[5] H. C. Sherman, *Chemistry of Food and Nutrition,* 1946.

RELATIVE COMPOSITIONS OF BLOOD PLASMA AND NORMAL URINE IN MAN

(Adapted from various authors, after Cushny)

Figures in milligrams per 100 cc

	Blood Plasma, %	*Urine,* %	*Number of Times Concentrated by Kidney*
Water	90–93	95	—
Proteins, fats, and other colloids	7–9	—	—
Glucose	0.1	—	—
Urea	0.03	2	60
Creatinine	0.001	0.075	75
Uric acid	0.004	0.05	12
Sodium	0.32	0.35	1
Potassium	0.02	0.15	7
Ammonium	0.001	0.04	40
Calcium	0.008	0.015	2
Magnesium	0.0025	0.006	2
Chloride	0.37	0.6	2
Phosphate	0.009	0.15	16
Sulphate	0.002	0.18	90

Note that the substances may be grouped roughly into three groups: (1) those not concentrated at all or only a few times, (2) those concentrated a greater number of times, and (3) those which are highly concentrated.

for substances that are concentrated the greatest number of times (in this table sulphate is concentrated 90 times) would mean that an enormous amount of fluid is filtered by the glomerulus and returned to the blood stream while passing through the tubule. It is known that certain cells in the kidney tubule have secretory ability which accounts for the high concentrations of certain substances in urine. It is evident that both the reabsorption and the secretion theory are needed to explain the varying concentrations of substances in urine and kidney activity. The other constituents diffuse back from the tubules into the blood in amounts sufficient to account for the smaller number of times they are concentrated in the urine, e.g., chloride two times. Substances such as glucose, sodium, calcium, etc., which diffuse back along with the water into the blood in considerable quantities, are sometimes called *high-threshold* substances. Those substances which are diffused back into the blood in only small quantities are known as *low-threshold* substances. Those substances which do not diffuse back into the blood are known as *no-threshold* substances.

The physiology of urine formation is related to the peculiarities in the kidney blood supply and structural differences in various parts of the renal tubule. The fact that the efferent vessel is smaller than the afferent makes blood pressure in the glomerulus high. Therefore water, salts, and noncolloidal materials in the blood (deproteinized plasma) are *filtered* through the capillary and the capsular epithelium into the renal capsule. The efferent vessel joins other, similar vessels which form plexuses about the tubules. Concentration of the urine is thought to take place chiefly in the descending loop of Henle, much water and certain percentages of salts passing through the walls of the tubule back into the blood stream by *diffusion*. Hence the blood plasma tends to regain its normal concentration of these substances, while the remainder of the salts, organic substances, and water form the now normally concentrated urine. It is evident that any factor which changes the blood pressure in the glomerulus will affect urinary production.

Hormonal control of kidney function. The reabsorption of water in the kidney tubules is an active process controlled by the antidiuretic hormone (ADH) of the posterior pituitary. The "salt hormone" of the adrenal cortex controls the reabsorption of potassium, sodium, and chloride in the tubule. In this way water and salt elimination by the kidney is rigorously controlled.

An understanding of the physiology of urine formation, with filtration of deproteinized blood plasma through the glomerular capsule and concentration in the renal tubule, helps to explain the varying symptoms of glomerular nephritis and tubular nephritis, respectively. When the glomerulus is impaired, proper filtration does not occur; nonprotein nitrogenous substances are retained in the blood, edema may be present, and a small amount of highly concentrated urine containing albumin may be passed. Acidosis is probable, owing to the retention of acid substances in the blood. In tubular nephritis, the usual dilute, plasma-like fluid is pressed through the epithelium of the capsule, but it cannot be concentrated because of tubular impairment and so is excreted as such. Acidosis, however, is likely to be present in this type of nephritis also, owing to the lack of ammonia production and absence of reabsorption of basic radicals by the tubules.

Diuretics are substances which increase the volume of urine excreted, causing a condition known as *diuresis*. Certain drugs, such as caffeine, act as diuretics by increasing the number of glomeruli functioning at a time. Others, "saline diuretics," when pressed through the renal capsule are unable to be reabsorbed by the tubules. They therefore so increase the concentration of salts within the tubule that no part of the urine can diffuse back into the blood stream. Still other drugs, as digitalis, cause diuresis by acting specifically upon the circulatory system.

Some constituents of normal urine are creatinine, urea, ammonia, hippuric acid, and purine bodies.

Creatinine ($C_4H_7N_3O$) is always present in the urine, and in amounts independent of the proteins of the diet. It is thought, therefore, to be an endogenous substance resulting from cellular metabolism of certain protoplasmic constituents. About 1 to 2 gm of creatinine are excreted in the urine daily.

The origin of creatinine is not clear, but it is thought to be made from creatine, which occurs in the metabolism of striped muscle tissue. It is not yet known whether this change from creatine to creatinine, which involves a loss of water, is accomplished in the blood or in the kidney. *Creatine* ($C_4H_9N_3O_2$) is not excreted as such in the urine of the adult male, but it is constantly present in the urine of children and perhaps in that of women after menstruation, during pregnancy, and in the puerperium. Creatine is also present in the urine during starvation or fever, probably because the body tissues are being utilized at a rate too high for all the creatine to be changed to creatinine.

Urea (CON_2H_4) constitutes about one-half (30 gm daily) of all the solids excreted in the urine. It is made by the liver cells from ammonium salts which are formed in the breakdown of proteins and are brought to the liver by the blood. Normally 27 to 28 mg of urea are contained in each 100 cc of blood. The kidneys constantly remove the urea as it is formed, keeping the amount in the blood stream at its proper level.

Ammonia (NH_3) in the urine is thought to be formed by the kidney either from amino acids or from urea brought to it in the blood. All of the ammonia so formed is not excreted; the blood of the renal vein contains more ammonia than that of the renal artery. The amount of ammonia produced by the kidney may depend upon the general need of the body for basic substances to offset acid substances in the blood and tissues. It is evident that the liver and kidneys work together in this respect, helping to keep the body fluids at their normal acid-base balance.

Hippuric acid ($C_9H_9NO_3$) is thought to be the means by which benzoic acid, a toxic substance occurring in food and from body processes, is eliminated from the body. A vegetable diet increases the quantity of hippuric acid excreted, probably because fruit and vegetables contain benzoic acid.

Purine bodies (*uric acid*, etc.)[6] are derived from foods containing nucleic acid (exogenous) and from the catabolism of the body proteins (endogenous). The exogenous purines excreted depend upon the quantity eaten of purine-containing foods such as meat; the endogenous purine waste depends upon the health and activities of the body and is normally fairly constant.

Some abnormal constituents appearing in urine are albumin, glucose, indican, acetone bodies, casts, calculi, pus, blood, and bile pigments.

Albumin. Serum albumin is a normal constituent of the blood plasma, but it is not usually pressed through the renal capsule. Its presence in the urine is spoken of as *albuminuria* and is usually a symptom of disease.

Certain conditions favoring albuminuria are:

Organic disease of the kidney, with injury to the renal capsule.

Increased blood pressure, which may be due to (1) stimulation of the vasoconstrictor mechanism, (2) vigorous muscular exercise, (3) inelastic arteries due to arteriosclerosis, (4) congestion of the kidneys because of interference with circulation through the renal vein, such as might result from pressure of tumors or a pregnant uterus.

Irritation of the kidney cells by poisons such as bacterial toxins, ether, turpentine, or heavy metals.

In abnormal conditions of the kidneys associated with albumin in the urine, there is usually retention of the inorganic salts and protein waste. In consequence, some of the salts may pass from the blood into the tissues and, by raising the osmotic pressure, promote excessive transudation of fluid into the tissues, thereby causing edema, or dropsy. Retained protein waste may undergo chemical changes which transform it into toxic substances and in this way cause the condition called uremia.

Glucose. Normal urine contains so little sugar that for clinical purposes it may be considered absent. In health the amount of glucose present in the blood varies from 0.08 to 0.18 per cent. A higher per cent is irritating to the tissues, so when the quantity of sugar eaten is greater than the system can promptly change to glycogen and fat, the kidneys excrete it. When glucose is found in the urine from this cause, the condition is called *temporary*

[6] For formulas, see Summary.

glycosuria. Frequent or continuous elimination of sugar shows that the body has not the usual power to oxidize sugar.

Indican. Indican (potassium indoxyl sulphate) is a potassium salt that is formed from indol. Indol results from the putrefaction of protein food in the large intestine. It is absorbed by the blood and carried to the liver, which it is thought changes the indol to indican, a less poisonous substance. Traces of indican are found in normal urine, but its presence in any amount is abnormal and denotes excessive putrefaction of protein food in the intestines, or putrefaction in the body itself, as in abscess formation. Excessive putrefaction may be due to a diseased condition of the intestine that interferes with absorption, to a diet containing too much protein food, or to constipation.

Acetone bodies. When there is excessive consumption of fat, as in starvation of a fat animal, and in diabetes mellitus, in which sugar is not oxidized and the body lives on its fats and proteins, acetone bodies are formed. Under these conditions the oxidation of fats is not complete. Acetone bodies are found in the urine of normal individuals when the oxidation of fats is not carried to completion, as during periods of fasting.

Casts. In some abnormal conditions the kidney tubules become lined with substances which harden and form a mold or cast inside the tube. Later these casts are washed out by the urine, and their presence in urine can be detected by the aid of a microscope. They are named either from the substances composing them or from their appearance. Thus there are pus casts, blood casts, epithelial casts from the walls of the tubes, granular casts from cells which have decomposed and form masses of granules, fatty casts from cells which have become fatty, and hyaline casts which are formed from coagulable elements of the blood.

Calculi. Mineral salts in the urine may precipitate and form calculi, or stones. Calculi may be formed in any part of the urinary tract from the tubules to the external orifice of the urethra. The causes which lead to their formation are an excessive amount of salts, a decrease in the amount of water, and abnormally acid or abnormally alkaline urine.

Pus. In suppurative conditions of any of the urinary organs, pus cells are present in the urine.

Blood. In cases of acute inflammation of any of the urinary organs, of tuberculosis, of cancer, and of renal stone, blood may be present in the urine, a condition known as *hematuria.* Blood imparts a smoky or reddish color to urine.

Bile pigments in the urine may be caused by obstructive jaundice, when bile has been pressed from the gallbladder into the blood stream, or by diseases in which an abnormal number of erythrocytes are destroyed. Bile pigments give the urine a greenish-yellow or golden-brown color.

Elimination of toxic substances. During illness it is the function of the kidneys to eliminate toxic substances that find their way into the blood, whether these substances result from defective metabolism, from bacterial activity, or from chemical poisons. This may account for the fact that after a severe illness the kidneys are often left in a damaged condition and suggests the desirability of a copious intake of water in order to decrease the concentration of toxic materials and thereby lessen the chances of injury to the tissues.

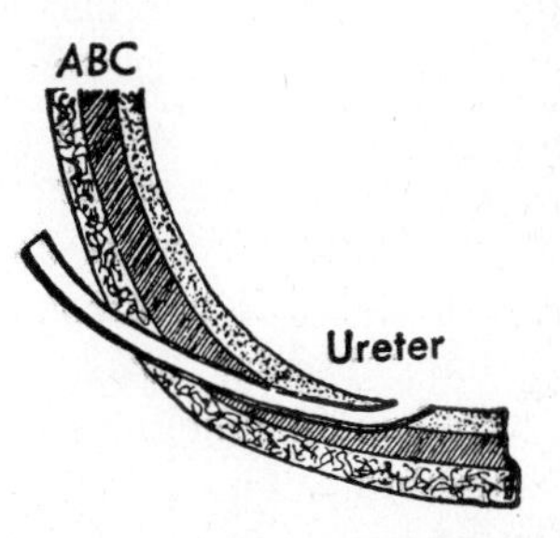

Fig. 334. Diagram showing entrance of ureter into bladder. *A,* fibrous coat; *B,* muscular coat; and *C,* mucous coat. The ureters run obliquely downward through the bladder wall for about 2 cm.

The ureters are two tubes which convey the urine from the kidneys to the bladder. Each ureter commences as a number of cuplike tubes, or *calyces,* which surround the renal papillae. The calyces (varying in number from 7 to 13) join and form two or three short tubes, which unite and form a funnel-shaped dilatation called the *renal pelvis.* From the pelvis the ureter proper, a cylindrical tube, passes to the fundus of the bladder. Each tube is about 25 to 30 cm long (10 to 12 in.) and about 4 to 5 mm ($\frac{1}{5}$ in.) in diameter. Each consists of three coats: an outer fibrous coat, a muscular coat, and an inner mucous lining. The contractions of the muscular coat produce peristaltic waves, which commence at the kidney end of the ureter and progress downward.

BLADDER

The bladder is a hollow muscular organ situated in the pelvic cavity behind the pubes, in front of the rectum in the male, and in front of the anterior wall of the vagina, and the neck of the uterus, in the female. It is a freely movable organ but is held in position by folds of peritoneum and fascia. During infancy it is conical in shape and projects above the upper border of the pubes into the hypogastric region. In the adult, when quite empty, it is placed deeply in the pelvis; when slightly distended it has a round form; but when greatly distended, it is ovoid in shape and rises to a considerable height in the abdominal cavity. It has four coats: (1) The *serous* coat is a reflection of the peritoneum and covers only the superior surface and the upper part of the lateral surfaces. (2) The *muscular* coat has three layers, an inner longitudinal, middle circular, and outer longitudinal. The circular fibers are collected into a layer of some thickness around the opening of the bladder into the urethra. These circular fibers form a sphincter muscle, which is normally in a state of contraction, relaxing only when the accumulation of urine within the bladder renders its expulsion necessary. (3) The *submucous* coat consists of areolar connective tissue and connects the mucous and muscular coats. (4) The *mucous* membrane of transitional epithelium lining the bladder is like that lining the ureters and the urethra. This coat is thrown into folds, or rugae, when the bladder is empty, with the exception of a small triangular area just above and behind the internal orifice of the urethra where the mucous membrane is firmly attached to the muscular coat. This area is called the *trigonum vesicae.*

There are three openings into the bladder. The two ureters open into the lower part about ½ in. from the median plane. The ureters take an oblique course through the wall of the bladder, downward and medialward. The urethra leads from the bladder, its vesical opening lying in the median plane below and in front of the openings of the ureters.

Nerve and blood supply. The bladder is supplied by the pelvic nerve of the craniosacral system and the thoracolumbar system from the inferior mesenteric plexus (Fig. 194).

The superior, middle, and inferior vesical arteries (branches of the hypogastric artery) supply the bladder. In the female, branches of the uterine and vaginal arteries also supply the bladder.

Function. The bladder serves as a reservoir for the reception of urine. Its capacity varies. When moderately distended, it holds about ½ liter (about 1 pt).

The urethra. In the female the urethra is a narrow membranous canal which extends from the bladder to the external orifice, the meatus urinarius. It is placed behind the symphysis pubis and is embedded in the anterior wall of the vagina. Its diameter, when undilated, is about 6 mm ($\frac{1}{4}$ in.), and its length is about 3.8 cm ($1\frac{1}{2}$ in.). Its direction is obliquely downward and forward, its course being slightly curved, with the concavity directed forward and upward. Its external orifice is the narrowest part and is located between the clitoris and the opening of the vagina.

The walls of the urethra consist of three coats: (1) an outer muscular coat, which is continuous with that of the bladder; (2) a thin layer of spongy tissue, containing a plexus of veins; and (3) a mucous coat, which is continuous internally with that lining the bladder and externally with that of the vulva.

The male urethra is about 20 cm (8 in.) long. It is divided into three portions: (1) the prostatic, which runs vertically through the prostate; (2) the membranous, which extends between the apex of the prostate and the bulb of the urethra; and (3) the cavernous portion, which extends from the membranous to the external orifice.

The male urethra is composed of (1) mucous membrane, which is continuous with the mucous membrane of the bladder and is prolonged into the ducts of the glands that open into the urethra; and (2) a submucous tissue, which connects it with the structures through which it passes.

Micturition. Urine is secreted continuously by the kidneys. It is carried to the bladder by the ureters and at intervals is expelled from the bladder through the urethra. The act by which the urine is expelled is called micturition. The desire to urinate is due to sensory stimulation in the bladder itself caused by pressure of urine, or reflex stimulation. The act is essentially a reflex through the central nervous system.

Involuntary micturition, or incontinence. Involuntary micturition may occur as the result of lack of consciousness or as the result of spinal injury involving the nerve centers which send nerves of control to the bladder. It may be due to a want of *tone* in the muscular walls, or it may result from some abnormal irritation due to irritant substances in the urine or to disease of the bladder (cystitis). Excessive nervousness may provoke the desire to urinate when there is only a small amount of urine in the bladder. This desire may also be aroused by visual and auditory impressions, such as the sight and sound of running water.

In young infants incontinence of urine is normal. The infant voids whenever the bladder is sufficiently distended to arouse a reflex stimulus.

Children vary markedly in the ease with which they learn to control micturition and defecation. During the first year, some children can be taught to associate the act with the proper time and place. By the second year, regular training in habit formation and proper feeding should enable the child to inhibit the normal stimulus and control micturition, at least during the day. Control of micturition at night is a habit requiring longer practice and is usually formed by the end of the second year. When involuntary voiding occurs at night with any degree of regularity after the third year, it is called enuresis. If this is not caused by nervous instability or irritation of the bladder, it will usually yield to proper training.

Retention of urine. Retention, or failure to void urine, may be due to: (1) some obstruction in the urethra or in the neck of the bladder, (2) nervous contraction of the urethra, or (3) dulling of the senses so that there is no desire to void. In the last two conditions retention is often overcome by measures which induce reflexes, i.e., pouring warm water over the vulva, or the sound of running water. If micturition does not occur and the bladder is not catheterized, distention of the organ may become extreme, and there is likely to be constant leakage, or involuntary voiding of small amounts of urine without emptying the bladder. This condition is described as retention with overflow.

Suppression of urine. A far more serious condition than retention is the failure of the kidneys to secrete urine. This is spoken of as suppression, or *anuria.* Unless suppression is relieved, a toxic condition known as uremia will develop. When the secretion of urine is decreased below the normal amount, the condition is spoken of as *oliguria.*

SUMMARY

Waste Products of Body
- Carbon dioxide
- Water
- Organic and inorganic salts
- Hair, nails, dead skin
- Indigestible and undigested portions of food

Wastes of Cell Metabolism
- (1) Liquid—water
- (2) Gas—carbon dioxide
- (3) Soluble salts
  - Organic salts, e.g., urea
  - Inorganic salts, e.g., sodium chloride
- (4) Others, e.g., heat

Excretory Organs
- Lungs
- Kidneys
- Alimentary canal
- Skin

Urinary System
- **Kidneys** (2)—secrete urine
- **Ureters** (2)—ducts which convey urine from kidneys to bladder

Urinary System
- **Bladder** (1)—reservoir for urine
- **Urethra** (1)—tube through which urine is voided

Kidneys

- **Location**
  - Posterior part of lumbar region, behind peritoneum
  - Placed on either side of spinal column and extend from upper border of twelfth thoracic to third lumbar vertebra
- **Covering and Support**
  - Embedded in a mass of fatty tissue, adipose capsule
  - Surrounded by fibrous tissue called renal fascia
  - Held in place by renal fascia which blends with fasciae on
    - quadratus lumborum,
    - psoas major, and
    - diaphragm
  - Also by pressure and counterpressure of neighboring organs
- **Size and Shape**
  - About 4.5 in. long, 2–3 in. broad, and 1 in. thick
  - Weight about 4.5 oz (135 gm)
  - Bean-shaped, tubular gland
  - Concave border directed toward median line of body
  - Hilum—fissure near center of concave side serves for vessels to enter and leave
- **Gross Structure**
  - **Pelvis**—upper expanded end of ureter
  - **Calyces**—cuplike cavities of the pelvis that receive papillae of pyramids
  - **Medulla**—inner striated portion, made up of cone-shaped masses
    - **Pyramids**—8–18
    - **Bases** directed toward circumference of kidney
    - **Papillae**—apices of pyramids, directed into pelvis
  - **Cortex**—outer portion of kidney
    - **Renal columns**—extensions of cortical substance between pyramids
- **Unit Pattern**
  - **Nephron**—consists of a renal tubule with its blood supply
  - **Renal tubules**—begin as capsules enclosing capillary tufts—glomeruli—in the cortex, and after a tortuous course open into straight collecting tubes which pour their contents into calyces of kidney pelvis
- **Blood Supply**
  - **Renal artery**—direct from aorta
  - Before entering kidney divides into several branches
  - **Arterial arches**
    - Lateral branches at the boundary zone between cortex and medulla
    - Send branches to cortex
  - **Venous arches**
    - Lateral branches at level of base of pyramids
    - Receive blood from cortex
    - Receive blood from medulla
  - Veins empty into renal vein, leave kidney at hilus, and empty into inferior vena cava

Kidneys

- **Nerves**
  - Nerves derived from renal plexus, and from the lesser and lowest splanchnic nerves
  - Vasomotor, by regulating size of blood vessels, influences blood pressure
- **Function of Kidneys**
  - To keep the body fluids at normal constancy by removing waste substances from the blood

Urine

- **Physical Characteristics**
  - An aqueous solution of organic and inorganic substances
  - **Color** and **transparency** depend upon concentration, diet, etc.
  - **Reaction,** usually slightly acid, *p*H 5–7
  - **Specific gravity** 1.010–1.030, depending upon proportions of solids and water
  - **Quantity**—1,200–1,500 cc daily. Affected by:
    - Amount of fluid ingested
    - Amount of fluid lost in other excretions than urine, e.g., perspiration
    - Ability of organs to function—heart, blood vessels, kidneys, etc.
    - Action of specific substances, as diuretics
- **Composition of Urine**
  - 95 per cent water
  - Remainder, about 3.7% organic and 1.3% inorganic wastes
    - May be exogenous or endogenous
  - **Exogenous,** e.g., urea
    - Amount excreted varies with diet
  - **Endogenous,** e.g., creatinine
    - Amount excreted does not vary with diet
- **Threshold substances**—valuable constituents, as glucose, are reabsorbed into blood stream
- **No-threshold substances**—those excreted in their entirety

Physiology of Urine Formation

- Efferent vessel is smaller than afferent, making blood pressure in glomerulus high
- Deproteinized plasma filters through capsule into tubule
- Efferent vessel, with others, forms plexuses about tubule
- Urine is concentrated in tubule by diffusion of water and some salts back into the blood

Some Constituents of Normal Urine

- **Inorganic**
  - Chlorides, Sulphates, Phosphates of Sodium, Potassium, Magnesium, Calcium
- **Organic**
  - **Creatinine** ($C_4H_7N_3O$)
    - 1–2 gm excreted daily
    - From endogenous metabolism, probably from creatine formed in metabolism of striped muscle tissue

- **Some Constituents of Normal Urine**
  - **Organic**
    - **Urea** [$CO(NH_2)_2$]
      - About 30 gm excreted daily
      - Constitutes one-half of all solids excreted in urine
      - Is made by liver cells from ammonium salts
    - **Ammonia** (NH_3)
      - 0.7 gm daily
      - Is thought to be made by the kidney from amino acids or urea. Amount made may depend upon body's need for basic radicals
    - **Hippuric acid** ($C_9H_9NO_3$)
      - About 1 gm daily
      - Amount increased on diet of foods containing benzoic acid
    - **Purine bodies**
      - Purine ($C_5H_4N_4$)
      - Adenine ($C_5H_3N_4NH_2$)
      - Guanine ($C_5H_3N_4ONH_2$)
      - Hypoxanthine ($C_5H_4N_4O$)
      - Xanthine ($C_5H_4N_4O_2$)
      - Uric acid ($C_5H_4N_4O_3$)
        - End products resulting from metabolism of nucleoproteins of food (exogenous) and tissues (endogenous)
        - Purine-free diet reduces amount of exogenous purine bodies in the urine
- **Some Abnormal Constituents of Urine**
  - **Albumin** (Albuminuria)
    - Caused by—
      - Disease of kidneys
      - Irritation of kidneys by poisons
      - Increased blood pressure
  - **Sugar** (Glycosuria)
    - Normal urine contains little sugar
    - Glycosuria may be temporary, as after ingestion of much sugar, or persistent, as in diabetes mellitus
  - **Indican**
    - Indican is potassium indoxyl sulphate

Some Abnormal Constituents of Urine

- **Indican**
  - **Caused by**
    - Excessive putrefaction of proteins in intestines, due to
      - Interference with absorption
      - Excess of protein food
      - Constipation
    - Putrefaction in body itself
- **Acetone bodies**—result of incomplete oxidation of fats
- **Casts**
  - Substances which harden and form a mold inside of tubules
  - Named from substances composing them or from their appearance
  - Varieties
    - Pus casts, blood casts, epithelial casts, granular casts, fatty casts, hyaline casts
- **Calculi**
  - Deposits of solid matter precipitated from the urine, vary in shape and size
  - **Causes**
    - Increase in slightly soluble constituents of urine
    - Decrease in amount of water secreted
    - Abnormally acid or abnormally alkaline urine
- **Pus**—due to suppurative conditions of urinary organs
- **Hematuria**
  - Blood in urine
  - Inflammation of urinary organs, tuberculosis, cancer, renal stone
- **Bile pigments**
  - In obstructive jaundice
  - In diseases in which many erythrocytes are destroyed
  - Give urine greenish-yellow or golden-brown color

Ureters

- Excretory ducts. Connect kidneys with bladder and serve as passageway for urine
- Commence as calyces which surround renal papillae. These join to form two or three short tubes, and these unite to form renal pelvis
- Duct is 25–30 cm long
- **Three coats**
  - (1) Mucous—lining
  - (2) Muscular
    - Inner, longitudinal layer
    - Outer, circular layer
  - (3) Fibrous—carries blood vessels and nerves

Bladder
- Hollow muscular organ
- Situated in pelvic cavity behind the pubes
  - In front of rectum in male
  - In front of anterior wall of vagina and neck of uterus in female
- Freely movable. Held in position by folds of peritoneum and fascia
- Size, shape, and position depend upon age, sex, and whether bladder is full or empty
- **Four coats**
  - (1) Mucous—lining
  - (2) Areolar—connects mucous and muscular
  - (3) Muscular
    - Inner layer—longitudinal
    - Middle layer—circular
    - Outer layer—longitudinal
  - (4) Serous—partial covering derived from peritoneum
- **Three openings**
  - Ureters run obliquely downward through the bladder wall opening into lower part, about ½ in. from median plane
  - Urethral opening is below and in front of the opening of the ureters
- **Function**
  - Serves as a reservoir for the reception of urine
  - When moderately distended, holds about 500 cc

Urethra
- Membranous canal, extends from the bladder to the meatus urinarius
- 3.8 cm long in female, about 20 cm long in male
- In female behind symphysis pubis, and embedded in the anterior wall of vagina
- **Three coats**
  - (1) Mucous—lining
  - (2) Submucous—supports network of veins
  - (3) Muscular
    - Inner—longitudinal
    - External—circular
- Meatus urinarius—external orifice located between clitoris and vagina

Micturition
- Act of expelling urine from bladder
- Reflex act—controlled by voluntary effort

Retention
- Failure to void urine
- Due to
  - (1) Obstruction in urethra or neck of bladder
  - (2) Nervous contraction of urethra
  - (3) Dulling of the senses
- May be accompanied by constant leakage, or involuntary voiding of small amounts

Suppression, or Anuria—Failure of the kidneys to secrete urine

Oliguria—Deficient secretion of urine

CHAPTER 22

THE SKIN AND APPENDAGES: HISTOLOGY REGULATION OF TEMPERATURE, VARIATIONS IN TEMPERATURE

Most of our contacts with the environment are through the skin. Since living cells must be surrounded with lymph, the contact of the body with the air is made by means of dead cells. These dead cells form a protective covering for the living cells. The living cells of the inner layers of the skin are constantly pushed to the outside, shrinking and undergoing progressive chemical changes which cement them firmly together and render them waterproof. In this way a tissue-fluid environment is maintained for living cells even though man lives in an air environment.

THE SKIN

The skin has many functions. It covers the body and protects the deeper tissues from drying and injury. It protects from invasion by infectious organisms. It is important in many ways in temperature regulation. It contains end organs of many of the sensory nerve fibers by means of which one becomes both physiologically and sometimes actually aware of items of the environment. It acts as an accessory mechanism for tactile and pressure corpuscles. In the skin fat, glucose, water, salts such as sodium chloride, etc., are stored temporarily by *inundation,* as Cannon has shown. It has excretory functions, eliminating water with the various salts which compose perspiration, and the dead cells themselves become an important way of eliminating many salts. It is an important light screen for the underlying living cells. It also has absorbing powers. It will absorb oily materials placed in contact with it. The skin doubtless has many other functions.

Skin area is especially important when an individual is burned. The following figures give the approximate areas of the body.

Head and neck 6 per cent; trunk 38 per cent (anterior trunk and genitals 20 per cent, posterior trunk 18 per cent). Upper extremities 18 per cent (arms 13.5 per cent hands 4.5 per cent). Lower extremities 38 per cent (thighs including buttocks 19 per cent, legs, 12.6 per cent, feet 6.4 per cent).

Structure. Skin consists of two distinct layers: (1) epidermis, cuticle, or scarf skin; (2) dermis, corium, or cutis vera.

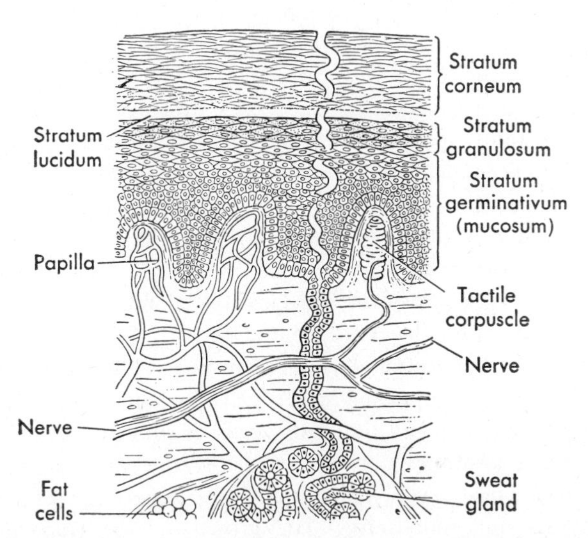

Fig. 335. Diagram of a section of the skin to show its structure. The epidermis consists of the strata corneum, ludicum, granulosum, and germinativum (mucosum). The corium lies below the epidermis. (See Fig. 337, p. 676.)

The epidermis is a stratified squamous epithelium, consisting of a variable number of layers of cells. It varies in thickness in different parts, being thickest on the palms of the hands and on the soles of the feet, where the skin is most exposed to friction, and thinnest on the ventral surface of the trunk and the inner surfaces of the limbs. It forms a protective covering over every part of the true skin and is closely molded on the papillary layer of the corium. The external surface of the epidermis is marked by a network of ridges caused by the size and arrangement of the papillae beneath. Some of these ridges are large and correspond to the folds produced by movements, e.g., at the joints; others are fine and intersect at various angles, e.g., upon the back of the hand. Upon the palmar surface of the fingers and hands and the soles of the feet, the ridges serve to increase resistance between contact surfaces and therefore prevent slipping. On the tips of the fingers and thumbs these ridges form distinct patterns which are peculiar to the individual and practically never change, hence the use of fingerprints for purposes of identification.

From without inward four regions of the epidermis are named: the *stratum corneum,* the *stratum lucidum,* the *stratum granulosum,* and the *stratum germinativum* (*mucosum*).

The three outer layers consist of cells which are practically dead and are constantly being shed and renewed from the cells of the stratum germinativum. In the stratum corneum the protoplasm of the cells has become changed into a protein substance called *keratin,* which acts as a waterproof covering. The reaction is acid; and many kinds of organisms, when placed upon the skin, are destroyed, presumably by the effect of the acidity. Underneath this is the stratum lucidum, a few layers of clear cells.

The stratum granulosum is formed by two or three layers of flattened cells that are supposed to be cells in transition between the stratum germinativum and the horny cells of the superficial layers.

The stratum germinativum consists of several layers of cells. The cells of the deepest layer are columnar in shape and are sometimes called the stratum mucosum. The growth of the epidermis takes place by multiplication of the cells of the germinative layer. As they multiply, they push upward toward the surface the cells previously formed. In their upward progress these cells undergo a chemical transformation, and the soft protoplasmic cells become converted into the flat scales which are constantly being rubbed off the surface of the skin. The pigment in the skin of the Negro, as well as that of the nipple in white races, is found in the cells of the stratum germinativum. No blood vessels pass into the epidermis, but it receives fine nerve fibers between the cells of the inner layers.

The corium is a highly sensitive and vascular layer of connective tissue. It contains numerous blood vessels, lymph vessels, nerves, glands, hair follicles, and papillae and is described as consisting of two layers: the *papillary,* or *superficial, layer,* and the *reticular,* or *deeper, layer.*

The surface of the *papillary,* or *superficial,* layer is increased by protrusions in the form of small conical elevations, called papillae, whence this layer derives its name. They project up into the epidermis, which is molded over them. The papillae consist of small bundles of fibrillated tissue, the fibrils being arranged parallel to the long axis of the papillae. Within this tissue is a loop of capillaries; and some papillae, especially those of the palmar surface of the hands and fingers, contain *tactile corpuscles,* which are numerous where the sense of touch is acute.

The *reticular,* or *deeper,* layer consists of strong bands of fibrous tissue and some fibers of elastic tissue. These bands interlace, and the tiny spaces left by their interlacement are occupied by adipose tissue and sweat glands. The reticular layer is attached to the parts beneath by a subcutaneous layer of areolar connective tissue, which, except in a few places, contains fat. In some parts, as on the front of the neck, the connection is loose and movable; in other parts, as on the palmar surface of the hands and the soles of the feet, the connection is close and firm. In youth the skin is both extensile and elastic, so that it can be stretched and wrinkled and return to its normal condition of smoothness. As age advances, the elasticity is lessened, and the wrinkles tend to become permanent.

Blood vessels and lymphatics. The arteries which supply the skin form a network in the subcutaneous connective tissue and send branches to the papillae, the hair follicles, and the sudoriferous glands. The capillaries of the skin are so numerous that when distended they are capable of holding a large proportion of the blood contained in the body. The amount of blood they contain is dependent on their caliber, which is regulated largely by the vasomotor nerve fibers.

There is a superficial and a deep network of lymphatics in the skin. These communicate with each other and with the lymphatics of the subcutaneous connective tissue.

Nerves. The skin contains the peripheral terminations of many nerve fibers. These fibers may be classified as follows:

1. Motor nerve fibers, including the vasoconstrictors and vasodilators distributed to the blood vessels, and motor nerve fibers distributed to the arrector muscles (arrectores pilorum).

2. Nerve fibers concerned with the temperature sense, which terminate in *cold receptors* (end organs of Krause, Fig. 157) and *receptors for warmth* (possibly the end organs of Ruffini).

3. Nerve fibers concerned with touch and pressure, which terminate in *touch* (Meissner's corpuscles and free nerve endings around hairs and skin) and *pressure receptors* (Pacinian corpuscles).

The nerve fibers concerned with touch and temperature are myelinated and belong to the group in the cranial and spinal nerves that is responsible for the beta and gamma waves of the A group of fibers.[1] The nerve fibers concerned with pain belong to the C group and are fine fibers, amyelinated or slightly myelinated. These fibers of the trunk and limbs run in the lateral spinothalamic tract of the spinal cord from cells in the dorsal root ganglia to the thalamus. The second neuron extends to a third neuron in the postcentral gyrus of the cerebral cortex. The fibers above the spinal cord are in the cranial nerves (p. 286).

4. Nerve fibers which are stimulated by pain. The number of *pain receptors* (free nerve-fiber ends) is estimated to be more than 2,000,000.

5. Secretory nerve fibers which are distributed to the glands. Because of the number of sensory nerve fibers which lead from the skin to centers in the brain and spinal cord, nearly every nerve center in the body may be affected by sensations arising in the skin. It is for this reason that hydrotherapeutic applications, heat, cold, and counterirritants excite so many and such varied reflexes.

THE APPENDAGES OF THE SKIN

The appendages of the skin are the nails, the hairs, the sebaceous glands, the sudoriferous, or sweat, glands, and their ducts.

The nails (ungues) are composed of clear, horny cells of the epidermis, joined so as to form a solid, continuous plate upon the dorsal surface of the terminal phalanges. Each nail is closely adherent to the underlying corium, which is modified to form what is called the bed, or *matrix*, of the nail. At the hinder part of the bed the skin forms a deep fold in which is lodged the root of the nail.

The nails grow in length by multiplication of the soft cells in the stratum germinativum at the root. The cells are transformed into

[1] J. Erlanger and H. S. Gasser, *American Journal of Physiology*, 1930.

hard, dry scales, which unite to form a solid plate; and the nail, constantly receiving additions, slides forward over its bed and projects beyond the end of the finger. When a nail is thrown off by suppuration or torn off by violence, a new one will grow in its place provided any of the cells of the stratum germinativum are left.

The hairs (**pili**) are growths of the epidermis, developed in the hair follicles, which extend downward into the subcutaneous tissue. The part which lies within the follicle is known as the root, and that portion which projects beyond the surface of the skin is called the shaft. The hair is composed of:

Cuticle, a single layer of scalelike cells, which overlap.

Cortex, a middle portion, which constitutes the chief part of the shaft, formed of elongated cells united to form flattened fibers which contain pigment granules in dark hair and air in white hair.

Medulla, an inner layer composed of rows of many-sided cells, which frequently contain air spaces. The fine hairs covering the surface of the body and the hairs of the head do not have this layer.

The root of the hair is enlarged at the bottom of the follicle into a bulb which is composed of growing cells and fits over a vascular papilla which projects into the follicle. Hair has no blood vessels but receives nourishment from the blood vessels of the papilla.

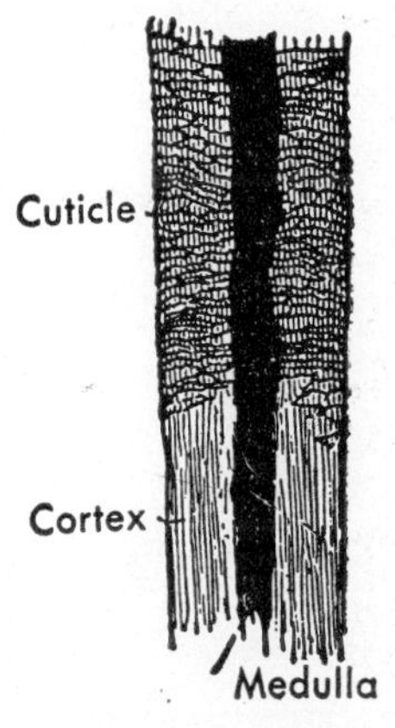

Fig. 336. Piece of human hair. (Highly magnified.)

Growth of hair. Hair grows from the papilla by multiplication of its cells (matrix cells). These cells become elongated to form the fibers of the fibrous portion, and as they are pushed to the surface, they become flattened and form the cuticle. If the scalp is thick, pliable, and moves freely over the skull, it is favorable to the growth of hair. A thin scalp that is drawn tightly over the skull tends to constrict the blood vessels, lessen the supply of blood, and cause atrophy of the roots of the hair by pressure; in such cases massage of the head loosens the scalp, improves the circulation of the blood, and usually stimulates the growth of the hair. The hairs are constantly falling out and constantly being replaced. In youth and early adult life not only may hairs be replaced, but there may be an increase in the number of hairs by development of new follicles. When the matrix cells lose their vitality, new hairs will not develop.

With the exceptions of the palms of the hands, the soles of the feet, and the last phalanges of the fingers and toes, the whole skin is studded with hairs. The hair of the scalp is long and coarse, but

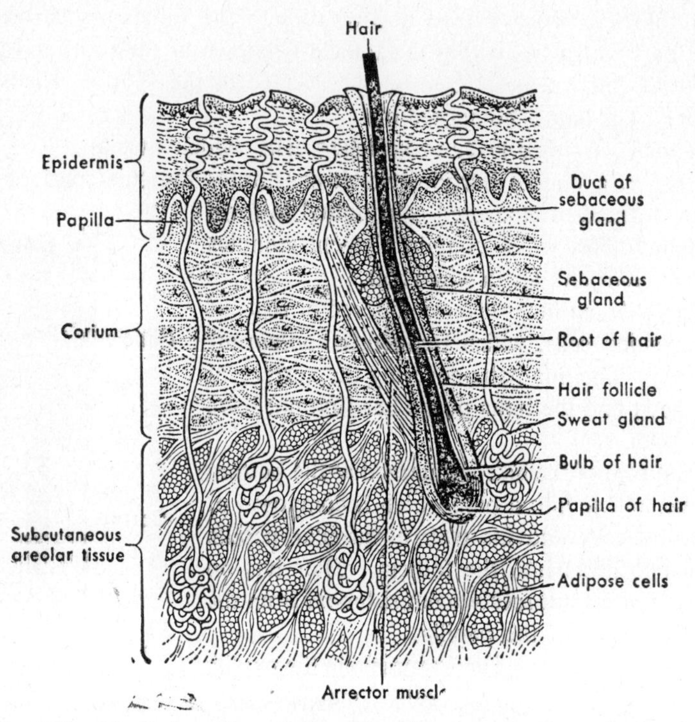

Fig. 337. Vertical section of the skin, showing sebaceous glands, sweat glands, hair, and follicle; also arrector muscle. (Gerrish.)

most of the hair is fine and extends only a little beyond the hair follicle.

Arrector (arrectores pilorum) muscles. The follicles containing the hairs are narrow pits which slant obliquely upward, so that the hairs they contain lie slanting on the surface of the body. Connected with each follicle are small bundles of involuntary muscle fibers called the *arrector muscles*. They arise from the papillary layer of the corium and are inserted into the hair follicle below the entrance of the duct of a sebaceous gland (Fig. 337). These muscles are situated on the side toward which the hairs slope, and when they contract, as they will under the influence of cold or fright, they straighten the follicles and elevate the hairs, producing the roughened condition of the skin known as "gooseflesh." Since the sebaceous gland is situated in the angle between the hair follicle and the muscle, contraction of the muscle squeezes the sebaceous secretion out from the duct of the gland. This secretion aids in preventing too great heat loss.

The sebaceous glands are compound alveolar glands lodged in the corium. They occur everywhere over the skin surface, with the exception of the palms of the hands and the soles of the feet. They are abundant in the scalp and face and are numerous around the apertures of the nose, mouth, external ears, and anus. Each alveolus of the gland is composed of a number of epithelial cells and is filled with larger cells containing fat. These cells are cast off bodily, their detritus going to form the secretion and new cells being continuously formed. Occasionally the ducts open upon the surface of the skin, but more frequently they open into the hair follicles. In the latter case the secretion from the gland passes out to the skin along the hair. Their size is not regulated by the length of the hair.

Some of the largest sebaceous glands are found on the nostrils and other parts of the face, where they may become enlarged with accumulated secretion. This retained secretion often becomes discolored, giving rise to the condition commonly known as blackheads. It also provides a medium for the growth of pus-producing organisms and consequently is a common source of pimples and boils.

Sebum is the secretion of the sebaceous glands. It contains fats, soaps, cholesterol, albuminous material, remnants of epithelial cells, and inorganic salts. It serves to protect the hairs from becoming too dry and brittle, as well as from becoming too easily saturated with moisture. Upon the surface of the skin it forms a thin protective layer, which serves to prevent undue absorption or evaporation of water from the skin. An accumulation of this sebaceous matter upon the skin of the fetus furnishes the thick, cheesy, oily substance called the *vernix caseosa*.

Sudoriferous, or sweat, glands are abundant over the whole skin but are largest and most numerous in the axillae, the palms of the hands, the soles of the feet, and the forehead. Each gland consists of a single tube, with a blind, coiled end which is lodged in the subcutaneous tissue. From the coiled end the tube is continued as the excretory duct of the gland up through the corium and epidermis and finally opens on the surface by a pore. Each tube is lined with secreting epithelium continuous with the epidermis. The coiled end is closely invested by capillaries, and the blood in the capillaries is separated from the cavity of the glandular tube by the thin membranes which form their respective walls.

Perspiration, or *sweat.* Pure sweat is very dilute and practically neutral. When gathered from the skin, it contains fragments of cells and sebum and has an average *p*H of 5.65. Perspiration contains the same inorganic constituents as the blood, but in lower concentration. The chief salt is sodium chloride. The organic constituents are similar to those found in the urine, namely, urea, uric acid, creatinine, sulphates of phenol, skatol, etc. Normally these are present in mere traces, but when sweating is profuse, the amount of urea eliminated may be considerable.

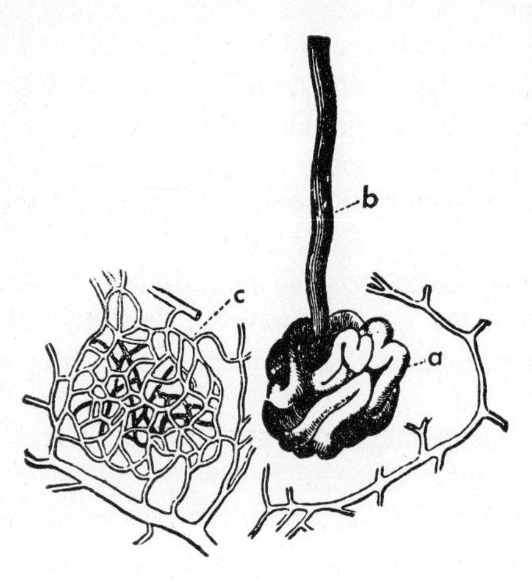

Fig. 338. Coiled end of a sweat gland. *a,* the coiled end; *b,* the duct; *c,* network of capillaries, inside which the sweat gland lies.

Quantity of perspiration. Under ordinary circumstances, the perspiration that the body is continually throwing off evaporates from the surface of the body without one's becoming sensible of it and is called *insensible perspiration.* When more sweat is poured upon the surface of the body than can be removed at once by evaporation, it appears on the skin in the form of drops, and is then spoken of as *sensible perspiration.*

The amount secreted during 24 hours varies greatly. It is estimated to average about 480 to 600 cc (16 to 20 oz), but may be increased to such an extent that even more than this may be secreted in an hour. It is *increased* by: increased temperature or humidity of the atmosphere; a dilute condition of the blood; exercise; pain; nausea; mental excitement or nervousness; dyspnea; use of diaphoretics, e.g., pilocarpine, physostigmine, strychnine, nicotine, camphor, etc.; and certain diseases, such as tuberculosis, acute rheumatism, and malaria.

The secretion of perspiration is *decreased* by: cold; voiding of a large quantity of urine; diarrhea; certain drugs such as atropine and morphine in large doses; and certain diseases, such as fevers, diabetes, and some cases of paralysis.

Activity of the sweat glands. Special secretory nerve fibers are supplied to the glandular epithelium of the sweat glands. The activity of these glands is supposed to be the result either of direct stimulation of the nerve endings in the glands or of indirect stimulation through the sensory fibers of the skin. The usual cause of profuse sweating is a high external temperature or muscular exercise. It is known that the high temperature acts upon the sensory cutaneous nerves, possibly the heat fibers, and stimulates the sweat fibers indirectly.

Functions of the sweat glands. While perspiration is an excretion, its value lies not so much in the elimination of waste matter as in the loss of body heat by the evaporation of water. Each gram of water requires about 0.5 Cal for evaporation, and this heat comes largely from the body. This loss of heat helps to balance the production of heat that is constantly taking place. When the kidneys are not functioning properly, and the blood contains an excessive amount of waste matter, the sweat glands will excrete some of the latter, particularly if their activity is stimulated. In the condition known as uremia, when the kidneys secrete little or no urine, the percentage of urea in perspiration rises.

Ceruminous glands. The skin lining the external auditory canal contains modified sweat glands called *ceruminous* glands. They secrete a yellow, pasty substance resembling wax, which is called cerumen. An accumulation of cerumen deep in the auditory canal may interfere with hearing.

BODY HEAT

From the standpoint of heat control, animals may be divided into two great classes:

Constant-temperature (homoiothermic) animals, or those whose temperature remains practically constant whether the surrounding air is hotter or cooler than the body. Birds and mammals (including human beings) are in this class.

Variable-temperature (poikilothermic) animals, or those whose temperature varies with that of the surrounding medium, e.g., reptiles, frogs, fishes, etc. In winter their temperature is low, and in summer their temperature approximates that of their surroundings. The human fetus is in this class.[2] Between these two groups are various classes of hibernating animals.

The great difference between these two classes of animals is in their reactions to external temperature. A cold environment reduces the temperature of the cold-blooded creature, reduces the metabolism of all its tissues, and thus reduces its heat production. The warm-blooded animal reacts in the opposite way. In a cold environment its temperature remains constant, its metabolism increases, and thus its heat production increases.

Production of heat. The heat of the body is derived by oxidation, which reduces complex substances of food to simple substances, thereby releasing energy. Every cell is a producer of heat. But tissues differ in their activities and in the quantity of heat they produce. The most important heat-generating organs are the skeletal muscles, which constitute about 50 per cent of the body weight. Next to these are the large glands, though it is thought that for equal weights of tissue, glands may produce more heat. Minor sources of heat are friction, i.e., that caused by the movements of the muscles and the circulation of the blood; hot substances ingested; and radiation from the sun and heating appliances.

Distribution of heat. The blood permeates all the tissues and serves as an absorbing medium for the heat. Wherever oxidation takes place and heat is generated, the temperature of the blood circulating in these tissues is raised. Wherever, on the other hand, the blood vessels are exposed to conditions which are cool because of heat lost by evaporation, etc., as in the moist membranes in the lungs or the more or less moist skin, the temperature of the blood is lowered. But these changes are not effected instantaneously, and

[2] At birth the heat-regulating mechanism is not "in working order," and infants are not able to regulate their body temperature; hence the importance of keeping them warm. Premature infants are even less able to regulate their body temperature; hence the need of special means to keep them warm.

consequently the temperature of some internal parts must always be higher than that of others. This is particularly true of the liver because its blood vessels are well protected against loss of heat. Because of this the temperature of the blood in different parts of the body varies slightly; but the circulation, now through warmer and again through cooler parts, tends to keep the average temperature of the blood at about 38° C.

Loss of heat. Heat is continually being produced in the body and continually leaving the body by the skin and the lungs, and by the urine and feces, which are at the temperature of the body when voided.

If weight is neither being lost nor gained, the calories lost per day match the calories taken in per day. One record of heat lost during a 24-hour day follows:

Lost from skin	2,156 Cal, or 87.5%
1,792 Cal, or 73.0%, by radiation and conduction	
364 Cal, or 14.5%, by evaporation of perspiration	
Lost in expired air	266 Cal, or 10.7%
182 Cal, or 7.2%, vaporization of water	
84 Cal, or 3.5%, warming air	
Lost in urine and feces	48 Cal, or 1.8%
Total heat loss per 24-hour day	*2,480 Cal, or 100.0%*

From these figures it is evident that the skin is the important factor in getting rid of body heat. This is due to the large surface offered for radiation, conduction, and evaporation and to the large amount of blood which constantly flows through the skin.

The temperature and humidity of the atmosphere may cause considerable difference in the percentages given above. A low temperature will increase the loss of heat by radiation, conduction, and convection and decrease that by evaporation. A high temperature will increase the relative amount of heat lost by evaporation, owing to the greater production of perspiration. Heat is lost by the evaporation of perspiration, by the warming of air taken into the lungs, and by the evaporation of the water which leaves the body by the lungs. It requires about 0.5 Cal for the evaporation of 1 gm of water, or about 250 Cal for the evaporation of 500 gm (1 pt) of water. It is estimated that under ordinary circumstances it requires about 250 Cal daily for the evaporation of the perspiration and about 250 Cal for the evaporation of the water (about 1 pt) lost through the lungs. In hot weather the perspiration may be greatly increased, and more heat than usual may be lost by evaporation of this water from the skin, particularly if the humidity is not excessive. Under these conditions there is less water lost as urine. High humidity interferes with the evaporation of perspiration, because air can take up only a certain amount of moisture. For this reason heat is more acutely felt when the humidity is high, even though the temperature is low, than when the humidity is low and the temperature high. Heat prostration and sunstroke are more likely to occur on hot days when the humidity is high than when it is low. Moving air favors evaporation because it tends to drive away evaporated moisture; hence the comfort derived from an electric fan, which keeps the air in motion.

THE REGULATION OF BODY TEMPERATURE

The constant temperature of the body is maintained by means of a balance between heat production (thermogenesis) and heat loss (thermolysis). The body must control the production and the loss of heat (thermotaxis). The part which the nervous system plays in this relationship is still an open question. The fact that infants cannot perform this function until some time after birth indicates that heat regulation follows a course parallel to the development of the nervous system. To many physiologists it seems reasonable to suppose that the accurate *balance* between heat production and heat loss is controlled by nerve fibers connected with a temperature-regulating center in the brain, and much experimental evidence indicates the existence of such a center in the thalamus.

Heat Production (thermogenic center in hypothalamus)	Amount and frequency of fuel intake (available calorific value)	Appetite, psychological states, plus hunger, which is related to the tone of the smooth muscle cells of the stomach
	Metabolic rate (rate of use of fuels)	Related to the tone of all cells of the body, but especially to tone of striped muscle cells and cells of large glands (great bulk of living human tissue)

Heat Loss (heat-loss center in hypothalamus)	*Radiation and conduction* related to quantity of blood through skin capillaries per unit of time	Concerned with tone of smooth muscle cells of arterioles and venules
	Evaporation of perspiration related to quantity of blood through skin capillaries per unit of time and, therefore, to temperature of skin and availability of raw materials for manufacture of perspiration	Concerned with tone of smooth muscle cells of arterioles and venules and with tone of cells of sweat glands

Heat production (chemical regulation) in the body is varied by increasing or decreasing the physiological oxidations. This end is effected in part by *taking food* and by *muscular exercise*. Chapter 20 describes how the oxidation of different foodstuffs produces varying amounts of heat and how the digestion products of foodstuffs, particularly proteins, stimulate the metabolism of the body and cause an increased production of heat. In this connection the action of enzymes and some of the internal secretions (e.g., thyroxin—and possibly adrenaline) is important. *Thyroxin* increases the oxidations of the body and stimulates all of the metabolic processes. During digestion heat is produced partly by the peristaltic action of the intestines and partly by the activity of the various digestive glands, particularly the liver. *Cold weather* stimulates the appetite; and an increased amount of food, usually accompanied by an increase of fats, increases heat production. Cold causes the muscles to contract and speeds up the processes of oxidation. Muscular contractions give rise to heat; therefore muscular activity counteracts the effects of external cold. On the other hand, muscular activity does not increase the temperature in warm weather to any marked extent. This is accounted for by the fact that when muscular exertion speeds up circulation, the blood vessels in the skin dilate, the sweat glands pour out more abundant secretion, and the heated blood passes in larger quantities through the cutaneous vessels, which are kept well cooled by the evaporation of the perspiration; the general average temperature of the body is thus maintained. Study Fig. 339 on p. 684 for centers that regulate body temperature.

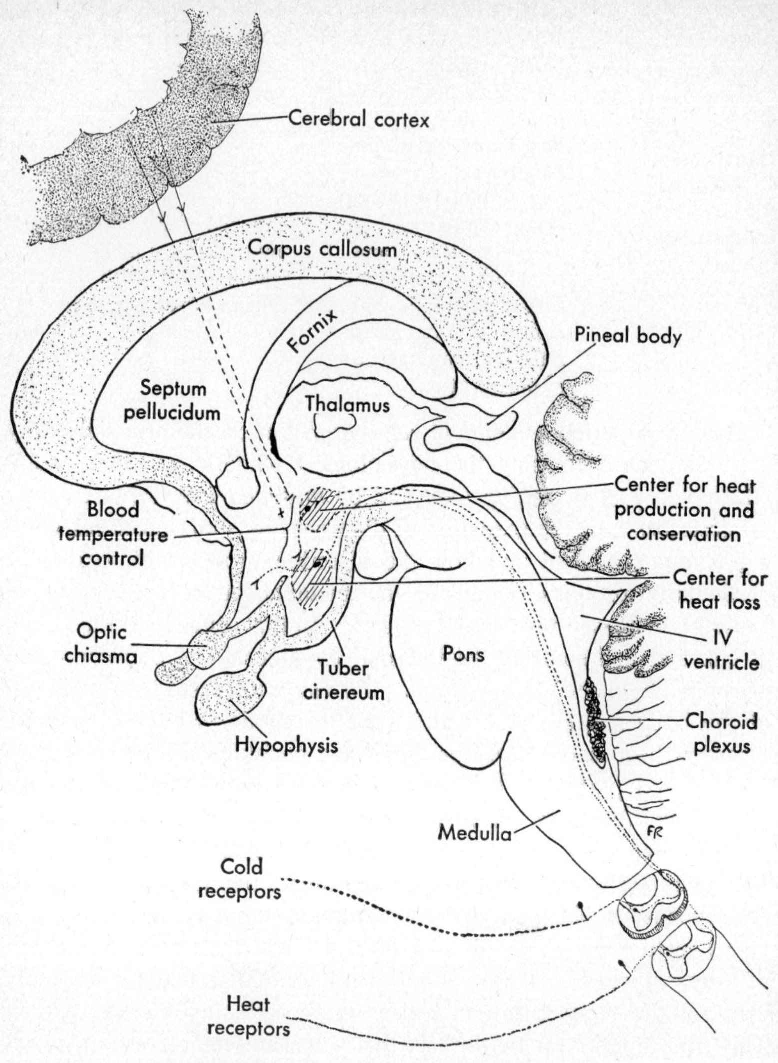

Fig. 339. Nervous control of body temperature.

Heat loss (physical regulation). To a small extent heat loss is controlled through an increase in the respiratory rate. The increased respirations associated with muscular activity aid somewhat in eliminating the excess heat produced, although this factor is not as important as sweating and flushing of the skin. In man respiration plays only a small part in temperature regulation; but in animals that do not perspire, respiration is an important means of regulating the temperature.

During muscular activity or when the external temperature is high, receptors for heat are stimulated, and impulses are transmitted over sensory fibers to the nerve centers controlling the motor fibers of the sweat glands. The motor fibers stimulate the activity of the sweat glands, and an increased amount of perspiration occurs. An increased amount of heat is required to vaporize this perspiration, and thus heat is lost. Excessive humidity interferes with the evaporation of water and thus interferes with the loss of heat; hence the discomfort experienced on hot, humid days.

The receptors for heat not only transmit impulses that stimulate the sweat glands to activity but at the same time transmit impulses that result in the depression of the vasoconstrictor center leading to the arterioles of the skin. In consequence the arterioles dilate, and more blood is sent to the surface to be cooled. This tends to increase the temperature of the skin and hence increases the heat lost by conduction and radiation. When the external temperature is low, the receptors for cold transmit impulses which result in stimulation of the vasoconstrictors and consequent contraction of the arterioles of the skin. This lessens the amount of blood in the skin arterioles, reduces the temperature of the skin, and so lessens the amount of heat loss.

The flow of blood through the skin tends to raise skin temperature, while the loss of heat to the environment tends to lower it. The nervous system adjusts the flow of blood in the skin in relation to environmental temperature at the skin surface in such a way that the body maintains a constant temperature. Temperature and heat-dissipating capacities of the environment at the skin surface are modified by clothing (amount, texture, style, color, etc.) and by ventilation (of rooms, etc.).

Thermal comfort seems to be closely related to skin temperature, and optimal average skin temperature seems to be about 33° C (92° F). This means apparently that metabolic rate, distribution of blood to skin, and sweat-gland activity are going on at optimal physiological levels when average skin temperature is about 33° C. If skin temperature rises or falls, physiological adjustments of heat production and heat loss can be made in order to maintain the normal body temperature, but they will not be made at the optimum physiological level. A sitting-resting person in thermal comfort is said to produce about 50 Cal per hour per square meter of surface area and, in a well-ventilated room (temperature 70° F, air movement at about 20 ft per minute, and humidity somewhat under 50 per cent), would lose less than 25 per cent (12 Cal) by evaporation of insensible perspiration and about 38 Cal per hour per square meter of surface area by radiation and conduction through the clothing to the environment. This gives a way of gauging (by means of heat-insulation value) the clothing needed in relation to activity, room temperature, etc., in order to keep skin temperature at approximately 33° C, or the temperature needed in relation to activity and clothing.

Recently it has been suggested that the concentration of the blood is a factor to be considered. The water of the body holds heat; and when the external temperature is low, water is withdrawn from the blood to the tissues, leaving the blood more concentrated. When the external temperature is high, water is withdrawn from the tissues to the blood. When the blood is dilute, an increased amount of water is brought to the surface, and an increased loss of heat results; but when the water is withheld in the interior of the body, less heat is lost. Other factors to be considered are size, age, and constitution.

Size. The quantity of heat produced by well-nourished animals, including man, is relatively constant; but the larger the surface of the body exposed to a cooler medium, the greater must be the loss of heat, since the heat lost is related to the *area of surface*. Small animals present a proportionately larger surface to the surrounding medium than larger animals; hence, the loss per unit of weight is greater, and this must be compensated for by a greater heat production. Skin secretions of different animals have varied effects in relation to heat insulation, and account must be taken of this also.

Age. In children the heat production is relatively large, because they are active and growing. Moreover, young children have not the constancy of temperature which is an evolved characteristic of adult life. On the contrary, they are subject to changes of body temperature which would be of grave import in an adult.

Constitution. Individuals differ greatly in their power of heat loss. Apart from differences in size and in the faculty of perspiration, there exist differences in compactness of shape, in the amount of adipose tissue protecting the viscera, etc.

Clothing aids the functions of the skin and the maintenance of heat, though, of course, clothes are not in themselves sources of heat. The kind of clothing to be worn should be determined by the necessity for diminishing the loss of heat, as in cold weather, or facilitating this loss, as in warm weather. Clothing of any kind captures a layer of warm and moist air between it and the skin and thus diminishes greatly the loss by evaporation, conduction, and radiation. In considering the heat value of clothing the important properties are: amount, quality, texture, style, color, etc.

Materials that are loosely woven are warmer than those that are tightly woven, because the meshes in a loosely woven material are capable of holding air, which is a poor conductor of heat. Two layers of thinner material are usually warmer than one layer of thicker material, because a layer of air is held between the two.

Thick material does not allow the warm air next to the body to penetrate to the outside.

Dark-colored materials absorb heat to some extent; hence they are warmer than light-colored textiles.

Thick, porous materials keep the body warm. Wool has an additional advantage, as evaporation takes place more slowly from it than from linen, cotton, or silk; and it has a greater capacity for absorbing moisture, so that the layer of air next to the skin does not become saturated with moisture. Thin and very porous materials help to keep the body cool, because they allow the air to penetrate to the skin, and thus assist the evaporation of sweat. Loose clothing facilitates heat loss.

Hot baths. The primary effect of a hot bath is to prevent loss of heat from the surface of the body, and some increase in temperature may result. If the bath is not continued for too long a time, this effect is counteracted by the increased perspiration that follows.

Cold, or tonic, baths. The primary effect of a cold bath is similar to the prolonged effect of cold air. The cold bath contracts the arterioles of the skin, drives the blood to the interior, and increases oxidation. If the bath is a short one and is followed by friction (contrast condition, cold followed by heat), the reaction is for the arterioles to dilate. The heated blood is sent to the surface, the circulation is quickened, and there is a consequent loss of heat. In health the gain in heat is usually balanced by the loss of heat, and the purpose of a cold bath is to exercise the arterioles and stimulate the circulation. If the bath is continued for some time, the temperature of the skin and of the muscles lying beneath is reduced, and either the heat-producing processes may be checked and a loss of temperature result, or shivering may intervene. In this case the muscular contractions and constriction of the blood vessels stimulate metabolism and heat production. When cold baths are given for the purpose of increasing heat elimination, friction is used during the bath to prevent shivering. Friction stimulates the sensory fibers of the skin, causes dilatation of the arterioles, and favors the flow of warm blood to the surface, thus decreasing the sensation of cold and increasing heat elimination. If properly given, cold baths stimulate the nervous system, improve the tone of the muscles, including the muscles of the heart and blood vessels, stimulate the circulation, and favor the elimination of heat.

VARIATIONS IN TEMPERATURE

The temperature of the human body is usually measured by a thermometer placed in the mouth, axilla, or rectum. Such measurements show slight variations. The normal temperature by mouth is about 37° C (98.6° F), by axilla the temperature is lower, and by rectum it is usually 1° higher.

Normal variations depend upon such factors as time of day, exercise, meals, age, sex, season, climate, and clothing. The temperature is usually lowest between 3 and 5 A.M. It rises slowly during the day, reaches its maximum at about 4 P.M., and falls again during the night. This corresponds to the usual temperature ranges in fever, when the minimum is in the early morning and the maximum is in the evening. Muscular activity and food cause a slight increase in temperature. This probably accounts for the increase in temperature during the day. In the case of night workers who sleep during the day, the increase in temperature occurs during the night, which is the period when food is eaten and work performed. Age has some influence. Infants and young children have a slightly higher temperature (about 1°) than adults. Their heat-regulating mechanism is more easily disturbed, and rise of temperature is caused by slight disturbances of digestion or metabolism and usually is less significant than the same increase in adults. Aged people show a tendency to revert to infantile conditions, and their temperature is usually slightly higher than in middle life. It is said that women have a slightly higher temperature than men. The effects of climate and clothing have been discussed.

Subnormal temperature. In order to carry on the activities essential to life, the body must maintain a normal temperature. If the temperature falls much below normal, to about 35° C (95° F), life can be maintained for only a short time. Subnormal temperature may be due to excessive loss of heat, profuse sweating, hemorrhage, and lessened heat production, as in starvation. In cases of starvation the fall of temperature is very marked, especially during the last days of life. The diminished activity of the tissues first affects the central nervous system; the patient becomes languid and drowsy, and finally unconscious; the heart beats more and more feebly, the breath comes more and more slowly, and the sleep of unconsciousness passes insensibly into the sleep of death.

Abnormal variations. Fever. The term *fever* is applied to an abnormal condition characterized by increased temperature, increased rate of heart action, increased respirations, increased tissue waste, faulty secretion, and various other symptoms such as thirst, weakness, and apathy, Some of the symptoms accompanying fever may be due to the substances or conditions causing the fever, some to the high temperature, and some to both.

Cause. The exact cause of fever is unknown. It may be due to diminished heat loss, increased heat production, or both. Fever is usually accompanied by increased catabolism of body tissue, as shown in the increased urea content of the urine, even though the dietary protein is not increased; in explanation of this it is suggested that the toxins which cause the fever produce some change in the cells which makes them susceptible to the action of oxidizing enzymes, so that oxidation and heat production are abnormally increased. Fever is accompanied also by retention of water in the tissues, resulting in a concentration of the blood. Usually the superficial blood vessels are contracted, and this is thought to be caused by the toxins acting on the vasoconstrictor centers. It is believed that fever and the conditions that accompany it are protective reactions to overcome the effect of toxins on the body. The reasons for this are based on various laboratory experiments. Animals have been inoculated with bacteria or bacterial toxins and then kept for a time at a temperature of about 40° C (105° F), with the result that they resisted the infection better than animals who were not subjected to this higher temperature. In connection with this, bacteriologists remind us that many organisms are killed at a temperature slightly above that of the body, and it may be that a high body temperature favors the formation of immune bodies. The contraction of the superficial blood vessels sends more blood to the interior of the body, thus providing an increased number of phagocytes and antibodies to fight the infection.

In fever therapy an artificial hyperpyrexia is brought about by the use of a hypertherm, or heating cabinet, in which the body temperature may be raised and held at some specific temperature, say 106 to 107° F, for a varying period of hours in order to render certain pathogenic organisms in the body nonviable. With such strenuous treatment care is taken to combat the decreased oxygen saturation in the body fluids by the use of oxygen and carbon dioxide mixture and to prevent excess alkalosis by the use of saline fluids.

SUMMARY

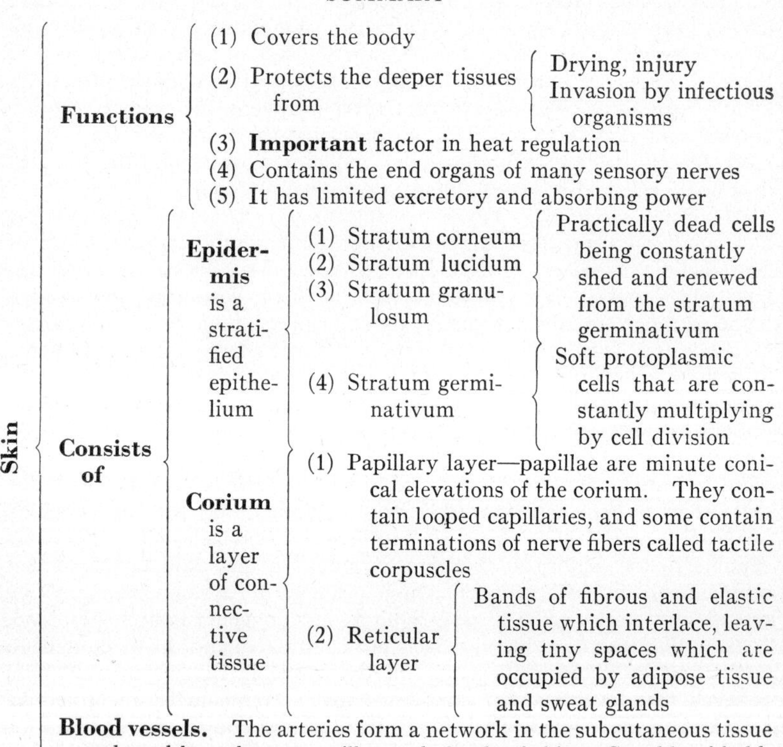

Skin

- **Functions**
  - (1) Covers the body
  - (2) Protects the deeper tissues from
    - Drying, injury
    - Invasion by infectious organisms
  - (3) **Important** factor in heat regulation
  - (4) Contains the end organs of many sensory nerves
  - (5) It has limited excretory and absorbing power
- **Consists of**
  - **Epidermis** is a stratified epithelium
    - (1) Stratum corneum
    - (2) Stratum lucidum
    - (3) Stratum granulosum
      - Practically dead cells being constantly shed and renewed from the stratum germinativum
    - (4) Stratum germinativum
      - Soft protoplasmic cells that are constantly multiplying by cell division
  - **Corium** is a layer of connective tissue
    - (1) Papillary layer—papillae are minute conical elevations of the corium. They contain looped capillaries, and some contain terminations of nerve fibers called tactile corpuscles
    - (2) Reticular layer
      - Bands of fibrous and elastic tissue which interlace, leaving tiny spaces which are occupied by adipose tissue and sweat glands
- **Blood vessels.** The arteries form a network in the subcutaneous tissue and send branches to papillae and glands of skin. Capable of holding a large proportion of total amount of blood in body
- **Lymphatics.** There is a superficial and a deep network of lymphatics in the skin

Skin

- **Nerve fibers**
  - (1) Motor fibers to blood vessels and arrector muscles
  - (2) Fibers concerned with temperature sense
  - (3) Fibers concerned with sense of touch and pressure
  - (4) Fibers stimulated by pain
  - (5) Secretory fibers which are distributed to the glands
- **Appendages**
  - Nails, hairs
  - Sebaceous glands, sudoriferous glands

Nails

- Consist of clear, horny cells of epidermis
- Corium forms a bed, or matrix, for nail
- Root of nail is lodged in a deep fold of the skin
- Nails grow in length from soft cells in stratum germinativum at root

Hairs (Pili)

- The hairs grow from the roots
- The roots are bulbs of soft, growing cells contained in the hair follicles
- Hair follicles are little pits developed in the corium
- Stems of hair extend beyond the surface of the skin, consist of three layers of cells: (1) cuticle, (2) cortex, and (3) medulla
- Found all over body, except
  - Palms of the hands
  - Soles of the feet
  - Last phalanges of the fingers and toes
- Arrector muscles are attached to corium and to each hair follicle. Contraction pulls hair up straight, drags follicles upward, forces secretion of sebaceous glands to surface, and forces blood to interior

Sebaceous Glands

- Compound alveolar glands, the ducts of which usually open into a hair follicle but may discharge separately on the surface of the skin
- Lie between arrector muscles and hairs
- Found over entire skin surface except
  - Palms of hands
  - Soles of feet
- Secrete *sebum*, a fatty, oily substance, which keeps the hair from becoming too dry and brittle, the skin flexible, forms a protective layer on surface of skin, and prevents undue absorption or evaporation of water from the skin

Sweat Glands

- Tubular glands, consist of single tubes with the blind ends coiled in balls, lodged in subcutaneous tissue, and surrounded by a capillary plexus. Secrete sweat and discharge it by means of ducts which open exteriorly

Sweat

- Watery, colorless, turbid liquid, salty taste, distinctive odor, and usually an acid reaction; *p*H is 5.65
- Contains the same inorganic constituents as the blood but in lower concentration
- Average quantity, about 16–20 oz in 24 hours

Sweat
- Amount increased by
  - (1) Increased temperature or humidity of the atmosphere
  - (2) Dilute condition of blood
  - (3) Exercise
  - (4) Pain
  - (5) Nausea
  - (6) Mental excitement or nervousness
  - (7) Dyspnea
  - (8) Use of diaphoretics, e.g., pilocarpine, physostigmine, strychnine, nicotine, camphor
  - (9) Various diseases, such as tuberculosis, acute rheumatism, and malaria
- Amount decreased by
  - (1) Cold
  - (2) Voiding a large quantity of urine
  - (3) Diarrhea
  - (4) Certain drugs, e.g., atropine and morphine
  - (5) Certain diseases

Activity of Sweat Glands Due to
- (1) Direct stimulation of nerve ending in sweat glands
- (2) Indirect stimulation through sensory nerves of the skin
- (3) Influenced by external heat, dyspnea, muscular exercise, strong emotions, and the action of various drugs

Function of Sweat
- Importance not in elimination of waste substances in perspiration, but elimination of **heat** needed to cause evaporation of perspiration
- When kidneys are not functioning properly, sweat glands will excrete waste substances, particularly if stimulated

Ceruminous Glands
- Modified sweat glands
- Found in skin of external auditory canal
- Secrete cerumen, a yellow, pasty substance, like wax

Body Heat
- Animals divided into two classes
  - (1) Homoiothermic, or those which have an almost constant temperature. Birds and mammals (including human beings) are in this class
  - (2) Poikilothermic, or those whose temperature varies with that of their environment, e.g., snakes, frogs, and fishes. The human fetus is in this class
- **Derived from**
  - (1) Food by a process of **oxidation,** which reduces complex substances to simple ones. Every body cell produces heat, but the most important heat-producing organs are the **muscles**

Body Heat

- **Derived from**
  - (2) Minor sources
    - Friction of muscles, blood, etc.
    - Hot substances ingested
    - Radiation from sun and heat appliances
- **Distributed**—by the blood circulating through the blood vessels
- **Lost by**
  - Skin—2,156 Cal, or 87.5%
    - Offers large surface for radiation, conduction, and evaporation of sweat
    - Contains large amount of blood
  - Lungs—266 Cal, or 10.7%, is lost in warming the inspired air and in the evaporation of the water of respiration
  - Urine and feces—48 Cal, or 1.8%, is lost in the urine and feces

Regulation of Body Heat

- Due to maintenance of *balance* between heat production and heat dissipation
- **Heat production**
  - By physiological oxidations, due to
    - Food
    - Muscular exercise
- **Heat loss**
  - (1) The respiratory center
  - (2) The sweat center and sweat nerves
  - (3) The vasomotor center and nerves
  - (4) The water content of the blood
- Coordinated by heat-regulating center in the brain
- **Other factors**
  - (1) Size
  - (2) Age
  - (3) Constitution
- **Aided by**
  - *a*) Use of suitable clothing
  - *b*) Use of hot and cold baths

Variations in Temperature

- The normal temperature by mouth is about 37° C (98.6° F)
- **Normal**
  - (1) Depends on where temperature is taken
    - Mouth
    - Axilla
    - Rectum
  - (2) Depends on time of day
    - Lowest in early morning, between 3 and 5 A.M.
    - Highest in late afternoon, about 4 P.M.
  - (3) Slightly increased by muscular activity and the digestive processes
  - (4) Age. Higher and more variable in
    - Infants, children, and the aged

Variations in Temperature

- **Normal**
  - (5) Sex
  - (6) Season
  - (7) Climate
  - (8) Clothing
- **Subnormal due to**
  - Excessive loss of heat
  - Profuse sweating and hemorrhage
  - Lessened heat production, as in starvation
- **Abnormal**
  - **Fever**
    - **Physiological effects**
      - Increased temperature
      - Increased pulse
      - Increased respirations
      - Increased tissue waste
      - Faulty secretion
      - Various other effects such as thirst, weakness, and apathy
    - **Cause**—mainly interference with heat loss
    - **Value**—may be a protective reaction

UNIT IV

Adaptive Response and the Special Senses

CHAPTER 23

THE SPECIAL SENSES: THE SENSORY UNIT, CUTANEOUS SENSATIONS, PAIN, THE TONGUE AND TASTE, THE NASAL EPITHELIUM AND SMELL, THE EAR AND HEARING, THE EYE AND SIGHT

The sensory nerve fibers have their peripheral endings in receptors, or sensory end organs. These receptors receive stimuli, and from them nerve impulses pass to the nerve fibers, which carry the impulses to centers in the central nervous system for interpretation or for linkage by means of synapses with other sensory and motor fibers for interpretation and control. All sensory impulses reach the thalamus either directly or indirectly by various pathways. Some reach the hypothalamus, which sends fibers to the preganglionic autonomic centers of the brain stem and cord. It is in this way that the individual responds to factors in the environment which bring about visceral changes. Through sight and hearing receptors, continuous streams of impulses reach the brain, some of which may change the heart rate, alter the secretion of digestive fluids, open capillaries, and cause blushing and many other visceral changes.

Sensory unit. Under the term *sensory unit* are included (1) a peripheral end organ, or receptor, (2) the sensory path through which the impulses are conveyed, and (3) a center in the cortex which interprets the sensation. It is through the agency of the sensory units that man derives information about himself and the world in which he lives. The limitations of the sense organs restrict knowledge of the many transformations of energy that are going on, except as man is able to devise means of bringing them artificially within the limitations of the sense organs. Thus knowledge of the microscopic forms of life depends upon the extension of sight by means of magnifying lenses. Radio waves circulated through the ether were unknown until the sense of hearing was extended by the use of sound-amplifying devices.

Sensations are the conscious results of processes which take place within the brain in consequence of nervous impulses derived from receptors. Many sensations are not followed by motor reactions but are stored as memory concepts and may be called into play at any time. The sensitiveness of the numerous receptors to stimulation varies. In some parts of the body the slightest pressure will arouse a sensation, while a similar degree of pressure in another part may fail to produce any sensation at all. The minimal stimulus necessary to arouse a sensation in any receptor is described as its *threshold* stimulus.

Sensations are felt and *interpreted in the brain.* The habit of *projecting sensations* to the part that is stimulated tends to obscure this fact. In reality individuals see and hear in the brain, because the eye and ear serve only as end organs to receive the stimuli which must reach the brain before sight and hearing take place.

Classification. One classification of sensations is based on the part of the body to which the sensation is projected. All sensations are aroused in the brain, and the changes which result in consciousness occur there. But the individual projects the sensation either to the exterior of the body or to some organ in the body. On this basis, sensations may be classified as external and internal. The external are those in which the sensations are projected to the exterior of the body; namely, sight, hearing, taste, smell, touch, pressure, and temperature (heat and cold). The internal are those in which the sensations are projected to the interior of the body, and include pain, muscle sense, sensations from the semicircular canals and vestibule of the internal ear, hunger, thirst, fatigue, and other less definite sensations from the viscera. Classification on this basis is not always satisfactory, because mental projections are not always the same.

The *receptors* themselves may be classified as (1) exteroceptors, stimulated by forces outside of the body; (2) proprioceptors, stimulated by activities that occur in connection with muscles and articulations; and (3) interoceptors, stimulated by substances or conditions within the viscera (Fig. 157).

Another classification of the receptors is based upon the kind of stimuli to which they are sensitive. This classification lists (1) chemoreceptors, such as those of smell and taste, which are excited by suitable concentrations of substances in solution; (2) mechanicoreceptors, such as those for sound, for pressure on the skin (touch), and receptors which are stimulated by mechanical pressure; and (3) radioreceptors, such as those for light, heat, and cold, which are excited by radiant energy. According to this classification, it is difficult to place the receptors for pain. Some receptors have not been

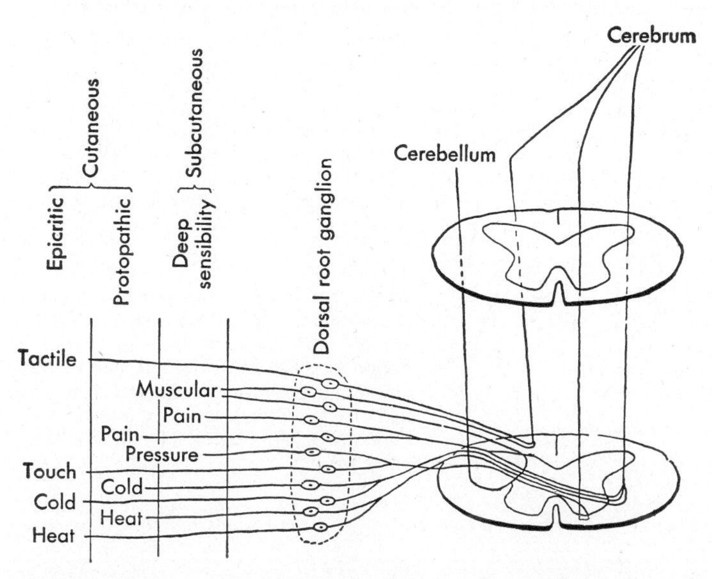

Fig. 340. Diagram of course of cutaneous fibers on reaching the cord.

identified anatomically, e.g., receptors for the senses of hunger, appetite, thirst, etc.; consequently, they cannot be classified satisfactorily.

Cutaneous sensation. The sensory nerves of the skin mediate different qualities of sensation, i.e., pressure, cold, heat, pain, etc. The surface of the skin is a mosaic of tiny sensory spots separated by relatively wide intervals. Each spot coincides with the location of some special end organ and serves a specific sense. These various spots are placed either singly or in clusters. In some locations one variety predominates, in others another. It is a matter of common knowledge that the sensitiveness of these varieties of cutaneous sensation differs in different parts of the body; e.g., the tip of the finger is more sensitive to pressure or contact than to alterations of temperature. The hot and cold spots and the pressure points can be located by passing a metallic point slowly over the skin. At certain points a feeling of contact or pressure will be experienced, and at other points a feeling of cold or heat, depending on whether the temperature of the instrument is higher or lower than that of the skin.

It is estimated that, for the whole cutaneous surface, there are something like 3,000,000 to 4,000,000 pain points, 500,000 pressure points, 150,000 cold points, and about 16,000 warm points.

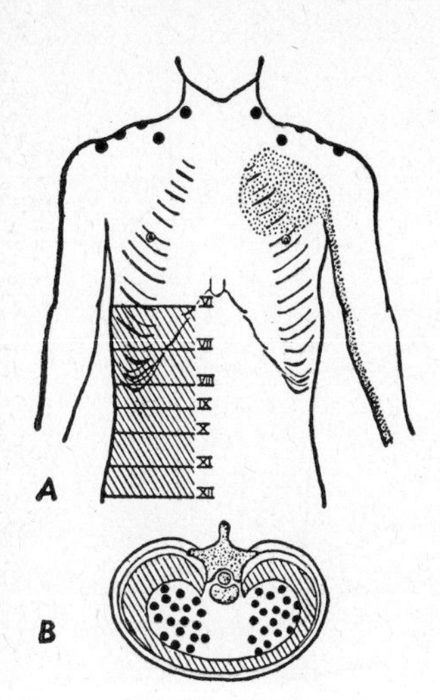

Fig. 341. Referred pain. *A,* cutaneous areas to which pain is referred. *B,* pleural surface of diaphragm. If center of diaphragm (dotted area) is stimulated, pain is felt in the region of the neck supplied by the third and fourth cervical nerves. If edge of diaphragm (striped area) is stimulated, pain is felt in the lower thorax and in the abdominal and lumbar regions supplied by the sixth to twelfth thoracic nerves. Pain of cardiac region is referred to the left side of the chest and down the inside of the left arm. (Modified from *Experimental Clinical Study of Pain,* by J. A. Capps, The Macmillan Company.)

Classification of cutaneous sensations. Some authorities suggest that the cutaneous senses be classified on the basis of the loss of sensation after division of the cutaneous nerves and the subsequent, gradual, and separate return of these sensations, after suture of the divided ends. The skin is supplied with two sets of nerve fibers which are named, respectively, protopathic and epicritic. The *protopathic* group is concerned with three qualities of sensation—pain, heat above 37° C (98.6° F), and cold below 26° C (78.8° F). This system conveys sensations of pain and of extreme changes of temperature, but the sensibility is low and the localization poor. This system is found in the viscera and from a functional standpoint may be considered as a defensive agency toward pathologic changes. The *epicritic* group contains separate fibers for heat, cold, light pressures, and tactile discriminations, which give us sensations of light touch and small differences of temperature between 26° C (78.8° F) and 37° C (98.6° F), i.e., the range of temperature to which the temperature fibers of the protopathic system are insensitive. It is through the sensations mediated by these fibers that we recognize the size and shape of objects. The epicritic group constitutes the special characteristic of the skin area and is not found in other organs.

In addition to the protopathic fibers the deeper tissues are supplied

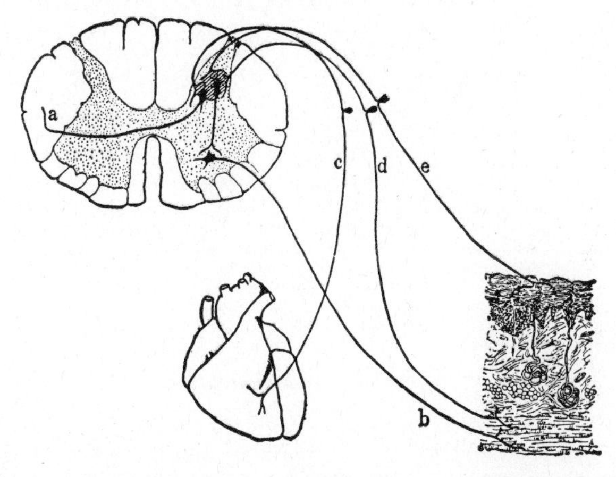

Fig. 342. Diagram to show a possible neural path for referred pain of cardiac origin. Nerve impulses *c* from disturbed heart bring about an "irritable area" in gray matter of cord. Nerve fibers from skin and muscle *d* and *e* enter this same region. Nerve fibers from this region carry impulses over path *a* to the spinothalamic tract and the cerebral cortex, and over path *b* to the chest muscles, which contract in an exaggerated manner.

with fibers which give a sense of pressure and, in the case of the muscles and joints, with fibers which give a knowledge of the position of the movable parts of the body. The paths which these various fibers take in their journey through the central nervous system may be studied in Fig. 340.

Pain. Excessive stimulation of any of the sense organs may give rise to unpleasant sensations, but it is thought that actual pain is caused by the stimulation of pain receptors. They are most abundant in superficial parts of the body, which are most exposed to injury. The meninges of the brain are sensitive to injury, but the brain itself is insensitive to pain. The deeper viscera of the thorax and abdomen are insensitive to mechanical, chemical, or thermal stimuli, though they are sensitive to pain arising from internal disorders (colic). If other receptors are stimulated simultaneously with those of pain, the sensation is modified, usually augmented; e.g., a burn causes stimulation of the hot spots and the pain receptors; hence, the resulting pain is interpreted as burning. Pain acts as a danger signal by means of which abnormal conditions may be detected. In many

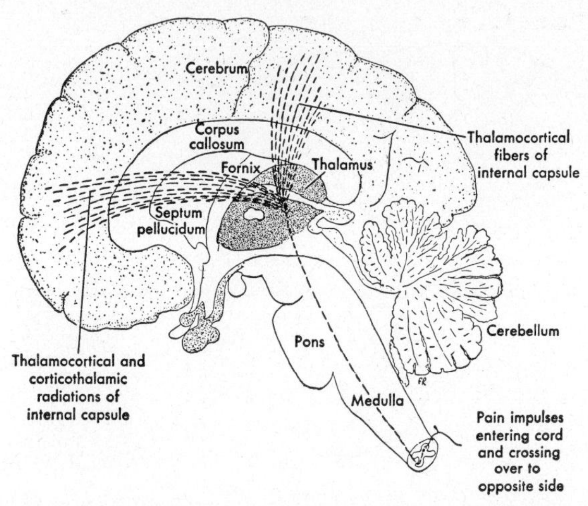

Fig. 343. Medial aspect of brain showing pathway for impulses of pain.

cases pain compels rest and thereby favors the healing of a diseased part.

Referred pains. Normally man is able accurately to localize pain arising in the skin. On the contrary, pain arising in the internal organs is often located very inaccurately. For example, the pain from a severe toothache may be projected quite diffusely to the side of the face. An interesting fact in this connection is that such pains are often referred to points on the skin and may be accompanied by skin areas of tenderness. Pains of this kind are spoken of as *referred pains.* It has been shown that the different visceral organs have a more or less definite relation to certain areas of the skin. Pains arising from stimuli acting upon the intestines are located in the skin of the back, loins, and abdomen, in the area supplied by the ninth, tenth, and eleventh thoracic nerves. Pains from irritations in the stomach are located in the skin over the ensiform cartilage, those from the heart in the scapular region. The explanation of this misreference is that the pain is referred to the skin region that is supplied from the spinal segment from which the organ in question receives its sensory fibers, the misreference being due to a diffusion in the nerve centers. Examples of referred pain are: in appendicitis the abdominal pains are often remote from the usual position of the appendix; in some pneumonia cases abdominal pain is the prominent symptom; in angina pectoris the pain radiates to the left shoulder and down the left arm, etc. (Figs. 341 and 342).

Muscular, or deep, sensibility. There are special sensory end organs in the muscles, the so-called *muscle spindles,* and in the attached tendons, the *tendon spindles,* or *tendon organs of Golgi.* From these end organs afferent fibers carry impulses to centers in the brain, which send out impulses along efferent fibers to the muscles. There is thus a circle of nerve fibers between the brain and the muscles, one fiber giving the sense of the condition of the muscle to the brain and another carrying the impulse from the brain to the muscle. This gives a certain consciousness of the condition of the muscles at all times and enables coordination of the contractions of harmonious groups in order to produce voluntary movements.

Hunger. The feeling that is commonly designated as hunger occurs normally at a certain time after meals and is usually projected to the region of the stomach. It is presumably due to contractions of the empty stomach, which stimulate the fibers distributed to the mucous membrane. If food is not taken, hunger increases in intensity for a time and is likely to cause fatigue and headache. Professional fasters state that after a few days the pangs of hunger diminish and sometimes disappear. In illness hunger contractions may not occur at all, even when the food taken is not sufficient. Probably this results from a lack of muscular tone in the stomach. On the other hand, hunger contractions may be frequent and severe even if an abundance of food is taken regularly, as in diabetes. This indicates malnutrition.

Appetite. Some authorities class appetite as a mild form of hunger. Others consider that hunger and appetite constitute two different sensations mediated by two different physiological mechanisms. The sensory apparatus for hunger lies in the walls of the stomach, probably in the muscular coats, and is stimulated by the contractions of the musculature. The sensations aroused are more or less disagreeable. Appetite, on the contrary, is an entirely pleasant sensation, aroused in part through the sensory nerves of taste and smell and associated with previous experiences.

Thirst. This sensation is projected to the pharynx, and the known facts indicate that the sensory fibers of this region have the important function of mediating this sense. Normal thirst sensations are designated as pharyngeal thirst to indicate the probable origin of the sensory stimuli. Local drying in this region, from dry or salty food or dry and dusty air, produces a sensation of thirst that may be appeased by moistening the membrane with a small amount of water not in itself sufficient to relieve a genuine water need of the body. Prolonged deprivation of water affects the water content of all the tissues and gives rise to sensations not of simple thirst alone but of actual pain and suffering. Under these conditions it is probable that sensory fibers are stimulated in many tissues, and in addition the metabolism of the nervous system is directly affected by loss of water.

Nausea. This sensation may be due to stimulation from the stomach, to substances in the blood, or to impulses coming from various parts of the body, e.g., the organs of sight, taste, and smell.

TASTE

The adequate stimulus for taste receptors is a savory substance in solution. The solution in the case of dry substances is effected by saliva. It is also necessary that the surface of the organs of taste shall be moist. The substances which excite the special sensation of *taste* act by producing a change in the taste buds, and this change furnishes the required stimulus.

Taste buds are ovoid bodies, with an external layer of supporting cells, and contain in the interior a number of elongated cells, which end in hairlike processes that project through the central taste pore. These cells are the sense cells, and the hairlike processes probably are the parts stimulated by the savory substances. The taste buds are found chiefly on the surface of the tongue, though some are scattered over the soft palate, fauces, and epiglottis.

The tongue is a freely movable muscular organ consisting of two distinct halves united in the center. The root of the tongue is directed backward and is attached to the hyoid bone by several muscles. It is connected with the epiglottis by three folds of mucous membrane, and with the soft palate by means of the glossopalatine arches.

Papillae of the tongue. The tongue is covered with mucous membrane, and the upper surface is studded with papillae. The papillae are projections of connective tissue covered with stratified squamous epithelium and contain a loop of capillaries, among which nerve fibers

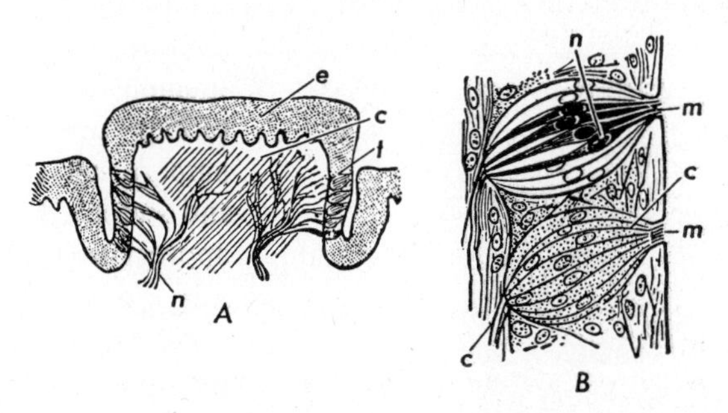

Fig. 344. *A,* a vallate papilla cut lengthwise; *c,* corium; *e,* epidermis; *n,* nerve fibers; *t,* taste buds. *B,* the two taste buds at *t* more highly magnified, the lower as seen from the outside showing *c,* the outer or supporting cells; the upper as seen in section showing *n,* four inner cells with processes *m,* projecting at the mouth of the bud.

are distributed. They give the tongue its characteristic rough appearance. Of these papillae there are four varieties:

Vallate (circumvallate) papillae are the largest, are circular in shape, and form a V-shaped row near the root of the tongue. They contain *taste buds.*

Fungiform papillae, so named because they resemble fungi in shape, are found principally on the tip and sides of the tongue.

Filiform papillae cover the anterior two-thirds of the tongue and bear delicate brushlike processes which seem to be specially connected with the sense of touch, which is very highly developed on the tip of the tongue.

Simple papillae similar to those of the skin cover the larger papillae and the whole of the mucous membrane of the dorsum of the tongue.

Nerve supply of the tongue. The nerve fibers which terminate in the tongue are: fibers of the lingual nerve, which is a sensory branch of the fifth, or trigeminal; fibers of the chorda tympani; a branch of the seventh, or facial; and fibers of the ninth, or glossopharyngeal, nerve.[1] Fibers from the seventh and ninth nerves are concerned with taste. The twelfth, or hypoglossal, nerve is distributed to the tongue. It is a motor nerve.

The sense of touch is very highly developed here, and with it the sense of temperature, pain, etc. Upon these tactile and muscular senses depends, to a great extent, the accuracy of the tongue in many of its important uses—speech, mastication, deglutition, sucking.

Classification of taste sensations. Taste sensations are very numerous, but four fundamental, or primary, sensations are recognized; namely, salty, bitter, acid, and sweet. Quantitative appreciation of taste is not great. All other taste sensations are combinations of these or combinations of one or more of them with sensations of odor or with sensations derived from stimulation of other nerves in the tongue. The seemingly great variety of taste sensations is due to the fact that they are confused or combined with simultaneous odor sensations. Thus the flavors in fruits are designated as tastes because they are experienced at the time these objects are eaten. If the nasal cavities are closed, as by holding the nose, the so-called taste often disappears in large measure. Very disagreeable tastes are usually due to unpleasant odor sensations, hence the practice of holding the nose when swallowing a nauseous dose. On the other hand, some volatile substances which enter the mouth through the nostrils and stimulate the taste buds are interpreted as

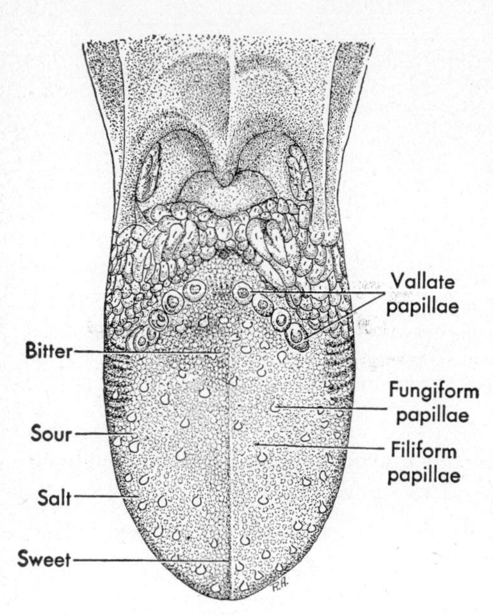

Fig. 345. The upper surface of the tongue showing kinds of papillae and areas for taste.

[1] This is a generally accepted view, but other statements may be found in various textbooks.

odors. The odor of chloroform is largely due to stimulation of the sweet taste in the tongue.

SMELL

The first essentials are a special nerve and nerve center, changes in the condition of which are perceived as sensations of odor. The special organs for this sense must be in their normal condition, and a stimulus (odoriferous substance) must be present to excite them.

The olfactory nerves are the special nerves of the sense of smell and are spread out in a fine network over the surface of the superior

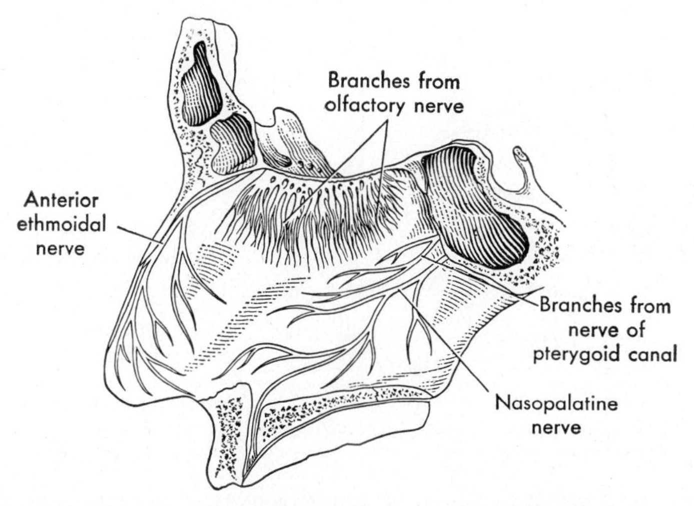

Fig. 346. The nerves of the left side of the septum of the nose. (After Gray's *Anatomy*.)

nasal conchae and the upper third of the septum. These nerves end in special organs known as olfactory cells, which lie among supporting epithelial cells of columnar shape. Each cell bears on its free end a tuft of six to eight hairlike processes. At the free edge of the cells there is a limiting membrane through which the olfactory hairs project. The basal ends of the olfactory cells are prolonged as nerve fibers, which pierce the cribriform plate of the ethmoid bone and end in a mass of gray matter called the olfactory bulb. In the olfactory bulb these fibers form synapses with the dendrites of the so-called mitral cells Through the axons of the mitral cells impulses are conducted to their various terminations in the olfactory lobe, either of the same or the opposite side.

The nerve fibers which ramify over the lower part of the lining membrane of the nasal cavity are branches of the fifth, or trigeminal, nerve. These fibers furnish the tactile sense and enable one to perceive, by the nose, the sensations of cold, heat, tickling, pain, and tension, or pressure. It is these nerve fibers which are affected by strong irritants, such as ammonia or pepper.

Odoriferous substances emit particles which usually are in gaseous form. These particles must penetrate into the upper part of the nasal chamber and, after solution in the moisture of the lining membrane, act chemically upon the sensitive hairs of the olfactory cells, which transmit impulses to the olfactory lobe and give rise to the sensation of smell.

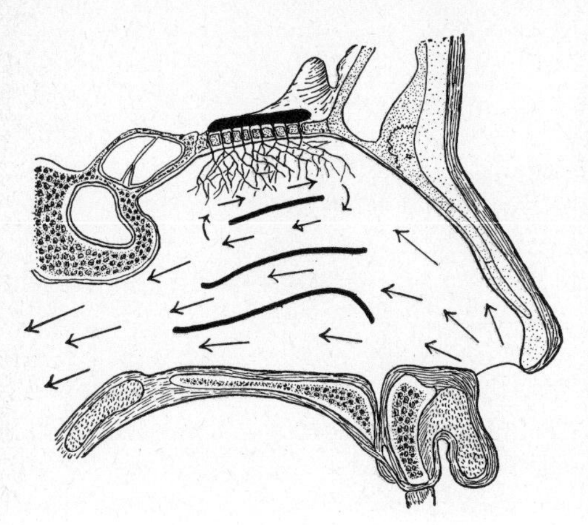

Fig. 347. Diagram of lateral wall of left nasal cavity. The three black lines represent the region of the inferior, middle, and superior conchae. Arrows indicate the direction of air flow. The olfactory lobe is shown with nerve fibers extending through orifices in the cribriform plate of the ethmoid bone. Olfactory nerve fibers are distributed in the mucosa above the superior conchae, the cell bodies lying in the nasal mucosa.

To smell anything particularly well, air is sniffed into the higher nasal chambers and thus brings the odoriferous particles in greater numbers into contact with the olfactory hairs. Odors can also reach the nose by way of the mouth. Many flavors of food are really odors rather than gustatory sensations, and one becomes aware of them just after swallowing. During swallowing the posterior nares are closed by the soft palate, which then opens, permitting odoriferous molecules to reach the sensory epithelium of the nose, through the wide posterior nares.

Each substance smelled causes its own particular sensation, and one is able not only to recognize a multitude of distinct odors but also to distinguish individual odors in a mixed smell. These odors are difficult to classify, i.e., it is not possible to pick out what might be called the fundamental odor sensations. One classification groups odors into pure odors, odors mixed with sensations from the mucous membrane of the nose, and odors mixed or confused with tastes. The pure odors are further subdivided into nine classes, namely, ethereal, aromatic, fragrant, ambrosial, garlic, burning odors, goat odors, repulsive odors, and nauseating or fetid odors. There is some evidence that the sense of odor is related to the radiant energy of the molecular vibrations of the substance giving the odor. This may lead to a more satisfactory classification.

The sensation of smell develops quickly after the contact of the odoriferous stimulus and may last a long time. When the stimulus is repeated, the sensation very soon dies out, the end organs of the sensory cells quickly becoming adapted. This accounts for the fact that one may easily become accustomed to foul odors and is an advantage when foul odors have to be endured. On the other hand, it emphasizes the importance of acting on the first sensation of a disagreeable odor, so as not to become accustomed to it.

The olfactory center in the uncus and the hippocampus of the brain is widely connected with other areas of the cerebrum. Olfactory memories are good. The sense of smell is widely and closely connected with the other senses and with many psychical activities.

HEARING

The auditory apparatus consists of the external ear; the middle ear, or tympanic cavity; the internal ear, or labyrinth; and the acoustic nerve and acoustic center.

The external ear consists of an expanded portion, named the pinna or auricula, and the external acoustic meatus, or auditory canal.

The *pinna* projects from the side of the head. It consists of a framework of cartilage, containing some adipose tissue and muscles; in the lobe, the cartilage is replaced by soft connective tissues. The pinna is covered with skin and joined to the surrounding parts by ligaments and muscles. It is very irregular in shape. The pinna serves to some extent to collect sound waves and direct them toward the external acoustic meatus.

The external acoustic meatus (**external auditory canal**) is a tubular passage, about 2.5 cm (1 in.) in length, which leads from the concha to the tympanic membrane. It forms an S-shaped curve and is directed inward, forward, and upward, then inward and backward. Lifting the pinna upward and backward tends to straighten the canal; but in children it is best straightened by drawing the pinna downward and backward. The external portion of this canal consists of cartilage, which is continuous with that of the pinna; the internal portion is hollowed out of the temporal bone. It is lined by a prolongation

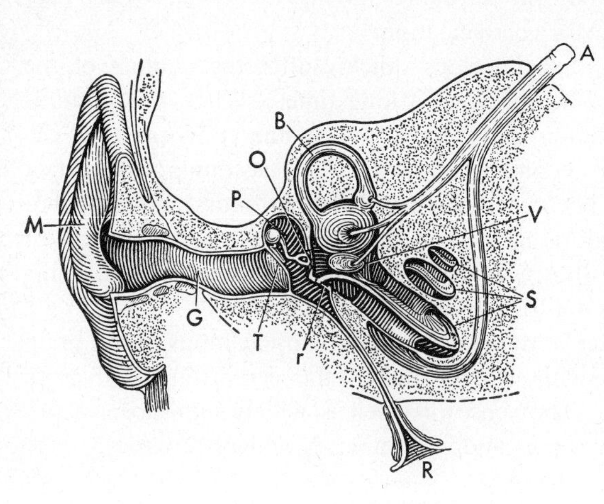

Fig. 348. Diagram of a section of the right ear. *M*, concha; *G*, external acoustic meatus; *T*, tympanic membrane; *P*, middle ear, or tympanic cavity; *O*, fenestra vestibuli; *r*, fenestra cochleae. The chain of tympanic ossicles extends from *T* to *O*. *R*, auditory tube; *V*, *B*, and *S*, bony labyrinth; *V*, vestibule; *B*, semicircular canal; *S*, cochlea; *A*, acoustic nerve dividing into branches for vestibule, semicircular canals, and cochleae.

of the skin, which in the outer half of the canal is very thick and not at all sensitive, and in the inner half is thin and highly sensitive. Near the orifice the skin is furnished with a few hairs and farther inward with modified sweat glands, and the ceruminous glands, which secrete the yellow, pasty cerumen, or earwax. The hairs and the cerumen protect the ear from the entrance of foreign substances.

The *tympanic membrane* (membrana tympani) separates the auditory canal from the tympanic cavity. It consists of a thin layer of fibrous tissue covered externally with skin and internally with mucous membrane.

The middle ear, or tympanic cavity, is a small, irregular bony cavity, situated in the petrous portion of the temporal bone. This air cavity is so small that probably five or six drops of water would fill it. It is separated from the external auditory canal by the tympanic membrane, and from the internal ear by a very thin bony wall ($\frac{1}{24}$ in.) in which there are two small openings: the *fenestra vestibuli* (*ovalis*) and the *fenestra cochleae* (*rotunda*). In the posterior, or mastoid, wall there is an opening into the mastoid antrum and mastoid cells; and because of this, infection of the middle ear may extend into the mastoid cells and cause mastoiditis.[2] In the anterior, or carotid, wall is an opening into the auditory tube, a small canal which leads to the nasopharynx. Thus, there are five openings in the middle ear; namely, the opening between it and the auditory canal; the fenestra vestibuli and the fenestra cochleae, which connect with the internal ear; the opening into the mastoid cells; and the opening into the auditory tube. The walls of the tympanic cavity are lined with mucous membrane, which is continuous anteriorly with the mucous membrane of the auditory tube and posteriorly with that of the mastoid antrum and mastoid cells.

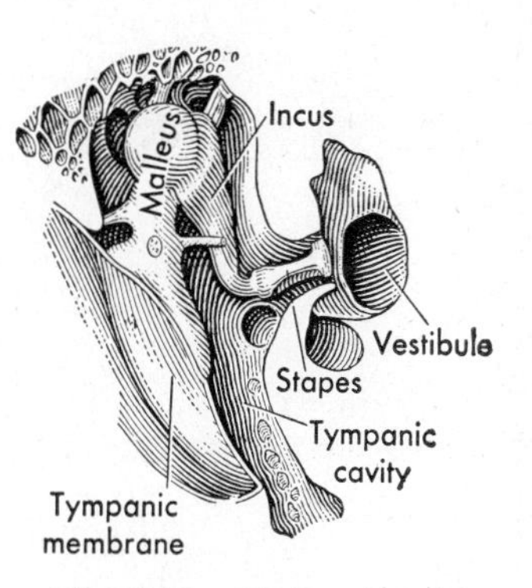

Fig. 349. Chain of ossicles and their ligaments, seen from the front.

Ossicles. Stretching across the cavity of the middle ear from the tympanic membrane to the fenestra vestibuli are three tiny, movable bones, named, because of their shapes, the *malleus,* or hammer, the *incus,* or anvil, and the *stapes,* or stirrup. The handle of the hammer is attached to the tympanic membrane, and the head is attached to the base of the anvil. The long process of the anvil is attached to the stapes, and the footpiece of the stapes occupies the fenestra vestibuli. These little bones are held in position, attached to the tympanic membrane, to each other, and to the edge of the fenestra vestibuli, by minute ligaments and muscles. They are set in motion with every movement of the tympanic membrane. Vibrations of the membrane are communicated to the malleus, taken up by the incus,

[2] The temporal bone at this point is very porous, and any suppurative process is exceedingly dangerous, for the infection may travel inward and invade the brain.

and transmitted to the stapes, which rocks in the fenestra vestibuli and is therefore capable of transmitting to the fluid in the cavity of the labyrinth the impulses which it receives. These bones form a series of levers, the effect of which is to magnify the force of the vibrations received at the tympanum about 10 times that at the oval window.

The auditory, or *Eustachian, tube* connects the cavity of the middle ear with the pharynx. It is about 36 mm ($1\frac{1}{2}$ in.) long and about 3 mm ($\frac{1}{8}$ in.) in diameter at its narrowest part. By means of this tube the pressure of the air on both sides of the tympanic membrane is equalized. In inflamed conditions, the auditory tube may become occluded, and this may prevent this equalization. Under such conditions, hearing is much impaired until the tube is opened. The pharyngeal opening of the tube is closed except when swallowing, yawning, and sneezing.

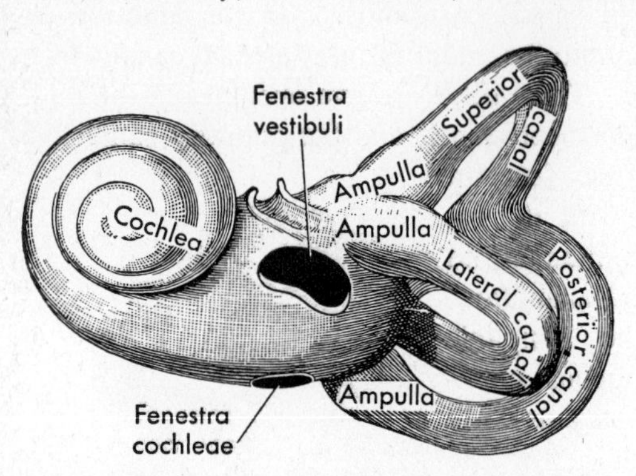

Fig. 350. Left osseous labyrinth, viewed from lateral side. (Cunningham.)

The internal ear, or labyrinth, receives the ultimate terminations of the auditory nerve. It consists of an *osseous labyrinth,* which is composed of a series of peculiarly shaped cavities, hollowed out of the petrous portion of the temporal bone and named from their shape:

Osseous labyrinth
1. The vestibule
2. The cochlea (snail shell)
3. The semicircular canals

Within the osseous labyrinth is a *membranous labyrinth,* having the same general form as the cavities in which it is contained, though considerably smaller, being separated from the bony walls by a quantity of fluid called the *perilymph.* It does not float loosely in this liquid but is attached to the bone by fibrous bands. The cavity of the membranous labyrinth contains fluid—the *endolymph*—and on its walls the ramifications of the acoustic nerve are distributed.

The *vestibule* is the central cavity of the osseous labyrinth; it is situated behind the cochlea and in front of the semicircular canals. It communicates with the middle ear by means of the fenestra vestibuli in its lateral or tympanic wall. The membranous labyrinth of the vestibule does not conform to the shape of the bony cavity but consists of two small sacs, called respectively the *saccule* and the *utricle.* The saccule is the smaller of the two and is situated near the opening of the scala vestibuli of the cochlea; the utricle is larger and occupies the upper and back part of the vestibule. These sacs are not directly connected with each other. From the posterior wall of the saccule, a canal, the *ductus endolymphaticus,* is given off. This duct is joined by a duct from the utricle and ends in a blind pouch on the posterior surface of the petrous portion of the temporal bone. The utricle, saccule, and ducts contain endolymph and are surrounded by perilymph. The inner wall of the saccule and utricle consists of two kinds of modified columnar cells on a basement membrane. One is a specialized nerve cell provided with stiff hairs, which project into the endolymph. Between the nerve cells are supporting cells, which are not ciliated and are not connected with nerve endings. The hair cells serve as end organs for fibers of the vestibular branch of the acoustic nerve, which arborize around the base of each hair cell.

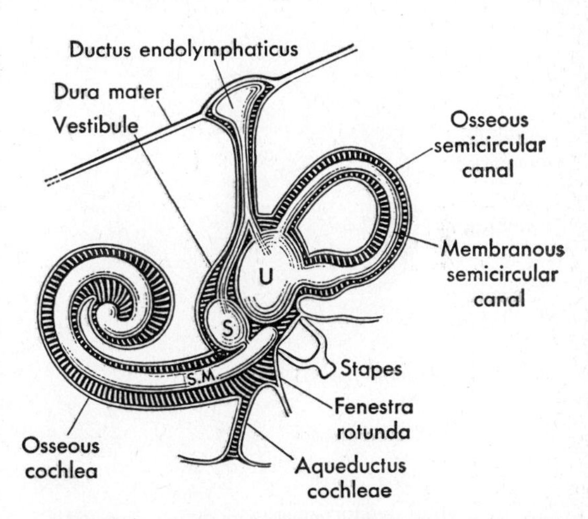

Fig. 351. Diagram of the membranous labyrinth of the inner ear enclosed in the bony labyrinth. *S,* saccule; *U,* utricle; *S.M.,* scala media.

The *cochlea* forms the anterior part of the bony labyrinth and is placed almost horizontally in front of the vestibule. It resembles a snail shell and consists of a spiral canal of $2\frac{3}{4}$ turns around a hollow, conical central pillar called the *modiolus,* from which a thin *lamina* of bone projects like a spiral shelf about halfway toward the outer wall of the canal. Within the bony cochlea is a *membranous cochlea,* which begins at the fenestra ovalis and duplicates the bony structure.

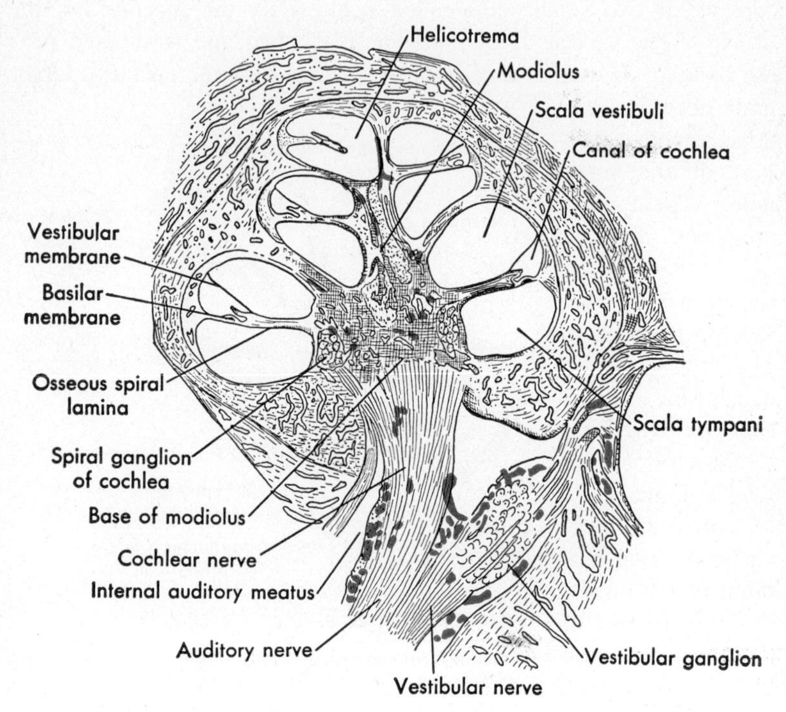

Fig. 352. Section of the cochlea, showing the scalae. (Toldt.)

The *basilar membrane* stretches from the free border of the lamina to the outer wall of the bony cochlea and completely divides its cavities into two passages, or *scalae,* which, however, communicate with each other at the apex of the modiolus by a small opening. The upper passage is the scala vestibuli, which terminates at the fenestra vestibuli; and the lower is the scala tympani, which terminates at the fenestra cochleae.

From the free border of the lamina, a second membrane, called the *vestibular membrane* (Reissner[3]), extends to the outer wall of the

[3] Ernst Reissner, German anatomist (1824–1878).

cochlea and is attached some distance above the basilar membrane. A triangular canal, called the *ductus cochlearis* or *scala media,* is thus formed between the scala vestibuli above and the scala tympani below.

On the basilar membrane the sound-sensitive epithelium of the *organ of Corti*[4] is located. This consists of a large number of *rod-shaped cells* and *hair cells,* extending into the endolymph of the scala media. The *tectorial membrane* projects from the spiral lamina over these cells of the organs of Corti. Some think the hairs of the hair cells are attached to the tectorial membrane so that as the basilar membrane vibrates the hair cells are alternately stretched and recoil. The fibers of the cochlear branch of the acoustic nerve arise in the

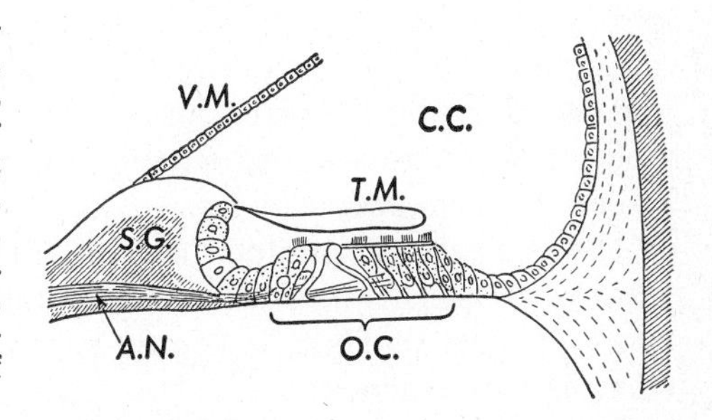

Fig. 353. Diagram, section of organ of Corti. (Highly magnified.) *A.N.*, acoustic nerve; *C.C.*, cochlear canal; *O.C.*, organ of Corti; *S.G.*, spiral ganglion; *T.M.*, tectorial membrane; *V.M.*, vestibular membrane. (T. H. Huxley and J. Barcroft, *Lessons in Elementary Physiology,* 6th ed. Courtesy of The Macmillan Company.)

nerve cells of the *spiral ganglion,* which is situated in the modiolus. These cells are bipolar and send one fiber toward the brain in the acoustic nerve and the other fiber to end in terminal arborizations around the hair cells of the organ of Corti.

The *semicircular canals* are three bony canals lying above and behind the vestibule and communicating with it by five openings, in one of which two tubes join. They are known as the *superior, posterior,* and *lateral* canals, and their position is such that each one is at right angles to the other two. One end of each tube is enlarged and forms what is known as the *ampulla.*

The *semicircular ducts* (membranous semicircular canals) are similar to the bony canals in number, shape, and general form, but their diameter is less. They open by five orifices into the utricle, one opening being common to the medial end of the superior and the

[4] Alfonso Corti, Italian histologist (1822–1876).

upper end of the posterior duct. In the ampullae the membranous canal is attached to the bony canal, and the epithelium is thrown into a ridge (the *crista ampullaris*) of cells with hairlike processes, which project into the endolymph. The hair cells are covered with a gelatinous substance which contains the otoliths. These are minute particles of calcium carbonate which pull and push on the hairs when the position of the head is changed. This forms the stimulus for initiation of reflexes that maintain balance against gravity. Some of the terminations of the vestibular branch of the acoustic nerve are distributed to these cells. Between the hair cells are supporting cells.

The acoustic center. Some investigators place the acoustic center in the temporal lobe of the cerebrum. Removal of both temporal lobes is followed by complete deafness. Removal of one temporal lobe is followed by impairment of hearing. Because of this it is thought that the connections of the acoustic center with the ear follow the scheme of the optic chiasma (Fig. 362); that is, some fibers from each ear cross to the opposite side of the cerebrum, and some end on the same side.

The acoustic, or **auditory, nerve** (VIII) is sensory and contains two sets of fibers, which differ in their function, origin, and destination. One set of fibers is known as the cochlear division and the other as the vestibular.

The *cochlear nerve* arises from bipolar cells in the spiral ganglion of the cochlea. The peripheral fibers pass to the cells of the organ of Corti, at which point the sound waves arouse the nerve impulses. The central fibers, forming part of the cochlear branch, pass into the lateral border of the medulla, terminating in the dorsal nucleus and ventral nucleus. From these nuclei the path is continued by secondary neurons to the auditory centers in the temporal lobes of the cerebrum.

The *vestibular nerve* arises from bipolar cells in the *vestibular ganglion* (ganglion of Scarpa), situated in the internal acoustic meatus. The peripheral fibers divide into three branches, which are distributed around the hair cells of the saccule, the utricle, and the ampullae of the semicircular canals. The central fibers, forming part of the vestibular branch, terminate in two nuclei[5] in the floor of the fourth ventricle. From these nuclei some fibers extend to the cere-

[5] Vestibular nucleus and Deiters' nucleus. Otto Friedrich Carl Deiters, German anatomist (1834–1863).

bellum, and others pass down the spinal cord to form connections with motor centers of the spinal nerves.

In the medulla most of the fibers from each ear cross to the opposite side; therefore injury to the temporal lobe of one side affects the hearing in the ear on the opposite side more than the hearing in the ear of the side injured. Destruction of the auditory centers in both lobes will result in defective hearing, but often the deafness is not complete or permanent. Because of this it is thought that the relay stations are capable of acting as subordinate auditory centers, i.e., of transferring auditory impulses to the association areas. The range of air vibrations for sound is from 40 to 20,000 vibrations per second.

Functions of inner ear and physiology of hearing. All bodies which produce sound are in a state of vibration and communicate their vibrations to the air with which they are in contact.

When these air waves, set in motion by sonorous bodies, enter the external auditory canal, they set the tympanic membrane vibrating. The stretched tympanic membrane takes up these vibrations from the air with great readiness. The vibrations of the tympanic membrane are then communicated by means of the auditory ossicles stretched across the middle ear to the perilymph and then to the endolymph of the inner ear. The movements of the fluids, in the rhythm of the air, stimulate the nerve endings in the organ of Corti; and from these, impulses are conveyed to the center of hearing in the cerebrum. Characteristics of sound are *loudness,* which varies with the *amplitude* of vibrations; *pitch,* which varies with the frequency of vibrations; and *timbre,* which is due to the pattern which the complex of vibrations makes. Various theories are held regarding the manner in which the organ of Corti is stimulated. Most recent work substantiates the *resonance* theory of Helmholtz. This theory postulates that the cochlea is the analyzer of sound. The theory makes use of the structure of the basilar membrane, which is said to have some 24,000 fibers running across it. These fibers vary gradually in length from around 130 μ at the base to about 275 μ at the apex of the cochlea. The short fibers are said to vibrate in response to vibrations at the tympanic membrane brought about by high notes; the long fibers at the apex vibrate in response to low tone vibrations. It is said that man is able to distinguish more than 10,000 pitches of tone. There are also more than 15,000 hair cells on the basilar membrane and more than 15,000 fibers in the cochlear nerve. This is thought to constitute an adequate physical mechanism for discrimination in auditory sensation. In all theories, the hair cells are involved. They are thought to be stimulated by vibrations of the cochlear fluids, tectorial membrane, basilar membrane, or combinations of these. Abnormal conditions in any part of the auditory mechanism stimulate the auditory nerve and give rise to noises that are described as rushing, roaring, humming, and ringing.

The sense of equilibrium. Among the various means (such as sight, touch, and muscular sense) whereby one is enabled to maintain equilibrium, coordinate movements, and become aware of position in space, one of the most important is the *action of the vestibule and semicircular canals.* These structures are found in the inner ear and communicate with the cochlea. As a result of much experimental work, many facts regarding the effects of injury to the semicircular canals have been accumulated; but it is difficult to interpret these facts, and several theories have been proposed. One theory is that movements of the head set up movements in the endolymph of the canals, and these act as stimuli to the nerve endings around the hair cells. These nerve endings serve as receptors and transmit the impulses to the cerebellum.

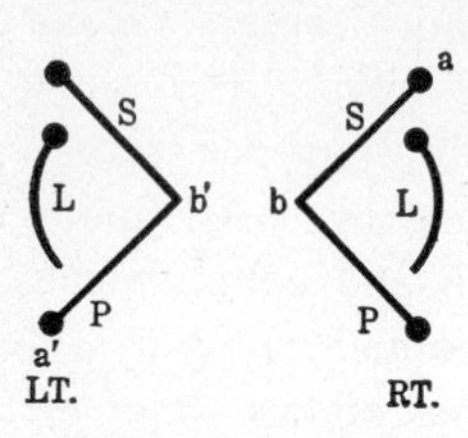

Fig. 354. Diagram showing relative position of the planes in which the semicircular canals lie. *Rt.*, right ear; *Lt.*, left ear; *S*, superior canal; *P*, posterior canal; *L*, lateral canal; *a*, ampulla of right superior canal; *a'*, ampulla of left posterior canal. The superior and posterior canals lie in nearly vertical planes. Each of the three canals lies in a plane practically at right angles to the other two. See text for explanation of fluid movement within the canals.

The canals are so arranged (Fig. 354) that any movement of the head causes an increase in the pressure of the endolymph in one ampulla and a corresponding diminution in the ampulla of the parallel canal on the opposite side. Thus, a nodding of the head to the right would cause a flow of the endolymph from *a* to *b* in the right superior canal but from *b'* to *a'* in the left posterior canal. Hence, the pressure upon the hairs is decreased in *a* but increased in *a'*. Such stimulations of the sensory hairs are transmitted by fibers of the vestibular nerve, through the cell bodies of the vestibular ganglion and the axons of the acoustic nerve, to the cerebellum. Impulses from receptors in the semicircular canals and from the otoliths, due to change in position of the head, initiate the righting reflex. Balance against gravity is thus maintained by the coordinated or effective response of the antigravity muscles. The cerebellum links the impulses that arise from stimulation of the sensory nerves of the semicircular canals, joints, etc., and sends the nerve impulses on to the motor centers of the cerebrum and spinal cord.

SIGHT

The visual apparatus consists of the bulb of the eye (eyeball), the optic nerve, and the visual center in the brain. In addition to these essential organs, there are accessory organs which are necessary for the protection and functioning of the eyeball.

Accessory organs of the eye. Under this heading are grouped eyebrows, eyelids, conjunctiva, lacrimal apparatus, muscles of the eyeball, and the fascia bulbi.

The eyebrow is a thickened ridge of skin, covered with short hairs. It is situated on the upper border of the orbit and protects the eye from too vivid light, perspiration, etc.

The eyelids (palpebrae) are two movable folds placed in front of the eye. They are covered externally with skin and internally with a mucous membrane, the conjunctiva, which is reflected from them over the bulb of the eye. They are composed of muscle fibers and dense fibrous tissue known as the *tarsal plates.* The upper lid is attached to a small muscle which is called the elevator of the upper lid (*levator palpebrae superioris*). Arranged as a sphincter around both lids is the *orbicularis oculi* muscle, which closes the eyelids.

The slit between the edges of the lids is called the palpebral fissure. It is the size of this fissure which causes the appearance of large and small eyes, as the size of the eyeball itself varies but little. The outer angle of this fissure is called the lateral palpebral commissure (*external canthus*)*;* the inner angle, the medial palpebral commissure (*internal canthus*).

The eyelids provide protection for the eye—movable shades which cover the eye during sleep, protect the eye from bright light and foreign objects, and spread the lubricating secretions of the eye over the surface of the eyeball.

Eyelashes and sebaceous glands. From the margin of each eyelid, a row of short thick hairs—the eyelashes—project. The follicles of the eyelashes receive a lubricating fluid from the sebaceous glands which open into them. If these glands become infected, a sty results. A *sty,* therefore, is comparable to a pimple or furuncle resulting from the infection of retained sebaceous fluid in other regions of the skin.

Lying between the conjunctiva and the tarsal cartilage of each eyelid is a row of elongated sebaceous glands—the tarsal, or Meibomian, glands—the ducts of which open on the edge of the eyelid. The secretion of these glands lubricates their edges and prevents adhesion of the eyelids. Distention of the gland is termed a *chalazion.*

The conjunctiva. The mucous membrane which lines the eyelids and is reflected over the forepart of the eyeball is called the conjunctiva. It is continuous with the lining membrane of the ducts of the tarsal glands, the lacrimal ducts, lacrimal sac, nasolacrimal duct, and nose.

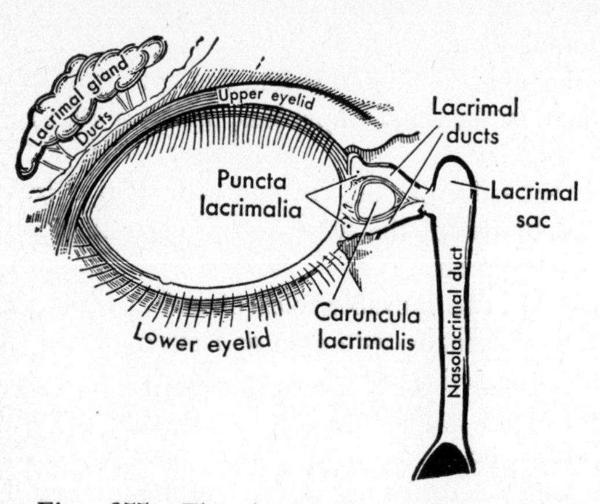

Fig. 355. The lacrimal apparatus. (After Gray's *Anatomy.*)

Lacrimal apparatus. This apparatus consists of the lacrimal gland, the lacrimal ducts, the lacrimal sac, and the nasolacrimal duct.

The *lacrimal gland* is a compound gland and is lodged in a depression of the frontal bone at the upper and outer angle of the orbit. It is about the size and shape of an almond and consists of two portions, a superior and inferior, which are partially separated by a fibrous septum.

Six to twelve minute *ducts* lead from the gland to the surface of the conjunctiva of the upper lid. The secretion (tears) is usually just enough to keep the eye moist and, after passing over the surface of the eyeball, is sucked through the *puncta* into two tiny *lacrimal ducts* and conveyed into the *lacrimal sac* at the inner angle of the eye. The *lacrimal sac* is the expanded upper end of the *nasolacrimal* duct, a small canal that opens into the nose. It is oval in shape and measures from 12 to 15 mm in length. The *caruncula lacrimalis* (caruncle) is a small reddish body situated at the medial commissure.

It contains sebaceous and sudoriferous glands and forms the whitish secretion which collects in this region.

The lacrimal gland secretes tears. This secretion is a dilute solution of various salts in water, which also contains small quantities of mucin. The ducts leading from the lacrimal gland carry the tears to the eyeball, and the lids spread it over the surface. Ordinarily this secretion is evaporated, or carried away by the nasolacrimal duct, as fast as formed; but under certain circumstances, as when the conjunctiva is irritated or when painful emotions arise in the mind, the secretion of the lacrimal gland exceeds the drainage power of the nasolacrimal duct, and the fluid, accumulating between the lids, at length overflows and runs down the cheeks. The purpose of the lacrimal secretion is to keep the surface of the eyes moist and to help remove foreign bodies, microorganisms, dust, etc. It is slightly antiseptic. The secretion is increased by foreign bodies (molecules of volatile gases, liquids, dust, microorganisms) that come in contact with the eyeball or lids, irritation of the nasal mucous membrane, bright light, and emotional states. Inflammation from the nose may spread to the nasolacrimal ducts, blocking them, and thus cause a slow dropping of tears from the inner angle of the eye.[6]

Muscles of the eye. For purposes of description the muscles of the eye are divided into groups—intrinsic and extrinsic. The intrinsic muscles are the ciliary muscle and the muscles of the iris. The extrinsic muscles hold the eyeball in place and control its movements. They include the four straight, or recti, and the two oblique, already described. The rectus muscles are so arranged that the superior and inferior oppose each other; the medial and lateral do likewise. This action is comparable to the opposed action of the flexors and extensors of the arms and legs and ensures accuracy of movement. Sometimes these muscles (particularly the medial and lateral recti) are unequal in length or strength, the equilibrium of the opposed muscles is upset, and the eye is then turned in the direction of the stronger muscle, producing a squint, or strabismus. There are various kinds of strabismus. Convergent strabismus is "walleye."

Fascia bulbi (*capsule of Tenon*).[7] Between the pad of fat and the

[6] The lacrimal glands do not develop sufficiently to secrete tears until about the fourth month of life; hence the need for protecting a baby's eyes from bright lights, dust, etc.

[7] Jacques René Tenon, French anatomist and surgeon (1724–1816).

eyeball is a serous sac—the fascia bulbi—which envelops the eyeball from the optic nerve to the ciliary region and forms a socket in which the eyeball rotates.

The nerves of the eye are: the optic nerve, concerned with vision only; the oculomotor nerve, which supplies (1) the medial, superior,

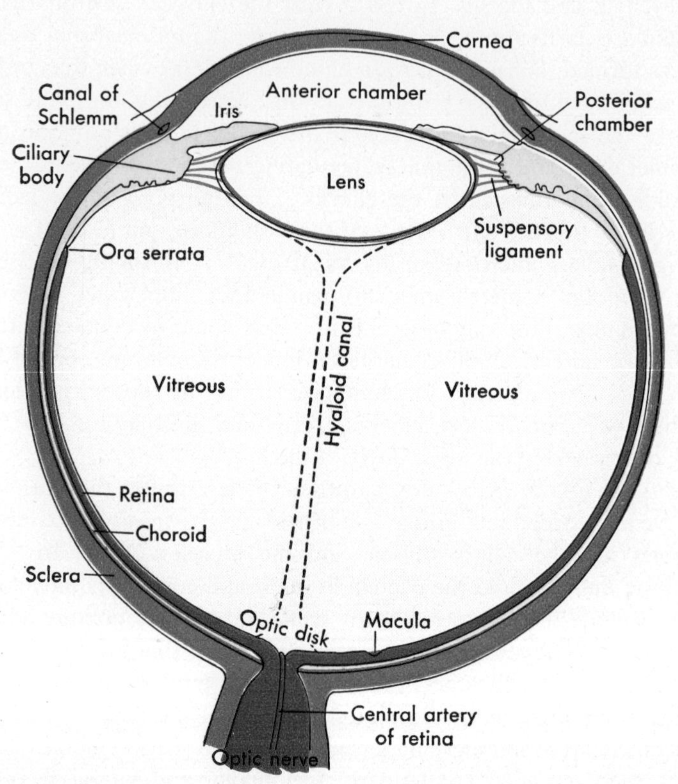

Fig. 356. Horizontal section of the eyeball. Retina red, choroid yellow between red and blue. (Charles H. May, *Manual of the Diseases of the Eye.*)

and inferior recti and the inferior oblique muscles, and (2) the ciliary muscle and the circular muscle of the iris (sphincter pupillae); the trochlear nerve, which controls the superior oblique muscle; the abducent, which controls the lateral rectus; and the ophthalmic, which is a branch of the trigeminal nerve, supplies general sensation, and sends branches to the cornea, ciliary body, and iris and to the lacrimal gland and conjunctiva. Autonomic fibers are supplied to the ciliary muscle and the radial muscle of the iris.

The orbits are the bony cavities in which the eyeballs are contained. Seven bones assist in the formation of each orbit, namely, the frontal, malar, maxilla, palatine, ethmoid, sphenoid, and lacrimal. As three of these bones are mesial (frontal, ethmoid, and sphenoid), there are only 11 bones forming both orbits.

Each orbit is shaped like a funnel; the large end, directed outward and forward, forms a strong bony edge which protects the eyeball. The small end is directed backward and inward and is pierced by a large opening—the optic foramen—through which the optic nerve and the ophthalmic artery pass from the cranial cavity to the eye. A larger opening to the outer side of the foramen—the superior orbital fissure—provides a passage for the orbital branches of the middle meningeal artery and the nerves which carry impulses to and from the muscles, i.e., the oculomotor, the trochlear, the abducent, and the ophthalmic. Each orbit contains the eyeball, muscles, nerves, vessels, lacrimal glands, fat, the fascia bulbi, and the fascia that holds these structures in place. The inner portion is lined with fibrous tissue and contains a pad of fat which serves as a cushion for the eyeball. During many forms of illness the body fat is oxidized at an unusual rate. Under such conditions the orbital fat becomes diminished, and the eyeballs sink in the orbits.

The bulb of the eye, or **eyeball,** is spherical in shape; but its transverse diameter is less than the anteroposterior so that it projects anteriorly and looks as if a section of a smaller sphere had been engrafted on the front of it.

The bulb of the eye is composed of three coats, or tunics. From the outside of the eyeball inward toward its center these are:

Fibrous: (1) sclera, (2) cornea

Vascular: (1) choroid, (2) ciliary body, (3) iris

Nervous: retina

It contains three refracting media. These are:

Aqueous humor

Crystalline lens and capsule

Vitreous body

The fibrous tunic is formed by the sclera and cornea.

1. The *sclera,* or *white of the eye,* covers the posterior five-sixths of the eyeball. It is composed of a firm, unyielding, fibrous mem-

brane, thicker behind than in front, and serves to maintain the shape of the eyeball and to protect the delicate structures contained within it. It is opaque, white, and smooth externally; behind, it is pierced by the optic nerve. Internally it is brown in color and is separated from the choroid by a lymph space. It is supplied with few blood vessels. A venous sinus—the canal of Schlemm[8]—encircles the cornea at the corneoscleral junction. Its nerves are derived from the ciliary nerve.

2. The *cornea* covers the anterior sixth of the eyeball. It is directly continuous with the sclera, which, however, overlaps it slightly above and below, as a watch crystal is overlapped by the case into which it is fitted. The cornea, like the sclera, is composed of fibrous tissue, which is firm and unyielding, but, unlike the sclera, it has no color and is perfectly transparent; it has been aptly termed the "window of the eye." The cornea is well supplied with nerves (derived from the ciliary) and lymph spaces but is destitute of blood vessels, so that it is dependent on the lymph contained in the lymph spaces for nutriment.

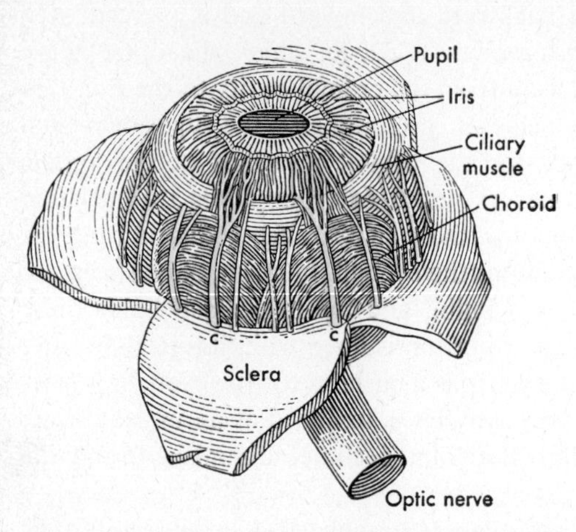

Fig. 357. Diagram of the eyeball, sclera cut and turned back. Note pupil, iris, ciliary muscle, choroid; *c*, ciliary nerves and optic nerve.

The vascular tunic (uvea, or uveal tract) consists, from behind forward, of the choroid, the ciliary body, and the iris.

1. The *choroid* is a thin, dark-brown membrane lining the inner surface of the sclera. It consists of a dense capillary plexus and small arteries and veins carrying blood to and from the plexus. Between these vessels are pigment cells which with other cells form a network, or stroma. The blood vessels and pigment cells render this mem-

[8] Friedrich Schlemm, German anatomist (1795–1858).

brane dark and opaque, so that it darkens the chamber of the eye by preventing the reflection of light. It extends to the ciliary body.

2. The *ciliary body* includes the orbicularis ciliaris, the ciliary processes, and the ciliaris muscle. The orbicularis ciliaris is a zone about 4 mm in width, which is directly continuous with the anterior part of the choroid.

Just behind the edge of the cornea, the choroid is folded inward and arranged in radiating folds, like a plaited ruffle, around the margin of the lens. There are from 60 to 80 of these folds, and they constitute the ciliary processes. They are well supplied with nerves and blood vessels and also support a muscle, the ciliaris (ciliary) muscle. The fibers of this muscle arise from the sclera near the cornea, and, extending backward, are inserted into the outer surface of the ciliary processes and the choroid. This muscle is the chief agent in accommodation. When it contracts, it draws forward the ciliary processes, relaxes the suspensory ligament of the lens, and allows the lens to become more convex.

3. The *iris* (*iris,* rainbow) is a circular, colored disk suspended in the aqueous humor in front of the lens and behind the cornea. It is attached at its circumference to the ciliary processes, with which it is practically continuous, and is also connected to the sclera and cornea at the point where they join one another. Except for this attachment at its circumference, it hangs free in the interior of the eyeball. In the middle of the iris is a circular hole, the *pupil,* through which light is admitted into the eye chamber. The iris is composed of connective tissue containing branched cells, numerous blood vessels, and nerves. The color of the eye is related to the number and size of pigment-bearing cells in the iris. If there is no pigment or very little, the eye is blue; with increasing amounts of pigment the eye is gray, brown, or black. It also contains two sets of muscles. One set is arranged like a sphincter with its fibers encircling the pupil and is called the *sphincter pupillae* (contractor of the pupil). The other set consists of fibers which radiate from the pupil to the outer circumference of the iris and is called the *dilator pupillae* (dilator of the pupil). The action of these muscles is antagonistic.

The posterior surface of the iris is covered by layers of pigmented epithelium of deep-purple tint, named *pars iridica retinae.* It is designed to prevent the entrance of light.

Function of the iris. The function of the iris is to regulate the amount of light entering the eye and thus assist in obtaining clear images. This regulation is accomplished by the action of the muscles described above, as their contraction or relaxation determines the size of the pupil. When the eye is accommodated for a near object or stimulated by a bright light, the sphincter muscle contracts and diminishes the size of the pupil. When, on the other hand, the eye is accommodated for a distant object or the light is dim, the dilator muscle contracts and increases the size of the pupil.

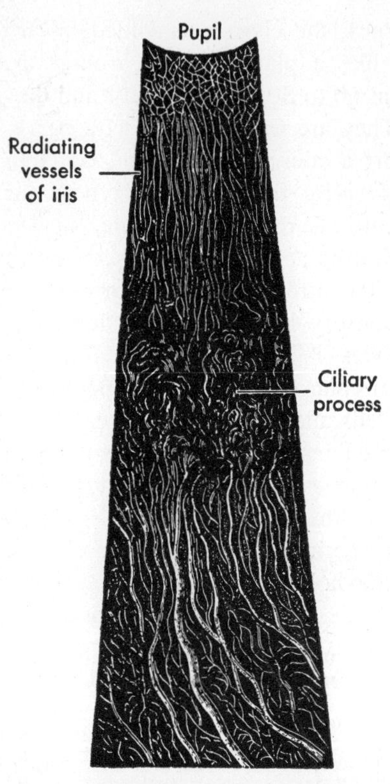

Fig. 358. Segment of the iris, ciliary body, and choroid, viewed from the internal surface. (Gerrish.)

The retina, or nervous tunic. The *retina,* the innermost coat of the eyeball, is a delicate nervous membrane which receives the images of external objects and transfers the impressions evoked by them to the center of sight in the cortex of the cerebrum. It occupies the space between the choroid coat and the hyaloid membrane of the vitreous body and extends forward almost to the posterior margin of the ciliary body, where it terminates in a jagged margin known as the *ora serrata.* It consists of three sets of neurons so arranged that the cell bodies and processes form seven layers. In addition there are two limiting membranes, one membrane (membrana limitans interna) in contact with the vitreous layer, and the second (membrana limitans externa) marking the internal limit of the rod-and-cone layer and a pigmented layer between the layer of rods and cones and the choroid coat.

The seventh layer is called the layer of rods and cones, and these act as end organs, or receptors, for the optic nerve. It is thought that the perception of color is the function of the cones, while the rods are sensitive to light and darkness and form the special apparatus for vision in dim lights. This layer is covered by pigmented epithelium. To reach this layer, rays of light pass through the cornea, lens, humors, and the entire thickness of the retina. The rods contain a pigment known as *rhodopsin,* or *visual purple,* which is bleached by

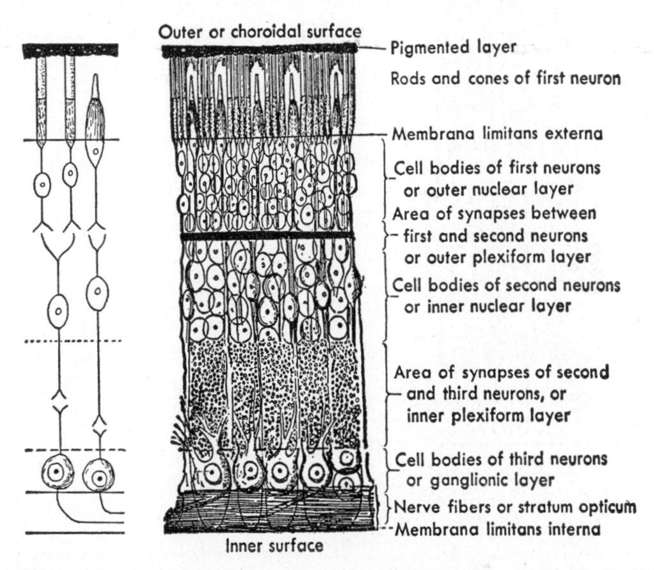

Fig. 359. Diagrammatic section of the human retina, showing individual connection of the cones with a single bipolar cell and multiple connections of the rods with a single bipolar cell.

light but is constantly regenerated, it is thought, by the pigmented epithelial cells that lie between the rods and the choroid coat.

Blind spot. The optic nerve pierces the eyeball not exactly at its most posterior point but a little to the inner side. This point is insensitive to light and is called the blind spot. There are no rods and cones at this spot, and rays of light falling upon it produce no sensation. The central artery of the retina[9] and its vein pass into the retina along with the optic nerve.

Macula lutea. One point of the retina is of great importance—the macula lutea, or yellow spot. It is situated about 2.08 mm

[9] Branch of the ophthalmic artery.

($\frac{1}{12}$ in.) to the outer side of the exit of the optic nerve and is the exact center of the retina. In its center is a tiny pit—*fovea centralis*—which is the center of direct vision. At this point there is an absence of rods but a great increase in the number of cones. This is the region of greatest visual acuity. In reading, the eyes move so as to bring the rays of light from word after word into the center of the fovea. Here each cone is connected to one ganglion cell from which one fiber leads to the brain. Elsewhere in the retina several cones or several rods are connected to one ganglion cell and one fiber of the optic nerve.

Fig. 360. Scheme of the structure of the primate retina, as shown by Polyak, 1941. *A*, rod; *B*, cone; *C*, horizontal cell; *d*, diffuse ganglionic cell. This diagram illustrates the very complex arrangement of retinal cells—a concept that the older diagrams do not give. (Courtesy of the University of Chicago Press.) For other identifications see S. R. Detwiler, *Vertebrate Photoreceptors*, The Macmillan Company, 1943.

Refracting media. The cornea and the aqueous humor form the first refracting medium. The *aqueous humor* fills the forward chamber; the latter is the space bounded by the cornea in front and by the lens, suspensory ligament, and ciliary body behind. This space is partially divided by the iris into an anterior and a posterior chamber. The aqueous humor is a clear, watery solution containing minute amounts of salts, mainly sodium chloride. It is believed that it is derived mainly from the capillaries by diffusion and that it drains away through the veins and through the spaces of Fontana[10] into the venous canal of Schlemm and then on into the larger veins of the eyeball.

The *crystalline lens* enclosed in its capsule is a transparent, refractive body, with convex anterior and posterior surfaces. It is placed directly behind the pupil, where it is retained in position by the counterbalancing pressure of the aqueous humor in front and the vitreous body behind, and by its own suspensory ligament, formed in

[10] Felice Fontana, Italian anatomist (1730–1805).

part by the hyaloid membrane and in part by fibers derived from the ciliary processes. The posterior surface is considerably more curved than the anterior, and the curvature of each varies with the period of life. In infancy, the lens is almost spherical; in the adult, of medium convexity; and in the aged, considerably flattened. The capsule surrounding the lens is elastic. Its reflective power is much greater than that of the aqueous or vitreous body.

In cataract the lens or its capsule becomes less transparent and blurs vision or causes blindness. It is thought that lack of sufficient vitamin B (riboflavin) and parathyroid hormone may be concerned.

The *vitreous body,* a semifluid albuminous tissue enclosed in a thin membrane, the hyaloid membrane, fills the posterior four-fifths of the bulb of the eye. The vitreous body distends the greater part of the sclera, supports the retina, which lies upon its surface, and preserves the spheroidal shape of the eyeball. Its refractive power, though slightly greater than that of the aqueous humor, does not differ much from that of water.

In glaucoma, intraocular pressure increases, cupping the optic disk and interfering with the proper distribution of blood to all the inner tissues of the eye. This increased presure may be due to increased blood pressure in the larger blood vessels of the eye, to altered osmotic conditions of blood and eye fluids, to rigidity of the eyeball, to improper functioning of intrinsic muscles of the eye, etc. It causes pain and interferes with vision.

Perception of light and color. Waves started by the motion of the molecules of bodies (especially hot bodies like the sun) cause ethereal vibrations, and these vibrations are of varying lengths. Vibrations from 0.0004 mm to 0.0008 mm long are called light and color waves. Those shorter than this are known as chemical or actinic rays; those much longer are electrical waves. When vibrations varying between 0.0004 and 0.0008 mm enter the eye, they cause chemical changes in the rods and cones which give rise to impulses that are carried by the optic nerve to the brain and result in sight. Just how this is accomplished is not known, but the rods contain a kind of pigment which is called rhodopsin, or visual purple, which is thought to function in these changes.

Refraction. Light rays may be refracted, or bent from their course. This is due to the fact that they travel at different rates in media of different density. For instance, light travels less rapidly in water than in air. For this reason where a ray of light in air strikes a body of water obliquely, it will be bent out of a straight line, as shown in Fig. 361. The light ray *AC,* instead of following the straight line *AB,* is bent on striking the surface of the denser medium, thereby being bent from its direct path toward *C.*

The bending is in proportion to the density of the medium. The cause of the refraction of oblique rays is that all the component rays do not reach the surface of the medium at the same time and those that enter first become retarded before those entering later. A ray of light which strikes a body of water perpendicularly will not be bent because all its component rays enter the water at the same time and hence are equally retarded. The central components of a light wave enter the eyes perpendicularly, and the sides obliquely. For clear vision the oblique rays must converge and come to a focus with the central rays on the retina. The aqueous humor, the crystalline lens, and the vitreous humor form a system of refractory devices. Rays of light are bent, or undergo refraction, chiefly on entering the cornea from the air, on entering the lens from the aqueous humor, and on leaving the lens and entering the vitreous fluid.

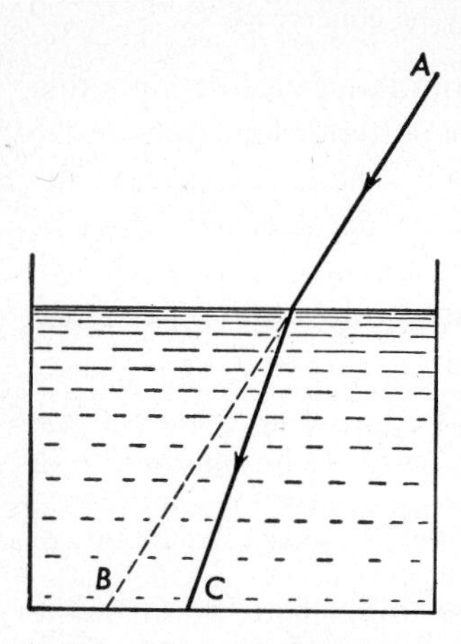

Fig. 361. Bending of a beam of light entering water.

Vision. Visible objects reflect light rays which fall upon them. These reflected rays are brought to a focus on the rods and cones of the retina, and the resulting nerve impulses are transmitted to the optic nerve and thence through various relay stations to the centers of vision in the occipital lobes of the cerebrum. From here it is believed the impulses are transmitted to the association areas, where they awaken memories that enable one to interpert their meaning. Several theories are advanced in explanation of the way in which rays of light stimulate the receptors of the optic nerve. One theory is that light rays bring about changes in the visual purple. This is questioned because the visual purple is absent from the cones of the fovea centralis, which is the place of most acute vision and the part on which the light rays are focused when the eyes are accommodated for near objects. Moreover, it has been found that the sensitiveness of the fovea centralis decreases in a dim light, while that of the peripheral portion of the retina, where the pigment is, increases. In a bright light the object is focused directly on the fovea, and the reflexes controlling accommodation help to bring this about. In a dim light the tendency is to diverge the eyes and thus bring the image into the peripheral part of the retina. These facts support the assumptions that in a bright light the cones of the fovea need no sensitizing agent but are directly affected by light; in a dim light the pupils are dilated and light spreads over other parts of the retina, the visual purple is acted upon by it and sensitizes the receptors in the peripheral portion so that they are stimulated by rays that otherwise would not affect them. For relations of vitamin A and visual purple, see p. 569.

Binocular vision. The value of two eyes instead of one is that true binocular vision is possible. This is distinctive in that it is stereoscopic. A stereoscopic picture consists of two views taken from slightly different angles. In stereoscopic vision, two optical images are made from slightly different angles. This gives the impression of distance and depth and is equivalent to adding a third dimension to the visual field. The processes necessary for binocular vision are convergence, or turning the eyes inward; change in the size of the pupil; accommodation; and refraction.

Convergence. In binocular vision it is necessary to turn the eyes inward, in order that the two images of a given object may lie upon what are called corresponding points of the two retinae. Excitation of two corresponding points causes only one sensation, which is the reason why binocular vision is not ordinarily double vision. Convergence of the eyes is brought about by innervation of the medial rectus muscles and is to some extent voluntary.

The optic chiasma. The correspondence of the two retinae and of the movements of the eyeballs is produced by a close connection of the nerve centers controlling these phenomena and by the arrangement of the nerve fibers in the optic nerves. The optic fibers from each retina pass backward through the optic foramen; and shortly after leaving the orbit the two nerves come together, and the fibers from the inner portion of each nerve cross. This is called the optic chiasma and is really an incomplete crossing of fibers, as the outer fibers do not cross. (See Fig. 362.)

Change in the size of the pupil. On looking at a near object in a bright light, the pupil contracts in order to direct the entering rays to the central parts of the lens, i.e., the part where the convexity and the consequent refractive power are greatest, and to the fovea centralis. In a dim light the pupil is dilated, causing a diffusion of the rays on the peripheral parts of the retina where there is sensitizing purple. The contraction of the pupil is brought about by the stimulation of the circular muscle of the iris by the oculomotor nerve; in a dim light the stimulation is lessened, and the pupil dilates. In excitement, fear, etc., its dilatation is due to stimulation of autonomic nerve fibers that arrive by way of the ophthalmic branch of the trigeminal nerve.

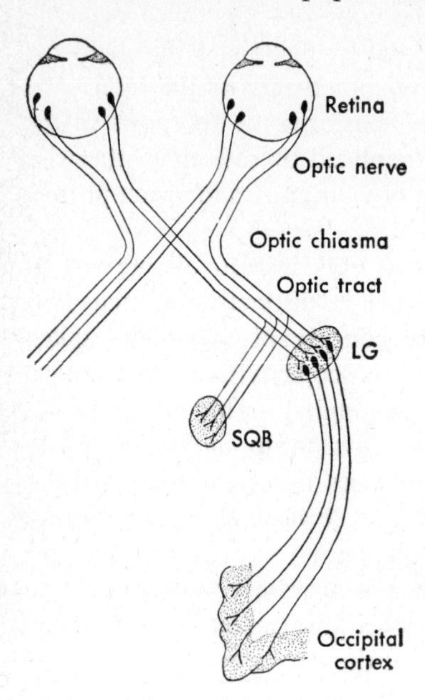

Fig. 362. Diagram of optic chiasma, optic nerves, and optic tracts. *LG,* lateral geniculate body; *SQB,* superior quadrigeminate body. Which fibers cross? Which do not cross?

Accommodation. Accommodation is the ability of the eye to adjust or focus for objects at different distances, because a sharply focused image must fall upon the retina in order to produce clear vision.

The generally accepted theory is that the ciliary muscle is the active agent in accommodation. When the eye is at rest or fixed upon distant objects, the suspensory ligament, which extends from the ciliary processes to the capsule of the lens, exerts a tension upon the capsule of the lens which keeps the lens flattened, particularly the anterior surface to which it is attached. When the eye becomes fixed on near objects, as in reading, sewing, etc., the ciliary muscle contracts and draws forward the choroid coat, which in turn releases the tension of the suspensory ligament upon the capsule of the lens and allows the anterior surface to become more convex. The accommodation for near objects is an active condition and is always more or less fatiguing. On the contrary, the accommodation for distant objects is a passive condition, in consequence of which the eye rests for an indefinite time upon remote objects without fatigue.

Refraction. Rays of light entering the eye are refracted, or bent, so that they come to a focus on the retina. It occurs because of the varying densities of the successive refractive media.

Inversion of images. Owing to refraction, light rays as they enter the eye cross each other and cause the image of external objects on

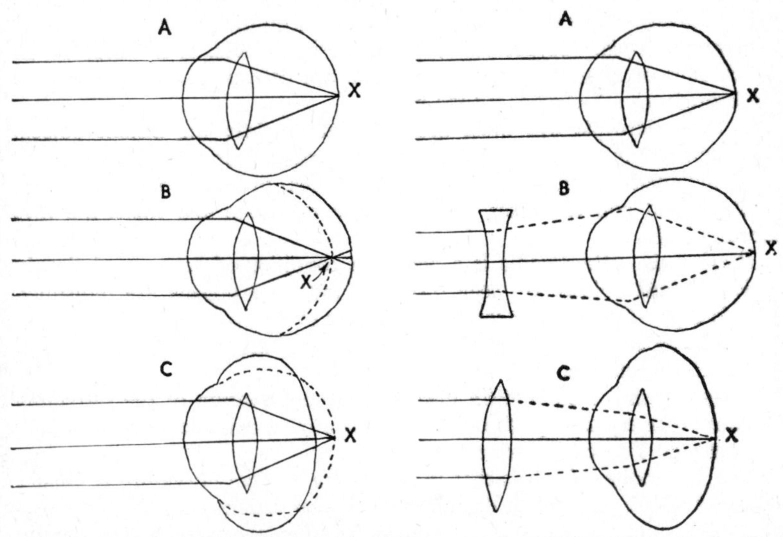

Fig. 363. *Left:* Diagram illustrating rays of light converging in *A*, normal eye; *B*, myopic eye; and *C*, hypermetropic eye. The parallel lines indicate light rays entering the eye; *x* is the point of convergence, or focus. In *A* the rays are brought to a focus (*x*) on the retina. In *B* they come to a focus in front of the retina. In *C* they would come to a focus behind the retina. *Right:* Diagrams illustrating the convergence of light rays in a normal eye (*A*), and the effects of concave lens (*B*) and convex lens (*C*) on rays of light.

the retina to be *inverted*. The question then arises, "Why is it that objects do not appear to be upside down?" This question is answered if it is remembered that actual visual sensations take place in the brain and that the projection of these senations to the interior is a secondary act that has been learned from experience.

Abnormal conditions that interfere with refraction. The normal eye is one in which at a distance of about 20 ft parallel rays of light focus on the retina when the eye is at rest. Such an eye is designated as emmetropic, or normal. Any abnormality in the refractive surfaces or the shape of the eyeball prevents this focusing of parallel rays and makes the eye ametropic, or abnormal.

The most common refractive troubles are: myopia, hypermetropia, presbyopia, and astigmatism.

Myopia. Myopia, or nearsightedness, is a condition in which rays of light converge too soon and are brought to a focus before reaching the retina. This is the opposite of hypermetropia and is caused by a cornea or lens that is too convex or an eyeball of too great depth. This condition is remedied by wearing concave lenses, which cause parallel rays of light to diverge before they converge and focus on the retina.

Hypermetropia. Hypermetropia, or farsightedness, is a condition in which rays of light from near objects do not converge soon enough and are brought to a focus behind the retina.

A hypermetropic eye must accommodate slightly for distant objects and overaccommodate for near objects. Hypermetropia is usually caused by a flattened condition of the lens or cornea, or an eyeball that is too shallow; and convex lenses are used to concentrate and focus the rays more quickly.

Presbyopia. Presbyopia is a defective condition of accommodation in which distant objects are seen distinctly but near objects are indistinct. This is a physiological process which affects every eye sooner or later. It is said to be caused by a loss of the elasticity of the lens and lack of tone of the ciliary muscle.

Astigmatism. Astigmatism means that the curvature of the refracting surfaces is unequal, e.g., the cornea is more curved vertically than it is in a horizontal direction or vice versa.

The commonest form is that in which the vertical curvature is greater than the horizontal, and is described as regular astigmatism "according to rule." Regular astigmatism is remedied by the use of cylindrical lenses, the focal lengths of which is different in two meridians at right angles to one another.

Color blindness refers to the inability to discriminate colors properly. There are many types, from complete insensitiveness to color to a weak power to discriminate. In the commonest type, known as dichromatism, the individual does not see red or green but only yellow and blue or various combinations of them. Some 4 per cent of all men are said to exhibit this type. Women, in general, do not show it. It is inherited but occurs practically only in males.

SUMMARY

Sensory Unit
- Peripheral end organ, or receptor
- Sensory path through which the impulses are conveyed
- A center in the nervous system for interpretation or linkage with motor nerves
- **Function**—enables one to derive information about the world

Sensation
- The conscious result of processes which take place within the brain due to impulses derived from receptors
- **Classifications**
  - Old classification
    - Special—sight, hearing, touch, taste, smell
    - General—all other sensations
  - Based on the part of the body to which the sensation is projected
    - External, or those in which the sensations are projected to the exterior of the body, i.e., sight, hearing, taste, smell, pressure, and temperature
    - Internal, or those in which the sensations are projected to the interior of the body, i.e., pain, muscle sense, sensations from the semicircular canals and vestibule of the internal ear, hunger, thirst, sexual sense, and fatigue

Receptors
- **Classification**
  - Based on location of stimuli
    - Exteroceptors stimulated by forces outside the body
    - Proprioceptors stimulated by activities that occur in muscles and articulations
    - Interoceptors stimulated by substances or conditions within the viscera
  - Based upon the kind of stimuli
    - Chemoreceptors stimulated by suitable concentration of substances in solution, e.g., taste, smell, etc.
    - Mechanicoreceptors stimulated by mechanical pressure or vibratory impacts, e.g., sound, touch, etc.
    - Radioreceptors stimulated by radiant energy, e.g., light, heat, cold, etc.

Cutaneous Sensations
- Surface of skin is a mosaic of sensory spots which coincide with location of special end organs for pressure, cold, heat, or pain
- **Classification**
  - Epicritic
    - (1) Fibers for heat, cold, light pressures, and tactile discriminations

Cutaneous Sensations
- **Classification**
  - Epicritic
    - (2) Fibers for small difference of temperature, i.e., between 26° C and 37° C (78.8° F and 98.6° F)
  - Protopathic
    - (1) Fibers for pain
    - (2) Fibers for heat above 37° C (98.6° F)
    - Fibers for cold below 26° C (78.8° F)
    - Sensibility low, localization poor

Pain
- Excessive stimulation of any of the sense organs causes unpleasant sensations
- Actual pain caused by stimulation of pain receptors, which are most abundant in the superficial parts of the body
- Pain and other receptors stimulated simultaneously, sensation modified, i.e., burning pain
- **Function**—serves as danger signal
- Referred Pain
  - Pain arising in viscera and referred to skin area supplied with sensory fibers from the same spinal segment that supplies organ in question

Muscular, or Deep, Sensibility
- (1) Sensory end organs in the muscles, i.e., muscle spindles
- (2) Sensory end organs situated in tendons, i.e., tendon spindles, or tendon organs of Golgi
- (3) Afferent fibers carry impulses to centers in brain
- (4) Efferent fibers carry impulses from brain to muscle

Hunger
- Normal gastric hunger due to contractions of empty stomach, acting on nerves distributed to mucous membrane
- Hunger contractions may be frequent and severe, even when food is taken regularly, as in diabetes

Appetite
- Aroused in part through sensory nerves of taste and smell, associated with previous experiences
- Thought of food associated with appetite induces flow of saliva and gastric fluid

Thirst
- Normal thirst sensations presumably due to stimulation of sensory nerves of pharynx
- Prolonged deprivation of water probably affects sensory nerves in many tissues and interferes with the metabolism of the nervous system

Nausea
- May be due to stimulation from the stomach, to substances in the blood, to impulses from organs of sight, taste, and smell

Taste
- Sensory apparatus
  - (1) Taste buds are end organs
  - (2) Nerve fibers of trigeminal, facial, and glossopharyngeal nerves
  - (3) Center in brain

- **Taste**
  - Solution of savory substances must come in contact with taste buds
  - Taste buds are distributed over
    - Surface of tongue
    - Soft palate and fauces
    - Tonsils and pharynx
- **Tongue**
  - Freely movable muscular organ
  - Attached to hyoid bone, epiglottis, and the glossopalatine arches
  - Surface covered by papillae containing capillaries and nerves
    - Vallate
    - Fungiform
    - Filiform
    - Simple
  - **Nerves**
    - Sensory
      - Lingual, branch of trigeminal
      - Chorda tympani, branch of the facial
      - Glossopharyngeal
    - Motor—Hypoglossal
  - **Sense of**
    - (1) Taste
    - (2) Temperature
    - (3) Pressure
    - (4) Pain
    - are all well developed
- **Classification of Taste Sensations**
  - **Four primary sensations**
    - Salty, bitter, acid, sweet
  - **All others are**
    - Combinations of primary sensations
    - Combinations of one or more plus odor
- **Smell**
  - Sensory apparatus
    - Olfactory nerve endings
    - Olfactory nerve fibers spread out in fine network over surface of superior nasal conchae and upper third of septum
    - Olfactory bulb and center in brain
  - Odors
    - Minute particles usually in gaseous form
      - Must be capable of solution in mucus
    - Classify
      - Pure odors
        - Ethereal, aromatic, fragrant, ambrosial, garlic, burning, goat, repulsive, fetid
      - Odors mixed with sensations
      - Odors mixed or confused with taste
  - Olfactory center in the brain is widely connected with other areas of the cerebrum
  - Branches of trigeminal nerve found in lining of lower part of nose (pressure)
- **Hearing**
  - Auditory apparatus
    - External ear
    - Middle ear, or tympanic cavity
    - Internal ear, or labyrinth
    - Acoustic, or auditory, nerve
    - Acoustic center in brain
  - Air waves enter meatus and cause vibrations of tympanic membrane. The vibrations are conveyed to nerve endings of organ of Corti and thence by the auditory nerve to the brain

EAR

- **External Ear**
  - Pinna, or auricle
    - **Structure**—cartilaginous framework, some fatty and muscular tissue, covered with skin
    - **Function**—collects sound waves
  - External acoustic meatus
    - 2.5 cm long, partly cartilage, partly bone
    - Leads from the concha to the tympanic membrane
    - Near orifice skin is furnished with hairs and ceruminous glands
    - Ceruminous glands secrete a yellow, pasty substance
- **Middle Ear**
  - An irregular cavity in the temporal bone
  - Five or six drops of water will fill it
  - Bones
    - Malleus (hammer)
    - Incus (anvil)
    - Stapes (stirrup)
  - Five openings
    - Opening between it and external auditory canal, covered by tympanic membrane
    - Connect with internal ear:
      - Fenestra vestibuli at end of scala vestibuli
      - Fenestra cochleae at end of scala tympani
    - Opening into mastoid antrum and mastoid cells
    - Eustachian (auditory) tube—connects with the pharynx; ventilates cavity
- **Internal Ear**
  - Osseous labyrinth
    - Vestibular branch of acoustic nerve distributed to vestibule and semicircular canals:
      - Vestibule behind the cochlea, in front of the semicircular canals
      - Semicircular canals
        - Three in number
        - Open into vestibule
    - Cochlea
      - A spiral canal $2\frac{3}{4}$ turns around modiolus
      - Cochlear branch of the acoustic nerve
  - Membranous labyrinth
    - Surrounded by perilymph
    - Contains endolymph
    - In the vestibule forms the
      - Saccule
      - Utricle
    - Lines the semicircular canals
    - Lines the cochlea
      - *Basilar membrane* extends from free border of lamina to outer wall of cochlea and separates the scala vestibuli and the scala tympani. Supports organ of Corti

EAR

- **Internal Ear**
  - Membranous labyrinth
    - Lines the cochlea
      - *Vestibular membrane* extends from free border of lamina to outer wall of cochlea and is attached above basilar membrane, forms *scala media*
  - Acoustic nerve
    - Cochlear arises from bipolar cells in spiral ganglion
      - Peripheral fibers from cells terminate in and around the cells of the organ of Corti
      - Central fibers pass into the medulla and terminate in two nuclei
    - Vestibular arises from bipolar cells in vestibular ganglion
      - Peripheral fibers terminate in hair cells of saccule, utricle, and ampullae of the semicircular canals
      - Central fibers pass into the medulla and terminate in two nuclei

Physiology of Hearing
- Sonorous bodies produce air waves
- Air waves enter external auditory canal, set tympanic membrane vibrating, vibrations communicated to ossicles, transmitted through fenestra vestibuli to perilymph, stimulate nerve endings in organ of Corti, impulses carried to center of hearing in brain

Sense of Equilibrium
- Function of the vestibule and semicircular canals
- Lining membrane supplied with sensory hairs which connect with vestibular nerve
- Flowing of the endolymph stimulates the sensory hairs; this is transmitted to the vestibular branch of the acoustic nerve, thence to cerebellum

Visual Apparatus
- Bulb of the eye
- Optic nerve
- Center in brain
- Accessory organs
  - Eyebrows
  - Eyelids
  - Conjunctiva
  - Lacrimal apparatus
  - Muscles of the eyeball
  - Fascia bulbi

Accessory Organs
- **Eyebrows**
  - Thickened ridges of skin furnished with short, thick hairs

- **Accessory Organs**
  - **Eyebrows**
    - Protect eyes from vivid light, perspiration, etc.
  - **Eyelids**
    - Folds of connective tissue covered with skin, lined with mucous membrane, conjunctiva, which is also reflected over the eyeball
    - Provided with lashes
    - Upper lid raised by levator palpebrae superioris
    - Both lids closed by orbicularis oculi muscle
    - Slit between lids called palpebral fissure
    - Inner angle of slit called medial palpebral commissure (internal canthus)
    - Outer angle of slit called lateral palpebral commissure (external canthus)
    - **Function**
      - (1) Cover the eyes
      - (2) Protect eyes from bright light and foreign objects
      - (3) Spread lubricating secretions over surface of eyeball
  - **Eyelashes and sebaceous glands**
    - Margin of each lid, a row of short hairs project
    - Sebaceous glands connected with lashes
    - Meibomian glands between conjunctiva and tarsal cartilage of each lid
    - Secretion lubricates edges, prevents adhesion of lids
  - **Conjunctiva**
    - Mucous membrane, lines eyelids and is reflected over eyeball. Continuous with mucous membrane of lacrimal ducts and nose
  - **Lacrimal apparatus**
    - Lacrimal gland—in the upper and outer part of the orbit. Secretes tears
    - Lacrimal ducts begin at puncta and open into lacrimal sac
    - Lacrimal sac, expansion of upper end of nasolacrimal duct. Between lateral ducts is the lacrimal caruncle
    - Nasolacrimal canal—extends from lacrimal sac to nose
    - Tears
      - Secretion constant, carried off by nasal duct
      - Dilute solution of various salts in water, also mucin
      - Function
        - Keep surface of eyes moist

- **Accessory Organs**
  - **Lacrimal apparatus**
    - Tears
      - Function
        - Help to remove foreign bodies, microorganisms, dust, etc.
  - **Muscles**
    - **Extrinsic**
      - Superior rectus
      - Inferior rectus
      - Medial, or internal, rectus
      - Lateral, or external, rectus
      - Superior oblique
      - Inferior oblique
    - **Intrinsic**
      - Ciliary muscle
        - Determines position of lens
      - Muscles of iris
        - Contractor of pupil
        - Dilator of pupil

- **Nerves of Eye**
  - (1) Optic nerve concerned with vision only
  - (2) Oculomotor controls
    - Medial rectus muscle
    - Superior rectus muscle
    - Inferior rectus muscle
    - Inferior oblique muscle
    - Ciliary muscle
    - Circular muscle of iris
  - (3) Trochlear controls superior oblique muscle
  - (4) Abducent controls lateral rectus muscle
  - (5) Ophthalmic supplies general sensation, such as pressure, muscle sense, and pain

- **Orbit**
  - A bony cavity formed by seven bones
    - Frontal, malar, maxilla, palatine, ethmoid, sphenoid, lacrimal
  - Contains eyeball, muscles, nerves, vessels, lacrimal glands, fat, fascia bulbi, and fascia holding structures in place
  - Lined by fibrous tissue
  - Pad of fat—supports eyeball
  - Fascia bulbi is a serous sac which envelops eyeball from optic nerve to ciliary region
  - Shaped like funnel
    - Large end directed outward and forward
    - Small end directed backward and inward
  - **Optic foramen**—opening for passage of optic nerve and ophthalmic artery
  - **Superior orbital fissure**—opening for passage of orbital branches of middle meningeal artery and oculomotor, trochlear, abducent, and ophthalmic nerves

- **Bulb of the Eye**
  - Spherical in shape, but it projects anteriorly
  - **Tunica**
    - (1) Fibrous—sclera and cornea
    - (2) Vascular—choroid, ciliary body, and iris
    - (3) Nervous—retina
  - **Media**
    - (1) Cornea and aqueous humor
    - (2) Crystalline lens and capsule
    - (3) Vitreous body

- **Protective Tunics**
  - **Sclera**
    - Tough, fibrous, opaque
    - Covers posterior five-sixths of eyeball
    - Opaque, white and smooth externally, brown internally
  - **Cornea**
    - Fibrous, transparent—covers one-sixth of eyeball
    - Well supplied with nerve fibers
- **Vascular Tunic**
  - **Choroid**
    - Composed of dense capillary network and stroma of cells, some of which are pigmented, lines the sclera
  - **Ciliary body**
    - Includes the orbicularis ciliaris, the ciliary processes, and the ciliaris muscle. The orbicularis ciliaris is a zone about 4 mm in width which is continuous with anterior part of choroid
    - Ciliary processes 60–80 radiating folds, arranged like a plaited ruffle around margin of lens
    - Support ciliaris muscle—action of this muscle determines position of lens
  - **Iris**
    - A circular colored disk suspended in front of lens and behind cornea. Hangs free except for attachment at circumference to the ciliary processes and choroid. Central perforation—pupil
    - Pupil contracted by circular, or sphincter, muscle
    - Pupil dilated by radial, or dilator, muscle
    - Composed of connective tissue, containing numerous blood vessels and nerves. Contains pigment cells
    - **Function**—regulates size of pupil and thereby amount of light entering eye
- **Nervous Tunic, or Retina**
  - Nervous layer—contains elements essential for reception of rays of light. Situated between the choroid coat and hyaloid membrane of the vitreous humor, extends forward and terminates in the *ora serrata*
  - Has three sets of neurons so arranged that seven layers are formed. These are held in place by neuroglia and two membranes. Counting from the hyaloid membrane outward: Membrana limitans interna
    - (1) Nerve fibers, or stratum opticum
    - (2) Cell bodies of third neurons, or ganglionic layer
    - (3) Area of synapses of second and third neurons, or inner plexiform layer
    - (4) Cell bodies of second neurons, or inner nuclear layer
    - (5) Area of synapses between first and second neurons, or outer plexiform layer

Nervous Tunic, or Retina
- (6) Cell bodies of first neurons, or outer nuclear layer
- Membrana limitans externa, marking internal limit of rods and cones
- (7) Rods and cones of first neurons are end organs, or receptors, for the optic nerve
- Pigmented layer between rods and cones and choroid
- **Blind spot**
  - Entrance of optic nerve and central artery and vein of the retina
  - There are no rods and cones
  - Totally insensitive to light
- **Macula lutea**
  - 2 mm outer side of blind spot
  - Central pit—fovea centralis—is the center of direct vision

Refracting Media
- **Aqueous humor**
  - Aqueous chamber is between cornea in front and lens, suspensory ligament, and ciliary body behind. Aqueous humor is a watery solution containing minute amounts of salts
  - Dialyzed from capillaries, drains away through canal of Schlemm
- **Crystalline lens**
  - Transparent, refractive body enclosed in an elastic capsule
  - Double convex in shape. Situated behind the pupil
  - Held in position by counterbalancing of aqueous humor, vitreous body, and suspensory ligament
- **Vitreous body**
  - Semifluid, albuminous tissue enclosed in hyaloid membrane
  - Fills posterior four-fifths of bulb of the eye, distends sclera, and supports retina

Perception of Light and Color
- Waves started by the motion of molecules of hot bodies cause vibrations in ether
- Waves vary in length
  - Electrical waves—may be miles long
  - Light waves—between 0.0004 and 0.0008 mm in length
  - Chemical waves—are shorter than light waves

Refraction—bending or deviation in the course of rays of light, in passing obliquely from one transparent medium into another of different density

Vision
- Visible objects reflect light waves which fall upon them
- These reflected rays are brought to focus on receptors (rods and cones) of retina, transmitted to optic nerve, and thence to centers of vision in occipital lobe of cerebrum, from here to association areas
- Various theories are suggested to account for vision
- Color blindness, kinds, inheritance of dichromatism

Processes Necessary for Binocular Vision

- (1) Convergence, or turning the eyes inward, in order to place the image on corresponding points of the two retinae
- (2) Change in size of pupil—contracts in a bright light—dilates in a dim light
- (3) Accommodation—ability of the eye to adjust itself so that it can see objects at varying distances
- (4) Refraction—bending of light rays entering the pupil so they come to a focus on the retina

Abnormal Conditions

- **Myopia**
  - Nearsightedness
  - **Cause**—rays of light converge too soon
- **Hypermetropia**
  - Farsightedness
  - **Cause**—rays of light do not converge soon enough
- **Presbyopia**
  - Defective condition of accommodation in which distant objects are seen distinctly but near objects are indistinct
- **Astigmatism**
  - Condition in which the curvature of the refracting surfaces is defective

UNIT V

The Structural and Functional Relationships for Human Reproduction and Development

CHAPTER 24

REPRODUCTION, EMBRYOLOGY, FETAL CIRCULATION, PARTURITION, AND INVOLUTION

The human organism begins life as a single-celled embryo, derived from the fusion of pre-existing parental cells, the ovum and the spermatozoon. This one-celled embryo undergoes repeated mitotic cell division, forming the multicellular organism.

During life, cell division continues throughout the body, growth and repair of the complex organism being accomplished by repeated cell divisions, the grouping of cells, the differentiation of cells, and the accumulation of more or less intercellular material between the cells. The purpose of this chapter is to present some of the phases of this development.

The life cycle of man may be divided into two periods of *prenatal life* and *postnatal life*. Prenatal life is generally divided into the following:

Period of the ovum—first two weeks of development.

Embryonic period—from the beginning of the third week to the end of the second month.

Fetal period—from the beginning of the third month to the time of birth.

Postnatal life may be divided into these periods:

Neonatal period—extending from birth to the end of the second week.

Infancy—from the second week to the end of the first year.

Childhood—from the end of infancy to puberty, which comes at about the twelfth to the sixteenth year.

Adolescence—extending from puberty into the early or middle twenties.

Maturity—early and middle maturity lasting from adolescence to the end of the reproductive period. These are followed by late maturity.

Puberty and adolescence. In the developing embryo the elements of the reproductive organs are among the first to be differentiated. In fact, many authorities consider them to be differentiated in the zygote. During fetal life, infancy, and childhood, these organs develop at somewhat the same rate as other parts of the body. They attain functional maturity at the time of puberty. In temperate climates the age at which boys usually attain puberty is about 15 or 16 years; in girls about 1 year younger, at 14. In southern countries it is somewhat earlier, and in arctic regions 1 or 2 years later. However, there is no fixed rule, as the time of puberty varies with individuals. It is thought that in the male the androgens may have an effect on the onset of puberty. This effect may be due either to an increased quantity of androgens or an increase in the sensitivity of the tissues to androgens.

Puberty is indicated in the male by production of functional spermatozoa, in the female by the onset of ovulation and menstruation. At birth the testes contain thousands of immature spermatocytes, and the ovaries contain thousands of primitive germ cells; but these do not mature until the onset of puberty. Puberty is also marked by the gradual appearance of the secondary sex characteristics. In the male the larynx increases in size and accentuates the prominence called "Adam's apple," the voice changes, the external genitals grow somewhat rapidly, and hair grows on the face, pubes, axillae, and other parts of the body. The girl undergoes a gradual change of figure; the pelvis widens, fat develops on the hips, the breasts develop, hair grows on the pubes and axillae, and menstruation begins.

These changes are not accomplished at once, but go on over a number of years, known as the period of adolescence, which extends from puberty until 17 to 20 years in the female and until 18 to 21 years in the male.

MALE ORGANS OF REPRODUCTION

The male organs of reproduction are: two testes, which produce the spermatozoa and internal secretions, two seminal ducts (sing. ductus deferens), two seminal vesicles, two ejaculatory ducts, two spermatic cords, the scrotum, the penis and urethra, the prostate gland, and two bulbourethral (Cowper's)[1] glands.

[1] William Cowper, English surgeon (1666–1709).

The testes are two glandular organs suspended from the inguinal region by the spermatic cord and surrounded and supported by the scrotum. Each gland weighs 10.5 to 14 gm and on the average is about 4 to 5 cm long, with an average breadth of about 3 cm, and it is somewhat greater in thickness. Each consists of two portions—the testis and the epididymis.

The *testis* is ovoid in shape, covered exteriorly by fibrous tissue which sends incomplete partitions into the central portion of the

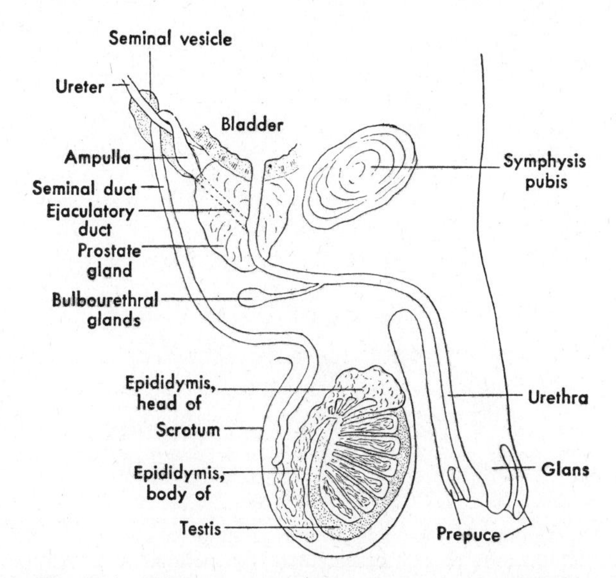

Fig. 364. Diagram of the male organs of reproduction.

gland, dividing it into communicating cavities. In these cavities are winding tubules (seminiferous tubules) surrounded by blood vessels and held together by interstitial tissue. These tubules inosculate in a sort of mesh (rete testis) and finally all unite in the epididymis.

The *epididymis* is a long, narrow body, which lies along the upper posterior portion of the testis and consists of a tortuous tubule some 5 m long, lined with mucous membrane and containing muscular tissue in its walls. It connects the testis with the seminal duct.

The functions of the testes are the production of spermatozoa and internal secretions containing hormones. If the testes are removed before puberty, sexual characteristics do not appear. The size of the thymus, pituitary, and cortex of the adrenals is increased, but the growth of the thyroid is diminished, and mental development tends to be retarded.

Descent of the testes. In early fetal life the testes are abdominal organs lying in front of and below the kidneys. During growth of the fetus they pass downward through the inguinal canal and shortly before birth are normally found in the scrotum. Sometimes, particularly in premature infants, a testis has not descended and is found

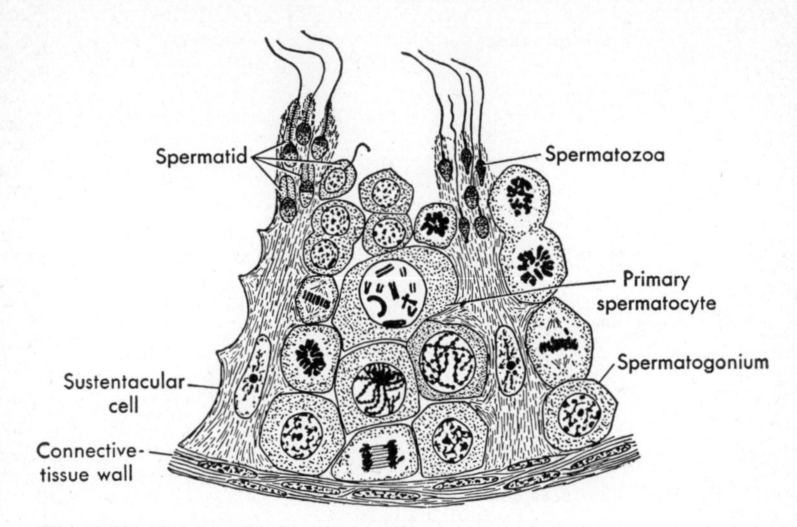

Fig. 365. Section of the wall of a portion of a seminiferous tubule to show spermatogenesis, or the development of spermatozoa. (From Walter, after Arey.)

in the inguinal canal or even in the abdominal cavity; as a rule it soon descends, but occasionally it does not descend. This nondescent is called *cryptorchism.* Cryptorchism may be either unilateral or bilateral.

Seminal duct (ductus deferens). Each duct is a continuation of the epididymis, and is the excretory duct of the testis. After a very devious course it reaches the prostate gland, which lies in front of the neck of the bladder. Here each duct ends by joining the duct from the corresponding seminal vesicle to form one of the ejaculatory ducts. Each consists of three coats: an external areolar, a middle muscular, and an internal mucous coat.

The seminal vesicles are two membranous pouches placed between the bladder and the rectum. They are pyramidal in form, with the broad ends directed backward and lateralward. The anterior portions converge, become narrowed, and unite on either side with the corresponding seminal duct to form the ejaculatory duct.

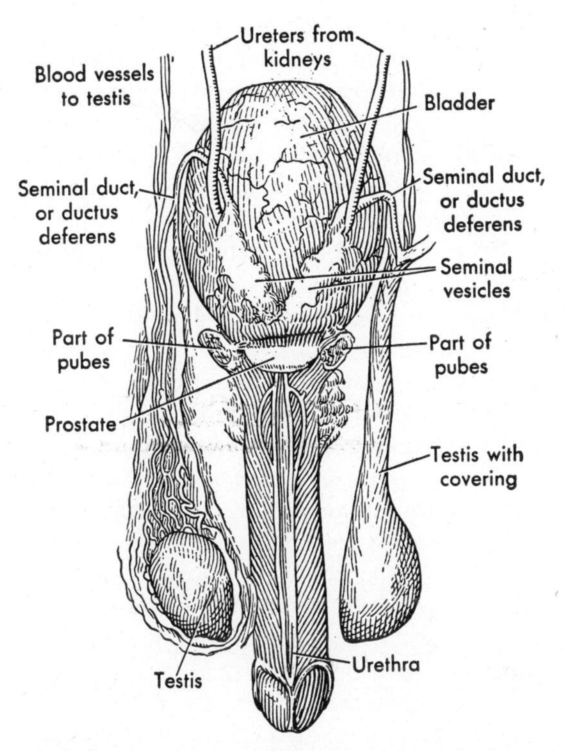

Fig. 366. Male reproductive organs, dorsal view.

Function. The seminal vesicles are glands which add a secretion to the semen aiding the motility of the spermatozoa.

The ejaculatory ducts are two in number, one on each side. They are formed by the union of the seminal vesicle with the seminal duct, descend, one on each side, and, passing between the lobes of the prostate gland, finally reach the urethra, into which they open and discharge their contents. Each has an external areolar, middle muscular, and internal mucous coat.

The spermatic cords. Each cord consists of a seminal duct, an artery, veins (forming a plexus called the pampiniform plexus), lymphatics, nerves, and connecting areolar tissue covered with fascia that is continuous with that covering the testis. These structures come together and form a cord just above the inguinal ring, through which the cord passes and descends into the scrotum, where it connects with the posterior surface of the testis.

The scrotum is a pouch consisting of the integument, a layer of thin, relatively dark skin, disposed in folds, or rugae, and in the adult covered with short hairs, and the dartos tunic, consisting of plain muscle fibers. The dartos sends in a fold of tissue which serves as a partition dividing the interior of the scrotum into two chambers. The scrotum contains and supports the testes and parts of the spermatic cords. The tissues of the scrotum are continuous with those of the groin and the perineum.

The penis, or organ of copulation of the male, is suspended from the front and sides of the pubic arch. It is composed of three cylindrical masses of cavernous tissue bound together by fibrous tissue and covered with skin. Two of the masses are lateral and are known as the *corpora cavernosa penis;* the third is median and is termed the *corpus cavernosum urethrae* because it contains the urethra. The term *cavernous tissue* is used because of the relatively large size of the venous spaces which exist in this tissue. It is also described as erectile tissue because the venous spaces may become distended with blood, during sexual excitement, and the penis thus becomes firm and erect. As the organ becomes more engorged, the process is known as erection. The skin covering the penis is continuous with that covering the scrotum, the perineum, and the pubes. At the end of the penis there is a slight enlargement known as the *glans penis,* in which the urethral orifice is situated. In the region of the glans, the loose integument of the penis becomes folded inward and then backward upon itself, forming the *prepuce,* or foreskin, which sometimes covers the glans too tightly, creating a condition known as *phimosis.*

The male urethra extends from the urethral orifice in the bladder to the external orifice at the end of the penis. The length is usually given as 17.5 to 20 cm (7 to 8 in.), a large part of which lies inside the pelvis. It is lined with mucous membrane and furnished with numerous muscular fibers.

The prostate gland is situated immediately below the internal urethral orifice. It is about the size of a chestnut and consists of a dense fibrous capsule containing glandular and muscular tissue. The glandular tissue consists of tubules which communicate with the urethra by minute orifices.

The function of the prostate gland is to secrete a viscid, alkaline fluid, the prostatic fluid, which aids in the motility of the spermatozoa. It is claimed that the vitality of the spermatozoa is enhanced

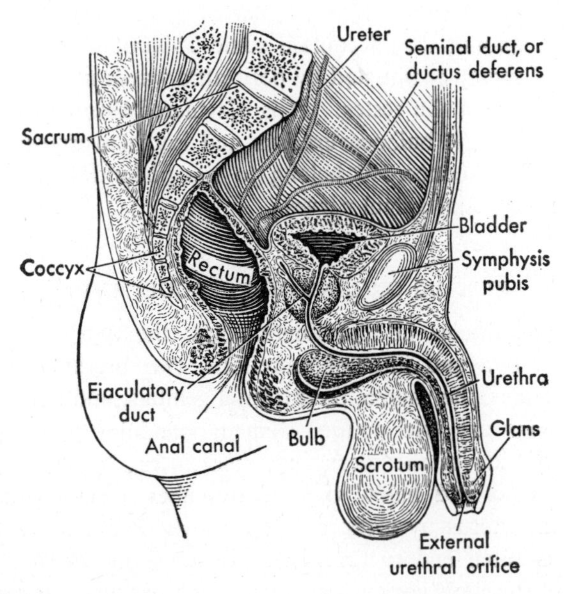

Fig. 367. Median section of male pelvis. (After Gray's *Anatomy*.)

by the prostate fluid and that the unfavorable acid environment which may exist in the urethra is neutralized through its alkaline reaction.

The bulbourethral glands (Cowper's glands) are two small bodies about the size of peas situated a little below the prostate gland one on either side of the membranous portion of the urethra. Each one is provided with a duct about 2.5 cm (1 in.) in length which terminates in a minute orifice in the wall of the urethra. These glands secrete a viscid fluid, which goes to form part of the seminal fluid.

The semen is a fluid derived from the testes, seminal vesicles, prostate gland, and the bulbourethral glands. It is a grayish-white, viscid fluid containing water, mucin, proteins, salts, and about 70,000,000 spermatozoa per cubic centimeter. Nonmotile spermatozoa are formed in the seminiferous tubules, are temporarily stored, and become physiologically mature in the epididymis. The motility of these cells is aided by the secretions of all the accessory glands of the reproductive tract. These secretions act as lubricants, give a practically neutral medium for spermatozoa, and probably have other functions. On the average, 3 to 5 cc leave the urethra at a time. This amount would contain some 250,000,000 spermatozoa. Discharge of semen is initiated by peristaltic waves moving along the tubes leading from the testes. It is ejaculated through the urethra under control of the bulbocavernosus and ischiocavernosus muscles. The nerve fibers concerned belong to the pudendal plexus, which arises from the sacral spinal nerves.

FEMALE ORGANS OF REPRODUCTION

The female organs of reproduction are: the two ovaries, which produce the ova and internal secretions, two uterine (Fallopian) tubes, the uterus, the vagina, the external genitals, and two breasts.

The ovaries are two almond-shaped, glandular bodies, situated one on each side of the uterus, attached to the back of the broad ligament behind and below the uterine tubes. Each ovary is attached to the uterus by a short ligament—the ligament of the ovary—and at its tubal end to the uterine tube by one of the fringelike processes of the fimbriated extremity. Each ovary weighs from 2 to 5 gm, and is about 4 cm long, 2 cm wide, and about 1 cm thick.

Structure. If the substance of an ovary is minutely examined, it is found to consist of a layer of columnar cells which constitutes the germinal epithelium; a stroma, or meshwork, of spindle-shaped cells with a small amount of connective tissue and an abundant supply of blood vessels; and the Graafian[2] follicles (vesicular ovarian follicles). In the cortical layer of the stroma, which is just beneath the germinal epithelium, a large number of vesicles of uniform size, about 0.25 mm in diameter, are found. These are the follicles in their earliest condition and are especially numerous in the ovary of a child. Be-

[2] Regnerus de Graaf, Dutch anatomist (1641–1673).

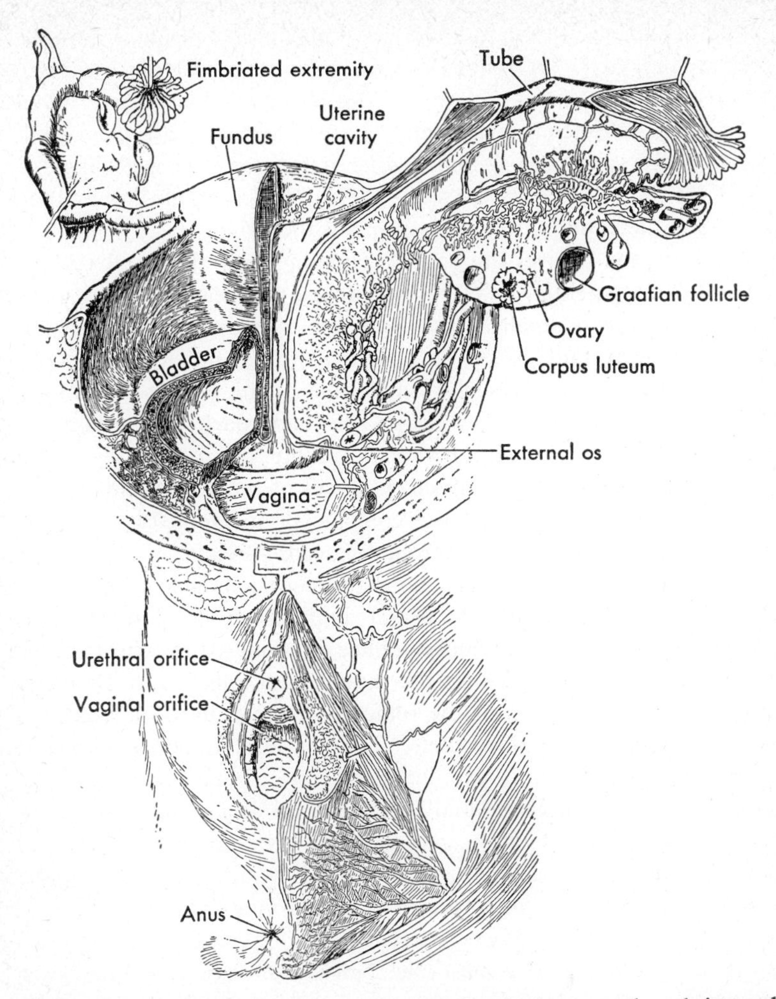

Fig. 368. Female generative tract, showing both external and internal organs. The internal organs are represented as laid out flat, parallel with the external organs.

tween puberty and the menopause large and mature follicles are found in the same layer, also corpora lutea, which are the remains of follicles that have ruptured and are undergoing atrophy and absorption.

The large ovarian follicles consist of an outer coat of fibrous tissue derived from the stroma, a middle coat also derived from the stroma, and an inner coat or lining, called the *membrana granulosa*, composed of several layers of cells. At one part of the mature follicle the cells of the membrana granulosa are collected into a mass called the *cumulus oöphorus* (discus proligerus), which projects into the cavity of the follicle. The cavity of the follicle contains albuminous fluid, and the cumulus oöphorus contains the ovum.

At birth the ovaries are said to contain many thousand (30,000 to 300,000) potential ova, closely packed near the periphery of the organ; but only a small number of these ever develop, as the great majority shrink and disappear. At the time of puberty the ovaries enlarge, become very vascular, and some of the follicles increase in size. As the follicles grow larger, they approach the surface and begin to form small protuberances on the outside of the ovary. When a follicle is fully matured, the wall of the ovary and the wall of the follicle rupture, and the contents of the follicle—the liquor folliculi, the ovum, and the surrounding cells—escape. This process of development, maturation, and rupture of a follicle is known as ovulation and continues at regular intervals from puberty to the menopause. A fully developed follicle is about 10 mm in diameter.

The corpus luteum. After the rupture of a follicle and the escape of the ovum, the lining of the follicle is thrown into folds, and the space is filled with a blood clot. The clot-filled follicle is known as *corpus hemorrhagicum.* Gradually luteum cells invade the clot; the clot is gradually absorbed and replaced by a scar of fibrous connective tissue known as the *corpus albicans.* Before puberty the surface of the ovary is smooth. During mature life it becomes pitted as the corpora albicantia are formed, leaving white scars. The corpus luteum continues to enlarge for about 2 weeks after ovulation, then disintegrates and is absorbed before the next follicle comes to maturity; but if fertilization of the ovum takes place, the corpus luteum continues to develop, reaching its maximum size in about 12 to 14 weeks. It persists in its fully developed state until the end of pregnancy. Sometimes this is called the true corpus luteum.

The functions of the ovaries are to produce, develop, and mature the ova and to discharge them when fully formed. The ovaries furnish at least two internal secretions (theelin and progesterone). (See p. 473.)

Comparative studies show that if the ovaries are removed before puberty or remain undeveloped, the uterus remains small, there is an absence of menstruation, the breasts do not develop, and the growth of hair on the body is greater than is usual in the female. If the ovaries are removed after puberty, the uterus and vagina atrophy, menstruation ceases, and various symptoms due to derangement of the nervous and metabolic processes may develop. These symptoms are similar to those which occur at the time when menstruation normally ceases, i.e., the menopause. The most typical are (1) vasomotor reactions, i.e., changes in the size of the blood vessels, which increase or diminish the amount of blood sent to the affected part, which is usually the skin of the face and neck. These changes

Fig. 369. Diagram of a longitudinal section of ovary and uterine tube opening into uterus at its narrow end. Broad ligament of uterus in background. Note longitudinally folded lining with fimbriae at open end. Cut surface of ovary shows corpus luteum (black) and ovarian follicles of various sizes.

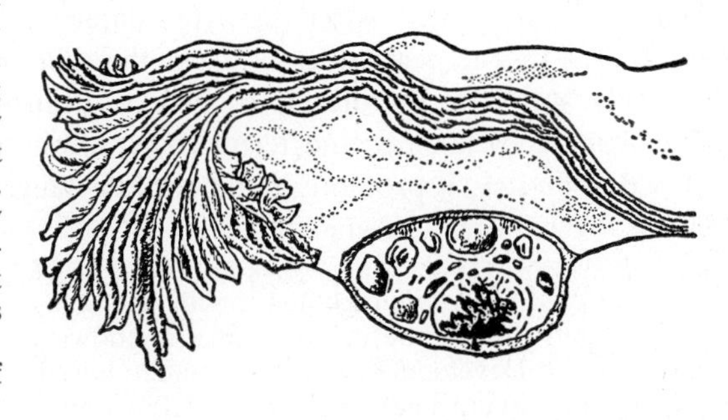

may give rise to sensations of alternating heat and cold, sweating, dizziness, muscular pains, headache, and sometimes abnormal mental conditions; (2) the amount of carbohydrates that can be oxidized is diminished, and obesity is common.

The uterine (Fallopian) tubes are two in number, one on each side, and pass from the upper angles of the uterus in a somewhat tortuous course between the folds and along the upper margin of the broad ligament toward the sides of the pelvis. They are about 7 to 14 cm (4 in.) long and at the point of attachment to the uterus are very narrow, but they gradually increase in size so that the distal end is larger. The margin of the distal end is surrounded by a number of fringelike processes called *fimbriae*. One of these fimbriae is attached to the ovary. The uterine opening of the tube is minute and will admit only a fine bristle; the abdominal opening is larger.

The uterine tube consists of three coats—the external, or *serous,* coat derived from the peritoneum; the middle, or *muscular,* coat having two layers—an external layer of muscular fibers longitudinally arranged and an internal layer of muscular fibers circularly arranged; the internal, or *mucous,* coat arranged in longitudinal folds and covered with ciliated epithelium. It is continuous at the inner end with the mucous lining of the uterus and at the distal end with the serous lining of the abdominal cavity. This is the only place in the body where a mucous and serous lining are continuous with one another (p. 61).

The function of the uterine tubes is to convey the ova from the ovaries to the uterus. It is thought that the movement of the cilia on the fimbriae and in the tubes produces a current which draws the ovum into the tube. After the ovum enters the tube, it is carried to the uterus by the peristaltic action of the tube and the movement of the cilia. It is in the uterine tube that fertilization takes place. If the ovum does not become fertilized within a short time (estimated at a day or so), it promptly undergoes disintegration and disappears in the secretions of the genital tract.

Occasionally an impregnated ovum remains in the tube, instead of passing into the uterine cavity. This is known as *tubal* or *tubal ectopic* pregnancy. Development may continue; but the erosive action of the impregnated ovum upon the tube usually results in hemorrhage, producing distention of the tube, which requires operation to forestall fatal bleeding. The ovum occasionally makes its way to the surface of the broad ligament or into the abdominal cavity, and development of the embryo may continue there, in rare cases to term. This is known as *abdominal* or *ectopic* pregnancy.

The uterus is a hollow, pear-shaped, muscular organ. It is situated in the pelvic cavity between the bladder and the rectum. Its length is estimated to be about 7.5 cm (3 in.), its width 5 cm (2 in.) at the upper part, and its thickness 2.5 cm (1 in.). During pregnancy the uterus becomes enormously enlarged, attains the length of 30 cm or more, extends into the epigastric region, and measures about 20 to 25 cm in width. After parturition the uterus returns to almost its original size but is always larger than before pregnancy. After the menopause, the uterus becomes smaller and atrophies.

The *fundus* is the convex part of the uterus above the entrance of the uterine tubes; the *body* is the part between the fundus and the cervix; the *cervix*, or neck, is the lower constricted part and extends from the body of the uterus into the vagina.

The cavity of the uterus is small because of the thickness of its walls. The part of the cavity within the body of the uterus is triangular in shape and has three openings, one at each upper angle communicating with the uterine tubes and one, the internal orifice, opening into the cavity of the cervix below. The cavity of the cervix, which is continuous with the cavity of the body of the uterus,

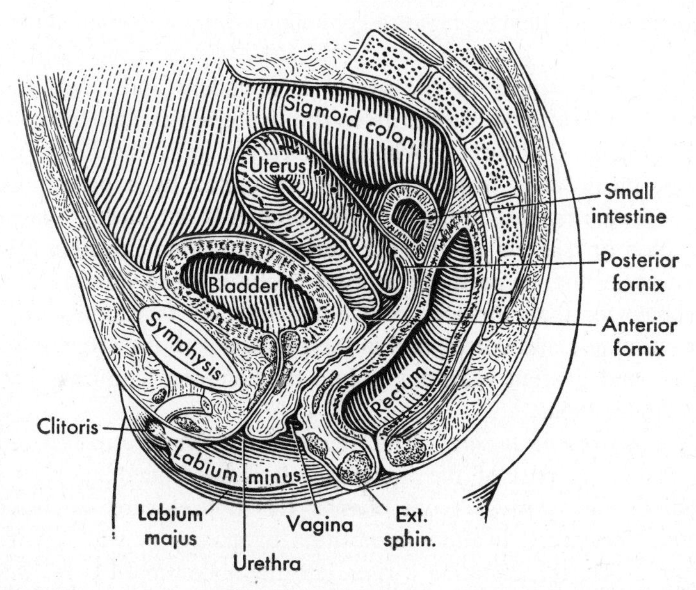

Fig. 370. Median section of female pelvis. (After Gray's *Anatomy*.)

is constricted above, where it opens into the body by means of the internal orifice (internal os), and below, where it opens into the vagina by means of the external orifice (external os). Between these two openings the canal of the cervix is somewhat enlarged.

Structure. The walls of the uterus are thick and consist of three coats:

An external *serous* coat derived from the peritoneum. It covers all of the fundus and the whole of the intestinal surface of the uterus but covers the anterior (vesical) surface only as far as the beginning of the cervix.

A middle *muscular* coat, which forms the bulk of the uterine walls. It consists of layers of plain muscular tissue intermixed with blood vessels, lymphatics, and nerves. The arrangement of the muscles is very complex, as they run longitudinally, spirally, and circularly, and cross and interlace in every direction.

An internal *mucous* membrane which is continuous with that lining the vagina and uterine tubes. It is highly vascular, provided with numerous uterine glands, and is covered with ciliated epithelium, except the lower third of the cervical canal, where it gradually changes to stratified squamous epithelium similar to that lining the vagina.

The blood supply of the uterus is abundant and reaches it by means of the uterine arteries from the hypogastrics and the ovarian arteries from the aorta. Where the cervix joins the body of the uterus, the arteries from both sides are united by a branch vessel called the circumflex artery. If this branch is cut during a surgical operation or if a tear of the neck during parturition extends so far as to sever it, the hemorrhage is very profuse. The arteries are remarkable for their tortuous course and frequent anastomoses. The veins are large and correspond in their behavior to the arteries. The uterine and ovarian veins empty into the common iliac veins via the hypogastric veins.

Position of the uterus. The uterus is not firmly attached or adherent to any part of the skeleton. It is suspended in the pelvic cavity by ligaments. A full bladder tilts it backward; a distended rectum, forward. It alters its position by gravity or with change of posture. During gestation it rises into the abdominal cavity.

The fundus of the uterus is inclined forward, and the external orifice is directed downward and backward.

Anteversion is the condition where the fundus turns too far forward. *Retroversion* is the condition where the fundus inclines backward. A bend may exist where the cervix joins the body. If the body is bent forward, it is described as *anteflexion;* if bent backward, *retroflexion.*

Ligaments. The uterus is supended by eight ligaments. Six are arranged in pairs.

The *broad,* or *lateral, ligaments,* two in number, are folds of peritoneum slung over the front and back of the uterus, extending laterally to the walls of the pelvis. They are composed of two opposed, serous layers, and between these layers are found the following structures: (1) uterine tubes, (2) the ovaries and their ligaments, (3) the round ligaments, (4) blood vessels and lymphatics, (5) nerves, and (6) some smooth muscle tissue.

The posterior fold covers the back of the uterus and extends far enough below to cover also the upper one-fifth of the back wall of the vagina, when it turns up and is reflected over the anterior wall of the rectum. Thus the uterus, with and between its two broad ligaments, forms a transverse partition in the pelvic cavity, the bladder, vagina, and urethra being in the front compartment and the rectum in the back compartment.

The smooth muscles of the broad ligaments are derived from the superficial muscular layer of the uterus. They pass out between the serous folds and become attached to the pelvic fascia.

Between the bladder and uterus the peritoneum forms a fold which is described as the *anterior ligament* of the uterus.

Behind the uterus the peritoneum forms a second and deeper fold, which is described as the *posterior ligament*.

The *round ligaments* are two rounded, fibromuscular cords, situated between the folds of the broad ligament. They are about 10 to 12 cm (4 to 5 in.) long, and take their origin from the upper angle of the uterus (on either side), in front of and below the attachment of the uterine tube. They extend forward, upward, and lateralward over the external iliac vessels. Each one passes through the external abdominal ring and along the inguinal canal to the labium majus, in which it terminates. The round ligaments are composed of muscles, areolar connective tissue, blood vessels, and nerves.

The *uterosacral ligaments* pass backward from the cervix, on either side of the rectum, to the posterior wall of the pelvis. They are partly serous, partly smooth muscular tissue.

The function of the uterus is to receive the ovum from the uterine tubes and, if it becomes fertilized, to retain it during its development. Later, when the ovum has developed into a mature fetus, it is expelled from the uterus, chiefly by the contractions of the uterine walls.

The vagina is a highly dilatable musculomembranous canal situated in front of the rectum and behind the bladder. It extends downward and forward from the uterus to the vulva.

The posterior wall is about 8 to 9 cm long, whereas the anterior wall is only 6 to 7 cm long. The upper portion of the vagina surrounds the vaginal portion of the cervix, forming a deep recess behind the cervix which is called the posterior fornix. The recesses at the front and sides are smaller and are called the anterior and lateral fornices.

Structure. The vagina consists of an internal mucous lining, continuous above with that of the uterus, and a muscular coat, with a layer of submucous areolar connective tissue between. The inner surface of the mucous membrane is thrown into two wide longitudinal folds, and from these transverse folds, or rugae, extend. The

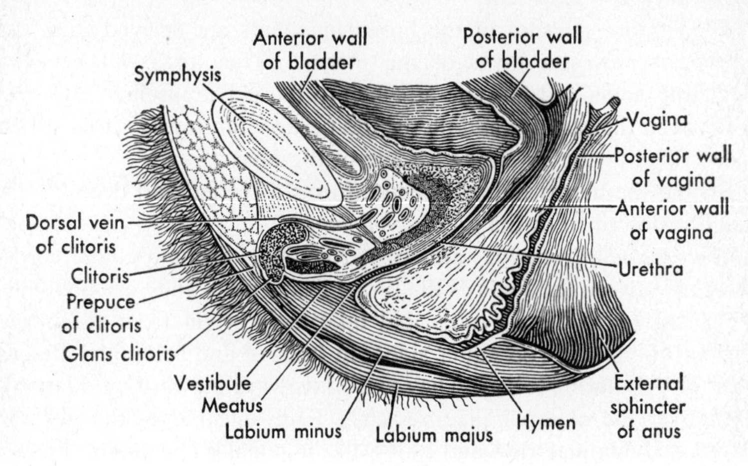

Fig. 371. Sagittal section of the vagina and neighboring parts. (Gerrish.)

muscular coat increases during pregnancy, and the rugae of the mucous coat allow for extreme dilatation of the canal during labor and birth.

The external genitals are grouped under the name of vulva or pudendum and include the following: the mons pubis, the labia majora, the labia minora, the clitoris, the vestibule of the vagina, and the greater vestibular glands.

The mons pubis is an eminence situated in front of the pubic symphysis. It consists of areolar, adipose, and fibrous connective tissue covered with skin and, after puberty, with hair.

The labia majora are two longitudinal folds of skin which are continuous with the mons pubis in front and extend to within 2.5 cm

(1 in.) of the anus behind. In the substance of the labia there are blood vessels, nerves, glands of the sebaceous type, and adipose and areolar connective tissue.

The labia minora are two longitudinal folds of stratified squamous epithelium containing numerous sebaceous glands. They are situated between the labia majora. They are joined anteriorly in the hood, or prepuce, of the clitoris and extend downward and backward for about 4 cm.

The clitoris is a small body situated at the apex of the triangle formed by the junction of the labia minora. It consists of erectile tissue, contains many vessels and nerves, and is almost completely covered by the hood, or prepuce.

The vestibule of the vagina is the cleft behind the clitoris and between the labia minora. The urethral and vaginal orifices and the openings of the ducts of the greater vestibular glands are in the vestibule. The hymen is a fold of mucous membrane which surrounds the lower part of the vagina and makes the orifice smaller. It is quite elastic and may remain intact even after childbirth, although it is usually ruptured. Occasionally it extends entirely across and closes the orifice altogether. This condition is spoken of as imperforate hymen.

The greater vestibular glands are two round, or oval, glands, situated on either side of the vagina. Their ducts open into the vestibule, one on either side, in the groove between the hymen and the labia minora. Their secretion lubricates the vulval canal.

The perineum is the external surface of the floor of the pelvis, from the pubic arch to the anus, with the underlying muscles and fascia. In the female it is perforated by the vagina. A wedge-shaped upward extension of the perineum, forming a septum between the vagina and the rectum, is called the perineal body. The perineum is distensible and stretches to a remarkable extent during labor. Nevertheless it is frequently torn, and when the tear is of any extent and is not repaired, the vagina and uterus lose the support afforded by it, and various abnormal conditions follow.

PHYSIOLOGY OF THE FEMALE REPRODUCTIVE ORGANS

The functions of the female reproductive organs are the formation and development of the ova, the retention and sustenance of the fecundated ovum until it develops into a mature fetus ready to live outside the body, the expulsion of the fetus, and the manufacture and liberation of endocrine secretions.

Puberty is the period at which the sexual organs become matured and functional.

Ovulation includes the process of the development and maturation of the follicle and its ovum, and the rupture of the follicle. Sometimes the term *ovulation* is used to refer simply to the extrusion of ova by the ovary.

The commonly accepted theory is that about or shortly before the age of puberty the Graafian follicles begin to discharge their ova and that this process continues until the menopause.

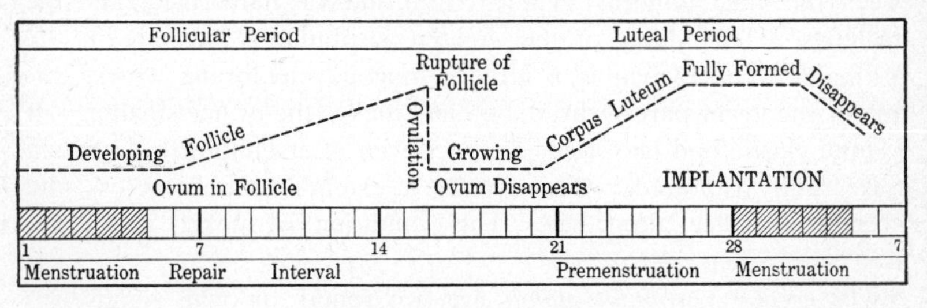

Fig. 372. Diagram to show sequence of events in ovary and in uterus in human menstrual cycle. (After G. W. Corner.)

Menstruation consists of the periodical discharge of bloody fluid from the uterine cavity. When once established, it occurs on the average every 28 days from the time of puberty to the menopause, with the exception of periods of pregnancy and lactation. The average duration is from 4 to 5 days. The amount of blood lost is subject to individual variations but is usually between 100 and 200 cc. The menstrual fluid consists of mucin, epithelial cells, and blood. The amount of destruction of the mucous membrane of the uterus varies a good deal, but the surface epithelium is always destroyed.

The mucous membrane of the uterus exhibits a constantly recurring cycle. For description this cycle may be divided into three or four periods. Three have been listed here.

Since ovulation and menstruation are directly related to each other it may be helpful to think through the sequence of events that lead up to menstruation and begin with the repair process.

1. Period of repair and proliferation (follicular phase). During this period the epithelium of the mucous membrane is restored to its usual thickness and new epithelium is formed. The uterus slightly enlarges, the stroma becomes more vascular, and the glands show proliferate changes. The follicle continues to develop and finally ruptures and ovulation takes place.

2. Premenstrual period or secretory stage (luteal phase). This usually begins 12–14 days before the first day of menstruation. The mucous membrane becomes thicker and congested with blood, its glands become elongated and more coiled, and secretion is markedly increased. It is evident that this stage is influenced by the developing corpus luteum. Toward the end of the secretory stage, vasoconstriction occurs.

3. Period of degeneration or menstruation, usually lasting about 4 or 5 days. In the secretory stage vasoconstriction occurred, which causes ischemia of the endometrium and results in necrosis of the superficial epithelium. There is dilation of the underlying blood vessels, capillary hemorrhage occurs, and the epithelium is cast off.

Reference to Fig. 372 will show that these periods can be considered, in relation to the ovary, as a follicular period of about 14 days and a luteal period of about 14 days.

If the body temperature is recorded daily during the cycle it will show a drop during the midmenstrual period. It then rises rapidly again, reaching a level of about 0.8° F above normal, and remains elevated until the menstrual flow begins, when it returns to normal.

Causes of menstruation. At the present time the generally accepted view is that menstruation is dependent upon the ovaries and that their influence is exerted through the medium of the blood. It is thought the luteal hormone is carried to the uterus by the blood and is responsible for the hypertrophy and congestion that precede menstruation. The fact that operations for the removal of the ovaries are followed by atrophy of the uterus and cessation of menstruation supports the theory that the ovaries are responsible for menstruation. Bard[3] puts it this way: "An ebb in the ovarian hormone tide serves to bring about augmented secretion of gonadotropic hormore. This causes increased ovarian activity which in turn

[3] Philip Bard, *Macleod's Physiology in Modern Medicine,* 1941.

brings about a depression in the anterior pituitary. The lag in these reciprocal responses gives time for successive waves of activity."

The menopause, or **climacteric,** is the physiological cessation of the menstrual flow, the end of the period during which the ovarian follicles develop in the ovaries, and consequently the end of the childbearing period. It is usually marked by atrophy of the breasts, uterus, tubes, and ovaries. Many individuals pass through the menopause without any untoward symptoms, while others may have vasomotor reactions, i.e., changes in the size of the blood vessels, which increase or diminish the amount of blood sent to the affected part, which is usually the skin of the face and neck. These changes give rise to sensations of alternating heat and cold, sweating, dizziness, muscular pains, headache, and sometimes abnormal mental conditions. Obesity may occur from a variety of causes. The age of menopause varies as does the age of puberty; in general, we may say the earlier the puberty, the later the menopause, and vice versa. In temperate climates the average period for the arrival of the menopause is from about 45 to 47 years.

Mammary glands. The two mammary glands, or breasts, secrete the milk which is needed for the nourishment of the young infant. Each breast covers a nearly circular space in front of the pectoral muscles, extending from the second rib above to the sixth rib below, and from the side of the sternum to the border of the armpit.

Structure. The breasts are convex in shape and are covered externally by skin. About the center of the convexity a papilla projects, which is called the *nipple.* The nipple contains the openings of the milk ducts and is surrounded by a small circular area of pink or dark-colored skin, which is called the *areola.* The breasts are compound glands and are divided by connective-tissue partitions into about 20 lobes, each of which possesses its own excretory duct, which, as it approaches the top of the breast, dilates and forms a small reservoir in which milk can be stored during the period when the gland is active. Each duct opens by a separate orifice upon the surface of the nipple. The lobes are subdivided, and the small lobes, or lobules, are made up of the terminal tubules of the duct, which lie in a mesh of fibrous areolar tissue containing considerable fat.

Blood vessels and nerves. The mammary glands are well supplied with blood brought to them by the thoracic branches of the axillary, internal mammary, and intercostal arteries. The nerve fibers are derived from the ventral and lateral cutaneous branches of the fourth, fifth, and sixth thoracic nerves.

Development of the mammary glands, or mammae. The increase in the size of the mammary glands at the time of puberty is due to an increased development of the connective tissue and fat. The glandular tissue remains undeveloped and does not function unless conception takes place. When conception occurs, the glandular tissue undergoes a process of gradual development that produces marked changes. The breasts become larger and firmer, the veins on the surface become more noticeable, the areola becomes enlarged and darkened, the nipple becomes more prominent, and toward the end of pregnancy a fluid called colostrum can be squeezed from the orifice of the ducts. After delivery the amount of colostrum increases for a day or two, and then the secretion changes to milk.

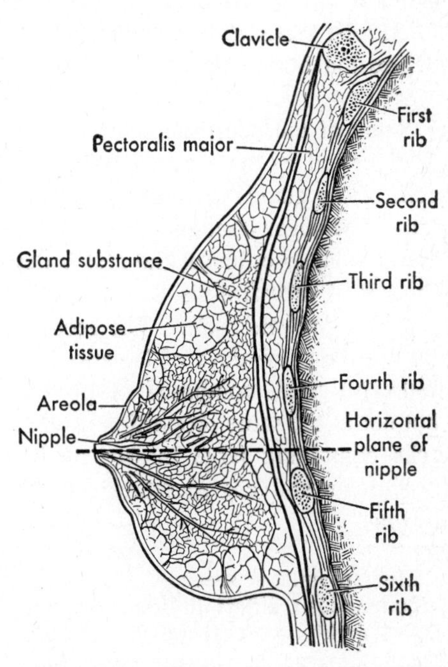

Fig. 373. Right breast in sagittal section, inner surface of outer segment. (Gerrish.)

The primary development and later functioning of the mammary glands suggest an intimate connection between these glands and the uterus, ovaries, and pituitary body. The present theory is that the increase in the size of the breasts at the time of puberty is influenced by estrin, the follicular hormone. If the ovaries are removed before puberty, the breasts do not develop; or if the ovaries are removed after puberty, the breasts are apt to atrophy.

It was formerly thought that the development of the glandular tissue following conception was due to hormones formed by the corpus luteum, but recent experiments on animals seem to indicate that the corpus luteum alone cannot cause growth and function of the mammary glands, while these effects are readily produced by extracts of the pituitary body (prolactin). To what extent the ovary also is necessary to growth of the glands and lactation is as yet uncertain. Secretion after childbirth is dependent on the emptying of the glands by suckling or by mechanical means.

The secretion of milk. The secretory portion of the mammary gland is composed of 15 to 20 lobes of alveolar glandular tissue. Each lobe sends a duct to the nipple. These ducts open in a slight depression at the tip of the nipple. The secretion of the gland is milk.

Some of the constituents of the milk, i.e., water, salts, and sugar, are secreted by these glands from the blood; but it is thought that the cells of the glands themselves disintegrate and form the proteins and fat. The sugar contained in the milk is lactose, and the sugar of the blood is glucose, and if the first is derived from the second, it is by a process of dehydrolysis.

Colostrum and milk. The secretion of the mammary glands during the first few days of lactation is called colostrum. It is a thin, yellowish fluid, composed of proteins, fat, sugar, salts, and water, but not in the same proportion as in milk. It also contains numerous cells containing large masses of fat. These are called *colostrum corpuscles* and are secreting cells that are not completely broken down.

Human milk is specially adapted to the requirements of the human infant and so differs in some respects from that of all other animals. Cow's milk is most frequently substituted for human milk. The relative composition of the two can be seen in the following table.

In substituting cow's milk for human milk, the differences that must be taken into consideration are not only the different relative proportions but also the following: The difference in the proteins; the protein of human milk is one-third caseinogen and two-thirds lactalbumin, and that of cow's milk is five-sixths caseinogen and one-sixth lactalbumin. The difference in the curds formed in the stomach; human milk curdles in small flocculi, and cow's milk curdles in large, heavy curds. The reaction of human milk is practically neutral, *p*H 7.0 to 7.2; cow's milk is slightly acid, *p*H 6.6 to 6.8 when first drawn, but the acidity increases on standing. Human milk is sterile, and cow's milk, due to the handling it undergoes, contains a large number of microorganisms. Pasteurization destroys the microorganisms usually found in milk, but unless it is done very carefully, it also destroys the vitamins. Human milk contains antitoxins and antibacterial substances that have been formed in the mother's blood; and as it is ingested directly from the breast, its germicidal power is at its height. Cow's milk may have germicidal value, but this soon deteriorates and usually is lost by the time it is given to the child.

	Human (Average) %	*Cow's (Average)* %
Water	88.4	87.1
Proteins	1.5	3.2
Fat	3.3	3.9
Lactose	6.5	4.9
Salts	0.3	0.9

EMBRYOLOGY

The individual starts life as a single microscopic cell. This cell, following a gradually advancing sequence of growth gradients (organizers,[4] etc.), develops in an orderly integrated way into an enormous number of cells, arranging them at the same time as the embryonic tissues, then forming the tissues into embryonic organs and developing the systems from these. The size of the individual and of its parts increases as the number of cells increases and as intercellular material secreted by the cells accumulates between the cells.

Embryonic tissues. In an embryo of about 6 weeks, the *germinal tissues,* which will develop into reproductive glands, may be distinguished from the *somatic tissues,* which will form the other parts of the body. The germinal tissues make their appearance as a pair of genital folds in the dorsal region of the embryo, and by about the tenth or eleventh week these are differentiated into the reproductive organs called *gonads.* The female gonads are called *ovaries;* the male gonads are called *testes.* During fetal life and infancy, these organs remain relatively inactive, undergoing cell division at somewhat the same rate as the other parts of the body but producing no functional *germ cells,* or *gametes.* At puberty the ovaries and testes (singular, *testis*) begin more rapid development, and functional germ cells (ova and spermatozoa) are produced.

[4] C. H. Waddington, *Organizers and Genes,* 1940, summarizes the theories of organizers and growth, evocation, competence, patterns as equilibria.

Germ cells. Egg cells, or ova, are thought to be derived from the germinal epithelium of the ovary. Each *ovum* comes to be surrounded by a cellular layer, which gradually enlarges to form the ovarian (*Graafian*) follicle (Fig. 374). Each follicle, when fully developed, forms a marked protuberance on the surface of the ovary and ruptures, liberating the contained ovum to the exterior of the ovary. This discharge of ova from the ovary is termed *ovulation;* and it is thought that, in general, ovulation takes place about midway between two menstrual periods, the ovaries possibly alternating each month in the liberation of a single ovum. The ovum is globular, about 0.2 mm ($\frac{1}{125}$ in.) in diameter. The fully grown ovarian

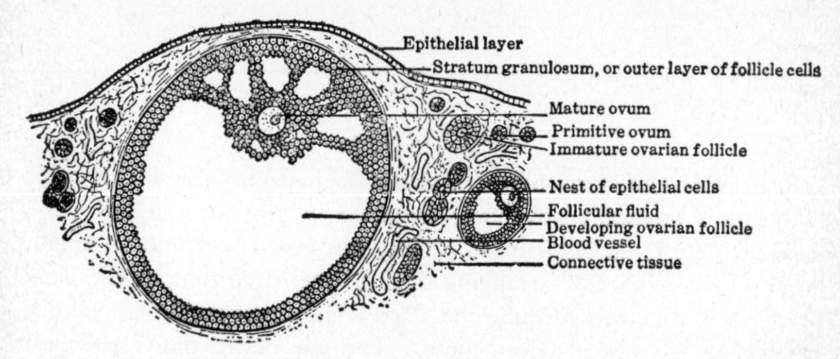

Fig. 374. Diagram of thin section of portion of an adult mammalian ovary. Shows one fully formed Graafian, or ovarian, follicle and others in various stages of development. (See Fig. 369 also.)

follicle is about 12 mm ($\frac{2}{5}$ in.) in diameter and is millions of times bulkier than a primary follicle. *Spermatozoa* (sperm cells) are derived by cell division from the epithelial cells which line the tubules of the testes. When fully formed, a spermatozoon is about 0.05 mm ($\frac{1}{500}$ in.) in length and is shaped much like a tadpole. It consists of an elliptical head about 0.003 mm long, a rod-shaped middle piece of about the same length, and a tail, which gradually tapers. The head contains the nuclear material of the cell and hence the chromatin; the middle piece contains the centrioles. When activated by the fluid constituents of the *semen,* the spermatozoa swim by means of lashing motions of the tail.

Maturation of the germ cells is the process by means of which the ovum and the spermatozoon are prepared for fertilization. During the process of formation of functional ova from the germinal epithelium of the ovary, the number of chromosomes in each ovum is reduced to one-half the original number. Human cells have 48 chromosomes (23 pairs of *ordinary chromosomes* and a pair of *sex chromosomes,* usually designated as *x* chromosomes). Each mature ovum has one member of each pair of chromosomes (23 chromosomes and an *x* chromosome). The details of the process are given in Fig. 375. In human individuals, it is thought that the cell divisions resulting in the reduction of the number of chromosomes are begun before ovulation, one polar body being produced, but that reduction is not completed by the formation of the second polar body until after fertilization has taken place. Human spermatozoa have 24 chromosomes, one member of each pair of chromosomes. Maturation of spermatozoa takes place before they are detached from the walls of the seminiferous tubules.

According to most authorities, the cells of male individuals have 48 chromosomes (23 pairs of autosomes and one pair of sex chromosomes, usually called *x* and *y*). Half of the spermatozoa have 23 chromosomes (abbreviation = chr) and an *x* chr, and half of them have 23 chr and a *y* chr, 24 chr in all, in each spermatozoon. If an ovum is fertilized by a spermatozoon with an *x* chr, it will then have 23 pairs of autosomes and a pair of *x* chr, and will, therefore, produce a female individual. If an ovum is fertilized by a spermatozoon having a *y* chr, it will then have 23 pairs of ordinary chr and a pair made up of an *x* and a *y* chr, and will produce a male individual. Since ova are all alike in the possession of an *x* chr, and since there are equal numbers of the two kinds of spermatozoa, there should be an equal number of male and female individuals produced, due to inheritance of sex through the chromosomes, as explained above.[5]

Fertilization. This term is applied to the penetration, or impregnation, of the ovum by the spermatozoon, and the *fusion of their nuclei.* An enzyme called *hyaluronidase* is present in spermatic fluid which is essential for removing the cumulus oöphorus from the ovum so that the sperm cell may enter. Fertilization is thought to take place in the upper portion of the uterine tube, probably within a few hours after ovulation and after *insemination*, or semination. During this time the ovum moves down into the tube from the ovary, and the spermatozoa move or are carried up through the

[5] For studies of selection of sperm cells, environmental factors, etc., see biological journals. L. B. Arey, *Developmental Anatomy,* 1954 edition, has a good bibliography at the close of the chapters.

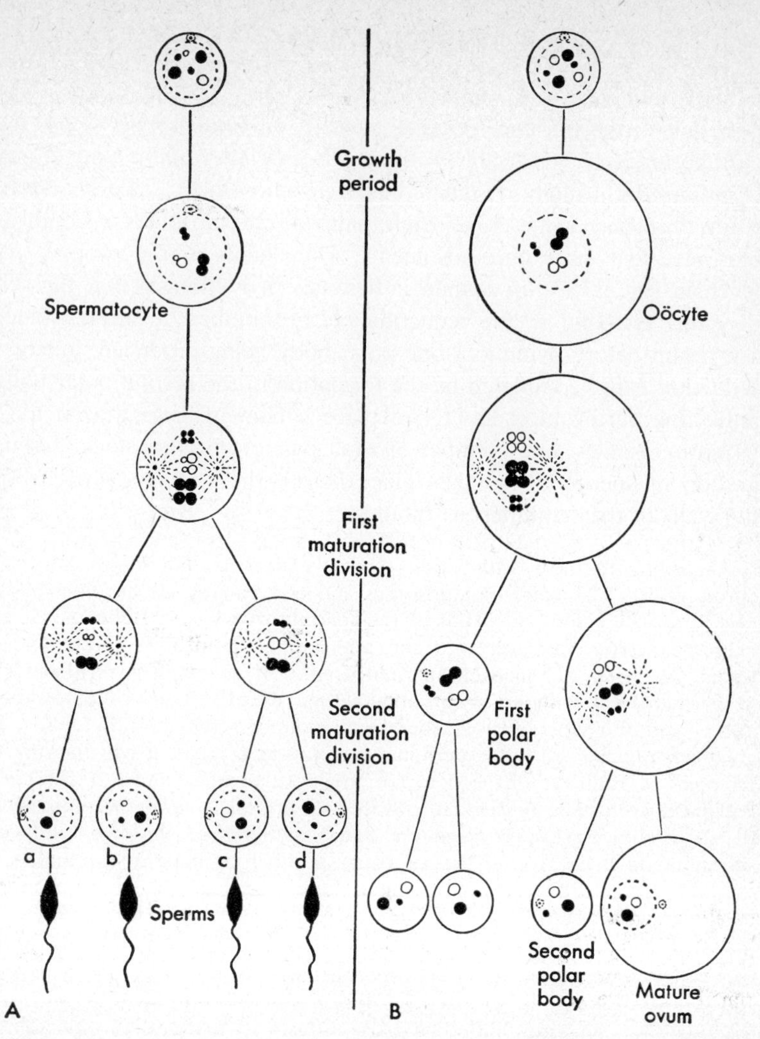

Fig. 375. Diagram to show maturation of reproductive cells. This process consists of meiosis (meiotic, or reduction, division) and mitosis. *A*, sperm cells; *B*, ova. In *A* each of these four cells is a functional sperm. In *B* only one of the cells is a functional ovum; the other three usually disappear. The spermatocyte and oöcyte are shown with two pairs of autosomes (black) and one pair of sex chromosomes (white). In the oöcyte the two sex chr are alike (white circles of same size)—the two *x* chr referred to in the text. It is evident that in *B* all possible ova will be alike in having one *x* chr. In the spermatocyte the two sex chr are not alike (a large white circle and a small white circle)—the *x* and *y* chr referred to. There are in *A*, therefore, two possible kinds of sperm cells—(*c*) and (*d*) containing an *x* chr and (*a*) and (*b*) containing a *y* chr. (Modified from *Essentials of Human Embryology*, by G. S. Dodds, John Wiley & Sons, Inc.)

uterus and tube. As soon as the head of the spermatozoon enters the ovum, the tail disappears, and the head takes on the appearance of a nucleus. The nucleus of the ovum and the nucleus of the spermatozoon approach each other and fuse more or less completely into a single nucleus known as the *cleavage nucleus.* The cell thus

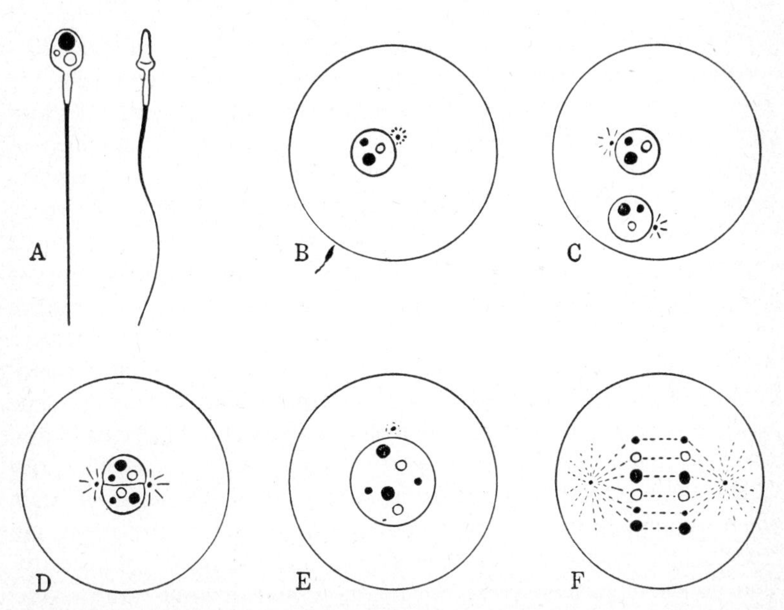

Fig. 376. Diagram to show fertilization. *A,* human sperm cell, face and side views. Note the head containing the chromosomes, the middle piece, and the tail. *B,* sperm entering ovum. *C,* sperm nucleus within ovum, showing chromosomes. *D,* the sperm nucleus approaches the ovum nucleus. *E,* the two nuclei fuse, and fertilization is complete; the zygote formed in *E* has six chromosomes—two pairs of autosomes (black) and one pair of sex chromosomes (white). Since the two sex chromosomes are alike (*x* chromosomes), the zygote would develop into a female. *F,* first embryonic cell division, which when complete would give a two-celled embryo. (Modified from *Essentials of Human Embryology,* by G. S. Dodds, John Wiley & Sons, Inc.)

resulting is a one-celled embryo (*zygote*) and contains the full number of chromosomes, 48, or 24 pairs, one member of each pair being derived from the ovum and one member from the spermatozoon, as shown in Fig. 376. The chromosomes are said to contain *genes.* A factor, or gene, is that portion of chromatin that influences some character in the offspring. One-half the *genes*, or inheritance determiners, which are in the chromosomes, are, therefore, derived from each parent, the germ cells forming the continuity link between succeeding generations.

Heredity. This term is applied to the transmission of the potential characteristics of parents to their children. The mechanism of transmission is thought to be the maturation of the germ cells and fertilization. It is evident that it is potentialities, rather than characteristics, that are transmitted. In different environments, during the growth of the individual and the development of the potentialities into characteristics, these developed characteristics may be modified to a greater or less extent. It is believed that the offspring receives one member of each pair of chromosomes from each parent. The genes of these chromosomes are similar; for instance, each of the paired chromosomes may carry a "gene for eye color," or for hair form, or for short fingers, etc. These genes may be the same or alternative in effect: for instance, in a given case (A), the offspring may receive a "gene for brown eyes" from each parent, in the paired chromosomes; or, in another case (B), the offspring may receive a "gene for brown eyes" from one parent and a "gene for blue eyes" from the other parent, in the paired chromosomes; or, in some other case (C), a "gene for blue eyes" from each parent.[6] Some traits seem to be dominant under usual conditions, so that if the genes are present, the traits are bound to appear. Some traits are recessive and do not appear, if the genes for the dominant traits are present. For instance, in (A), the individual will have brown eyes; in (B), since brown is dominant to blue, the offspring will have brown eyes, although the alternative "blue gene" is present: in (C), the eyes will be blue, there being no "gene for brown eyes" present. Dominance may be complete, partial, or absent. A gene which is dominant to a certain gene may be recessive to some other gene. Genes for recessive traits may be carried through generations, along with genes for dominant traits, the recessive traits only appearing as characteristics in individuals lacking the dominant genes.[7]

Cleavage and the formation of germ layers. One of the results of fertilization is the activation of the fertilized ovum, so that

[6] Actually, the inheritance of these characteristics is much more complex.

[7] For description of experiments in genetics and their cytological interpretation, see books on heredity, such as H. E. Walter, *Genetics;* E. W. Sinnott and L. C. Dunn, *Principles of Genetics.*

it begins to undergo cell division. It divides and subdivides, until an enormous number of cells called *blastomeres* is formed; typically, the single cell becomes two, then four, then eight, etc., though this uniform rate is not maintained. The first few blastomeres form a mulberry-shaped group of cells called the *morula*. As the blastomeres increase in number, they come to form a hollow, fluid-filled ball of cells, with a disk of cells near its periphery. This is a modi-

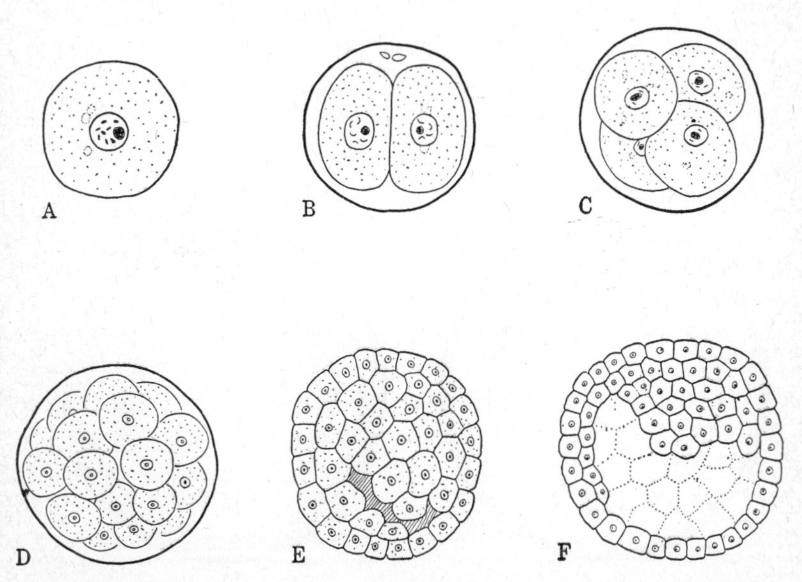

Fig. 377. Diagram to show very early stages of mammalian development. *A*, one-celled embryo; *B*, two-celled embryo; *C*, four-celled embryo; *D*, berry-like ball of cells, or *morula; E,* beginning the formation of the blastocyst; *F*, well-developed blastocyst, consisting of a hollow ball of *trophoblast* cells and an inner mass of cells known as *blastomeres.*

fied blastula (hollow ball of cells) and is known as the *blastocyst,* or blastodermic vesicle. The cells of the disk (blastoderm) develop into the embryo, while the outer cells later assist in the formation of the fetal membranes. The cells in the blastoderm increase in number and come to be arranged in two layers, an outer layer of cells called the *ectoderm* and an inner layer called the *entoderm.*[8] This embryo, composed of two cell layers, is called the *gastrula.* Between the ectoderm and entoderm the *mesoderm* is formed. The

[8] Also spelled endoderm.

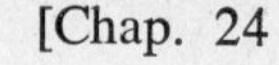

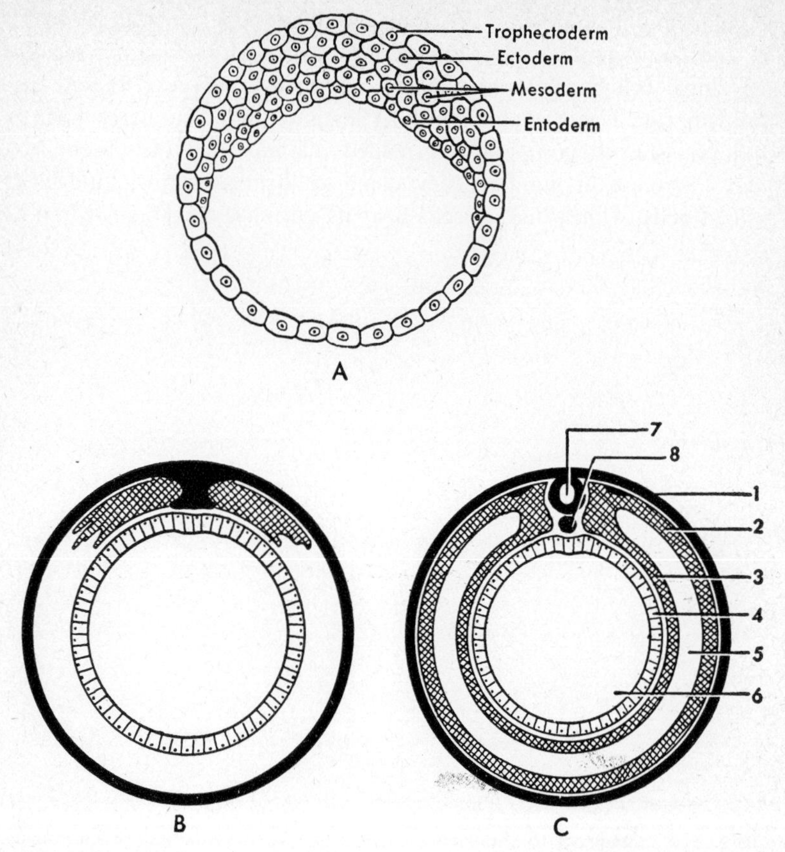

Fig. 378. Diagram of a section of an embryo, showing the beginning of tissue formation. *A,* the embryonic disk shows the cells arranged in layers. *B, C,* two diagrams of sections of embryos showing later stages of tissue formation. *Ectoderm* is shown in black, *entoderm* as cells, the two layers of *mesoderm* are cross-hatched. *1,* ectoderm; *2,* parietal mesoderm; *3,* visceral mesoderm; *4,* entoderm; *5,* future body cavity; *6,* enteron; *7,* neural tube; *8,* notochord.

mesoderm separates into two layers, the somatic, or parietal, layer lying next to the ectoderm, and the splanchnic, or visceral, layer lying next to the entoderm. These three layers of cells are called *germ layers* and also *embryonic* or *primary tissues.*[9] The ectoderm and the somatic mesoderm form the *somatopleure*, and from this the

[9] G. S. Dodd's *Essentials of Human Embryology.*

body wall is developed. The entoderm and the splanchnic mesoderm form the *splanchnopleure,* and from this the viscera are developed. Between the two layers of mesoderm is a cavity, the *celom,* which develops into the body cavity and which later becomes divided into the peritoneal cavity, the pleural cavities, and the pericardial cavity. Development of the germ layers into the differential tissues is called *histogenesis.* In this way, the general plan of vertebrate structure, that is, a tubular body wall enclosing a cavity in which lie the more or less tubular viscera, comes to be formed.

Histogenesis. The tissues derived from these germ layers are as follows:

Ectoderm	Epidermis and its appendages Epithelium of the beginning of gastropulmonary tract (nose, mouth) and the termination (anus) Tissues of the nervous system
Mesoderm	The connective and supporting tissues The muscular tissues Tissues of the vascular and lymphatic systems, including endothelium and circulating cells. Endothelium lines closed cavities of body Epithelium of a part of the genitourinary system
Entoderm	Epithelium of the gastropulmonary tract, except those portions of ectodermic origin at the beginning (mouth, nose) and the termination (anus)

From the layers of the mesoderm in its early history, the *mesenchyme* is formed. This is a network of irregular cells, loosely joined together. The spaces in this network contain a fluid. From the mesenchyme the supporting and circulatory tissues of the body are gradually developed. The ectoderm and entoderm probably contribute to the formation of the mesenchyme.

Organ formation. As the tissues are gradually differentiated, they become grouped into organs, mainly by local unequal rates of growth and the consequent production of pockets or folds, or, in other words, by invagination or evagination. This process is called *organogenesis.* As the embryo grows, the embryonic organs increase in size and become differentiated locally in various ways and organized into systems of organs. The primitive alimentary canal is folded into tubular form from the splanchnopleure; and the pancreas, liver, and respiratory system develop as pockets growing out from this tube. They are therefore lined with entoderm. The *neural tube* is infolded from the ectoderm and gives rise to the brain, the spinal cord, ganglia, and nerves. The *notochord*, or embryonic backbone, develops between the alimentary tube and the neural tube and later becomes replaced by the segmented vertebral column. Blood tubes develop in the mesenchyme as a network of channels lined with endothelium, which later differentiates into the blood and lymph vascular systems. The urogenital system develops from the intermediate mass of mesoderm in the dorsal region of the body.

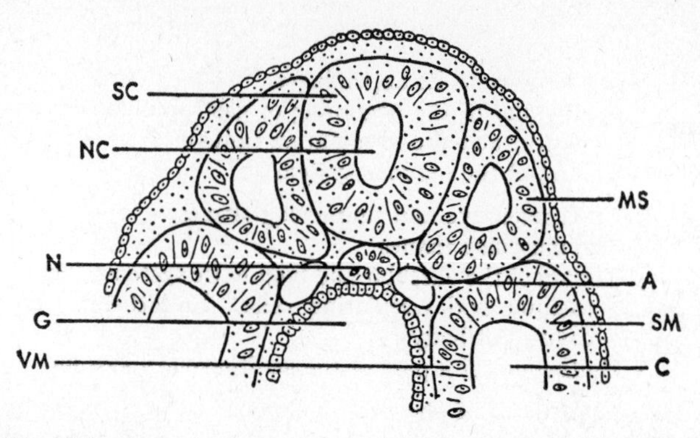

Fig. 379. Diagram of a portion of a cross section of a human embryo about 2 weeks of age. *A*, aorta; *C*, celom; *G*, midgut; *MS*, myotome, or mesoblastic somite; *N*, notochord; *NC*, central canal of neural tube; *SC*, wall of neural tube; *SM*, somatic mesoderm; *VM*, splanchnic mesoderm.

The association of the systems working harmoniously together produces the organism.

Implantation. During the early stages of the cleavage period, the embryo descends through the uterine tube into the uterus. About the eighth or tenth day after fertilization, it becomes embedded or implanted in the uterine mucosa, which grows over it. Certain membranes—maternal and fetal—are formed for the protection and nourishment of the developing fetus.

Maternal, or decidual, membranes. The mucosa, or lining membrane, of the uterus becomes differentiated into the *decidua basalis* (*serotina*), against which the embryo rests and which forms the maternal part of the placenta, the *decidua capsularis* (*reflexa*), which grows over the embryo, and the *decidua vera* (*parietalis*), the remainder of the uterine mucosa (Fig. 380). The decidual membranes, together with the fetal membranes, are discarded as the *afterbirth* soon after the extrusion of the child from the uterus.

The fetal membranes are the *chorion,* the *amnion,* the *yolk sac,* and the *allantois.* The *yolk sac* and *allantois* are pocketlike extensions of the ventral side of the embryo and are to be seen in the umbilical cord of a *young* embryo.[10] The *amnion* is a thin, transparent sac filled with fluid, which surrounds the embryo, keeping it moist and equalizing the varying pressures around it. The *chorion* is developed from the outer layer of the blastocyst, together with a layer of mesoderm, and surrounds the embryo. It develops villi, which erode the mucosa of the uterus, forming pits into which they grow. The villi extending into the decidua basalis become numerous, long, and highly branched. This portion of the chorion is called *chorion frondosum* and becomes the fetal part of the placenta.

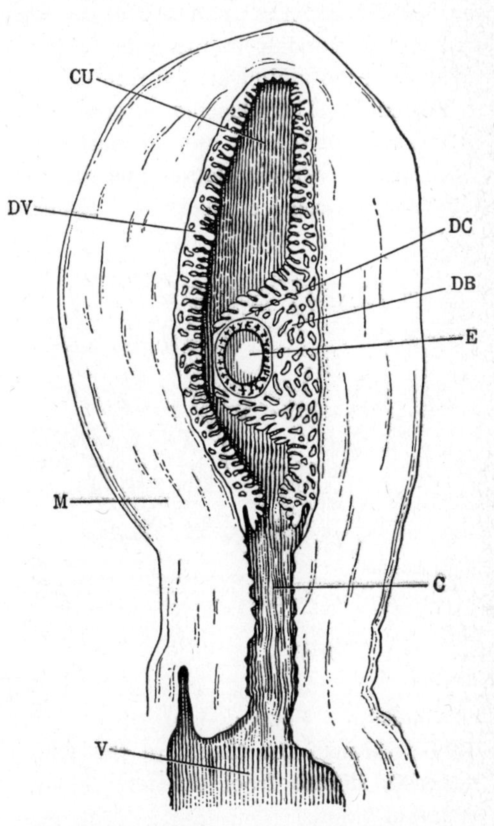

Fig. 380. Diagram of a longitudinal section of the gravid uterus, showing the relations of the implanted embryo and the uterine mucosa. *C,* cervix of uterus; *CU,* cavity of uterus; *DB,* decidua basalis; *DC,* decidua capsularis; *DV,* decidua vera; *E,* embryo about 2 weeks of age in chorion showing villi; *M,* muscle layers of wall of uterus; *V,* vagina.

[10] The yolk sac and allantois, though serving in a certain sense as fetal membranes in some species, can scarcely be called fetal membranes in man.

The villi on the other portions of the chorion do not usually continue to grow with the embryo and may even disappear, leaving this surface of the chorion smooth (*chorion laeve*). Through the *umbilical cord*, which attaches the fetus[11] to the placenta, two umbilical arteries carry fetal blood from the fetus to the capillaries of the fetal placenta, and the umbilical vein returns this blood to the fetus.

Placenta. The placenta is composed of an embryonic portion (*chorion frondosum*) and a maternal portion (*decidua basalis*). Each has its own circulation, and there is no connection between the fetal and maternal circulations. Exchange of materials takes place

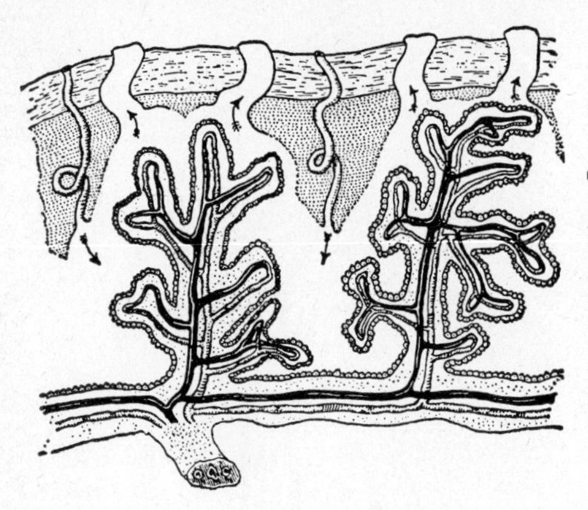

Fig. 381. Placenta with its blood supply. Cellular line separates maternal and embryonic portions. The embryonic portion is shown as two much-lobed villi enclosing blood vessels. Note umbilical cord containing umbilical vein and two arteries. Two uterine arteries and two uterine veins are indicated by the arrows. White area would contain maternal blood.

by diffusion. Food and oxygen in solution in the mother's blood diffuse through the tissues of the villi into the fetal blood; waste materials diffuse from the fetal blood through the villi into the maternal blood.

It is estimated that the total area of villi through which this diffusion can take place is about 6.5 sq m.

Early embryos. Figures 383 and 384 show human embryos of different ages. It will be noted that development takes place most rapidly in the head region and with increasing slowness toward the caudal end of the embryo, more rapidly also in the dorsal than in the ventral region. The embryo, at first a disk of cells, by local

[11] The term *embryo* is applied to the organism during the first 2 months of development. During the third month, the embryo begins to take on human appearance, and is called a *fetus*.

differential rates of growth, which result in folding, comes to have a roughly cylindrical form and is attached to the embryonic membranes by the body stalk. Among the earliest visible external structures are the *primitive groove*, an important area of rapid cell division and tissue differentiation lasting for a short time only; the *neural groove*, which by growth and infolding becomes the *neural tube*, from which

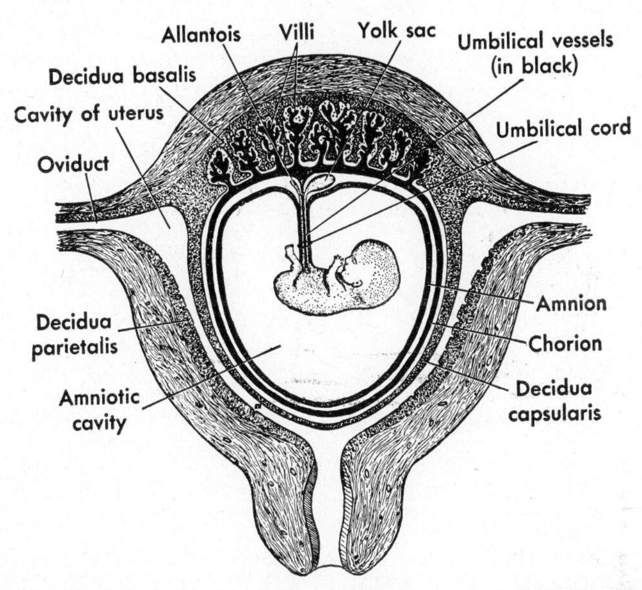

Fig. 382. Sectional view of human uterus with fetal membranes and their relationship to the uterus and the embryo. (A. F. Huettner, *Fundamentals of Comparative Embryology of the Vertebrates,* 1st ed. Courtesy of The Macmillan Company.)

the nervous system is developed; and the *somites*, paired masses of mesodermal tissue, showing through the ectoderm. The somites develop into the vertebrae, and the ribs, muscles, and connective tissues organized with them. For the development of other external features reference must be made to textbooks of embryology.

Intrauterine growth. During the period of intrauterine life growth takes place rapidly. From the union of the ovum, which is 0.2 mm ($\frac{1}{125}$ in.) in diameter, and the spermatozoon, which is much smaller, there is developed in 2 weeks' time an embryo which

(Text continued on p. 784.)

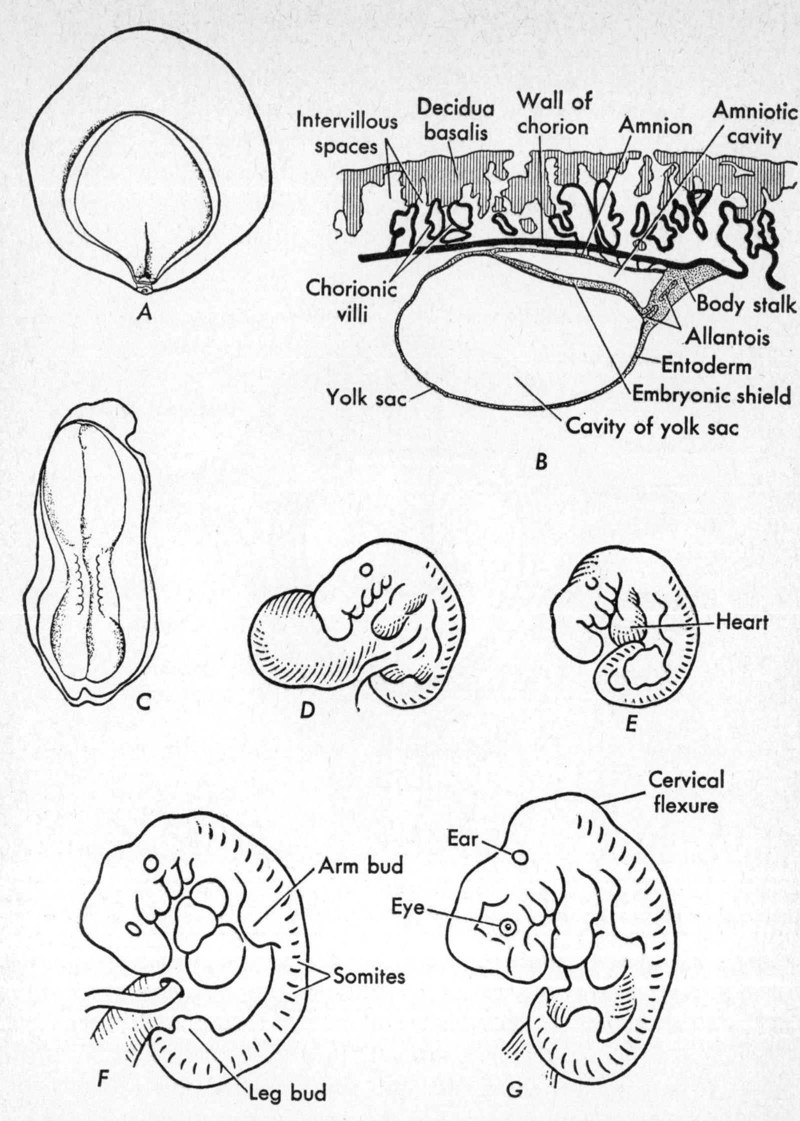

Fig. 383. Human embryos. *A, C,* dorsal view. *A,* shows *primitive* streak. *B,* human embryo (Mateer) described by Streeter, estimated to be 17 days old and 0.92 mm long. Formation of the body stalk. The amnion is detaching itself from the chorion. (A. F. Huettner, *Fundamentals of Comparative Embryology of the Vertebrates,* 1st ed. Courtesy of The Macmillan Company.) *C* (1.38 mm long) shows primitive streak, *neural groove,* and the beginning of *somites.* Amnion has been cut away to show embryo. *D, E, F, G,* ages from 2 to 5 weeks.

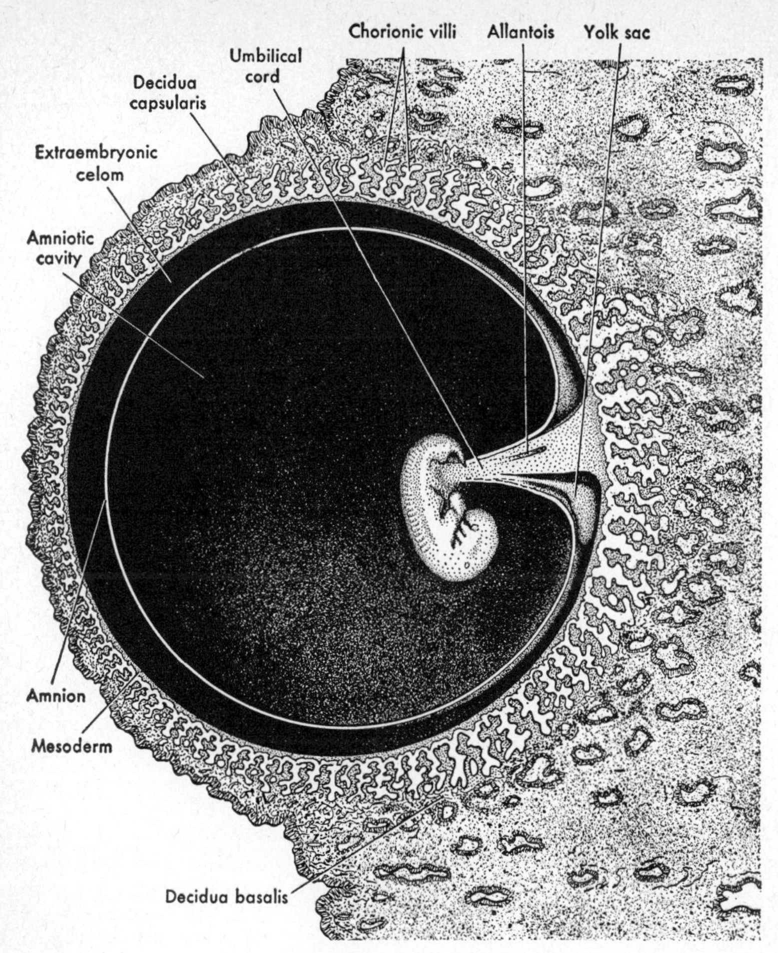

Fig. 384. Stereogram of a blastocyst containing a human embryo, 33 days old and 5 mm long. The blastocyst has formed an elevation of the surface of the uterine mucosa. The extraembryonic celom has been almost entirely replaced by the amniotic cavity. The body stalk has been converted into the umbilical cord by being covered with the amnion. The rudimentary umbilical cord contains the allantois, yolk sac, and the umbilical vessels. The latter are not shown. The chorionic villi in the decidua capsularis begin to atrophy and those in the decidua basalis are growing larger and show a greater branching. (A. F. Huettner, *Fundamentals of Comparative Embryology of the Vertebrates,* 1st ed. Courtesy of The Macmillan Company.)

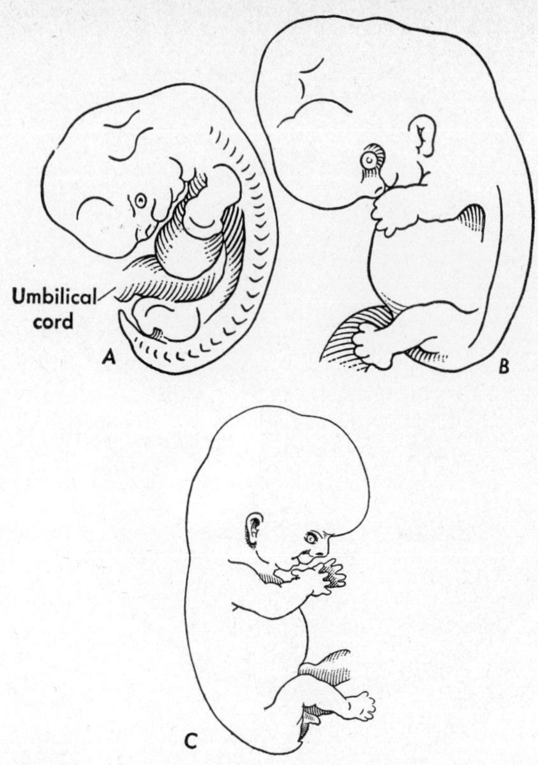

Fig. 385. Human embryos. Approximate age of *A* and *B*, 6 to 7 weeks. *C*, age about 2 months, length about 25 mm. Human form established, fetal period of development begun. Cervical flexure nearly gone; head, though still relatively large, lifted; body straightened. Neck well marked though short. Face developing; eyelids, nose, lips, external ears, cheeks appear. Ventral body wall formed, prominence due to the heart less conspicuous. Fore limbs have appearance of arms, thumbs well marked off from fingers, shoulders begin to appear. Legs smaller than arms; soles of feet close together.

is about 6.25 mm ($\frac{1}{4}$ in.). At the end of 4 weeks it is about 12.5 mm ($\frac{1}{2}$ in.) long, and at 2 months it is 25 mm long and is called a fetus, because it begins to have the appearance of a human being. The usual duration of pregnancy is about 10 lunar, or 9 calendar, months, estimated from the first day of the last menstrual period. At the end of 6 months the fetus is sufficiently developed to live outside the mother's body, but it is frail and requires a great deal of care.

This table shows growth in weight and in height during intrauterine life.

Age of Embryo	*Weight*	*Crown-heel Length*
8 weeks	3 gm (about $\frac{1}{10}$ oz)	2.5 cm (about 1 in.)
12 weeks	36 gm (about 1 oz)	9 cm (about 3.6 in.)
20 weeks	330 gm (about 11 oz)	25 cm (about 10 in.)
28 weeks	1,000 gm (about 2 lb)	35 cm (about 14 in.)
36 weeks	2,400 gm (about 5 lb)	45 cm (about 18 in.)
40 weeks	3,200 gm (about $7\frac{1}{4}$ lb)	50 cm (about 20 in.)

With growth of the fetus the uterus comes to occupy more and more abdominal space.

This table shows the position of the top of the uterus at different ages of the fetus.

By the end of the	*The top of the uterus has reached a level*
fourth month	halfway between the symphysis and umbilicus
sixth month	even with the umbilicus
ninth month	about even with the end of the sternum
tenth month	about that of the eighth month—a little below the sternum

FETAL CIRCULATION

Certain structures are necessary to the performance of fetal circulation but cease to be of use after birth. They are as follows:

Foramen ovale. An opening between the two atria. It furnishes direct communication between them.

Ductus arteriosus. A blood vessel connecting the pulmonary artery with the aorta.

Ductus venosus. A blood vessel connecting the umbilical vein and the inferior vena cava.

The placenta and umbilical cord. The fetal placenta is a mass of tissue, rich in blood vessels, which is in close contact with the lining of the uterus. The umbilical cord unites the placenta with the navel of the child. The cord is made up of two arteries and one large vein protected by Wharton's jelly. The arteries are branches of the arterial system of the fetus and carry blood to the fetal placenta, where it is separated by the thinnest of walls from the maternal blood in the blood vessels of the uterus. The usual distinctions between arterial and venous blood cannot be recognized, as the blood of the fetus is never up to the arterial standard of the mother. The best blood is that which has been improved by effecting exchanges with the blood in the uterine vessels and is carried from the placenta to the fetus by the umbilical vein. By means of the placenta the fetus obtains oxygen; the placenta is also the seat of the absorption of food and the unloading of wastes. Consequently it represents a combination of organs and serves as the respiratory, digestive, and excretory mechanism for the fetus.

Course of the blood. The blood is carried from the placenta along the umbilical cord by the umbilical vein. Entering the fetus, it is conveyed into the ascending vena cava partly through the liver but chiefly through the ductus venosus, which connects these two vessels. From the ascending vena cava it enters the right atrium, passes through the foramen ovale into the left atrium, thence into the left ventricle, and out through the aorta, which distributes it principally to the head and upper extremities. The blood from the head and upper extremities returns by the descending vena cava to the right atrium, then passes into the right ventricle and out through the pulmonary artery to the lungs. As the lungs in the fetus are solid, they require very little blood (only for nutrition); the greater part of the blood passes through the ductus arteriosus into the descending aorta, where, mixing with the blood delivered to the aorta by the left ventricle, it descends to supply the viscera of the abdomen and pelvis and the lower extremities. The greater amount of this blood is carried to the placenta by the two umbilical arteries, but a small amount passes back into the ascending vena cava and mixes with the blood from the placenta.

From this description of the fetal circulation, it follows:

That the placenta serves the purpose of a respiratory, nutritive, and excretory organ.

That the liver receives blood directly from the placenta; hence the large size of the liver at birth.

That the blood from the placenta passes almost directly into the arch of the aorta and is distributed by its branches to the head and upper extremities.

That the blood in the descending aorta is chiefly derived from that which has already circulated in the upper extremities and, mixed with only a small quantity from the left ventricle, is distributed to the lower extremities.

Changes in the vascular system at birth. From the foregoing description it is obvious that at birth very important changes take place:

With the first inspiration the lungs expand, the pulmonary vessels are dilated, and blood flows into the pulmonary arteries from the right ventricle.

The ductus arteriosus becomes obliterated, eventually forming fibrous cords.

The blood clots in the umbilical vein, which eventually becomes obliterated, forming the ligamentum teres of the liver.

The ductus venosus becomes obliterated and eventually forms the ligamentum venosum of the liver.

Pressure in the right atrium decreases, and pressure in the left atrium and ventricle increases. This results in the closure of the foramen ovale, which occurs immediately after birth. Complete anatomical obliteration occurs gradually. Occasionally the foramen ovale does not become completely obliterated, venous blood enters the arterial system, and a "blue baby" is the result.

The branches of the arteries which extend to the umbilicus become obliterated and form fibrous cords, the lateral umbilical ligaments. The pelvic branches of the arteries remain open and constitute the hypogastric arteries and the first part of the superior vesical arteries of the adult.

Parturition takes place, normally, at about 280 days from the beginning of the last menstrual period and results in the birth of the child. It is brought about by the periodic contractions of the muscles of the wall of the uterus, aided by contractions of the muscles of the wall of the abdomen.

Various investigations into the causes of the onset of parturition at the end of the tenth lunar month and into the process of *labor* have been carried on. At present, however, no single factor or sequence of groups of factors is certainly known to initiate and control the processes. There is widespread belief in the theory of hormonal control of parturition, but this theory is not without its opponents. Immediately preceding delivery, there is an increase in excretion of free theelin and an abrupt decrease in the amount of combined theelin excreted. This may indicate that the presence of theelin controls parturition. It is known that injection of theelin

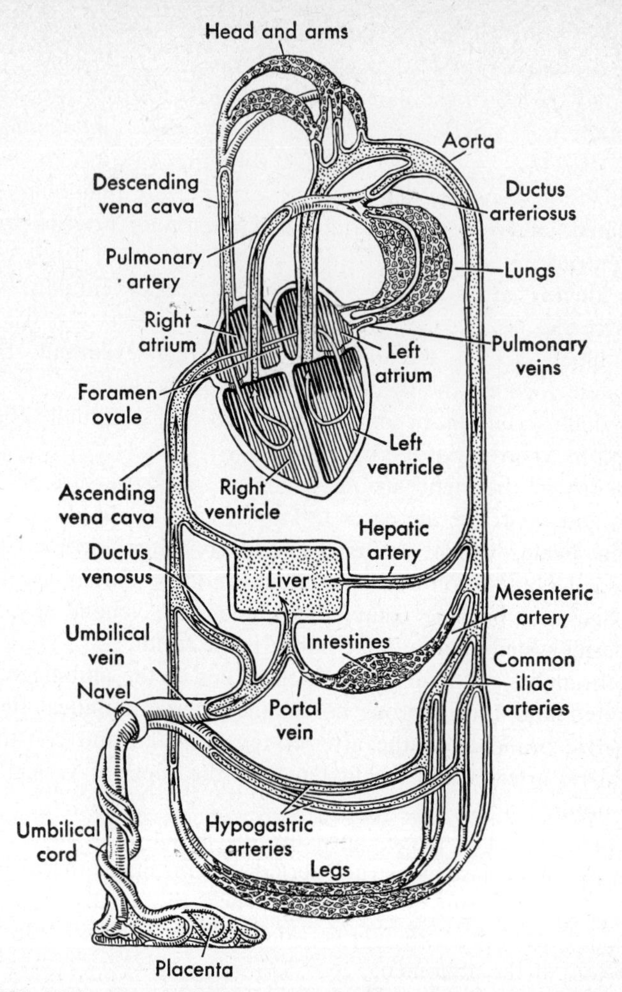

Fig. 386. Diagram showing course of fetal circulation through hypogastric arteries, ductus venosus, ductus arteriosus, and the foramen ovale. (Modified from *The American Textbook on Obstetrics.*)

may lead to abortion in pregnant animals. Although the oxytocic factor of pituitary extract is widely used to bring on rhythmic contractions of the uterus in labor, removal of the pituitary in laboratory animals interferes in no way with the onset of normal labor and parturition. Therefore it is doubtful if the pituitary *normally* plays a part in parturition in the human.

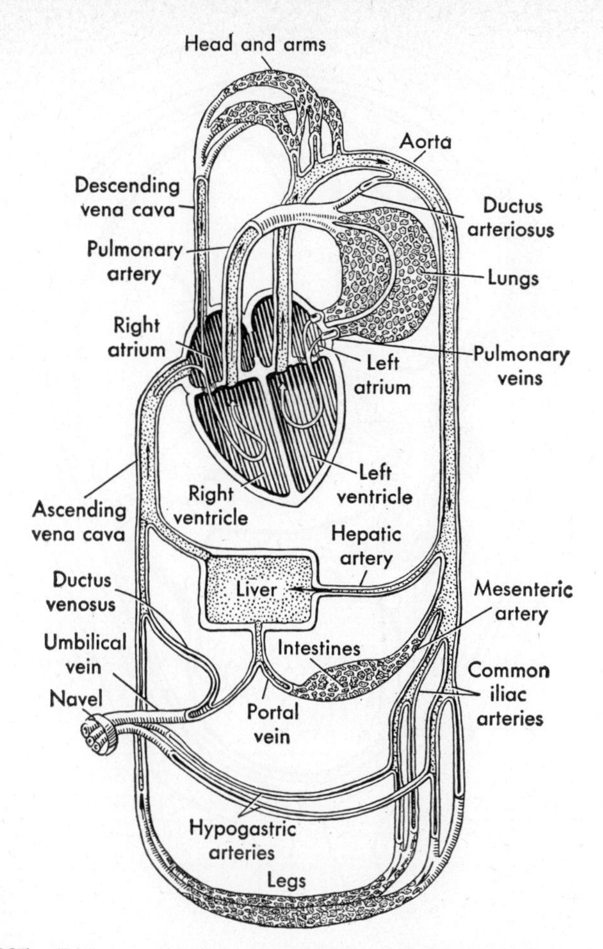

Fig. 387. Diagram showing circulation of the blood after birth, with hypogastric arteries, ductus venosus, ductus arteriosus, and foramen ovale *in process of obliteration* and pulmonary circulation greatly increased. The ductus arteriosus and the ductus venosus become the ligamentum arteriosus and the ligamentum venosus; the umbilical vein becomes the ligamentum teres. (Modified from *The American Textbook on Obstetrics.*)

The duration of labor is variable, but the average length of time is about 12 to 18 hours. During the first stage, or *stage of dilation,* the cervix of the uterus is dilated, and as a rule the amnion is ruptured and the amniotic fluid expelled; during the second stage, or *stage of descent,* the child descends through the vagina and is ex-

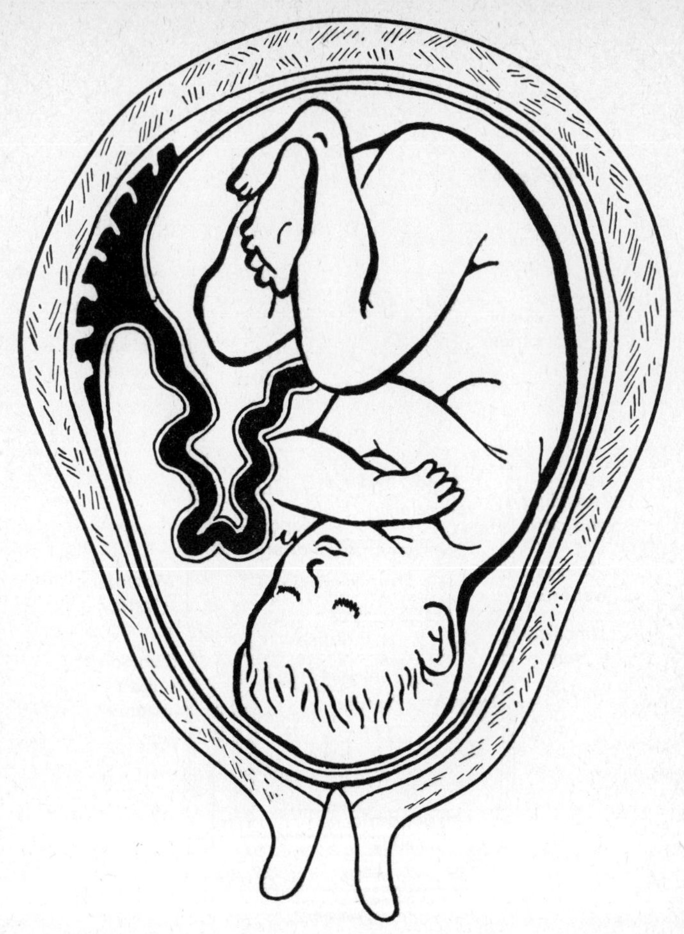

Fig. 388. Diagram of a section of uterus with well-developed fetus. The amnion is represented by a thin black line bounding the amniotic fluid in which the fetus lies. The heavy black line is the chorion. The fetal piacenta and umbilical cord are shown at the left.

pelled; during the third stage, or *placental stage,* the fetal membranes are expelled.

Involution is the process of rapid decrease in the size of the uterus. It is brought about by a gradual autolysis, or "self-digestion," of the uterine wall and requires from 6 to 8 weeks. During this time the uterus resumes its original position in the pelvic cavity and approximately its original size.

SUMMARY

Life Processes
- Phenomena of activity and growth
- Early life—anabolic processes in ascendancy, individual gains in weight and stature
- Middle life—anabolic and catabolic changes balance each other, and individual holds his own
- Old age—catabolic changes in ascendancy; final result is death

REPRODUCTION

Reproduction
- Means by which new life is brought into existence
- Sexual type depends upon the union of two cells, one of which is produced by the male and one by the female organism

Puberty
- Age at which sex organs become matured and functional
- Boy begins to develop into man—in temperate climates about 15 years of age—essential feature is production of mature spermatozoa
- Period marked by gradual appearance of secondary sex characteristics
- Girl begins to develop into woman—in temperate climates about 14 years of age—essential feature is production of mature ova
- Period marked by gradual appearance of secondary sex characteristics. Function of menstruation starts

Adolescence—period from puberty to early or middle twenties

Male Organs of Generation
- 2 testes which produce spermatozoa and an internal secretion
- 2 seminal ducts (ducti deferentes, or vasa deferentia)
- 2 seminal vesicles
- 2 ejaculatory ducts
- 2 spermatic cords
- Scrotum
- Penis and urethra
- Prostate gland
- 2 bulbourethral glands (Cowper's)

Testes
- Two glandular organs which produce the spermatozoa
- **Structure**
  - **Testis**—ovoid body covered by fibrous tissue. Central portion consists of irregular cavities filled with seminiferous tubules and blood vessels
  - **Epididymis**—tortuous tubule, forms long, narrow body which lies along upper posterior portion of testis

Testes
- **Location**
  - In early fetal life in abdomen. Before birth are normally drawn downward to scrotum and are suspended by spermatic cord
- **Function**
  - Production of spermatozoa
  - Production of internal secretion

Seminal Ducts
- Each duct is a continuation of epididymis and is excretory duct of testis. Serves to connect epididymis and seminal vesicle of each side

Seminal Vesicles
- Two membranous pouches located between bladder and rectum. Connect seminal ducts with ejaculatory ducts
- **Function**—add a secretion to semen

Ejaculatory Ducts
- Formed by union of seminal vesicle and seminal duct of each side
- They descend, converge, pass between lobes of prostate gland, and open into urethra

Spermatic Cords
- Each cord consists of a seminal duct, an artery, veins forming a plexus called the pampiniform plexus, lymphatics, nerves, and connecting areolar tissue covered with fascia
- Extend from inguinal ring to back of testes

Scrotum
- Pouch which contains testes and part of each spermatic cord
- **Structure**
  - Thin, dark skin, disposed in folds, or rugae
  - Dartos—consists of plain muscle fibers and numerous blood vessels

Penis
- Organ suspended from front and sides of pubic arch
- Consists of three cylindrical bodies of cavernous tissue
  - Two corpora cavernosa penis
  - A corpus cavernosum urethrae
- Bound together by fibrous tissue
- Covered with skin that is continuous with that covering scrotum, perineum, and the pubes
- Urethra—extends from urethral orifice in bladder through corpus cavernosum urethrae—17.5–20 cm long
- Expansion at lower extremity—glans penis—urethral orifice—covered by foreskin, or prepuce

The Prostate
- Situated immediately below internal urethral orifice
- Size of chestnut
- **Consists of**
  - Fibrous capsule containing glandular and muscular tissue. Glandular tissue consists of tubules which empty into urethra
- **Function**—secretion of prostatic fluid

Bulbo-urethral Glands
- Located one on each side of prostate gland—about the size of a pea—1-in. duct terminates in wall of urethra
- **Function**—secretion of a fluid which forms part of seminal fluid

Semen
- Fluid derived from the various sex glands in the male
- 70,000,000 spermatozoa per cu cm. Nerve fibers from pudendal plexus control discharge

Female Organs of Generation
- 2 ovaries, which produce ova and internal secretions
- 2 uterine tubes
- Uterus
- Vagina
- External genitals
- 2 breasts

Ovaries
- 2 almond-shaped glandular bodies
- **Attached**
  - To back of broad ligament
  - To uterus—by ligament of ovary
  - To tubes—by fimbriae
- **Size about**
  - $1\frac{1}{2}$ in. long
  - $\frac{3}{4}$ in. wide
  - $\frac{1}{3}$ in. thick
- **Weight**—2–3.5 gm
- **Structure**
  - Layer of germinal epithelium
  - Stroma, or meshwork, of cells
  - Graafian follicles
    - In early life—vesicles about 0.25 mm in diameter
    - Between puberty and menopause—mature follicles and corpora lutea
    - Outer coat fibrous tissue derived from stroma
    - Middle coat—theca interna derived from stroma
    - An inner coat called *membrana granulosa*
- **Function**
  - Produce, develop, mature, and discharge ova
  - Form internal secretions
  - One secreted by Graafian follicles—called theelin, related to menstrual cycle
  - One secreted by corpus luteum—called luteal secretion. Essential to implantation

Uterine, or Fallopian, Tubes
- **Location**
  - Enclosed in layers of broad ligament
  - Extend from upper angles of uterus to sides of pelvis
- **Divisions**
  - Isthmus—or inner constricted portion near uterus
  - Ampulla—dilated portion which curves over ovary
  - Infundibulum—trumpet-shaped extremity—fimbriae

- **Uterine, or Fallopian, Tubes**
  - **Three coats**
    - External, or serous
    - Middle, or muscular
    - Internal, or mucous, arranged in longitudinal folds and covered with cilia
  - **Function**—convey ova to uterus
- **Uterus**
  - Hollow, pear-shaped, muscular organ, placed in pelvis between bladder and rectum
  - **Divisions**
    - Fundus—rounded upper portion, above entrance of tubes
    - Body—portion below fundus, above isthmus
    - Cervix—lower and smaller portion which extends into vagina
  - **Three coats**
    - External, or serous, derived from peritoneum, covers intestinal surface and anterior surface to beginning of cervix
    - Muscular
      - Circular layer
      - Longitudinal layer
      - Spiral layer
      - Interlaced in every direction
    - Mucous membrane, lines the uterus
  - **Blood vessels**
    - Uterine arteries from hypogastrics
    - Ovarian arteries from aorta
    - Remarkable for tortuous course and frequent anastomoses
  - **Ligaments**
    - Broad, or lateral—two layers of serous membrane
    - Anterior—peritoneal fold between bladder and uterus
    - Posterior—peritoneal fold behind uterus
    - Round—two fibromuscular cords
    - Uterosacral—two partly serous, partly muscular ligaments
  - **Function**—to receive ovum and, if it becomes fertilized, to retain it until developed and then to expel it
- **Vagina**
  - Extends from uterus to vulva
  - **Coats**
    - Internal mucous lining arranged in rugae
    - Layer of submucous tissue
    - Muscular coat
  - **Location**—placed in front of rectum, behind bladder
- **External Genitals**
  - **Mons pubis**—a cushion of areolar, fibrous, and adipose tissue, in front of pubic symphysis, covered with skin and after puberty with hair
  - **Labia majora**—two folds that extend from the mons pubis to within an inch of the anus
  - **Labia minora**—two folds situated between the labia majora

- **External Genitals**
  - **Clitoris**—small body, situated at apex of the triangle formed by junction of labia minora. Well supplied with nerves and blood vessels
  - **Vestibule**—cleft between labia minora
  - **Hymen**—fold of mucous membrane that surrounds vaginal orifice
  - **Glands**—greater vestibular—oval bodies situated on either side of vagina
- **Physiology of Generative Organs**
  - **Function**
    - Formation and development of ovum
    - Retention and sustenance of fecundated ovum until able to live outside of body.
    - Expulsion of fetus
  - **Ovulation**
    - Process of development and maturation of follicle and ovum, and discharge of ovum
  - **Menstruation**
    - A flow of blood from the uterus. Occurs on an average every 28 days. Extends from puberty (14 years) to the menopause, or climacteric (about 45 years). This represents the childbearing period of a woman's life
    - Changes in connection with menstrual cycle
      - (1) Period of repair and proliferation
      - (2) Premenstrual period (secretory stage)
      - (3) Period of degeneration or menstruation
  - **Menopause**—physiological cessation of menstrual flow
- **Mammary Glands**
  - **Function**—to secrete milk to nourish infant
  - **Location**
    - Extend from second to sixth rib and from sternum to armpit
  - **Structure**
    - Outer surface convex—papilla projects from center, called nipple—contains openings of milk ducts. Nipple surrounded by areola
    - Consists of connective-tissue framework which divides the gland into about 20 lobes
    - Lobes are subdivided into lobules
    - Lobules are made up of terminal tubules of the duct
    - Each lobe possesses its own excretory duct, which is called lactiferous and is sacculated
  - **Blood vessels**
    - Axillary
    - Internal mammary
    - Intercostal
  - **Nerves**—derived from cutaneous branches of fourth, fifth, and sixth thoracic nerves

Mammary Glands
- **Development**
  - Primary development and later functioning suggest an intimate connection between these glands, the uterus, ovaries, and pituitary body
  - Formerly thought functional development that follows conception was probably due to hormone produced by mature corpus luteum. Recent experiments on animals seem to indicate that these effects are readily produced by pituitary hormone
- **Secretion of milk**
  - Water, Salts, Sugar — Secreted from blood
  - Proteins, Fat — Formed by disintegration of cells lining lactiferous tubules

Colostrum—thin, yellowish fluid secreted during first few days of lactation

Milk
- **Composition**

	Human, %	*Cow's, %*
Water	88.4	87.1
Proteins	1.5	3.2
Fat	3.3	3.9
Lactose	6.5	4.9
Salts	0.3	0.9

- **Differences**
  - Different relative proportions
  - Difference in proteins
    - Human: Caseinogen, one-third; Lactalbumin, two-thirds
    - Cow's: Caseinogen, five-sixths; Lactalbumin, one-sixth
  - Difference in curds
    - Human—small flocculi
    - Cow's—heavy curds
  - Difference in reaction
    - Human—practically neutral, pH 7.0–7.2
    - Cow's—slightly acid, pH 6.6–6.8
  - Bacterial count
    - Human—sterile
    - Cow's—sterile, but due to handling, it contains large number of bacteria
  - Germicidal power
    - Human—antitoxins and antibacterial substances found in mother's blood
    - Cow's—may have germicidal value, but it is lost by time milk reaches infant

EMBRYOLOGY

The new organism arises by fusion of parental cells—ova and spermatozoa. Growth is accomplished by the multiplication of cells, differentiation of cells, and accumulation of material between them.

Germinal Tissue
- *Somatic tissues*—all tissues except reproductive tissues
- *Germinal tissues*
  - *Ovaries*—produce ova
  - *Testes*—produce spermatozoa

Germ Cells, or Gametes
- *Ova*, or *egg cells*, $\frac{1}{125}$ in. in diameter. Surrounded by Graafian follicles
- *Ovulation*—rupture of Graafian follicles and discharge of ova from ovary. Probably occurs about midway between menstrual periods
- *Spermatozoa*, $\frac{1}{500}$ in. long. Head is nucleus

Maturation of Germ Cells
- Human cells have 48 chromosomes (or 24 pairs)
- During maturation the number of chromosomes in the germ cells is reduced to one-half that number
- Each mature ovum or spermatozoon has, therefore, 24 chromosomes—one member of each of the 24 pairs
- Maturation of ova probably not complete until after fertilization
- Maturation of spermatozoa complete before they are detached from the walls of the seminiferous tubules

Fertilization
- Term applied to penetration of ovum by spermatozoon and fusion of their nuclei
- Restores original number of chromosomes—48. One-half the chromosomes contributed by each germ cell
- Activates the ovum. Cell division begins

Heredity
- Transmission of potential characteristics from parents to children
- One member of each pair of chromosomes from each parent
- Paired chromosomes contain factors for similar or alternative traits
- Dominance may be complete, partial, or absent
- Factors for recessive characteristics may be carried through generations, developing into characteristics only when the dominant factors are not present

Embryonic Development
- **Cleavage and formation of germ layers**
  - *Fertilization* activates the ovum—cell multiplication begun
  - *Morula*—a group of cells resulting from cell multiplication
  - *Blastocyst*, or modified blastula—a hollow ball of cells, with disk of cells near the periphery at one point—the *blastoderm*
  - *Gastrula*—an embryo of two layers, ectoderm and entoderm. *Ectoderm* is the outer layer of blastoderm; *entoderm* is the inner layer of blastoderm

- **Embryonic Development**
  - **Cleavage and formation of germ layers**
    - *Mesoderm* develops between ectoderm and entoderm. Somatic layer of mesoderm and the ectoderm give rise to the body wall (*somatopleure*). Splanchnic layer and entoderm give rise to viscera (*splanchnopleure*)
    - *Celom*—the cavity between the layers of mesoderm becomes divided into the peritoneal, pleural, and pericardial cavities
  - **Tissue formation**
    - **Ectoderm**
      - Epidermis and its appendages
      - Epithelium of beginning of gastropulmonary tract (nose, mouth) and termination (anus)
      - Tissues of nervous system
    - **Mesoderm**
      - Connective and supporting tissues
      - Muscular tissues
      - Tissues of vascular and lymphatic systems, including endothelium and circulating cells. Endothelium lines closed cavities of body
      - Epithelium of a part of genitourinary system
    - **Entoderm**
      - Epithelium of gastropulmonary tract, except those portions of ectodermic origin at beginning (mouth, nose) and at termination (anus)
    - **Mesenchyme**
      - Loosely arranged network of cells derived from layers of mesoderm in dorsal region where tissue is not completely separated into layers
      - Develops into supporting and circulatory tissue

- **Embryonic Development**
  - **Organ formation**
    - Tissues become pocketed or folded by unequal rates of growth, forming organs
    - Embryonic organs grow and become differentiated into systems
    - Association of systems produces organism
    - Primitive alimentary canal formed from entoderm. Pancreas, liver, and respiratory organs develop as pockets growing out from it
    - Neural tube developed from ectoderm. Gives rise to brain, spinal cord, and nerves
    - *Notochord*, or embryonic backbone, a rod of cells between neural tube and alimentary canal. It is later replaced by segmented vertebral column
    - Blood and lymph vascular systems develop from mesenchyme
- **Implantation**
  - Embryo embedded in uterine mucosa about 8–10 days after fertilization
  - **Maternal, or decidual, membranes**
    - *Decidua basalis*—uterine mucosa, in which embryo lies. It forms maternal portion of placenta
    - *Decidua capsularis*—uterine mucosa grown over embryo
    - *Decidua vera*—the other portions of the uterine mucosa
    - Deciduae discarded with fetal membranes after birth of child, as the *afterbirth*
  - **Fetal membranes**
    - *Yolk sac* and *allantois* are pocketlike extensions of ventral side of embryo. They are to be seen in umbilical cord of a *young* embryo
    - *Amnion*—a thin, fluid-filled sac surrounding embryo
    - *Chorion*—developed from outer layer of blastocyst and mesoderm. It forms villi which grow into *decidua basalis*. *Chorion frondosum* is that portion with numerous, long, much-branched villi. It forms fetal portion of placenta. *Chorion laeve* is the other portion of chorion

- **Implantation**
  - **Placenta**
    - *Chorion frondosum*, the fetal portion.
    - *Decidua basalis*, the maternal portion
    - There is no connection between fetal and maternal blood. Exchange of materials is by diffusion and absorption. Food and oxygen diffuse through tissues of villi from maternal blood into fetal blood. Wastes diffuse through tissues of villi from fetal to maternal blood
    - *Umbilical cord* attaches fetus to placenta. It contains two umbilical arteries and an umbilical vein
- **Development of Form**
  - Embryos develop more rapidly at head end and with increasing slowness toward caudal end—more rapidly in dorsal than in ventral region
  - **External structures of young embryos**
    - *Primitive groove*—area of rapid cell multiplication and differentiation lasts for a short time only
    - *Neural groove*—becomes neural tube and develops into brain, spinal cord, and nerves
    - *Somites*—paired masses of mesoderm, which develop into vertebrae and the ribs, muscles, and connective tissues organized with them
- **Fetal Circulation**
  - Direct communication between right and left atrium by means of foramen ovale
  - Direct communication between pulmonary artery and aorta through the ductus arteriosus
  - Direct communication between umbilical vein and inferior vena cava through the ductus venosus
  - Oxygen and nutritive substances obtained from placenta
  - Waste eliminated by placenta
- **Changes in Vascular System at Birth**
  - Umbilical vein and ductus venosus become fibrous cords
  - Respiration stimulates pulmonary circulation; this raises the blood pressure in left atrium and closes foramen ovale soon after birth
  - Ductus arteriosus becomes a fibrous cord
  - Branches of the hypogastric arteries become fibrous cords

Parturition—about 280 days from beginning of last menses

- Stages of labor
  - Stage of dilation
  - Stage of descent
  - Placental stage

Involution—return of uterus to approximately normal size following parturition

REFERENCE BOOKS AND BOOKS FOR FURTHER STUDY

Anatomy

Buchanon, A. R. *Functional Neuro-Anatomy.* Lea and Febiger, Philadelphia.

Callander, C. Latimer. *Surgical Anatomy.* W. B. Saunders Company, Philadelphia.

Cunningham, D. J. *Textbook of Anatomy.* Edited by J. C. Brash and E. B. Jamieson. Oxford University Press, New York.

Drinker, C. K., and Joffey, J. M. *Lymphatics, Lymph, and Lymphoid Tissue.* Harvard University Press, Cambridge, Mass.

Elliott, H. C. *Textbook of the Nervous System.* J. B. Lippincott Company, Philadelphia.

Eycleshymer, A. C., and Jones, T. S. *Hand-Atlas of Clinical Anatomy.* Lea and Febiger, Philadelphia.

Eycleshymer, A. C., and Shoemaker, D. M. *A Cross-Section Anatomy.* D. Appleton and Company, Inc., New York.

Gardner, E. *Fundamentals of Neurology.* W. B. Saunders Company, Philadelphia.

Grant, J. C. *An Atlas of Anatomy.* The Williams and Wilkins Company, Baltimore.

Grant, J. C. *A Method of Anatomy.* William Wood and Company, Baltimore.

Gray's Anatomy. Edited by W. H. Lewis. Lea and Febiger, Philadelphia.

Howell, A. B. *Gross Anatomy.* D. Appleton-Century Company, New York.

Krieg, W. J. S. *Functional Neuroanatomy.* The Blakiston Company, Philadelphia.

Krogh, A. *Anatomy and Physiology of the Capillaries.* Yale University Press, New Haven.

Kuntz, A. *The Autonomic Nervous System.* Lea and Febiger, Philadelphia.

Kuntz, A. *Textbook of Neuro-Anatomy.* Lea and Febiger, Philadelphia.

Larsell, O. *Anatomy of the Nervous System.* D. Appleton-Century Company, New York.

Mettler, F. A. *Neuroanatomy.* C. V. Mosby Company, St. Louis.

Netter, F. H. *The Nervous System,* Volume I. Ciba Pharmaceutical Products Company, Summit, N. J.

Penfield, W., and Rasmussen, T. *The Cerebral Cortex of Man.* The Macmillan Company, New York.

Piersol, G. A. *Human Anatomy*. Edited by G. C. Huber. J. B. Lippincott Company, Philadelphia.

Quiring, D. P. *Collateral Circulation*. Lea and Febiger, Philadelphia.

Quiring, D. P. *The Head, Neck and Trunk*. Lea and Febiger, Philadelphia.

Quiring, D. P., and Boroush, E. L. *The Extremities*. Lea and Febiger, Philadelphia.

Ranson, S. W., and Clark, S. L. *Anatomy of the Nervous System*. W. B. Saunders Company, Philadelphia.

Rasmussen, A. T. *Principal Nervous Pathways*. The Macmillan Company, New York.

Schaeffer, J. P. *Morris' Human Anatomy*. The Blakiston Company, Philadelphia.

Sobotta, J., and McMurrich, J. P. *Atlas of Human Anatomy* (3 vols.). G. E. Stechert and Company, New York.

Spalteholz, W. *Hand Atlas of Human Anatomy*. J. B. Lippincott Company, Philadelphia.

Strong, O. S., and Elwyn, A. *Human Neuroanatomy*. The Williams and Wilkins Company, Baltimore.

Toldt, C. *An Atlas of Human Anatomy* (2 vols.). The Macmillan Company, New York.

Physiology

Alvarez, W. C. *The Mechanics of the Digestive Tract*. Paul B. Hoeber, Inc., New York.

Bard, P. *Macleod's Physiology in Modern Medicine*. The Blakiston Company, Philadelphia.

Barnes, T. C. *Textbook of General Physiology*. The Blakiston Company, Philadelphia.

Best, C. H., and Taylor, N. B. *The Physiological Basis of Medical Practice*. William Wood and Company, Baltimore.

Cannon, W. B. *Bodily Changes in Pain, Hunger, Fear, and Rage*. D. Appleton-Century Company, New York.

Cannon, W. B. *The Wisdom of the Body*. W. W. Norton and Company, New York.

Carlson, A. J., and Johnson, V. *The Machinery of the Body*. University of Chicago Press, Chicago.

Fulton, J. F. *Functional Localization in Relation to Frontal Lobotomy*, Oxford University Press, New York.

Fulton, J. F. *Howell's Textbook of Physiology*. W. B. Saunders Company, Philadelphia.

Fulton, J. F. *Physiology of the Nervous System*. Oxford University Press, New York.

Gregg, D. E. *Coronary Circulation in Health and Disease*. Lea and Febiger, Philadelphia.

Heilbrunn, L. V. *An Outline of General Physiology.* W. B. Saunders Company, Philadelphia.

Henderson, L. J. *Blood: A Study in General Physiology.* Yale University Press, New Haven.

Livingston, W. K. *Pain Mechanisms; A Physiologic Interpretation of Causalgia and Its Related States.* The Macmillan Company, New York.

McDowell, R. J. S. *Handbook of Physiology and Biochemistry.* The Blakiston Company, Philadelphia.

Martin, E. G. *The Human Body.* Henry Holt and Company, Inc., New York.

Mitchell, P. H. *A Textbook of General Physiology.* McGraw-Hill Book Company, Inc., New York.

Rogers, C. G. *Textbook of Comparative Physiology.* McGraw-Hill Book Company, Inc., New York.

Schneider, E. C., and Karpovich, P. *Physiology of Muscular Exercise.* W. B. Saunders Company, Philadelphia.

Schoenheimer, R. *Dynamic State of Body Constituents.* Harvard University Press, Cambridge, Mass.

Starling, E. H. *Principles of Human Physiology.* Lea and Febiger, Philadelphia.

Wiggers, C. J. *Circulatory Dynamics.* Grune and Stratton, New York.

Wiggers, C. J. *Physiology in Health and Disease.* Lea and Febiger, Philadelphia.

Winton, F. P., and Bayliss, L. E. *Human Physiology.* The Blakiston Company, Philadelphia.

Wolf, S. G., and Wolff, H. G. *Human Gastric Function; An Experimental Study of a Man and His Stomach.* Oxford University Press, New York.

Wright, S. *Applied Physiology.* Oxford University Press, New York.

Zoethout, W. D., and Tuttle, W. W. *Textbook of Physiology.* C. V. Mosby Company, St. Louis.

Histology and Cells

American Association for the Advancement of Science. *Cell and Protoplasm.* Edited by F. R. Moulton. Science Press, Publication No. 14, New York.

Andrew, W. *Cellular Changes with Age.* Charles C Thomas, Springfield.

Bailey's Textbook of Histology. Edited by P. E. Smith William Wood and Company, Baltimore.

Bremer, J. L., and Weatherford, H. L. *A Textbook of Histology.* The Blakiston Company, Philadelphia.

Clark, W. E. L. *Tissues of the Body: An Introduction to the Study of Anatomy.* Oxford University Press, New York.

Cowdry, E. V. *A Textbook of Histology.* Lea and Febiger, Philadelphia.

Dawson, H. L. *Lambert's Histology.* The Blakiston Company, Philadelphia.

Gerard, R. W. *Unresting Cells.* Harper & Brothers, New York.

Ham, A. *Histology.* J. B. Lippincott Company, Philadelphia.

Hewer, E. E. *Textbook of Histology for Medical Students.* C. V. Mosby Company, St. Louis.

Jordan, H. A. *A Textbook of Histology.* D. Appleton-Century Company, New York.

Lee, D. H. K. *The Physiology of Tissues and Organs.* Charles C Thomas, Springfield, Ill.

Maximow, A. A., and Bloom, W. *A Textbook of Histology.* W. B. Saunders Company, Philadelphia.

Nonidez, J. F., and Windle, W. F. *Textbook of Histology.* McGraw-Hill Book Company, Inc., New York.

Schafer, E. S. *The Essentials of Histology.* Lea and Febiger, Philadelphia.

Seifriz, W. E. *Protoplasm.* McGraw-Hill Book Company, Inc., New York.

Sharp, L. *Introduction to Cytology.* McGraw-Hill Book Company, Inc., New York.

Whipple, G. H. *Hemoglobin, Plasma Protein and Cell Protein; Their Production and Interchange.* Charles C Thomas, Springfield, Ill.

Chemistry and Biochemistry

Bell, G. H., Davidson, J. N., and Scarborough, H. *Textbook of Physiology and Biochemistry.* The Williams and Wilkins Company, Baltimore.

Bull, H. B. *Physical Biochemistry.* John Wiley and Sons, Inc., New York.

Cameron, A. T. *Textbook of Biochemistry.* The Macmillan Company, New York.

Everett, M. K. *Medical Biochemistry.* Paul B. Hoeber, Inc., New York.

Mathews, A. P. *Physiological Chemistry.* The Williams and Wilkins Company, Baltimore.

Mathews, A. P. *Principles of Biochemistry.* William Wood and Company, Baltimore.

Mitchell, P. H. *A Textbook of Biochemistry.* McGraw-Hill Book Company, Inc., New York.

Sherman, H. C. *Chemistry of Food and Nutrition.* The Macmillan Company, New York.

West, E. S., and Todd, W. R. *Textbook of Biochemistry.* The Macmillan Company, New York.

Special Problems

Adrian, E. D. *The Basis of Sensation.* W. W. Norton and Company, New York.

Alvarez, W. C. *Nervousness, Indigestion and Pain.* Paul B. Hoeber, Inc., New York.

Arey, L. B. *Developmental Anatomy.* W. B. Saunders Company, Philadelphia.

Barcroft, Sir J. *The Brain and Its Environment.* Yale University Press, New Haven.

Buchsbaum, R. M. *Animals without Backbones; An Introduction to the Invertebrates.* University of Chicago Press, Chicago.

Colin, E. C. *Elements of Genetics: Mendel's Laws of Heredity with Special Application to Man.* The Blakiston Company, Philadelphia.

Davson, H., and Danielli, J. F. *Permeability of Natural Membranes.* The Macmillan Company, New York.

Dodds, G. S. *Essentials of Human Embryology.* John Wiley and Sons, Inc., New York.

Fulton, J. F. *History of Physiology and Medicine.* Charles C Thomas, Springfield, Ill.

Gamble, J. L. *Extracellular Fluid.* Harvard University Press, Cambridge, Mass.

Glasser, O. *Medical Physics.* The Year Book Publishers, Chicago.

Grollman, A. *Essentials of Endocrinology.* J. B. Lippincott Company, Philadelphia.

Hamilton, W. J., and others. *Human Embryology (Prenatal Development of Form and Function.)* The Williams and Wilkins Company, Baltimore.

Harrison, R. J. *The Child Unborn.* The Macmillan Company, New York.

Hegner, R. W. *College Zoology.* The Macmillan Company, New York.

Huettner, A. F. *Fundamentals of Comparative Embryology of Vertebrates.* The Macmillan Company, New York.

Macleod, G., and Taylor, C. M. *Rose's Foundations of Nutrition.* The Macmillan Company, New York.

Marshall, A. M. *The Frog.* The Macmillan Company, New York.

Mavor, J. W. *General Biology.* The Macmillan Company, New York.

Osgood, E. E. *Laboratory Diagnosis.* The Blakiston Company, Philadelphia.

Patten, B. M. *Human Embryology.* The Blakiston Company, Philadelphia.

Selye, H. *The Story of the Adapatation Syndrome.* Acta, Inc., Montreal.

Selye, H. *Stress.* Acta, Inc., Montreal.

Stitt, E. R. *Practical Bacteriology, Blood Work, and Animal Parasitology.* The Blakiston Company, Philadelphia.

Todd, J. G., and Sanford, A. H. *Clinical Diagnosis by Laboratory Methods.* W. B. Saunders Company, Philadelphia.

Turner, C. D. *General Endocrinology.* W. B. Saunders Company, Philadelphia.

Waddington, C. H. *Organizers and Genes.* Cambridge University Press, London.

Walter, H. E. *Biology of the Vertebrates.* The Macmillan Company, New York.

Walter, H. E. *Genetics.* The Macmillan Company, New York.

Wilder, H. H. *History of the Human Body.* Henry Holt and Company, Inc., New York.

Laboratory Manuals

Baumgartner, W. J. *Laboratory Manual of the Foetal Pig.* The Macmillan Company, New York.

Bensley, B. A. *Practical Anatomy of the Rabbit.* The Blakiston Company, Philadelphia.

Booth, E. S. *Laboratory Anatomy of the Cat.* The author, Walla Walla College, College Place, Wash.

Greene, E. C. *Anatomy of the Rat.* American Philosophical Society, Philadelphia.

Reighard, J., and Jennings, H. S. *Anatomy of the Cat.* Henry Holt and Company, Inc., New York.

Stackpole, C. E., and Leavell, L. C. *Laboratory Manual in Anatomy and Physiology.* The Macmillan Company, New York.

Visscher, M. B., and Smith, P. W. *Experimental Physiology.* Lea and Febiger, Philadelphia.

Zoethout, W. D. *Laboratory Experiments in Physiology.* The C. V. Mosby Company, Medical Publishers, St. Louis.

Dictionaries

The American Illustrated Medical Dictionary
The Macmillan Medical Dictionary
New Gould Medical Dictionary
Stedman's Practical Medical Dictionary
Chamber's Technical Dictionary

METRIC SYSTEM

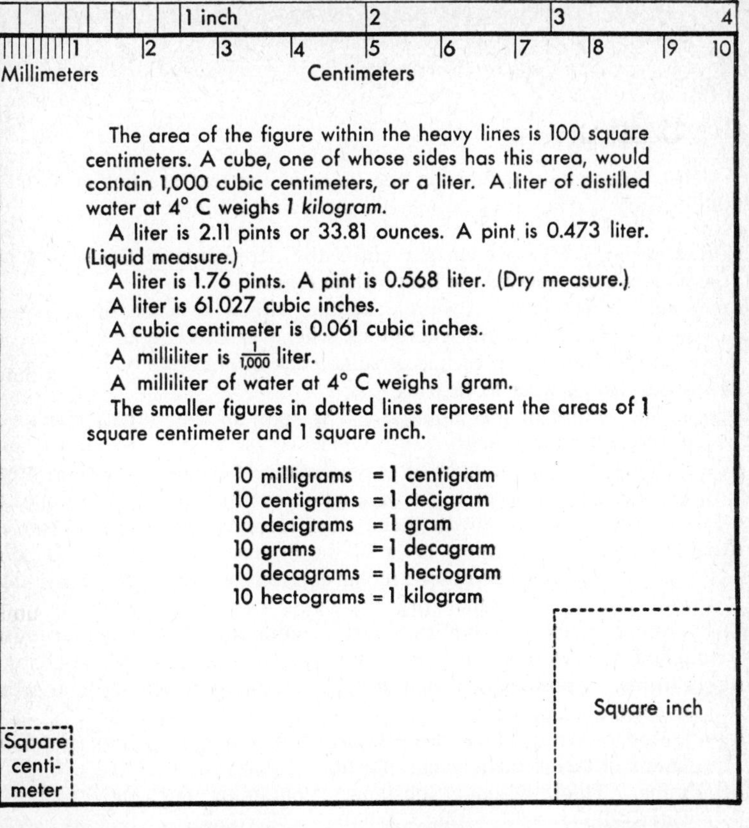

The area of the figure within the heavy lines is 100 square centimeters. A cube, one of whose sides has this area, would contain 1,000 cubic centimeters, or a liter. A liter of distilled water at 4° C weighs *1 kilogram.*

A liter is 2.11 pints or 33.81 ounces. A pint is 0.473 liter. (Liquid measure.)

A liter is 1.76 pints. A pint is 0.568 liter. (Dry measure.)

A liter is 61.027 cubic inches.

A cubic centimeter is 0.061 cubic inches.

A milliliter is $\frac{1}{1,000}$ liter.

A milliliter of water at 4° C weighs 1 gram.

The smaller figures in dotted lines represent the areas of 1 square centimeter and 1 square inch.

10 milligrams = 1 centigram
10 centigrams = 1 decigram
10 decigrams = 1 gram
10 grams = 1 decagram
10 decagrams = 1 hectogram
10 hectograms = 1 kilogram

1 millimeter	=	0.039 inch	1 decigram	=	1.543 grains
1 centimeter	=	0.393 inch	1 gram	=	15.432 grains
1 decimeter	=	3.937 inches	1 decagram	=	154.323 grains
1 meter	=	39.370 inches	1 hectogram	=	1,543.235 grains
			1 kilogram	=	15,432.350 grains
1 milligram	=	0.015 grain	1 kilogram	=	35.274 ounces
1 centigram	=	0.154 grain	1 kilogram	=	2.204 pounds

1 micron (μ) = 0.001 millimeter
1 millimicron, mμ or $\mu\mu$ = 0.000001 millimeter
1 Å, or Angstrom unit = 0.0001 μ, or 10^{-7} millimeters

Avoirdupois weights are used in weighing the organs of the body. One ounce avoirdupois = 28.35 grams. For the sake of simplicity in converting figures in the text from one system to the other, we have assumed

1 in. to equal 25 mm
1 in. to equal 25,000 microns (μ)
1 in. to equal 2.5 cm
1 cm to equal 0.4 in.
1 cc to equal 15 minims
30 cc to equal 1 oz

GLOSSARY

Terms adequately defined in the body of the text are not included in the Glossary. They may be located through the Index.

Acetab'ulum. "Shallow vinegar cup"; the depression in the innominate bone which receives the head of the femur

Ac'etone. $(CH_3)_2CO$. A simple ketone with a sweetish odor, present in blood and in body excretions whenever fats are used in metabolism without the presence of sufficient carbohydrate

Acous'tic. Pertaining to hearing

Acro'mion. Point of the shoulder. The spine of the scapula terminates in the acromion process

Adsorption. The attachment of one substance to the surface of another. The concentration of a gas or a substance in solution in a liquid on a surface in contact with them. See technical dictionaries also

Adventi'tia. The outermost layer of the organs which are not bounded by a serous coat, the outer areolar connective tissue being continuous with that of the other organs; e.g., blood vessels have no outer limiting membrane but lie in the common areolar tissue. Often called the *externa*

Agglu'tinins. Substances which induce adhesion or clumping together of cells

Ag'minated. Arranged in clusters, grouped together, as the agminated glands of Peyer in the small intestine

Albu'mins. Thick, viscous substances containing nitrogen; soluble in water, dilute acids, dilute salines, and concentrated solutions of magnesium sulphate and sodium chloride. They are coagulated by heat and strong acids. Examples: egg albumin and serum albumin of blood.

Al'kalies. Strong bases; that is, the hydroxides of active metals such as sodium, potassium, magnesium, etc.

Amphib'ia. A class of vertebrate animals that is able to live both on land and in water, as the frog

Ampul'la. A flasklike dilatation of a canal

Ana'erobe. Any microorganism which is able to live without free air or oxygen

An'tigens. Name given to foreign proteins and certain other substances which upon entering the blood stream cause the formation of protective substances (antibodies) in the serum

A'pex, pl. Ap'ices. The top or pointed extremity of a body

Arach'noid. Resembling a cobweb

Arboriza'tion. A branching distribution of veinlets or of nerve filaments, especially the branched, terminal ramifications of neurofibrils

A'trium. A hall or antechamber. The upper chamber of each half of the heart

Au'ricle. Ear. The *pinna* of the ear which with the *external meatus* constitutes the external ear

Au'tacoid. An early name suggested by Sir Edward Albert Sharpey-Schaefer (1850–1935) for hormones which were then divided into two groups—hormones stimulating and chalones inhibiting autacoids. Terms not now in general use

Autonom'ic. Performed without the will; automatic

Auton'omy. Independence; the condition of having independent functions, limitations, and laws

Az'ygos. Without a fellow; single; unpaired

Bas'al metab'olism. Rate of energy metabolism of a person at rest 12 to 18 hours after eating, as measured by the calorimeter

Bas'ilar. Pertaining to the base of an object; e.g., basilar artery located at the base of the brain

Bicip'ital. Referring to the biceps

Blas'tula. A hollow sphere of embryonic cells; the last stage in development before the embryo divides into two layers

B.N.A. Basle Nomina Anatomica, 1895. Anatomical terms in prevalent use. Doubtless there will be a further study of this list with some possible changes, e.g., as shown in the Jena Nomina Anatomica

Brachiocephal'ic. Pertaining to both the upper arm and head, as the brachiocephalic (innominate) artery and veins

Butyr'ic acid. C_3H_7COOH. A rancid, sticky fluid occurring in butter, cod-liver oil, etc., and as a product of protein putrefaction in body excretions

Cal'cify. Harden by deposit of salts of calcium; petrify

Cal'culus, pl. **Cal'culi.** A stone

Cal'orie. The calorie is the amount of heat required to raise the temperature of 1 kg of water 1° C or that of 1 lb of water 4° F. The small calorie is the amount of heat required to raise the temperature of 1 gm of water 1° C

Canalic'ulus, pl. **Canalic'uli.** A minute channel or vessel

Car'dia. The heart. The upper orifice of the stomach

Car'diograph. Instrument used to obtain tracing showing the force and form of the heart's movements

Castra'tion. Removal of the testes in the male or the ovaries in the female

Catacrot'ic. Referring to an irregularity of the pulse in which the heat is marked by two or more expansions of the artery. A tracing of this pulse shows one or more abnormal elevations on the downward stroke

Catal′ysis. A changing of the speed of a reaction, produced by the presence of a substance which does not itself enter the final product

Ce′liac. Pertaining to the abdominal cavity. A hollow

Celluli′tis. Inflammation of connective tissue, most commonly of superficial fascia

Chi′asm. An X-shaped crossing or decussation, especially that of the fibers of the optic nerve

Choa′na. Any funnel-shaped cavity, such as the posterior nares

Choles′terol. A sterol, $C_{27}H_{43}OH$, found in small quantities in the protoplasm of all cells, especially in nerve tissue, blood cells, and bile. Same as cholesterin

Chor′da tym′pani. The tympanic cord, a branch of the facial (seventh cranial) nerve, which traverses the tympanic cavity and joins the gustatory (lingual) nerve

Chro′maffin. Certain cells occurring in the adrenal, coccygeal, and carotid glands, along the sympathetic nerves and in various organs, stain deeply with chrome salts; hence the name chromaffin. The whole system of such tissue throughout the body is named the chromaffin or chromaphil system

Chro′matin. Portions of the nucleus which stain deeply with basic dyes, e.g., methylene blue

Chro′mosomes. Segments of chromatin in the nucleus of cells, concerned with the transmission of hereditary characteristics and the direction of the embryo's development

Cica′trix. The mark or scar left after the healing of a wound, due to the formation of embryonal tissue in repair

Cister′na mag′na, or cister′na cerebel′lo med′ullaris. That portion of the subarachnoid cavity between the cerebellum and the medulla

Clea′vage. The process of division of the fertilized ovum before differentiation into layers occurs. Same as segmentation

Col′loid. A state of matter in which particles having a diameter of 0.1 μ–0.001 μ are suspended in a dispersing medium, water, etc.

Com′missure. A joining. A bundle of nerve fibers passing from one side of the brain or spinal cord to the other side. The corner or angle of the eyes or lips

Compo′nent. A constituent part. A series of neurons forming a special system (afferent or efferent)

Congen′ital. Existing from or before birth

Cor′acoid. Shaped like a crow's beak; a process of the scapula

Cor′pus. A body or mass

Correla′tion. The interdependence of organs or functions; the reciprocal relations of organs

Cor′tex. The bark or outer layer; the outer portion of an organ

Cre′nated. Notched on the edge

Crepita'tion. A grating or crackling sound or sensation, like that produced by fragments of a fractured bone rubbing together
Crib'riform. Perforated like a sieve
Cu'mulus oöphor'us. Term applied to a mass of cells clinging to the ovum where it is set free from the ovary. Same as discus proligerus and ovarian mound
Cu'neiform. Wedge-shaped
Cyano'sis. Blueness of the skin, resulting from insufficient oxygenation of the blood
Cyclo'sis. Circulatory movement of protoplasm in cells—seen well in plant cells and in protozoa. Often spoken of as "streaming of protoplasm"
Cysti'tis. Inflammation of the bladder

Dea'minization. Liberation of ammonia from an amide
Dec'ibel. Unit of loudness, tenth of a bel. Used for measuring loudness differences of sounds
Decid'uous. That which falls off; not permanent
Dehydra'tion. Removal of water from a substance
Delir'ium cor'dis. Violent, tremulous beating of the heart
Del'toid. Triangular; resembling in shape the Greek letter Δ, delta
Diapede'sis. Passing of the erythrocytes through vessel walls without rupture
Diath'esis. A predisposition to certain kinds of disease
Dichot'omous. Divided into two; consisting of a pair or pairs
Dis'cus prolig'erus. See cumulus oöphorus
Distilla'tion. A process which consists of volatilizing a liquid by heat and then condensing the vapor, to separate the more volatile and less volatile parts of the liquid
Dura mater. "Hard mother"; the tough outer membrane enveloping the brain and spinal cord
Dyne. Unit of force which, when acting on a mass of 1 gm for 1 second, will cause an acceleration of 1 cm per second
Dyspha'gia. Difficulty in swallowing

Ectop'ic. Out of place. *Ectopic gestation* refers to pregnancy when the fecundated ovum, instead of entering the uterus, remains in either a Fallopian tube or the abdominal cavity
Ede'ma. Swelling due to abnormal effusion of serous fluid into the tissues
Elastic'ity. Quality of being easily deformed and quick in recovering
Empir'ical. Founded on experience; relating to the treatment of disease according to symptoms alone, without regard for scientific knowledge
Emul'sion. A mixture of two fluids insoluble in each other, where one is dispersed through the other in the form of finely divided globules
Endem'ic. Occurring frequently within a certain district

Endochon'dral ossifica'tion. Ossification in which cartilage is formed first and then is gradually replaced with bone

Endog'enous. Originating within the organism. Opposite of exogenous

En'ergy. Capacity or ability to do work

Equilib'rium. The balanced condition resulting when opposing forces are exactly equal. The term may refer to the maintenance of correct concentration of constituents of the body fluids or to the harmonious action of the organs of the body as in standing, etc.

Erg. A unit of work. The work done in moving a body 1 cm against a force of 1 dyne

Eth'moid. Resembling a sieve

Evagina'tion. Protrusion of some part or organ

Evapora'tion. The changing of a liquid into a vapor. Heat is necessary for evaporation, and if not otherwise supplied it is taken from near objects. Thus, the heat necessary for the evaporation of perspiration is taken from the body

Exog'enous. Originating outside of the organism; opposite of endogenous

Exophthal'mic. Pertaining to abnormal protrusion of the eyeball

Exten'sile. Capable of being stretched

Ex'udate. A fluid or semifluid which has oozed through the tissues into a cavity or upon the body surface

Fal'ciform. Sickle-shaped

Fascic'ulus, pl. Fascic'uli. A bundle of close-set fibers, usually muscle or nerve fibers

Fecunda'tion. Fertilization; impregnation

Fenes'trated. Having windowlike openings; perforated

Fibril'la, pl. Fibril'lae. A small fiber or filament

Fim'bria, pl. Fim'briae. A fringe

Flat'ulence. Distention due to generation of gases in the stomach and intestine

Flu'id. Includes liquids and gases. Not rigid, not solid. Takes shape of container

Fontanel'. "Little fountain"; the rise and fall of the pulse can be observed through the membranous interspaces or fontanels in the infant's cranium

Funic'ulus, pl. Funic'uli. The umbilical cord; the spermatic cord; one of the bundle of nerve fibers of which a nerve trunk is made up

Gastrocne'mius. "Belly of the leg"; one of the calf muscles

Gas'troepiplo'ic. Pertaining to the stomach and greater omentum

Gen'erative. Having the power or function of reproduction

Genes. Theoretical components of the chromosomes which determine hereditary traits

Gesta'tion. Pregnancy

Gli'a. Neuroglia

Glob'ulins. Protein substances (myosin, fibrinogen, etc.) similar to albumins but insoluble in water and soluble in dilute solutions of neutral salts

Glycogen'esis. The production of glycogen, animal starch

Glycogenol'ysis. Splitting of glycogen into glucose

Glycol'ysis. Splitting of glucose into carbon dioxide and water

Gon'ad. Gamete-producing gland; e.g., testis, ovary

Granula'tions. Grainlike bodies made up of large, soft vegetative cells which form upon the surface of wounds and ulcers; upon contraction they produce the scar

Hem'atin. A nonprotein substance containing iron; the colored constitutent of hemoglobin

Hemorrhoi'dal. Pertaining to hemorrhoids, small tumors caused by dilation of the veins of the anal region

Hi'lus or **Hi'lum.** The depression, usually on the concave surface of a gland, where vessels and ducts enter or leave

Homeos'tasis. Constancy of the internal environment

Homoge'neous. Of the same kind or quality throughout; uniform in nature; the reverse of heterogeneous

Hy'aline. Glassy; translucent

Hydroceph'alus. An excess of fluid in the brain, accompanied by an enlarged head and mental weakness

Hydrostatic pressure. A pressure exerted uniformly and perpendicularly to all surfaces as by a homogeneous liquid

Hyperglyce'mia. An abnormal amount of sugar in the blood

Immis'cible. Immiscible liquids are liquids which do not mix. Emulsions are mixtures of immiscible liquids

In'guinal. Pertaining to the groin

Inos'culate. To join by direct union or anastomosis

Interme'diary meta'bolism. Refers to metabolism taking place after absorption and before excretion from the excretory organs, that is, either in cells or in the body fluids—tissue fluid, lymph, or blood

Internun'cial. Acting as a medium between two organs or nerve centers

Intersti'tial. In the interspaces of a tissue; refers to the connective-tissue framework of glands

Ionize. To separate into ions which are atoms or groups of atoms bearing electric charges

Ische'mia. Local anemia due to mechanical obstruction (mainly arterial narrowing) to the blood supply.

Is'chium, pl. **Is'chia.** The lower portion of the os innominatum; that upon which the body is supported in a sitting posture

Isoagglu'tinin. A substance present in the blood serum which can agglutinate or clump together the erythrocytes of other individuals of the same species

Isoagglutin'ogen. A substance in blood cells which stimulates the action of agglutinins

Isohemol'ysin. A substance present in the blood serum which can dissolve the erythrocytes of other individuals of the same species

I'sotope. Isotopes are atoms that have different numbers of neutrons in their nuclei, but have the same number of free protons. They weigh differently but behave alike chemically. Some of them are radioactive and can be used as tracers in the body; for instance, iron, iodine, etc.

Ketogen'ic. Tending to produce acetone bodies

Lacta'tion. The secretion of milk

Lacu'na, pl. Lacu'nae. A minute, hollow space

Lambdoi'dal. Resembling the Greek letter Λ, lambda

Lamel'la, pl. Lamel'lae. A thin plate, or layer

Lam'ina. A thin plate; a germinal layer

Laryn'goscope. The instrument by which the larynx may be examined in the living subject

Lens. A transparent substance, usually glass, bounded by two curved surfaces or by one curved and one plane. There are two general classes of lenses: concave, which are thinner at the center than at the edges; and convex, which are thicker at the center than at the edges

Lig'ature. A thread or wire for tying a vessel. The act of applying a ligature

Lin'ea as'pera. A rough, longitudinal line on the back of the femur

Lymphangi'tis. Inflammation of a lymphatic vessel

Ly'sin. Lysis, "to dissolve." An antibody which can dissolve cells, etc.

Macera'tion. The softening of the parts of a tissue by soaking

Macroscop'ic. That which can be viewed with the naked eye

Manom'eter. An instrument for measuring the pressure or tension of liquids or gases

Maras'mus. Progressive wasting and emaciation, especially in young infants

Matura'tion. Cell division in which the number of chromosomes in the germ cells is reduced to one-half the number usual for the species

Medul'la. The central portions of an organ. Marrow. The medulla oblongata of the spinal cord

Mesoco'lon. A process of the peritoneum by which the colon is attached to the posterior abdominal wall

Microceph'alus. An idiot or fetus with a very small head

Mononu'clear. Having but one nucleus

Mo'tor. Producing or subserving motion. A muscle, nerve, or center that affects or produces movement

Mu′cin. A glycoprotein, a constituent of mucus
My′elocyte. A bone-marrow cell giving rise to granulocytes
Myogen′ic. Originating in muscular tissue
Myoneu′ral. Pertaining to both muscle and nerve
My′osin. A globulin, chief protein substance of muscle

Na′ris, pl. **Na′res.** A nostril
Navic′ular. Boatlike
Ner′vus er′igens, pl. **Ner′vi erigen′tes.** A nerve fiber supplying the bladder, genitals, and rectum; derived from the second and third sacral nerves
Neurasthen′ic. The name for a group of symptoms resulting from some functional disorder of the nervous system, with severe depression of the vital forces. It is usually due to prolonged and excessive expenditure of energy and is marked by tendency to fatigue, lack of energy, pain in the back, loss of memory, insomnia, constipation, loss of appetite, etc.
Neurogen′ic. Originating in nerve tissue
Nic′otine. An acid, colorless, fluid base obtained from tobacco; exceedingly poisonous
No′tochord. The primitive backbone in the embryo
Nu′cleus cunea′tus. A group of nerve cells in the medulla in which the fibers of the fasciculus cuneatus of the spinal cord terminate
Nu′cleus gra′cilis. A group of nerve cells in the medulla, in which the fibers of the fasciculus gracilis of the spinal cord terminate

Odon′toid. Toothlike
O′öcyte. The primitive ovum in the cortex of the ovary, before maturation takes place
Organ′ic. Pertaining to an organ or organs. Having an organized structure. Arising from an organism. In chemistry, a compound containing carbon
Os, pl. **O′ra.** A mouth
Os, pl. **Os′sa.** A bone
Osmo′sis. The passage of fluids and solutions, separated by a membrane or other porous septum, through the partition, so as to become mixed or diffused through each other
Os′sa innomina′ta, pl. of **Os innomina′tum.** "Unnamed bones." The irregular bones of the pelvis, unnamed on account of their nonresemblance to any known object
Os′teoblasts. The cells forming or developing into bone
Os′teoclast. A large cell found in the bone marrow, believed to be capable of absorbing bone
Osteogenet′ic. Pertaining to the formation of bone
O′tic. Pertaining to the ear
O′toliths. Particles of calcium carbonate and phosphate found in the internal ear on the hair cells

Papil'la, pl. **Papil'lae.** A small eminence; a nipplelike process
Paranas'al si'nuses. Sinuses which communicate with the cavity of the nose. They are often called air sinuses of the head
Patel'la. A small pan; the kneecap
Ped'icle. A stalk
Pedun'cle. A narrow part acting as a support
Pet'rous. Stonelike
Phlebot'omy. The surgical opening of a vein; venesection
Phren'ic. Pertaining to the diaphragm
Phys'ics. The science of the laws and phenomena of nature, especially of the forces and general properties of matter
Pi'a ma'ter. "Tender mother"; the innermost membrane closely enveloping the brain and spinal cord
Pir'iform. Pear-shaped
Pis'iform. Pea-shaped
Plas'tids. Bodies in cytoplasm thought to form areas for specific chemical activity, e.g., chromoplasts and leucoplasts
Polar'ity. Tendency of a body to exhibit opposite properties in opposite directions; referring to the possession of positive and negative poles
Poles. Points having opposite properties, occurring at the opposite extremities of an axis. Either end of a spindle in mitosis
Precip'itins. Antibodies in the blood serum which are capable of precipitating foreign proteins
Psy'chical. Pertaining to the mind
Ptery'goid: Wing-shaped
Pyogen'ic. Producing pus
Pyrex'ia. Elevation of temperature; fever

Quadrigem'inal. In four parts

Radia'tion. The act of spreading outward from a central point. The diffusion of rays of light
Recep'tor. This word is used with various meanings. One dictionary defines it as a sense organ; another as "nerve endings in organs of sense." Others would define it as the "ends of an afferent nerve fiber." It should be used and interpreted with care
Rec'tus, pl. **Rec'ti.** Straight. Name given to certain muscles of the eye and abdomen
Reflec'tion. The return of rays, beams, sounds, or the like from a surface. Reflection of light is of two kinds, *regular* and *diffused.* When a beam of light enters a darkened room through a small opening and strikes a mirror, a reflected beam will be seen traveling along some definite path. This is called *regular reflection.* Should the light, however, fall on a piece of white paper, it would be reflected and scattered in all directions. This is called *diffused reflection* and is caused by the inequalities of the reflecting surface. All rough surfaces, as well as dust and moisture in the atmosphere, serve to

diffuse light. If this were not the case, it would be dark everywhere except in the direct path of light from some luminous body

Regurgita'tion. The casting-up of undigested food from the stomach. A backward flowing of blood through the left atrioventricular opening because of imperfect closure of the bicuspid valve

Rhe'obase. Minimal electric current required to excite a tissue, e.g., nerve or muscle

Rhom'boid. A quadrilateral figure whose opposite sides and angles are equal but which is neither equilateral nor equiangular

Saliva'tion. An excessive secretion of saliva

Saponifica'tion. Conversion into soap

Sciat'ic. Pertaining to the ischium

Segmenta'tion. The process of division of the fertilized ovum before differentiation into layers occurs. Same as cleavage

Sig'moid. Shaped like the letter S

Ska'tol or **Sca'tol.** A strong-smelling crystalline substance from human feces, produced by decomposition of proteins in the intestine

Sol'ute. A dissolved substance

Sol'vent. A substance, usually liquid, which is capable of dissolving something

Somat'ic. Pertaining to the body, especially the body wall

Specif'ic grav'ity. A comparison between the weight of a substance and the weight of an equal volume of some other substance taken as a standard. The standards usually referred to are air for gases, and water for liquids and solids. For instance, the specific gravity (sp. gr.) of carbon dioxide (air standard) is 1.5, meaning that it is 1.5 times as heavy as an equal volume of air. Again, the specific gravity of mercury (water standard) is 13.6, meaning that mercury is 13.6 times as heavy as an equal volume of water. The specific gravity of solutions, as a salt solution, will necessarily vary with the concentration

Sphe'noid. Wedge-shaped

Sphinc'ter. A circular muscle which contracts the aperture to which it is attached

Splanch'nic. Pertaining to the viscera

Summa'tion. Addition; finding of total or sum

Sur'face ten'sion. The force which exists in the surface film of liquids which tends to bring the contained volume into a form having the least superficial area. It is due to the fact that the particles in the film are not equally acted on from all sides but instead are attracted inward by the pull of molecules below them

Su'ture. That which is sewn together, a seam; the synarthrosis between two cranial bones

Syner'gic or **Synerget'ic.** Acting in harmonious cooperation, said especially of certain muscles

Ten′do Achil′lis. "Tendon of Achilles." The tendon attached to the heel, so named because Achilles is supposed to have been held by the heel when his mother dipped him in the river Styx to render him invulnerable

Tet′any. A disease characterized by painful tonic and symmetric spasm of the muscles of the extremities

The′nar. Mound at base of thumb

Thermogen′esis. The production of heat, especially the process of generating heat within the animal body

Thermol′ysis. Dissociation by means of heat. The dissipation of bodily heat

Thermotax′is. The normal adjustment of the bodily temperature. The movement of organisms in response to heat

Trabec′ula, pl. Trabec′ulae. A supporting fiber; a prolongation of fibrous membrane which forms septa or partitions

Troch′lear. Pertaining to a pulley. The trochlear nerve supplies the superior oblique muscle

Unicel′lular. Composed of a single cell

U′vula. "Little grape"; the soft mass which projects downward from the posterior middle of the soft palate

Vac′uole. A space or cavity within the protoplasm of a cell, containing nutritive or waste substances

Vas′cular. Latin, *vasculum*, a small tube. Refers to tubes conveying liquids, as blood vascular and lymph vascular systems

Ve′na co′mes, pl. Ve′nae com′ites. A deep vein following the same course as the corresponding artery

Ve′na com′itans, pl. Ve′nae comitan′tes. Same as vena comes

Ver′miform. Worm-shaped

Vo′lar. Pertaining to the palm of the hand or the sole of the foot

INDEX